NONDESTRUCTIVE TESTING

HANDBOOK

Third Edition

Volume 5

Electromagnetic Testing

Technical Editor
Satish S. Udpa

Editor
Patrick O. Moore

American Society for Nondestructive Testing

ASNT exists to create a safer world by promoting the profession and technologies of nondestructive testing.

American Society for Nondestructive Testing, Incorporated
1711 Arlingate Lane
PO Box 28518
Columbus, OH 43228-0518
(614) 274-6003; fax (614) 274-6899
www.asnt.org

Errata

Errata if available for this printing may be obtained from ASNT's Web site, www.asnt.org, or as hard copy by mail, free on request from ASNT at the address above.

Library of Congress Cataloging-in-Publication Data

Electromagnetic Testing / technical editor, Satish S. Udpa; editor,
Patrick O. Moore.
p. cm. — (Nondestructive testing handbook (3rd ed.); v. 5).
Includes bibliographic references and index.
ISBN 1-57117-046-4 (alk. paper)
1. Magnetic testing. 2. Electric testing. I. Udpa, Satish S. II. Moore,
Patrick O. III. Series: Nondestructive testing handbook (3rd ed.) ; v. 5.
TA417.3.E525 2004
620.1'1278--dc22

200400448

Published by the American Society for Nondestructive Testing
PRINTED IN THE UNITED STATES OF AMERICA

President's Foreword

It was 1955 when the Board of Directors resolved to begin work on the *Nondestructive Testing Handbook* with Robert C. McMaster serving as editor. The first edition of the series began and was published in 1959. A review of the credits at the beginning of the first edition makes it clear that this documentation of our body of knowledge was a collaboration of a large number of distinguished volunteers. Since the beginning of this society, volunteers have been our greatest asset. They do not appear on the financial balance sheet but they make this society great.

At the time of publication of the first edition, President Maurice J. Curtis wrote that the *NDT Handbook* "will be significantly revised in future editions to keep pace with progress. Similarly, the functions and services of the Society for Nondestructive Testing will change as it strives to provide the utmost service of its membership to science, industry, and mankind."

In that spirit, the society published the second edition in ten volumes and is in the process of publishing the third edition. *Electromagnetic Testing* is the fifth volume to be published in this edition. The corresponding volume in the second edition was published in 1986. With the changes in signal processing capabilities and advances in science, this new volume is necessary if we are to "keep pace with progress," as Curtis said.

The *NDT Handbook* continues to be one of the finest examples of what society volunteers can accomplish. We have the support of a full time editor but the text is provided and peer reviewed by volunteers. The contributors and reviewers are listed in the preface to this volume. We all owe our appreciation these individuals for volunteering their time, energy and resources to document the science and practice of electromagnetic testing.

Three individuals gave conspicuously of their time and I would like to provide a personal thanks on behalf of the Society to Satish S. Udpa, technical editor; Mani Mina, coordinator; and James E. Cox, Electromagnetic Testing Committee chair. Thank you for your efforts.

I challenge each nondestructive testing professional to get involved in making our professional organization better, especially if you feel that important information is missing from any society publication. We each have a unique knowledge. The volunteers who worked on this *NDT Handbook* were willing to share their knowledge. In the future, we need your contribution to have comprehensive volumes.

Again thanks to all who contributed.

Joseph L. Mackin
ASNT President, 2003-2004

Foreword

Aims of a Handbook

The volume you are holding in your hand is the fifth in the third edition of the *Nondestructive Testing Handbook*. In the beginning of each volume, it has been useful to state the purposes and nature of the *NDT Handbook* series.

Handbooks exist in many disciplines of science and technology, and certain features set them apart from other reference works. A handbook should ideally give the basic knowledge necessary for an understanding of the technology, including both scientific principles and means of application.

The typical reader may be assumed to have completed three years of college toward a degree in mechanical engineering or materials science and hence has the background of an elementary physics or mechanics course. Additionally, this volume provides a positive reinforcement for the use of computer based media that enhances its educational value and enlightens all levels of education and training.

Standards, specifications, recommended practices and inspection procedures may be discussed in a handbook for instructional purposes, but at a level of generalization that is illustrative rather than comprehensive. Standards writing bodies take great pains to ensure that their documents are definitive in wording and technical accuracy. People writing contracts or procedures should consult the actual standards when appropriate.

Those who design qualifying examinations or study for them draw on handbooks as a quick and convenient way of approximating the body of knowledge. Committees and individuals who write or anticipate questions are selective in what they draw from any source. The parts of a handbook that give scientific background, for instance, may have little bearing on a practical examination except to provide the physical foundation to assist handling of more challenging tasks. Other parts of a handbook are specific to a certain industry. This handbook provides a collection of perspectives on its subject to broaden its value and convenience to the nondestructive testing community.

The present volume is a worthy addition to the third edition. The editors, technical editors, ASNT staff, many contributors and reviewers worked together to bring the project to completion. For their scholarship and dedication, I thank them all.

Gary L. Workman
Handbook Development Director

Preface

Electromagnetic nondestructive test methods have come a long way since the work reported by S.M. Saxby concerning magnetic techniques for inspecting gun barrels in 1868 and the early eddy current experiments conducted by D.E. Hughes in 1879. The level of sophistication of these methods has grown considerably with time as has the diversity of applications. Progress in the fields of electronics, sensor technology and computer engineering has facilitated the development of new and interesting approaches for improving the sensitivity and resolution of such techniques. The ready availability of microprocessors has contributed greatly to improving both the quality and quantity of information made available to the user. The emergence of new techniques for modeling has not only given insight into the underlying physical process but also given tools for the design of new sensors, the prediction of test performance and the development of new and improved techniques. Major sections of this *NDT Handbook* had to be rewritten to provide readers with information about a number of these exciting developments. The resulting volume represents the collective wisdom of many volunteers who have chosen to share their expertise.

All electromagnetic test techniques are described by Maxwell's equations. These elegant equations describe magnetic particle testing just as effectively as eddy current and microwave techniques. An important factor that contributes to the differences in the underlying physical processes associated with each technique is the excitation frequency. The excitation frequency associated with magnetostatic techniques such as magnetic particle testing is either zero (direct current) or very low. Eddy current techniques use excitation frequencies in the quasistatic range (100 Hz to 10 MHz). Microwave testing uses much higher frequencies. Thus, these electromagnetic methods can be viewed as a continuum of techniques governed by the common thread of Maxwell's equations, with the character of the underlying physical process dictated by the choice of excitation frequency.

Techniques for modeling magnetic particle techniques, for example, have much in common with techniques for simulating other tests described in this volume. We have therefore included some discussion on magnetic particle testing despite the fact that a separate *NDT Handbook* volume covers that method.

A handbook, by definition, is a source of fundamental information. Its readers come from diverse backgrounds. Practitioners who wish to seek additional information about a technique are just as welcome to use the book as a student seeking basic information about a test technique. Many readers may be interested in knowing about advanced modeling techniques; others are interested in modern applications of electromagnetic testing. This volume covers both theory and practice. In some ways, the coverage is different from that in previous editions. The references at the end of each chapter will also allow the reader to explore this fascinating subject.

The volume represents the work of many in the field. I am immensely thankful to all the contributors. I am equally grateful to all the reviewers who took the time to review and comment on material that was sent to them on short notice. I am particularly thankful to Mani Mina, of Iowa State University, who went to extraordinary lengths to contact many of the contributors, to John Bowler, also of Iowa State University, who invested a lot of time to make sure that technical details presented in the book are correct and to Patrick Moore at ASNT for making this a labor of love. Finally, I would be remiss if I did not acknowledge the help, support and advice that I have received over the years from William Lord. He was instrumental in infecting me with the love for the subject and helped shape many of the ideas in the book.

Satish S. Udpa
Technical Editor

Editor's Preface

The first edition of the *Nondestructive Testing Handbook,* published in 1959, introduced eddy current testing to test technicians in the United States and the world.

The second edition started slowly. Two drafts of the electromagnetic volume, in the 1960s and 1970s, did not get into print. With the help of ASNT staff, the second edition volume was finally published in 1986.

The second edition, like the first, featured the analytical approach of Friedrich Förster. In addition to its extensive treatment of applications, the second edition also featured sections on magnetic flux leakage and microwave testing.

In 1999, work on the third edition of *Electromagnetic Testing* began in the Technical and Education Council's Handbook Development Committee and Electromagnetic Testing Committee. Production began in 2001, when Satish Udpa assumed the duties of technical editor.

Since 1990, numerical techniques have increasingly replaced analytical ones in research and applications and are well represented in this volume.

In this edition, an attempt is made to standardize the ways that technology is referred to. One is that conductivity measurements are given in siemens per meter in addition to the more familiar percentages of the International Annealed Copper Standard. Another is that, wherever practical, alloys are referred to by their numerical designations in the Unified Numbering System. It is hoped that these changes will help to harmonize nondestructive testing with sister disciplines and make the volume more useful to posterity.

I would personally like to thank members of ASNT staff who helped to make this book better. Hollis Humphries and Joy Grimm produced many excellent graphics and laid out the chapters.

Senior Manager of Publications Paul McIntire believed strongly in the value of this book. He provided many valuable suggestions, read every chapter in galley and reviewed the book again before publication. McIntire's personal attention prevented more than a few errors and has helped readability throughout the volume. He supported the project at every stage of development and production.

People listed as contributors in the acknowledgments below were also reviewers but are listed once, as contributors.

Patrick O. Moore
NDT Handbook Editor

Acknowledgments

Thanks are extended to the Institute of Electrical and Electronics Engineers and its Center for the History of Electrical Engineering, for help in obtaining illustrations.

The infrastructure chapter is indebted to Bernard Mayton of the Virginia Department of Transportation and Cesar Apusen of the Virginia Transportation Research Council for their assistance in conducting field tests.

The following people are thanked for their contributions to the discussion of offshore weld testing: Fred Cella and the Shaw Group, Stone and Webster; Dennis Donovan, Coastal Inspection Services; Helen Goldberg, Sea Test Services; and Mike Wallace, Core Technical Services.

Sources of illustrations are acknowledged in a section at the end of this book.

Handbook Development Committee

Gary L. Workman, University of Alabama, Huntsville
Michael W. Allgaier, Allgaier Consulting and Training
Albert S. Birks, Naval Surface Warfare Center
Richard H. Bossi, Boeing Aerospace Company
Lisa J.H. Brasche, Iowa State University
William C. Chedister, Circle Systems
James E. Cox, Zetec, Incorporated
James L. Doyle, Jr., NorthWest Research Associates
Nat Y. Faransso, KBR
François Gagnon, Vibra-K Consultants
Matthew J. Golis
Robert E. Green, Jr.
Gerard K. Hacker, Teledyne Brown Engineering
Harb S. Hayre, Ceie Specs

Eric v.K. Hill, Embry-Riddle Aeronautical University
Frank A. Iddings
Charles N. Jackson, Jr.
Morteza K. Jafari, Fugro South
John K. Keve, DynCorp Tri-Cities Services
Doron Kishoni, Business Solutions USA
Lloyd P. Lemle, Jr., ConocoPhillips
Xavier P.V. Maldague, University Laval
George A. Matzkanin, Nondestructive Testing Information and Analysis Center
Paul McIntire, American Society for Nondestructive Testing
Michael L. Mester
Ron K. Miller, Physical Acoustics Corporation
Scott D. Miller, Saudi Aramco, Saudi Arabia
Mani Mina, Technology Resource Group
Patrick O. Moore, American Society for Nondestructive Testing
Stanley Ness
Louis G. Pagliaro, Engineered Sintered Components Company
Emmanuel E. Papadakis, Quality Systems Concepts
Stanislav I. Rokhlin, Ohio State University
Frank J. Sattler
Fred Seppi, Williams International
Kermit A. Skeie
Roderic K. Stanley, NDE Information Consultants
Holger H. Streckert, General Atomics
Stuart A. Tison, Millipore Corporation
Noel A. Tracy, Universal Technology Corporation
Satish S. Udpa, Michigan State University
James L. Walker, University of Alabama, Huntsville
Mark F.A. Warchol, Alcoa, Incorporated
Glenn A. Washer, Turner-Fairbank Highway Research Center
George C. Wheeler

Contributors

David M. Amos, MFE Enterprises
Donald M. Bailey
David R. Bajula, Longview Inspection
Craig W. Benson, General Dynamics
Thomas U. Bipes, Progress Energy
Anmol S. Birring, NDE Associates
Ronald J. Botsko
David J. Brown, Zetec, Incorporated
John R. Bowler, Iowa State University
James E. Cox, Zetec, Incorporated
Gerald L. Fitzpatrick, PRI Research and Development Corporation
Paul M. Gammell
Lawrence O. Goldberg, Sea Test Services
Donald J. Hagemaier
Johann H. Hinken, Hochschule Magdeburg-Stendal, Germany
Tatsuo Hiroshima, Marktec Corporation, Japan
Nathan Ida, University of Akron
Timothy E. Kinsella, Dassault Falcon Jet
Margarit G. Lozev, Edison Welding Institute
Martin C. Lugg, TSC Inspection Systems, United Kingdom
David D. Mackintosh, Queens University, Canada
Walter R. Matulewicz, Huddleston Technical Services
Joseph M. Mandula
Michael L. Mester
Masashi Mizuno, Tohoku Steel, Japan
George Mordwinkin, Sensor Corporation
Nasser Qaddoumi, American University of Sharjah, United Arab Emirates
Pradeep Ramuhalli, Michigan State University
Gary P. Rogles, AmerenUE
David E. Russell, Russell NDE Systems, Canada
Takehide Sakamoto, Sumitomo Metal Technology, Japan
Ram P. Samy, The Timken Company
Michael L. Scott, Wiss, Janney, Elstner Associates
Ad A. Shatat, Russell NDE Systems, Canada
William C.L. Shih, PRI Research and Development Corporation
Jack C. Spanner, Jr., Electric Power Research Institute
Roderic K. Stanley, NDE Information Consultants
Eric J. Strauts, TEEM Electronics
Hal Streckert, General Atomics
Wolfgang Stumm
Theodoros Theodoulidis, Aristotle University of Thessaloniki, Greece
Marvin W. Trimm, Westinghouse Savannah River Company
Lalita Udpa, Michigan State University
Satish S. Udpa, Michigan State University
Andrew P. Washabaugh, Jentek Sensors
Glenn A. Washer, Turner-Fairbank Highway Research Center
Herbert R. Weischedel, NDT Technologies, Incorporated
Markus Zahn, Massachusetts Institute of Technology
Reza Zoughi, University of Missouri, Rolla

Reviewers

R. David Adamson, Pacific Gas and Electric
David L. Atherton, Queens University, Canada
Alfred J. Bahr, SRI International
Narendra K. Batra, Naval Research Laboratory
Mark J. Bergander, Scientific Technologies
Albert S. Birks, Naval Surface Warfare Center
Lisa J.H. Brasche, Iowa State University
Timothy L. Brown, Foerster Instruments
Donald Bugden, Magnetic Analysis Corporation
Val S. Cecco, R/D Tech, Canada

Robert L. Crane, Air Force Research Laboratory
Alfred E. Crouch, Southwest Research Institute
James M. Daughton, NVE Corporation
Claude D. Davis, Unified Testing Services
Thomas J. Davis, HD Laboratories
John C. Deboo, Boeing Commercial Airplane Company
Louis M. de la Pintiere
Hitoshi Domon, Tokyo Engineering University, Japan
John J. Flaherty, Flare Technology
Michael J. Gilkey, General Dynamics
Mark H. Gehlen, UniWest
Neil J. Goldfine, Jentek Sensors
Tad S. Golosinski, University of Missouri, Rolla
David P. Harvey, Oremet-Wah Chang
Rudolf G. Hentschel
James W. Houf, American Society for Nondestructive Testing
Morteza K. Jafari, Fugro South
Marvin Johnson, Owens Services Corporation
Marty J. Klatt, R/D Tech, Canada
Gary G. Korkala, Security Defense Systems
Karl E. Kraft, Tinker Air Force Base
Kenji J. Krzywosz, Electric Power Research Institute
Lloyd P. Lemle, Jr., ConocoPhillips
Raymond J. Leseck, US Airways
Clayton D. Luce, Northeast NDE Company
Joseph L. Mackin, International Pipe Inspectors Association
Shreekanth Mandayam, Rowan University
John F. Manning, MTS, Incorporated
Michael J. Maulucci
Paul McIntire, American Society for Nondestructive Testing
Scott M. Metzger, Stupp Pipeline Services
Scott D. Miller, Saudi Aramco, Saudi Arabia
Mani Mina, Iowa State University
William Arthur Mullan, Boeing Aerospace Company
Shridhar C. Nath, General Electric Global Research Center
Monty O'Connor, Eddy Current Technology
Allan F. Pardini
Ronald B. Peoples, Foerster Instruments
Dennis N. Poffenroth
G. Alan Raine, TSC Inspection Systems, United Kingdom
Paul Ristuccia, Boeing Commercial Airplane Company
Harold A. Sabbagh, Victor Technologies
E.H. Sabbagh, Victor Technologies
Robert Saglio, Commissariat à l'Energie Atomique, France
Frank J. Sattler
Thomas R. Schmidt
Ujjal Sen, Spectronics Corporation
Richard L. Skaugset, PRI Research and Development Corporation
Thierry Sollier, Commissariat à l'Energie Atomique, France
Tadeusz Stepinski, Uppsala University, Sweden
Douglas J. Strand, Army Research Laboratory, Aberdeen Proving Ground
Yushi Sun, Innovative Materials Testing Technologies
Richard L. Tiefenauer, Saint Louis Testing Laboratories
Orestes J. Varonis
Mark F.A. Warchol, Alcoa, Incorporated
Gary Zimmerman, Zetec, Incorporated

Contents

1

CHAPTER

Introduction to Electromagnetic Testing

Marvin W. Trimm, Westinghouse Savannah River Company, Aiken, South Carolina (Parts 1 and 2)

Holger H. Streckert, General Atomics, San Diego, California (Part 3)

Andrew P. Washabaugh, Jentek Sensors, Waltham, Massachusetts (Part 2)

PART 1. Nondestructive Testing

Definition

Nondestructive testing (NDT) has been defined as comprising those methods used to test a part or material or system without impairing its future usefulness.[1] The term is generally applied to nonmedical investigations of material integrity.

Strictly speaking, this definition of nondestructive testing includes noninvasive medical diagnostics. Ultrasound, X-rays and endoscopes are used by both medical and industrial nondestructive testing. Medical nondestructive testing, however, has come to be treated by a body of learning so separate from industrial nondestructive testing that today most physicians do not use the word *nondestructive.*

Nondestructive testing is used to investigate specifically the material integrity or properties of the test object. A number of other technologies — for instance, radio astronomy, voltage and amperage measurement and rheometry (flow measurement) — are nondestructive but are not used specifically to evaluate material properties. Radar and sonar are classified as nondestructive testing when used to inspect dams, for instance, but not when they are used to chart a river bottom.

Nondestructive testing asks "Is there something wrong with this material?" In contrast, performance and proof tests ask "Does this component work?" It is not considered nondestructive testing when an inspector checks a circuit by running electric current through it. Hydrostatic pressure testing is another form of proof testing, one that sometimes destroys the test object.

Another gray area that invites various interpretations in defining nondestructive testing is *future usefulness*. Some material investigations involve taking a sample of the tested part for a test that is inherently destructive. A noncritical part of a pressure vessel may be scraped or shaved to get a sample for electron microscopy, for example. Although future usefulness of the vessel is not impaired by the loss of material, the procedure is inherently destructive and the shaving itself — in one sense the true *test object* — has been removed from service permanently.

The idea of future usefulness is relevant to the quality control practice of sampling. Sampling (that is, less than 100 percent testing to draw inferences about the unsampled lots) *is* nondestructive testing if the tested sample is returned to service. If the steel is tested to verify the alloy in some bolts that can then be returned to service, then the test is nondestructive. In contrast, even if spectroscopy used in the chemical testing of many fluids is inherently nondestructive, the testing is destructive if the samples are poured down the drain after testing.

Nondestructive testing is not confined to crack detection. Other discontinuities include porosity, wall thinning from corrosion and many sorts of disbonds. Nondestructive material characterization is a growing field concerned with material properties including material identification and microstructural characteristics — such as resin curing, case hardening and stress — that have a direct influence on the service life of the test object.

Methods and Techniques

Nondestructive testing has also been defined by listing or classifying the various techniques.[1-3] This sense of *nondestructive testing* is practical in that it typically highlights methods in use by industry.

In the *Nondestructive Testing Handbook,* the word *method* is used for a group of test techniques that share a form of probing energy. Ultrasonic test methods, for example, use acoustic waves faster than sound. Infrared and thermal testing and radiographic testing both use electromagnetic radiation, each in a defined wavelength range. A technique, in contrast, has features that adapt the method to the application. Through-transmission immersion testing is a technique of the ultrasonic method, for example.

Purposes of Nondestructive Testing

Since the 1920s, the art of testing without destroying the test object has developed from a laboratory curiosity to an indispensable tool of fabrication, construction, manufacturing and maintenance processes. No longer is visual testing of materials, parts and complete products the principal means of determining adequate quality. Nondestructive tests in great variety are in worldwide use to detect variations in structure, minute changes in surface finish, the presence of cracks or other physical discontinuities, to measure the thickness of materials and coatings and to determine other characteristics of industrial products. Scientists and engineers of many countries have contributed greatly to nondestructive test development and applications.

The various nondestructive testing methods are covered in detail in the literature but it is always wise to consider objectives before details. How is nondestructive testing useful? Why do thousands of industrial concerns buy the testing equipment, pay the subsequent operating costs of the testing and even reshape manufacturing processes to fit the needs and findings of nondestructive testing?

Modern nondestructive tests are used by manufacturers (1) to ensure product integrity and in turn reliability, (2) to avoid failures, prevent accidents and save human life (see Figs. 1 and 2), (3) to make a profit for the user, (4) to ensure customer satisfaction and maintain the manufacturer's reputation, (5) to aid in better product design, (6) to control manufacturing processes, (7) to lower manufacturing costs, (8) to maintain uniform quality level and (9) to ensure operational readiness.

These reasons for widespread and profitable nondestructive testing are sufficient in themselves but parallel developments have contributed to its growth and acceptance.

Increased Demand on Machines

In the interest of greater performance and reduced cost for materials, the design engineer is often under pressure to reduce weight. This can sometimes be done by substituting aluminum alloys, magnesium alloys or composite materials for steel or iron but such light parts may not be the same size or design as those they replace. The tendency is also to reduce the size. These pressures on the designer have subjected parts of all sorts to increased stress levels. Even such commonplace objects as sewing machines, sauce pans and luggage are also lighter and more heavily loaded than ever before. The stress to be supported is seldom static. It often fluctuates and reverses at low or high frequencies. Frequency of stress reversals increases with the speeds of modern machines and thus parts tend to fatigue and fail more rapidly.

Another cause of increased stress on modern products is a reduction in the safety factor. An engineer designs with certain known loads in mind. On the supposition that materials and workmanship are never perfect, a safety factor of 2, 3, 5 or 10 is applied. However, a lower factor is often used that depends on considerations such as cost or weight.

New demands on machinery have also stimulated the development and use of new materials whose operating characteristics and performance are not completely known. These new materials

FIGURE 1. Fatigue cracks contributed to damage to aircraft fuselage in flight (April 1988).

FIGURE 2. Boilers operate with high internal steam pressure. Material discontinuities can lead to sudden, violent failure with possible injury to people and property.

could create greater and potentially dangerous problems. For example, an aircraft part was built from an alloy whose work hardening, notch resistance and fatigue life were not well known. After relatively short periods of service, some of the aircraft using these parts suffered disastrous failures. Sufficient and proper nondestructive tests could have saved many lives.

As technology improves and as service requirements increase, machines are subjected to greater variations and to wider extremes of all kinds of stress, creating an increasing demand for stronger or more damage tolerant materials.

Engineering Demands for Sounder Materials

Another justification for nondestructive tests is the designer's demand for sounder materials. As size and weight decrease and the factor of safety is lowered, more emphasis is placed on better raw material control and higher quality of materials, manufacturing processes and workmanship.

An interesting fact is that a producer of raw material or of a finished product sometimes does not improve quality or performance until that improvement is demanded by the customer. The pressure of the customer is transferred to implementation of improved design or manufacturing. Nondestructive testing is frequently called on to deliver this new quality level.

Public Demands for Greater Safety

The demands and expectations of the public for greater safety are apparent everywhere. Review the record of the courts in granting high awards to injured persons. Consider the outcry for greater automobile safety, as evidenced by the required automotive safety belts and the demand for air bags, blowout proof tires and antilock braking systems. The publicly supported activities of the National Safety Council, Underwriters Laboratories, the Occupational Safety and Health Administration and the Federal Aviation Administration in the United States, as well as the work of similar agencies abroad, are only a few of the ways in which this demand for safety is expressed. It has been expressed directly by passengers who cancel reservations following a serious aircraft accident. This demand for personal safety has been another strong force in the development of nondestructive tests.

Rising Costs of Failure

Aside from awards to the injured or to estates of the deceased and aside from costs to the public (because of evacuation occasioned by chemical leaks) consider briefly other factors in the rising costs of mechanical failure. These costs are increasing for many reasons. Some important ones are (1) greater costs of materials and labor, (2) greater costs of complex parts, (3) greater costs because of the complexity of assemblies, (4) greater probability that failure of one part will cause failure of others because of overloads, (5) trend to lower factors of safety, (6) probability that the failure of one part will damage other parts of high value and (7) part failure in an integrated automatic production machine, shutting down an entire high speed production line. When production was carried out on many separate machines, the broken one could be bypassed until repaired. Today one machine is tied into the production of several others. Loss of such production is one of the greatest losses resulting from part failure.

Applications of Nondestructive Testing

Nondestructive testing is a branch of the materials sciences that is concerned with all aspects of the uniformity, quality and serviceability of materials and structures. The science of nondestructive testing incorporates all the technology for detection and measurement of significant properties, including discontinuities, in items ranging from research specimens to finished hardware and products in service. By definition nondestructive test methods provide a means for examining materials and structures without disruption or impairment of serviceability. Nondestructive testing makes it possible for internal properties or hidden discontinuities to be revealed or inferred.

Nondestructive testing is becoming increasingly vital in the effective conduct of research, development, design and manufacturing programs. Only with appropriate nondestructive testing methods can the benefits of advanced materials science be fully realized. The information required for appreciating the broad scope of nondestructive testing is available in many publications and reports.

Classification of Methods

The National Materials Advisory Board (NMAB) Ad Hoc Committee on Nondestructive Evaluation adopted a system that classified techniques into six major method categories: visual, penetrating radiation, magnetic-electrical, mechanical vibration, thermal and chemical/electrochemical.[3] A modified version is presented in Table 1.[1]

Each method can be completely characterized in terms of five principal factors: (1) energy source or medium used to probe the object (such as X-rays, ultrasonic waves or thermal radiation); (2) nature of the signals, image or signature resulting from interaction with the object (attenuation of X-rays or reflection of ultrasound, for example); (3) means of detecting or sensing resultant signals (photoemulsion, piezoelectric crystal or inductance coil); (4) means of indicating or recording signals (meter deflection, oscilloscope trace or radiograph); and (5) basis for interpreting the results (direct or indirect indication, qualitative or quantitative and pertinent dependencies).

The objective of each method is to provide information about one or more of the following material parameters: (1) discontinuities and separations (cracks, voids, inclusions, delaminations and others); (2) structure or malstructure (crystalline structure, grain size, segregation, misalignment and others); (3) dimensions and metrology (thickness, diameter, gap size, discontinuity size and others); (4) physical and mechanical properties (reflectivity, conductivity, elastic modulus, sonic velocity and others); (5) composition and chemical analysis (alloy identification, impurities, elemental distributions and others); (6) stress and dynamic response (residual stress, crack growth, wear, vibration and others); (7) signature analysis (image content, frequency spectrum, field configuration and others); and (8) abnormal sources of heat.

Material characteristics in Table 1 are further defined in Table 2 with respect to specific objectives and specific attributes to be measured, detected and defined.

The limitations of a method include conditions to be met for method application (access, physical contact, preparation and others) and requirements to adapt the probe or probe medium to the object examined. Other factors limit the detection or characterization of discontinuities, properties and other attributes and limit interpretation of signals or images generated.

Classification Relative to Test Object

Nondestructive testing techniques may be classified according to how they detect indications relative to the surface of a test object. Surface methods include liquid

Table 1. Nondestructive testing method categories.

Categories	Objectives
Basic Categories	
Mechanical and optical	color, cracks, dimensions, film thickness, gaging, reflectivity, strain distribution and magnitude, surface finish, surface flaws, through-cracks
Penetrating radiation	cracks; density and chemistry variations; elemental distribution; foreign objects; inclusions; microporosity; misalignment; missing parts; segregation; service degradation; shrinkage; thickness; voids
Electromagnetic and electronic	alloy content; anisotropy; cavities; cold work; local strain, hardness; composition; contamination; corrosion; cracks; crack depth; crystal structure; electrical conductivities; flakes; heat treatment; hot tears; inclusions; ion concentrations; laps; lattice strain; layer thickness; moisture content; polarization; seams; segregation; shrinkage; state of cure; tensile strength; thickness; disbonds; voids
Sonic and ultrasonic	crack initiation and propagation; cracks, voids; damping factor; degree of cure; degree of impregnation; degree of sintering; delaminations; density; dimensions; elastic moduli; grain size; inclusions; mechanical degradation; misalignment; porosity; radiation degradation; structure of composites; surface stress; tensile, shear and compressive strength; disbonds; wear
Infrared and thermal	anisotropy, bonding; composition; emissivity; heat contours; plating thickness; porosity; reflectivity; stress; thermal conductivity; thickness; voids; cracks; delaminations; heat treatment; state of cure; moisture; corrosion
Chemical and analytical	alloy identification; composition; cracks; elemental analysis and distribution; grain size; inclusions; macrostructure; porosity; segregation; surface anomalies
Auxiliary Categories	
Image generation	dimensional variations; dynamic performance; anomaly characterization and definition; anomaly distribution; anomaly propagation; magnetic field configurations
Signal image analysis	data selection, processing and display; anomaly mapping, correlation and identification; image enhancement; separation of multiple variables; signature analysis

penetrant testing, visual testing, grid testing and moiré testing. Surface/near-surface methods include tap, potential drop, holographic, shearographic, magnetic particle and electromagnetic testing. When surface or near-surface methods are applied during intermediate manufacturing processes, they provide preliminary assurance that volumetric methods performed on the completed object or component will reveal few rejectable discontinuities. Volumetric methods include radiography, ultrasonic testing, acoustic emission testing and less widely used methods such as acoustoultrasonic testing and magnetic resonance imaging. Through-boundary techniques include leak testing, some

TABLE 2. Objectives of nondestructive testing methods.

Objectives	Attributes Measured or Detected
Discontinuities and Separations	
Surface anomalies	roughness, scratches, gouges, crazing, pitting, imbedded foreign material
Surface connected anomalies	cracks, porosity, pinholes, laps, seams, folds, inclusions
Internal anomalies	cracks, separations, hot tears, cold shuts, shrinkage, voids, lack of fusion, pores, cavities, delaminations, disbonds, poor bonds, inclusions, segregations
Structure	
Microstructure	molecular structure; crystalline structure and/or strain; lattice structure; strain; dislocation; vacancy; deformation
Matrix structure	grain structure, size, orientation and phase; sinter and porosity; impregnation; filler and/or reinforcement distribution; anisotropy; heterogeneity; segregation
Small structural anomalies	leaks (lack of seal or through-holes), poor fit, poor contact, loose parts, loose particles, foreign objects
Gross structural anomalies	assembly errors; misalignment; poor spacing or ordering; deformation; malformation; missing parts
Dimensions and Metrology	
Displacement; position	linear measurement; separation; gap size; discontinuity size, depth, location and orientation
Dimensional variations	unevenness; nonuniformity; eccentricity; shape and contour; size and mass variations
Thickness; density	film, coating, layer, plating, wall and sheet thickness; density or thickness variations
Physical and Mechanical Properties	
Electrical properties	resistivity; conductivity; dielectric constant and dissipation factor
Magnetic properties	polarization; permeability; ferromagnetism; cohesive force, susceptibility
Thermal properties	conductivity; thermal time constant and thermoelectric potential; diffusivity; effusivity; specific heat
Mechanical properties	compressive, shear and tensile strength (and moduli); Poisson's ratio; sonic velocity; hardness; temper and embrittlement
Surface properties	color, reflectivity, refraction index, emissivity
Chemical Composition and Analysis	
Elemental analysis	detection, identification, distribution and/or profile
Impurity concentrations	contamination, depletion, doping and diffusants
Metallurgical content	variation; alloy identification, verification and sorting
Physiochemical state	moisture content; degree of cure; ion concentrations and corrosion; reaction products
Stress and Dynamic Response	
Stress, strain, fatigue	heat treatment, annealing and cold work effects; stress and strain; fatigue damage and life (residual)
Mechanical damage	wear, spalling, erosion, friction effects
Chemical damage	corrosion, stress corrosion, phase transformation
Other damage	radiation damage and high frequency voltage breakdown
Dynamic performance	crack initiation, crack propagation, plastic deformation, creep, excessive motion, vibration, damping, timing of events, any anomalous behavior
Signature Analysis	
Electromagnetic field	potential; intensity; field distribution and pattern
Thermal field	isotherms, heat contours, temperatures, heat flow, temperature distribution, heat leaks, hot spots, contrast
Acoustic signature	noise, vibration characteristics, frequency amplitude, harmonic spectrum, harmonic analysis, sonic emissions, ultrasonic emissions
Radioactive signature	distribution and diffusion of isotopes and tracers
Signal or image analysis	image enhancement and quantization; pattern recognition; densitometry; signal classification, separation; and correlation; discontinuity identification, definition (size and shape) and distribution analysis; discontinuity mapping and display

infrared thermographic techniques, airborne ultrasonic testing and certain techniques of acoustic emission testing. Other less easily classified methods are material identification, vibration analysis and strain gaging.

No one nondestructive testing method is all revealing. In some cases, one method or technique may be adequate for testing a specific object or component. However, in most cases, it takes a series of test methods to do a complete nondestructive test of an object or component. For example, if surface cracks must be detected and eliminated and if the object or component is made of ferromagnetic material, then magnetic particle testing would be the appropriate choice. If the material is aluminum or titanium, then the choice would be liquid penetrant or electromagnetic testing. However, if internal discontinuities are to be detected, then ultrasonic testing or radiography would be the selection. The exact technique in each case would depend on the thickness and nature of the material and the types of discontinuities that must be detected.

Value of Nondestructive Testing

The contribution of nondestructive testing to profits has been acknowledged in the medical field and computer and aerospace industries. However, in industries such as heavy metals, although nondestructive testing may be reluctantly accepted, its contribution to profits may not be obvious to management. Nondestructive testing is sometimes thought of only as a cost item and can be curtailed by industry downsizing. When a company cuts costs, two vulnerable areas are quality and safety. When bidding contract work, companies add profit margin to all cost items, including nondestructive testing, so a profit should be made on the nondestructive testing. The attitude toward nondestructive testing is positive when management understands its value.

Nondestructive testing should be used as a control mechanism to ensure that manufacturing processes are within design performance requirements. When used properly, nondestructive testing saves money for the manufacturer. Rather than costing the manufacturer money, nondestructive testing should add profits to the manufacturing process.

Overview of Other Nondestructive Testing Methods

To optimize the use of nondestructive testing it is necessary first to understand the principles and applications of all the methods. This volume features electromagnetic testing (Fig. 3) — only one of the nondestructive test methods. The following section briefly describes several other methods and the applications associated with them.

FIGURE 3. Electromagnetic testing: (a) representative setup for eddy current test; (b) in-service detection of discontinuities.

(a)

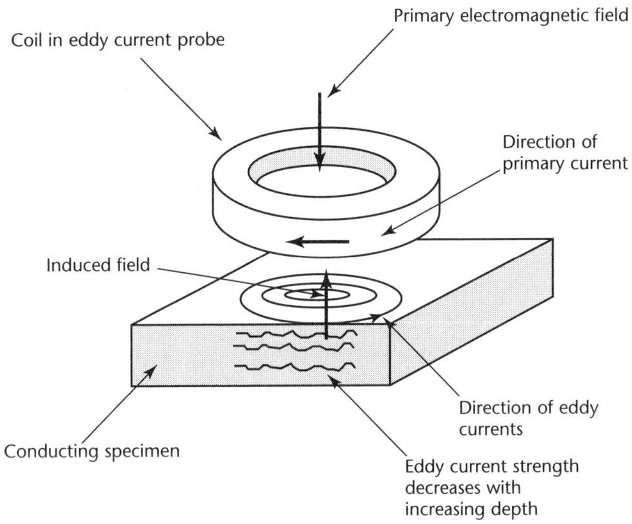

(b)

Visual Testing

Principles. Visual testing (Fig. 4) is the observation of a test object, either directly with the eyes or indirectly using optical instruments, by an inspector to evaluate the presence of surface anomalies and the object's conformance to specification. Visual testing should be the first nondestructive testing method applied to an item. The test procedure is to clean the surface, provide adequate illumination and observe. A prerequisite necessary for competent visual testing of an object is knowledge of the manufacturing processes by which it was made, of its service history and of its potential failure modes, as well as related industry experience.

Applications. Visual testing provides a means of detecting and examining a variety of surface discontinuities. It is the most widely used method for detecting and examining for surface discontinuities associated with various structural failure mechanisms. Even when other nondestructive tests are performed, visual tests often provide a useful supplement. When the eddy current testing of process tubing is performed, for example, visual testing is often performed to verify and more closely examine the surface condition. The following discontinuities may be detected by a simple visual test: surface discontinuities, cracks, misalignment, warping, corrosion, wear and physical damage.

Liquid Penetrant Testing

Principles. Liquid penetrant testing (Fig. 5) reveals discontinuities open to the surfaces of solid and nonporous materials. Indications of a wide variety of discontinuity sizes can be found regardless of the configuration of the work piece and regardless of discontinuity orientations. Liquid penetrants seep into various types of minute surface openings by capillary action. The cavities of interest can be very small, often invisible to the unaided eye. The ability of a given liquid to flow over a surface and enter surface cavities depends principally on the following: cleanliness of the surface, surface tension of the liquid, configuration of the cavity, contact angle of the liquid, ability of the liquid to wet the surface, cleanliness of the cavity and size of surface opening of the cavity.

Applications. The principal industrial uses of liquid penetrant testing include postfabrication testing, receiving testing, in-process testing and quality control, maintenance and overhaul in the transportation industries, in-plant and machinery maintenance and in testing of large components. The following are some of the typically detected discontinuities: surface discontinuities, seams, cracks, laps, porosity and leak paths.

Magnetic Particle Testing

Principles. Magnetic particle testing (Fig. 6) is a method of locating surface and slightly subsurface discontinuities in ferromagnetic materials. It depends on the fact that when the material or part under

FIGURE 4. Visual test using borescope to view interior of cylinder.

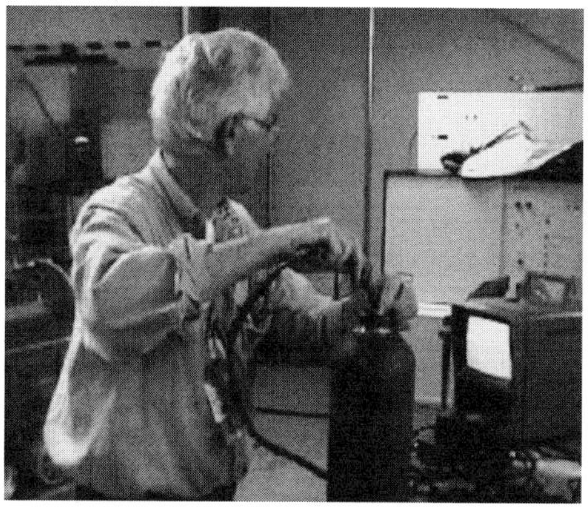

FIGURE 5. Liquid penetrant indication of cracking.

FIGURE 6. In magnetic particle testing, particles gather where lines of magnetic force leak from discontinuity.

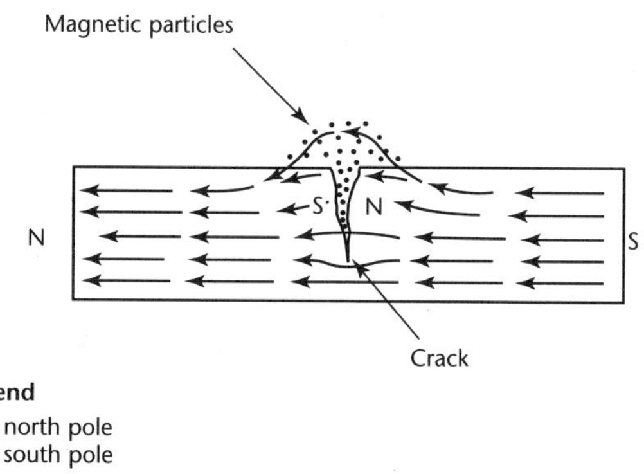

test is magnetized, discontinuities that lie in a direction generally transverse to the direction of the magnetic field will cause a leakage field to be formed at and above the surface of the part. The presence of this leakage field and therefore the presence of the discontinuity is detected with fine ferromagnetic particles applied over the surface, with some of the particles being gathered and held to form an outline of the discontinuity. This generally indicates its location, size, shape and extent. Magnetic particles are applied over a surface as dry particles or as wet particles in a liquid carrier such as water or oil.

Applications. The principal industrial uses of magnetic particle testing include final, receiving and in-process testing; for quality control; for maintenance and overhaul in the transportation industries; for plant and machinery maintenance; and for testing of large components. Some of the typically detected discontinuities are surface discontinuities, seams, cracks and laps.

Radiographic Testing

Principles. Radiographic testing (Fig. 7) is based on the differential absorption of penetrating radiation — either electromagnetic radiation of very short wavelength or particulate radiation (X-rays, gamma rays and neutron rays) — by the part or object being tested. Different portions of an object absorb different amounts of penetrating radiation because of differences in density and variations in thickness of the part or differences in absorption characteristics caused by variation in composition. These variations in the absorption of the penetrating radiation can be monitored by detecting the unabsorbed radiation that passes through the object. This monitoring may be in different forms. The traditional form is through radiation sensitive film. Radioscopic sensors provide digital images. X-ray computed tomography is a radiographic technique.

FIGURE 7. Representative setup for radiographic testing.

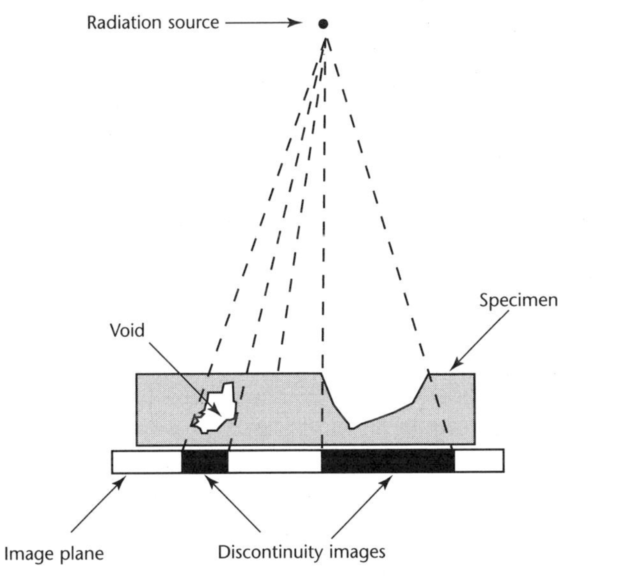

Applications. The principal industrial uses of radiographic testing involve testing of castings and weldments, particularly where there is a critical need to ensure freedom from internal discontinuities. For example radiography is often specified for thick wall castings and weldments for steam power equipment (boiler and turbine components and assemblies). Radiography can also be used on forgings and mechanical assemblies, although with mechanical assemblies radiography is usually limited to testing for conditions and proper placement of components. Typically detected discontinuities and conditions include inclusions, lack of fusion, cracks, corrosion, porosity, leak paths, missing or incomplete components and debris.

Ultrasonic Testing

Principles. Ultrasonic testing (Fig. 8) is a nondestructive method in which beams of sound waves at a frequency too high to hear are introduced into materials for the detection of surface and subsurface discontinuities in the material. These acoustic waves travel through the material with some attendant loss of energy (attenuation) and are reflected at interfaces. The echoes are then analyzed to define the presence and locations of discontinuities.

Applications. Ultrasonic testing of metals is widely used, principally for the detection of discontinuities. This method can be used to detect internal discontinuities in most engineering metals and alloys. Bonds produced by welding, brazing, soldering and adhesives can also be ultrasonically examined. Inline techniques have been developed for monitoring and classifying materials as acceptable, salvageable or scrap and for process control. Other applications include testing of piping and pressure vessels, nuclear systems, motor vehicles, machinery, structures, railroad rolling stock and bridges and thickness measurement.

Leak Testing

Principles. Leak testing is concerned with the flow of liquids or gases from pressurized or into evacuated components. The principles of leak testing involve the physics of fluid (liquids or gases) flowing through a barrier where a pressure differential or capillary action exists. Leaking fluids (liquid or gas) can propagate from inside a component or assembly to the outside, or vice versa, as a result of a pressure differential between the two regions or as a result of permeation through a barrier.

Leak testing encompasses procedures that fall into these basic functions: leak location, leakage measurement and leakage monitoring. There are several subsidiary methods of leak testing, entailing tracer gas detection (Fig. 9), pressure change measurement, observation of bubble formation and other means.

Applications. Like other forms of nondestructive testing leak testing has an impact on the safety and performance of a product. Reliable leak testing decreases costs by reducing the number of reworked products, warranty repairs and liability claims. The most common reasons for performing a leak test are to prevent the loss of costly materials or energy; to prevent contamination of the environment; to ensure component or system reliability; and to prevent the potential for an explosion or fire.

FIGURE 8. Classic setups for ultrasonic testing: (a) longitudinal wave technique; (b) shear wave technique.

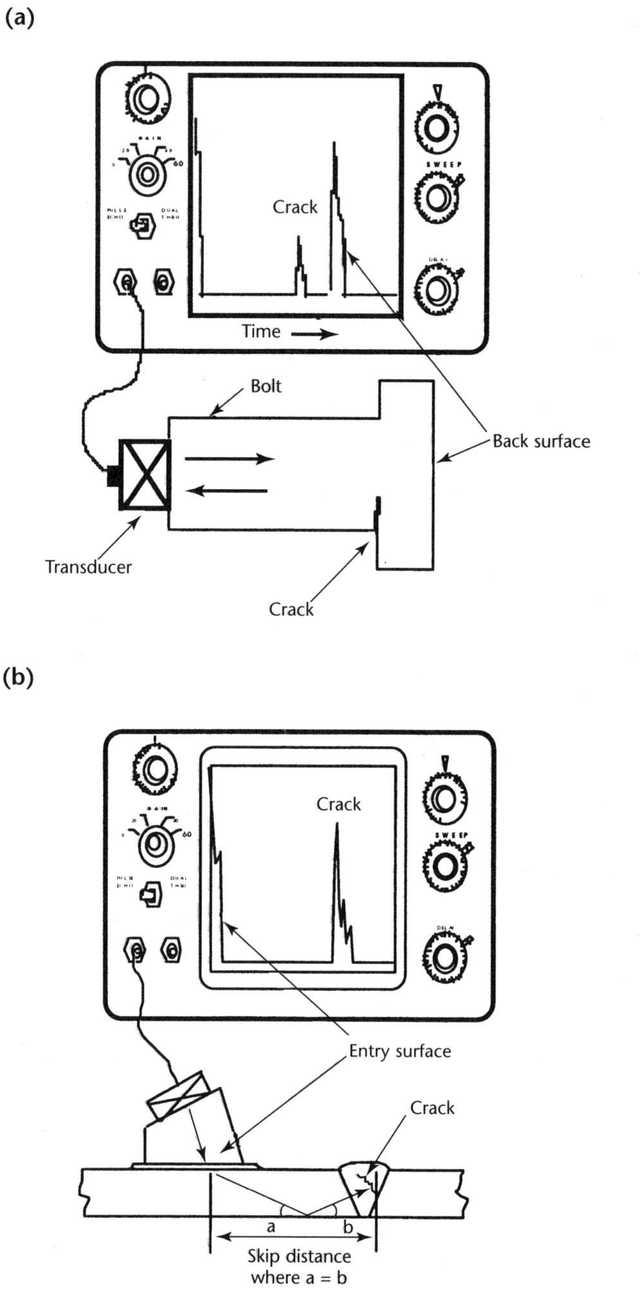

Acoustic Emission Testing

Principles. Acoustic emissions are mechanical waves produced by sudden movement in stressed materials. The classic sources of acoustic emission are discontinuity related deformation processes such as crack growth and plastic deformation. Sudden movement at the source produces a stress wave that radiates out into the structure and excites a sensitive piezoelectric sensor. As the stress in the material is raised, emissions are generated. The signals from one or more sensors are amplified and measured to produce data for display and interpretation.

FIGURE 9. Leakage measurement dynamic leak testing using vacuum pumping: (a) pressurized system mode for leak testing of smaller components; (b) pressurized envelope mode for leak testing of larger volume systems.

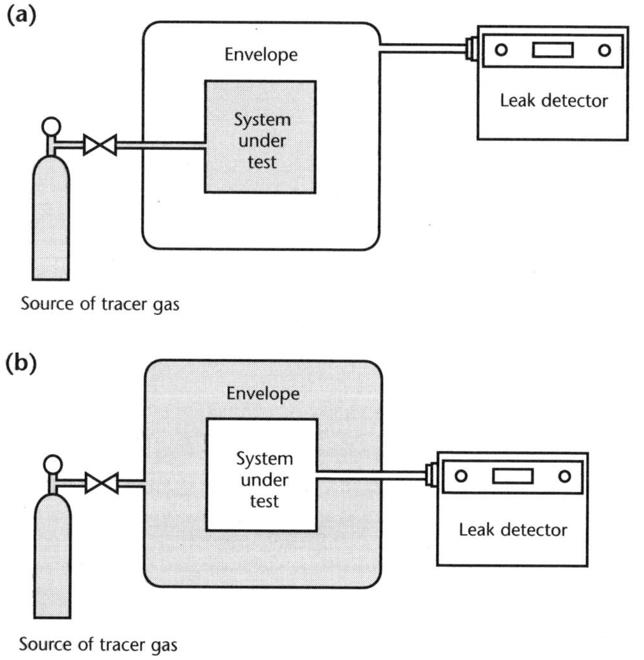

The source of acoustic emission energy is the elastic stress field in the material. Without stress there is no emission. Therefore, an acoustic emission test (Fig. 10) is usually carried out during a controlled loading of the structure. This can be a proof load before service; a controlled variation of load while the structure is in service; a fatigue, pressure or creep test; or a complex loading program. Often a structure is going to be loaded hydrostatically anyway during service and acoustic emission testing is used because it gives valuable additional information about the expected performance of the structure under load. Other times, acoustic emission testing is selected for reasons of economy or safety and a special loading procedure is arranged to meet the needs of the acoustic emission test.

Applications. Acoustic emission is a natural phenomenon occurring in a wide range of materials, structures and processes. The largest scale events observed with acoustic emission testing are seismic and the smallest are small dislocations in stressed metals.

The equipment used is highly sensitive to any kind of movement in its operating frequency (typically 20 to 1200 kHz). The equipment can detect not only crack growth and material deformation but also such processes as solidification, friction, impact, flow and phase transformations. Therefore acoustic emission testing is also used for in-process weld monitoring; detecting tool contact and tool wear during automatic machining; detecting wear and loss of lubrication in rotating equipment; detecting loose parts and loose particles; detecting and monitoring leaks, cavitation and flow; preservice proof testing; inservice weld monitoring; and leak testing.

Infrared and Thermal Testing

Principles. Conduction and convection are the primary mechanisms of heat transfer in an object or system. However, electromagnetic radiation is emitted from a heated body when electrons in that body change to a lower energy state. Thermal testing involves the measurement or mapping of surface temperatures when heat flows from, to or through a test object. Temperature differentials on a surface, or changes in surface temperature with time, are related to heat flow patterns and can be used to detect discontinuities or to determine the heat transfer characteristics of an object. For example, during the operation of an electrical breaker, a hot spot detected at an electrical termination may be caused by a loose or corroded connection (Fig. 11). The resistance to electrical flow

Figure 10. Acoustic emission testing setup in which eight sensors permit computer to calculate location of crack propagation.

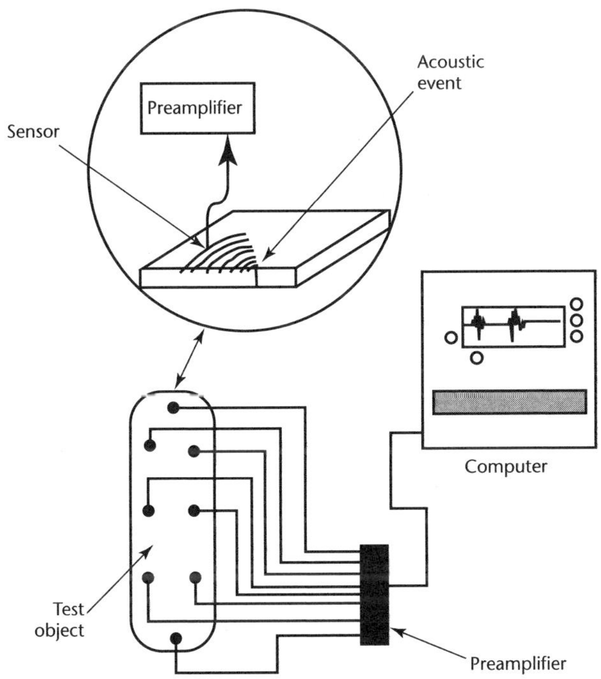

Figure 11. Infrared thermography of automatic transfer switches of emergency diesel generator. Hot spots appear bright in thermogram (inset).

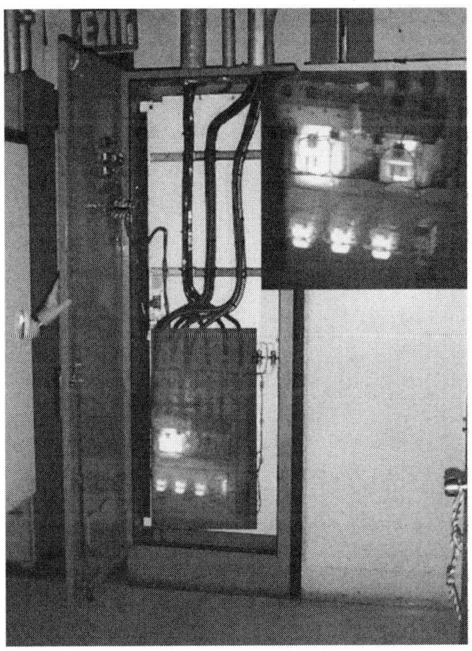

through the connection produces an increase in surface temperature of the connection.

Applications. There are two basic categories of infrared and thermal test applications: electrical and mechanical. The specific applications within these two categories are numerous. Electrical applications include transmission and distribution lines, transformers, disconnects, switches, fuses, relays, breakers, motor windings, capacitor banks, cable trays, bus taps and other components and subsystems. Mechanical applications include insulation (in boilers, furnaces, kilns, piping, ducts, vessels, refrigerated trucks and systems, tank cars and elsewhere), friction in rotating equipment (bearings, couplings, gears, gearboxes, conveyor belts, pumps, compressors and other components) and fluid flow (steam lines; heat exchangers; tank fluid levels; exothermic reactions; heating, ventilation and air conditioning systems; leaks above and below ground; cooling and heating; tube blockages; environmental assessment of thermal discharge; boiler or furnace air leakage; condenser or turbine system leakage; pumps; compressors; and other system applications).

Other Methods

There are many other methods of nondestructive testing, including optical methods such as holography, shearography and moiré imaging; material identification methods such as chemical spot testing, spark testing and spectroscopy; strain gaging; and acoustic methods such as vibration analysis and tapping.

PART 2. Management of Electromagnetic Testing

Selection of Electromagnetic Testing[4]

Electromagnetic tests are an important and widely used method within the broad field of nondestructive materials testing. The electromagnetic test method includes several subsidiary methods, sometimes called *submethods* or *techniques:* eddy current testing, remote field testing, flux leakage testing, alternating current field measurement testing and microwave testing. Of these several submethods, conventional eddy current testing is the most widely used. Magnetic particle testing is an electromagnetic test that industry administers as a separate method.

Applications of eddy current tests in industry are numerous and widespread. The total number of test measurements made annually by this nondestructive test method may exceed that of all other methods combined. Eddy current testing is used for the following:

1. noncontacting measurement of the thickness of metallic foils, sheets, plates, tube walls and machined parts from one side only;
2. measurement of the thickness of coatings over base materials where the coating and base material have significantly different electrical or magnetic properties;
3. identifying or separating materials by composition or structure;
4. detecting material discontinuities that lie in planes transverse to the eddy currents, such as cracks, seams, laps, score marks or plug cuts, drilled and other holes and laminations at cut edges of sheet or plate;
5. identifying and controlling heat treatment conditions and evaluation of fire damage to metallic structures;
6. determining depths of case hardening of steels and some ferrous alloys;
7. locating hidden metallic objects such as underground pipes, buried bombs or ore bodies, or metallic objects accidentally packaged in foodstuffs;
8. timing or locating the motions of hidden parts of mechanisms, counting metallic objects on conveyor lines or detecting metallic missiles in flight; and
9. precise dimensional measurement of symmetric, machined or ground and polished metallic parts, such as bearings and bearing races, small mechanism components and others.

Advantages of Electromagnetic Testing

Modern eddy current and electromagnetic test techniques offer low cost means for high speed, large scale testing of metallic materials such as those used in nuclear, aerospace, marine, high pressure, high temperature and high speed engineering systems where premature failures could represent economic disasters or the endangering of human life. The method's special suitability for testing of automobiles, engines, machine parts and consumer products has long been recognized.

Like other nondestructive methods, eddy current tests permit measurements of material properties and dimensions and detection of discontinuities. In general, electromagnetic tests provide nearly instantaneous measurements. The test speed and modern signal analysis permit such analysis to be performed in real time. Consequently, the method can be used in production lines to test swiftly moving bars, tubes, sheets, plates, welds and other symmetric parts. These parts either pass through test coils or are scanned by moving test probes. The automation of eddy current testing and test data evaluation permits mass testing of similar parts at high rates, with economies not attainable by other commonly used nondestructive tests. The results can be optimized for automation of test systems, for sorting of test parts, for control of manufacturing processes and for automatic documentation for process control and statistical quality control.

Small, portable forms of eddy current test instrumentation provide simple and rapid means for (1) manual quality tests by individual operators and (2) mechanized test systems to sort mixed lots of materials, to follow deterioration of materials and equipment in service and to verify process quality.

Limitations of Eddy Current Tests

Limitations of eddy current tests are a direct consequence of the specific nature of the test and of the response of electrically conductive test materials to the externally applied, time varying magnetic fields used to excite eddy current flow. In general, eddy current tests are applicable only to test materials with significant electrical conductivity, such as metals and alloys and composites with conducting layers or reinforcing fibers. They can be used, however, to measure thicknesses of nonconducting layers on the surface of conducting metallic materials by the liftoff effect in which the coating separates the test probe from the conducting material by the thickness of the nonconductive coating or sheet material.

Eddy current tests provide maximum test sensitivity for the surface and near surface layers of the test material adjacent to the source of excitation. In some cases it may be difficult or impossible to penetrate to the center of thick specimens because of skin effect and attenuation of the electromagnetic field at certain depths below the surface. Eddy current tests tend to be insensitive to laminar discontinuities, which lie parallel to the induced eddy currents. They do tend to respond, however, to discontinuities that lie transverse to the flow of eddy currents within test materials, where these discontinuities interrupt, lengthen or distort the current flow paths.

Management of Electromagnetic Testing Programs

Management of an electromagnetic testing program will require consideration of many items before a program can produce the desired results. Six basic questions must be answered before a true direction can be charted. They are as follows.

1. Are regulatory requirements in place that mandate program characteristics?
2. What is the magnitude of the program that will provide desired results?
3. What provisions must be made for personnel safety and for compliance with environmental regulations?
4. What is the performance date for a program to be fully implemented?
5. Is there a cost benefit of electromagnetic testing?
6. What are the available resources in personnel and money?

Once these questions are answered, then a recommendation can be made to determine the best path forward. Three primary paths are (1) service companies, (2) consultants and (3) in-house programs.

Although these are the primary paths, some programs may, routinely or as needed, require support personnel from a combination of two or more of these sources. Before a final decision is made, advantages and disadvantages of each path must be considered. Therefore the following details must be considered.

Service Companies

1. Who will identify the components within the facility to be examined?
2. Will the contract be for *time and materials* or have a specific *scope of work*?
3. If a *time and materials* contract is awarded, who will monitor the time and materials charged?
4. If a *scope of work* is required, who is technically qualified to develop and approve it?
5. What products or documents (test reports, trending, recommendations, root cause analysis and others) will be provided once the tests are completed?
6. Who will evaluate and accept the product (test reports, trending, recommendations, root cause analysis and others) within the service company?
7. Do the service company workers possess qualifications and certifications required by contract and by applicable regulations?
8. Do the service company workers require site specific training (confined space entry, electrical safety, hazardous materials and others) or clearance to enter and work in the facility?
9. Does the service company retain any liability for test results?

Consultants

1. Will the contract be for *time and materials* or have a specific *scope of work*?
2. If a *scope of work* is required, who is technically qualified to develop and approve it?
3. Who will identify the required qualifications of the consultant?
4. Is the purpose of the consultant to develop or update a program or is it to oversee and evaluate the performance of an existing program?
5 Will the consultant have oversight responsibility for tests performed?
6. What products (trending, recommendations, root cause analysis and others) are provided once the tests are completed?

7. Who will evaluate the consultant's performance (test reports, trending, recommendations, root cause analysis and other functions) within the sponsoring company?
8. Does the consultant possess qualifications and certifications required by contract and by applicable regulations?
9. Does the consultant require site specific training (confined space entry, electrical safety, hazardous materials and others) or clearance to enter and work in the facility?
10. Does the consultant retain any liability for test results?

In-House Programs

1. Who will determine the scope of the program, such as which techniques will be used (eddy current, flux leakage and others)?
2. What are the regulatory requirements (codes and standards) associated with program development and implementation?
3. Who will develop a *cost benefit* analysis for the program?
4. How much time and resources are available to establish the program?
5. What are the qualification requirements (education, training, experience and others) for personnel?
6. Do program personnel require additional training (safety, confined space entry or others) or qualifications?
7. Are subject matter experts required to provide technical guidance during personnel development?
8. Are procedures required to perform work in the facility?
9. If procedures are required, who will develop, review and approve them?
10. Who will determine the technical specifications for test equipment?

Test Procedures for Electromagnetic Testing

The conduct of facility operations (in-house or contracted) should be performed in accordance with specific instructions from an expert. This is typically accomplished using written instructions in the form of a technical procedure. In many cases, codes and specifications will require a technical procedure to perform required tests.

The procedure process can take many forms, including general instructions that address only major aspects of test techniques. Or a procedure may be written as a step-by-step process requiring a supervisor's initial or signature after each step. The following is a typical format for an industrial procedure.

1. The *purpose* identifies the intent of the procedure.
2. The *scope* establishes the latitude of items, tests and techniques covered and not covered by the procedure.
3. *References* are specific documents from which criteria are extracted or documents satisfied by implementation of the procedure.
4. *Definitions* are needed for terms and abbreviations that are not common knowledge to people who will read the procedure.
5. Statements about *personnel requirements* address specific requirements to perform tasks in accordance with the procedure — issues such as personnel qualification, certification, access clearance and others.
6. *Equipment* characteristics, calibration requirements and model numbers of qualified equipment must be specified.
7. The test *procedure* provides a sequential process to be used to conduct test activities.
8. *Acceptance criteria* establish component characteristics that will identify the items suitable for service.
9. *Reports* (records) provide the means to document specific test techniques, equipment used, personnel performing activity, date performed and test results.
10. *Attachments* may include (if required) items such as report forms, instrument calibration forms, qualified equipment matrix, schedules and others.

Once the procedure is completed, typically an expert in the subject matter evaluates it. If the procedure is judged to meet identified requirements, the expert will approve it for use. Some codes and standards also require the procedure to be qualified — that is, demonstrated to the satisfaction of a representative of a regulatory body or jurisdictional authority.

Test Specifications for Electromagnetic Testing[4]

An electromagnetic specification must anticipate a number of issues that arise during testing.

Means of Induction and Detection of Magnetic Fields

Electromagnetic nondestructive test methods use either static or time varying electromagnetic fields as a probing medium (1) to explore the properties of test materials, (2) to locate discontinuities or (3) to detect variations in geometry and dimensions of test materials. The magnitudes, time lags, phase angles and flow patterns of the resulting fields are sensed by using probes such as sensing coils or solid state magnetic field detectors (such as hall effect devices).

Eddy Current Test Frequencies

A single electromagnetic test system can be used for many different measurements through the selection of test frequencies. These frequencies are those of the excitation current applied to the coils of the electromagnetic test probes. Frequency is measured in hertz (Hz), where 1 Hz = 1 cycle per second. Most industrial electromagnetic tests are made in the frequency range between 5 Hz and 10 MHz.

Most types of electromagnetic test equipment provide either variable frequency oscillators or several fixed frequency steps. Thus, appropriate test frequencies can be readily selected by the user to meet special test requirements. Low excitation frequencies are used to penetrate deeper within a conducting test material. High test frequencies can be used for selective examination of near surface regions, testing of thin materials and for testing of materials that have low electrical conductivities.

Interpretation

Interpretation may be complex, especially before a procedure has been established. The interpreter must have a knowledge of the following: (1) the underlying physical process, (2) techniques and equipment used to obtain the data and displays, (3) details about the item being examined (configuration, material characteristics, fabrication process, potential discontinuities and intended service conditions) and (4) acceptance criteria.

Ensuring Reliability of Test Results

When a test is performed, there are four possible outcomes: (1) a rejectable discontinuity can be found when one is present; (2) a rejectable discontinuity can be missed even when one is present; (3) a rejectable discontinuity can be indicated when none is present and (4) no rejectable discontinuity is found when none is present. A reliable testing process and a qualified inspector should find all discontinuities of concern with no discontinuities missed (no errors as in case 2 above) and no false calls (case 3 above).

To achieve this goal, the probability of finding a rejectable discontinuity must be high and the inspector must be both proficient in the testing process and motivated to perform with maximum efficiency. A reckless inspector may accept parts that contain discontinuities, with the result of possible inservice part failure. A conservative inspector may reject parts that contain rejectable discontinuities but the inspector also may reject parts that do not contain rejectable discontinuities, with the result of unnecessary scrap and repair. Neither scenario is desirable.

Electromagnetic Testing Standards

Traditionally, the purpose of specifications and standards has been to define the requirements that goods or services must meet. As such, they are intended to be incorporated into contracts so that both the buyer and provider have a well defined description of what one will receive and the other will provide.

Standards have undergone a process of peer review in industry and can be invoked with the force of law by contract or by government regulation. In contrast, a specification represents an employer's instructions to employees and is specific to a contract or work place. Specifications may form the basis of standards through a review process. Standards and specifications exist in three basic areas: equipment, processes and personnel.

1. Standards for equipment include criteria that address probes, artificial discontinuities and test results. Reference standards are work pieces that contain artificial discontinuities for instrument calibration and test procedure verification.
2. ASTM International and other organizations publish standards for test techniques. Some other standards are for quality assurance procedures and are not specific to a test method or even to testing in general. Tables 3 and 4 list some standards used in electromagnetic testing. The United States Department of Defense has replaced most military specifications and standards with industry consensus specifications and standards. A source for nondestructive testing standards is the *Annual Book of ASTM Standards.*[5]
3. Qualification and certification of test personnel are discussed below with specific reference to recommendations of *ASNT Recommended Practice No. SNT-TC-1A.*[6]

Personnel Qualification and Certification

One of the most critical aspects of the test process is the qualification of test personnel. Nondestructive testing is sometimes referred to as a *special process.* The term simply means that it is very difficult to determine the adequacy of a test by merely observing the process or the documentation generated at its conclusion. The quality of the test is largely dependent on the skills and knowledge of the inspector.

The American Society for Nondestructive Testing (ASNT) has been a world leader in the qualification and certification of nondestructive testing personnel for many years. By 1999, the American Society for Nondestructive Testing had instituted three major programs for the qualification and certification of nondestructive testing personnel.

TABLE 3. Electromagnetic testing standards published by ASTM International.

Miscellaneous

E 543, *Standard Practice for Agencies Performing Nondestructive Testing*

E 1004, *Standard Practice for Determining Electrical Conductivity Using the Electromagnetic (Eddy-Current) Method*

E 1312, *Standard Practice for Electromagnetic (Eddy-Current) Examination of Ferromagnetic Cylindrical Bar Product above the Curie Temperature*

E 1316, *Standard Terminology for Nondestructive Examinations:* Section C, *Electromagnetic Testing*

E 1571, *Standard Practice for Electromagnetic Examination of Ferromagnetic Steel Wire Rope*

E 1606, *Standard Practice for Electromagnetic (Eddy-Current) Examination of Copper Redraw Rod for Electrical Purposes*

E 1629, *Standard Practice for Determining the Impedance of Absolute Eddy Current Probes*

F 673, *Standard Test Methods for Measuring Resistivity of Semiconductor Slices or Sheet Resistance of Semiconductor Films with a Noncontact Eddy-Current Gage*

Coating Thickness

B 244, *Standard Test Method for Measurement of Thickness of Anodic Coatings on Aluminum and of Other Nonconductive Coatings on Nonmagnetic Basis Metals with Eddy-Current Instruments*

B 499, *Standard Test Method for Measurement of Coating Thicknesses by the Magnetic Method: Nonmagnetic Coatings on Magnetic Basis Metals*

B 659, *Standard Guide for Measuring Thickness of Metallic and Inorganic Coatings*

E 376, *Standard Practice for Measuring Coating Thickness by Magnetic-Field or Eddy-Current (Electromagnetic) Test Methods*

Geophysical Measurements

D 4748, *Standard Test Method for Determining the Thickness of Bound Pavement Layers Using Short-Pulse Radar*

D 6429, *Standard Guide for Selecting Surface Geophysical Methods*

D 6432, *Standard Guide for Using the Surface Ground Penetrating Radar Method for Subsurface Investigation*

D 6565, *Standard Test Method for Determination of Water (Moisture) Content of Soil by the Time-Domain Reflectometry (TDR) Method*

D 6639, *Standard Guide for Using the Frequency Domain Electromagnetic Method for Subsurface Investigations*

D 6726, *Standard Guide for Conducting Borehole Geophysical Logging — Electromagnetic Induction*

Material Identification

E 566, *Standard Practice for Electromagnetic (Eddy-Current) Sorting of Ferrous Metals*

E 703, *Standard Practice for Electromagnetic (Eddy-Current) Sorting of Non-Ferrous Metals*

E 1476, *Standard Guide for Metals Identification, Grade Verification, and Sorting*

Tubular Products

A 135, *Standard Specification for Electric-Resistance-Welded Steel Pipe*

E 215, *Standard Practice for Standardizing Equipment for Electromagnetic Examination of Seamless Aluminum-Alloy Tube*

E 243, *Standard Practice for Electromagnetic (Eddy-Current) Examination of Copper-Alloy Tubes*

E 309, *Standard Practice for Eddy-Current Examination of Steel Tubular Products Using Magnetic Saturation*

E 426, *Standard Practice for Electromagnetic (Eddy-Current) Examination of Seamless and Welded Tubular Products, Austenitic Stainless Steel and Similar Alloys*

E 570, *Standard Practice for Flux Leakage Examination of Ferromagnetic Steel Tubular Products*

E 571, *Standard Practice for Electromagnetic (Eddy-Current) Examination of Nickel and Nickel-Alloy Tubular Products*

E 690, *Standard Practice for In Situ Electromagnetic (Eddy-Current) Examination of Nonmagnetic Heat Exchanger Tubes*

E 1033, *Standard Practice for Electromagnetic (Eddy-Current) Examination of Type F — Continuously Welded (CW) Ferromagnetic Pipe and Tubing above the Curie temperature*

E 2096, *Standard Practice for In Situ Examination of Ferromagnetic Heat-Exchanger Tubes Using Remote Field Testing*

1. *Recommended Practice No. SNT-TC-1A* provides guidelines for personnel qualification and certification in nondestructive testing. This recommended practice identifies the specific attributes that should be considered when qualifying nondestructive testing personnel. It requires the employer to develop and implement a *written practice* (procedure) that details the specific process and any limitation in the qualification and certification of nondestructive testing personnel.[6]
2. ANSI/ASNT CP-189, *Standard for Qualification and Certification of Nondestructive Testing Personnel* resembles *SNT-TC-1A* but also establishes specific attributes for the qualification and certification of nondestructive testing personnel. However, CP-189 is a consensus standard as defined by the American National Standards Institute (ANSI). It is recognized as the American standard for nondestructive testing. It is not considered a *recommended practice;* it is a national standard.[7]

TABLE 4. Some standards for electromagnetic testing.

Issuing Organization	Representative Standards and Related Documents
American National Standards Institute	ANSI B3.1, *Rolling Element Bearings — Aircraft Engine, Engine Gearbox, and Accessory Applications — Eddy Current Inspection*
American Petroleum Institute	API 510, *Pressure Vessel Inspection Code: Maintenance Inspection, Rating, Repair and Alteration*
	API 570, *Piping Inspection Code: Inspection, Repair, Alteration, and Rerating of In-Service Piping Systems*
	API 650, *Welded Steel Tanks for Oil Storage*
	API 1104, *Welding, Pipelines and Related Facilities*
American Society for Nondestructive Testing	*ASNT Recommended Practice No. SNT-TC-1A*
	ANSI/ASNT CP-189, *ASNT Standard for Qualification and Certification of Nondestructive Testing Personnel*
American Society of Mechanical Engineers	ANSI/ASME B31.1, *Power Piping*
	ANSI/ASME B31.3, *Process Piping*
	ASME *Boiler and Pressure Vessel Code:* Section V — *Power Boilers:* Article 8, *Eddy Current Examination of Tubular Products*
	ASME *Boiler and Pressure Vessel Code:* Section XI — *Inservice Inspection of Nuclear Vessels.* N-553-1, *Eddy Current Surface Examination Section XI, Division 1*
	ASME PTC 19-1, *Performance Test Codes, Supplement on Instruction and Apparatus*
American Welding Society	AWS D1.1, *Structural Welding Code — Steel*
ASTM International	See Table 3
Canadian General Standards Board	CAN/CGSB-48.9712, *Non-Destructive Testing — Qualification and Certification of Personnel*
	48.14-M86-CAN/CGSB, *Advanced Manual for: Eddy Current Test Method Amendment No. 1 May 1997 R(1997)*
Chinese National Standards	Z8005100, *General Rules for Eddy Current Testing*
Deutsche Institut für Normung	DIN 54141-3, *Non-Destructive Testing; Eddy Current Testing of Pipes and Tubes; Procedure*
European Association of Aerospace Industries	AECMA PREN 2002-20, *Aerospace Series Test Methods for Metallic Materials:* Part 20: *Eddy Current Testing of Circular Cross-Section Tubes,* Edition P 1
European Committee for Standardization	EN 12084, *Non-Destructive Testing — Eddy Current Testing — General Principles and Guidelines*
International Organization for Standardization	ISO 9712, *Nondestructive Testing – Qualification and Certification of Personnel*
Japanese Standards Association	JIS Z 2314, *Test Methods for Performance Characteristics of Eddy Current Testing Instruments*
Society of Automotive Engineers	SAE ARP 891A, *Determination of Aluminum Alloy Tempers through Electrical Conductivity Measurement (Eddy Current) (R 1988)*
	SAE ARP 1926, *Cure Monitor, Electrical Methods*
	SAE ARP 4402, *Eddy Current Inspection of Open Fastener Holes in Aluminum Aircraft Structure*
	SAE ARP 4462, *Barkhausen Noise Inspection for Detecting Grinding Burns*
	SAE AS 4787, *Eddy Current Inspection of Circular Holes in Nonferrous Metallic Aircraft Engine Hardware*
	SAE DFT K-89AW, *Eddy Current Inspection of Circular Holes in Nonferrous Metallic*
	SAE J 425, *Electromagnetic Testing by Eddy Current Methods, Information Report; March 1991*
United States Department of Defense	MIL-P-85585, *Probes, Eddy Current, Unshielded, Single Coil, Absolute*
	MIL-STD-1537B, *Electrical Conductivity Test for Verification of Heat Treatment of Aluminum Alloys, Eddy Current Method*
	MIL-STD-2032, *Eddy Current Inspection of Heat Exchanger Tubing on Ships of the United States Navy*
	MIL-STD-2195, *Inspection Procedure for Detection and Measurement of Dealloying Corrosion on Aluminum Bronze and Nickel-Aluminum Bronze Components*

3. The *ASNT Central Certification Program (ACCP),* unlike *SNT-TC-1A* and CP-189, is a third party certification process that identifies qualification and certification attributes for Level II and Level III nondestructive testing personnel. The American Society for Nondestructive Testing certifies that the individual has the skills and knowledge for many nondestructive test method applications. It does not remove the responsibility for the final determination of personnel qualifications from the employer. The employer evaluates an individual's skills and knowledge for application of company procedures using designated techniques and equipment identified for specific tests.[8]

Selections from *Recommended Practice No. SNT-TC-1A*

To give a general idea of the contents of these documents, the following items are specified in the 2001 edition of *Recommended Practice No. SNT-TC-1A.* (The following text has been excerpted and adapted. The original text is arranged in outline format and includes recommendations that are not specific to electromagnetic testing.)

Scope. This recommended practice has been prepared to establish guidelines for the qualification and certification of nondestructive test personnel whose specific jobs require appropriate knowledge of the technical principles underlying the nondestructive tests they perform, witness, monitor or evaluate. This document provides guidelines for the establishment of a qualification and certification program.

Written Practice. The employer shall establish a written practice for the control and administration of nondestructive testing personnel training, examination and certification. The employer's written practice should describe the responsibility of each level of certification for determining the acceptability of materials or components in accordance with the applicable codes, standards, specifications and procedures.

Education, Training, Experience Requirements for Initial Qualification. Candidates for certification in nondestructive testing should have sufficient education, training and experience to ensure qualification in those nondestructive testing methods for which they are being considered for certification. Table 6.3.1A [Table 5 in this volume, for electromagnetic testing] lists the recommended training and experience factors to be considered by the employer in establishing written practices for initial qualification of Level I and II individuals.

Training Programs. Personnel being considered for initial certification should complete sufficient organized training to become thoroughly familiar with the principles and practices of the specified nondestructive test method related to the level of certification desired and applicable to the processes to be used and the products to be tested.

Examinations. For Level I and II personnel, a composite grade should be determined by simple averaging of the results of the general, specific and practical examinations described below. Examinations administered for qualification should result in a passing composite grade of at least 80 percent, with no individual examination having a passing grade less than 70 percent. The examination for near vision acuity should ensure natural or corrected near distance acuity in at least one eye such that the applicant can read a minimum of jaeger size 2 or equivalent type and size letter at a distance of not less than 305 mm (12 in.) on a standard jaeger test chart. This test should be administered annually.

Practical Examination for NDT Level I and II. The candidate should demonstrate ability to operate the necessary nondestructive test equipment and to record and analyze the resultant information to the degree required. At least one selected specimen should be tested and the results of the nondestructive test analyzed by the candidate.

Certification. Certification of all levels of nondestructive test personnel is the responsibility of the employer. Certification of nondestructive test personnel shall be based on demonstration of satisfactory qualification [in accordance with sections on education, training, experience and examinations] as described in the employer's written practice. Personnel certification records shall be maintained on file by the employer.

Table 5. Recommended training and experience for electromagnetic testing personnel according to *Recommended Practice No. SNT-TC-1A.*[6]

	Level I	Level II
High school graduate[a]	40 h	40 h
Two years of college[b]	24 h	40 h
Work experience[c]	210 h	630 h

a. Or equivalent.

b. Completion with a passing grade of at least two years of engineering or science study in a university, college or technical school.

c. Minimum work experience per level. Note: for Level II certification, the experience shall consist of time as Level I or equivalent. If a person is being qualified directly to Level II with no time at Level I, the required experience shall consist of the sum of the times required for Level I and Level II and the required training shall consist of the sum of the hours required for Level I and Level II.

Recertification. All levels of nondestructive testing personnel shall be recertified periodically in accordance with the following: evidence of continuing satisfactory performance; and reexamination in those portions of the examination deemed necessary by the employer's NDT Level III. Recommended maximum recertification intervals are three years for Level I and II and five years for Level III.

The minimum number of questions that should be administered in the written examination for eddy current test personnel is as follows: 40 questions in the general examination and 20 questions in the specific examination. The number of questions is the same for Level I and Level II personnel.

These recommendations from *Recommended Practice No. SNT-TC-1A* are cited only to provide a general idea of the specific items that must be considered in the development of an in-house nondestructive testing program. Because the items are paraphrased, those developing a personnel qualification program should consult the complete text of *SNT-TC-1A* and other applicable procedures and practices. If an outside agency is contracted for electromagnetic test services, then the contractor must have a qualification and certification program to satisfy most codes and standards.

Central Certification

Another standard that may be a source for compliance is contained in the requirements of the International Organization for Standardization (ISO). The work of preparing international standards is normally carried out through technical committees of the International Organization for Standardization, a worldwide federation of national standards bodies. Each ISO member body interested in a subject for which a technical committee has been established has the right to be represented on that committee. International organizations, governmental and nongovernmental, in liaison with the International Organization for Standardization, also take part in the work.

Technical Committee ISO/TC 135, Non-Destructive Testing Subcommittee SC 7, Personnel Qualification, prepared international standard ISO 9712, *Nondestructive Testing – Qualification and Certification of Personnel.*[9] In its statement of scope, ISO 9712 states that it "establishes a system for the qualification and certification, by a certification body, of personnel to perform industrial nondestructive testing (NDT) using any of the following methods: (a) eddy current testing; (b) liquid penetrant testing; (c) magnetic particle testing; (d) radiographic testing; (e) ultrasonic testing" and that the "system described in this International Standard may also apply to visual testing (VT), leak testing (LT), neutron radiography (NR), acoustic emission (AE) and other nondestructive test methods where independent certification programs exist." The applicability of ISO 9712 to electromagnetic testing therefore depends on activity of the national certifying body.

Safety in Electromagnetic Testing

To manage an electromagnetic testing program, as with any testing program, the first obligation is to ensure safe working conditions. The following are components of a safety program that may be required or at least deserve serious consideration.

1. Before work is to begin, identify the safety and operational rules and codes applicable to the areas, equipment and systems to be tested.
2. Provide proper safety equipment (protective barriers, hard hat, safety harnesses, steel toed shoes, hearing protection and others).
3. Before the test, perform a thorough visual survey to determine all the hazards and to identify necessary safeguards to protect test personnel and equipment.
4. Notify operative personnel to identify the location and specific material, equipment or systems to be tested. In addition, it must be determined whether signs or locks restrict access by personnel. Be aware of equipment that may be operated remotely or may be started by time delay.
5. Be aware of any potentially explosive atmosphere. Determine whether it is safe to take test equipment into the area.
6. Do not enter any *roped off* or *no entry* areas without permission and approval.
7. When working on or around moving or electrical equipment, the inspector should remove pens, watches, rings or objects in pockets that may touch (or fall into) energized equipment.
8. Know interplant communication and evacuation systems.
9. Never let unqualified personnel operate equipment independently from qualified supervision.

10. Keep a safe distance between the inspector and any energized equipment. In the United States, these distances can be found in documents from the Occupational Safety and Health Administration, the National Fire Prevention Association (*National Electric Code*),[10] the Institute of Electrical and Electronics Engineers (*National Electrical Safety Code*)[11] and other organizations.
11. Be aware of the personnel responsibilities before entering a *confined space*. All such areas must be tested satisfactorily for gas and oxygen levels before entry and periodically thereafter. If odors are noticed or if unusual sensations such as ear aches, dizziness or difficulty in breathing are experienced, leave the area immediately.

Most facilities in the United States are required by law to follow the requirements in the applicable standard. Two Occupational Safety and Health Standards in the United States that should be reviewed are *Occupational Safety and Health Standards* for general industry[12] and the *Occupational Safety and Health Standards for the Construction Industry.*[13]

Personnel safety is always the first consideration for every job.

PART 3. Units of Measure for Electromagnetic Testing

Origin and Use of SI System

In 1960, the General Conference on Weights and Measures established the International System of Units. *Le Systéme International d'Unités* (SI) was designed so that a single set of measurement units could be used by all branches of science, engineering and the general public. Without SI, this *Nondestructive Testing Handbook* volume could have contained a confusing mix of obsolete centimeter-gram-second (CGS) units, imperial units and the units preferred by certain localities or scientific specialties.

SI is the modern version of the metric system and ends the division between metric units used by scientists and metric units used by engineers and the public. Scientists have given up their units based on centimeter and gram and engineers made a fundamental change in abandoning the kilogram-force in favor of the newton. Electrical engineers have retained their ampere, volt and ohm but changed all units related to magnetism.

Table 6 lists the seven SI base units. Table 7 lists derived units with special names. Table 8 gives examples of conversions to SI units. In SI, the unit of time is the second (s) but hour (h) is recognized for use with SI.

For more information, the reader is referred to the information available through national standards organizations and specialized information compiled by technical organizations.[14-17]

Multipliers

In science and engineering, very large or very small numbers with units are expressed by using the SI multipliers, prefixes of 10^3 intervals (Table 9). The multiplier becomes a property of the SI unit. For example a millimeter (mm) is 0.001 meter (m). The volume unit cubic centimeter (cm^3) is $(0.01\ m)^3$ or $10^{-6}\ m^3$. Unit submultiples such as the centimeter, decimeter, dekameter (or decameter) and hectometer are often avoided in scientific and technical uses of SI because of their variance from the 10^3 interval. However, dm^3 and cm^3 are commonly used. Note that 1 cm^3 is not equal to 0.01 m^3. Nevertheless, in equations, submultiples

TABLE 6. SI base units.

Quantity	Unit	Symbol
Length	meter	m
Mass	kilogram	kg
Time	second	s
Electric current	ampere	A
Temperature	kelvin	K
Amount of substance	mole	mol
Luminous intensity	candela	cd

TABLE 7. SI derived units with special names.[a]

Quantity	Units	Symbol	Relation to Other SI Units[b]
Capacitance	farad	F	$C{\cdot}V^{-1}$
Catalytic activity	katal	kat	$s^{-1}{\cdot}mol$
Conductance	siemens	S	$A{\cdot}V^{-1}$
Energy	joule	J	$N{\cdot}m$
Frequency (periodic)	hertz	Hz	$1{\cdot}s^{-1}$
Force	newton	N	$kg{\cdot}m{\cdot}s^{-2}$
Inductance	henry	H	$Wb{\cdot}A^{-1}$
Illuminance	lux	lx	$lm{\cdot}m^{-2}$
Luminous flux	lumen	lm	$cd{\cdot}sr$
Electric charge	coulomb	C	$A{\cdot}s$
Electric potential[c]	volt	V	$W{\cdot}A^{-1}$
Electric resistance	ohm	Ω	$V{\cdot}A^{-1}$
Magnetic flux	weber	Wb	$V{\cdot}s$
Magnetic flux density	tesla	T	$Wb{\cdot}m^{-2}$
Plane angle	radian	rad	1
Power	watt	W	$J{\cdot}s^{-1}$
Pressure (stress)	pascal	Pa	$N{\cdot}m^{-2}$
Radiation absorbed dose	gray	Gy	$J{\cdot}kg^{-1}$
Radiation dose equivalent	sievert	Sv	$J{\cdot}kg^{-1}$
Radioactivity	becquerel	Bq	$1{\cdot}s^{-1}$
Solid angle	steradian	sr	1
Temperature, celsius	degree celsius	°C	K
Time[a]	hour	h	3600 s
Volume[a]	liter	L	dm^3

a. Hour and liter are not SI units but are accepted for use with the SI.
b. Number one (1) expresses dimensionless relationship.
c. Electromotive force.

Table 8. Examples of conversions to SI units.

Quantity	Measurement in Non-SI Unit	Multiply by	To Get Measurement in SI Unit
Angle	minute (min)	$2.908\,882 \times 10^{-4}$	radian (rad)
	degree (deg)	$1.745\,329 \times 10^{-2}$	radian (rad)
Area	square inch (in.^{2})	645	square millimeter (mm^2)
Distance	angstrom (Å)	0.1	nanometer (nm)
	inch (in.)	25.4	millimeter (mm)
Energy	British thermal unit (BTU)	1.055	kilojoule (kJ)
	calorie (cal), thermochemical	4.184	joule (J)
Power	British thermal unit per hour (BTU·h^{-1})	0.293	watt (W)
Specific heat	British thermal unit per pound degree fahrenheit (BTU·lb$_m^{-1}$·°F^{-1})	4.19	kilojoule per kilogram per kelvin (kJ·kg^{-1}·K^{-1})
Force (torque, couple)	foot-pound (ft-lb$_f$)	1.36	joule (J)
Pressure	pound force per square inch (lb$_f$·in.$^{-2}$)	6.89	kilopascal (kPa)
Frequency (cycle)	cycle per minute	60^{-1}	hertz (Hz)
Illuminance	footcandle (ftc)	10.76	lux (lx)
	phot (ph)	10 000	lux (lx)
Luminance	candela per square foot (cd·ft^{-2})	10.76	candela per square meter (cd·m^{-2})
	candela per square inch (cd·in.$^{-2}$)	1 550	candela per square meter (cd·m^{-2})
	footlambert (ftl)	3.426	candela per square meter (cd·m^{-2})
	lambert	3 183 (= 10 000/π)	candela per square meter (cd·m^{-2})
	nit (nt)	1	candela per square meter (cd·m^{-2})
	stilb (sb)	10 000	candela per square meter (cd·m^{-2})
Radioactivity	curie (Ci)	37	gigabecquerel (GBq)
Ionizing radiation exposure	roentgen (R)	0.258	millicoulomb per kilogram (mC·kg^{-1})
Mass	pound (lb$_m$)	0.454	kilogram (kg)
Temperature (difference)	degree fahrenheit (°F)	0.556	kelvin (K) or degree celsius (°C)
Temperature (scale)	degree fahrenheit (°F)	(°F – 32)/1.8	degree celsius (°C)
Temperature (scale)	degree fahrenheit (°F)	((°F – 32)/1.8) + 273.15	kelvin (K)

such as centimeter (cm) or decimeter (dm) are often avoided because they disturb the convenient 10^3 or 10^{-3} intervals that make equations easy to manipulate.

In SI, the distinction between upper and lower case letters is meaningful and should be observed. For example, the meanings of the prefix *m* (milli) and the prefix *M* (mega) differ by nine orders of magnitude.

Table 9. SI prefixes and multipliers.

Prefix	Symbol	Multiplier
yotta	Y	10^{24}
zetta	Z	10^{21}
exa	E	10^{18}
peta	P	10^{15}
tera	T	10^{12}
giga	G	10^{9}
mega	M	10^{6}
kilo	k	10^{3}
hecto[a]	h	10^{2}
deka (or deca)[a]	da	10
deci[a]	d	10^{-1}
centi[a]	c	10^{-2}
milli	m	10^{-3}
micro	µ	10^{-6}
nano	n	10^{-9}
pico	p	10^{-12}
femto	f	10^{-15}
atto	a	10^{-18}
zepto	z	10^{-21}
yocto	y	10^{-24}

a. Avoid these prefixes (except in dm^3 and cm^3) for science and engineering.

SI Units for Electromagnetic Testing

CGS Units

Table 10 gives examples of centimeter-gram-second (CGS) units. These units are not accepted for use with the SI. Furthermore, no other units of the various CGS systems of units, which includes the CGS electrostatic, CGS electromagnetic and CGS gaussian systems, are accepted for use with SI except such units as the centimeter (cm), gram (g) and second (s) that are also defined in SI.

The oersted, gauss and maxwell are part of the electromagnetic

Table 10. Units from the centimeter-gram-second (CGS) system of units and not accepted for use with SI. Factor to convert each CGS unit to SI unit is given.

Physical Quantity	CGS Unit	Multiply by	SI Unit	SI Symbol
Basic CGS Units				
Magnetic field intensity	oersted (Oe)	$10^3{\cdot}(4\pi)^{-1}$	ampere per meter	$A{\cdot}m^{-1}$
Magnetic flux	maxwell (Mx)	10^{-8}	weber	Wb
Magnetic flux density	gauss (G)	10^{-4}	tesla	T
Magnetic potential difference	gilbert (Gb)	$10{\cdot}(4\pi)^{-1}$	ampere	A
Electromagnetic CGS Units				
Capacitance	abfarad	10^9	farad	F
Charge	abcoulomb	10	coulomb	C
Conductance	abmho	10^9	siemens	S
Current	abampere	10	ampere	A
Inductance	abhenry	10^{-9}	henry	H
Magnetic field intensity	abampere per centimeter	10^3	ampere per meter	$A{\cdot}m^{-1}$
Potential	abvolt	10^{-8}	volt	V
Resistance	abohm	10^{-9}	ohm	Ω
Electrostatic CGS Units				
Capacitance	statfarad	$1.112\,650 \times 10^{-12}$	farad	F
Charge	statcoulomb	3.3356×10^{-10}	coulomb	C
Conductance	statmho	$1.112\,65 \times 10^{-12}$	siemens	S
Current	statampere	$3.335\,641 \times 10^{-11}$	ampere	A
Inductance	stathenry	$8.987\,552 \times 10^{11}$	henry	H
Potential	statvolt	$2.997\,925 \times 10^{2}$	volt	V
Resistance	statohm	$8.987\,55 \times 10^{11}$	ohm	Ω

three-dimensional CGS system. When only mechanical and electric quantities are considered, these three units cannot strictly speaking be compared each to the corresponding unit of SI, which has four dimensions.

Ampere per Meter Replaces Oersted. One ampere per meter ($A{\cdot}m^{-1}$) equals about one eightieth of an oersted (Oe). The relationship is $1\ Oe = 1000{\cdot}(4\pi)^{-1}\ A{\cdot}m^{-1} = 79.57747\ A{\cdot}m^{-1}$.

Tesla Replaces Gauss. One tesla (T) equals ten thousand gauss (G). $1\ G = 10^{-4}\ T = 0.1\ mT$.

Weber Replaces Maxwell. One weber (Wb) equals 10^8 maxwell (Mx). $1\ Mx = 10^{-8}\ Wb = 0.01\ \mu Wb = 10\ nWb$.

Conductivity and Resistivity

In the twentieth century, the conductivity of a given metal was conventionally expressed as a percentage of pure copper's conductivity with reference to the International Annealed Copper Standard (IACS).[18] In SI, conductivity is expressed in siemens per meter ($S{\cdot}m^{-1}$). The conductivity of pure copper (100 percent IACS) is $58\ MS{\cdot}m^{-1}$.

Resistivity is the inverse of conductivity and is expressed in ohm meter. Table 11 gives the formulas for conversion to and from units for conductivity and resistivity.

Table 11. Conversion of Units for Conductivity σ and Resistivity ρ.

From Unit	Operation	To Get Unit
Conductivity Unit to Conductivity Unit[a]		
$S{\cdot}m^{-1}$	$(S{\cdot}m^{-1}) \times 10^{-6}$	$MS{\cdot}m^{-1}$
$S{\cdot}m^{-1}$	$(S{\cdot}m^{-1}) \times (1.724 \times 10^{-6})$	%IACS
$MS{\cdot}m^{-1}$	$(MS{\cdot}m^{-1}) \times 10^{6}$	$S{\cdot}m^{-1}$
$MS{\cdot}m^{-1}$	$(MS{\cdot}m^{-1}) \times 1.724$	%IACS
%IACS	$\%IACS \times (5.800 \times 10^{5})$	$S{\cdot}m^{-1}$
%IACS	$\%IACS \times 0.580$	$MS{\cdot}m^{-1}$
Conductivity Unit[a] to Resistivity Unit		
$S{\cdot}m^{-1}$	$1 \div (S{\cdot}m^{-1})$	$\Omega{\cdot}m$
$S{\cdot}m^{-1}$	$(1 \times 10^{8}) \div (S{\cdot}m^{-1})$	$\mu\Omega{\cdot}cm$
$MS{\cdot}m^{-1}$	$(1 \times 10^{-6}) \div (MS{\cdot}m^{-1})$	$\Omega{\cdot}m$
$MS{\cdot}m^{-1}$	$(1 \times 10^{2}) \div (MS{\cdot}m^{-1})$	$\mu\Omega{\cdot}cm$
%IACS	$(1.724 \times 10^{-6}) \div \%IACS$	$\Omega{\cdot}m$
%IACS	$172.4 \div \%IACS$	$\mu\Omega{\cdot}cm$
Resistivity Unit to Conductivity Unit[a]		
$\Omega{\cdot}m$	$1 \div (\Omega{\cdot}m)$	$S{\cdot}m^{-1}$
$\Omega{\cdot}m$	$(1 \times 10^{-6}) \div (\Omega{\cdot}m)$	$MS{\cdot}m^{-1}$
$\Omega{\cdot}m$	$(1.724 \times 10^{-6}) \div (\Omega{\cdot}m)$	%IACS
$\mu\Omega{\cdot}cm$	$(1 \times 10^{8}) \div (\mu\Omega{\cdot}cm)$	$S{\cdot}m^{-1}$
$\mu\Omega{\cdot}cm$	$(1 \times 10^{2}) \div (\mu\Omega{\cdot}cm)$	$MS{\cdot}m^{-1}$
$\mu\Omega{\cdot}cm$	$172.4 \div (\mu\Omega{\cdot}cm)$	%IACS[a]
Resistivity Unit to Resistivity Unit		
$\Omega{\cdot}m$	$(\Omega{\cdot}m) \times 10^{8}$	$\mu\Omega{\cdot}cm$
$\mu\Omega{\cdot}m$	$(\mu\Omega{\cdot}m) \times 10^{-8}$	$\Omega{\cdot}m$

a. %IACS: percentage of International Annealed Copper Standard.[18]

References

1. *Nondestructive Testing Handbook,* second edition: Vol. 10, *Nondestructive Testing Overview.* Columbus, OH: American Society for Nondestructive Testing (1996).
2. Wenk, S.A. and R.C. McMaster. *Choosing NDT: Applications, Costs and Benefits of Nondestructive Testing in Your Quality Assurance Program.* Columbus, OH: American Society for Nondestructive Testing (1987).
3. *Nondestructive Testing Methods.* TO33B-1-1 (NAVAIR 01-1A-16) TM43-0103. Washington, DC: Department of Defense (June 1984).
4. *Nondestructive Testing Handbook,* second edition: Vol. 4, *Electromagnetic Testing.* Columbus, OH: American Society for Nondestructive Testing (1986).
5. *Annual Book of ASTM Standards:* Section 3, *Metals Test Methods and Analytical Procedures.* Vol. 03.03, *Nondestructive Testing.* West Conshohocken, PA: ASTM International (2001).
6. *Recommended Practice No. SNT-TC-1A.* Columbus, OH: American Society for Nondestructive Testing (2001).
7. ANSI/ASNT CP-189, *Standard for Qualification and Certification of Nondestructive Testing Personnel.* Columbus, OH: American Society for Nondestructive Testing (2001).
8. *ASNT Central Certification Program (ACCP),* Revision 3 (November 1997). Columbus, OH: American Society for Nondestructive Testing (1998).
9. ISO 9712, *Nondestructive Testing — Qualification and Certification of Personnel.* Geneva, Switzerland: International Organization for Standardization.
10. NFPA 70, *National Electric Code,* 2002 edition. Quincy, MA: National Fire Prevention Association (2001).
11. *National Electrical Safety Code,* 2002 edition. New York, NY: Institute of Electrical and Electronics Engineers (2001).
12. 29 CFR 1910, *Occupational Safety and Health Standards* [*Code of Federal Regulations:* Title 29, *Labor.*] Washington, DC: United States Department of Labor, Occupational Safety and Health Administration; United States Government Printing Office.
13. 29 CFR 1926, *Occupational Safety and Health Standards for the Construction Industry* [*Code of Federal Regulations:* Title 29, *Labor*]. Washington, DC: United States Department of Labor, Occupational Safety and Health Administration; United States Government Printing Office.
14. IEEE/ASTM SI 10-1997, *Standard for Use of the International System of Units (SI): The Modernized Metric System.* West Conshohocken, PA: ASTM International (1996).
15. Taylor, B.N. *Guide for the Use of the International System of Units (SI).* National Institute of Standards and Technology Special Publication 811, 1995 edition. Washington, DC: United States Government Printing Office (1995).
16. Taylor, B.N., ed. *Interpretation of the SI for the United States and Federal Government and Metric Conversion Policy.* NIST Special Publication 814, 1998 Edition. Washington, DC: United States Government Printing Office (1998).
17. Taylor, B.N., ed. *The International System of Units (SI),* 2001 edition. NIST Special Publication 330. Washington DC: United States Government Printing Office (2001).
18. IEC 60028, *International Standard of Resistance for Copper.* Geneva, Switzerland: International Electrotechnical Commission (2001).

2

CHAPTER

History of Electromagnetic Testing

PART 1. Electromagnetic Theory

This chapter previously appeared as an article by Robert McMaster and in the second edition of the *Nondestructive Testing Handbook*.[1,2] This chapter covers electromagnetic induction developments before 1960 and closes with a brief discussion of microwave testing before 1980.

Early Observations of Magnetic Attraction

It is probable that no other form of nondestructive testing has a history of scientific creativity and practical development that compares with electromagnetic induction and eddy current testing.

Electromagnetic testing has the most ancient name of all nondestructive testing methods. Thales of Miletus (sixth century B.C.) first recorded that rubbing amber induced a state in which the amber would attract other light objects. The Greek word for amber is *electron*. Thales also mentioned the remarkable powers of the lodestone (iron oxide), also known as *magnetite* from the place where it was found: Magnesia in Thessaly.[3]

Democritus (about 400 B.C.) provided concepts of an atomic structure of matter. His six principles were listed by John Tyndall and quoted by Robert A. Millikan. The fifth principle states that the "varieties of all things depend upon the varieties of their atoms, in number, size and aggregation."[3] Many electromagnetic tests are intended to identify the specific atoms in materials under test and the discontinuities that occur in structures when needed atoms are missing or separated from their neighbors.

By A.D. 1200, the use of the magnetic compass was reported in China. At about the same time, Alexander Neckam, an Englishman, also reported the use of the compass in navigation.[4] In the year 1600, William Gilbert, physician to England's Queen Elizabeth I, wrote in his book *De Magnete* a comprehensive description of his 18 years of experiments and his theory of magnetism.[5]

Benjamin Franklin

Robert A. Millikan, in his *Early Views of Electricity*, states that there were "no electrical theories of any kind" before Benjamin Franklin, who around 1747 observed that "electrical matter consists of particles extremely subtle, since it can permeate common matter, even the densest, with such freedom and ease as not to receive any appreciable resistance."[3]

Franklin "recognized two kinds of electrification and introduced the terms positive and negative to distinguish them. He arbitrarily called any body *positively electrified* if it was repelled by a glass rod that had been rubbed with silk and negatively electrified if it was repelled by sealing wax that had been rubbed with cat's fur. These are today our definitions of positive and negative electrical charges."[3]

19th Century Development of Induced Currents

Electromagnetic induction was not observed and explained before the 19th century. James Clerk Maxwell (see Fig. 1) in his remarkable two-volume work

FIGURE 1. James Clerk Maxwell.

A Treatise on Electricity and Magnetism[6] summarized the first 50 years of this development.

Örsted Discovers Electric Current's Magnetic Field

Maxwell explains that "conjectures of various kinds had been made as to the relation between magnetism and electricity, but the laws of these phenomena, and the form of these relations, remained entirely unknown till Hans Christian Örsted [Fig. 2a], at a private lecture to a few advanced students at Copenhagen, observed that a wire connecting the ends of a voltaic battery affected a magnet in its vicinity." Örsted's published account in 1820 observes that "the current itself ... was the cause of the action, and that the 'electric conflict acts in a revolving manner,' that is, that a magnet placed near a wire transmitting an electric current tends to set itself perpendicular to the wire, and with the same end always pointing forwards as the magnet is moved round the wire. ... The space in which these forces act may therefore be considered as a magnetic field" (Fig. 2b). Örsted's discovery meant that the "lines of magnetic force are everywhere at right angles to planes drawn through the wire, and are therefore circles each in a plane perpendicular to the wire" passing through the plane's center.[6]

FIGURE 2. Hans Christian Örsted: (a) with students Örsted discovers electric current's magnetic effect on compass when circuit is completed; (b) Örsted's observation that compass needle near electric current moves to position perpendicular to direction of current.

(a)

(b)

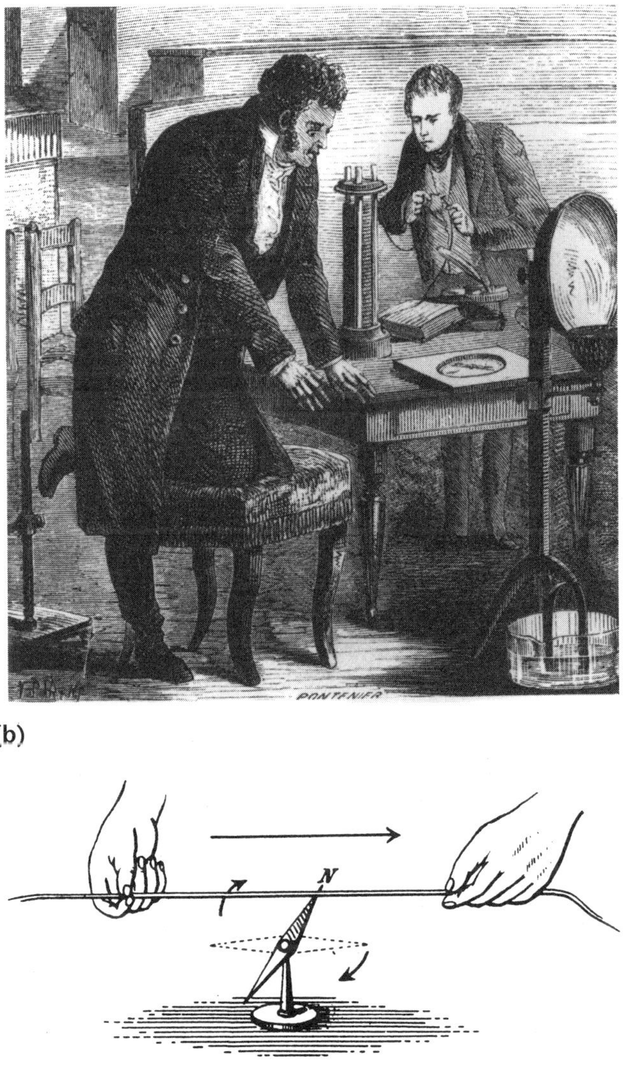

Ampere's Experiments

In his first experiment, André Marie Ampere (Fig. 3a) showed that two equivalent currents close together and flowing in opposite directions neutralize each other (Fig. 3b). Maxwell explains that an insulated wire may be looped back on itself so as to have no effect on the astatic balance: "This principle is of great importance in the construction of electric apparatus, as it affords the means of conveying the current to and from any galvanometer or other instrument in such a way that no electromagnetic effect is produced by the current on its passage to and from the instrument."[6] Techniques like this are commonly used to connect instruments to sensing coils or semiconductor detectors for detecting eddy current magnetic field test signals. At higher frequencies, shielding by concentric conductors, usually grounded at one end, aids in avoidance of interfering signals from ambient electromagnetic fields or moving ferromagnetic machine parts or test objects.

Ampere's second experiment concerned crooked paths of currents. Maxwell explains that "one of the wires is bent and crooked with a number of small sinuosities, but so that in every part of its course it remains very near the straight wire. ... A current flowing through the crooked wire and back again through the straight wire, is found to be without influence on the astatic balance. This proves that the effect of the current running through any crooked part of the wire is equivalent to the same current running in the straight line joining its extremities, provided the crooked line is in no part of its course far from the straight one. Hence any small element of a circuit is equivalent to two or more component elements, the relation between the component elements and the resultant element being the same as that between component and resultant displacements or velocities."[6] This basic principle has been generally ignored with respect to its significance in detection of very small discontinuities that locally distort eddy current flow paths. A circular test coil, for example, produces a mirror image circular flow path of eddy currents in the adjacent test material. Small diversions and excursions of eddy currents from a truly circular path will have little effect on relatively large pickup coils but small semiconductor detectors can have far greater sensitivity to small distortions of the eddy current magnetic field.

Ampere's third experiment demonstrated that external currents or magnets had no tendency to move a straight current carrying conductor in the direction of its length. The fourth experiment showed that the force acting between two adjacent current carrying loops varies as the square of the distance between the two loops.[6]

FIGURE 3. André Marie Ampere: (a) portrait; (b) Maxwell's sketch illustrating Ampere's basic test arrangement with astatic balance coil arrangement.

(a)

(b)

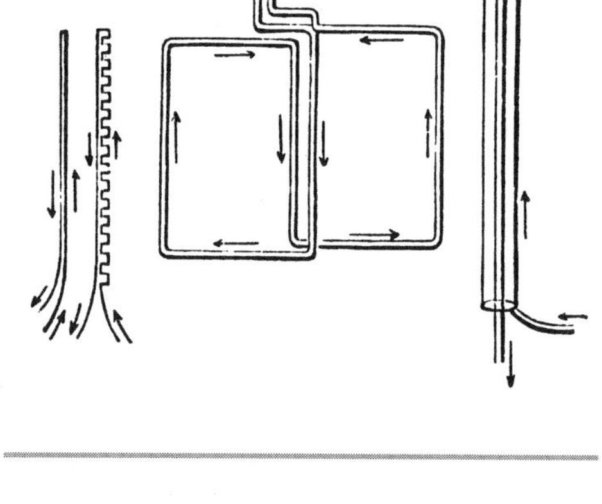

Faraday's Law of Electromagnetic Induction

In 1831, both Joseph Henry in the United States and Michael Faraday (Fig. 4) in England discovered electromagnetic induction. Maxwell notes that "Faraday, who had been for some time endeavoring to produce electric currents by magnetic or electric action, discovered the conditions of magnetoelectric induction. The method that Faraday used in his researches consisted of a constant appeal to experiment as a means of testing the truth of his ideas, and a constant cultivation of ideas under the direct influence of experiment." Because Faraday discusses "his unsuccessful as well as his successful experiments, and his crude ideas as well as his developed ones," the reader may feel "sympathy even more than admiration, and is tempted to believe that, if he had the opportunity, he too would be a discoverer. Every student ... should study Faraday for the cultivation of a scientific spirit, by means of the action and reaction which will take place between the newly discovered facts as introduced to him by Faraday and the nascent ideas of his own mind."[6]

FIGURE 4. Michael Faraday, evidently holding bar magnet.

The method of Faraday seems to be intimately related to the method of partial differential equations and integrations throughout all space. "He never considers bodies as existing with nothing between them but their distance, and acting on one another according to some function of that distance. He conceives all space as a field of force, the lines of force being in general curved, and those due to any body extending from it on all sides, their directions being modified by the presence of other bodies. He even speaks of the lines of force belonging to a body as in some sense part of itself, so that in its action on distant bodies it cannot be said to act where it is not. This, however, is not a dominant idea with Faraday. He would probably have said that the field of space is full of lines of force, whose arrangement depends on that of the bodies in the field, and that the mechanical and electrical action on each body is determined by the lines which abut on it."[6]

Maxwell describes the first form of Faraday's law: "The primary circuit is connected with a voltaic battery by which the primary current may be produced, maintained, stopped, or reversed. The secondary circuit includes a galvanometer," which is placed so that the primary current does not affect it. Parts of the primary and secondary currents are straight wires placed parallel and near to each other.

When a current is suddenly sent through the primary circuit, Maxwell explains, "the galvanometer of the secondary circuit indicates a current in the secondary straight wire in the *opposite* direction. This is called the induced current. If the primary current is maintained constant, the induced current soon disappears, and the primary current appears to produce no effect on the secondary circuit. If now the primary current is stopped, a secondary current is observed, which is in the *same* direction as the primary current. Every variation of the primary current produces electromotive force in the secondary circuit. When the primary current increases, the electromotive force is in the opposite direction to the current. When it diminishes, the electromotive force is in the same direction as the current. ... These effects of induction are increased by bringing the two wires nearer together. They are also increased by forming them into two circular or spiral coils placed close together, and still more by placing an iron rod or a bundle of iron wires inside the coils."[6]

This experiment demonstrates the fundamental principles for using magnetizing coils in eddy current testing. The need for a time varying primary current is clearly indicated. The advantage of close coupling or spacing between the magnetizing coil and test metal surface is also shown. This translates into control of liftoff of probe coils and preference for high coil fill factors with encircling coil eddy current tests. The need for pulsating or alternating primary current is also now evident. Finally, the advantages of using ferrite or iron cores in eddy current probe coils are suggested. Eddy current test systems at the beginning of the twenty-first century make full use of each of these principles, enunciated clearly by Faraday in 1831.

Induction by Moving the Circuits

Faraday found that, by moving the primary circuit toward the secondary circuit, current could be induced in the secondary current in a direction *opposite* to the primary current. Similarly, Faraday found that moving the secondary circuit toward the primary induces a current opposite to the primary current. Also, moving the secondary circuit *away* from the primary induces a current in the *same* direction as the primary current. Maxwell explains that "the direction of the secondary current is such that the mechanical action between the two conductors is opposite to the direction of motion, being a repulsion when the wires are approaching, and an attraction when they are receding."[6] This electromotive force was observed by Faraday but was given more systematic treatment by H.F. Lenz (see below).

Three principles are implied by the concept of induction by motion of the primary circuit. The first is that polarized and directional secondary currents can be induced by moving a straight line primary current over a conducting test surface. Secondly, alternating current could be induced in a conducting secondary circuit or test material when a constant current primary coil is moved cyclically up and down or side to side over a secondary coil or conducting test surface. A third concept implied by the technique of induction from a moving primary circuit would be that of using direct current magnetic field detectors to measure the magnitude of secondary current or eddy currents in a conducting material, under or lagging behind the moving primary coil.

A practical example of testing by moving the secondary circuit would be the rapid movement of conductive test material, such as sheet metal in a rolling mill, past a stationary direct current test coil, inducing a flow of current in material both approaching and leaving the area of this local magnetization. Detectors of the eddy current field in either location can respond to local discontinuities or material property variations that influence the amplitude and distribution of the eddy currents.

Faraday also found that current could be induced by the relative motion of a magnet and the secondary circuit. Maxwell explains that "if we substitute for the primary circuit a magnetic shell, whose edge coincides with the circuit, whose strength is numerically equal to that of the current in the circuit, and whose austral face corresponds to the positive face of the circuit, then the phenomena produced by the relative motion of this shell and the secondary circuit are the same as those observed in the case of the primary circuit."[6] The coil of the preceding examples can be replaced by a permanent magnet when relative motion exists between the magnet and test material in eddy current tests, provided that adequate secondary current magnitude and speed of motion can be attained.

Faraday's Legacy

Maxwell finally states the "true law of magneto-electric induction" in the following terms: "The total electromotive force acting around a circuit at any instant is measured by the rate of decrease of the number of lines of magnetic force which pass through it. ... The time integral of the total electromotive force acting round any circuit, together with the number of lines of magnetic force which pass through the circuit, is a constant quantity." This quantity "may even be called the fundamental quantity in the theory of electromagnetism." Faraday recognized "in the secondary circuit, when in the electromagnetic field, a 'peculiar electrical condition of matter,' to which he gave the name of the Electrotonic State."[6] This quantity appears to be similar to the concept of flux linkage, measured by the product of the number of winding turns and the total magnetic flux enclosed in the winding.

Michael Faraday's two-volume work *Experimental Researches in Electricity* influenced numerous investigators and inventors in Europe and the United States from the 1830s to the end of the nineteenth century. This led many others to experiment with electromagnetic effects and to develop many basic inventions such as Morse's telegraph, Bell's telephone and Edison's many improvements on telegraphic, telephonic, fire alarm and stock ticker communication systems. In 1831, Faraday also showed before the Royal Society a homopolar generator (a disk rotating between the poles of a large horseshoe magnet) for converting mechanical energy into electric energy.

Faraday's influence on inventors with little or no scientific training was very great, for Faraday's accounts of his experiments did not use any complicated mathematical formulas. To inventors like Thomas Edison, Faraday appeared to be the master experimenter, whose laboratory notes communicated the highest intellectual excitement — and hope as well. Faraday's explanations were simple, steeped in the spirit of truthfulness and humility before Nature. For Faraday, the natural laws were revealed through experiment. To American inventors, Faraday, poor and self-educated, indifferent to money or titles, exemplified the ethics of a true man of science, whom others could emulate. Thus, during the period from 1831 to about 1875, the inventions made on the basis of Faraday's research were often developed by trial and error, empirically and step by step.

Lenz, Neumann and Helmholtz

In 1834, Heinrich Friedrich Lenz described electromotive force — the relationship, in Maxwell's words, "between the phenomena of mechanical action of electric currents, as defined by Ampere's formula, and the induction of electric currents by the relative motion of conductors."[6] More generally, Lenz's law states that the electromagnetic field will act so as to oppose or resist any effort made to change its intensity or configuration. Where mechanical motion causes the change, mechanical force developed within the system will oppose the change. If mechanical motion is absent, electromotive forces will be induced that tend to maintain the status quo, namely to maintain the total flux linkages in the system.

On the basis of Lenz's law, Franz E. Neumann in 1845 formulated his mathematical theory of induction, in effect, as Maxwell says, "completing for the induction of currents the mathematical treatment which Ampere had applied to their mechanical action."

In Maxwell's opinion, "a step of still greater scientific importance" was Hermann L. von Helmholtz's derivation in 1847 of the laws of induction from the laws of conservation of energy. He and William Thompson, working independently, showed "that the induction of electric currents discovered by Faraday could be mathematically deduced from the electromagnetic actions discovered by Örsted and Ampere by the application of the principle of the Conservation of Energy."[6]

Maxwell's Equations

James Clerk Maxwell conceived and published the comprehensive group of relations for the electromagnetic field known as *Maxwell's equations*,[6] which mathematically represent almost the entire present knowledge of this subject. Maxwell's remarkable achievement of integrating the available knowledge concerning electromagnetic circuits and fields provides the basis for analysis of all basic eddy current and electromagnetic induction problems — and for most of modern electromagnetic theory.

These simple equations in both integral and differential form were derived by the methods of Lagrange, using relationships from the calculus of variations. Solutions for alternating fields are also available for many configurations of the fields. It is of interest that simpler techniques using an operational map have been devised for presenting these types of equations and their derivations in simple form for use by second-year engineering students. The equations are available in nearly all basic textbooks on the electromagnetic field. Kelvin devised the solutions of Bessel's equation for the cases of probe coils and provided the *kelvin functions* from which simple cases can be readily calculated by hand or with computer.

Since 1900, physicists and researchers in electricity and magnetism have occupied themselves with applications of Maxwell's theory. However, no one has conceived any significant new law to be added to Maxwell's principles, with the possible exception of Einstein's theory of relativity, which extends the theory of the three-dimensional electromagnetic field to a four-dimensional framework, including time.

PART 2. Industrial Development of Electromagnetic Tests

Electromagnetic testing since 1880 has evolved from relatively simple devices for metal characterization to microwave testing and sophisticated systems with quadrature phase analysis. Much of this development has been chronicled in patents and summarized elsewhere.[7]

In 1868, a British engineering publication reported that discontinuities were being located in gun barrels by using a magnetic compass to register the flux.[8]

Hughes' Eddy Current Test

Alexander Graham Bell invented and patented the first practical telephone in 1876. In 1879, David E. Hughes used the telephone as an indicating device to detect imbalance between two pairs of induction coils with which he performed eddy current comparison tests of coins. In his demonstration and report to the Physical Society, he stated that "if we introduce into one pair of induction coils any conducting body ... there are set up in these bodies electric currents which react both upon the primary and secondary coils, producing extra currents whose forces will be proportional to the mass and its specific conducting power." Two identical shillings "will be completely balanced" if one is put in the center of each of the coils. "If, however, these shillings are in the slightest degree worn, or have a different temperature, we at once perceive this difference." Hughes called his apparatus "a rapid and perfect coin-detector" that could "test any alloy, giving instantly its electrical value."[9]

Hughes then measured the electrical conductivity of different metals, using copper as a reference value of 100, producing a series of values like the conductivity values expressed in the late twentieth century as percentages of the International Annealed Copper Standard (IACS). He also made tests on ferromagnetic materials that differentiated between soft iron and hard steel. Finally, Hughes provided curves showing the effects of varying percentages of alloying elements (silver-gold, copper-tin and tin-lead). He thus established the basic principles of testing and of interpretation of modern eddy current and magnetic induction tests.

Bell's Electromagnetic Induction Metal Detector

After consultation with Hughes, Bell used an induction sensing device to look for a bullet in United States President James A. Garfield after he was shot in 1881 (Fig. 5). The attempt was a failure, probably because signals from bedsprings interfered with the test.[10]

Early Tests for Eddy Current and Hysteresis Losses in Electrical Steel Sheets

Active practical interest in electromagnetic techniques for sorting metals and detecting discontinuities did not result in many useful test devices before the twentieth century. However, numerous developments (including alternating current electric power systems, transformers and other induction machines) provided a base of practical design and a need to investigate the losses occurring in magnetic core materials used in these devices. From 1890 to 1925, much effort was devoted to reducing eddy current and magnetic hysteresis losses in laminated steel sheets, particularly (1) by

FIGURE 5. Alexander Graham Bell and assistant look for bullet in President James A. Garfield.

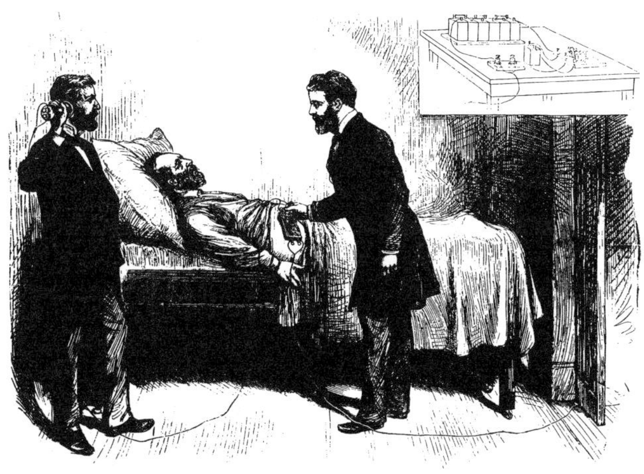

addition of silicon and other alloying elements that lowered their electrical conductivity and (2) by using purer iron alloys with, in some cases, directional rolling to attain maximum permeability and minimum hysteresis losses.

To a first approximation, in cores formed of thin magnetic laminations, it was shown that eddy current losses tended to increase in proportion with the square of the frequency and that hysteresis losses tended to increase in accordance with the 1.6th power of the frequency of alternation of the magnetic field intensity. Numerous laboratories, including those of electrical equipment manufacturers (such as Westinghouse and the General Electric Company) and electrical steel sheet manufacturers (such as Allegheny Ludlum and Armco Steel Company) established measurement laboratories to monitor properties of production steel sheets and ensure specified electromagnetic loss factors for electrical steel sheets. The Epstein test and many others were used for these material tests.

Many improvements resulted, including (1) thinner sheets, (2) oriented steel sheets and (3) insulating coatings between sheets to limit eddy current flow paths. Also discovered during these magnetic core improvements were the undesirable effects of mechanical clamping stresses and stresses resulting from punching and shearing of laminations, which tended to increase core losses under alternating current excitation. Hydrogen annealing and other techniques, such as those developed by Trigvie Yensen of Westinghouse Research Laboratories, led to magnetic sheet alloys with superior properties. Control of other alloying elements, additions of up to 50 percent nickel and orientation of grain structures and magnetic domains were used to develop special steels with rectangular hysteresis loops. These steels are used in magnetic switching of electrical currents, saturable reactors, magnetic amplifiers and many novel electromagnetic devices.

These developments illustrated the variations in electrical conductivity, magnetic permeability, grain orientation, anisotropy, mechanical stresses, alloy contents and impurity contents that, in turn, influenced the electromagnetic response of ferromagnetic materials and changed the apparent inductance and resistive losses measured by their magnetizing coils. Direct current bias to adjust the apparent inductance in saturable reactors and transductors for power control purposes also illustrated a means for reducing magnetic permeability and incremental inductance or inductive reactance. It was also observed that many magnetic core materials introduced odd harmonics into the magnetizing currents or voltages across inductances of their magnetizing coils (or into unloaded secondary windings on the cores). The high sensitivity of the harmonic signals to material conditions and mechanical stressing were known and purposely avoided where possible.

FIGURE 6. Charles Proteus Steinmetz: (a) portrait; (b) Steinmetz and Thomas Alva Edison examine porcelain insulators broken by current from Steinmetz's high voltage generator.

(a)

(b)

These various effects, well known to electrical designers at the turn of the century, have since become possible techniques for control or readout of eddy current nondestructive test signals. In general, however, the highly permeable electrical steel sheets now commercially available are not ideal for eddy current tests because their eddy current losses are so low. For their evaluation, electromagnetic induction tests responsive primarily to hysteresis effects, including higher harmonic effects, may prove more useful.

Steinmetz's Vectors

By the 1890s, Charles Proteus Steinmetz had come to the United States and begun work for General Electric (Fig. 6). Steinmetz had a colorful, outspoken character and was a hard working industrial researcher in the modern sense.

In the late 19th century, the sinusoidal oscillations of alternating current electric power system voltages and currents introduced new complexities in analysis of circuit performance, as compared with analyses for Thomas A. Edison's earlier direct current electric power systems. Detailed solutions of Maxwell's equations required vector calculus.

Steinmetz developed much simplified techniques of analysis using rotating line segments that he called *vectors* (later called *sinors*) to represent sinusoidal quantities. As such line segments rotated about one end (at the origin of coordinates), their vertical projections mapped out the ordinates of the sinusoidal waves when these vertical projections were plotted as functions of time. Together with the technique of representing impedances on a complex plane, these phasor quantities reduced the solutions for steady state alternating current to simple algebra and trigonometry rather than to integral calculus.

Later, after World War II, these techniques of signal analysis on the complex plane were to become widely used in analysis of eddy current tests

FIGURE 7. Charles W. Burrows' eddy current test arrangement with comparator circuit from patent filed in 1923 (United States Patent 1 686 679, *Apparatus for Testing Magnetizable Objects*).[12]

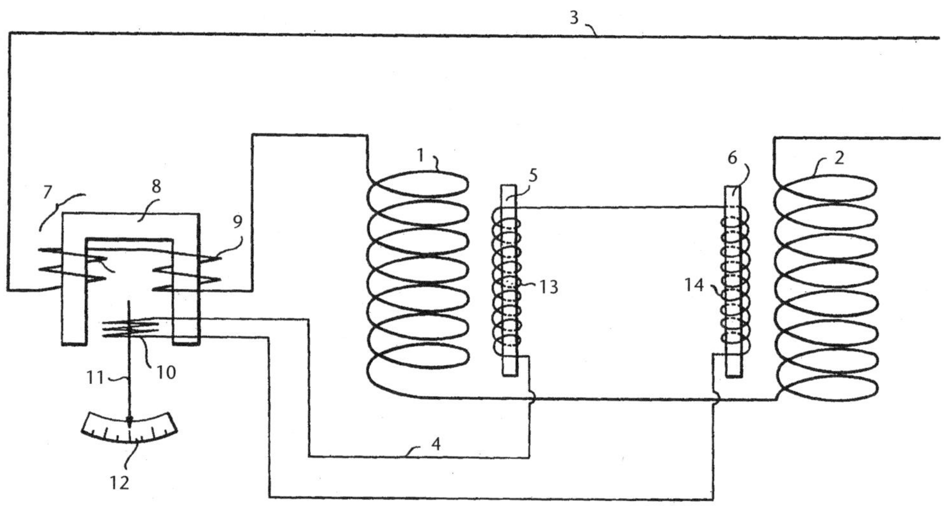

Legend

1. Primary coil to energize reference circuit.
2. Primary coil to energize test specimen circuit.
3. Primary circuit.
4. Closed secondary circuit.
5. Standard reference specimen.
6. Test specimen.
7. Dynamometer.
8. Core.
9. Stationary coil.
10. Moving coil.
11. Pointer.
12. Scale or dial.
13. Test coil for reference circuit.
14. Test coil for test specimen circuit.

following their clear enunciation by Friedrich Förster.[11] The corresponding impedance diagrams on the complex plane and oscilloscope displays provide direct means for interpreting many of the changes observed in eddy current nondestructive testing. These two-dimensional impedance diagrams, with the inductive reactance as the ordinate and resistive (energy loss) values as the abscissa, permit mapping of numerous different test conditions and prediction of various effects observed in single-frequency alternating current electromagnetic tests by technicians and test operators who do not know calculus.

Early Industrial Development of Electromagnetic Induction Comparators

Numerous electromagnetic induction or eddy current comparators were patented in the United States in the period from 1925 until the end of World War II in 1945. Innumerable examples of comparator tests were reported in the literature and in patents. Many provided simple comparator coils into which round bars or other test objects were placed, producing simple changes in amplitudes of test signals or unbalancing simple bridge circuits (see Fig. 7).[7,8,12]

In nearly all cases, particularly where ferromagnetic test materials were involved, no quantitative analyses of test object dimensions, properties or discontinuities were possible with such instruments. Often, difficulties were encountered in reproducing test results: some test circuits were adjusted or balanced to optimize signal differences between a known sound test object and a known anomalous test object for each group of objects to be tested. Little or no correlation could then be obtained between various types of specimens, each type having been compared to an arbitrarily selected specimen of the same specific type.

Many simple comparators operated at 60 Hz from 110 V alternating current circuits, using conventional instruments such as volt meters, ampere meters, watt meters and occasionally phase meters. Such meters typically absorbed energy from the test circuits and had accuracies and reproducibilities often of only one or two percent of full scale readings. In other cases, wheatstone bridge circuits were used to balance comparison test arrangements and to provide greater sensitivity to signal differences. For the most part, many of these early comparator systems were short lived and received little acceptance in industry. By comparison, a few such developments, sponsored by major industries or persistent creative inventors who sought support and formed their own companies, survived and have been used in modernized form by industry in the United States.

American Developments of Electromagnetic Tests for Steel Products

Development continued for electromagnetic induction tests for round bars, tubes, billets and other products of the steel industry in the United States. Advances at Magnetic Analysis Corporation and at Republic Steel and Tubes were based on the continuing efforts of a few dedicated individuals who passed their skills and enthusiasm along to successors in the same organizations. Charles W. Burrows (Fig. 7), Carl Kinsley and Theodore W. Zuschlag were among the pioneers at Magnetic Analysis Corporation.[7,8,12,13] Horace G. Knerr, Cecil Farrow and Alfred R. Sharples received basic patents for Republic Steel and Tubes (Fig. 8). Their developments were extended and continued in the Electromechanical Research Center of Republic Steel (later to become LTV Steel), Cleveland, Ohio, by Cecil Farrow, William Archibald Black, William C. Harmon and I.G. Orellana.[7,8,14] Automated electromagnetic testing was applied to the large scale, automated, production line

FIGURE 8. Cecil Farrow (right) watches operation of system for electromagnetic testing of longitudinally welded steel tubes.

testing of tubes, bars and billets. Other companies had early inventors and developers of electromagnetic tests but in many cases management did not support their developments long enough to achieve practical applications.

Within the General Electric Company, an early sequence of inventive development was pioneered by James A. Sams, Charles D. Moriarty and H.D. Roop.[7,8] Ross Gunn of the United States Naval Research Laboratory designed a new form of probe coil magnetizing system with two small diameter pickup coils displaced symmetrically along a diameter of the magnetizing coil. This was an early example of using one coil size for magnetization and much differently sized pickup coils in nonconcentric positions.

Developments in Electromagnetic Induction Tests

Rapid technological developments in many fields before and during World War II (1939 to 1945) contributed both to the demand for nondestructive testing and to the development of advanced test techniques. Radar and sonar systems made acceptable the viewing of test data on the screens of cathode ray tubes and oscilloscopes. Developments in electronic instrumentation, and in magnetic sensors used both for degaussing ships and for actuating magnetic mines, brought a resurgence of activity. After the war, developments such as Floyd Firestone's Supersonic Reflectoscope for ultrasonic testing and Förster's advanced eddy current and magnetometer systems became available as industrial nondestructive test systems. These systems offered new dimensions for nondestructive measurement of material properties, the locations and the relative sizes of discontinuities. The ten-year lag (from 1945 to about 1955) in industry's acceptance of novel developments was uniquely short in the case of these instruments.

Electronic instrumentation based on vacuum and gas filled electron tubes was approaching the peak of its development. These developments permitted easy construction of variable frequency oscillators and power supplies for the magnetizing coils of eddy current test systems. They also permitted minute voltage or current signals to be amplified linearly to levels adequate for display systems, graphic and permanent recording systems and for operation of sorting gates, automation of scanning and mechanization of materials handling during tests.

The aerospace and nuclear power industries were developing rapidly and made unique demands for sensitivity and consistency of instruments used in materials evaluation and inservice reliability assessment. These industries (and government agencies related to them) were the primary sponsors of research to advance all forms of nondestructive testing. However, governmental support of eddy current instrumentation remained significantly less than for other fields of nondestructive testing until Friedrich Förster's technology was introduced to this country.

Friedrich Förster

The introduction by Förster of sophisticated, stable quantitative test equipment and of practical techniques for analysis of quantitative test signals on the complex plane were important factors contributing to the rapid development and acceptance of electromagnetic induction and eddy current tests from 1950 to 1965 in the United States.

Förster's experience before World War II included advanced university education in physics and, in German research institutes, a significant introduction to electromagnetic measurements related to the metallurgy and structure of steels and nonferrous metals. During World War II, his knowledge was used in naval warfare, particularly with respect to magnetic mines. At the conclusion of the war, after a period of imprisonment by the French, Förster retrieved his technical reports and, "with the aid of a screwdriver and a technician," began further development of electromagnetic test instruments in the upper story of an old inn just a few kilometers from Reutlingen — the place where he later established the Institut Dr. Förster.

By 1950, he had developed a precise theory for many basic types of eddy current tests, including both absolute and differential or comparator test systems and probe or fork coil systems used with thin sheets and extended surfaces.[11] Painstaking calibration tests were made with these coil systems and with mercury models (in which discontinuities could be simulated by insertion of small pieces of insulators). Each test was confirmed by precise solution of Maxwell's differential equations for the various boundary conditions involved with coils and test objects, at least for symmetrical cases such as round bars, tubes and flat sheets where such mathematical integrations were feasible.

Further studies were made of the nonlinear response characteristics of ferromagnetic test objects. Techniques using very low test frequencies (5 Hz), harmonic signal analysis, comparators at various levels of magnetization and precise bridge circuits were developed. In most instances, Förster replaced measurements of the inductance or impedance of magnetizing coils with the more precise technique of measuring response with unloaded secondary coils coupled to the test materials (the secondary coil's coupling with the test material is almost identical to that of the magnetizing coils).

The extent and depth of these scientific studies were not matched by any United States laboratory, either government sponsored or operated independently. By extensive publications (not initially in the form of United States patents but in the open literature), Förster made the results of this research available to the world of technical personnel. His contribution of almost the entire theory and technology of electromagnetic induction and eddy current test techniques in the first edition of the *Nondestructive Testing Handbook* of the American Society for Nondestructive Testing (ASNT) provided the means for educating thousands of nondestructive test personnel in the theory, techniques, equipment and interpretation of eddy current tests.[15] This integrated presentation was then used throughout the world to update eddy current test technology.

The unique developments in Förster's new laboratory in Reutlingen, Federal Republic of Germany, were made known in the United States not only by those who read his German publications before 1950 but also through missions in which American personnel were sent to Förster's laboratory for education and experience with these new forms of test instrumentation. Richard Hochschild, for example, made a visit of perhaps six months to Reutlingen. Upon his return, he prepared summary reports that were distributed by the Atomic Energy Commission, the sponsors of his visit.[16]

In the United States, numerous facilities began research to test these new concepts and instrumentation, including significant efforts at Oak Ridge, Hanford and other facilities. The creative work of Hugo L. Libby at Hanford and others at Oak Ridge may well have been sponsored in response to the original work done by Förster.

Even more significant was the transfer of Förster's technology to American firms manufacturing and distributing nondestructive test equipment since 1952. Förster made his first presentation before an ASNT audience in the early 1950s after learning a very little English aboard ship. Agreements for licensing under Förster patents were later concluded. The nondestructive testing staff at Battelle Memorial Institute in Columbus, Ohio, modified the basic Förster instruments for use with United States components and electron tubes.

During the next few years, increasing amounts of Förster's technology were transferred to Magnaflux, whose staff under Glenn L. McClurg became qualified in design and production of Förster's various instruments and then marketed these electromagnetic induction test systems throughout the United States. The collaboration between Förster and the Magnaflux Corporation lasted perhaps ten years, during which rapid progress was made both in the German laboratory and in the United States.

Proliferation of Eddy Current Equipment

Upon termination in the 1960s of the arrangement with Magnaflux, Förster marketed his instruments through the Foerster-Hoover organization in Ann Arbor, Michigan. Rudolf G. Hentschel, who was trained in Reutlingen at Institut Dr. Förster, transferred information to this new organization. After a few years, the licensing of Foerster instruments to Automation Industries resulted in further transfer of advanced technology and marketing through a new organization. A later arrangement with Krautkramer Branson repeated this unique educational process.

Organizations manufacturing many types of nondestructive test equipment and marketing their services widely in the United States made advances in test technologies and broadened the range of applications. Most of these instruments have been updated to semiconductor circuit elements and integrated circuits. Many instruments in the twenty-first century operate with absolute or differential probe coils, encircling coils, internal bobbin coils and various special coil and circuit arrangements — many of which Förster described in the first edition of the *Nondestructive Testing Handbook.*

Self-balancing or self-adjusting instruments, which establish reference points by placing probes on standard test materials or specimens, are available in several cases, using developments by Hugo Libby and other innovators. Designs of probes based on digital computer analyses of eddy current distributions in single-layer or multiple-layer sheet materials have been made feasible through the pioneering work at Oak Ridge National Laboratory. Special probes with split coils, internal magnetic shields and other complexities have also been developed for crack detection and special applications. Digital displays of test signals are also being used.

Microwave Nondestructive Testing

At very high frequencies, electromagnetic fields can be concentrated into beams and propagated through space. When such a beam pulse strikes a conducting metallic surface, for example, it is reflected and may return as an echo to the site of the original pulse transmitter, or to other detectors, as in radar detection. In dielectric materials, microwaves can be subject to rotations and phase shifts, as well as to attenuation due to dielectric hysteresis losses. In many ways, microwave nondestructive test systems are analogous in performance applications to immersion ultrasonic test systems. By Maxwell's theory of the electromagnetic field, microwaves are reflected like light waves by eddy currents induced in the surface layers of highly conducting metallic materials. Thus, microwaves appear to have the capacity to apply high frequency eddy current tests to a metallic surface from a distance and perhaps to scan such surfaces to detect discontinuities that change the pulse reflection patterns.

When his eddy current systems were sold to the Budd Company, Richard Hochschild turned his attention to formation and development of Microwave Instruments Company, Corona del Mar, California. Soon a series of instrument systems had been developed and the long task of educating industrial and scientific users in the capabilities and applications of electromagnetic tests had to be done all over again for these new higher frequencies.

The theory and design of microwave generators, horns, antennas, detectors and display systems had been developed for long distance ranging in radar. Many textbooks presented the electromagnetic theory of microwaves in terms readily used by electrical engineers. Microwave system components and electron tubes were commercially available. However, electrical engineers were rarely aware of the needs of nondestructive testing engineers, and nondestructive testing engineers had little familiarity with microwaves. In fact, many nondestructive test personnel were still just beginning to use and understand eddy current testing at the lower frequencies.

After several years of diligent development, continued application research and marketing efforts by Richard Hochschild with the assistance of Ronald Botsko, the pioneer organization Microwave Instruments Company was sold and its proprietor moved to the area of medical services. A few other organizations built simple microwave test systems but the development of industrial microwave nondestructive testing languished during the 1970s. Limited research sponsored by government agencies resulted in possibilities for crack detection from a distance.

The theory of microwave antennas and of time domain reflectometry of microwaves in tubes, passing along wires and reflecting and refracting in dielectric layers, promises the possibility of valuable nondestructive testing applications. Because microwaves can be focused, microwave systems could be designed to operate in a manner analogous to optical instruments and ultrasonic systems.

A large scale example of microwave exploration of test objects at great distances is occurring in radio astronomy laboratories throughout the world. Many radio signals from objects billions of kilometers away have been confirmed by films from optical telescopes and the locations of others have been predicted. Emissions are detected from galaxies, black holes and other astronomical features. J.D. Kraus has recognized this as a form of *nondestructive testing of outer space* and has written a biographical book called *The Big Ear,*[17] which clearly and simply summarizes a lifetime of study and applications of Maxwell's theory of electromagnetic fields.

References

1. McMaster, R.C. "The Origins of Electromagnetic Testing." *Materials Evaluation*. Vol. 43, No. 8. Columbus, OH: American Society for Nondestructive Testing (July 1986): p 946-956.
2. McMaster, R.C. Section 1, "Introduction to Electromagnetic Testing." *Nondestructive Testing Handbook*, second edition: Vol. 4, *Electromagnetic Testing*. Columbus, OH: American Society for Nondestructive Testing (1986): p 2-12.
3. Millikan, R.A. "Early Views of Electricity." *Electrons (+ and –), Protons, Photons, Neutrons, and Cosmic Rays*. Chicago, IL: University of Chicago Press (1935-36).
4. Holmes, U.T., Jr. *Daily Living in the Twelfth Century*. Madison, WI: University of Wisconsin Press (1952): p 49-50.
5. Gilbert, W. *De Magnete* (translated by P.F. Mottelay, 1892). New York, NY: Dover Press (1958).
6. Maxwell, J.C. *A Treatise on Electricity and Magnetism*, third edition (1891). Vol. 2. New, York, NY: Dover Press (1954): p 138-262.
7. McMaster, R.C. and S.A. Wenk. "A Basic Guide for Management's Choice of Non-Destructive Tests." *Symposium on the Role of Non-Destructive Testing in the Economics of Production*. Special Technical Publication 112. West Conshohocken, PA: ASTM International (1951).
8. Saxby, S.M. "Magnetic Testing of Iron." *Engineering*. Vol. 5. London, United Kingdom: Office for Advertisements and Publication (1868): p 297.
9. Hughes, D.E. "Induction-Balance and Experimental Researches Therewith." *Philosophical Magazine*. Series 5, Vol. 8. London, United Kingdom: Taylor and Francis, Limited (1879): p 50-57.
10. Davis, R.S. "Bell's Use of Induction Balance: Searching for a Bullet in President Garfield." *Materials Evaluation*. Vol. 46, No. 12. Columbus, OH: American Society for Nondestructive Testing (November 1988): p 1528, 1530, 1532, 1560.
11. Förster, F. "The First Picture: A Review of the Initial Steps in the Development of Eight Branches of Nondestructive Material Testing." *Materials Evaluation*. Vol. 41, No. 3 (December 1983): p 1477-1488.
12. Burrows, C.W. United States Patent 1 686 679, *Apparatus for Testing Magnetizable Objects* (October 1928).
13. Zuschlag, T. "Magnetic Analysis Inspection in the Steel Industry." *Symposium on Magnetic Testing, 1948* [Detroit, Michigan, June 1948]. Special Technical Publication 85. West Conshohocken, PA: ASTM International (1949): p 113-122.
14. Black, W.A. "Eddy Current Testing of Steel Tubing, 1929-60." *Materials Evaluation*. Vol. 43, No. 12. Columbus, OH: American Society for Nondestructive Testing (November 1985): p 1490, 1492-1493, 1495-1498.
15. Förster, F. Sections 36-42. *Nondestructive Testing Handbook*, first edition: Vol. 2. Columbus, OH: American Society for Nondestructive Testing (1959).
16. Hochschild, R. "Eddy Current Testing by Impedance Analysis." *Nondestructive Testing*. Vol. 12, No. 3. Columbus, OH: American Society for Nondestructive Testing (May-June 1954): p 35-44.
17. Kraus, J.D. *The Big Ear*. Powell, OH: Cygnus-Quasar Books (1976).

Bibliography

Electromagnetic Induction Techniques

Albin, J. "Salvaging and Process Control with the Cyclograph." *The Iron Age*. Vol. 155. Newton, MA: Cahners Business Information, Division of Reed Elsevier (17 May 1945): p 62-64.

Brenner, A. and E. Kellogg. "An Electric Gage for Measuring the Inside Diameter of Tubes." *Journal of Research*. Vol. 42, No. 5. Gaithersburg, MD: National Institute of Standards and Technology (May 1949): p 461-464.

Brenner, A. and E. Kellogg. "Magnetic Measurement of the Thickness of Composite Copper and Nickel Coatings on Steel." *Journal of Research*. Vol. 40, No. 4. Gaithersburg, MD: National Institute of Standards and Technology (April 1948): p 295-299.

Carside, J.E. "Metallic Materials Inspection." *Metal Treatment*. Vol. 13. London, United Kingdom: Fuel and Metallurgical Journals Limited (Spring 1946): p 3-18.

Cavanagh, P.E. "A Method for Predicting Failure of Metals." *ASTM Bulletin*. No. 143. West Conshohocken, PA: ASTM International (December 1946): p 30.

Cavanagh, P.E. "The Progress of Failure in Metals As Traced by Changes in Magnetic and Electrical Properties." *Proceedings*. Vol. 47. West Conshohocken, PA: ASTM International (1947): p 639.

Cavanagh, R.L. "Nondestructive Testing of Drill Pipe." *Oil Weekly*. Vol. 125. Houston, TX: Gulf Publishing Company (10 March 1947): p 42-44.

Cavanagh, R.L. "Nondestructive Testing of Metal Parts." *Steel Processing*. Vol. 32, No. 7. Pittsburgh, PA: Steel Publications, for the American Drop Forge Association (July 1946): p 436-440.

"Electronic Comparators." *Automobile Engineer*. Vol. 37. London, United Kingdom: IPC Transport Press Limited, for the Institution of Automobile Engineers (July 1947): p 271-272.

Ford, L.H. and C.E. Webb. "Nondestructive Testing of Welds." *The Engineer*. Vol. 165. London, United Kingdom: Office of "The Engineer" (8 April 1938): p 400-401.

Förster, F. and H. Breitfeld. "Nondestructive Test by an Electrical Method." *Aluminum*. Vol. 25. Berlin, Germany: Aluminum-Zentrale GmbH (March 1943): p 130.

Förster, F. and H. Breitfeld. "Nondestructive Testing of Light Metals Using a Testing Coil." *Light Metals Bulletin*. Vol. 7. London, United Kingdom: British Aluminum Company (28 April 1944): p 442-443.

Gunn, R. "Eddy-Current Method for Flaw Detection in Nonmagnetic Metals." *Journal of Applied Mechanics*. Vol. 8, No. 1. New York, NY: American Society of Mechanical Engineers (March 1941): p A22-A26.

Hastings, C.H. "Recording Magnetic Detector Locates Flaws in Ferrous Metals." *Product Engineering*. Vol. 18. New York, NY: Morgan-Grampian Publishing (April 1947): p 110-112.

Hastings, C.H. "A New Type of Magnetic Flaw Detector." *Proceedings*. Vol. 47. West Conshohocken, PA: ASTM International (1947): p 651.

Henry, E.B. "The Role of Nondestructive Testing in the Production of Pipe and Tubing." *Materials Evaluation*. Vol. 47, No. 6. Columbus, OH: American Society for Nondestructive Testing (June 1989): p 714-715, 718, 720, 722-724.

Jellinghaus, W. and F. Stablein. "Nondestructive Testing to Detect Internal Seams in Sheets." *Technische Mitteilungen Krupp*, Ausgabe A: *Forschungsbericht*, Vol. 4. Essen, Germany: Friedrich-Krupp-GmbH, Technische Werksleitung (April 1941): p 31-36.

Jupe, J.H. "Crack Detector for Production Testing." *Electronics*. Vol. 18, No. 10. New York, NY: McGraw-Hill (October 1945): p 114-115.

Lichy, C.M. "Determination of Seams in Steel by Magnetic Analysis." *Electronic Methods of Inspection of Metals*. Materials Park, OH: ASM International (1947): p 97-106.

Mader, H. "Magneto-Inductive Testing." *Metal Industry*. Vol. 68. New York, NY: Metal Industry Publishing Company (18 January 1946): p 46-48.

Matthaes, K. "Magnetinduktive Stahlprüfung" [Magneto-Inductive Testing of Steel]. *Zeitschrift für Metalkunde*. Vol. 39. Stuttgart, Germany: Dr. Riederer-Verlag, for Deutsche Gesellschaft für Metallkunde (September 1948): p 257-272.

McMaster, R.C. "The History, Present Status, and Future Development of Eddy Current Tests." *Eddy Current Nondestructive Testing*. Special Technical Publication 589. West Conshohocken, PA: ASTM International (1981): p 1-32.

Nelson, G.A. "The Probolog, for Inspecting Nonmagnetic Tubing." *Metal Progress*. Vol. 56. Materials Park, OH: ASM International (July 1949): p 81-85.

"Nondestructive Testing." *Automobile Engineering*. Vol. 34. Chicago, IL: American Technical Society (May 1944): p 181.

Polgreen, G.R. and G.M. Tomlin. "Electrical Nondestructive Testing of Materials." *Electronic Engineering*. Vol. 18, No. 218. London, United Kingdom: Morgan-Grampian Publishing (April 1946): p 100-105.

Robinson, I.R. "Magnetic and Inductive Nondestructive Testing of Metals." *Metal Treatment and Drop Forging*. Vol. 16. London, United Kingdom: Fuel and Metallurgical Journals Limited (Spring 1949): p 12-24.

Schmidt, T.R. "History of the Remote-Field Eddy Current Inspection Technique." *Materials Evaluation.* Vol. 47, No. 1. Columbus, OH: American Society for Nondestructive Testing (January 1989): p 14, 17-18, 20-22.

Schneider, P. "Measuring the Wall Thickness of Light-Metal Cast Parts with Dr. Förster's 'Sondenkawimeter'." *Metall.* Vol. 3. Frankfurt, Germany: IG Metall (October 1949): p 321-324.

Segsworth, R.S. "Uses of the DuMont Cyclograph for Testing of Metals." *Electronic Methods of Inspection of Metals.* Materials Park, OH: ASM International (1947): p 54-70.

Trost, A. "Testing Non-Ferrous Pipes, Bars and Shapes with Eddy Currents." *Metallwirtschaft, Metallwissenschaft, Metalltechnik.* Vol 20. Berlin, Germany: G. Lüttke Verlag (1941): p 697-699.

Vosskuhler, G.H. "Zerstörungsfreie Prüfung der Al-Mg-Zn Legierung Hy 43 auf Magnetinduktivem Wege" [Nondestructive Testing of the Al-Mg-Zn Alloy Hy43 by Magnetoinductive Means]. *Metall.* Vol. 3. Frankfurt, Germany: IG Metall (August-September 1949): p 247-251, 292-295.

Zeluff, V. "Electronic Inspection." *Scientific American.* Vol. 174, No. 2. New York, NY: Scientific American Publishing Company (February 1946): p 59-61.

Zijlstra, P. "An Apparatus for Detecting Superficial Cracks in Wires." *Philips Technical Review.* Vol. 11. Eindhoven, Netherlands: Philips Research Laboratory (July 1949): p 12-15.

Rail Testing

Clarke, J.G. and C.F. Spitzer. "Electronic Locator for Salvaging Trolley Rails." *Electronics.* Vol. 17. New York, NY: McGraw-Hill (January 1944): p 129.

Davis, R.S. "Harcourt C. Drake, Henry W. Keevil, and the Development of Induction-Based Rail Testing." *Materials Evaluation.* Vol. 48, No. 9. Columbus, OH: American Society for Nondestructive Testing (September 1990): p 1165-1168, 1171. See also *Materials Evaluation,* Vol. 48, No. 12 (December 1990): p 1440.

Keevil, W.R. "History and Development of Rail Flaw Detector Cars." *Materials Evaluation.* Vol. 49, No. 1. Columbus, OH: American Society for Nondestructive Testing (January 1991): p 71-76.

"Rail Testing Cars, 1928-49." *Materials Evaluation.* Vol. 50, No. 2. Columbus, OH: American Society for Nondestructive Testing (February 1992): p 307-310.

Wickre, J.M. " Fishing for Fissures: Sources for the History of Rail Testing Cars, 1927-60." *Materials Evaluation.* Vol. 43, No. 4. Columbus, OH: American Society for Nondestructive Testing (March 1985): p 372-379.

Wire Rope

Cavanagh, P.E. "Some Changes in Physical Properties of Steels and Wire Rope during Fatigue Failure." *Transactions.* Montreal, Quebec, Canada: Canadian Institute of Mining and Metallurgy (July 1947): p 401-411.

Cavanagh, P.E. and R.S. Segsworth. "Nondestructive Inspection of Mine Hoist Cable." *Transactions.* Vol. 38. Materials Park, OH: ASM International (1947): p 517-550.

Gee, J. "Testing and Inspection of Wire Ropes." *Mine and Quarry Engineering.* Vol. 14. London, United Kingdom: Electrical Press, Limited (December 1948): p 375.

Weischedel, H.R. "Electromagnetic Wire Rope Inspection in Germany, 1925-40." *Materials Evaluation.* Vol. 46, No. 6. Columbus, OH: American Society for Nondestructive Testing (May 1988): p 734-736.

3

CHAPTER

Principles of Electromagnetic Testing

Lalita Udpa, Michigan State University, East Lansing, Michigan

Satish S. Udpa, Michigan State University, East Lansing, Michigan

PART 1. Introduction to Principles of Electromagnetic Testing

The electromagnetic test method of nondestructive testing, like other nondestructive test methods, involves the application of electromagnetic energy to evaluate the condition of test objects. The energy interacts with the material and a snapshot of the interaction process is analyzed to ascertain the condition of the material. Although electromagnetic methods, in principle, cover a wide range of techniques, the term *electromagnetic testing* is generally used to denote several techniques, including magnetic flux leakage testing, eddy current testing and microwave testing. Radiation methods such as infrared and thermal testing and radiographic testing are often not thought of as electromagnetic methods even though they are governed by the same physical laws.

Although all electromagnetic methods are governed by Maxwell's equations, the distinctive nature of each method stems from differences in excitation frequencies, the nature of the transducers used and the signal analysis techniques for characterizing the state of the test object. As an example, magnetic flux leakage techniques typically use excitation frequencies near 0 Hz whereas eddy current techniques use excitation frequencies from about 100 Hz to about 10 MHz. Microwave testing uses excitation sources usually in excess of 100 MHz.

As the excitation frequency increases from zero, the underlying physical process gradually changes. Below about 10 MHz, the field is said to be quasistatic, which means that displacement current is negligible. As the frequency finally increases beyond quasistatic values, the energy propagates in the form of waves into the tested material. Differences in the underlying processes associated with each frequency make it possible for electromagnetic techniques to test a wide range of materials.

The principles underlying three techniques correspond to the three frequency ranges discussed below: (1) magnetic flux leakage testing (low frequency), (2) eddy current testing (middle frequency) and (3) microwave testing (high frequency).

Part 2. Magnetic Flux Leakage Testing

Magnetic flux leakage testing is used extensively in industry for testing ferromagnetic parts and components. The magnetic flux leakage technique involves magnetization of the test object by a permanent magnet or by passing an excitation current directly through an electromagnet. The presence of a discontinuity on or near the surface of the sample disturbs the magnetic flux lines and results in a local leakage field around the discontinuity. The magnetic flux leakage field can be detected using a variety of techniques. In magnetic particle testing, the leakage field is imaged by dusting the surface of the test object with magnetic particles coated with fluorescent dye.[1] Force exerted by the magnetic leakage field around a crack attracts the particles to line up along surface cracks. The magnetic flux leakage field can also be detected using noncontact sensors such as a hall effect probe or a simple induction coil. A hall probe using an element oriented parallel to the sample surface is sensitive to the normal component of the magnetic flux leakage field and generates a typical signal, as shown in Fig. 1 for a rectangular notch.[2]

To understand the operation of magnetic flux leakage, it is useful to consider the physics of permanent magnets. A permanent magnet can be considered an agglomeration of domains that can be thought of as elementary magnets obtained as a result of the fact that the dipole moments of uncompensated electron spins contained within the domain are held parallel.[3] In the demagnetized state, the domains orient themselves randomly (Fig. 2a) so that closed paths for the magnetic flux

Figure 1. Typical leakage field signal.[2]

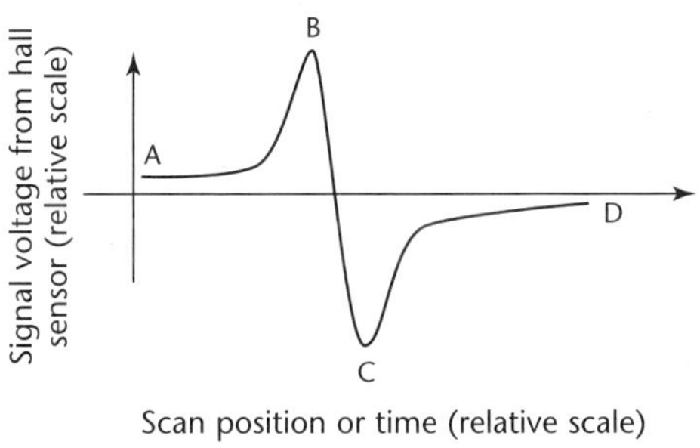

Figure 2. Physics of permanent magnets: (a) random orientation of domains in unmagnetized state; (b) domains aligned in direction of applied field; (c) relaxation of parallel alignment of domains when magnetic field is removed; (d) self-demagnetization of material after magnetic field is removed; (e) reversion of domains to random orientation when gap is removed.[3] See Fig. 3 for characteristic curve.

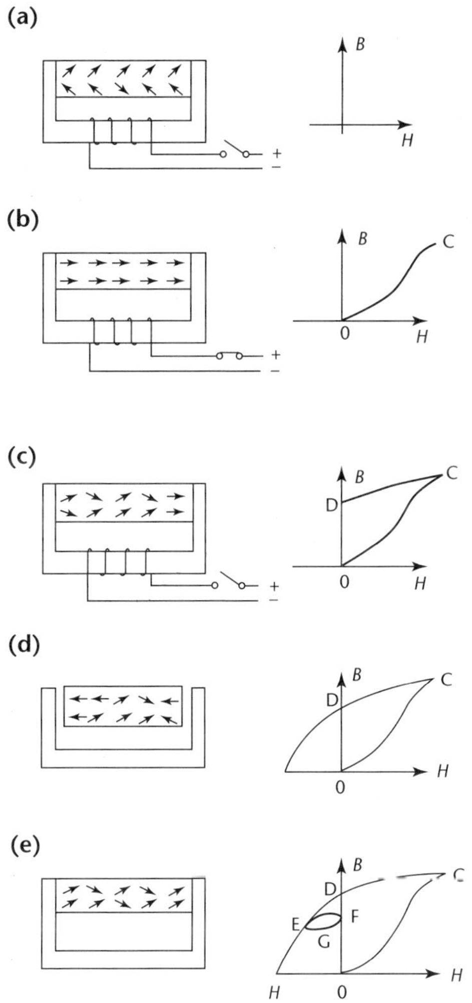

Legend

B = magnetic flux density (relative scale)
H = magnetic field intensity (relative scale)

exist in the material. The magnetostatic energy under this condition is a minimum. This state is indicated by point 0 on the characteristic curve of magnetic flux density B versus magnetic field intensity H (Fig. 3). When an external magnetic field is applied, the domains tend to align with the direction of the applied field, thereby increasing B. The operating point now moves into region 0A in Fig. 3. The size and orientation of the domains are affected by the potential energy (1) arising out of the interaction between neighboring atoms, (2) associated with the anisotropy energy and (3) associated with the external field energy.

As the external magnetic field is increased, the operating point moves into region AB of Fig. 3. The domain walls start shifting and ultimately reach a state when each crystal represents a single domain. Further increases in the magnetic field intensity result in magnetic saturation, a state in which the domains rotate against the forces of anisotropy until all the domains get aligned in the direction of the applied field (Fig. 2b).[3] This state is represented by the region BC on the curve of B versus H (Fig. 3).

If the applied magnetic field is then withdrawn, the domains relax. As a result, the parallel alignment of the domains is disturbed (Fig. 2c). The residual flux B_r represents a new minimum energy, at point D where magnetization $H = 0$.

If a gap is then introduced as shown in Fig. 2d, the material self-demagnetizes. The imbalance created by the gap results in a realignment of the domains closest to the gap.

FIGURE 3. Typical characteristic curve of magnetic flux density B versus magnetic field intensity H.[3]

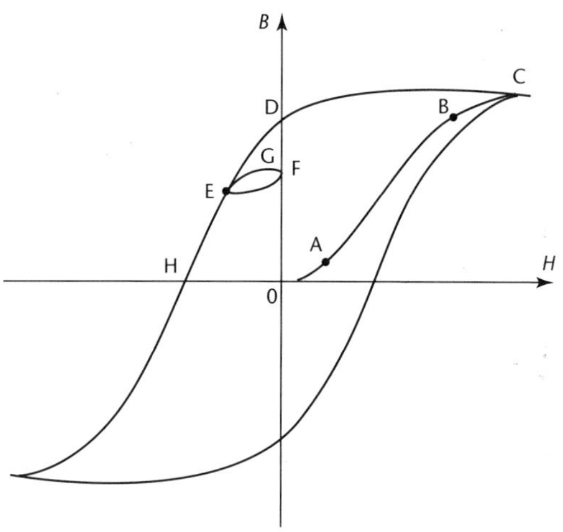

Legend

B = magnetic flux density (relative scale)
H = magnetic field intensity (relative scale)

These domains take up orientations that are 180 degrees from the original orientation. The mechanical energy injected into the system to introduce the gaps is used to transfer the operating point from D to E. If the air gap is then reduced to zero as shown in Fig. 2e, the operating point moves along the minor or recoil loop to F and the domains revert back very nearly to the same orientation as before. If the gap is once again restored, the operating point then moves toward point E along the recoil loop FGE. Repeated cycles of opening and closing the gap cause the minor recoil loop to be traced.

The presence of a discontinuity causes a reduction in the cross sectional area of the test object, thereby resulting in a local increase in the magnetic flux density.[4,5] A reduction in the permeability together with an increase in the magnetic flux density would cause the flux to leak into the surrounding medium. Magnetic leakage fields can be subdivided further into active or residual leakage fields. To understand the origin of the leakage fields and choice of initial magnetization for the active leakage field technique, consider an unmagnetized steel billet with a surface discontinuity, as shown in Fig. 4a. Let A represent the cross sectional area of the billet and let a represent the cross sectional area of the discontinuity. The cross sectional area of the sound portion of the billet in the vicinity of the discontinuity is reduced to $(A - a)$ units (Fig. 4b).

Magnetic field $\boldsymbol{H}$ is a vector quantity because it has both magnitude and direction. In the characteristic curve of Figs. 2 and 3, for isotropic materials, H is the magnitude component and so is a scalar quantity.

Then, place the billet in a uniform magnetic field $\boldsymbol{H}$ and represent the induced flux density in the sound portion of the billet by B_1 (weber per square meter). This magnetic flux density corresponds to a point P to the right of μ_{max} on the permeability curve of the material, as illustrated in Fig. 4c. The corresponding point on the initial magnetization curve in Fig. 4c is point Q. The magnetic flux density passing through the sound part of the billet is B_1 (tesla). Now, if it is assumed that this same magnetic flux is to pass through the reduced billet area in the vicinity of the discontinuity, then the flux density present in this section is greater than B_1 and is equal to $B_1 A \cdot (A-a)^{-1}$, namely B_2.

This local increase of magnetic flux density results in a change of the operating point on the magnetization curve from Q to Q´ and a corresponding decrease of local permeability from P to P´. However, this results in conflicting demands in the vicinity of the discontinuity. The magnetic flux density must increase with a reduction of cross sectional area but this change drives the permeability in the restricted region of the billet to a value less than that present in the sound regions. Consequently, some of the flux *leaks* into the surrounding medium near the discontinuity and is called a *leakage field* (Fig. 4d). The detection of this leakage field is the basis of magnetic flux leakage testing.

Subsurface Discontinuities

If a discontinuity is farther below the surface, the difficulty of detecting these magnetic leakage fields is much greater. The reason for this difficulty is that the surrounding material tends to smooth out the field distortion due to the subsurface discontinuity, thus resulting in a small field disturbance on the surface of the billet.[4] Because most detectors used to monitor the magnetic leakage fields rely on a sharp change of field gradient to record the presence of the field, it is naturally difficult to sense the location of subsurface discontinuities, as illustrated in Fig. 5.

Degree of Initial Magnetization

The initial operating point on the permeability characteristic of the material is very important.[4] For example, if this point should lie to the left of μ_{max}, as illustrated by the point T in Fig. 4c, an increase of magnetic flux density with an area reduction due to a discontinuity would drive the local permeability higher than the permeability of a material free of discontinuities. Thus, there is a possibility

FIGURE 4. Billet with discontinuity: (a) view of billet; (b) cross section through discontinuity; (c) magnetic characteristics of billet material; (d) billet in magnetic field, showing discontinuity leakage field.[4]

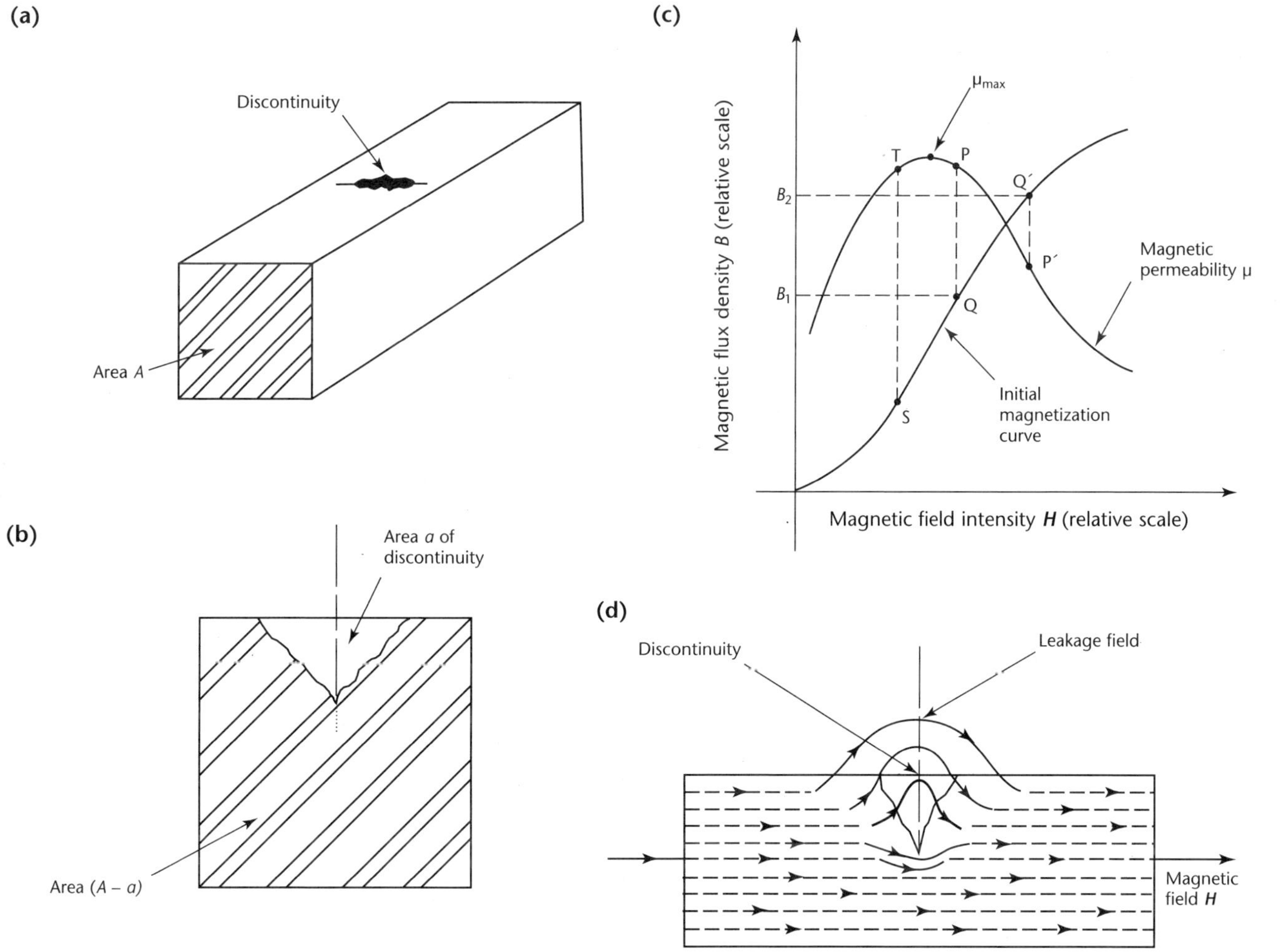

that the discontinuity may go undetected in these circumstances.

Moreover, if the initial magnetization of the material should locate the operating point near saturation, then the difference between the magnetic flux density in the material and the leakage magnetic field in the surrounding medium decreases with increasing discontinuity cross sectional area. Therefore, the problem of quantitatively detecting the discontinuities is magnified because it becomes increasingly more difficult to discriminate between the severity of the various heterogeneities.

Also, because the degree of magnetization is so great, the surface roughness is easily mistaken for actual discontinuities and results in unwarranted rejection of test objects.

Therefore, there exists an upper and lower limit of magnetization to which a test object should be subjected if the magnetic leakage field technique of nondestructive testing is to be most successful. Magnetization of the test object lies on the linear part of the magnetization curve in such a way that the material permeability is maximum. Magnetization should not approach saturation but should have a value of flux density that locates the initial operating point of the material on the steepest part of the initial induction curve.

If the degree of magnetization is too low, discontinuities may go unnoticed and, if the magnetization level is too high, a lack of discontinuity discrimination may result in false indications.

FIGURE 5. Billet with subsurface discontinuity, showing resultant leakage field.[4]

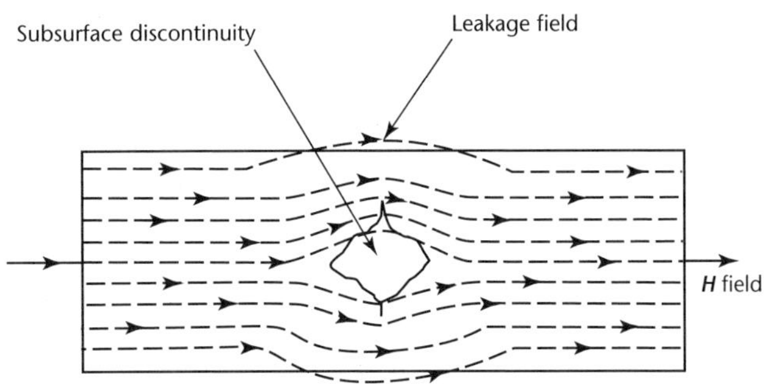

PART 3. Eddy Current Testing

Eddy current techniques of nondestructive testing rely on the principles of magnetic induction to interrogate the materials under test. A complete understanding of the underlying physical process can only be gained through Maxwell's equations. However, the physical basis of the technique can also be understood qualitatively. Eddy current testing is based on the fact that, when a coil excited by an alternating current is brought close to a material, the terminal impedance of the coil changes.[6] The change is associated with the fact that the primary field set up by the eddy current coil induces eddy currents within the electrically conducting specimen.

In conformity with Lenz's law, the direction of the induced eddy currents and consequently the secondary field generated by these currents oppose the change in the primary field (Fig. 6).[7,8] If the test object is nonferromagnetic, the magnetic flux leakage associated with the coil decreases because of the opposing nature of the primary and secondary fields. Because the self-inductance of the coil is defined as *flux linkages per ampere*, the inductance of the coil decreases. Accompanying the decrease in inductance is an increase in resistance, owing to the fact that the eddy current losses incurred within the specimen have to be met by the source of primary excitation. This loss manifests itself as a change in coil resistance.

The presence of a discontinuity or heterogeneity in the test object causes a reduction as well as a redistribution of the eddy currents. Consequently, the changes in the inductance and resistance of the excitation coil are correspondingly less. Figure 7a shows how the impedance of a coil changes as it comes in contact with nonferromagnetic conducting specimens with and without discontinuities. It should be noted that Fig. 7 greatly exaggerates these changes.[7,8]

The underlying process is more complicated when the test object is ferromagnetic. Counteracting the decrease in inductance (due to the influence of eddy currents induced in the test object) is an increase in inductance attributable to the higher permeability of the material. The latter effect generally predominates,

FIGURE 6. Alternating current coil over conducting test object, showing opposite direction of primary and induced currents.[7]

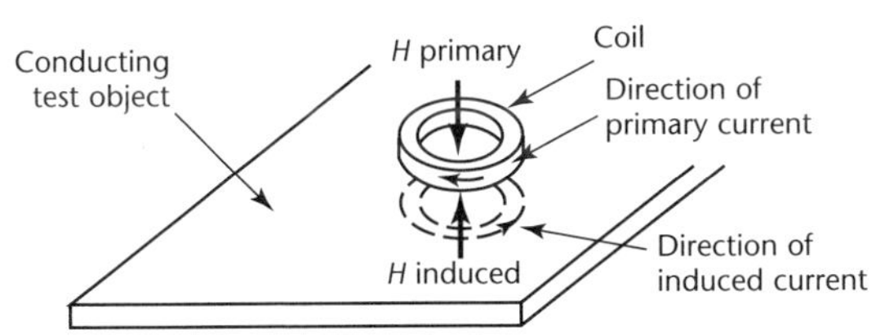

FIGURE 7. Impedance plane trajectories of coil over specimens: (a) over nonferromagnetic specimen; (b) over ferromagnetic specimen. Changes are exaggerated for clarity.[7,8]

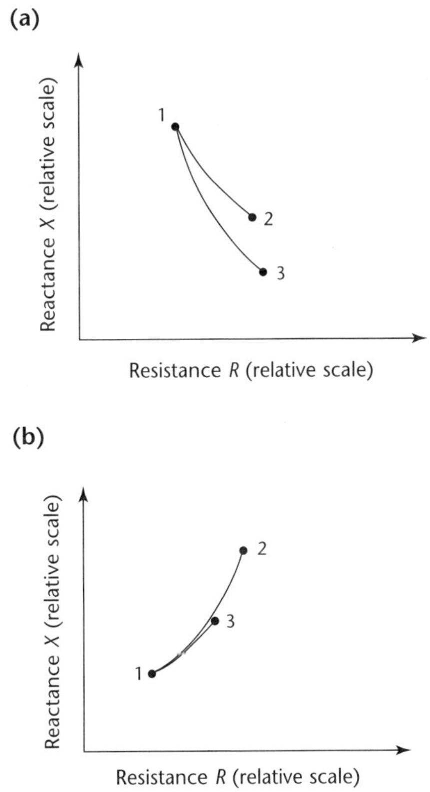

Legend
1. Coil in air.
2. Coil over specimen with discontinuity.
3. Coil over specimen without discontinuity.

so the inductance of the coil increases when the coil comes in contact with a ferromagnetic specimen (Fig. 7). The change in inductance is also accompanied by an increase in resistance attributable to the eddy current and hysteresis losses.[7-9]

The variations in coil impedance caused by discontinuities in the test object are often very small in comparison with the quiescent value of the coil impedance. Detection and measurement of these small changes is often accomplished using bridge circuits.

In this regard, there is a useful distinction between two kinds of probe coils: (1) absolute coils and (2) differential coils. An absolute coil responds to the electromagnetic properties of the test object in the magnetic field of the coil without comparison to the response of a second coil. Differential coils are two or more coils connected in such a way that electromagnetic differences in the regions beneath the coils will cause an imbalance between them to be signaled.

A problem of the absolute eddy current probe is the difficulty of detecting small changes in impedance, which are superimposed over the value in air. In addition, changes in the coil parameters because of environmental factors and liftoff can often mask changes due to discontinuities, making signal interpretation very difficult.

An alternative to the absolute eddy current probe is the differential eddy current probe. Figure 8 shows a differential eddy current probe designed for testing tubes.[9] The probe consists of two identical coils mounted on the same axis as the tube but spaced apart by a small distance. The two coils form two arms of a bridge circuit as illustrated in Fig. 9.[9] The bridge imbalance signal is the voltage difference across the impedance of two coils. When the probe is moved past a discontinuity, the change in impedance of the leading coil when it scans a discontinuity results in an imbalance voltage. The differential impedance traces a trajectory 0A0 in the impedance plane shown in Fig. 10.[8] Similarly, when the trailing coil scans the discontinuity, the differential impedance traces the

FIGURE 8. Differential eddy current probe for inspecting tubes from the inside.[7]

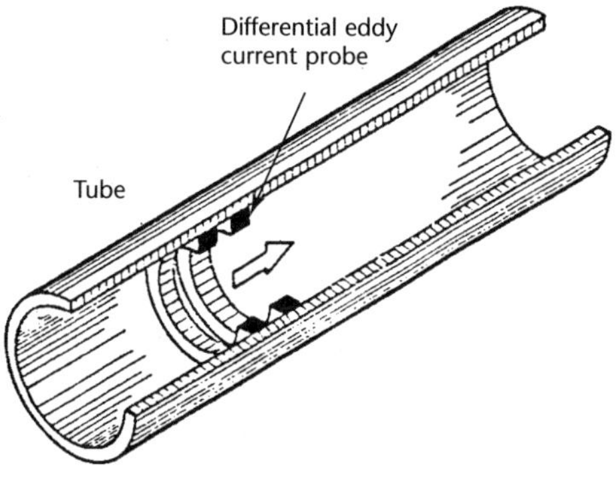

FIGURE 9. Alternating current bridge for measuring changes in impedance.

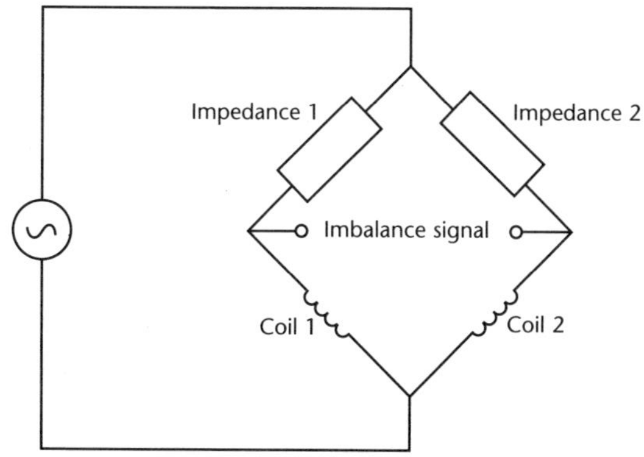

FIGURE 10. Narrow axisymmetric outside diameter groove having width less than spacing of differential coil: (a) diagram; (b) impedance plane trajectory obtained for groove.[7]

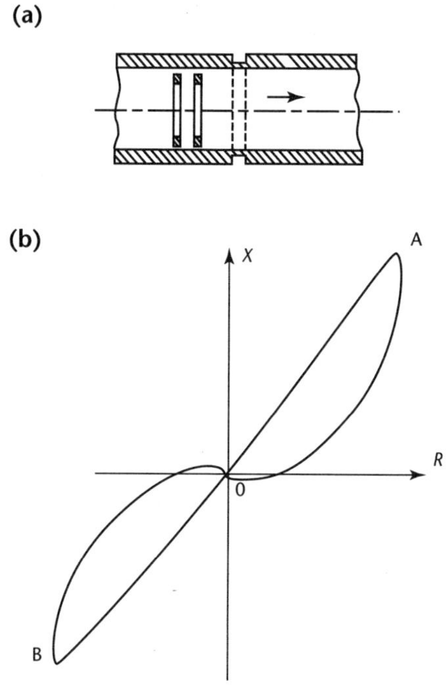

Legend
R = resistance (relative scale)
X = reactance (relative scale)

trajectory 0B0 in the opposite direction. The shape of the impedance plane trajectory is a function of the nature of the discontinuity. This information is used in inverting the measured eddy current probe signal to determine the shape and size of the discontinuity.

Electrical Conductivity and Resistivity

In eddy current testing, instead of describing conductivity in absolute terms, an arbitrary unit has been widely adopted. Because the relative conductivities of metals and alloys vary over a wide range, a conductivity benchmark has been widely used. In 1913, the International Electrochemical Commission established that a specified grade of high purity copper, fully annealed — measuring 1 m long, having a uniform section of 1 mm^2 and having a resistance of 17.241 mΩ at 20 °C — would be arbitrarily considered 100 percent conductive. The symbol for conductivity is σ and the unit is siemens per meter. Conductivity is also often expressed as a percentage of the International Annealed Copper Standard (IACS).[10]

Table 1 lists the conductivity σ and the resistivity ρ of selected materials. Note that conductance and resistance are direct reciprocals: a good conductor is a poor resistor. Resistivity ρ is expressed in absolute terms of ohm meter or of microohm centimeter. To convert, simply follow Eq. 1:

$$\text{(1)} \quad 1\,(\text{percent IACS}) = 0.58\ \text{MS}\cdot\text{m}^{-1} = \frac{1.724 \times 10^{-6}}{\rho\,(\Omega\cdot\text{m})} = \frac{172.4}{\rho\,(\mu\Omega\cdot\text{cm})}$$

The impedance of a test coil varies with the conductivity of a nearby material. Figure 11 shows how the magnitude of impedance decreases with increasing conductivity.

The coil's inductive reactance is plotted on the Y axis; coil resistance is plotted on the X axis. The 0 percent conductivity point, or air point, is when the coil's empty reactance is maximum.

Conductivity is influenced by many factors. Figure 11 represents a measured conductivity locus. Table 1 lists conductivities of materials with different chemical compositions.[11-13]

TABLE 1. Electrical resistivity and conductivity of selected metals.[11]

Metal	Conductivity MS·m^{-1}	Conductivity (percent IACS[a])	Resistivity Ω·m	Resistivity (μΩ·cm)
Aluminum, pure	35.38	(61.00)	2.83×10^{-8}	(2.83)
Aluminum (99.99 percent)	37.67	(64.94)	2.65×10^{-8}	(2.65)
Antimony	2.55	(4.40)	3.92×10^{-7}	(39.18)
Bronze, commercial annealed	25.52	(44.00)	3.92×10^{-8}	(3.92)
Cadmium	14.62	(25.20)	6.84×10^{-8}	(6.84)
Calcium	28.25	(48.70)	3.54×10^{-8}	(3.54)
Chromium	5.10	(8.80)	1.96×10^{-7}	(19.59)
Cobalt	16.01	(27.60)	6.25×10^{-8}	(6.25)
Copper	58.00	(100.00)	1.72×10^{-8}	(1.72)
Gold	40.60	(70.00)	2.46×10^{-8}	(2.46)
Iron, pure	10.44	(18.00)	9.58×10^{-8}	(9.58)
Iron ingot (99.9 percent)	9.05	(15.60)	1.11×10^{-7}	(11.05)
Magnesium, pure	22.39	(38.60)	4.47×10^{-8}	(4.47)
Molybdenum	19.14	(33.00)	5.22×10^{-8}	(5.22)
Nickel	14.62	(25.20)	6.84×10^{-8}	(6.84)
Selenium	8.35	(14.40)	1.20×10^{-7}	(11.97)
Silver, tin solder	9.63	(16.60)	1.04×10^{-7}	(10.39)
Steel, high alloy	1.68	(2.90)	5.94×10^{-7}	(59.45)
Tin, pure	8.70	(15.00)	1.15×10^{-7}	(11.49)
Tin foil	2.44	(4.20)	4.10×10^{-7}	(41.05)
Tungsten	18.21	(31.40)	5.49×10^{-8}	(5.49)
Zinc, commercial rolled	16.24	(28.00)	6.16×10^{-8}	(6.16)

a. International Annealed Copper Standard.

FIGURE 11. Measured conductivity locus, with conductivity expressed in siemens per meter (percentages of International Annealed Copper Standard).[6,10,12]

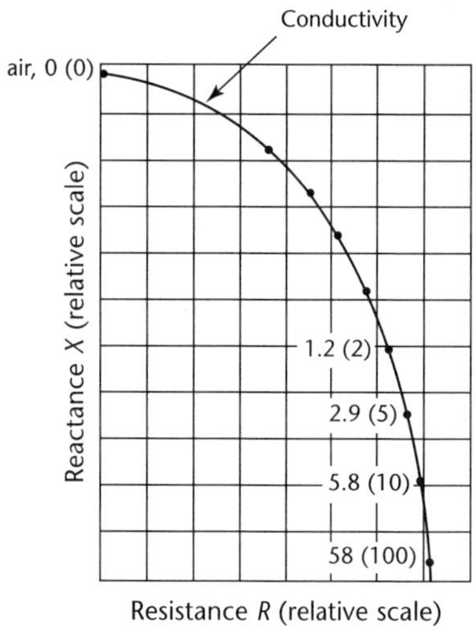

Impedance

The vector sum of the reactive and resistive components is impedance. Impedance is a quantity with magnitude and direction directly proportional to frequency. To construct a universal impedance diagram valid for all frequencies, the impedance must be normalized.[6] Figure 12 shows a typical normalized impedance diagram.[6]

Primary impedance Z_P is affected by changes in frequency ($\omega = 2\pi f$). Figure 12a represents primary impedance without a secondary circuit or test object.

Figure 12b illustrates the effect of frequency on primary impedance with a secondary circuit or test object present. The primary resistance R_1 shown in Fig. 12a has been omitted from Fig. 12b because resistance has a relatively small effect on frequency. The term $\omega L_S G$ in Fig. 12b represents a reference quantity for the secondary impedance, where G is

FIGURE 12. Effect of frequency change: (a) primary impedance without secondary circuit; (b) primary impedance with secondary circuit.[6,12]

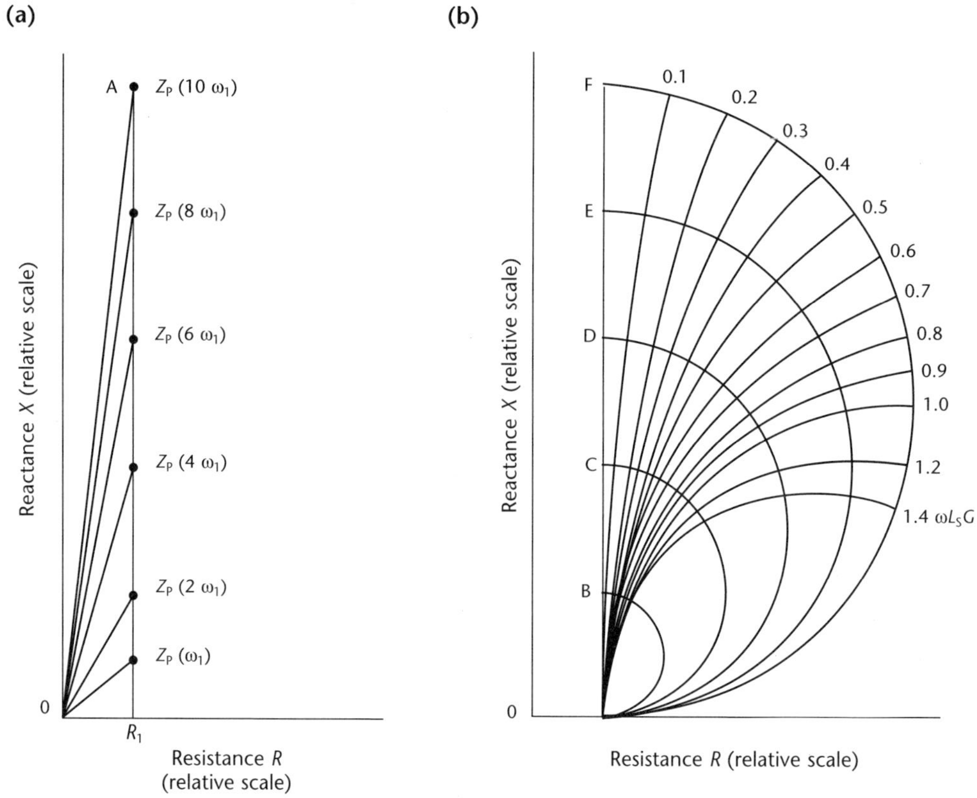

Legend

B, C, D, E, F = loci for selected values of Z_P
G = secondary conductance
Z_P = primary impedance
ω = angular frequency = $2\pi f$ where f = frequency (Hz)
ωL_S = secondary reactance

secondary conductance (siemens) and ωL_S is secondary reactance (ohm).

Further normalization is accomplished by dividing the reactive and resistive components by the primary inductive reactance ωL_0 without a secondary circuit present. In Fig. 13, the terms $\omega L \cdot (\omega L_0)^{-1}$ and $R \cdot (\omega L_0)^{-1}$ represent the relative impedance of the test coil as affected by the test object.

Signals generated by changes in ωL or R caused by test object conditions such as surface and subsurface discontinuities (as well as variations in liftoff, material thickness and conductivity) may be noted by $\Delta(\omega L)$ or ΔR to indicate a change in the impedance.

FIGURE 13. Normalized impedance diagram for long coil encircling solid cylindrical nonferromagnetic bar and for thin wall tube. Coil fill factor = 1.0.[12,13]

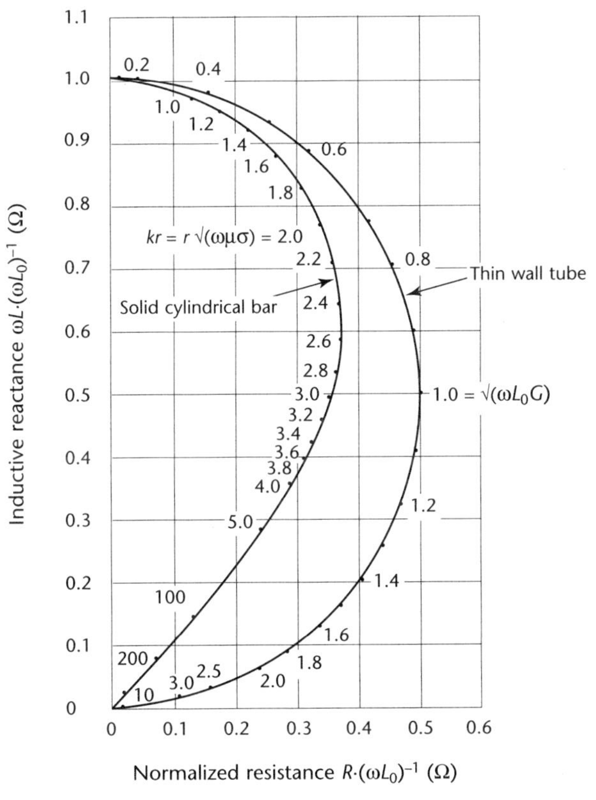

Legend

k = $\sqrt{(\omega\mu\sigma)}$ = electromagnetic wave propagation constant for conducting material
r = radius of conducting cylinder (m)
μ = magnetic permeability of bar ($4\pi \times 10^{-7}$ H·m^{-1} if bar is nonmagnetic)
σ = electrical conductivity of bar (S·m^{-1})
ω = angular frequency = $2\pi f$ where f = frequency (Hz)
$\sqrt{(\omega L_0 G)}$ = equivalent of $\sqrt{(\omega\mu\sigma)}$ for simplified electrical circuits, where G = conductance (S) and L_0 = inductance in air (H)

PART 4. Microwave Testing

Microwave Radiation[14]

The term *microwave* is used to denote all electromagnetic radiation waves whose frequencies lie between 0.3 and 300 GHz. These frequencies correspond to a range of free space wavelengths in vacuum from about 1 m (39 in.) to 1 mm (0.04 in.). In vacuum or air, microwaves travel at the velocity of light, about 2.998×10^8 m·s^{-1} (671 million mi·h^{-1}).

Microwaves in the frequency range above about 40 GHz are generally referred to as *millimeter waves* because their wavelengths in free space are conveniently measured in millimeters.

As seen in Fig. 14, microwaves occupy that portion of the electromagnetic spectrum between radio waves and infrared radiation.[14,15] Microwaves are common in daily life. The public first became familiar with them as the form of energy used for radar. Microwave ovens are commonly used both to cook and to dry foods. Telephone and communication circuits use microwave relay stations to transmit signals over distances of many miles. Television signals are often transmitted by means of microwaves and are sent and received by dish antennas, which are used in larger sizes for space communications and for radio astronomy. Guidance, tracking and control of spacecraft are made possible by microwaves. Microwaves are also used for nondestructive testing and spectroscopy.

Basic Experimental Approach

Microwaves propagate readily through most nonmetallic materials. In contrast, microwaves reflect almost completely from metal surfaces, penetrating only microscopic distances below the surface.

Figure 15 illustrates a typical experimental setup for the microwave transmission technique.[16] The basic idea is that a nonmetallic (dielectric) test object is irradiated by microwave energy from a transmitting horn antenna; the signal then travels through the sample and is received by a receiving horn antenna. The phase difference between the incident and the received signals is directly related to the slab thickness and its relative permittivity ε_r, which in general is a complex parameter:

$$(2) \quad \varepsilon_r = \varepsilon_r' - j\varepsilon_r''$$

Relative dielectric permittivity ε_r'' is related to the attenuation experienced by the signal while traveling through the slab. (Absolute permittivity is measured in farad per meter; relative permittivity is a ratio, nondimensional.)

A microwave sweep oscillator is used to generate a swept frequency signal, which is passed through an isolator and is then split into a test signal and a reference

FIGURE 14. Electromagnetic spectrum wavelengths and frequencies. Microwaves are between infrared and radio waves.[14]

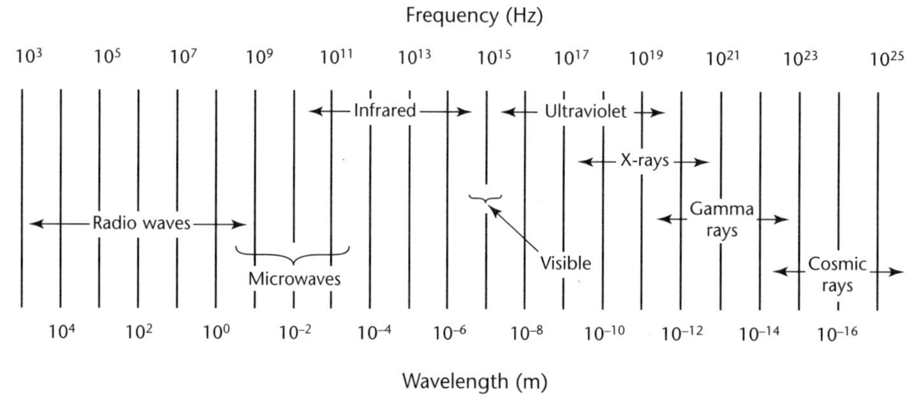

signal. The reference signal becomes the input signal to the reference channel of a microwave network analyzer. The test signal is fed through another isolator, which prevents reflections from corrupting the reference signal. After passing through a frequency meter, the signal irradiates the sample under test through a small transmitting horn antenna. The signal that propagates through the sample is then picked up by a small receiving horn antenna and is subsequently directed to the test channel of the network analyzer. The network analyzer compares the amplitude and the phase of the test signal with those of the reference signal.

A lossless dielectric material ($\varepsilon_r'' = 0$) has a relative permittivity ε_r that is real and greater than 1. The wavelength λ (meter) and the phase constant β (radian per meter) for an electromagnetic wave propagating in such a dielectric are:

$$(3) \quad \lambda = \frac{\lambda_0}{\sqrt{\varepsilon_r}}$$

and:

$$(4) \quad \beta = \frac{2\pi}{\lambda}$$

where:

$$(5) \quad \lambda_0 = \frac{c}{f}$$

and where c is the speed of light (about 2.998×10^8 m·s^{-1}) and f is frequency (hertz). It is evident that the wavelength λ in a dielectric material is shorter than wavelength λ_0 in free space. Hence, a dielectric slab has a longer electrical length than a column of air of equal thickness. The electrical length is defined as the number of wavelengths between two points. This reduction in wavelength causes a greater phase shift per unit length for a wave propagating in the dielectric, which is used to determine both the relative permittivity and the thickness of dielectric slabs.

Figure 16 shows the side and axial section views of an open ended coaxial aperture for the microwave reflection technique.[17] This technique is used for evaluation of surface cracks in metals.

The dominant mode of operation for coaxial lines is the transverse electromagnetic mode. The word *transverse* refers to the fact that the directions of the electric (radial) lines and magnetic field (concentric) lines are orthogonal to each other and both are orthogonal to the direction of energy propagation (along the coaxial line), as shown in Fig. 17.

When a coaxial line is cut at one end and is terminated by a metal plate, it is said to be *short circuited.* As a result, the electric field at the plate (the short circuit) is totally reflected with a phase shift of 180 degrees. When a crack is introduced in the metal plate, nearly all of the incident signal is reflected. However, the phase of the reflected signal depends on the frequency of operation, coaxial aperture dimensions, the dimensions of the crack and its location in the open ended coaxial aperture. The length of the crack exposed to the probing aperture

FIGURE 15. Measurement apparatus for transmission technique of microwave testing.[15]

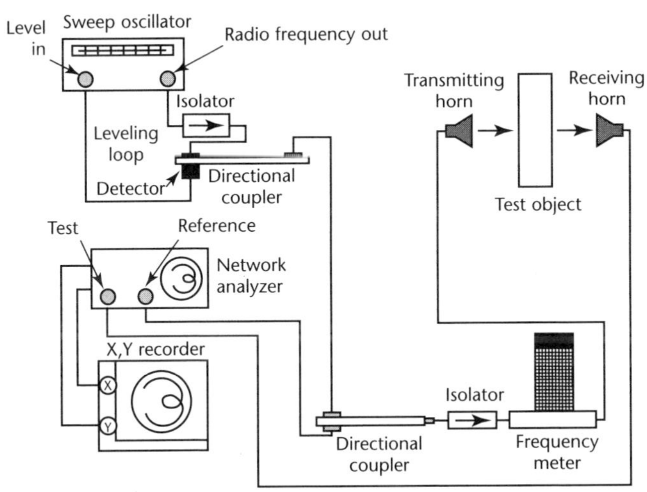

FIGURE 16. Open ended coaxial aperture for microwave reflection technique for evaluation of surface cracks in metals: (a) side view; (b) axial view.[16]

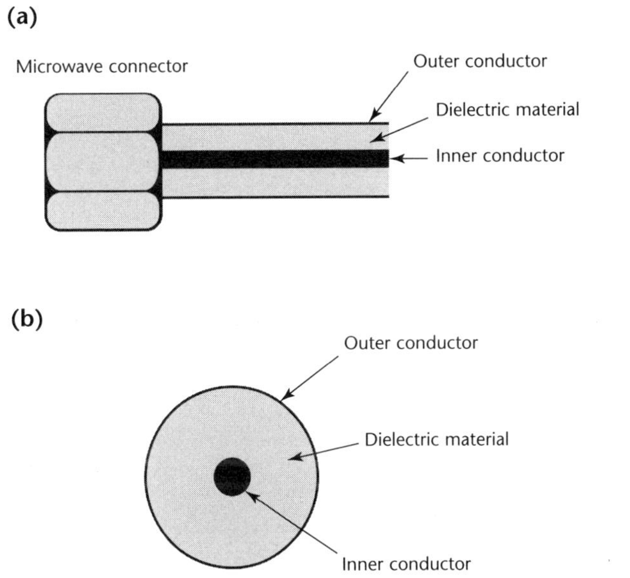

FIGURE 17. Field distributions and relative crack geometry at coaxial probe aperture, illustrating arrangement of transverse electromagnetic mode.[16]

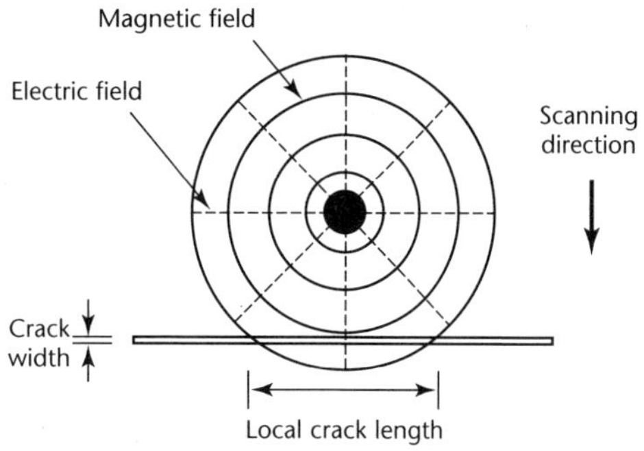

changes as a function of the scanning distance, as can be seen in Fig. 17.

One may use a vector network analyzer and measure the change in the magnitude and phase of the reflection coefficient. However, a relatively small, simple and inexpensive reflectometer or phase detector may be designed and constructed using discrete microwave components. In this way, a direct current voltage proportional to the magnitude or phase of the reflection coefficient can be measured and recorded to indicate the presence and the properties of a crack.

References

1. *Nondestructive Testing Handbook,* second edition: Vol. 6, *Magnetic Particle Testing.* Columbus, OH: American Society for Nondestructive Testing (1989).
2. Udpa, L. *Imaging of Electromagnetic NDE Phenomena.* Ph.D. dissertation. Fort Collins, CO: Colorado State University (1986).
3. Udpa, S.S. *Finite Element Modeling of Residual Magnetic Phenomenon.* M.S. thesis. Fort Collins, CO: Colorado State University (1981).
4. Lord, W. and D.J. Oswald. "Leakage Field Methods of Defect Detection." *International Journal of NDT.* Vol. 4. Kidlington, United Kingdom: Elsevier Science Limited (1972): p 249-274.
5. Förster, F. "On the Way from the 'Know-How' to the 'Know-Why' in the Magnetic Leakage Field Method of Nondestructive Testing (Part One)." *Materials Evaluation.* Vol. 43, No. 9. Columbus, OH: American Society for Nondestructive Testing (September 1985): p 1154-1162.
6. Libby, H.L. *Introduction to Electromagnetic Nondestructive Test Methods.* New York, NY: John Wiley and Sons (1971).
7. McMaster, R.C. and S.S. Udpa. Section 2, "Basic Concepts and Theory of Eddy Current Testing." *Nondestructive Testing Handbook,* second edition: Vol. 4, *Electromagnetic Testing.* Columbus, OH: American Society for Nondestructive Testing (1986): p 25-51.
8. Udpa, S.S. and L. Udpa. "Eddy Current Nondestructive Evaluation." *Wiley Encyclopedia of Electrical and Electronics Engineering.* New York, NY: John Wiley and Sons (1999): p 149-163.
9. Satish, S.R. *Parametric Signal Processing for Eddy Current NDT.* Ph.D. dissertation. Fort Collins, CO: Colorado State University (1983).
10. IEC 60028, *International Standard of Resistance for Copper.* Geneva, Switzerland: International Electrotechnical Commission (2001).
11. *Electrical Conductivity of Materials — Report ECT R8418-R1.* Virginia Beach, VA: Eddy Current Technology (2003).
12. *ASNT Level III Study Guide: Eddy Current Testing Method.* Columbus, OH: American Society for Nondestructive Testing (1983).
13. "Eddy-Current Inspection." *ASM Handbook,* ninth edition: Vol. 17, *Nondestructive Inspection and Quality Control.* Materials Park, OH: ASM International (1992): p 164-194.
14. Botsco, R.J. and R.C. McMaster. Section 16, "Microwave Theory." *Nondestructive Testing Handbook,* second edition: Vol. 4, *Electromagnetic Testing.* Columbus, OH: American Society for Nondestructive Testing (1986): p 461-488.
15. Wang, Y. and R. Zoughi. "Interaction of Surface Cracks in Metals with Open Ended Coaxial Probes at Microwave Frequencies." *Materials Evaluation.* Vol. 58, No. 10. Columbus, OH: American Society for Nondestructive Testing (October 2000): p 1228-1234.
16. Zoughi, R. and M. Lujan. "Nondestructive Microwave Thickness Measurements of Dielectric Slabs." *Materials Evaluation.* Vol. 48, No. 9. Columbus, OH: American Society for Nondestructive Testing (September 1990): p 1100-1105.
17. Zoughi, R. *Microwave Non-Destructive Testing and Evaluation.* Boston, MA: Kluwer Academic Publishers (2000).

Bibliography

Blitz, J. *Electrical and Magnetic Methods of Nondestructive Testing,* second edition. London, United Kingdom: Chapman and Hall (1997).

Cecco, V.S., G. Van Drunen and F.L. Sharp. *Eddy Current Testing,* U.S. edition. Columbia, MD: GP Courseware (1987).

Davis, J.M. and M. King. *Mathematic Formulas and References for Nondestructive Testing — Eddy Current.* Las Vegas, NV: Art Room Corporation (2001).

Hagemaier, D.J. *Fundamentals of Eddy Current Testing.* Columbus, OH: American Society for Nondestructive Testing (1990).

Harvey, E.D. *Eddy Current Testing Theory and Practice.* Columbus, OH: American Society for Nondestructive Testing (1995).

Lord, W., ed. *Electromagnetic Methods of Nondestructive Testing.* New York, NY: Gordon and Breach Science Publishers (1985).

4

CHAPTER

Modeling of Electromagnetic Testing

Lalita S. Udpa, Michigan State University, East Lansing, Michigan

Nathan Ida, University of Akron, Akron, Ohio (Parts 1 and 4)

John R. Bowler, Iowa State University, Ames, Iowa (Part 3)

Theodoros Theodoulidis, Aristotle University of Thessaloniki, Thessaloniki, Greece (Part 2)

PART 1. Modeling of Phenomena for Electromagnetic Testing

Introduction

Mathematical models are used to simulate the eddy current phenomenon and its applications in nondestructive testing. Models typically simulate an eddy current test and predict the probe signal associated with a specific discontinuity (a region where conductivity or permeability changes abruptly) under different experimental conditions. Results of such parametric studies are useful in designing probes, visualizing the interaction of the field with discontinuities, optimizing the test setup and generating discontinuity signatures that can be used to develop signal interpretation algorithms. Simulation models are relatively inexpensive compared to data acquired experimentally from artificial discontinuities.

All electromagnetic phenomena, including those relating magnetic flux leakage and eddy current testing, are governed by differential equations.[1]

Basic Differential Equations for Electromagnetic Fields[2]

The differential equations governing general, time varying electromagnetic fields at low frequencies, in regions that include magnetic and conducting materials and applied current densities, are derived from Maxwell's equations:[1]

$$\nabla \times \boldsymbol{E} = -\frac{\partial \boldsymbol{B}}{\partial t} \tag{1}$$

$$\nabla \times \boldsymbol{H} = \boldsymbol{J} \tag{2}$$

$$\nabla \cdot \boldsymbol{B} = 0 \tag{3}$$

$$\nabla \cdot \boldsymbol{D} = \rho \tag{4}$$

where $\boldsymbol{B}$ is magnetic flux density (tesla), $\boldsymbol{D}$ is electric flux density (coulomb per meter squared), $\boldsymbol{E}$ is electric field intensity (volt per meter), $\boldsymbol{H}$ is magnetic field intensity (ampere per meter), $\boldsymbol{J}$ is current density (ampere per square meter), t is time (second) and ρ is charge density (coulomb per cubic meter).

Equation 2 depends on the quasistatic approximation, which neglects displacement current. The microwave technique needs the displacement current but its omission is justifiable in the eddy current technique because the highest frequencies encountered are only on the order of a few megahertz. At such frequencies the conduction current in metals is typically many orders of magnitude greater than the displacement current. Charge can accumulate on discontinuity boundaries and on the surface of conductors, giving rise to a jump in the normal component of the electric field. However, Eq. 2 implies that $\nabla \cdot \boldsymbol{J} = 0$, which means for example that the current normal to a surface that acquires a charge is negligible. Although the charging current can be neglected, the effect of the charge on the electric field cannot be ignored. If the boundary is not abrupt, the incident charge is distributed over a volume.

Note that by setting all time derivatives to zero, these equations can be used to describe magnetic flux leakage phenomena. The same numerical model used for eddy current testing can be applied to magnetic flux leakage testing by equating the frequency of the source current to zero.

In addition to Maxwell's equations, the following relations describe linear, isotropic media:

$$\boldsymbol{B} = \mu \boldsymbol{H} \tag{5}$$

$$\boldsymbol{D} = \varepsilon \boldsymbol{E} \tag{6}$$

$$\boldsymbol{J} = \sigma \boldsymbol{E} \tag{7}$$

The permittivity or dielectric constant ε (farad per meter), the magnetic permeability μ (henry per meter) and the electric conductivity σ (siemens per meter) are treated here as scalar constants. In anisotropic media, each becomes a 3×3 tensor. Nonlinear behavior for any of the three properties may exist in a given situation. Although nonlinearity in conductivity and permittivity are rarely encountered in eddy current problems, nonlinearity of magnetic materials is common and is expressed as the field dependence of the permeability. For practical eddy current applications, the excitation levels are usually low enough to justify the linearity assumption for magnetic materials.

Using this assumption and substituting Eq. 5, Eq. 2 becomes:

$$(8)\quad \frac{1}{\mu}\nabla \times \boldsymbol{B} = \boldsymbol{J}$$

This, however, is not sufficient to completely specify the fields within the solution region because the current density $\boldsymbol{J}$ contains two different sources. The first and most obvious is the applied current density J_s. A second component is the induced eddy current density J_e.

Thus, Eq. 8 becomes:

$$(9)\quad \frac{1}{\mu}\nabla \times \boldsymbol{B} = J_s + J_e$$

At this point, it is useful to introduce the magnetic vector potential $\boldsymbol{A}$, which is defined as follows:

$$(10)\quad \boldsymbol{B} = \nabla \times \boldsymbol{A}$$

Substituting this in Eq. 8 and Eq. 1 results in Eqs. 11 and 12 for a source free region:

$$(11)\quad \frac{1}{\mu}\nabla \times \left(\nabla \times \boldsymbol{A}\right) = \sigma \boldsymbol{E}$$

$$(12)\quad \nabla \times \boldsymbol{E} = -\nabla \times \frac{\partial \boldsymbol{A}}{\partial t}$$

The electric field in Eq. 12 is:

$$(13)\quad \boldsymbol{E} = -\frac{\partial \boldsymbol{A}}{\partial t} - \nabla\phi$$

Equation 13 shows that the electric field can be partitioned into a magnetic vector potential term and a contribution written as the gradient of a scalar potential ϕ. The gradient of the potential is included to express the electric field as a general form that satisfies Eq. 12. The scalar potential is eliminated when Eq. 13 is substituted into Eq. 1 because the curl of the gradient is identically zero.

Therefore, the electromagnetic field is defined for any particular physical problem but $\boldsymbol{A}$ and ϕ are not yet defined. For example, a different potential gradient could be added to the vector potential term instead of the original $\nabla\phi$ and $\boldsymbol{A}$ could be adjusted to give the correct electric field. The resulting expression would satisfy Eq. 12 and yield the same magnetic flux from Eq. 1. Therefore, there is flexibility in the choice of $\boldsymbol{A}$ and ϕ. To ensure that the potentials are uniquely defined, the partition of the field must be fixed in some way. This is usually done by completing the definition of $\boldsymbol{A}$.

A vector field may be defined, apart from an arbitrary constant, by specifying its curl and its divergence. In the case of the magnetic vector potential, the curl is given by Eq. 10. It is necessary only to decide on the divergence to have it fully specified. The specification of the divergence is called the gage condition.

Substituting Eq. 13 into Eq. 12 gives:

$$(14)\quad \frac{1}{\mu}\nabla \times \left(\nabla \times \boldsymbol{A}\right) = J_s - \sigma\frac{\partial \boldsymbol{A}}{\partial t} - \sigma\nabla\phi$$

Expanding the left side with the vector identity $\nabla \times \nabla \times = \nabla\nabla - \nabla$ gives:

$$(15)\quad \frac{1}{\mu}\nabla^2 \boldsymbol{A} = -J_s + \sigma\frac{\partial \boldsymbol{A}}{\partial t} + \sigma\nabla\phi + \frac{1}{\mu}\nabla\nabla\cdot\boldsymbol{A}$$

The divergence of $\boldsymbol{A}$ is commonly defined as zero (coulomb gage) but this would not, in general, separate the scalar and vector potentials. Instead, the gage condition is chosen:

$$(16)\quad \sigma\phi + \frac{1}{\mu}\nabla\cdot\boldsymbol{A} = 0$$

which eliminates the last two terms in Eq. 15 to give:

$$(17)\quad \frac{1}{\mu}\nabla^2 \boldsymbol{A} = -J_s + \sigma\frac{\partial \boldsymbol{A}}{\partial t}$$

Equation 17 resembles the diffusion equation for heat flow and has similar time domain solutions.

Most eddy current testing, however, is performed with alternating current, for which time dependence is simply a harmonic oscillation in time. The harmonic oscillation is characterized by an amplitude and a phase, which can be conveniently represented in phasor form: $\boldsymbol{A}(\boldsymbol{r},t) = \Re\{\boldsymbol{A}(\boldsymbol{r})\,e^{j\omega t}\}$, where $\boldsymbol{A}(\boldsymbol{r})$ is a complex vector representing the amplitude and phase of the components of the magnetic vector potential and where $j = \sqrt{(-1)}$, $\Re$ denotes the operation of taking the real part and ω is angular frequency (radian per second). Note that the same symbol is used here to represent both the time dependent, real quantity $\boldsymbol{A}(\boldsymbol{r},t)$ and the complex quantity $\boldsymbol{A}(\boldsymbol{r})$ but they are distinguished by their arguments. Elsewhere, the arguments will not be given and the distinction between the two must be recognized by the context. The time derivative gives:

$$\text{(18)} \quad \frac{\partial A(r,t)}{\partial t} = \Re\left\{ j\omega\, A(r)\, e^{j\omega t} \right\}$$

Hence, for time harmonic theory, $j\omega$ is substituted for $\partial \cdot (\partial t)^{-1}$ in Eq. 17 and the vector potential can be viewed as a complex phasor. In this way, Eq. 17 becomes Eq. 19:

$$\text{(19)} \quad \frac{1}{\mu} \nabla^2 A = - J_s + j\omega\sigma A$$

Analytical and Numerical Models

There are different kinds of models. Some kinds are analytical and some are numerical. Analytical models are more computationally efficient than numerical models. However, numerical models are far more flexible and can be used to model complex discontinuity geometries, material nonlinearity and other complexities associated with real test scenarios.

Described below are analytical models that characterize eddy current behavior in homogeneous conducting media free from discontinuities, particularly the model proposed by Dodd and Deeds[10] and its extensions. Analytical and integral solutions, numerical techniques that cover discontinuities in materials, are also described below, as well as numerical techniques based on finite difference and finite element analysis techniques.

PART 2. Modeling of Homogeneous Conducting Media

Background

Model based quantitative eddy current testing has evolved steadily with improvements in computing power. A focus on accurate modeling has led to a thorough understanding of eddy current testing and to full automation of field tests.[2-7] Modeling is performed by solving Maxwell's equations and the solutions can be expressed either analytically or numerically. Analytical solutions provide closed form expressions for the parameters of interest in eddy current testing and are the subject of the present discussion.

Eddy current testing models can be used for coil design, test frequency selection and interpretation of test data. Important quantities to be calculated are the eddy current distribution induced in the specimen undergoing testing, as well as the resulting impedance change of the coil. Calculation and visualization of the eddy current flow pattern can be used to assess the true depth of penetration into the material and the interaction with particular discontinuities. In this way, the coil configuration can be optimized to ensure maximum interaction with given discontinuity types, properly taking into account frequency and material parameters. Calculation and visualization of impedance plane loci can be used for comparison with actual test signals. This comparison provides a better understanding of impedance variations from known discontinuities of particular size and orientation as well as from particular material and spatial features of the test object.

Problems concerning eddy current induction are formulated by means of differential equations, which determine the magnetic field and related quantities at a certain point in terms of an existing source current density. The flow of eddy currents is calculated by using the diffusion differential equation, which is conveniently expressed in terms of the magnetic vector potential. There are two ways of solving this differential equation: analytical techniques and numerical ones.

Analytically, the equation is solved by the technique of separation of variables within a region of the geometry. The influence of sources outside the region is accounted for by imposing appropriate boundary conditions. Analytical solutions may handle two-dimensional problems, axisymmetric problems and in certain cases three-dimensional ones, as long as the corresponding equations are linear and the geometry of boundaries and sources are relatively simple. Because the class of geometries that can be treated is usually restricted to problems with canonical boundaries (planar, cylindrical and spherical regions), these techniques allow only for an approximation to problems with noncanonical boundaries or discontinuities. The solutions from analytical techniques are general and exact and they provide deeper insight into the problem. They are obtained normally in the form of a mathematical relationship, which can then be used for analysis, parametric studies and calibration of test systems. An important aspect of analytical models is that closed form expressions are easily coded, either with higher programming languages or with commercial mathematical packages, and therefore require minimal effort by the developer. When the solutions are coded, they are much faster than numerical techniques, which require significantly longer computing times.

Analytical solutions are also used for validation of solutions from more complex numerical techniques. The latter produce numerical results rather than closed form expressions and their accuracy can be confirmed independently by analytical models, which provide an inexpensive alternative to experimental verification of numerical results.

Models for problems having canonical boundaries are described below, beginning with the well established models developed by Dodd and Deeds. Extensions of these models as well as three-dimensional models and semianalytical models for problems involving canonical boundaries are presented. Approximate solutions with application to discontinuity modeling are presented elsewhere, below.

Analytical Models

In the case of a two-dimensional axisymmetric geometry with rotational symmetry about the Z axis, Eq. 17 in a source free region becomes:

$$(20)\quad j\omega\mu\sigma A_\phi = \frac{\partial^2 A_\phi}{\partial r^2} + \frac{1}{r}\frac{\partial A_\phi}{\partial r} + \frac{\partial^2 A_\phi}{\partial z^2} - \frac{A_\phi}{r^2}$$

The above equation is solved by adopting the technique of separation of variables. Although many applications can be modeled with an axisymmetric geometry, many applications are described by a three-dimensional geometry that exhibits special difficulties. These difficulties arise when using curvilinear coordinates for the description of the problem because the components of $\boldsymbol{A}$ (Eq. 17) are coupled together in the resulting scalar differential equations. In this case, the technique of separating variables cannot be applied. The inconvenience is avoided by using the second order vector potential $\boldsymbol{W}$, which was introduced by Smythe.[8] For the case of a solenoidal $\boldsymbol{A}$, having a zero divergence as in Eq. 17, $\boldsymbol{W}$ is defined as:

$$(21)\quad \boldsymbol{A} = \nabla \times \boldsymbol{W} = \nabla \times \left[W_a \boldsymbol{u} + \left(\boldsymbol{u} \times \nabla W_b\right)\right]$$

where $\boldsymbol{u}$ is a fixed unit vector and W_a and W_b are two orthogonal scalar functions satisfying the scalar equation:

$$(22)\quad \nabla^2 W_{a,b} = j\omega\mu\sigma W_{a,b}$$

Because the above equation is separable in a number of coordinate systems, formulations based on $\boldsymbol{W}$ can be used effectively for the separation of the vector differential equation of Eq. 17.

Analytical models suitable for eddy current testing have been developed over the years by workers in nondestructive testing and in geophysics and by designers of magnets, motors and accelerators. Initially, the basic problem studied was that of a filamentary current source beside a conducting test object. A review and a list of solutions are presented by Tegopoulos and Kriezis[9] for a variety of configurations with regard to the shape of the sources and the geometry of the conducting media. The two-dimensional problems are studied by using the magnetic vector potential $\boldsymbol{A}$, whereas the three-dimensional problems are treated by using the second order vector potential $\boldsymbol{W}$.

Dodd and Deeds Models

In the theory of eddy current testing, the work of Dodd and Deeds[10] has provided the basis for one of the most popular models. Building on a range of earlier work, they presented solutions for eddy current distributions, in the form of fourier-bessel integrals, for a number of axisymmetric coil configurations often encountered in eddy current test applications. These solutions have been applied to the calculation of eddy currents produced by cylindrical coils in planar and cylindrical conductors, in the analysis of coil impedance changes caused by the presence of such conductors and in the prediction of impedance changes caused by subsurface discontinuities.[11,12] An

FIGURE 1. Test object geometries for models of Dodd and Deeds: (a) layered half space; (b) layered bore hole; (c) layered sphere.

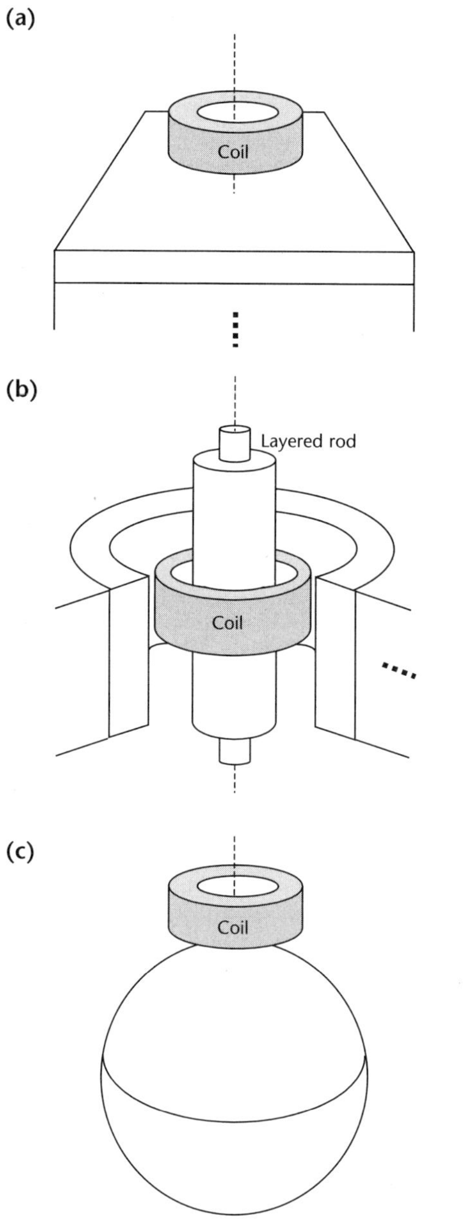

essential feature of Dodd and Deeds[3] analysis is that at typical eddy current frequencies, a multiple-turn coil wound with round insulated wire can be approximated by a current sheet, obtaining the electromagnetic field by superposition.

The differential equation solved was Eq. 20 and the impedance of the coil was calculated from the following expression for axial symmetry:

$$\text{(23)} \quad Z = j\omega 2\pi \frac{N}{A_{cs}} \iint\limits_{A_{cs}} r A_\phi \, dr \, dz$$

where A_{cs} is the cross sectional area (square meter) and N is the number of turns in the coil. The superposition principle is applied by integrating the magnetic vector potential over the cross sectional area of the coil.

Closed form expressions for the electromagnetic field and the coil impedance were obtained for a variety of common test object geometries (Fig. 1): for a cylindrical coil of rectangular cross section above a layered plane, encircling a layered rod or inside a cylindrically layered bore hole. The spherical configuration of Fig. 1c was also considered but the particular case of a rectangular cross section coil was analyzed by Nikitin.[13,14] Once the calculations are performed using a single coil, the analysis can be extended to multiple coil configurations simply by superimposing the solutions.[11,15] Dodd's models were also extended to an arbitrary number of layers, by using the matrix technique proposed by Cheng, Dodd and Deeds.[16-18]

For the case of a coil over a homogeneous conducting half space (Fig. 2a), the analytical expression for the coil impedance is given:

$$\text{(24)} \quad Z = \frac{j\omega\pi\mu_0 N^2}{l^2 \left(r_2 - r_1\right)^2} \int\limits_0^\infty \frac{J^2\left(r_1, r_2\right)}{a^6} \times \left[2\left(al + e^{-al} - 1\right) + \left(e^{-a(l + l_0)} - e^{-a l_0}\right)^2 \times \frac{a\mu_r - a_1}{a\mu_r + a_1} \right] da$$

where:

$$\text{(25)} \quad J\left(r_1, r_2\right) = \int\limits_{a \cdot r_1}^{a \cdot r_2} x J_1(x) \, dx$$

and:

FIGURE 2. Coil above metal plate: (a) geometric configuration; (b) normalized impedance plane display.

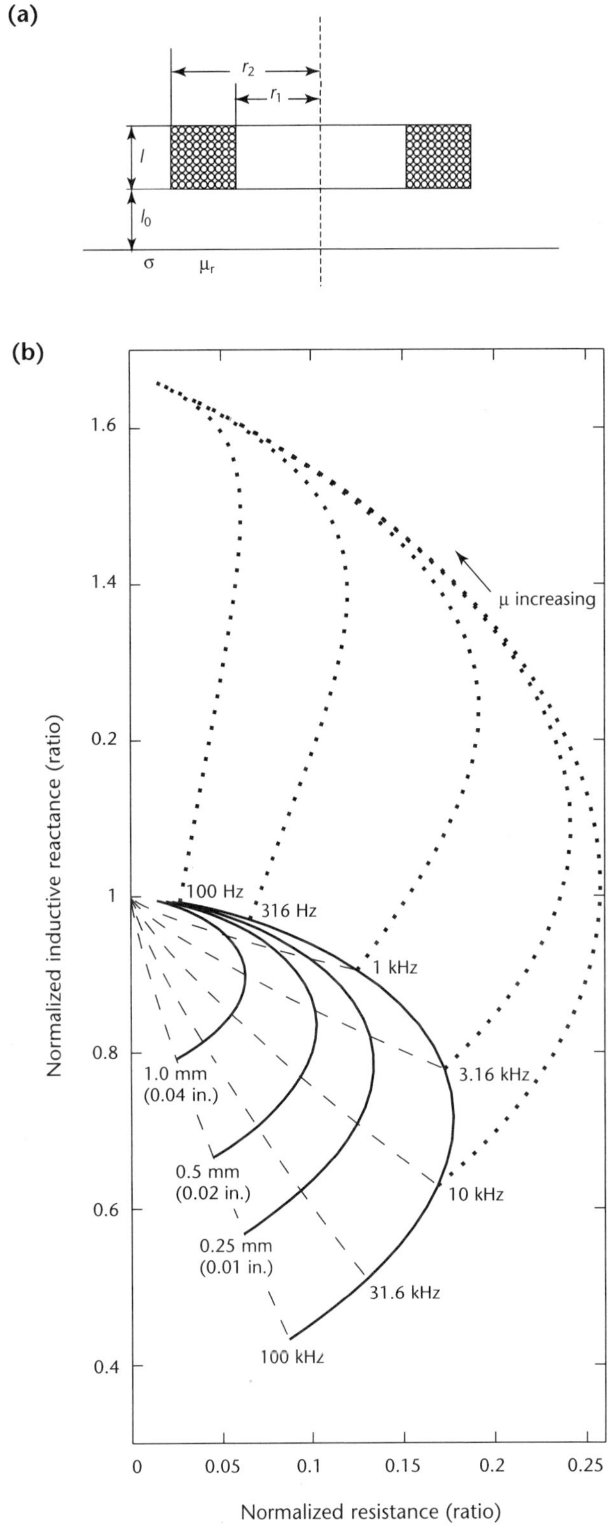

Legend

r_1 = coil inner radius = 2 mm (0.08 in.)
r_2 = coil outer radius = 4 mm (0.16 in.)
l = coil width = 1 mm (0.04 in.)
μ_r = relative magnetic permeability of half space (ratio) = 1
σ = 35.4 MS·m^{-1} (61 percent International Annealed Copper Standard)

$$\textbf{(26)} \quad a_1 = \sqrt{a^2 + j\omega\mu_r\mu_0\sigma}$$

where a is the integration variable, $J_1(x)$ is the bessel function of the first kind and first order, l is the width of the coil (meter), l_0 is the liftoff (meter), r_1 is the inner radius of the coil (meter), r_2 is the outer radius of the coil (meter), μ_r is relative magnetic permeability (dimensionless), μ_0 is magnetic permeability (henry per meter) of free space and σ is conductivity (siemens per meter).

The eddy current density is calculated from the magnetic vector potential:

$$\textbf{(27)} \quad J_\phi = -j\omega\sigma A_\phi$$

In the case of a normal coil above a half-space conductor (Fig. 2a), the induced current density is as follows:

$$\textbf{(28)} \quad J_\phi(r,z) = \frac{-j\omega\sigma\mu_r\mu_0 N}{l(r_2 - r_1)} \int_0^\infty \frac{J(r_1,r_2)}{a^2} \times \left(e^{-a(l+l_0)} - e^{-al_0}\right) \times \frac{e^{a_1 z}}{a\mu_r + a_1}\, da$$

Equations 24 and 28 involve the numerical computation of an infinite integral. Numerical integration techniques available in most numerical analysis software packages can be used to calculate the integrals.

Fig. 2b is an example of computer generated impedance display for a surface coil. The impedance is depicted normalized, using the inductive reactance of the coil in air as the normalizing factor. (This quantity can also be computed from Eq. 24 by setting conductivity to zero, $a_1 = a$). The main purpose of such impedance displays is that they demonstrate the optimum frequency for a specific test. This frequency is usually the one that produces the best phase difference between the loci of two parameters. The conducting half-space material is aluminum and the solid curve represents the locus produced by varying the excitation frequency. Because the conductivity and frequency always appear as a product in Eq. 22, the same curve would have been produced for a constant excitation frequency and a varying conductivity. The dashed lines are the liftoff curves and represent the impedance variation with coil liftoff. The dotted curves show the impedance variation with frequency for different magnetic permeabilities of the half-space material.

Figure 3 is an example of a computer generated display of eddy current contours induced by a surface coil at various frequencies. As expected, the higher frequencies result in a smaller penetration of the eddy currents in the conducting object. Using Eq. 28 for a variety of coils reveals that peak eddy current densities associated with larger coils fall off more slowly with depth than those produced by smaller coils. A similar investigation conducted by Mottl[19] showed that the standard depth of penetration and linear-with-depth phase delay, obtained as solutions for the plane wave case, very rarely approximate the eddy current distribution in conducting samples beneath a real coil. The standard depth of penetration remains a material parameter rather than a real measure of penetration.

The Dodd and Deeds models have been proven very useful because they were successful in predicting experimental data from eddy current measurements. Since the 1970s, they have been widely used by the nondestructive testing community in

FIGURE 3. Contours of eddy currents induced by surface coil at various frequencies: (a) 1 kHz; (b) 10 kHz; (c) 100 kHz.

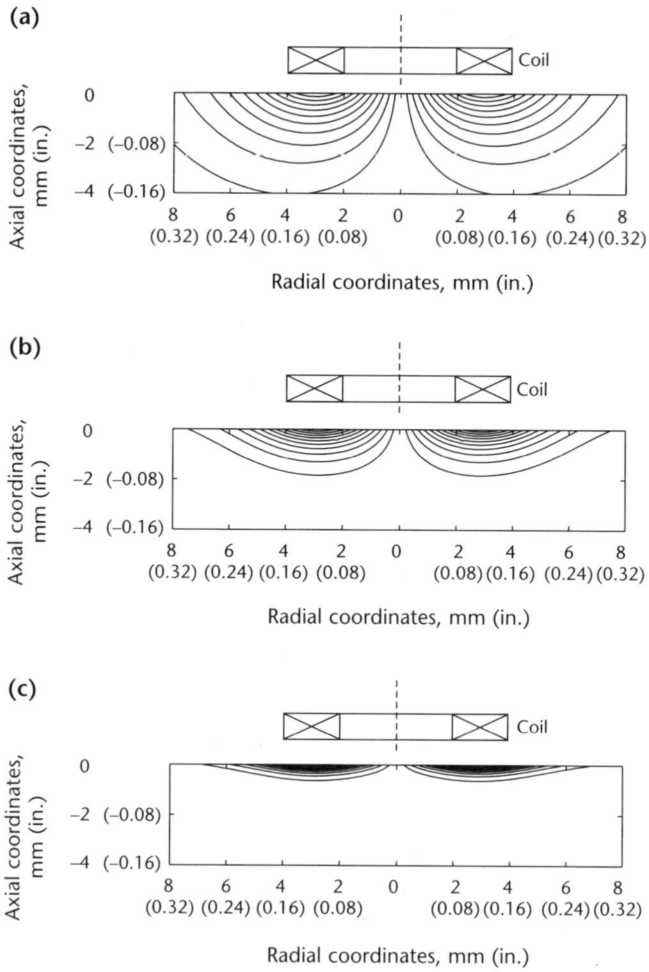

the design of eddy current tests. More specifically, they have been used to optimize general types of eddy current tests such as thickness and conductivity measurements, to optimize specific tests for specific problems and to help design general induction instrumentation for process control.

Extensions of Dodd and Deeds Models

The Dodd and Deeds models assume a harmonic time variation for the solution of the diffusion equation. Similar modeling techniques can be used in the case of transient coil excitations, such as step time functions or rectangular pulses. These current excitations are used in the pulsed eddy current technique, which is applied to either metal loss or crack detection at greater depths.

In addition to coil superposition, different frequencies can also be superimposed to obtain the response of a transient eddy current system. A simple technique of evaluating transient fields is to obtain, through a fourier transform, the frequency spectrum of the excitation current pulse and to calculate the voltage response at each frequency, thus acquiring the voltage frequency spectrum. The transient voltage response is then obtained by an inverse fourier transform. A distinct advantage of this technique is that it can be approximated numerically using the fast fourier transform. Bowler[20] uses this approach for a pulsed excitation having the form of a step function with the coil located above a layered system consisting of two slabs. The configuration mimics geometries encountered in the detection and identification of the metal loss in lap joints of aircraft.

Another technique of evaluating transient fields is to compute the laplace transform of the field equations, solve the transformed equations and recover the time domain behavior through an inverse laplace transform. This approach is followed by Waidelich,[21] Ludwig,[22] Sapunov[23] and Bowler[24] to obtain the voltage response of a coil situated above a layered conducting plane. In the case of a homogeneous conducting half space or for simple thin plate systems,[25] the inverse laplace transform can be obtained analytically but in the case of a layered half space this is not possible and numerical techniques are needed to obtain the response as a function of time. In the above situation, a robust numerical routine should be used for computing the inverse laplace transform. In other situations, it is preferable to work with the frequency domain solution, as already described, using the fourier transform.

Figures 4 to 6 show voltage responses derived for the case described by Bowler.[20] The voltage response is computed by numerically evaluating the inverse laplace transform. It is observed that certain features of the pulse, such as the amplitude of the pulse, the time of arrival of the maximum and the cross point, are sensitive to different geometry characteristics, thus making possible the estimation of metal loss.

Other extensions of Dodd's modeling technique concern the conductivity and permeability profiles of the test objects. Applications include case hardening, heat treatment, ion bombardment or chemical processes, which produce smoothly varying near surface conductivity and permeability profiles. In these cases, where for example the conductivity $\sigma(z)$ in Eq. 20 is a continuous function of depth, the electromagnetic field and the

FIGURE 4. Top plate metal loss in system of two plates: (a) setup; (b) transient electric potential. Depicted signal is coil voltage subtracted from response of same coil due to conducting half space. Percentage of parameter variation is in terms of thickness of one slab.

(a)

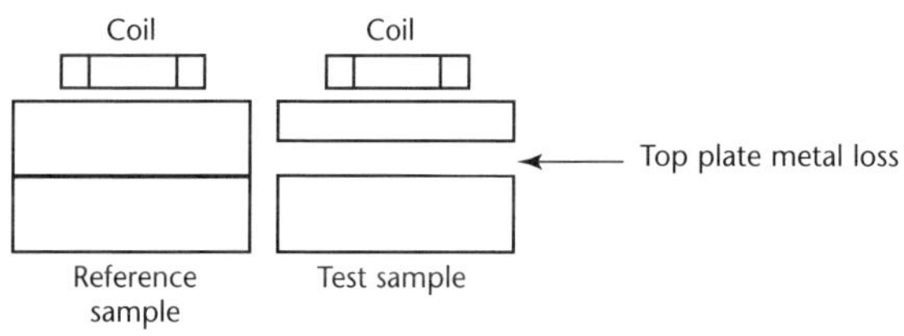

(b)

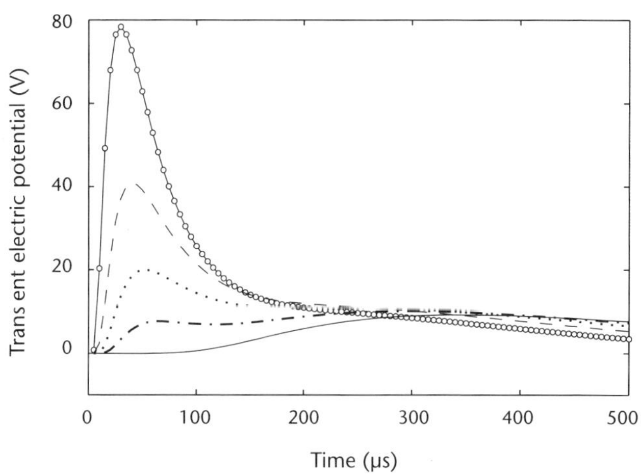

Legend

- —○— = 50 percent
- — · = 37.5 percent
- . . . = 25 percent
- . — = 12.5 percent
- —— = 0 percent

impedance of the coil can be evaluated in two ways.

The first is to solve Eq. 20 analytically for special forms of conductivity variations. Such solutions that result in closed form expressions involving higher transcendental functions have been derived by many researchers for specific functions not only of conductivity but also of magnetic permeability profiles.[26-29] This approach is much faster than the more general piecewise approach described next.

As discussed above, Cheng[17] extended Dodd and Deed's models to layered regions with an arbitrary number of layers. If continuous conductivity and permeability profiles are replaced with piecewise constant profiles, then it is possible to approximate numerically the coil impedance by implementing the above technique. The greater the number of layers, the better the approximation. Using this technique, Uzal[26] studied the problem of a coated conductor whose coating conductivity varied continuously with depth and permeability. Although this technique is slower than the one based on the analytical solution for each specific profile, it is more general and particularly useful when it is desired to solve the inverse problem, that is, to evaluate the profile from variable frequency measurements. The piecewise approach was also extended to cylindrical and spherical test objects by Uzal and Theodoulidis, respectively.[30,31]

Three-Dimensional Models

The models described so far are two-dimensional and axisymmetric. Their simplicity lies in the fact that the magnetic vector potential has only one component and the technique of separation of variables is applicable.

A significant amount of work concerns models of coils that have shapes other than the classical cylindrical coil or positions that destroy the axisymmetry. A problem of great interest is the evaluation of the three-dimensional electromagnetic field for a coil with an arbitrary shape and orientation above a conducting half space.

FIGURE 5. Plate separation in system of two plates: (a) setup; (b) transient electric potential. Depicted signal is coil voltage subtracted from response of same coil due to conducting half space. Percentage of parameter variation is in terms of thickness of one slab.

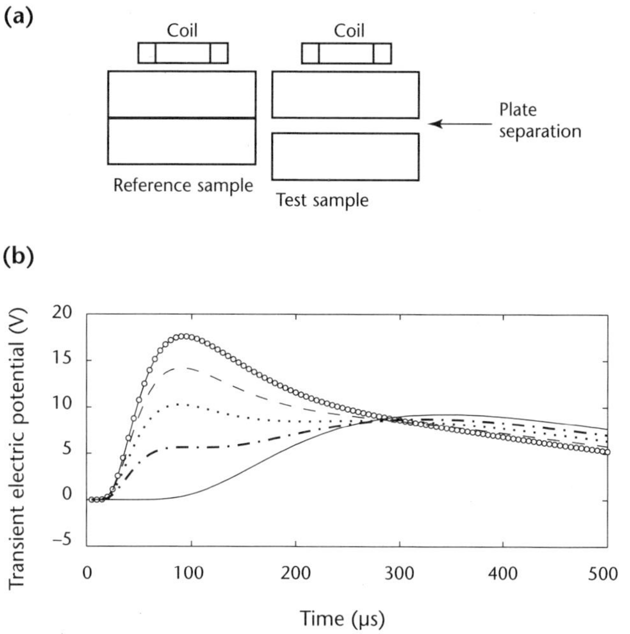

Legend
- —o— = 50 percent
- — · = 37.5 percent
- · · · = 25 percent
- · — = 12.5 percent
- —— = 0 percent

FIGURE 6. Bottom plate metal loss above system of two plates: (a) setup; (b) transient electric potential. Depicted signal is coil voltage subtracted from response of same coil due to conducting half space. Percentage of parameter variation is in terms of thickness of one slab.

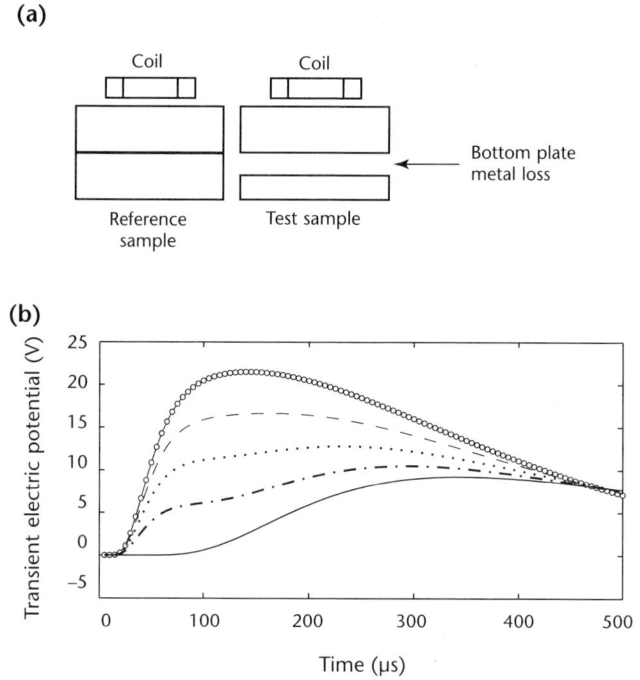

Legend
- —o— = 50 percent
- — · = 37.5 percent
- · · · = 25 percent
- · — = 12.5 percent
- —— = 0 percent

Weaver[32] presented a general theory of electromagnetic induction in a conducting half space by an external magnetic source using the electric and magnetic hertz vectors whereas Hannakam[33] provided solutions for a filamentary coil using the similar second order vector potential formulation. Based on the latter formulation, Kriezis[34] evaluated the eddy current density induced in a conducting half space by a filamentary coil whose axis is parallel to the surface.

Other researchers like Beissner[35] and Bowler[36] have favored Green's dyadic functions in solving the problem. Bowler was able to present analytical expressions for the eddy current density of a vertically oriented cylindrical coil over a conducting half space, thus extending the results of Kriezis to an eddy current probe coil of finite cross section. Beissner[37] and Tsaknakis[38] presented formulas for the eddy current distribution from cylindrically symmetric sources inclined at an arbitrary angle with respect to the surface normal. The general solution for a nonsymmetric source is in the form of a two-dimensional fourier integral. Numerical computations for the nonsymmetric case are therefore more demanding than those needed to evaluate fields from Dodd and Deeds formulas, where the integrals are one-dimensional. A semianalytical model was also presented by Juillard[39] for the same problem where the coil is divided in a number of elements called *point current sources*. The problem is solved for each point current source and superposition is applied to compute the electromagnetic field from the whole coil. Another technique for computing the magnetic field, based on the fourier transform, was presented by Panas[40] and Sadeghi,[41] who solved the problem of an elliptical coil and a rectangular coil in an inclined position, respectively.

An important conclusion of all these studies is that the eddy currents induced in the conductor flow parallel to the surface of the conductor, irrespective of the shape of the inducing coil. Figures 7 and 8 show the eddy currents induced on the surface of a conducting metal half space from a rectangular coil when the coil is parallel and perpendicular to the metal.

The problem of an arbitrarily shaped coil beside a cylindrical conducting system was studied by Hannakam[42] with the second order vector potential and by Grimberg[43,44] with dyadic Green's functions. Hannakam,[45] Theodoulidis[46] and Mrozynski[47] extended the second order vector potential formulation in the spherical coordinate system to solve for an arbitrarily shaped coil beside a conducting sphere. An important conclusion was that the eddy currents flow in spherical surfaces concentric with the conductor's surface.

All of the above analytical solutions concern the electromagnetic field with emphasis on the induced eddy current density. The impedance change of the coil, on the other hand, is calculated in two steps: (1) first the three-dimensional problem of evaluating the electromagnetic field is solved analytically and (2) then the general expression of the impedance change of a coil is applied. An impedance change expression was derived by Auld.[48] It was shown, through the lorenz reciprocity theorem, that the change in the impedance of an eddy current probe in the presence of a discontinuity is expressed in terms of an integral

FIGURE 7. Eddy current testing with rectangular coil parallel to test object: (a) setup; (b) eddy current pattern.

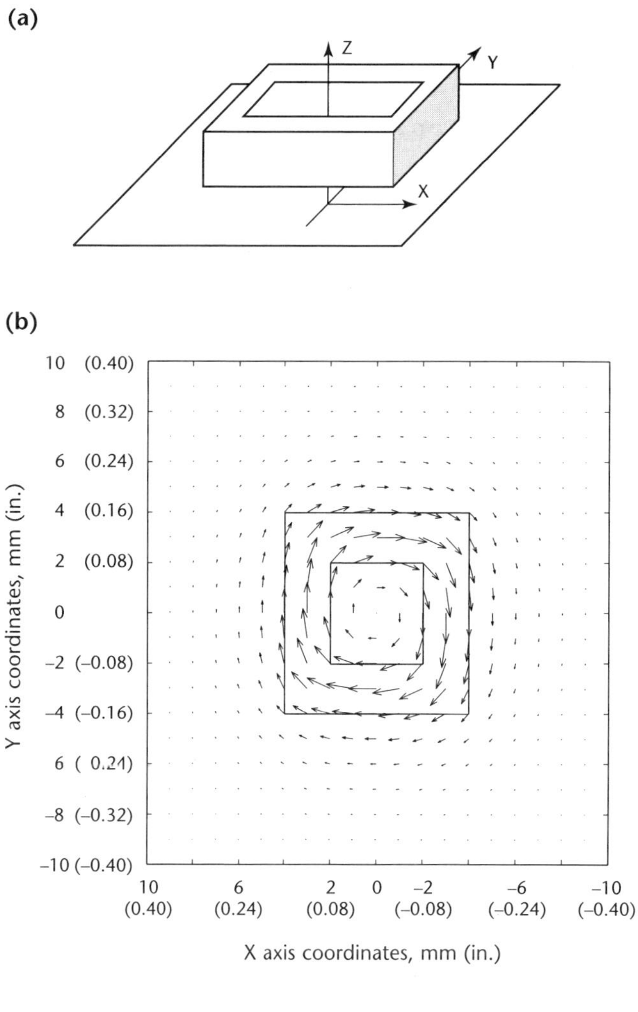

evaluated over any closed surface S containing the discontinuity:

$$(29)\quad \Delta Z = \frac{1}{I^2} \times \oint \left(\boldsymbol{E'} \times \boldsymbol{H} - \boldsymbol{E} \times \boldsymbol{H'} \right) \boldsymbol{n} \, dS$$

where $\boldsymbol{n}$ is the unit vector normal to the surface and where $\boldsymbol{E}$ and $\boldsymbol{H}$ are the electric and magnetic field intensities; the primed symbols denote the fields in the presence of the discontinuity and the unprimed symbols denote the fields in the absence of the discontinuity. The ΔZ formula is well suited to derivation of general expressions and can also be used effectively to compute the impedance change of a coil in canonical problems.[5] This development is significant because the coil geometry does not appear explicitly (no integrals appear over the volume of the coil) and allows the choice of planar, cylindrical and spherical boundaries in keeping with the symmetry of the problem.

In the particular case of a coil with arbitrary shape and orientation, above a conducting half space, the surface of integration coincides with the surface of the half space, closed by a surface at infinity, which makes no contribution.

Following this approach and solving analytically for the three-dimensional electromagnetic field, Burke[49,50] presented the following general expression for the impedance of any coil over a conducting half space:

$$(30)\quad \Delta Z = \frac{2 j \omega}{\mu_0 I^2} \int_{-\infty}^{\infty} du \int_{-\infty}^{\infty} dv \, \hat{B}_Z^S(u,v) \times \hat{B}_Z^S(-u,-v) \, \frac{a \mu_r - a_1}{a \left(a \mu_r + a_1 \right)}$$

where u and v are integration variables,

$$(31)\quad a = \sqrt{u^2 + v^2}$$

and:

$$(32)\quad a_1 = \sqrt{a^2 + j \omega \mu_r \mu_0 \sigma}$$

The term $\hat{B}_Z^S(u,v)$ denotes the double fourier transform of the normal component of the source magnetic field on the surface of the metal plane. For simple coil shapes, it has an analytical expression in terms of u and v. For more complex shapes, it has to be calculated numerically using the Biot-Savart law. The same approach was followed by Theodoulidis[51,52] for evaluating the impedance of a rectangular coil over a conducting half space and was further extended to cylindrical coordinates for evaluating the impedance of a bobbin coil in an offset position to a tube, thus simulating the wobble signal present during tube tests.

Perturbation and Eigenfunction Expansion

The class of problems that can be solved analytically can be extended with the aid of perturbation techniques, which are often used to provide solutions to physical problems that would otherwise be difficult or time consuming to treat. Perturbation techniques are inherently approximate and their main applicability is in the modeling of discontinuities. Such techniques can be used by assuming that the conductivities of the discontinuity and the surrounding medium do not differ very much or by considering limiting cases such as a high frequency limit.[53]

Nevertheless, perturbation techniques have also been applied to models of canonical problems. A technique called the *layer approximation,* based on the analytic transfer matrix solution for the electric field in a layered metal, was used by Satveli[54] to calculate the impedance change in a number of canonical problems. Burke[55] also has presented a perturbation technique, which enables the impedance computation in the high frequency limit when the conducting region is canonical. The technique was applied to the cases of a two-dimensional conducting wedge and a slot in a conducting half space.

Eigenfunction expansions can also be used to further extend the class of problems that can be solved analytically.[56-58] The problem is again solved using separation of variables; because the region of interest is finite, however, extra boundary conditions limit the domain of the solution. As a result, the solution involves series instead of integrals. The coefficients of the series are computed by solving a matrix system, which is formed by imposing the interface and boundary conditions of the problem. The numerical computation of the coefficients classifies the technique as semianalytical. The technique was effectively used by Theodoulidis[59] to derive an expression for the impedance of a ferrite cored probe coil over a conducting layered half space.

Conclusions

Analytical solutions in eddy current testing, although restricted to certain geometries as compared to the more general numerical solutions, have an explicit and closed form. The models are not computationally intensive and offer accurate solutions. They have limited scope but not limited value.

Whenever plausible, analytical solutions are preferable to numerical ones because they are easier to apply, are less expensive to compute, are more accurate and finally allow for easy parametric studies of the test geometry.

PART 3. Analytical and Integral Models for Simulating Cracks

Introduction

Eddy current nondestructive testing uses inductive probes to excite currents in electrical conductors. The simple fact that the coil carrying an alternating current can sense a discontinuity in a metal is intuitively easy to understand but evaluating the signal for a given configuration of coil and discontinuity is not always easy. The present discussion describes calculations of probe signals from cracks, starting with a review of the basic theoretical concepts and moving on to a number of related techniques for evaluating probe response.

Early investigators applied concepts from other fields of electromagnetism to problems in eddy current testing. The researcher in relatively unexplored areas of electromagnetic theory inevitably brings concepts from the parent discipline and adapts them for the new field of investigation. As advances in the new area begin to mature, the new discipline adopts distinct themes and approaches that are successful and rewarding. At the end of the twentieth century, eddy current nondestructive testing was at a point of early maturity. Basic problems had been solved satisfactorily yet many problems remained open and relatively underdeveloped.

This discussion of crack theory briefly reviews a few significant early developments relevant to the treatment of crack problems in eddy current testing, including the analysis of the spherical inclusion and the penny shaped crack. Recent advanced developments in the evaluation of crack signals are then briefly outlined. Two approaches are described: (1) integral techniques that represent the effect of a discontinuity in terms of dipole distribution and (2) approaches valid at high frequencies that use small approximations of standard depth of penetration.

Elements of Crack Theory

The pioneering achievements of Friedrich Förster and his colleagues in eddy current nondestructive testing resulted from extensive theoretical and experimental investigations,[60] laying the foundations on which others have built over the intervening half century. Early uses of eddy current testing investigated by Förster are metal sorting, hardness measurement and the evaluation of heat treatments through the effects of electrical resistivity variations. In developing instruments for these measurements, Förster recorded the impedance change of a solenoid when it was near an electrically conducting material. In the initial investigations, the solenoid impedance changes due to the cylindrical rods acting as cores were measured using an inductance bridge. It soon became apparent that the measurements yielded results dependent on the dimensions of the rod. Consequently, much effort would be devoted to the problem of separating the effects of variations in the sample dimensions and the variation in resistivity. Förster's ultimate success was made possible by his willingness and ability to analyze the problem theoretically.[61]

Förster used analytical expressions for the impedance of an infinite solenoid in the presence of a conducting rod to account for the effects of variations in rod diameter and material properties. Later Dodd and Deeds derived closed form integral expressions for the field and impedance of an axial coil of finite length encircling an infinitely long rod.[10] In addition, they derived integral expressions for the impedance and field of a normal coil above a layered half-space conductor, a normal coil being one whose axis is normal to the surface of the conductor. Although ferrite cored probes may be preferred for discontinuity detection because of their enhanced sensitivity, air cored coils have been widely used in calculations because of the ease of evaluating the field using the formulas of Dodd and Deeds. Usually numerical techniques are needed to calculate the fields of probes with ferrite cores.[1,62] However, Theodoulidis has shown that solutions satisfying Maxwell's equations for axially symmetric ferrite cored probes can be found.[59,63] Other significant and interesting results using the analytical solutions of Maxwell's equations are described elsewhere in this chapter.

More than a decade after Förster's work became widely known, an embryonic discontinuity theory was given in the dissertation of Michael Burrows.[64] Central to the thesis is the idea that a small discontinuity in a conductor, such as a tiny spherical cavity, produces a perturbed field that is the same as that of a suitably chosen dipole. Because the discontinuity is small compared with the standard depth of penetration and small on the scale of other spatial variations of the unperturbed field, the field can be approximated as locally uniform and the polarization of a spheroidal discontinuity can be found by using standard textbook theory.[65] Having determined the dipole intensity, Burrows found the induced electromotive force in a pickup coil due to the discontinuity by using an expression derived from reciprocity principles.[66] Because key elements of this approach arise in more advanced treatments of discontinuities, the dipole analysis will be summarized later.

The small discontinuity analysis is itself of limited practical application but the principle of representing the effect of a discontinuity by an equivalent electromagnetic source distribution can be applied to arbitrary discontinuities using either multipole expansions or a dipole distribution. Multipole techniques for representing the field have not been pursued[67] extensively in nondestructive testing although they may be fruitful. However, numerous approaches have been developed based on the representation of a discontinuity in terms of a current dipole distribution. Early developments in which a volume dipole density was expanded in terms of volume elements were made by the geophysics community,[68-70] followed by an adaption of the technique by McKirdy[71] and by Bowler, Jenkins, Sabbagh and Sabbagh[72,73] to the solution of problems in eddy current testing. An account of the volume element technique is given in this handbook and elsewhere.

Although the equivalent source representation is a common feature of a number of crack response calculations, a seminal article by Kahn, Spal and Feldman[74] can be seen as a significant initial step for developments that have taken a different path. In the essentially two-dimensional problem, the field is uniform along the length of a crack of constant depth and negligible opening. If the standard depth of penetration is small compared with the crack depth, the current flow follows stream lines parallel to the crack faces except at the corners where the crack meets the surface of the conductor and in the region of the crack edge. Kahn gives local solutions for the corner, face and edge field, each of which contribute to the impedance change. An interesting feature of the edge field is that it has the same mathematical form as that given by Sommerfeld[75] for the diffraction of a plane wave by a half-plane barrier.

The diffraction of an electromagnetic wave at a thin conducting barrier and the flow of eddy currents around the edge of a crack are physically distinct phenomena but both are governed by Maxwell's equations and are subject to comparable boundary conditions. For a time harmonic field, the physical difference between the two cases is manifest in the wave number, a number that is real in a lossless medium but complex in a conductor. Hence the solutions are essentially the same, differing only in the nature of the wave number.

Before describing in more detail the implications of the equivalent source approaches, typical examples of the outcomes of such calculations in the form of probe signals due to cracks are reviewed.

Impedance Plane

The impedance of an eddy current probe varies with frequency and with its proximity to the conductor as measured by a liftoff parameter, defined here as the distance from the surface of the conductor to the base of the coil. Energy dissipated by induced current is related to an increase in the resistive part of the driving point impedance whereas the reactive component of impedance is reduced by the induced current as a consequence of Lenz's law. Following Förster, the probe impedance $Z_n = R_n + jX_n$, normalized with respect to the magnitude of the free space coil reactance X_0, varies with frequency as shown on the impedance plane diagram (Fig. 9),[76] where the normalized reactance, $X_n = X{\cdot}X_0^{-1}$ is plotted against the normalized resistance $R_n = (R - R_0){\cdot}(X_0)^{-1}$, R_0 being the free space coil resistance. In the low frequency limit, $X_n = 1$ and $R_n = 0$, as represented by a point at the top of the main curve. In the high frequency limit, the curve intersects the reactance axis at a value of X_n, about 0.68 in this case, which depends on the coil geometry. For flat pancake coils with a small liftoff, the limiting value of the normalized reactance has a lower value than for a longer solenoidal coil with larger liftoff. Thus, the high frequency intersection point is a measure of the coupling between the probe and the work piece, having a low normalized reactance for greater coupling.

The data for the main curve in Fig. 9 were calculated from a Dodd and Deeds formula[10] and hence represent results of an idealization that neglects interwinding capacitance and the effects of a finite penetration depth in the windings. The parameters of a coil are taken from a benchmark experiment on simulated cracks in aluminum.[76] Superimposed on the diagram are two signals calculated using the lowest and highest frequencies of the experiments, 250 Hz and 50 kHz respectively, for the same simulated planar crack.

In Fig. 10, the calculated discontinuity signals are displayed with the background coil impedance removed. The response is for a normal coil whose axis is in the plane of the crack. Taking the crack plane to be the $x = 0$ plane, then the impedance variation shown occurs as the coil is moved in the horizontal Y direction from one end of the crack to the other. The numerical techniques used for calculating these impedance variations are described elsewhere in this chapter. First some general comments are in order, concerning the nature of numerical schemes.

The discontinuity impedance is calculated from the electromagnetic field in the presence of the discontinuity. Simple cases that can be dealt with analytically are discussed first. For more complicated geometries, numerical techniques are needed. Numerical techniques for solving electromagnetic field problems are traditionally categorized as differential or integral techniques. Finite element and finite difference techniques are the most common in the differential category whereas the integral techniques can be classified as boundary element and volume element techniques.

Most numerical schemes introduce a set of localized functions defined with respect to a grid or mesh. Often these functions are low order polynomials, which interpolate between nodal points or the edges of cells. Typically, they do not satisfy Maxwell's equations (or the integral equivalent) nor does a linear superposition of them form a solution. Nevertheless, it is postulated that a superposition of such functions gives a

FIGURE 9. Calculated normalized impedance variation with frequency of normal coil. Two discontinuity signals from semielliptical simulated crack are also shown. Discontinuity responses were calculated for excitation frequencies of 250 Hz, upper trace, and 50 kHz, lower trace, for same simulated crack. Details of coil parameters and simulated crack are given by Harrison and Burke.[76]

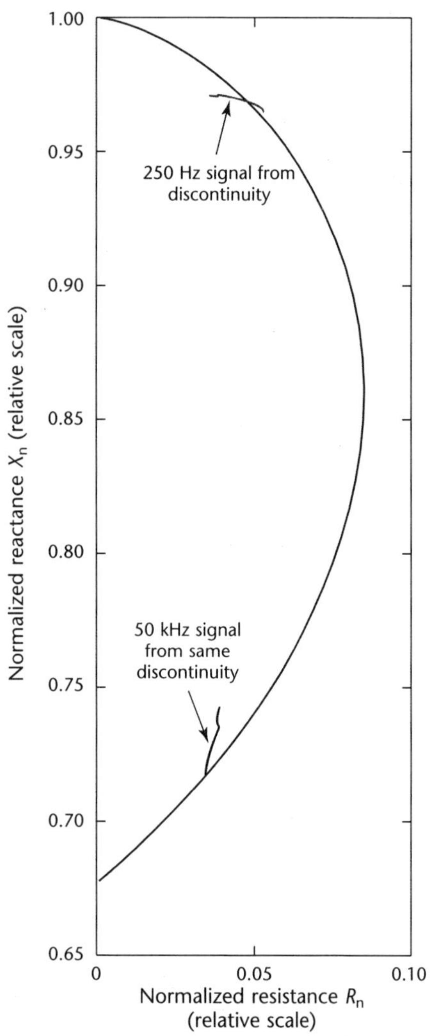

FIGURE 10. Normalized impedance due to semielliptical simulated crack shown as impedance plane locus. Trace is obtained from impedance variation as coil position is varied along crack. Note that impedance of discontinuity has been normalized by dividing by free space coil reactance at designated frequency. Details of coil parameters and simulated crack are given by Harrison and Burke.[76]

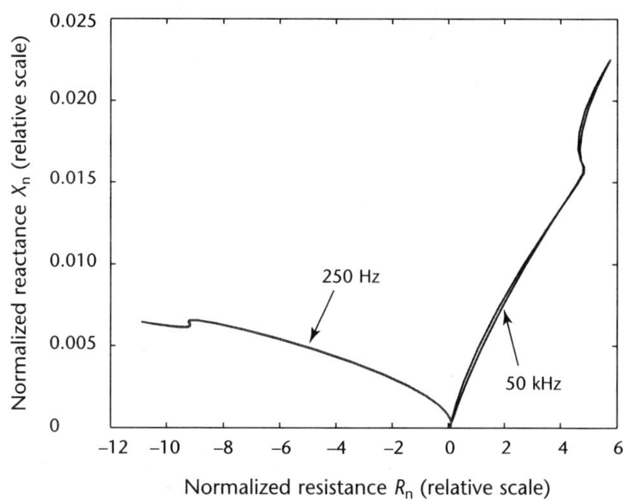

reasonably accurate numerical approximation of a solution.

The numerical results rarely come with a guarantee of accuracy. Because of the way in which a solution is constructed, the results are dependent on a mesh or grid. In the absence of error estimates, and these are rarely given, it is important that code is validated because, even if it is bug free, the onus is on the author to demonstrate that the results are reliable.

Elementary techniques, on the other hand, provide a means of predicting limited results. A number of simple formulas for evaluating discontinuity signals are given below, preceded by a summary of the basic expressions for a current dipole field. The dipole theory is presented in a way that anticipates the more advanced numerical techniques for homogeneous conducting media, in which integral formulations are used.

Current Dipole

Static Current Monopole

The current dipole is formed from two monopoles of opposite polarity adjacent to one another. A current monopole is a point source of current with intensity I. In an unbounded homogeneous region, the current spreads uniformly in all directions from the source. Hence, the current density obeys an inverse square law and is directed radially from the source. Suppose a current monopole located at a position represented by the vector $\mathbf{r}'$ gives rise to a current density $\mathbf{J}$ at some other point whose coordinate is $\mathbf{r}$. Then:

$$\text{(33)} \quad \mathbf{J} = \frac{1}{4\pi}\frac{I}{R^2}\hat{\mathbf{R}}$$

where $R = |\mathbf{r}-\mathbf{r}'|$ and $\hat{\mathbf{R}}$ is a radial unit vector. Expressing the electric field as $\mathbf{E} = -\nabla\Phi$, then the current monopole potential is:

$$\text{(34)} \quad \Phi = \frac{1}{\sigma_0}\frac{I}{4\pi R}$$

where σ_0 is the electrical conductivity of the medium. The potential satisfies the laplace equation, $\nabla^2\Phi = 0$, except at the singular point where the point source is located. The $(4\pi R)^{-1}$ dependence of a static potential due to a point source is identified as a scalar Green's function for a laplacian problem in three dimensions.

Static Dipole Field

Let two current monopoles of opposite polarity approach one another while keeping constant the product of their source intensity and their separation. With initial separation $\delta\mathbf{r}$, the dipole potential is:

$$\text{(35)} \quad \Phi = \frac{1}{4\pi\sigma_0} \times \lim_{\delta\mathbf{r}' \to 0}\left[\frac{I}{\left|\mathbf{r} - \left(\mathbf{r}' + \delta\mathbf{r}'\right)\right|} - \frac{I}{\left|\mathbf{r} - \mathbf{r}'\right|}\right]$$

In general, the limit of $f(\mathbf{r}' + \delta\mathbf{r}) - f(\mathbf{r}')$ as the separation $\delta\mathbf{r}'$ tends to zero can be written as $\delta\mathbf{r}'\cdot\nabla' f(\mathbf{r}')$. Hence the limit above can be related to the gradient of R^{-1} with respect to the primed source coordinates. The gradient may be written in terms of the unprimed field coordinates with a reversal of sign. Also expressing the dipole moment as the (finite) limit of $\mathbf{p} = I\delta\mathbf{r}'$ gives the static current dipole potential:

$$\text{(36)} \quad \Phi = -\frac{1}{\sigma_0}\nabla\left(\frac{1}{4\pi R}\right)\cdot\mathbf{p}$$

where $\mathbf{p}$ is the dipole moment (ampere meter).

By taking the negative gradient to find the electric field and multiplying by the conductivity, the current density can be written:

$$\text{(37)} \quad \mathbf{J} = \nabla\nabla\left(\frac{1}{4\pi R}\right)\cdot\mathbf{p}$$

Although the scalar product here can be seen as producing a scalar function on which the first gradient acts, the above expression can also be interpreted as a dyadic operator, $\nabla\nabla(4\pi R)^{-1}$, acting on the vector $\mathbf{p}$. The final result is the same but the second viewpoint prompts the idea that the dyad may be detached from the vector on which it acts and given a separate mathematical life. Studying the properties of dyadic Green's functions[77] leads to distinct ways of finding solutions of Maxwell's equations as outlined below.

Before returning to the role of the dyadic Green's functions, a simple illustration of the fundamental utility of the current dipole is given. The dipole field is used to express the solution of a problem in which a uniform current in an otherwise homogeneous conductor of electrical conductivity σ_0 encounters a spherical inclusion of uniform conductivity σ.

Small Spherical Inclusion

The spherical inclusion problem, usually found in textbooks as a problem in electrostatics involving a dielectric sphere, has a solution that satisfies the laplace equation inside and outside the sphere. Interface conditions on its surface ensure that the normal current and tangential electric field are continuous. Given a uniform field E_0 in the Z direction, which is also the polar direction of a spherical coordinate system ($z = R \cos \theta$) and defining the parameter s as the conductivity ratio $s = \sigma \cdot \sigma_0^{-1}$, the internal potential (volt) is:[65]

$$(38) \quad \Phi_{in} = -\frac{3}{s+2} E_0 R \cos\theta$$

whereas outside the sphere the potential is:

$$(39) \quad \Phi_{out} = -E_0 R \cos\theta + \frac{s-1}{s+2} E_0 \frac{a^3}{R^2} \cos\theta$$

where θ is the polar angle (radian). The external potential can also be written:

$$(40) \quad \Phi_{out} = -E_0 R \cos\theta - \frac{1}{4\pi\sigma_0} \nabla\left(\frac{1}{R}\right) \cdot \boldsymbol{p}$$

where the dipole intensity and direction are given by:

$$(41) \quad \boldsymbol{p} = 4\pi \frac{s-1}{s+2} \sigma_0 E_0 a^3 \hat{z}$$

Perhaps of greater interest here is the fact that the external electric field can be written:

$$(42) \quad E_{out} = E_0 + \frac{1}{\sigma_0} \nabla\nabla\left(\frac{1}{4\pi R}\right) \cdot \boldsymbol{p}$$

where $E_0 = E_0\hat{z}$. This goes beyond the basic textbook account by expressing the field of the dipole in terms of a dyadic Green's function, $\sigma_0^{-1}\nabla\nabla(4\pi R)^{-1}$.

Equation 42 can apply to a dipole of arbitrary orientation. Figure 11 shows the current associated with the perturbed field that when added to the unperturbed current $\sigma_0 E_0 \hat{z}$ gives the total current density.

An additional point of interest is that the dipole intensity can be related to a uniform current dipole density $\boldsymbol{P}$ distributed in the spherical region. By putting $\boldsymbol{p} = 4 \cdot 3^{-1} \times \pi a^3 \boldsymbol{P}$, it is found that:

$$(43) \quad \boldsymbol{P} = (\sigma - \sigma_0) \boldsymbol{E}$$

where $\boldsymbol{E}$ is the electric field in the sphere given by taking the negative gradient of Eq. 38.

Dynamic Current Dipole

In eddy current testing, the fields are dynamic rather than static. Therefore, the dynamic current dipole has a more significant elemental discontinuity field than does the field of static current dipole. The dynamic current dipole for a time harmonic field is described by essentially the same equations as those used for the textbook treatment of the hertzian dipole.[78] The difference arises from the fact that in eddy current applications the host medium is a conductor, not air. In a good conductor such as a metal, the charge current is much larger than the displacement current. Consequently, the latter can be neglected. This means that Ampère's law (Eq. 2), $\nabla \times \boldsymbol{H} = \boldsymbol{J}$, is adequate and Maxwell's addition of the displacement current $j\omega\boldsymbol{D}$ to the right hand side of this relationship is not needed. Here, the field is expressed in terms of complex phasors, which means, for example, that the magnetic field varies in time as the real part of $\boldsymbol{H}e^{j\omega t}$, ω being the angular frequency (radian per second) of the excitation. The neglect of displacement current means that solutions are sought in the quasistatic limit. As a short cut from the description of waves in air to fields in a conductor, the displacement current $j\omega\varepsilon_0\boldsymbol{E}$, which appears in standard hertzian

FIGURE 11. Perturbed current at small spherical inclusion in metal.

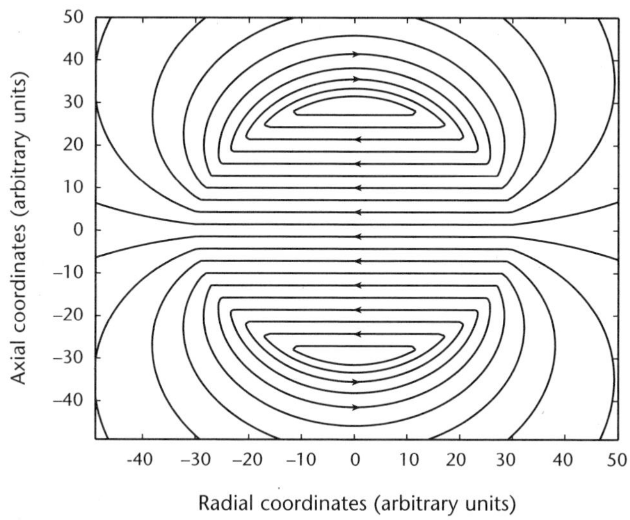

dipole theory,[78] can usually be replaced with the charge current $\sigma_0 \boldsymbol{E}$.

It is convenient to express the dynamic field in terms of a magnetic vector potential $\boldsymbol{A}$, related to the magnetic flux density:

$$\text{(44)} \quad \nabla \times \boldsymbol{A} = \boldsymbol{B}$$

and having a gage condition:

$$\text{(45)} \quad \nabla \cdot \boldsymbol{A} = -\mu_0 \sigma_0 \Phi$$

replacing the usual lorenz condition. For a current dipole in an unbounded conductor of conductivity σ_0 and permeability of vacuum, the magnetic vector potential is:

$$\text{(46)} \quad \boldsymbol{A} = \mu_0 \frac{e^{-jkR}}{4\pi R} \boldsymbol{I} \cdot \boldsymbol{p}$$

where $k = \sqrt{(-j\omega_0\mu_0\sigma)}$, taking the root with a positive real part and $\boldsymbol{I} \cdot \boldsymbol{p} = \boldsymbol{p}$. The parameter k is related to the standard depth of penetration δ (meter):

$$\text{(47)} \quad k = \frac{1 - j}{\delta}$$

where:

$$\text{(48)} \quad \delta = \sqrt{\frac{2}{\omega \mu_0 \sigma_0}}$$

The identity dyad $\boldsymbol{I}$ in Eq. 46 has been inserted to express the magnetic vector potential as a dot product of a dyadic operator acting on a vector source, this being the appropriate general form for the relationship between a vector source and a vector field $\boldsymbol{A}$. The magnetic field due to the current dipole is found from:

$$\text{(49)} \quad \boldsymbol{H} = \frac{1}{\mu_0} \nabla \times \boldsymbol{A}$$

The electric field is found from Ampère's law in the form:

$$\text{(50)} \quad \nabla \times \boldsymbol{H} = \sigma_0 \boldsymbol{E}$$

Combining Eqs. 46, 49 and 50 gives:

$$\text{(51)} \quad \boldsymbol{E} = \frac{1}{\sigma_0 \mu_0} \nabla \times \nabla \times \boldsymbol{A} = -j\omega\mu_0 \left\{ \left(\boldsymbol{I} + \frac{1}{k^2} \nabla\nabla \right) \frac{e^{-jkR}}{4\pi R} \right\} \cdot \boldsymbol{p}$$

Equation 51 has been derived from the identity $\nabla \times \nabla \times = \nabla\nabla \cdot - \nabla^2$ and from the fact that the vector potential satisfies Helmholtz's equation.[79] A discussion of the dyadic form between the braces has been given by Tai.[77,80]

Clearly, the dynamic dipole field reduces to the static case, Eq. 42 in the limit, as angular frequency ω goes to zero. It also reduces to the static case in the near field, where the first term of the dyadic operator is negligible. This is a reminder of the fact that a local field on a scale small compared with the standard depth of penetration δ can often be analyzed using electrostatic or magnetostatic theory.

Equation 51 may be generalized to give the perturbed field due to a volumetric discontinuity by representing the effect of such a discontinuity as a general dipole distribution $\boldsymbol{P}(\boldsymbol{r}')$. Then the perturbed electric field is found by replacing the point dipole $\boldsymbol{p}$ in Eq. 51 by $\boldsymbol{P}(\boldsymbol{r}')$ and integrating with respect to the source coordinate $\boldsymbol{r}'$ over the region of the dipole density. This field representation is used in volume integral formulations and is a preliminary step toward a volume element calculation of the dipole density.[72] Similarly, the effects of a thin crack can be represented by a surface dipole layer and form the basis of a boundary element formulation.[94] In either case, the dipole density is determined by an integral equation. Having found a solution, the probe signal due to the discontinuity can be calculated from the probe response formulas below.

Probe Response

An eddy current probe senses discontinuities through changes of impedance. There are a number of techniques for calculating the discontinuity response depending on the details of the approach used. For example, Kahn and others used the integration of the poynting vector over a surface.[74] Auld uses a reciprocity relationship attributed to Lorenz[48] whereas others use a reciprocal relationship associated with Rumsey.[66,81] Rumsey's relationship is used next.

The coil current density can be represented by a function $\boldsymbol{J}$. With $\boldsymbol{E}^{(P)}$ defined as the perturbed field due to the discontinuity, the probe impedance change due to the discontinuity is:

$$\text{(52)} \quad I^2 \Delta Z = -\int_{\Omega_J} \boldsymbol{E}^{(P)} \cdot \boldsymbol{J} \, d\boldsymbol{r}$$

where the integration is over the coil region denoted by Ω_J.

The coil current can be used as a phase reference and taken to be real. Although the coil current is confined to the coil windings, these are usually on such a small scale that the current density can be approximated as a smooth function, usually a constant, over the coil cross section. In a calculation in which the effects of the discontinuity are represented by a dipole volume distribution $\boldsymbol{P}$, Rumsey's reciprocal relations may be invoked to write the impedance change in terms of the unperturbed electric field at the discontinuity $\boldsymbol{E}^{(0)}$:[72]

$$(53)\quad I^2 \Delta Z = -\int_{\Omega_P} \boldsymbol{E}^{(0)} \cdot \boldsymbol{P} d\boldsymbol{r}$$

where the integration is now over the region Ω_P where the discontinuity conductivity differs from that of the host. Equation 53 is advantageous because $\boldsymbol{P}$ is usually calculated directly by an integral equation technique whereas the evaluation of $\boldsymbol{E}^{(P)}$ for Eq. 52 requires an additional step once the dipole density has been found. In general:

$$(54)\quad \boldsymbol{P} = \left[\sigma(\boldsymbol{r}) - \sigma_0\right] \boldsymbol{E}(\boldsymbol{r})$$

which defines $\boldsymbol{P}(f)$ for an arbitrary discontinuity whose conductivity $\sigma(\boldsymbol{r})$ differs from that of the host conductivity σ_0. For the special case of the small spherical region with constant conductivity, a similar relationship (Eq. 43) is used.

Small Discontinuities

For a small spherical discontinuity, such as a gas bubble or spherical inclusion in a conductor, the impedance change sensed by a probe is given by:

$$(55)\quad I^2 Z_B = -\boldsymbol{E}^{(0)} \cdot \boldsymbol{p}$$

An explicit expression for the response can be found using a suitable unperturbed field, for example the normal coil field in a half-space conductor.[10] A simpler case is one where the field at the surface of the conductor is uniform. This approximation may in practice be reasonable if the probe dimensions are larger than the standard depth of penetration. With H_0 as the tangential magnetic field in the (horizontal) Y direction and the Z direction normal to the surface of the conductor, the unperturbed electromagnetic field in a conductor below the plane $z = 0$ is given by:

$$(56)\quad \boldsymbol{H}^{(0)} = H_0 e^{jkz} \hat{y}$$

and:

$$(57)\quad \boldsymbol{E}^{(0)} = \frac{jk}{\sigma_0} H_0 e^{jkz} \hat{x}$$

By substituting the expression for the dipole density of a small spherical inclusion given in Eq. 41 into the relation Eq. 55 with $\boldsymbol{E}^{(0)}$ given by Eq. 57, it is found that, for a small spherical cavity ($\sigma = 0$), centered at $\boldsymbol{r} = \boldsymbol{r}_0$ and of radius a (meter), the impedance is:

$$(58)\quad Z_B = -\frac{2\pi}{\sigma_0} k^2 a^3 \left(\frac{H_0}{I}\right)^2 e^{2jkz_0}$$

The impedance change is proportional to a^3 simply because the dipole intensity varies in proportion to the volume of the sphere. Note that the ratio $H_0 \cdot I^{-1}$ is real for a magnetic field uniform at the surface. However, it may be useful to estimate the small sphere response for a nonuniform field, for which Eq. 55 applies if the unperturbed field is known. Note that the maximum value of the ratio $H_0 \cdot I^{-1}$ can be regarded as a figure of merit for the probe because the signal intensity depends on its square. Note also that the factor $2jkz_0$ in the exponential of Eq. 58 indicates that the signal is attenuated over a path of length $2z_0$, representing the round trip distance from the surface to the discontinuity and back.

Another small discontinuity result that can be found by elementary means is the response due to a semicircular surface crack of negligible opening whose radius is smaller than the standard depth of penetration. The assumption of a relatively large standard depth of penetration means that the local field can be treated as static in the sense that it may be described by a potential satisfying the laplace equation. The surface of the conductor acts as a plane of reflection, allowing a conversion of the semicircular crack problem to a circular crack problem by appealing to the technique of images. The problem can then be solved as if the crack were a thin disk in a uniform stream of incompressible fluid. With a uniform applied electromagnetic field given by Eqs. 56 and 57, the impedance change due to such a crack is:[82,83]

$$(59)\quad Z_S = -\frac{4}{3\sigma_0} k^2 a^3 \left(\frac{H_0}{I}\right)^2$$

Of practical importance is the question of what limits the detection of small cracks. Equation 59 yields insight and significant basic information in this regard. First, note that the response depends on the third power of the crack radius. Second, the impedance change increases in proportion to the frequency because $k^2 = -j\omega_0\mu_0\sigma_0$:

Thirdly note that for a strictly uniform field, the change in impedance is purely inductive (imaginary), $H_0 \cdot I^{-1}$ being real. Even if the assumptions that went into the derivation of the simple relation given by Eq. 59 are not precisely satisfied, the equation can provide an approximate answer. If the accuracy is inadequate, improvements may be made by extending the results to higher order terms by using perturbation theory or by taking into account nonuniformities in the field by an extension of the basic analytical technique.[83]

Long Crack

A long crack of constant depth d (meter) may be treated as a two-dimensional problem provided that the unperturbed field does not vary along its length. Such a configuration does not relate directly to most practical problems but its solution has had an impact on the understanding of crack fields. The problem can be solved analytically in the low and high frequency regimes that correspond to small and large standard depths of penetration compared with the crack depth.

According to the thin penetration approach of Kahn, assuming the crack is in the plane $x = 0$, the field on the crack faces has the form:

$$\textbf{(60)} \quad \boldsymbol{H} = H_0 e^{-jk|x|} \hat{y}$$

and:

$$\textbf{(61)} \quad \boldsymbol{E} = \mp \eta H_0 e^{-jk|x|} \hat{z}$$

where η is the characteristic impedance of the medium:

$$\textbf{(62)} \quad \eta = \frac{jk}{\sigma_0}$$

This field can be used to evaluate the complex time average poynting vector $\boldsymbol{P}$ (not to be confused with dipole density $\boldsymbol{P}$) at the crack faces from:

$$\textbf{(63)} \quad \boldsymbol{P} = \frac{1}{2} \boldsymbol{E} \times \boldsymbol{H}^*$$

where the asterisk (*) denotes the complex conjugate. The uniform face field means that:

$$\textbf{(64)} \quad \boldsymbol{P} = \pm \frac{1}{2} \eta H_0^2 \hat{x}$$

where the characteristic impedance of the medium is $\eta = jk \cdot (\sigma_0)^{-1}$. The upper and lower signs on the right side refer to the positive and negative sides of the crack, respectively. Integrating the poynting vector over the crack surface and equating the result to the energy transferred at the drive point of the probe gives an impedance:

$$\textbf{(65)} \quad Z_S = \frac{2}{\sigma_0} jkd \left(\frac{H_0}{I} \right)^2$$

per unit length of the crack. To Eq. 65 must be added the corner and edge effects that together with Z_S give rise to a combined impedance:

$$\textbf{(66)} \quad Z_K = \frac{1}{\sigma_0} \left(2jkd + 1 - \frac{8}{\pi} \right) \left(\frac{H_0}{I} \right)^2$$

The three contributions to the impedance per unit length include the field at the edge (represented by the 1 in parentheses) and the corner field (the $8 \cdot \pi^{-1}$ term). A complete analysis of the above expression is given elsewhere.[84] The impedance in this problem therefore contains a dominant face term that varies as the square root of frequency, is proportional to the crack depth and has a phase angle of $\pi \cdot 4^{-1}$ with respect to the drive current. The additional terms due to the edge and corner are resistive.

Complementary to the kahn thin penetration result is a formula valid in the low frequency regime that can be found from a solution valid in the static, direct current limit. In this regime, ikd is a small parameter; this fact can be exploited to find a field solution in the form of an ordered series using Rayleigh-Ritz perturbation theory. Likewise, the impedance can be expressed as an ordered series:[83]

$$\textbf{(67)} \quad Z_L = \frac{1}{\sigma_0} \left(\frac{H_0}{I} \right)^2 \times \left[\frac{\pi}{2} (jkd)^2 - \frac{4}{3} (jkd)^3 + \ldots \right]$$

For a uniform excitation field, the leading term at low frequency is purely inductive and increases linearly with frequency and as the square of the crack depth.

The long crack theory is readily extended in range from the high frequency limit to lower frequencies by accounting for the interaction between the edge and corner fields through the Weiner-Hopf technique[85] and by applying the perturbation technique to extend the range of validity of the low frequency approximation.[83] The impedance results of these extensions, shown in Fig. 12, have been compared with numerical results of a boundary element code.[83] In these figures, the impedance is normalized by writing:

$$(68)\quad Z_n = \frac{\sigma_0 I^2 \Delta Z}{H_0^2}$$

Hence, the kahn impedance (Eq. 66) is written in terms of the normalized impedance:

$$(69)\quad Z_n = 2jkd + 1 - \frac{8}{\pi}$$

The main benefit of the study of the two dimension problem is that it provides a simple test bed for new techniques, including an adaption of the geometrical theory of diffraction,[86] to problems in eddy current crack interaction.

FIGURE 12. Analytical and numerical results of change in normalized impedance Z_n due to long surface breaking crack: (a) for inductive, or imaginary, component; (b) for resistive, or real, component.

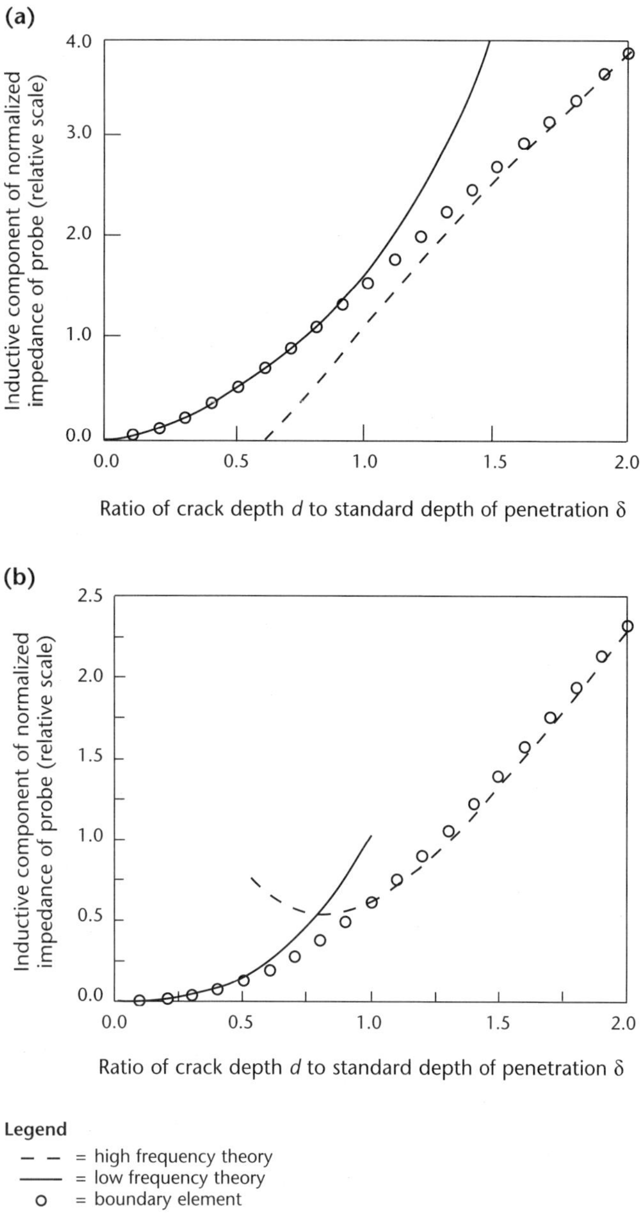

Advanced Techniques

Two types of advanced techniques for evaluating probe signals due to cracks are considered next. First, equivalent source techniques are discussed, of which the Burrows small discontinuity theory[64] is an elementary precursor. Second, the thin penetration approaches, prototyped by Kahn and others[74] and applicable to both ferromagnetic and nonferromagnetic materials, are described.

The equivalent source techniques cover all frequencies and are closely linked with field formulations based on integral equations. They can be used to evaluate fields at cracks in ferromagnetic material but here the description will be limited to materials with the permeability of a vacuum.

Finding a numerical solution from integral equations can be more demanding in the thin penetration regime because a large number of volumetric cells or boundary elements may be needed to give an accurate result. Usually, a grid containing several cells per standard depth of penetration is required, so the number of unknowns and the computational cost are usually high in the thin penetration regime. Because this cost is avoided in approaches that explicitly take advantage of small penetration depth approximations, the techniques described here are complementary. To understand dipole and thin skin techniques, it is helpful to consider the behavior of the electric field near the crack mouth and the properties of the field at the crack face.

Electric Field at Crack Opening

The crack opening is typically much smaller than the standard depth of penetration. Therefore, a local field theory for this region can be based on Maxwell's equations in the static limit. Because the electric field varies relatively slowly along the crack mouth away from the ends, a two-dimensional solution in a plane perpendicular to the mouth direction adequately captures the significant features. This approach implies that the solution of the laplace equation in two dimensions is suitable for the task.

The geometry of the problem (Fig. 13) lends itself to the Schwarz-Christoffel theory,[87,88] which yields a conformal transformation to map the domain of the crack and the adjoining half plane above it into a half plane. An elementary solution for the half plane will lead to a fixed potential difference across the crack. Then, an inverse transform can be applied to produce a representation of the electric field at the crack mouth. In this case, a suitable analytic inverse transform is apparently lacking and the mapping must be done numerically by using, possibly, the newton-raphson iterative technique or the brent algorithm.[89]

Förster[90] and others[91] have used conformal mapping to determine the magnetic flux leakage at the crack mouth. In fact, the mapping is used widely to find the magnetic field at the gap between two pole pieces such as the field at the gap between the poles of a magnetic recording head.[92] In eddy current problems, the electric field is needed rather than the magnetic field but the solution is essentially the same (Fig. 13).

At the corners, the electric field is singular, varying in magnitude in air close to the corner as $(r_{corner})^{-1/3}$, where (r_{corner}) is the radial distance from the apex of the corner. This behavior is characteristic of the field in the vicinity of a right angled wedge.[93] Between the crack faces, the field tends to become more uniform deeper into the crack. The magnitude of the field between the faces depends on how deep and wide the crack is. If the crack is made narrower while the potential across the crack remains the same, then the magnitude of the electric field increases. In the limit of closure without contact, the electric field forms a singular layer, infinitely strong, of infinitesimal thickness. It is this limiting case that will be explored here because the singular layer has a simple mathematical representation.

Impenetrable Crack

In calculations of the field perturbation due to a crack, it is usual and convenient to apply a boundary condition that states that the normal component of the current density in the conductor at the crack face is zero. Although the surface of the crack supports a distribution of electrical charge and the charge must get there somehow, in the quasistatic approximation the charging current is neglected. In a conductor, the displacement current $j\omega\varepsilon_0 E$ is neglected because it is very much smaller than the charge current $\sigma_0 E$. Even at high eddy current test frequencies, ~10 MHz, where the magnitude of displacement current is greater than at lower frequencies, the ratio $\varepsilon_0\omega\cdot\sigma_0^{-1}$ is on the order of 10^{-9} for a low conductivity metal, 0.58 MS·m^{-1} (1 percent of the International Annealed Copper Standard). However, the accuracy of a boundary condition that neglects the charging current at the crack face is dependent on crack width. Therefore, it is necessary to seek a justification for the quasistatic approximation in this context.

The normal component of the true current, to use Maxwell's term for the sum of the displacement and charge current, is continuous across an interface. Therefore, the displacement current between the faces and directed across the crack is equal to the charging current at the conducting side of the crack face. Hence, the boundary condition is justified if the displacement current $j\omega e_0 E_n$ across the crack is negligible compared with the tangential charge current $\sigma_0 E_t$ at the crack

Figure 13. Electric field at crack opening.

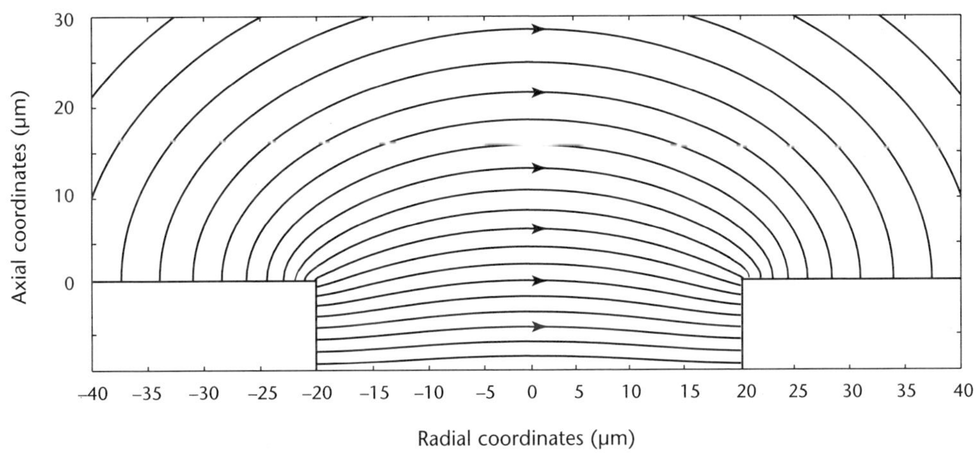

face. In the following argument, these currents are estimated and compared.

Applying Stokes' theorem to Faraday's induction law in differential form gives an integral form of the induction law in which the line integral of the electric field around a closed path is equated to the rate of change of magnetic flux through the surface S bounded by the path. If it happens that the rate of change of magnetic flux through S can be neglected. Then the line integral is approximately zero:

$$\oint_C \mathbf{E} \cdot d\mathbf{s} \;\cong\; 0 \tag{70}$$

where C is the path bounding S and ds is an incremental displacement along the path.

For this case, the path links points ABCD (Fig. 14) in the limit as the points approach the crack surface. By considering an exponential field at the crack face, it can be shown that the magnetic flux through S is less than the path integral of $\mathbf{E}$ over a crack face by a factor on the order $w \cdot \delta^{-1}$, where w is crack width (meter) and δ is standard depth of penetration (meter). Hence, if w is small compared with the standard depth of penetration, as it usually is, then Eq. 70 is a reasonable approximation. This equation indicates that the following are of roughly comparable order of magnitude: $2E_0 d \cong E_n w$, where E_n is the normal component (volt per meter) of the electric field in the crack and E_0 is the tangential field (volt per meter) at the outer surface. That being the case, the ratio of the displacement current across the crack $j\omega\varepsilon_0 E_n$ to the tangential face current $\sigma_0 E_0$ is small if:

$$\frac{\omega \varepsilon_0 d}{\sigma_0 w} \;<<\; 1 \tag{71}$$

This condition for the validity of the quasistatic approximation at the crack is usually satisfied. For example, if $d \cdot w^{-1} \cong 10^4$, then $\omega\varepsilon_0 d \cdot (\sigma_0 w)^{-1} \cong 10^{-5}$ at 10 MHz in a conductor with a low conductivity, 0.5 MS·m^{-1} (1 percent International Annealed Copper Standard). Assuming the quasistatic approximation for a nonconducting crack, the zero normal current at the crack face is written:

$$J_n^{\pm} \;=\; 0 \tag{72}$$

where the ± sign denotes points on one or the other crack face approached from the interior of the conductor.

FIGURE 14. Integration path C, crossing crack.

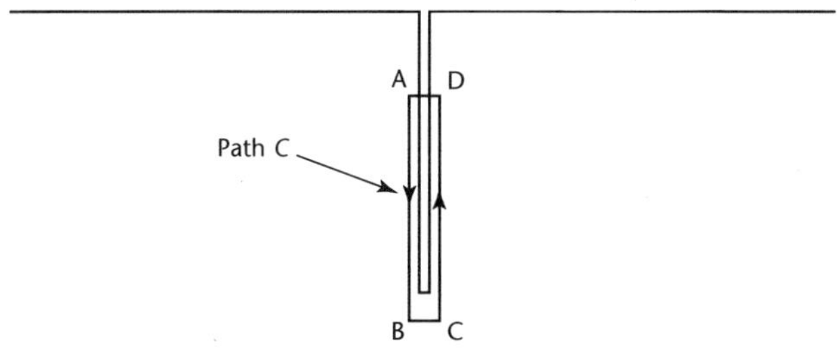

Surface Current Dipole Distribution

A basic problem to be considered initially is the probe response to an ideal crack, defined as having negligible opening compared with the standard depth of penetration but satisfying the requirement in Eq. 72 for the validity of the quasistatic approximation. The crack is therefore impenetrable to the flow of electric current. The ideal crack is defined, for example, with respect to an open surface S_0 bounded by the crack edge and by the intersection of the crack with the surface of the conductor (Fig. 15).

Eddy currents flow around the buried crack edge such that the current density is different at points adjacent to one another on opposite faces. The fact that the crack opening is neglected means that the ideal crack gives rise to a discontinuity in the tangential current density at S_0 and, consequently, a discontinuity in the tangential electric field. The solution of the ideal crack problem can be found by evaluating the discontinuity in the field

FIGURE 15. Side view of coil and crack, showing crack in Y,Z plane of coordinate system. Surface S_0 is part of the Y,Z plane occupied by the crack.

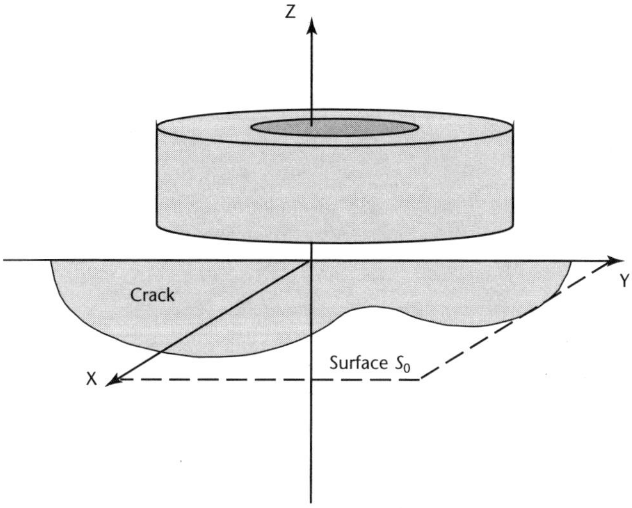

directly or by expressing the jump in the field in terms of an equivalent dipole source distribution, either electric [94] or magnetic.[95] The relationship between the field and the equivalent current dipole source is described next.

For an open crack, the volume dipole density $\boldsymbol{P}$ is defined by Eq. 54 and, like the electric field in the crack, is larger for cracks of narrower opening. However, the integral of $\boldsymbol{P}$ along a path C_n across the crack is expected to tend to a finite value in the limit as the crack opening becomes infinitesimal. With w as the width (meter) of the crack opening, the limit is written:

$$(73)\quad \lim_{w\to 0} \int_{C_n} \boldsymbol{P}\cdot d\boldsymbol{n} = p$$

where p is the surface dipole density having the vector representation $\boldsymbol{p} = \hat{n}p$. For a crack whose interior has zero conductivity, it can be seen from the definition (Eq. 54) that $\boldsymbol{P} = -\sigma_0\boldsymbol{E}$. Therefore:

$$(74)\quad \lim_{w\to 0} \int_{C_n} \boldsymbol{E}\cdot d\boldsymbol{n} = \frac{1}{\sigma_0}p$$

This relation can be used in formulas for the line integral (Eq. 70) along a path C_0 around a segment of the surface of an ideal crack (Fig. 16) to give:

$$(75)\quad \left(\boldsymbol{E}_t^+ - \boldsymbol{E}_t^-\right)\cdot\delta_S + \frac{p_A - p_B}{\sigma_0} = 0$$

where the subscript t denotes components tangential to S_0 and where δ_S is a displacement vector between points A and B on the surface S_0 (Fig. 16):

$$(76)\quad p_A - p_B = \nabla_t p\cdot\delta_S$$

Because δ_S is arbitrary, it can be seen that:

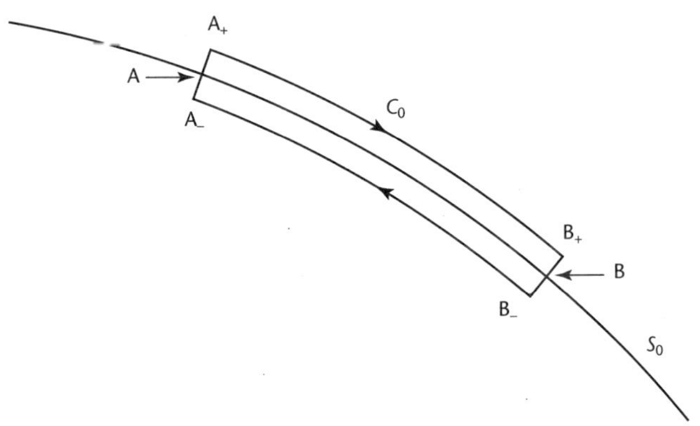

FIGURE 16. Integration path C_0 crosses crack at points A and B and is formed in limit as A± and B± approach surface S_0.

$$(77)\quad \boldsymbol{E}_t^+ - \boldsymbol{E}_t^- = -\frac{1}{\sigma_0}\nabla_t p$$

A similar relationship between the jump in the electric field at a surface and the gradient of the surface dipole density exists for the electrostatic charge dipole layer.[65] Here it relates the discontinuity in the dynamic tangential electric field at an ideal crack surface S_0 to the surface distribution of dynamic current dipoles whose orientation is normal to S_0.

Two properties of the dipole density are worthy of note at this point. Firstly, it tends to zero at the buried crack edge. Secondly, the derivative $\partial P\cdot(\partial z)^{-1}$ is zero at the crack mouth, z being the coordinate whose axis is normal to the surface of the conductor (Fig. 15). These properties are written as:

$$(78)\quad p(\boldsymbol{r}_e) = 0$$

and:

$$(79)\quad \left.\frac{\partial p}{\partial z}\right|_{\boldsymbol{r}_m} = 0$$

where $\boldsymbol{r}_e$ is the coordinate of an edge point and $\boldsymbol{r}_m$ is the coordinate of a point at the crack mouth. For example, Eq. 80 gives the dipole density for a long straight crack of depth d in a uniform unperturbed field E_0:[83]

$$(80)\quad p(z) = -2\sigma_0 E_0\sqrt{d^2 - z^2}$$

Note that $p(z)$ vanishes at $z = -d$ and that the derivative with respect to z vanishes at $z = 0$ in keeping with the general properties in Eqs. 78 and 79. In addition, it is important to be aware that the electric field has a half-power singularity at the edge of an ideal crack varying locally as:[96]

$$(81)\quad \boldsymbol{E}(\rho,\phi) \cong \frac{1}{\sqrt{\rho}} \times \left[\hat{\rho}\cos\left(\frac{\phi}{2}\right) - \hat{\phi}\sin\left(\frac{\phi}{2}\right)\right]$$

where ρ is the perpendicular distance (meter) of a point from the edge and ϕ is an angle (radian) measured from the surface S_0 in a plane perpendicular to the edge. This means that, in general, the dipole density varies as:

$$(82)\quad p \propto \sqrt{\rho}$$

near the edge.

The solution of the eddy current ideal crack problem has been reduced to one of finding the surface dipole density p. Thus a scalar replaces a two-component vector, the jump in the tangential electric field. Consequently fewer unknowns are needed for a numerical solution. To calculate p, it is necessary to know the continuity conditions that apply to the magnetic field at the crack surface S_0 because these conditions will be needed in the derivation of an equation from which the dipole density can be calculated. Although the details of these derivations will not be given here, it is useful to understand the continuity conditions that apply to the magnetic field at the ideal crack surface.

The jump in the tangential electric field at the ideal crack is inseparable from the singular property of the electric field between the crack faces, as expressed here in terms of a current dipole layer. However, no such singular behavior occurs in the magnetic field. The truth of this can be demonstrated by following an argument like the one for the electric field but applying Stokes' theorem to Ampère's law rather than to the induction law, thereby forming the line integral of $\boldsymbol{H}$ around the path C_0. Following this parallel argument, it can be deduced that the line integral of $\boldsymbol{H}$ vanishes as the closed path $A_-A_+B_+B_-$ (Fig. 16) collapses onto the crack but no singular behavior of the magnetic field in the crack could lead to a discontinuity in the tangential magnetic field. It is concluded that:

$$\textbf{(83)} \quad \boldsymbol{H}_t^+ - \boldsymbol{H}_t^- = 0$$

at S_0. In addition, it may be recalled that the normal magnetic flux B (tesla) is continuous at an interface.[65] At a crack, which is in fact a double interface, the same relationship holds:

$$\textbf{(84)} \quad B_n^+ - B_n^- = 0$$

To confirm the consistency of the continuity conditions at S_0, note that Faraday's induction law implies that the normal magnetic flux density at S_0 is:

$$\textbf{(85)} \quad B_n^\pm = -\frac{1}{j\omega}\hat{n}\cdot\nabla_t \times \boldsymbol{E}_t^\pm$$

By using this relationship to express the difference $B_{n+} - B_{n-}$ in terms of the jump in the tangential electric field and substituting for the jump in the transverse electric field using Eq. 77, the transverse curl acts on the transverse gradient of the dipole density to give zero. Thus, the continuity of the normal flux density is ensured by the fact that the jump in the tangential electric field is expressed as the tangential gradient of a scalar function. Having now defined the continuity conditions at the surface S_0, one is equipped for the task of finding a governing equation for the dipole density p.

Integral Formulation

The most common approach to the solution of electromagnetic field problems at low frequencies, such as the modeling of electrical machines, electromagnets and eddy current discontinuity detection, is to use a differential formulation as the basis of a finite element solution. However, in the area of antennas and electromagnetic wave propagation, integral techniques are used more commonly than the finite element scheme. In the approaches described here, the aim is to compute solutions for simple but realistic geometries using relatively few unknowns and adapt the forward problem solver for the task of iterative inversion. Integral equation techniques are better suited to this strategy, particularly if the region of the required solution can be confined to the discontinuity. The implication is that the number of unknowns is small and the forward solver is fast.

In antenna theory, the hertzian dipole is used as a fundamental solution from which the field of a wire antenna is found by integration over the wire structure, a step that is justified by the principle of superposition. The elementary current dipole field (Eq. 51) like the hertzian dipole, plays the role of a fundamental solution in a conductor. It allows the field of an extensive discontinuity in a conductor to be expressed as an integral over a discontinuity region. The fundamental solution is written here as:

$$\textbf{(86)} \quad \boldsymbol{E}(\boldsymbol{r},\boldsymbol{r}') = -j\omega\mu_0 \boldsymbol{G}(\boldsymbol{r}|\boldsymbol{r}')\cdot \boldsymbol{p}(\boldsymbol{r}')$$

where $\boldsymbol{G}(\boldsymbol{r}|\boldsymbol{r}')$ is a dyadic Green's function transforming the current dipole source $\boldsymbol{p}$ into the electric field. For a dipole embedded in an unbounded domain, the dyadic Green's function is given in the braces of Eq. 51. A more representative configuration in eddy current testing is one in which a probe in air interacts with a discontinuity in a conducting plate. If the standard depth of penetration is smaller than the plate thickness, the conductor can be considered as occupying a half space (Fig. 15). The dyadic Green's function for a half space, like the fundamental solution, is known in explicit analytical form.[68,94] Hence the discontinuity field can be written as an integral over the discontinuity in the knowledge that the integral kernel will ensure that the correct continuity conditions will be satisfied automatically at the interface of air and conductor.

For a crack in a half-space conductor ($z = 0$), the electric field is written as the sum of the unperturbed probe field $\boldsymbol{E}^{(0)}$ and the discontinuity field:

$$(87)\quad \boldsymbol{E}(\boldsymbol{r}) = \boldsymbol{E}^{(0)}(\boldsymbol{r}) - j\omega\mu_0 \int_{S_0} \boldsymbol{G}(\boldsymbol{r}|\boldsymbol{r}') \cdot \boldsymbol{p}(\boldsymbol{r}') dS'$$

Here the field due to the crack is expressed in terms of its equivalent source $\boldsymbol{p}$ as superposition of dipole fields written as an integral over the crack surface S_0. It should be noted that, rather than simply invoking the principle of superposition, the formal techniques of deriving integral equations for the field are based on Green's second theorem.[77] Equation 87 is multiplied by the conductivity σ_0 and the condition (Eq. 72) is applied so that the normal component of the current density at a point at the crack surface is zero:

$$(88)\quad J_n^{(0)}(\boldsymbol{r}^{\pm}) = -k^2 \times \int_{S_0} G^{nn}(\boldsymbol{r}^{\pm}|\boldsymbol{r}')\, p(\boldsymbol{r}') dS'$$

where:

$$(89)\quad J_n^{(0)}(\boldsymbol{r}^{\pm}) = \sigma_0 \hat{n} \cdot E^{(0)}(\boldsymbol{r}^{\pm})$$

and:

$$(90)\quad G^{nn}(\boldsymbol{r}|\boldsymbol{r}') = \hat{n} \cdot \boldsymbol{G}(\boldsymbol{r}|\boldsymbol{r}') \cdot \hat{n}$$

It is to be understood that the field point whose coordinate is $\boldsymbol{r}$ approaches a point $\boldsymbol{r}^{\pm}$ on the crack and that this limiting process takes place after the integration has been performed. Equation 88 determines the current dipole density on the surface S_0.

Rather than seeking a solution of the integral equation itself, an approximation is constructed by expanding the unknown $p(\boldsymbol{r})$ as a linear superposition of a set of N basis functions and the expansion coefficients determined by using the moment technique.[97] By this approximation procedure, a matrix equation replaces the integral equation as the means of finding the field. The solution of the matrix equation can then be found by standard numerical techniques.[89] The classic text on the moment technique in electromagnetism is by Harrington[97] and a more recent volume on the subject, which includes the treatment of dyadic Green's functions, is by Wang.[98]

Having calculated a discrete estimate of the dipole density $p(\boldsymbol{r}_0)$, $\boldsymbol{r}_0 \in S_0$, the coil impedance change due to the discontinuity is determined from a variant of Eq. 53:

$$(91)\quad I^2 \Delta Z = -\int_{S_0} \boldsymbol{E}^{(0)} \cdot \boldsymbol{p}\, dS$$

where the integration is over the surface S_0. In applying the moment technique to the ideal crack problem,[94] the discrete approximation of the dipole density converts the impedance integral to a summation.

Boundary Element Results

Results have been calculated using a version of the moment technique in which the dipole density is approximated as a piecewise constant with respect to a regular grid of rectangular boundary elements. For a piecewise constant solution, it is necessary to find the value of the constant coefficient for each of, say, N cells. This value is obtained by expressing the dipole density as a linear superposition of N rectangular pulse functions, substituting the expansion into Eq. 88 and demanding that the resulting equation is satisfied at the center of each and every rectangular cell, a step known as point matching or collocation. The procedure leads to an $N \times N$ matrix equation for the coefficients of the piecewise constant approximation.

In general, the moment technique proceeds by expanding the unknown function in terms of suitable set basis functions defined with respect to a grid or a set of nodal points subdividing the domain of the solution. Therefore, the dipole density can be approximated by using a set of basis functions that lead to a smoother representation of the solution than does the piecewise constant approximation. This approximation certainly leads to improved results.[99] However, despite the relatively crude approximation of the piecewise constant solution, the results (Fig. 17) agree reasonably well with experiment on a semielliptical artificial crack.[76] Incidentally, note that the theoretical predictions computed with a grid of 16×8 elements are also used to generate the 250 Hz impedance plane plot in Fig. 10. The computed results in Fig. 17 are plotted for three different rectangular cell sizes showing the dependence of the results on the number of unknowns. A reasonably accurate result can be achieved with only 128 unknowns and the finer grid results are consistent with each other.

Figure 18 shows similar low frequency (250 Hz) results for a simulated crack whose shape is shown in Fig. 15. At intermediate frequencies, the crack opening must be taken into account[100] and at high frequencies, the number of boundary elements must be increased. However, in the thin penetration regime, boundary elements can be avoided altogether as discussed in the following section.

FIGURE 17. Variation with probe position for coil whose axis is in plane of semielliptical simulated crack in aluminum: (a) resistance change; (b) reactance change.[76]

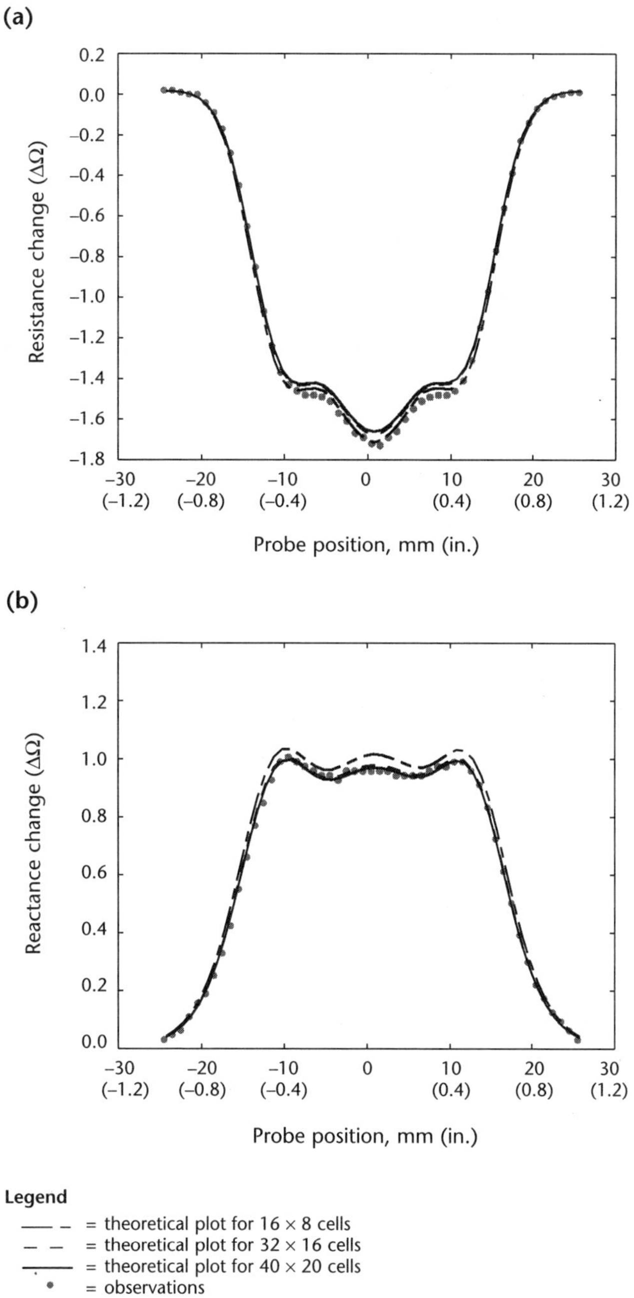

Thin Penetration Crack Theory

A number of approaches have been used to determine the electromagnetic field at a crack for the thin penetration regime. In this regime, in which the standard depth of penetration is very much smaller than the length and depth of the crack, eddy currents are confined to a region close to the conductor and to the crack surface. It is found that their distribution over the crack is governed by the solution of the laplace equation in the domain of the crack face. The reduction to a two-dimensional laplace problem is theoretically attractive because a number of standard techniques can be adopted to solve such problems. From the practical point of view, it is often desirable to carry out eddy current testing and experiments in the thin penetration regime because the sensitivity to discontinuities is greater at high frequencies. In testing ferromagnetic materials for cracks, the standard depth of penetration is usually much smaller than the overall discontinuity dimensions. Hence, the high frequency limit has important practical significance.

The main theoretical question to be faced in seeking a solution of the two-dimensional laplace problem is, "What are the boundary conditions?" Beginning in the early 1980s, a research group at University College London in the United Kingdom produced a series of articles on the alternating current potential drop technique for measuring cracks. A number of these articles were based on the *unfolding model*.[101,102] This model was successfully applied to the problem of finding the depth of cracks in ferromagnetic steel in the thin penetration regime. The problem domain can be divided into two equal parts, each consisting of a half plane at the surface of the conductor and a crack face at right angles to it. The line adjoining the half plane and the crack face is called the *fold line.* By unfolding the crack face into the surface plane of the conductor, a modified problem domain is formed. A scalar potential representing the electromagnetic field in the plane was deemed to be continuous and have continuous normal gradient at the fold line. At the crack edge, a boundary condition on the potential was deduced from the fact that the electric field tangential to the tip is zero. These constraints are sufficient to form a well posed, two-dimensional laplace problem that was solved to give results in agreement with experiment. Estimates of crack depth in steel components using alternating current potential drop were improved as a result of this work.

The unfolding model is not valid for nonferrous material but an alternative thin penetration theory was developed for eddy current testing in such materials by Auld and others, who considered cracks in aluminum alloys.[48,103] Auld's boundary condition assumes that the external magnetic field tangential to the conductor surface is not perturbed by the crack. The assumption may have been inspired by Kahn's two-dimensional long crack problem[74] because it is exact when the magnetic field is uniform along the length of a crack of uniform depth but, for a nonuniform probe field at a finite crack, it is approximate. The approximation is reasonable provided the coil diameter is large compared with the crack size but this limitation leaves room for improvement in the predictions.

It became evident in the late 1980s that the differences between the London group's model and Auld's approach ought to be reconcilable in a unified theory that would be valid for arbitrary permeability. In seeking the unified approach, the perturbation in the magnetic field at the crack mouth was taken into account by Lewis, Michael, Lugg and Collins,[104,105] who derived a boundary condition using a flux conservation argument applied to a region around the opening. The resulting theory is applicable to materials of arbitrary relative permeability and corroborates the unfolding model in the high permeability limit.

Alternative Formulations

A more formal approach to obtaining the unified theory is to start with a technique valid at an arbitrary frequency and specialize it systematically for the thin penetration regime. A suitable

FIGURE 18. Variation with probe position for coil whose axis is in plane of semielliptical simulated crack in aluminum: (a) resistance; (b) reactance. Calculations were performed at 250 Hz and by using 32 × 16 grid. See Harrison and Burke for details of coil parameters and simulated crack.[76]

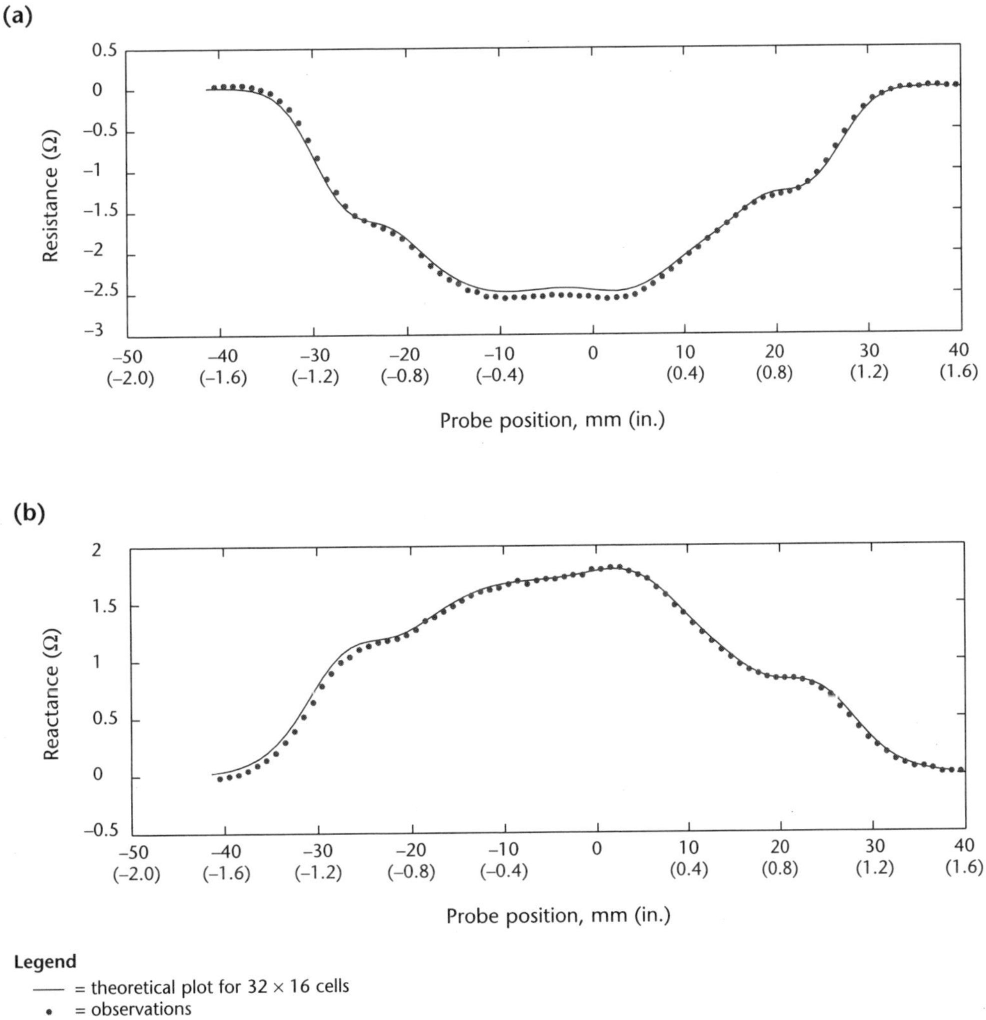

formulation for this strategy is one where the electromagnetic field in the conductor is expressed in terms of transverse electric and transverse magnetic hertz potentials,[106] ψ and ψ' respectively. Then, the electric and magnetic fields take the forms:

$$(92)\quad E(r) = i\omega\mu_0\mu_r\left[\nabla \times \hat{x}\psi'(r) - \nabla \times \nabla \times \hat{x}\psi(r)\right]$$

and:

$$(93)\quad H(r) = \nabla \times \nabla \times \hat{x}\psi'(r) - k^2\nabla \times \hat{x}\psi(r)$$

where $z < 0$ and where the preferred direction $\hat{x}$ is normal to the crack plane.

In a half-space problem formulated using hertz potentials, it is usual to choose the preferred direction as the normal to the interface between the air and the conductor. In this way the two potentials are decoupled at the interface. Although the present choice of preferred direction leads to coupled interface conditions, the chosen modes are decoupled at the crack surface. In fact, the transverse electric mode does not interact directly with an ideal crack at all. Instead, it is perturbed indirectly through its coupling with the transverse magnetic mode at the surface of the conductor. Because direct transverse electric interaction with the crack is absent, the transverse electric potential and its gradients are continuous at the ideal crack plane. In contrast, the transverse magnetic potential is subject to a direct interaction of the crack with the field and therefore exhibits a discontinuity at the crack.

To examine the discontinuity of the transverse magnetic hertz potential, it is necessary to reconsider the properties of the electromagnetic field at the crack.

First, the fact that the normal component of the electric field at the surface of the crack is zero means that:

$$(94)\quad \nabla_t^2\psi\left(r^{\pm}\right) = 0$$

at the crack surface, as can be seen by taking the x component of the electric field as expressed in Eq. 92 and applying the quasistatic condition for the nonconducting crack given in Eq. 72. Thus the transverse magnetic potential satisfies the two-dimensional laplace equation on the crack for an arbitrary excitation frequency and standard depth of penetration.

Second, in the absence of direct transverse electric interaction with the crack, the continuity of the tangential magnetic field (Eq. 83) implies that:

$$(95)\quad \psi\left(r^{+}\right) - \psi\left(r^{-}\right) = 0$$

as can be deduced from Eqs. 93 and 83. Thus the transverse magnetic potential itself is continuous at the crack surface S_0.

Third, noting that the jump in the electric field is due solely to the transverse magnetic mode, it can be seen from the form of the transverse magnetic contribution in Eq. 92 combined with Eq. 77 that:

$$(96)\quad \frac{\partial\psi}{\partial x}\bigg|_{r+} - \frac{\partial\psi}{\partial x}\bigg|_{r-} = \frac{1}{k^2}\,p(r_0)$$

It is concluded that the transverse magnetic potential has a discontinuity in its normal gradient at the crack surface S_0.

It can be shown that the transverse electric hertz potential ψ, expressed as the sum of unperturbed and perturbed components, is given by:

$$(97)\quad \psi(r) = \psi^{(0)}(r) - \int_{S_0} G(r,r') \times \left[\frac{\partial\psi}{\partial x}\bigg|_{r+} - \frac{\partial\psi}{\partial x}\bigg|_{r-}\right] dS'$$

for arbitrary frequency and standard depth of penetration. The Green's function $G(\boldsymbol{r},\boldsymbol{r}')$ accounts for the cross coupling between transverse electric and transverse magnetic modes.[107]

Several approaches for finding a solution to the ideal crack problem follow immediately from Eq. 97, both at an arbitrary frequency and for the thin penetration regime. For example, without restricting the frequency, one can use a symmetry argument to write the jump in the derivative of the potential at the crack as $2(\partial\psi)\cdot(\partial x)^{-1}$. Differentiating Eq. 97 with respect to x and assigning the field coordinate to a point at the crack face denoted by $\boldsymbol{r}^{\pm}$ will give an equation for the normal derivative of ψ. From the solution, p can be found from Eq. 96 and the probe impedance due to the discontinuity found from Eq. 91. The following approach has appeared in the literature.

Setting the field coordinate in Eq. 97 to $\boldsymbol{r}^{\pm}$ and using Eq. 96 gives:

$$(98)\quad \psi\left(\boldsymbol{r}^{\pm}\right) = \psi^{(0)}\left(\boldsymbol{r}^{\pm}\right) - \frac{1}{k^2}\int_{S_0} G\left(\boldsymbol{r}^{\pm},\boldsymbol{r}'\right)p\left(\boldsymbol{r}'\right)dS'$$

Essentially the same equation as Eq. 98 is found using a magnetic vector potential formulation.[108] Finding a solution relies on the fact that the unknown potential $\psi(r^{\pm})$ satisfies the laplace equation on S_0 (Eq. 94) and must be determined simultaneously with $p(f)$. These two unknown functions can indeed be found from the same equation simultaneously by imposing further constraints. The additional constraints are not the boundary conditions on the laplace problem for ψ at the crack face, because these are not defined. Instead, the boundary conditions at an arbitrary frequency (Eqs. 78 and 79) are imposed on p.

In finding a solution using the moment technique using N equations for N unknowns, a reduction in the unknowns needed to approximate p can be made because the prior knowledge derived from Eqs. 78 and 79 restricts its behavior at the perimeter of the crack. This technique releases some degrees of freedom that can be used to represent $\psi(r^{\pm})$ as a solution of the laplace equation on the crack face. By management of the unknown coefficients in this way, a solution can be found that agrees with experiment.[108]

FIGURE 19. Inductance and resistance variation with probe position for coil whose axis is in plane of semielliptical artificial crack in aluminum: (a) inductance plot; (b) resistance plot. Theory (solid line) is compared with experimental results (points) acquired at 50 kHz. See Harrison and Burke for details of coil parameters and simulated crack.[76]

(a)

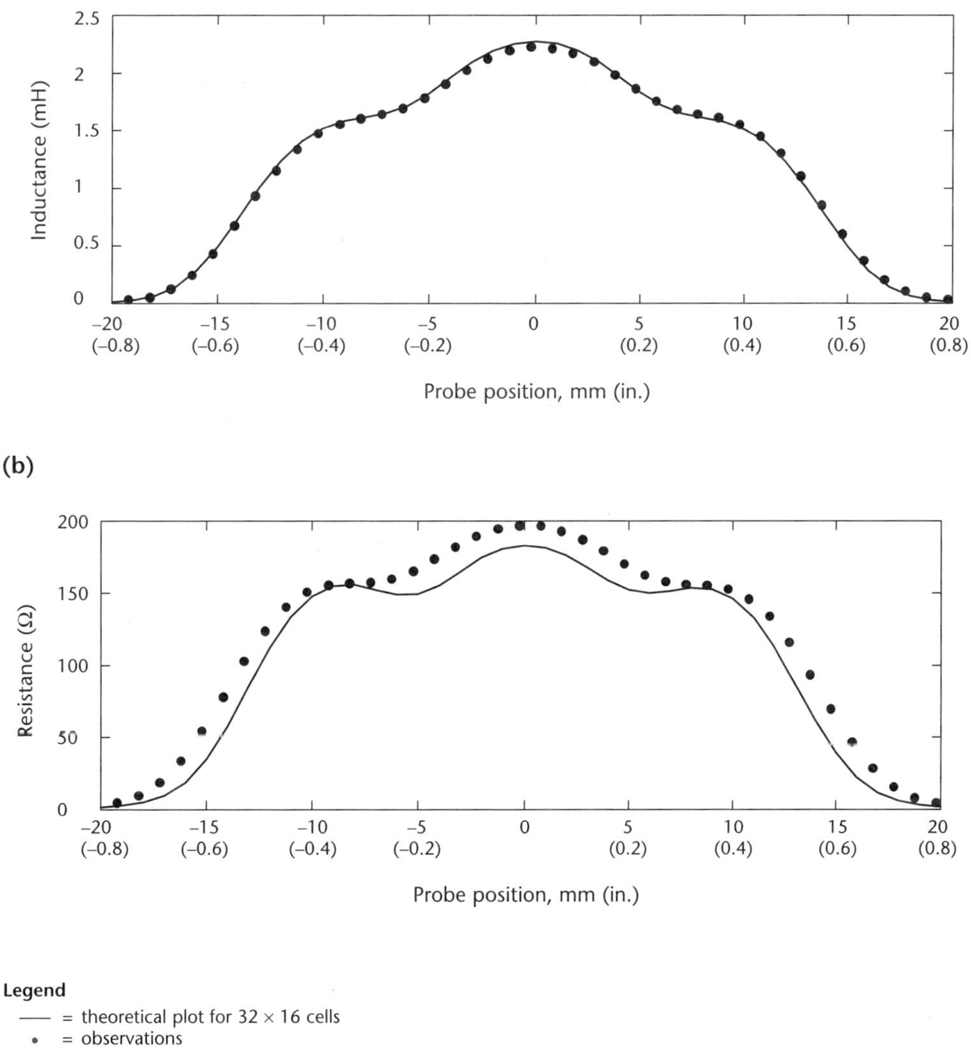

Thin Penetration Regime

As Auld has shown, a suitable boundary condition for formulating a well defined laplace problem on S_0 in the thin penetration regime can be derived from the magnetic field at the crack mouth. The transverse magnetic component of the magnetic field in the Y direction can be written:

$$\text{(99)} \quad \left.\frac{\partial \psi}{\partial z}\right|_{r_m} = \left.\frac{\partial \psi^{(0)}}{\partial z}\right|_{r_m} + \left.\frac{\partial \psi^{(c)}}{\partial z}\right|_{r_m}$$

where $\psi^{(c)}$ is the perturbed potential (volt) due to the crack. As it stands, Eq. 99 cannot be used immediately as a boundary condition because the perturbed field at the mouth is not known in advance. Auld got around this problem by neglecting the perturbation of the magnetic field at the crack mouth, a reasonable approximation because it can be small for nonferromagnetic materials. Taking the field perturbation into account increases the complexity of the problem[107] but improves the accuracy of the results for nonferrous alloys and gives results valid for ferromagnetic materials.[109]

Results of impedance predictions[110] and measurements for a semielliptical artificial crack are shown in Fig. 19. The experimental data are taken from a series of measurements made at 16 frequencies.[76] For a comparison with thin penetration theory, results at the highest frequency (50 kHz) are shown. Calculations were performed with conformal mapping.[110] At this frequency, the depth of the simulated crack, 8.61 mm (0.339 in.), is more than 18 times the standard depth of penetration, 0.47 mm (0.019 in.). Note that the theory underpredicts the resistive component by about 10 percent. However, this component, is small compared with the inductive reactance, which has a maximum value over 600 Ω. The overall accuracy of the predictions is good.

Part 4. Computer Modeling of Eddy Current Fields[2]

Mathematical Basis of Modeling

Computer modeling is used to simulate reality. In the case of eddy current testing, the computer can be programmed to replicate (1) the physics of the testing media, (2) the characteristics of the test object and (3) the geometry of the test — and to then display a dynamic visual image of all three during testing.

Modeling may be used for computer aided design of eddy current test components, as well as research on specific or general test applications. Both of these potential uses reduce the need for trial and error manufacturing of sample components and provide an alternative to actual tests that are difficult, hazardous or costly.

By definition, modeling is the design of a mathematical system that obeys certain fact based conditions. The behavior of the model is then used to understand an analogous physical system. The value of the model relies directly on its ability accurately to duplicate the behavior of the system for which it is the analog.

All electromagnetic phenomena, including those related to eddy current testing, are described by Maxwell's equations (Eqs. 1 to 4). In performing an eddy current test, these relations are used even if this is not explicitly known to the inspector. The designs of eddy current tests and equipment are based on Maxwell's equations, regardless of the actual technique used in the design process. In modeling electromagnetic phenomena, it is natural to rely on the solution of Maxwell's equations. The more accurately these equations can be modeled, the better the resulting general model. The usefulness of eddy current tests and the information available from interpretation of their test signals can only be as good as the understanding of the underlying principles. Authoritative decisions regarding a test signal cannot be made if all aspects of the magnetic field's interaction with materials and material discontinuities are not fully and uniquely understood.

The solution of Maxwell's equations is at the heart of any eddy current discontinuity characterization scheme. The ability to interpret signals, design tests and equipment and ultimately to solve the inverse problem in eddy current testing is directly related to the ability to solve these equations within realistic testing geometries.

Although the solution of field problems has preoccupied scientists and engineers since the publication of Maxwell's *Treatise*[1] in 1873, the complexity of field relations and interactions has limited such attempts to well behaved, simple problems. Such an extensive effort has yielded few specific results and not one general model capable of describing all electromagnetic field phenomena. Models that could be applied generally could be developed only after the introduction of digital computers.

This is not surprising considering the complexity of the interactions involved. Eddy current techniques of nondestructive testing rely on alternating current excitation that induces secondary currents and fields in the test material. Discontinuities in the test object cause changes in the induced fields, which are usually monitored by measurable changes in coil impedances. Thus, the technique requires indirect measurement (through impedance changes) of secondary (induced) fields and currents.

The nature of field problems, nondestructive testing applications in particular, leads to three-dimensional, nonlinear, partial differential equations within awkward boundary conditions. The solution domain often includes complicated discontinuity shapes. For the case of moving probe applications, the solution is a function of time and position. In addition, the solution domain is unbounded: the field only decays to zero at infinity. These very general requirements encompass the whole spectrum of possible difficulties (with the exception of high frequency problems).

These complications have led to a definite reliance on experimental techniques[111-113] and on analytical models wherever such models could be found.

Types of Modeling

Experimental Modeling. Experimental techniques are derived from measurements in eddy current tests, either actual or simulated. The value of such models is limited because of their empirical nature and the fact that their extension to other geometries is more or less speculative. This limitation does not mean that accurate, controlled experimental data are not valuable in modeling. Both analytical and especially numerical modeling rely on such data for confirmation.

Experimental techniques have many limitations and, although very useful at times, are not always reliable for modeling. The generality required for modeling does not exist with experimental techniques.

Analytical Modeling. Analytical models are derived from elementary field and circuit theory relations. At the very basis of this approach is the fact that some simplifying assumptions must be made with regard to the test environment. These assumptions include (1) those that are satisfied with little or no errors, such as linearity in eddy current calculations, and (2) those that imply large errors or even a modification of the geometry, relying on the hope that, by doing so, the solution is still an approximation of the actual problem modeled. In this category are symmetry considerations, boundary conditions and discontinuity shape approximations. In spite of extensive simplification, analytical models are extremely complicated and the results tend to be limited to a single geometry or class of problems.

Numerical Modeling. Numerical modeling in nondestructive testing is an outgrowth of the failure of analytical models reliably to predict the necessary field interactions with any degree of generality. A numerical model uses a digital computer to solve the governing equations directly, with few simplifying assumptions. This in itself is enough to explain the value of such models. Numerical modeling allows the solution of very complex problems and, at the same time, does not require the user to know the intricacies of electromagnetic theory or differential calculus. All that the user is required to do is input the problem variables and, if necessary, verify the results experimentally.

General Overview of Analytical and Numerical Modeling

The existence of models for eddy current testing phenomena depends entirely on the ability to solve Maxwell's equations with or without approximations. The value of such models in solving the inversion problem satisfactorily is beyond dispute. Solution of this important problem is possible only with the development of good theoretical models, capable of predicting the complex interactions of a multitude of factors in the test object. A good, reliable theoretical model for nondestructive testing should be able to satisfy the following conditions.

1. The model should describe the physics of interaction between the applied alternating current field, induced currents and discontinuities in the test object.
2. The model should serve as a theoretical test bed for situations difficult or impossible to replicate experimentally.
3. The model should generate eddy current output signals for a wide variety of discontinuities and specimen shapes, avoiding costly sample preparation and helping to determine discontinuity characterization parameters.
4. The model should provide training data for automated discontinuity characterization systems and equipment.
5. The model should aid in the design of eddy current probes for specific applications.

For the purpose of deriving such a model, two main avenues are available: the analytical approach and the numerical approach.

Analytical Modeling

Analytical models are derived from basic field and circuit theory considerations. In effect, an attempt is made to solve Maxwell's equations directly. These equations are generally three-dimensional, nonlinear, partial differential equations within complex boundaries and discontinuity shapes. In addition, for moving probe problems, the solution is both time and position dependent. It is therefore not surprising to find that such solutions are only possible for the most elementary of test geometries, with simplifying assumptions in terms of geometry, dimensionality, discontinuity shapes and sources. This oversimplification accounts for the fact that analytical models are limited in scope, applicable only to selected problems and not easily extended to other geometries. On the other hand, the solution to problems for which an analytical model applies is relatively simple and allows parameter change studies.

Analytical modeling has its roots in the pioneering work of Ampère, Örsted, Faraday, Lenz, Gauss, Helmholtz, Henry and Foucault, culminating in Maxwell's *Treatise*,[1] in which many practical problems are addressed and solved, including many that are directly related to nondestructive testing. In view of this background and the work of Hughes,[114] it is surprising that the first serious attempt at modeling was undertaken only after Steinmetz introduced the complex notation for field quantities,[115] thus paving the way for the early modeling work of Förster and the introduction of impedance plane diagrams as an accepted technique of presenting eddy current test data.

The phenomena associated with eddy current testing can be examined only after Maxwell's equations have been manipulated and simplified into a form suitable for solution by one of the techniques for solving partial differential equations. Such techniques as separation of variables, bessel functions, power series and the various transform techniques, especially the fourier transform, are used.

Förster and Stambke[115] used bessel functions to find the complex effective permeability of a metal rod encircled by a secondary search coil and an alternating excitation coil. The effective permeability concept is used to directly connect the dissipative and inductive quantities in the specimen with the resistive and inductive terms in the impedance plane plot. Hochschild[111] solves a fairly simple problem (a cylindrical sample and a concentric coil), using bessel functions directly for the magnetic flux density B (tesla) within the conductor, in terms of the flux density at the conductor's surface, using the following expression:

$$\frac{\partial^2 B_z}{\partial r^2} + \frac{1}{r}\frac{\partial B_z}{\partial r} - \gamma^2 B_z = 0 \tag{100}$$

where r is radial distance (meter) and γ is a function of the standard depth of penetration.

This leads to an expression for the flux linkage with the encircling coil from which the induced voltage and the real and imaginary parts of the coil impedance are found. The well known comma shaped curves can be derived by plotting the impedance for various frequencies and conductivities.

The same geometry is solved by Libby[116] by using the definition of magnetic vector potential and substituting it for the flux density. To obtain the coil's impedance, the magnetic vector potential is related to the induced voltage A_θ (volt) in a current loop and then the impedance of the coil is obtained as a closed form expression in a source free region:

$$\frac{\partial^2 A_\phi}{\partial r^2} + \frac{1}{r}\frac{\partial A_\phi}{\partial r} - \frac{A_\phi}{r^2} + \omega^2 \mu \varepsilon A_\phi - j\omega\sigma A_\phi = 0 \tag{101}$$

Waidelich and Renken[117] used the image coil concept to calculate the impedance of a coil in the vicinity of conducting media and compared the results with experimental data. Vine[118] shows this to be the limiting case of a single current loop above a conducting plate of finite thickness. Cheng[119] examines the same situation and, using the magnetic vector potential in cylindrical coordinates, obtains an integral expression for the coil impedance.

Expanding on this idea, Dodd[120] obtains expressions for the coil impedance of a rectangular cross section coil above a two-conductor plane and encircling a two-conductor rod.

To avoid the complexities of Maxwell's integral equations, Graneau and Swann[121] and Graneau[122] replace conducting media with an infinite number of filamentary circuits corresponding to streamlines of current flow. This substitution leads to a coupled circuit model and a power series representation for the induced currents.

Burrows[64] introduces magnetic and electric dipoles to represent small discontinuities. Dodd and others[11] predict the induced voltage in a circular coil due to small discontinuities.

Many other theoretical and experimental concepts[74,123-127] have been used to approximate the solution to nondestructive testing problems with varying degrees of simplification and success. The underlying assumptions made in deriving these models are of such a restrictive nature that their application to other problems, in more realistic geometries, is all but impossible.

Integral Solution Technique

To obtain a closed form solution for simple geometries, analytical techniques require solution of integral equations and, if the integration can be carried out, a solution may be obtained for a particular geometry. Beyond the simplification of the actual geometry and a lack of generality, a new consideration should be introduced: integration processes that may or may not be done analytically.

In such cases, numerical integration is used. Because of this need to numerically integrate analytical expressions, the technique is considered a hybrid that bridges the gap between the two techniques. Inherent in it are all the initial approximations of an analytical technique combined with the flexibility of numerical techniques.

The names *integral solution technique* and *boundary value solution* should not be confused with similar or identical terms used for finite element solution of field problems in integral form. This later usage refers to the formulation of Maxwell's equations (or any other physical system) as a boundary integral problem, usually solved by surface discretization of a volumetric region. The term *boundary value solution* is here much more restricted and reflects the fact that an integral expression is obtained and its solution is in terms of orthogonal functions along the boundaries of the solution region.[123,128]

Dodd[15,128,129] obtained such expressions for a variety of testing geometries that fall into two major categories: (1) multilayered conductors and (2) multiple coaxial cylindrical layers. These two categories include many of the more practical and useful testing geometries.

The solutions are in the following form:[15]

$$(102)\quad A(r,z) = \frac{1}{2}\eta_c \mu I \int_0^{\infty} \left\{ \frac{e^{(-\alpha \ell_1)}}{\alpha^3 V_{22}} \right. \times \left[1 - e^{-\alpha(\ell_2 - \ell_1)}\right] J(R_2 R_1) \cdot \left[V_{12}(\eta,1) e^{-\alpha \eta z} + V_{22}(\eta,1) e^{\alpha \eta z}\right]\Bigg\} J_1(\alpha r) d\alpha$$

where $A(r,z)$ is the magnetic vector potential in the nth plane for a coil above a number of planar conductors, d indicates a differential, I is the current per turn, J is a bessel function of the first kind and first order, α is the separation constant of the differential equation, η_c is the turn density of the coil, μ is the magnetic permeability (henry per meter), $V_{22}, V_{12}(\eta,1)$ and $V_{22}(\eta,1)$ are transformation matrices and where:

$$(103)\quad J(R_2,R_1) = \int_{\alpha R_1}^{\alpha R_2} \chi J_1(\chi) d\chi$$

where R_1 and R_2 are the inner and outer coil radii (meter).

Although these expressions are complicated and very difficult to evaluate on paper, they are easily integrated numerically on computers.

Once the magnetic vector potential has been evaluated, any magnetic or related quantity may be calculated. These are derivable either directly from Eq. 102 or through other known relations.

For example, if the driving coil is coaxial with the pickup coil in an axisymmetric geometry, the induced voltage in the pickup coil can be written:

$$(104)\quad V = \frac{j\omega 2\pi n}{a} \iint_a rA\, dr\, dz$$

where a is the cross sectional area (square meter) of the coil winding and n is the number of turns in the coil.

The integration is performed over the coil cross section and is particularly convenient for coils with rectangular cross sections (as is usually the case with eddy current coils).

The impedance $\mathbf{Z}$ of a coil may be found from Eq. 104 by dividing by I:

$$(105)\quad Z = \frac{V}{I} = \frac{j\omega 2\pi n}{Ia} \iint_a rA\, dr\, dz$$

Similarly, eddy current densities, flux densities, stored and dissipated energies or forces due to eddy currents may be calculated.

The technique described above has been programmed for minicomputers as well as mainframes and the computer programs are available in the open literature,[128-130] accounting for the technique's success.

The integral solution technique has several advantages over purely analytical techniques, the most important being the wider range of application afforded by numerical integration. Because of this, the integral solution technique has been applied to eddy current problems ranging from single-coil, single-conductor situations to situations using multiple coils and multilayered materials for simulation of such important problems as discontinuity detection and measurement of cladding thickness, conductivity, permeability and liftoff.

The technique still suffers from the problems associated with analytical techniques: (1) it is not general and (2) it requires the evaluation of an integral expression for each class of problems. In addition, because of its reliance on orthogonal functions on the boundaries, it can only take into account relatively simple geometries and discontinuity shapes.

Lastly, the technique assumes linear material properties and superposition of solutions, a difficulty found in all boundary integral techniques and not easily extendable to nonlinear problems. This assumption is acceptable for eddy current testing applications but is not general enough to form the basis of an all-purpose solution technique, as is the case with the more general numerical techniques.

Numerical Modeling

Numerical modeling is different from analytical modeling. For the purposes of nondestructive testing, the most important aspect of numerical modeling is the fact that none of the simplifying assumptions made in the analytical approach are necessary to reach a satisfactory solution.

Starting with Maxwell's equations, there are many different ways to proceed and each requires the formulation of the original equations in some particular form.[131-133] Assumptions are made for the sake of simplicity and economics of the solution and not to render the equations solvable. Thus, a two-dimensional or axisymmetric solution is assumed if the geometry is approximately two-dimensional or has axial symmetry. A two-dimensional solution is likewise a special case of the more general three-dimensional solution. Similarly, an axisymmetric solution is the solution of a three-dimensional problem in cylindrical coordinates. Other assumptions help to shorten the solution process or to obtain insight into the problem before a more general solution is attempted. A linear formulation may be used as a first approximation or in cases where the problem is indeed linear.

The application of a numerical process for the purpose of obtaining a model is, to a large extent, a matter of choosing the numerical technique to be used, a matter of making correct assumptions for geometry and the nature of the solution and, perhaps most important, a matter of economics. Within these constraints, any accuracy can be obtained regardless of geometry, linearity, nonlinearity or dimensionality of the problem.

Numerical techniques in general are far more powerful than analytical techniques. At the same time, the solution is obtained as numerical data rather than a closed form solution. As such, the solution to a particular problem may not be usable for the analysis of a different, perhaps similar problem. Thus, parameter study requires repetitive solution and it is not always possible to deduce parameters from the solution, as is the case with analytical solutions. This disadvantage is hardly significant: the same repetitive process allows study of parameters for which the analytical approach cannot be applied (such as changes in the geometry of an arbitrarily shaped discontinuity).

Finite Difference Technique

The finite difference technique has been used as a general means to solve partial differential equations. The reasons for its widespread application are many. The technique is relatively easy to apply, as well as being general. It is equally applicable to direct current fields, to quasistatic or transient fields and to linear and nonlinear problems. In its simplest form, the formulation of the field equations consists of simply replacing the partial derivatives by appropriate difference formulas. A solution can then be obtained for the dependent variable at discrete points within the solution region either by an iterative process or by the solution of a system of algebraic equations, depending on which finite difference formula is applied.

The application of the finite difference technique is complicated by problems of convergence and stability of the solution as well as by restrictions on the discretization process. Although regular sets of discretization points (grids) are easy to handle, irregular grids are not. Discretization of complex geometries into regular grids is not practical and irregular grids may in some cases make the solution nonconvergent. In field problems, the inability to properly discretize small areas (such as air gaps or discontinuities) is detrimental to the finite difference technique. In addition, the technique is a nodal technique and cannot take into account distributed parameters such as current densities, conductivities and permeability. These have to be described as equivalent nodal quantities with all the associated errors. The obtained solution is valid only at the nodal points.

Finite Difference Representation

If an attempt is made to solve a partial differential equation such as Eq. 100 or 101, it should be possible either to integrate the equation or to represent the partial derivatives in terms of the unknowns themselves at discrete points in space. The finite difference algorithm is an implementation of the second approach.

Considering Fig. 20, where a general function is described, the true derivative $dy \cdot (dx)^{-1}$ at a point x_i is the tangent to the curve at this point. An approximation to the derivative can be found by taking two

FIGURE 20. General function and finite difference approximation to true derivative.

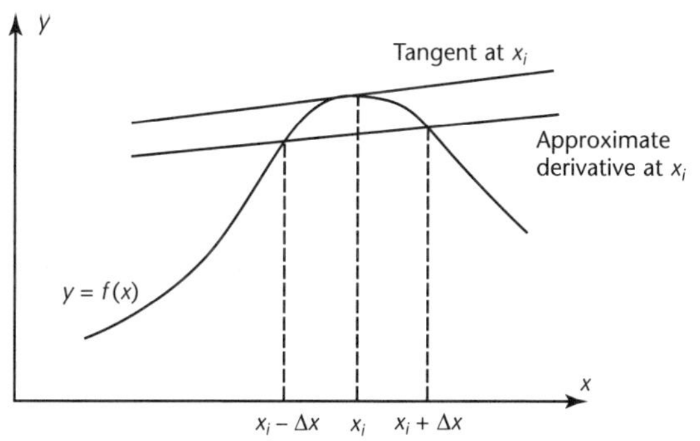

points, (one point on each side of x_i) and passing a straight line through them. If the two points are chosen to be equally spaced about point x_i (as in Fig. 20), the following expression for the slope of the line can be obtained:

$$\text{(106)} \quad y'_{(x_i)} = \frac{y_{(x_i+\Delta x)} - y_{(x_i-\Delta x)}}{2\Delta x}$$

By denoting in short form $y'_{(x_i)}$ as y'_i, $y_{(x_i+dx)}$ as $y_{(i+1)}$ and $y_{(x_i-dx)}$ as $y_{(i-1)}$, a simpler expression linking the approximation $y'_{(x_i)}$ to the function value at $x_{(i-1)}$ and $x_{(i+1)}$ can be written:

$$\text{(107)} \quad y'_i = \frac{y_{(i+1)} - y_{(i-1)}}{2\Delta x}$$

The same result can be obtained formally by using a taylor series expansion. By expanding the function $y = f(x)$ about the point x_i for $x = x_i - \Delta x$ and $x = x_i + \Delta x$, the following results are obtained:

$$\text{(108)} \quad y_{(x_1+\Delta x)} = y_i + y'_i \Delta x + \frac{y''_i (\Delta x)^2}{2!} + \frac{y'''_i (\Delta x)^3}{3!} + \cdots$$

$$\text{(109)} \quad y_{(x_1-\Delta x)} = y_i - y' \Delta x + \frac{y''_i (\Delta x)^2}{2!} - \frac{y'''_i (\Delta x)^3}{3!} + \cdots$$

By subtracting Eq. 109 from Eq. 108 and rearranging the terms, the first derivative can be written as:

$$\text{(110)} \quad y'_i = \frac{y_{(x_i + \Delta x)} - y_{(x_i - \Delta x)}}{2\Delta x} - \left[\frac{1}{6} y'''_i (\Delta x)^2\right] + \cdots$$

Then, neglecting terms with $(\Delta x)^2$ or higher powers of Δx, the expression in Eq. 107 is obtained. This technique is less intuitive than the one used to derive Eq. 107 but shows two important points.

1. The error produced by the approximation is on the order of $(\Delta x)^2$. Thus, there is a simple way of estimating the error and, at the same time, the solution can be improved by reducing the size of Δx.
2. The finite difference formula in Eq. 110 was obtained by choosing an appropriate expansion to cancel specific terms of the expansions. This indicates that higher order derivatives and different approximations to the same derivatives can be obtained. Indeed, many useful difference formulas have been derived.[134]

An approximation for the second derivative is obtained by adding the two expansions in Eqs. 108 and 109 and rearranging the terms:

$$\text{(111)} \quad y''_i = \frac{y_{(i+1)} - 2y_i + y_{(i-1)}}{(\Delta x)^2} - \left[\frac{1}{12} y''''_i (\Delta x)^2 + \cdots\right]$$

This particular approximation also introduces an error on the order of $(\Delta x)^2$.

The finite difference expressions derived here use points on both sides of the point at which the derivatives are calculated. They are therefore called *central difference expressions*. Backward and forward difference formulas may also be used.[134]

Finite Difference Formulation for Two-Dimensional and Axisymmetric Field Problems

The first step in the field formulation consists of replacing the partial derivatives by a difference equation. Referring to the grid in Fig. 21, Eq. 19 reduces in the case of an axisymmetric geometry in cylindrical coordinates to:

$$\text{(112)} \quad \frac{1}{\mu} \times \left[\frac{A_{(i-1),j} - 2A_{i,j} + A_{(i+1),j}}{(\Delta r)^2} + \frac{1}{r}\left(\frac{A_{(i+1),j} - A_{i,j}}{\Delta r}\right) + \frac{A_{i,(j-1)} - 2A_{i,j} + A_{i,(j+1)}}{(\Delta z)^2} - \frac{A_{i,j}}{r^2}\right] = -I_{i,j} + j\omega\sigma A_{i,j}$$

Here the current density J_s was replaced by an equivalent nodal current $I_{i,j}$. A similar expression may be written for the two-dimensional field equation.

If the grid is equally spaced in both directions ($W_r = W_z = h$) and a constant permeability can be assumed for all points, Eq. 112 is further simplified as:

$$(113)\quad \frac{1}{\mu} \times \left(\left\{ \left[A_{(i+1),j} + A_{(i-1),j} + A_{i,(i+1)} + A_{i,(j-1)} - 4A_{i,j} \right] \times \left(h^2\right)^{-1} \right\} + \frac{1}{rh}\left[A_{(i+1),j} - A_{(i-1),j} \right] - \frac{A_{i,j}}{r^2} \right) - j\omega\sigma A_{i,j} = -I_{i,j}$$

Equation 112 forms the basis for solution of eddy current problems in axisymmetric geometries. In its present form, the equation is of little use because, besides applying only to regular grids, it can be used to describe only nonmagnetic media. This limitation can be seen by testing where the interface between two materials is shown (Fig. 22). When calculating the value of the magnetic vector potential for points on the interface, some points lie in areas of different permeability. Equation 113 requires a single value for all points forming the expression. Similar problems are encountered with description of conductivities and current densities because all three properties are volumetric properties rather than point values.

To circumvent these problems, some special techniques were derived by several researchers. One such technique came from Erdelyi and Fuchs[135,136] and was later refined by Demerdash[137,138] for problems in electrical machines. If the five nodes associated with the calculation of the magnetic vector potential are considered (as in Fig. 22), four distinct regions are observed. Assuming that nodes must coincide with material boundaries, each such domain may have different material properties and current densities associated with it.

Under the present assumptions, an equivalent material property that is a weighted average of the properties of the four domains is produced. The current at the central node then becomes:

$$(114)\quad I_i = \frac{1}{8}\left(S_1 J_1 + S_2 J_2 + S_3 J_3 + S_4 J_4\right)$$

The weights S_1 to S_4 are functions of the four domains around node i. These areas need not be equal or rectangular.

Similarly, average values for conductivity σ and reluctivity ν are defined as:

$$(115)\quad \sigma_1 = \frac{1}{8}\left(S_1\sigma_1 + S_2\sigma_2 + S_3\sigma_3 + S_4\sigma_4\right)$$

and:

$$(116)\quad \nu_1 = \frac{1}{\mu_1} = \frac{1}{8}\left(S_1\nu_1 + S_2\nu_2 + S_3\nu_3 + S_4\nu_4\right)$$

The net effect of these approximations is to change the material properties at the edges of material discontinuities, assuming that the errors introduced by doing so are small. This approximation may or may not be good, depending on

FIGURE 21. Simple finite difference grid.

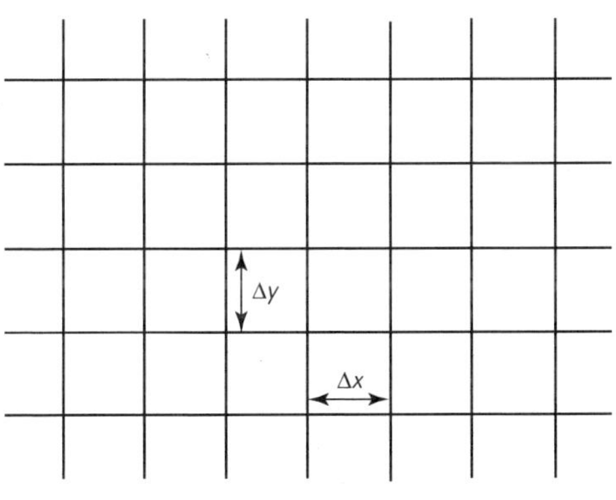

FIGURE 22. Section of finite difference grid showing domain of node *i*. Five nodes shown are needed to approximate second derivatives of magnetic vector potential with respect to *x* and *y* or to *r* and *z* in axisymmetric geometries.

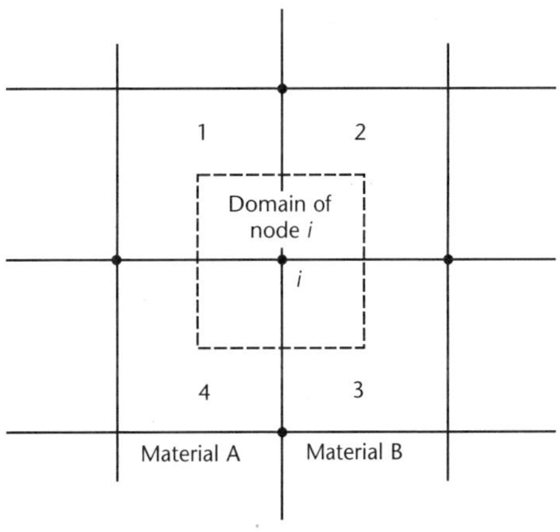

the grid. The same technique is used to calculate the equivalent material properties in nonregular and nonrectangular grids but the weighting is more complicated.[137]

Boundaries and Boundary Conditions

The first aspect to be considered is the structure of the grid itself. If the grid is kept uniform, the boundaries can be located only along lines connecting nodes. A curved boundary becomes a jagged line, as shown in Fig. 23. This approximation may be good for very fine grids where outer boundaries are involved. Curved material boundaries where the field gradients are high require better fitting to reduce errors.

This fitting can be done as shown in Fig. 24. The technique consists of calculating scale factors that effectively move nodes from their original location to the boundary. By doing so, the regularity of the grid is lost and, in the solution process, Eq. 113 will have to be modified to account for the different grid spacings used.

The boundary conditions most often encountered in eddy current field problems are dirichlet boundary conditions (prescribed values of the magnetic vector potential) and, in most cases relevant to nondestructive test applications, the values are zero. These are taken into account automatically by Eq. 113 when the five-node stencil of Fig. 22 hits the boundary. Other types of boundary conditions may be used with the finite difference technique.

FIGURE 23. Boundary representation in regular grid.

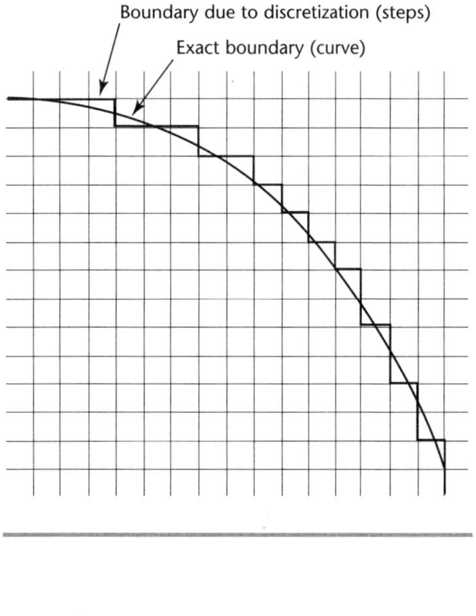

Nonuniform and Nonrectangular Meshes

The idea of scaling the distance between nodes to fit the grid to the boundary of the solution domain or material interfaces may be used with nonuniform grids. There is no requirement that the grid be rectangular. If nonuniform grids are used, Eq. 111 will have to be modified because the definition of the second derivative is based on a uniform grid. One approach is to rewrite the difference expression in Eq. 111 in the following form:

$$(117)\quad -I_{i,j} = W_1\left[A_{i,j} - A_{(i+1),j}\right] + W_2\left[A_{i,j} - A_{(i-1),j}\right] + W_3\left[A_{i,j} - A_{i,(j+1)}\right] + W_4\left[A_{i,j} - A_{i,(j-1)}\right] - j\omega\sigma A_{i,j}$$

The four weights are then calculated as weighted averages[137] that depend on the grid structure and material permeabilities. The calculation of the current at nodal points follows a technique similar to the one described above.

Solution of System of Equations

After a satisfactory description of the geometry has been found, the finite difference equation is applied to all interior nodes in the solution region. This step results in a relation between the

FIGURE 24. More accurate representation of boundaries by using irregular spacings at boundary.

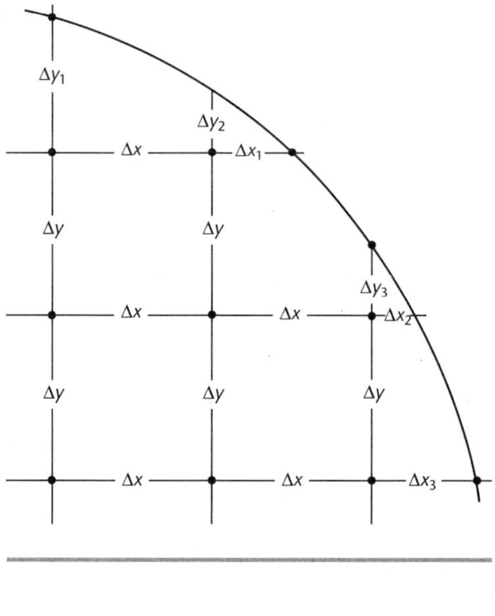

unknown at the node being assembled and four neighboring nodes, according to Eq. 113. For each node in the solution region, a linear algebraic equation is assembled.

To find the unknown values of the magnetic vector potential, there are two basic techniques available: (1) the iterative solution and (2) the matrix inversion solution.

Iterative Solution

The iterative solution is the simplest way to reach a solution and is often used with the finite difference technique. It consists of assuming an initial solution throughout the solution region (either an approximation or, if this is not possible, zero). The correct current distribution together with the boundary conditions are next applied. Then, the finite difference equation is applied to each interior node and the value at each node is updated in turn until a new solution is obtained. The process is repeated until the change in the solution is smaller than a predetermined value. The number of iterations required to obtain a good solution may be quite large. In some cases the solution may not converge to the final solution. This technique is called the *explicit technique* and because of its simplicity is often used in nonlinear and time dependent problems.[139] There are ways to refine the basic technique to make it more stable[140] or to accelerate convergence.[141] It is also possible to estimate beforehand if the solution is convergent, based on the grid spacing and the type of finite difference formulas used.

Matrix Inversion Solution

Instead of assuming an initial solution, the unknown values may be entered in the finite difference equation (Eq. 113) and a system of N equations with N unknowns is assembled (N is the number of interior grid points excluding boundary nodes). The final result is a system of complex algebraic equations:

$$\text{(118)} \quad \big[[G] + j[R]\big]\{A\} = \{I\}$$

where $[G]$ = the coefficient matrix resulting from the finite difference description of the partial derivatives, $\{I\}$ = a vector of equivalent applied currents at the nodes in the solution region and $[R]$ = a matrix resulting from the eddy current distribution.

The imaginary part of the matrix is due to eddy currents alone and will disappear for direct current applications. The technique outlined above is an implicit technique, absolutely stable, and therefore does not require iteration. The matrix in Eq. 118 can be solved using any standard technique such as the gauss elimination or the conjugate gradient technique.[142] The derivation presented refers to the axisymmetric field equation. An identical procedure can be used for two-dimensional geometries, replacing the necessary partial derivatives by appropriate finite difference expressions.

Finite Element Technique

The finite element technique has a briefer history than the finite difference technique. It evolved in the late 1950s as a numerical technique in structural analysis[143] but has spread quickly to become a major analysis tool in diverse areas of engineering[144-147] as well as in the physical sciences[148,149] and medical research.[150]

Because of its success in modeling intricate geometries efficiently and accurately, the potential for its application to electrical and magnetic fields was recognized in the early 1970s and has been applied with great success to the study of direct current and low frequency electromagnetic fields in electrical machines,[151-153] large magnet structures[154] and permanent magnet design.[155]

The finite element technique has considerable advantages over the finite difference technique, including the ease of handling boundary conditions and the ability to follow awkwardly shaped boundaries.[156] The technique is by definition a volumetric technique where various parameters are associated with the volume (or the surface, in the case of two-dimensional and axisymmetric formulations). Therefore, it is naturally suited to the modeling of continuum problems.

The finite element technique is also quite flexible in terms of the discretization process. Being a discrete technique, it requires discretization of the solution region but no restrictions are imposed on the shape, size and number of finite elements.[143] The solution process as well as the formulation is not affected by the size and shape of the elements used. Furthermore, although the finite difference technique assumes linear relations between the unknowns, the finite element technique can handle higher order relations as well.[143,157] Problems with convergence have no meaning in the context of finite elements.

These factors are of particular importance for the simulation of electromagnetic test techniques and the technique has received considerable attention. Numerical models based on the finite element technique have been developed for two-dimensional[158,159] and three-dimensional[160,161] eddy current applications.

When compared to finite difference techniques, problems solved by the finite element technique generally require larger computer resources, especially for nonlinear and time dependent problems. The technique does not lend itself well to the solution of transient problems because it cannot efficiently handle time discretization.

The two techniques are complementary, each being suited to the solution of different situations. Time integration in finite element computer codes is usually handled by various forms of finite difference schemes.[162]

In the following discussions, the finite element formulation of the electromagnetic field equations is outlined with reference to a particular element shape. The technique of formulation is completely general, however, and any other element shape can be used with relatively minor changes in the formulation.

Finite Element Formulation for Two-Dimensional and Axisymmetric Geometries

The finite element technique does not provide a direct solution to electromagnetic field equations. Rather, the solution is obtained by first formulating these equations into a suitable form for finite element solution and then solving the resulting set of simultaneous algebraic equations for the magnetic vector potential at discrete points in the solution region.

The formulation of the two-dimensional and axisymmetric field equations is presented here using the magnetic vector potential and an energy functional equivalent to the original equations. The following assumptions are made throughout this derivation.

1. The source current density J_s and the magnetic vector potential $\boldsymbol{A}$ vary sinusoidally with time. Harmonics in both the source and induced fields are absent.
2. The source medium is assumed to be infinitely conducting, thus effectively neglecting eddy currents in the source. In the case of eddy current probes, this is equivalent to subtracting the coil's direct current resistance from the resulting impedance, a common practice in interpreting eddy current signals.
3. Electrical conductivity σ and magnetic permeability μ of materials in the solution region are single-valued within each element. Each element is therefore a linear region but spatial variations between neighboring elements is allowed both in μ and σ. Also, these values can be different in each direction within the element, thus allowing modeling of anisotropic media. The linearity assumption provides satisfactory results because the applied and induced current densities in practical testing applications are very low.
4. The model is based on two-dimensional or axisymmetric assumptions. Thus, only one component of the magnetic vector potential is present, either in the Z direction (two-dimensional) or in the θ direction (axisymmetric) and is perpendicular to the cross section of the geometry modeled.

Among the many techniques available for finite element formulation of general problems, the weighted residual technique (Galerkin's technique) and the use of an energy functional stemming from a global energy balance concept,[143] in conjunction with variational techniques, are the most commonly used. Both procedures allow a direct formulation based on the original equations and both are satisfactory in terms of the resulting solution. Galerkin's technique is more convenient when only the differential equations and their boundary conditions are available.[157] When physical interpretation of the problem is important, the energy balance formulation introduced by Oden[163,164] in 1969 offers an attractive alternative.

A general energy functional for electromagnetic field problems can be written as:

$$\textbf{(119)} \quad F(A) = \int_v \left(E_s - E_i + E_d\right) dv$$

where E_s is the stored energy (joule) due to the magnetic field, E_i is the input energy (joule) derived from impressed current densities, E_d is the energy (joule) dissipated through eddy current densities in the conducting parts of the geometry, excluding sources, and v is volume (cubic meter).

The derivation that follows is limited to two-dimensional and axisymmetric geometries. It uses a particular type of isoparametric element (four-node quadrilateral element) but it is completely general and applies to other types of elements as well. A complete derivation for two-dimensional and axisymmetric problems in terms of triangular elements has been published.[158] Researchers describe a three-dimensional derivation in terms of tetrahedral elements[165] and a three-dimensional derivation in terms of eight-node and twenty-node hexahedral isoparametric elements.[160]

Energy Functional for Eddy Current Problems

The general energy functional in Eq. 119 applies to any field situation regardless of dimensionality, because it is a statement of energy balance in the system. In terms of the magnetic vector potential $\boldsymbol{A}$, the functional for two-dimensional or axisymmetric geometries can be written:

$$(120)\quad F(A) = \int_v \left[\frac{1}{2\mu} \left(\left| \frac{\partial A}{\partial x} \right|^2 + \left| \frac{\partial A}{\partial y} \right|^2 \right) + \frac{j\omega\sigma}{2} |A|^2 - J_S \cdot A \right] dv$$

where J_s is source current density (ampere per square meter) and where, for the axisymmetric case, y is replaced by z and $[\partial A \cdot (\partial x)^{-1}]$ is replaced by $[\partial A \cdot (\partial r)^{-1} + A \cdot r^{-1}]$. In the two-dimensional case, $dv = dx dy$ and in the axisymmetric case $dv = 2\pi r dr dz$.

From the principles of variational calculus, it can be shown that a correct solution to the governing partial differential equation is obtained by minimizing the energy functional throughout the solution region. Although proof of this statement[165] is not included here, it is a very important step because it also defines the natural boundary conditions that are implicit in the formulation and that need not be explicitly applied. The most important of these conditions are on the interfaces between different materials.

Finite Element Discretization

The solution of the variational formulation of the eddy current equations presented above is performed by finding a set of functions that minimize the functional $F(A)$. Because it is not possible to minimize the functional everywhere, it is minimized at discrete points (nodes) in a bounded region (solution region). The discretization of the solution region is therefore a very important step in the finite element technique because the number of nodes as well as their location in the solution region has an impact on the solution. A discretization (mesh) with few nodes in regions of high gradients in the solution will introduce errors whereas too many nodes will unnecessarily complicate and lengthen the solution. It is at this stage in the finite element process where efficient discretization techniques combined with correct judgment can have a significant impact on the efficiency and accuracy of the solution.

A large variety of volumetric finite elements can be used to discretize the region but, depending on the equations to be solved, some are more useful than others. The most common finite elements used for two-dimensional field calculations are the triangular and quadrilateral elements.[157] Linear or first order elements are defined as having a node at each vertex of the element as in Fig. 25a. Parabolic or second order elements are defined by including a node between each pair of adjacent vertices:

$$(121)\quad \begin{aligned} N_1 &= (0.25)\ (1 - \xi)\ (1 - \eta) \\ N_2 &= (0.25)\ (1 + \xi)\ (1 - \eta) \\ N_3 &= (0.25)\ (1 + \xi)\ (1 + \eta) \\ N_4 &= (0.25)\ (1 - \xi)\ (1 + \eta) \end{aligned}$$

FIGURE 25. Quadrilateral finite elements and shape functions: (a) element in local coordinate system; (b) mapped, curvilinear element in global coordinate system.

(a)

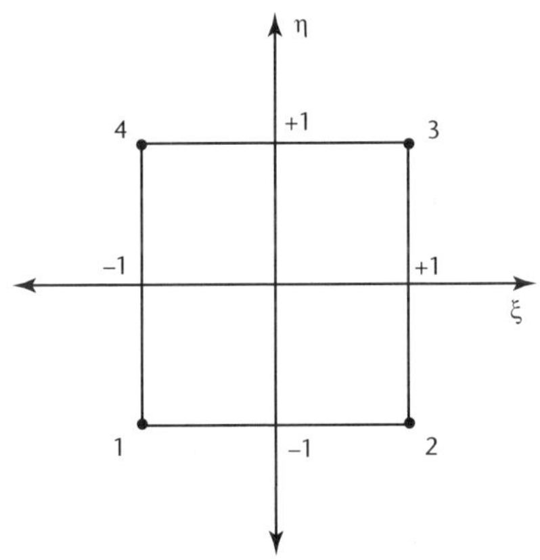

(b)

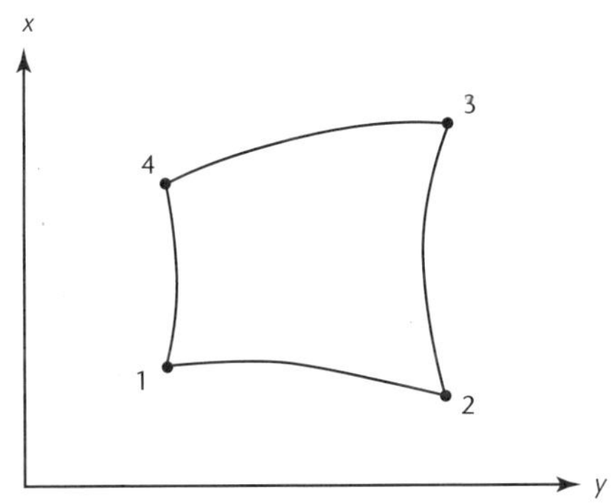

where ξ and η are finite element coordinates. Higher order elements can be defined similarly.

Automatic mesh generators for a variety of element shapes are available and offer a simple and efficient way to define the necessary input data.[166,167]

The details of the discretization process are discussed in the literature.[157] The following steps and assumptions are general and applicable to the discretization of a region into finite elements.

1. The solution region is subdivided into finite elements. The number and shape of the elements are not restricted in any way. The element density must be chosen for the geometry of the region and expected gradients in the solution. Small, dense elements must be used in regions of high curvature or high gradients.
2. Material interfaces within the solution region and on the boundaries must coincide with element boundaries. An element cannot cover more than one material.
3. The current density and each conductivity and permeability component are assumed to be constant within the element. Calculated quantities are either nodal values, as in the case for the magnetic vector potential, or quantities associated with the element such as flux densities or energy. In this case, the calculated value is associated with the volume (energy) or with the centroid of the element (flux density).
4. At the outer boundaries of the solution region, the magnetic vector potential is either zero (by ensuring that the discretized region extends far enough to have negligible flux density on the boundary) or is otherwise prescribed from known or calculated conditions.
5. The discretizations for two-dimensional and axisymmetric geometries are identical. The difference between the two formulations manifests itself in the integration over the element volume and in the form of the functional itself.

Finite Element Formulation

The discretization of the solution region is a geometrical procedure and by itself is not sufficient to ensure that the chosen elements can be used for finite element solution. A set N_i of special functions, called *interpolating* or *shape functions,* must be chosen for the element:[143]

$$A = \sum_{i=1}^{n} N_i A_i \tag{122}$$

where A_i is the nodal vector potential (weber per meter) and n is the number of nodes in the element.

These must meet two conditions to ensure convergence of the solution as the size of the elements decreases: (1) at element interfaces C_r, continuity must be maintained (compatibility requirement); and (2) within an element C_{r+1}, continuity must be met (completeness requirement).

In these conditions, r is the necessary continuity order of the function at the element boundaries and C_r continuity means that the function and its first r derivatives are continuous. For field problems formulated in terms of the magnetic vector potential, only C_0 continuity is necessary, meaning that only the function is continuous. Its first derivatives define the flux density components, which are not necessarily continuous.

The shape functions for a given element can be obtained by a variety of techniques, each leading to a separate set of functions. One common technique is to define polynomials to fit the sides of the element based on a natural system of coordinates.[143,157] The technique used here is to define the shape functions in a convenient, local system of coordinates and then to map the functions into the cartesian or cylindrical system (isoparametric elements) in which the solution is required.

Quadrilateral Isoparametric Elements

The interpolation functions for quadrilateral elements are those of the serendipity family.[143,157] The elements are created in a local system ξ,η where the functions are found by testing. Considering Fig. 25, the shape functions for the four nodes are written in the following form:

$$N_i = K\left(1 + \xi\xi_i\right)\left(1 + \eta\eta_i\right) \tag{123}$$

where ξ_i, η_i = +1 or –1 and where K is a constant.

The shape functions for quadrilateral elements are shown in Eq. 121 for a particular choice of local coordinates.

These shape functions are mapped into the global coordinate system using the shape functions themselves for the mapping (isoparametric mapping). The elements then become curvilinear as shown schematically in Fig. 25b.

The variation of the magnetic vector potential within the element can be written in terms of the shape functions and the nodal unknowns as:

$$(124)\quad A(\xi,\eta) = \sum_{i=1}^{4} N_i(\xi,\eta) A_i$$

Thus, a complete description of the magnetic vector potential within the finite element has been obtained in terms of (1) the shape functions and (2) the unknown values of the magnetic vector potential at the nodes of the element. The energy functional in Eq. 120 requires the definition of the first derivatives of A with respect to x and y. These can be written as:

$$(125)\quad \frac{\partial A}{\partial x} = \sum_{i=1}^{4} \frac{\partial N_i}{\partial x} A_i$$

and:

$$(126)\quad \frac{\partial A}{\partial y} = \sum_{i=1}^{4} \frac{\partial N_i}{\partial y} A_i$$

The shape functions are derived in local coordinates. Their derivatives are obtained in the local system using the chain rule of differentiation:

$$(127)\quad \frac{\partial N_i}{\partial \xi} = \frac{\partial N_i}{\partial x}\frac{\partial x}{\partial \xi} + \frac{\partial N_i}{\partial y}\frac{\partial y}{\partial \xi}$$

and:

$$(128)\quad \frac{\partial N_i}{\partial \eta} = \frac{\partial N_i}{\partial x}\frac{\partial x}{\partial \eta} + \frac{\partial N_i}{\partial y}\frac{\partial y}{\partial \eta}$$

Rewriting these equations in matrix form and inverting the system to obtain the derivatives of the shape functions in the global system, the following is obtained:[143]

$$(129)\quad \begin{Bmatrix} \dfrac{\partial N_i}{\partial x} \\ \dfrac{\partial N_i}{\partial y} \end{Bmatrix} = [J]^{-1} \begin{Bmatrix} \dfrac{\partial N_i}{\partial \xi} \\ \dfrac{\partial N_i}{\partial \eta} \end{Bmatrix}$$

where $[J]$ is the jacobian matrix calculated for Eqs. 127 and 128. Similarly, the following expression may be used for two-dimensional formulation:

$$(130)\quad dv = dx\,dy = \left|[J]\right| d\xi\, d\eta$$

and Eq. 131 applies to the axisymmetric formulation:

$$(131)\quad dv = 2\pi r\,dr\,dz = \left|[J]\right| 2\pi\xi\, d\eta\, d$$

To evaluate the elemental contribution for the two-dimensional formulation, Eqs. 124 to 126 must be integrated over the element:

$$(132)\quad \int_{-1}^{+1}\int_{-1}^{+1} f'(\xi,\eta)\, d\xi\, d\eta$$

or for the axisymmetric formulation:

$$(133)\quad 2\pi r \int_{-1}^{+1}\int_{-1}^{+1} f'(\xi,\eta)\, d\xi\, d\eta$$

where $f'(\xi,\eta)$ is the transformed $f(x,y)$. The equations above can only be integrated numerically using a technique like the gaussian quadrature.

Functional Minimization

Energy balance in the solution region is achieved by minimizing the energy functional in Eq. 120 at every node of the region. The first partial derivative of $F(A)$ with respect to each nodal value A is set to zero:

$$(134)\quad \frac{\partial F(A)}{\partial A_i} = 0$$

Instead of performing this operation over the entire region, it is convenient to do it element by element and then to sum the contribution of individual elements in order to obtain N simultaneous linear algebraic equations in N unknown values of the magnetic vector potential for the entire solution region. In this way, a repeatable process is performed on each element, a process well adapted for automatic assembly on a computer. The size of the elemental matrix is 4×4 (for a four-node element) and the individual contributions to the elemental matrix are summarized below:

$$(135)\quad s_{i,j} = \frac{1}{\mu}\int_V \left(\frac{\partial N_i}{\partial x}\frac{\partial N_j}{\partial x} + \frac{\partial N_i}{\partial y}\frac{\partial N_j}{\partial y} \right) dv$$

$$(136)\quad r_{i,j} = \omega\sigma \int_V N_i N_j\, dv$$

$$q_i = -J_s \int_v N_i \, dv \quad (137)$$

where the indices *i,j* vary from 1 to 4.

Each coefficient is numerically integrated over the volume of the element using gaussian quadrature[157] and then summed into a complex elemental matrix of the form:

$$\left[[s]_e + j[r]_e\right]\{a\}_e = \{q\}_e \quad (138)$$

where $\{a\}$ is the 4×1 vector of unknowns, $\{q\}$ is the source vector, $[r]$ is the imaginary part and $[s]$ is the real part of the elemental matrix.

The elemental matrices of all the elements in the solution region are summed into a global system of the form:

$$[G]\{A\} = \{Q\} \quad (139)$$

where there are a total of N equations in N unknowns, N being the total number of nodes in the solution region. This matrix is symmetric and banded. The bandwidth depends on the number of elements, the number of nodes per element and especially on the way the nodes are numbered.

The system in Eq. 139 can be solved by any standard solution technique (such as gauss elimination or the conjugate gradient technique) to yield the nodal values of the magnetic vector potential.

Boundary Conditions

The field equations (formulated in terms of finite elements in Eq. 139 for eddy current problems) can only be solved provided a correct set of boundary conditions is specified. Either dirichlet boundary conditions (for which the function A is known on the boundary) or neumann boundary conditions (for which the first derivative of A is known) can be specified. In the finite element analysis of magnetic field problems, it is more convenient to specify dirichlet boundary conditions because the global matrix in Eq. 139 can accommodate the function value A but not its derivative. Moreover, the neumann boundary conditions are implicit in the formulation in Eq. 139 and need not be specified.

Calculations with Magnetic Vector Potential

Although the developed formulation yields a correct solution to the problem at hand, this solution is in terms of the magnetic vector potential. Being an auxiliary function, it is not measurable by itself and therefore is of little value for comparison with measurements and other calculations. From the definition of the magnetic vector potential as $\boldsymbol{B} = \nabla \times \boldsymbol{A}$, the flux density is immediately defined. Other quantities are calculable, including coil impedances, stored and dissipated energy and eddy current density. The derivation of quantities calculated for the magnetic vector potential can be found in a number of sources.[15,159,161]

The solution (either by finite differences or finite elements) is correct in terms of the magnetic vector potential. The magnetic vector potential is an auxiliary function used to simplify the solution and is not by itself a measurable quantity. It is necessary to calculate other quantities such as flux densities, eddy current densities and coil impedances. Because coil impedance is of greater importance in eddy current testing, its derivation is outlined below. The literature may be consulted for the calculation of other quantities.[158-161]

In axisymmetric geometries, the impedance of a coil is calculated from the value of the magnetic vector potential in the coil's cross section, starting with the general formula for the impedance of a loop of wire carrying an alternating current I_s:

$$Z = \frac{j\omega}{I_s} \oint A \, d\ell \quad (140)$$

Because the magnetic vector potential has only one component in the direction of the current for one turn of a coil having a radius r, the impedance is:

$$Z_i = \frac{j\omega 2\pi r_i A_i}{I_s} \quad (141)$$

where A_i is the magnetic vector potential (volt) at r_i and I_s is the root mean square value of the current (ampere).

By integrating this over an elemental area with N_s turns per unit area (square meter), the expression becomes:

$$Z_i = \frac{j\omega \, 2\pi r_{ci} \, A_{ci} \, N_s \Delta_i}{I_s} \quad (142)$$

where A_{ci} is the magnetic vector potential at the same point, r_{ci} is the average distance (meter) of the area chosen from the axis of symmetry and Δ_i is the elemental area (square meter).

In a finite element calculation, r_{ci}, A_{ci} and Δ_i are the distance from the axis to the centroid of the element, the magnetic vector potential at the centroid of the element and the area of the element, respectively. In the case of finite differences, the same quantities can be used by imagining the area between four nodes as being an elemental area. The centroidal value of the magnetic vector potential can be calculated as the average of the four nodal values of the element.

When this value is summed over the elements in the coil's cross section and it is noted that $J_s = N_s I_s$, the coil impedance becomes:

$$(143)\quad Z_{coil} = \frac{j\omega\, 2\pi J_s}{I_s^2} \sum_{i=1}^{N} r_{ci}\, \Delta_i\, A_{ci}$$

For differential probes, the impedance is calculated separately for each coil and the two impedances are added:

$$(144)\quad Z_{probe} = Z_a + Z_b = \frac{j\omega 2\pi J_s}{I_s^2} \left[\sum_{i=1}^{N_b} r_{ci}\, \Delta_i\, A_{ci} - \sum_{j=1}^{N_a} r_{cj} \Delta_j A_{cj} \right]$$

In three-dimensional geometries, this technique cannot be used because it assumes that the magnetic vector potential is constant along the circumference of the coils. The impedance can be determined by calculating the stored and dissipated energies, then finding the resistance of the coils from the dissipated energy and the inductance from the stored energy.[168,169]

Modeling of Physics of Eddy Current Testing

The interaction of electromagnetic fields with material discontinuities, the basis of all electromagnetic testing methods, is a complicated phenomenon. Attempts at solving Maxwell's equations are in effect attempts to describe these interactions in some detail.

If the basic transformer equivalent model associated with an eddy current test were to be used, the various equivalent circuit parameters can be determined and such questions as the coil impedance for a certain, simplified geometry may be answered. It is quite a different task to analyze with any degree of generality the details of magnetic field interaction with the material. To do this, a continuous impedance plane trajectory is needed and the details of field distribution in materials and discontinuities must be calculable.

The numerical approach to this problem provides these data as an integral part of the calculation. Field distributions, eddy current densities and impedance values are calculated for any or all probe positions. The user may choose to test all or part of these data, either personally or in a computerized testing procedure.

To demonstrate the value of a numerical model for understanding field interactions with materials, the geometry in Fig. 26 can be analyzed. It comprises a conducting nonmagnetic tube inside a carbon steel plate with a gap 0.4 mm (0.015 in.) between the two. The tube is 1.3 mm (0.05 in.) thick, 22 mm (0.88 in.) in diameter and the plate is 19 mm (0.75 in.) thick. The response of the differential eddy current probe is obtained by moving the coils from a point where, because of the distance from the steel plate, no eddy currents are induced. The probe is moved a very short distance toward the plate and the impedance as well as the field distribution are recalculated. Repeating this process, a large number of probe positions are calculated, resulting in a smooth, continuous impedance plane trajectory.

To perform this calculation, the geometry in Fig. 26 is discretized into some 3000 quadrilateral elements (6000 triangular elements could be used to produce an identical mesh with identical results in half the geometry because of symmetry). This produces a mesh with 3146 nodes and a system of equations with 3146 unknowns and a semibandwidth of 27. The mesh allows movement of the probe in 140 probe positions to produce a curve composed of 140 impedance points. In general, 30 to 50 probe positions are sufficient but more positions may be needed to model complex or composite discontinuities.

Figure 27 shows a series of five probe positions, the respective impedance plane trajectory in the upper left corner and the flux distribution around the coils. In this sequence, a small inside diameter

FIGURE 26. Steam generator geometry.

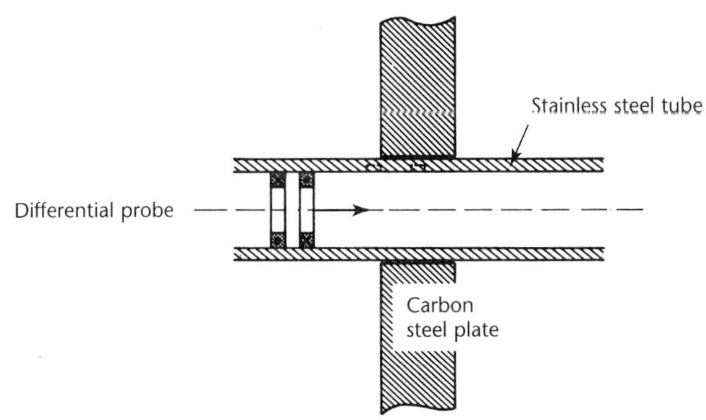

axisymmetric slot is simulated in the tube at the center of the support plate.

In Fig. 27a, the coils are both well away from the steel plate, so both have essentially the same field distribution. The single point to the left represents the impedance of the probe at this particular probe position (zero impedance point).

In Fig. 27b, the leading coil is close to the edge of the plate and a dramatic change in the flux distribution has taken place compared with Fig. 27a. The trailing coil, however, has a distribution that has changed very little. The change in the impedance is clearly noticeable. It should be remembered, however, that there are some 35 probe positions between those shown in Figs. 27a and 27b.

Figure 27c represents the situation where the coils are centered with the middle of the support plate. Because the two coils are differentially connected, this situation is identical to the one in Fig. 27a in terms of impedance and the impedance trajectory now describes a close curve. The effect of the plate (large lobe) and the discontinuity are shown.

As the leading coil leaves the support plate region, the process is repeated and Figs. 27d and 27e are in effect, a reflection of Figs. 27b and 27a. Thus, a full symmetric impedance plane trajectory has been described. The signal in Fig. 27 was obtained as a combined signal of the steel plate and the slot. The two signals were quite distinct because the discontinuity is relatively far from the edge of the steel plate.

In a sequence of this type, not only is it possible to obtain a full description of the impedance plane trajectory but also to test the details of subtle changes in flux and eddy current distribution, the effects of conductivity and permeability on flux lines and the progression of impedance changes.

As a second example of a similar situation, consider the sequence in Fig. 28. It repeats the sequence of Fig. 27 but this time the discontinuity is directly under the edge of the steel plate.

Starting with Fig. 28a, far from the plate and discontinuity, the situation is identical to that in Fig. 27a. As the probe approaches the plate, the difference between the two geometries becomes apparent. The flux lines are affected by the plate and the discontinuity, producing a pronounced composite signal. This first lobe of the composite signal is completed with Fig. 28c. The rest of the curve repeats Figs. 27d and 27e and the second part of the curve is identical to that in Fig. 27 because the discontinuity has no influence when the probe leaves the plate region.

By calculating the impedance at a large number of probe positions in the sequence and by photographing these as an animation sequence, a new dimension in field modeling is achieved. Not only does the sequence reveal all the necessary details on the measurement but also produces a startling and unique view of the changes in the magnetic field as they occur in real time. The figures presented for Fig. 28 are in fact still frames from an animated movie that describes this and other simple testing geometries.[170]

Figure 27. Finite element modeling of circumferential inside diameter slot under center of steel support plate: (a) probe far from plate; (b) probe approaches support plate and leading coil experiences large change in field and produces impedance plot; (c) coils are centered with support plate and closed contour has been described; (d) probe leaves support plate and trailing coil experiences large change; (e) probe is far from support plate and complete impedance plane trajectory has been described.

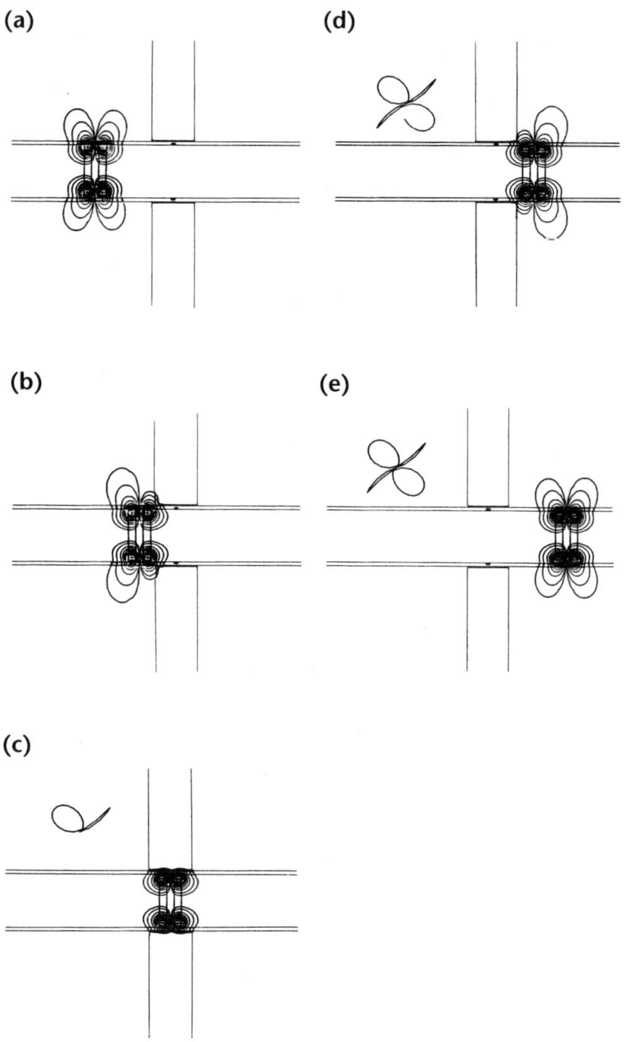

Modeling for Probe Design

Another very important aspect of eddy current modeling is the probe itself. It is safe to assume that the quality of probes used in eddy current testing has more effect on the results than all other factors, yet their design has been based in the past on empirical considerations and on experience. This implies a trial and error procedure by which a probe is designed and built, then tested. The process is repeated until a satisfactory result is obtained. A numerical approach allows a more detailed design and a potentially better product can be obtained at much lower expense and in shorter design times. Both general purpose probes[171] and specialized probes[172] can be designed. Detailed studies can be performed about the various parameters of the probe — its frequency response and its response to various discontinuities and material properties. These data may be used either to evaluate existing probes or to design new ones. In addition, a numerical model evaluates parameters impossible to evaluate in any other way. A simple example is the calculation of flux densities in any part of the tested material. Knowledge of flux densities can reveal saturation effects and possible nonlinear behavior of the tested material.

FIGURE 28. Finite element modeling of circumferential inside diameter slot under edge of steel support plate: (a) probe is far from plate; (b) probe approaches support plate and leading coil experiences large change in field and produces impedance plot; (c) coils are centered with support plate and closed contour has been described; (d) probe leaves support plate and trailing coil experiences large change; (e) probe is far from support plate and complete impedance plane trajectory has been described and, because of asymmetry in geometry, signal is asymmetric.

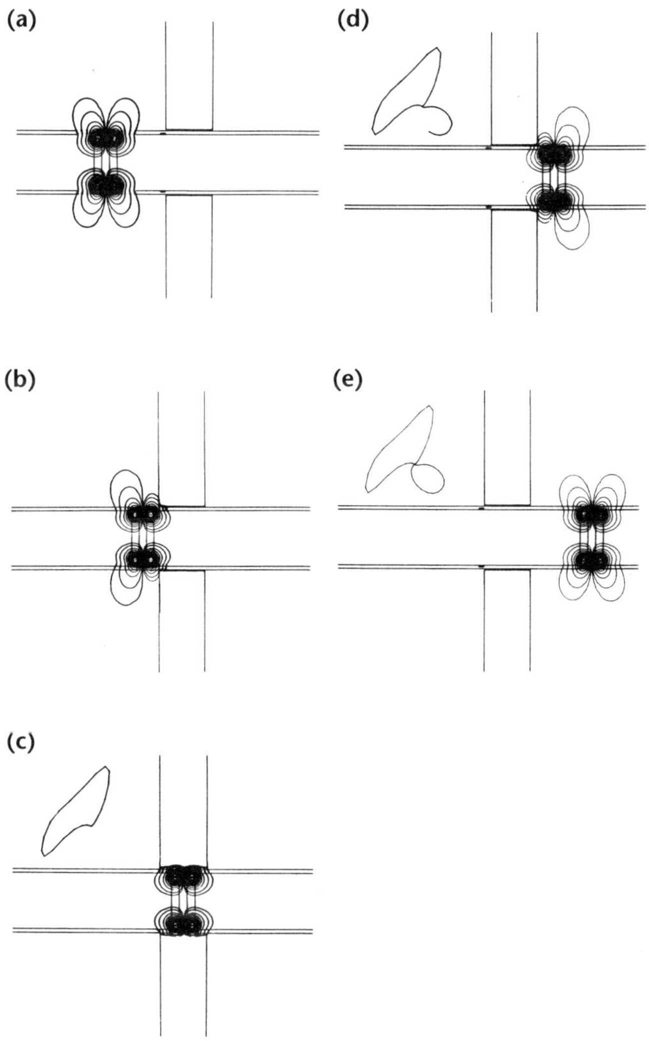

Finite Element Design of Simple Differential and Absolute Eddy Current Probes

The geometry being modeled is shown in Fig. 29a, where a differential eddy current probe is placed inside a tube of heat resistant nickel chromium alloy (Unified Numbering System N06600) with a simulated discontinuity. The geometry is

FIGURE 29. Differential eddy current probe inside tube with outside diameter axisymmetric slot: (a) geometry; (b) finite element mesh (half region).

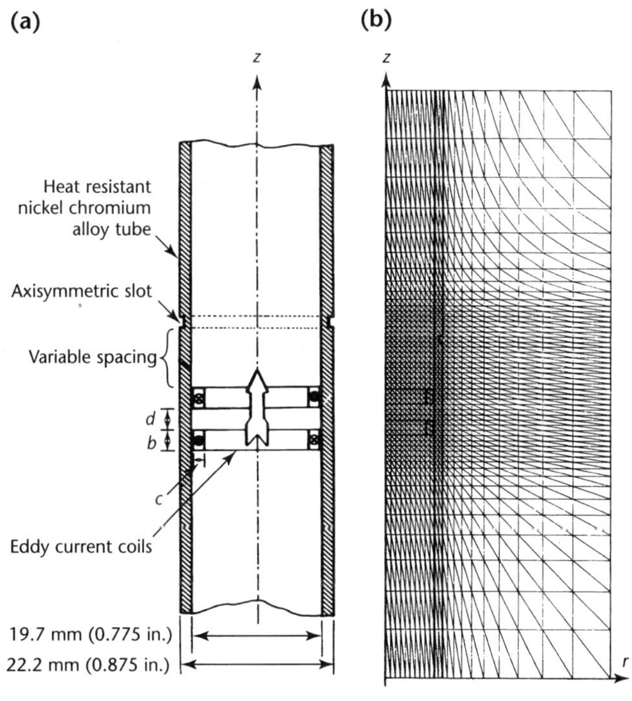

Legend

r = radial coordinates
z = axial coordinates

discretized by using triangular finite elements.[171] In this situation, both the spacing and the width of the coils can be varied.

The geometry in Fig. 29a is discretized into a large number of triangular elements as shown in Fig. 29b. The finite element technique is applied to solve for the magnetic vector potential at each node of the mesh in Fig. 29b. From these values, the impedance of the coil is calculated at discrete probe positions to form the impedance plane trajectory caused by the discontinuity.

Symmetry exists about the Z axis and only half of the geometry is analyzed using the axisymmetric formulation. Also, because symmetry exists about the center of the discontinuity, the probe is allowed to move up to the point where it is centered with the discontinuity and the calculated impedance values are reflected to form a full impedance plane trajectory.

Figure 30 compares the experimental results (Fig. 30a) and finite element results (Fig. 30b) from a differential probe with coils 2 mm (0.08 in.) wide at spacings from 1 to 8.5 mm (0.04 to 0.34 in.). The indication is from a slot (shown in Fig. 29a) measuring 1 mm (0.04 in.) wide and 0.4 mm (0.015 in.) deep on the outer surface of a 22 mm (0.87 in.) tube made of heat resistant nickel chromium alloy (Unified Numbering System N06600).

The experimental results were obtained using a specially designed eddy current probe with interchangeable coils and variable spacing between the coils. These results show clearly that as the spacing of the coils increases the resulting impedance plane trajectory loses its differential nature and the probe behaves increasingly as two distinct absolute probes. On the other hand, decreasing the spacing widens the loops but also reduces the amplitude of the trajectories.

Figure 31 compares different sized coils at a constant spacing for the same discontinuity as in Fig. 30. The spacing is 2.5 mm (0.1 in.) and the coil width varies from 0.5 to 7.5 mm (0.02 to 0.3 in.). In

FIGURE 30. Impedance plane trajectories for outside diameter axisymmetric slot and distance *d* between two coils of probe: (a) experimental element; (b) finite element.

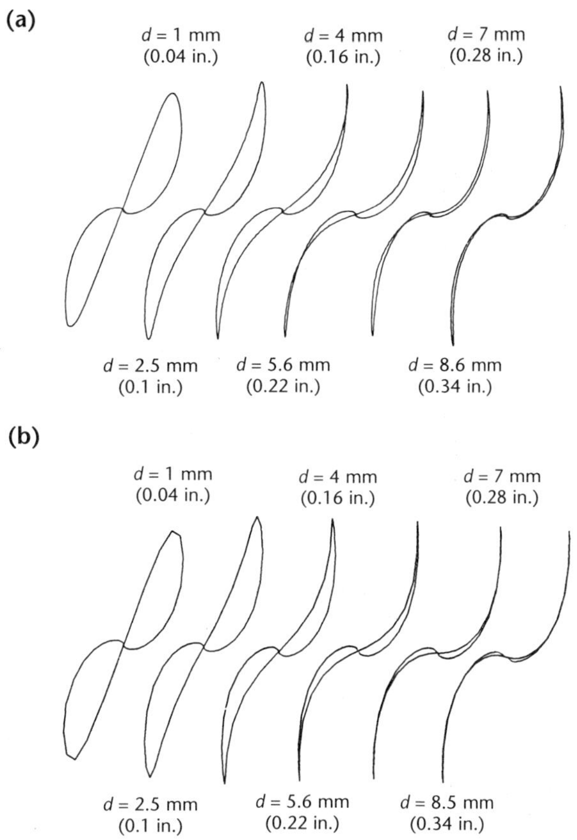

FIGURE 31. Impedance plane trajectories for different coil sizes at 100 kHz and constant spacing between coils for slot in Fig. 30: (a) experimental; (b) finite element.

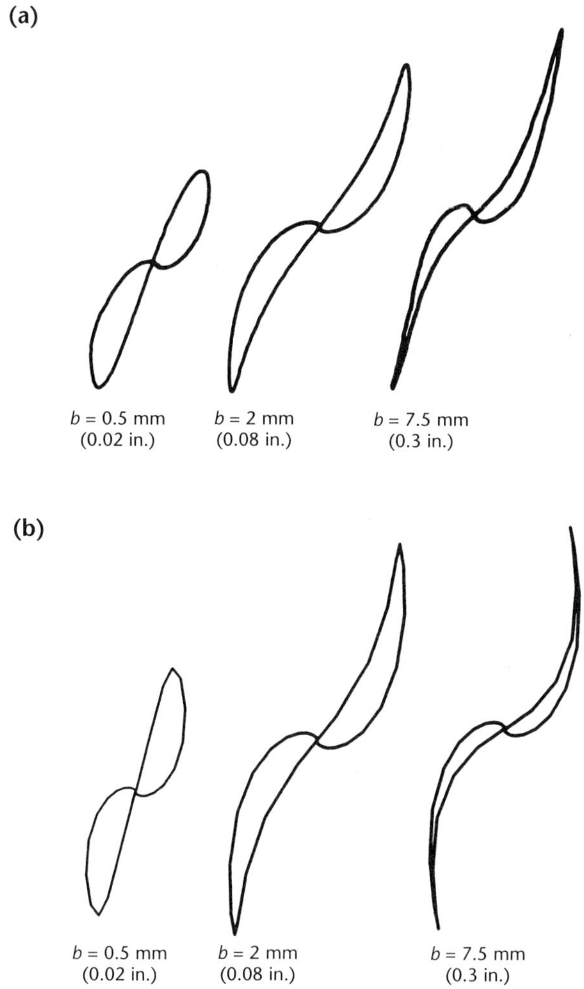

this case, as the coil becomes wider, the amplitude increases and the shape becomes narrower. From these calculations and experiments, it is clear that a good compromise is achieved by choosing a probe whose coil width and spacing is comparable to the width of the discontinuity. Further finite element predictions were made to investigate this model as a design tool.

Impedance plane trajectories were calculated and plotted by varying the following parameters in a given probe.

1. For frequency, impedance plane trajectories were calculated at 50, 100 and 150 kHz (Fig. 32).
2. For discontinuity geometry, two different discontinuities were simulated: a 1 mm (0.04 in.) wide, 0.4 mm (0.016 in.) deep outside diameter discontinuity and a 1 mm (0.04 in.) wide, 0.76 mm (0.030 in.) deep outside diameter discontinuity.
3. For coil spacing, calculations were performed at coil spacings of 1.0, 2.5, 4.0, 5.6, 7.0 and 8.6 mm (0.04, 0.10, 0.16, 0.22, 0.28 and 0.34 in.). The plots in Fig. 30 show the relation in the amplitude for smaller and larger discontinuities and the importance of choosing the correct spacing for the probe if meaningful signals are to be obtained.
4. For signal rotation with frequency due to discontinuities, rotation of the signal from two probes at 1 mm (0.4 in.) and 6.3 mm (0.25 in.) spacings were modeled at various frequencies for a 10 mm (0.4 in.) wide, 0.76 mm (0.03 in.) deep slot and for a 19 mm (0.75 in.) carbon steel support plate.

FIGURE 32. Finite element predicted impedance plane trajectories for different discontinuities, frequencies and coil spacings. Small signals are for outside diameter axisymmetric slot; larger signals are for a deeper slot (*d* = coil spacing).

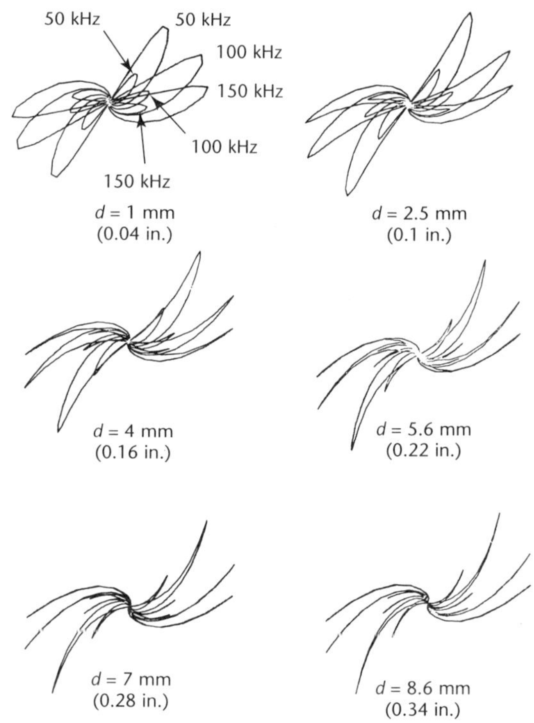

Figure 33 indicates that the rotation is more or less linear at higher frequencies (the optimal frequency for this particular probe is about 125 kHz) but is nonlinear at lower frequencies. The same phenomena are observed experimentally and must be taken into account when relating measured parameters to signal rotation.

As a second example of the application of the finite element model to probe design, the geometry in Fig. 34 was studied. The component is a section of a steam generator tube of heat resistant nickel chromium alloy (Unified Numbering System N08800) inside the tube sheet region. The steam generator contains rolled tubes where the rolling

FIGURE 33. Signal rotation versus frequency for one probe with different coil spacings.

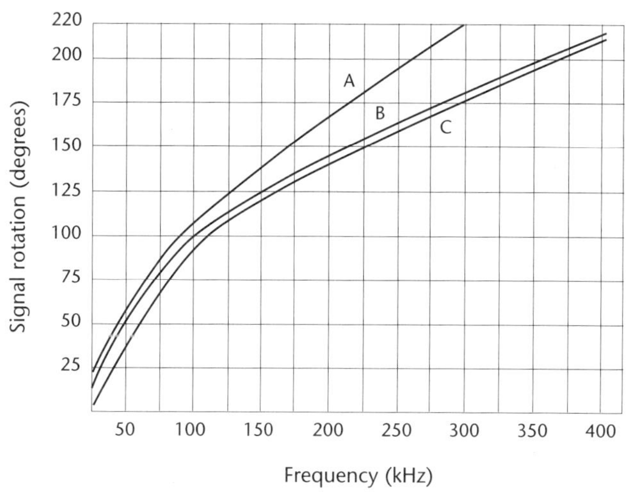

Legend

A. 1 mm (0.04 in.) spacing for probe 1 and axisymmetric slot.
B. 6.4 mm (0.25 in.) spacing for probe 1 and axisymmetric slot.
C. 4.5 mm (0.18 in.) spacing for probe 1 and support plate.

region can be at varying distances from the tube sheet inner surface. The absolute coil is 1 mm (0.04 in.) thick and has a length of 1 mm (0.04 in.), which needs to be optimized for the particular application.

In addition, the signal from the rolling region is to be modeled for identification of the tube condition.[173]

To determine the probe length needed to obtain the best signal for different locations of the rolling region relative to the tube sheet surface, three coil lengths — 1, 3 and 9 mm (0.04, 0.12 and 0.36 in.) — were modeled for the same three distances. The finite element results for these nine situations are plotted in Fig. 35.

The longer the coil, in comparison with the distance between the two factors that cause the signal change (tube sheet and rolling region), the less distinct the phenomena are in the signal. Thus a coil 9 mm (0.36 in.) long, testing for the rolling region that is only 1 mm (0.04 in.) away from the tube sheet surface, produces a flat composite signal in which the rolling and the tube sheet cannot be distinguished as in Fig. 35g. The other extreme is when the coil is much smaller than the distance as in Fig. 35c. Here the two signals are simply superimposed and one signal does not affect the other.

The curves in Fig. 35 are generated at 100 kHz and are, in general, a composite signal. The lower, comma shaped parts of the curves are due to the effect of the tube sheet. The upper part is due to the rolling region. These curves compare very well with experimental results, such as the curve in Fig. 36 taken at 100 kHz. The choice of coil size and shape might be complicated by additional factors, such as the minimum number of required turns, but as can be seen from these results the coil should be of the same general length as the effect it is measuring.

These results show that the numerical model is a powerful tool in probe design. It is less expensive and more accurate than empirical techniques and can be used beyond the restricting approximations of an analytical model. This versatility is more important for complex, multicoil probes where the interaction between the coils, and possibly ferrite cores and shields, complicates the design. The responses due to three-dimensional and subsurface discontinuities as well as changes in material properties can also be modeled and used as valuable input in the design process.

Figure 34. Geometry of steam generator section showing absolute coil, tube sheet and corrosion resistant nickel alloy tube.

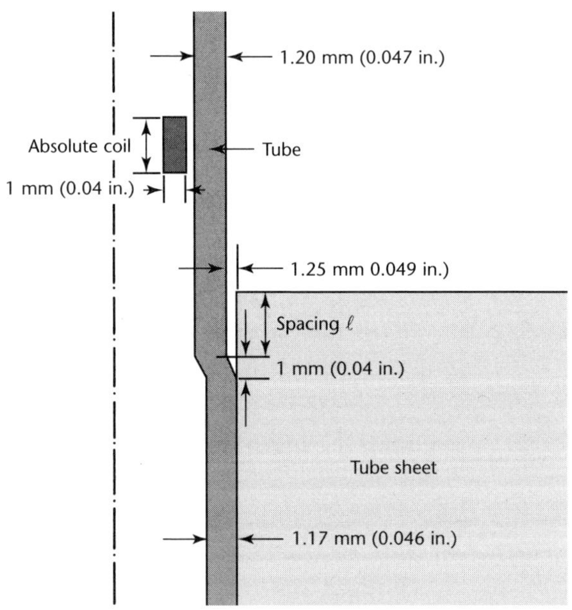

Modeling for Simulation

A third application of the numerical models is in the area of simulation of test geometries that are difficult, expensive or impossible to simulate experimentally. In this case, a numerical model is not an option but rather a necessity. Examples of these difficult geometries include subsurface discontinuities, arbitrarily shaped geometries and discontinuities in nuclear power plant structures, where confirmation of the discontinuity shape cannot be obtained by visual testing. Information regarding these discontinuities is of great importance but cannot be obtained reliably by any other means. Even if the cost of producing reference standards to calibrate testing equipment could be justified, these would be approximations of real discontinuities and would be limited to the particular discontinuity prepared.

The numerical model, once its versatility and accuracy have been demonstrated, can handle the task of producing training data in an economical and convenient way. It can produce the data necessary for many testing configurations (two-dimensional, axisymmetric and three-dimensional) and produce the output in a form that is directly compatible with computer analysis of raw data.

One situation where measurements are not generally possible is the testing for buildup of corrosion products such as magnetite. Producing sample test objects by packing such small gaps with magnetite is very difficult and the numerical model presents the only reasonable alternative.

The model described above was applied to a detailed numerical study of magnetite buildup in the crevice gap of pressurized water reactor steam generators.[169] There is uncertainty about how the magnetite accumulates in the crevices. Similarly, in the chemical flushing process, the magnetite is removed but the signal obtained while monitoring the process depends on the way the magnetite is flushed. Several possibilities have been modeled numerically: (1) radial or axial buildup (or flushing), (2) axial buildup from one side of the support plate and (3) flushing in the presence of tube denting. Because of the large number of impedance plane trajectories obtained, only representative data are presented here.

The geometry is presented in Fig. 37. It consists of a tube of high temperature nickel chromium alloy (Unified Numbering System N06600), 22 mm (0.87 in.) in diameter and 1.3 mm (0.05 in.) in wall thickness inside a 19 mm (0.75 in.) carbon steel support plate. The crevice gap between the tube and support plate is nominally 0.4 mm

FIGURE 35. Impedance plane trajectories for coils of length a and for spacing ℓ, where ℓ is distance between tube sheet and rolling distance to inner surface: (a) a = 1 mm (0.04 in.), ℓ = 1 mm (0.04 in.); (b) a = 1 mm (0.04 in.), ℓ = 3 mm (0.12 in.); (c) a = 1 mm (0.04 in.), ℓ = 9 mm (0.36 in.); (d) a = 3 mm (0.12 in.), ℓ = 1 mm (0.04 in.); (e) a = 3 mm (0.12 in.), ℓ = 3 mm (0.12 in.); (f) a = 3 mm (0.12 in.), ℓ = 9 mm (0.36 in.); (g) a = 9 mm (0.36 in.), ℓ = 1 mm (0.04 in.); (h) a = 9 mm (0.36 in.), ℓ = 3 mm (0.12 in.); (i) a = 9 mm (0.36 in.), ℓ = 9 mm (0.36 in.).

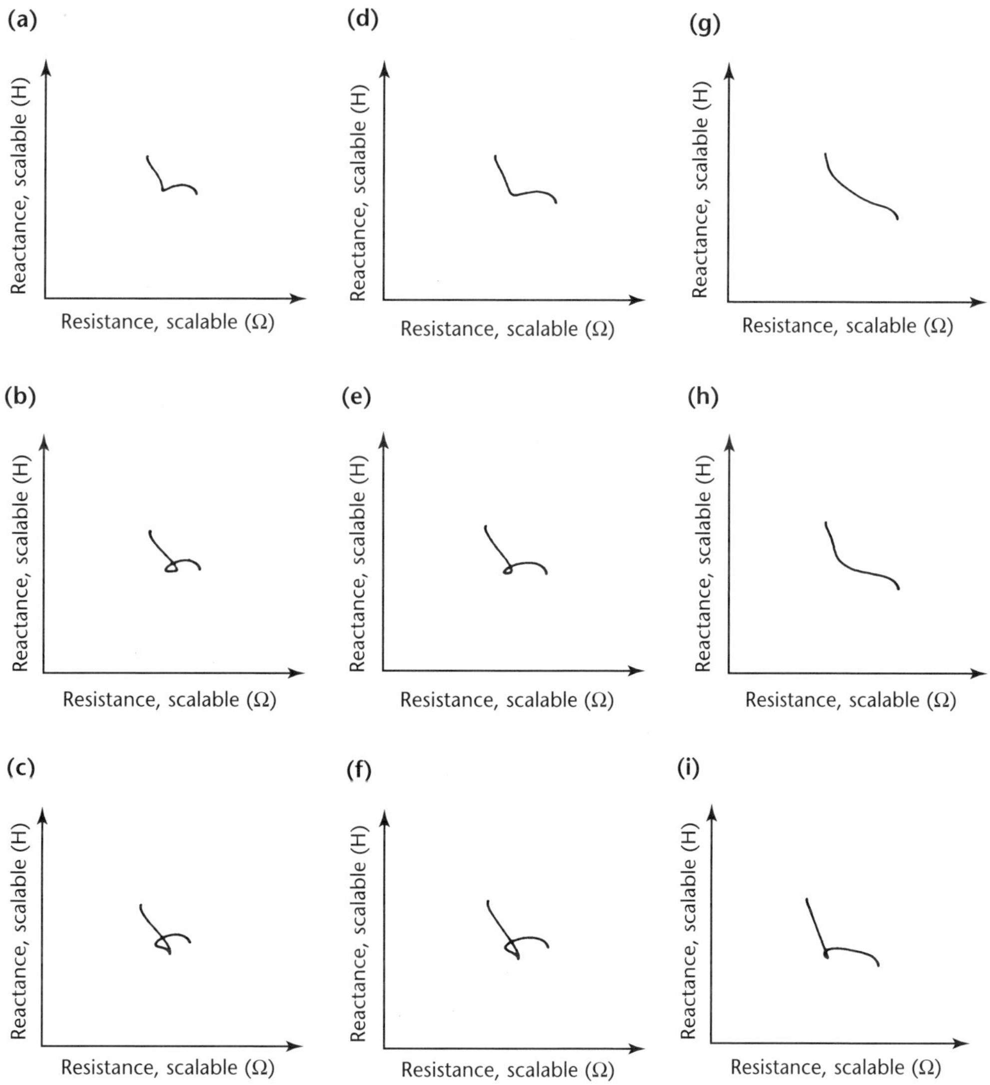

(0.015 in.) and the signals were calculated for a differential eddy current probe at 100 kHz. The dashed lines in Fig. 37 schematically represent the area in which denting of the tubes was modeled.

The first part of the study dealt with radial buildup of magnetite from the support plate toward the tube, or flushing of magnetite from the tube toward the support plate. Figure 38 shows the geometry involved and the impedance plane trajectories for magnetite buildup in layers of 0.08 mm (0.003 in.). The change in signal from a clean gap to one full of magnetite is quite dramatic, both in shape and amplitude.

The second part of the study assumed an axial buildup of magnetite from the center of the support plate outward. Although this direction is not very likely during buildup, it is representative of chemical flushing, where the chemical agents attack the magnetite from both sides. Figure 39 shows this situation: the changes between a full gap and partially filled gap are dramatic.

Another aspect of testing steam generator tubing is that of denting due to magnetite buildup. Similarly, monitoring the flushing of magnetite in this situation is more important to ensure complete cleaning of the magnetite in the gap.

Figure 40 shows the impedance plane trajectories of axial magnetite buildup (or flushing) in various amounts from a clean

FIGURE 36. Experimental impedance plane trajectory at 100 kHz from 3 mm (0.12 in.) long coil at nominal spacing of tube sheet and rolling.

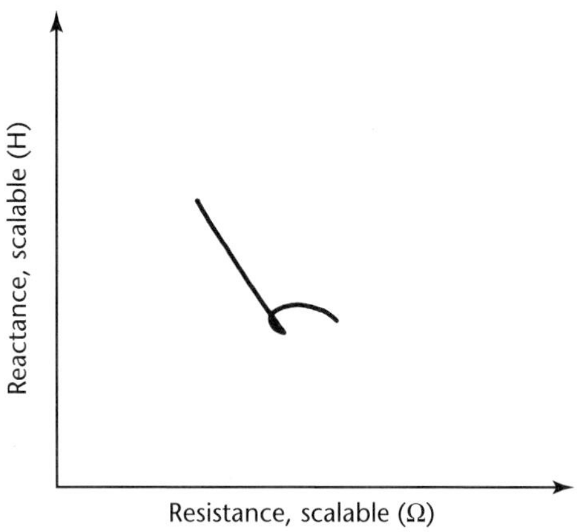

FIGURE 37. Geometry used to study effect of magnetite accumulation in crevice gap between tube and support plate. Dashed line represents area in which denting is modeled.

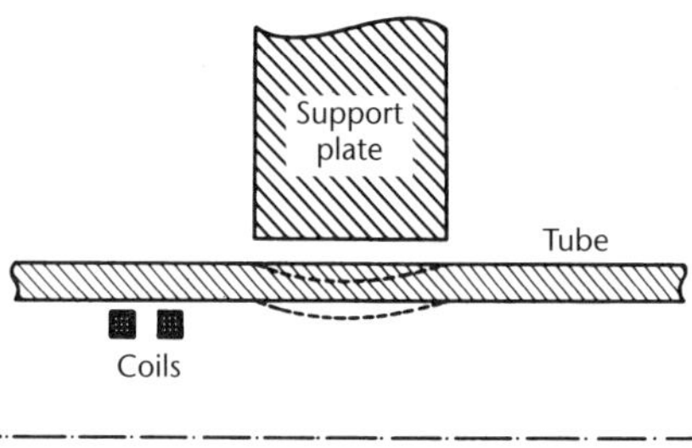

FIGURE 38. Modeling of radial buildup of magnetite of thickness ω.

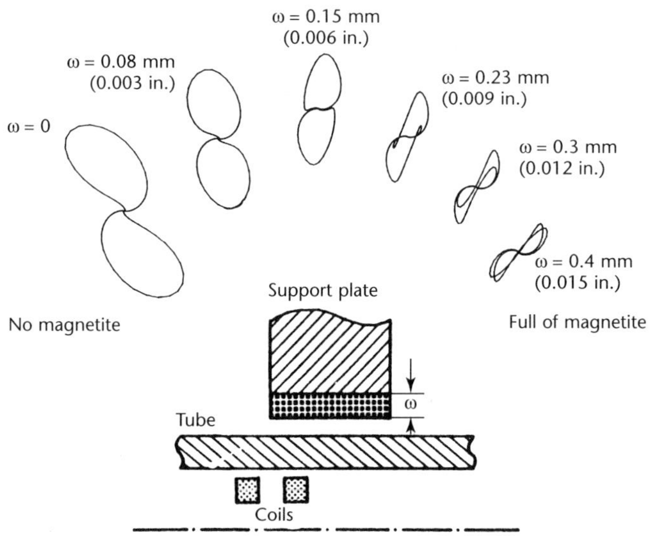

FIGURE 39. Modeling of axial buildup of magnetite (of extent ω) outward from center.

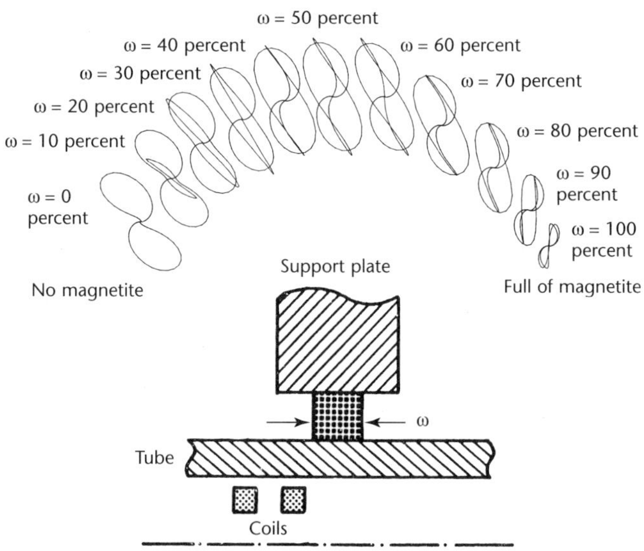

gap to a gap full of magnetite. From these plots, it is evident that even small amounts of magnetite in the gap affect the impedance plane trajectory.

In addition to the data presented above, studies were carried out (1) for axial buildup from one side of the support plate (Fig. 41), (2) for axial buildup from one side of the support plate with a portion of the gap clean on both sides of the magnetite band (Fig. 42) and (3) for denting of the tube without the presence of magnetite (Fig. 43). For comparison, the clean crevice gap trajectory is given again in Fig. 43.

FIGURE 40. Modeling of magnetite buildup in presence of denting.

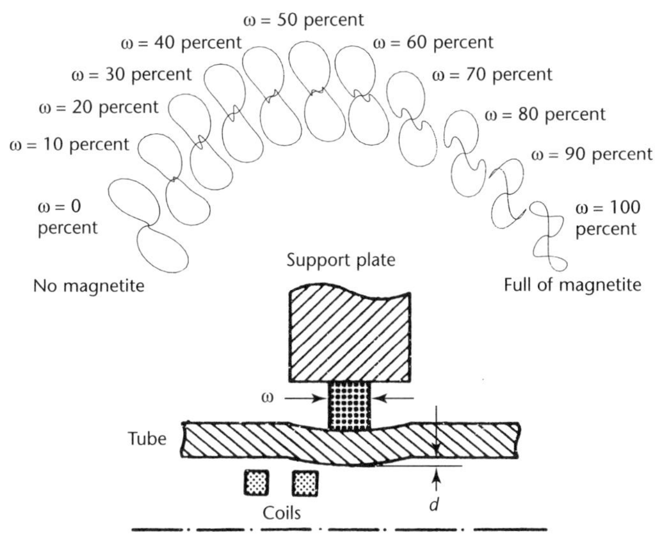

Legend
d = 0.1 mm (0.004 in.)
ω = magnetite

FIGURE 41. Modeling of buildup of magnetite of extent ω from one side of support plate.

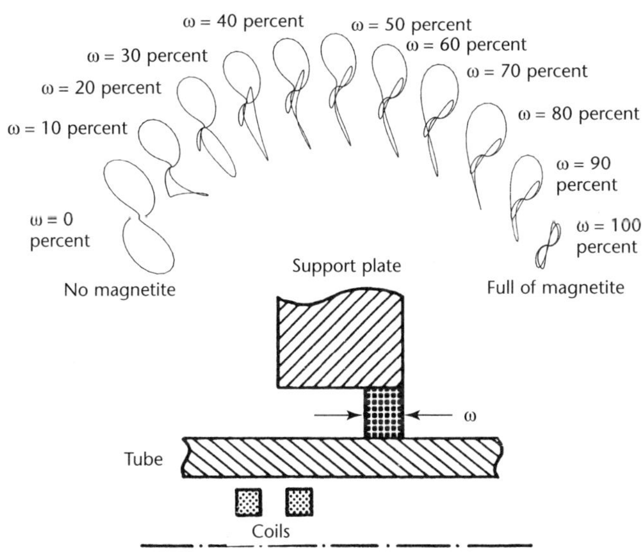

FIGURE 42. Modeling of buildup from one side of support plate with portion of gap clean.

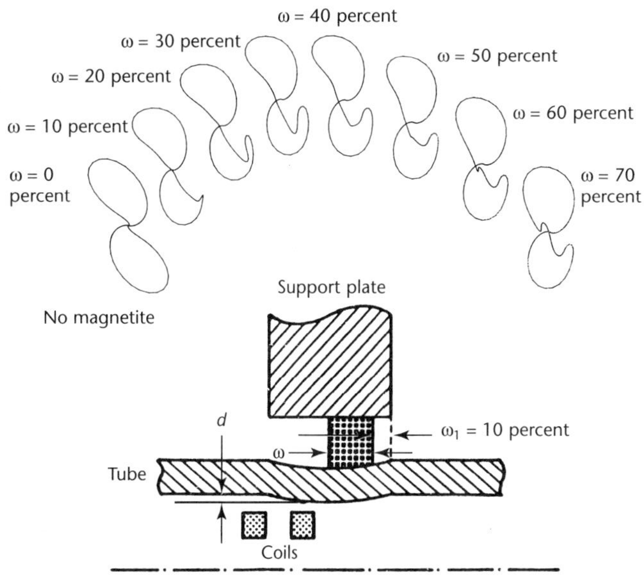

Legend
d = 0.1 mm (0.004 in.)
ω = magnetite

FIGURE 43. Modeling of denting in tube without presence of magnetite, representative of flushed gap d.

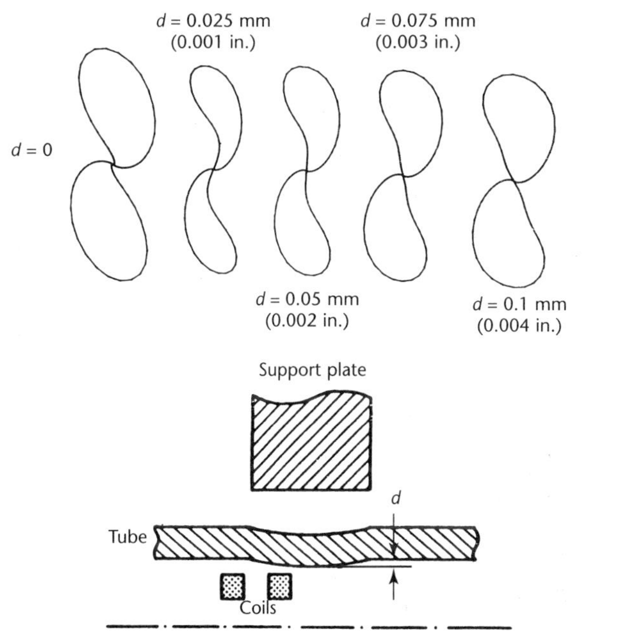

The geometries modeled above indicate the extent and versatility of the numerical model. There still remains the question of comparison with real, known data. The experimental measurement of a support plate signal in the presence of a clean support plate is only part of the answer. Most of the trajectories in Figs. 38 to 43 cannot be reproduced experimentally because of the difficulty of building experimental setups that reflect test conditions.

To partly answer the question of comparisons with known data, the experiment in Fig. 44 was carried out. Here, a 25 mm (1 in.) thick support plate was drilled to provide a 1.5 mm (0.06 in.) gap, which was then packed axially with magnetite powder at about 30 percent (Fig. 44a), 60 percent (Fig. 44b) and 90 percent (Fig. 44c) of the gap length. The resemblance to corresponding numerical trajectories is immediately evident.

More convincing evidence of the ability, accuracy and usefulness of numerical modeling is provided in Fig. 45. These impedance plane trajectories were obtained during the chemical flushing of a model boiler. The data are from a tube with a 0.1 mm (0.004 in.) radial dent at various stages of flushing. Figure 45a was taken before flushing began and shows an identifiable dent filled with magnetite. As the flushing progresses, the gap shows various stages of cleaning. Thus, for example, in Fig. 45b, one side of the gap is almost completely clean, as indicated by the large lower lobe. The rest is still packed with magnetite, indicating an uneven flushing from both sides of the support plate. Figure 45c shows a clean gap while the dent in the tube is visible.

This particular experiment provides a convincing experimental confirmation of the numerical model and demonstrates its value in interpreting data and in monitoring the flushing process.

FIGURE 44. Experimental impedance plane trajectories from crevice gaps of 1.5 mm (0.06 in.) width in 25 mm (1 in.) support plate hole packed with magnetite in various amounts: (a) 30 percent full; (b) 60 percent full; (c) 90 percent full.

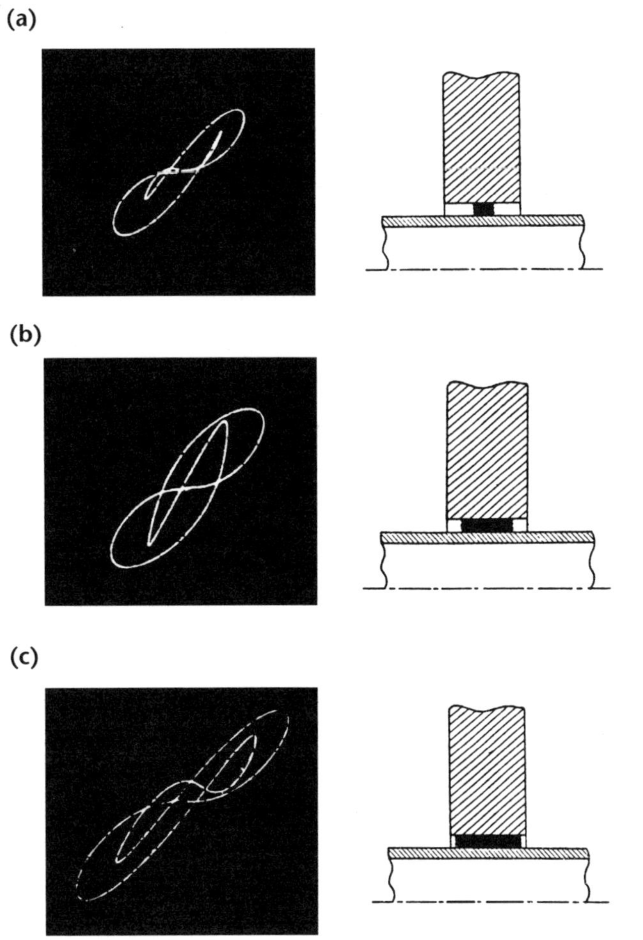

FIGURE 45. Experimental data taken during flushing process in model boiler: (a) impedance plane trajectory of support plate before flushing; (b) part of gap cleaned, indicating unequal flushing from both sides; (c) clean gap after flushing.

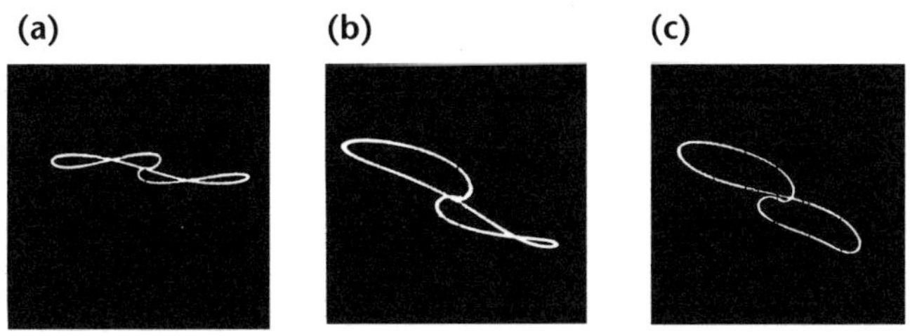

Conclusions

Computer modeling of eddy current testing phenomena has evolved from an experimental state through more sophisticated analytical techniques into general numerical models. The three techniques of modeling each have advantages and shortcomings but the numerical technique possesses the generality needed to model the intricacies of interactions between field and discontinuity.

As with almost any computer application, the numerical solution of eddy current problems is based on known physical relations. Analysis is part of the process of acquiring knowledge.

References

1. Maxwell, J.C. *A Treatise on Electricity and Magnetism,* third edition. New York, NY: Dover Publications (1891).
2. Ida, N. Section 19, "Computer Modeling of Eddy Current Fields." *Nondestructive Testing Handbook,* second edition: Vol. 4, *Electromagnetic Testing.* Columbus, OH: American Society for Nondestructive Testing (1986): p 562-590.
3. McNab, A. "A Review of Eddy Current System Technology." *British Journal of Non-Destructive Testing.* Vol. 30, No. 7. Northampton, United Kingdom: British Institute of Non-Destructive Testing (July 1988): p 249-255.
4. Auld, B.A. "John Moulder and the Evolution of Model-Based Quantitative Eddy Current NDE." *Review of Progress in Quantitative Nondestructive Evaluation.* Vol. 18A. New York, NY: Kluwer/Plenum (1999): p 441-448.
5. Auld, B.A. and J.C. Moulder. "Review of Advances in Quantitative Eddy Current Nondestructive Evaluation." *Journal of Nondestructive Evaluation.* Vol. 18, No. 1. New York, NY: Plenum (1999): p 3-36.
6. Becker, R., K. Betzold, K.D. Boness, R. Collins, C.C. Holt and J. Simkin. "The Modeling of Electrical Current NDT Methods and Its Applications to Weld Testing." Vol. 28. *British Journal of Non-Destructive Testing.* Northampton, United Kingdom: British Institute of Non-Destructive Testing. "Part 1" (September 1986): p 286-294. "Part 2" (November 1986): p 361-370.
7. Ida, N. *Numerical Modeling for Electromagnetic Non-Destructive Evaluation.* London, United Kingdom: Chapman and Hall (1995).
8. Smythe, W.R. *Static and Dynamic Electricity.* New York, NY: McGraw-Hill (1968).
9. Tegopoulos, J.A. and E.E. Kriezis. *Eddy Currents in Linear Conducting Media.* Amsterdam, Netherlands: Elsevier Science Publishers (1985).
10. Dodd, C.V. and W.E. Deeds. "Analytical Solutions to Eddy-Current Probe-Coil Problems." *Journal of Applied Physics.* Vol. 39, No. 6. Melville, NY: American Institute of Physics (1968): p 2829-2838.
11. Dodd, C.V., W.E. Deeds and J.W. Luquire. "Integral Solutions to Some Eddy Current Problems." *International Journal of Nondestructive Testing.* Vol. 1. Kidlington, United Kingdom: Elsevier Science Limited (1969): p 29-90.
12. Luquire, J.W., W.E. Deeds and C.V. Dodd. "Axially Symmetric Eddy Currents in a Spherical Conductor." *Journal of Applied Physics.* Vol. 41, No. 10. Melville, NY: American Institute of Physics (1970): p 3976-3982.
13. Nikitin, A.I. "Effect of a Spherical Conducting Body on the Parameters of Eddy-Current Probes." *Russian Journal of Nondestructive Testing.* New York, NY: Plenum/Consultants Bureau (1969): p 144-151.
14. Nikitin, A.I. and L.V. Babushkina. "Solution of the Problem of Eddy Currents in a Conducting Sphere Situated in the Field of a Superposed Transducer." *Russian Journal of Nondestructive Testing.* New York, NY: Plenum/Consultants Bureau (1989): p 863-869.
15. Dodd, C.V. "The Use of Computer Modeling for Eddy-Current Testing." *Research Techniques in Nondestructive Testing.* Vol. 3. New York, NY: Academic Press (1977): p 429-479.
16. Luquire, J.W., W.E. Deeds and C.V. Dodd. "Alternating Current Distribution between Planar Conductors." *Journal of Applied Physics.* Vol. 41, No. 10. Melville, NY: American Institute of Physics (1970): p 3983-3991.
17. Cheng, C.C., C.V. Dodd and W.E. Deeds. "General Analysis of Probe Coils near Stratified Conductors." *International Journal of Nondestructive Testing.* Vol. 3. Kidlington, United Kingdom: Elsevier Science Limited (1971): p 109-130.

18. Dodd, C.V., C.C. Cheng and W.E. Deeds. "Induction Coils Coaxial with an Arbitrary Number of Cylindrical Conductors." *Journal of Applied Physics*. Vol. 45, No. 2. Melville, NY: American Institute of Physics (1974): p 638-647.
19. Mottl, Z. "The Quantitative Relations between True and Standard Depth of Penetration for Air-Cored Probe Coils in Eddy Current Testing." *NDT International*. Vol. 23, No. 1. Kidlington, United Kingdom: Elsevier Science Limited (1990): p 11-18.
20. Bowler, J.R. "Transient Eddy Currents in Layered Media As a Model of Corrosion Detection." *Nondestructive Testing and Evaluation*. Vol. 6. New York, NY: Gordon and Breach Science Publishers (1992): p 315-322.
21. Waidelich, D.L. "Pulsed Eddy-Current Testing of Steel Sheets." *Eddy Current Characterization of Materials and Structures*. Special Technical Publication 722. West Conshohocken, PA: ASTM International (1981): p 367-373.
22. Ludwig, R. and X.-W. Dai. "Numerical and Analytical Modeling of Pulsed Eddy Currents in a Conducting Half-Space." *IEEE Transactions on Magnetics*. Vol. 26, No. 1. New York, NY: Institute of Electrical and Electronics Engineers (1990): p 299-307.
23. Sapunov, V.M. and P.I. Beda. "Eddy-Current Inspection of Sheet of Nonmagnetic Material by Superposed Transducer Excited by Pulsed Current with Nonideal Shape." *Russian Journal of Nondestructive Testing*. Vol. 27, No. 10. New York, NY: Plenum/Consultants Bureau (1991): p 743-750.
24. Bowler, J. and M. Johnson. "Pulsed Eddy-Current Response to a Conducting Half-Space." *IEEE Transactions on Magnetics*. Vol. 33, No. 3. New York, NY: Institute of Electrical and Electronics Engineers (1997): p 2258-2264.
25. Burke, S.K., G.R. Hugo and D.J. Harrison. "Transient Eddy-Current NDE for Hidden Corrosion in Multilayer Structures." *Review of Progress in Quantitative Nondestructive Evaluation* [San Diego, CA, July-August 1997]. Vol. 17A. New York, NY: Plenum (1998): p 307-314.
26. Uzal, E., J.C. Moulder, S. Mitra and J.H. Rose. "Impedance of Coils over Layered Metals with Continuously Variable Conductivity and Permeability: Theory and Experiment." *Journal of Applied Physics*. Vol. 74, No. 3. Melville, NY: American Institute of Physics (1993): p 2076-2089.
27. Kolyshkin, A.A. and R. Vaillancourt. "Analytical Solutions to Eddy-Current Testing Problems for a Layered Medium with Varying Properties." *IEEE Transactions on Magnetics*. Vol. 33, No. 4. New York, NY: Institute of Electrical and Electronics Engineers (1997): p 2473-2477.
28. Kolyshkin, A.A. and R. Vaillancourt. "Series Solution of an Eddy-Current Problem for a Sphere with Varying Conductivity and Permeability Profiles." *IEEE Transactions on Magnetics*. Vol. 35, No. 6. New York, NY: Institute of Electrical and Electronics Engineers (1999): p 4445-4451.
29. Theodoulidis, T.P., T.D. Tsiboukis and E.E. Kriezis. "Analytical Solutions in Eddy Current Testing of Layered Metals with Continuous Conductivity Profiles." *IEEE Transactions on Magnetics*. Vol. 31, No. 3. New York, NY: Institute of Electrical and Electronics Engineers (1995): p 2254-2260.
30. Uzal, E., I. Ozkol and M.O. Kaya. "Impedance of a Coil Surrounding an Infinite Cylinder with an Arbitrary Radial Conductivity Profile." *IEEE Transactions on Magnetics*. Vol. 34, No. 1. New York, NY: Institute of Electrical and Electronics Engineers (1998): p 213-217.
31. Theodoulidis, T.P. and E.E. Kriezis. "Coil Impedance Due to a Sphere of Arbitrary Radial Conductivity and Permeability Profiles." *IEEE Transactions on Magnetics*. Vol. 38, No. 3. New York, NY: Institute of Electrical and Electronics Engineers (2002): p 1452-1460.
32. Weaver, J.T. "The General Theory of Electromagnetic Induction in a Conducting Half-Space." *Geophysical Journal of the Royal Astronomical Society*. Vol. 22. Oxford, United Kingdom: Blackwell Scientific Publications, for the Royal Astronomical Society (1970): p 83-100.
33. Hannakam, L. "Wirbelstrome in Leitenden Halbraum bei Beliebiger Form der Erregenden Leiterschleife." *Archiv für Elektrotechnik*. Vol. 54. Berlin, Germany: Springer-Verlag (1972): p 251-261.
34. Kriezis, E.E. and I.E. Xypteras. "Eddy Current Distribution and Loss in a Semi-Infinite Conducting Space Due to Vertical Current Loop." *ETZ Archiv*. Berlin, Germany: VDE-Verlag (1979): p 201-207.

35. Beissner, R.E. "Analytic Green's Dyads for an Electrically Conducting Half-Space." *Journal of Applied Physics*. Vol. 60, No. 3. Melville, NY: American Institute of Physics (1986): p 855-858.
36. Bowler, J.R. "Eddy Current Calculations Using Half-Space Green's Functions." *Journal of Applied Physics*. Vol. 61, No. 3. Melville, NY: American Institute of Physics (1987): p 833-839.
37. Beissner, R.E. and M.J. Sablik. "Theory of Eddy Currents Induced by a Nonsymmetric Coil above a Conducting Half-Space." *Journal of Applied Physics*. Vol. 56, No. 2. Melville, NY: American Institute of Physics (1984): p 448-454.
38. Tsaknakis, H.J. and E.E. Kriezis. "Field Distribution Due to a Circular Current Loop Placed in an Arbitrary Position above a Conducting Plate" *IEEE Transactions on Geoscience and Remote Sensing*. Vol. 23, No. 4. New York, NY: Institute of Electrical and Electronics Engineers (1985): p 834-840.
39. Juillard, J., B. Barmon and G. Berthiau. "Simple Analytical Three-Dimensional Eddy-Current Model." *IEEE Transactions on Magnetics*. Vol. 36, No. 1. New York, NY: Institute of Electrical and Electronics Engineers (2000): p 258-266.
40. Panas, S.M. and A.G. Papayiannakis. "Eddy Currents in an Infinite Slab Due to an Elliptic Current Excitation." *IEEE Transactions on Magnetics*. Vol. 27, No. 5. New York, NY: Institute of Electrical and Electronics Engineers (1991): p 4328-4337.
41. Sadeghi, S.H.H. and A.H. Salemi. "Electromagnetic Field Distributions around Conducting Slabs, Produced by Eddy-Current Probes with Arbitrary Shape Current-Carrying Excitation Loops." *IEE Proceedings: Science, Measurement and Technology*. Vol. 148, No. 4. London, United Kingdom: Institution of Electrical Engineers (2001): p 187-192.
42. Hannakam, L. "Wirbelstrome in einem Massiven Zylinder bei Beliebig Geformter Erregender Leiterschleife." *Archiv für Elektrotechnik*. Vol. 55. Berlin, Germany: Springer-Verlag (1973): p 207-215.
43. Grimberg, R., E. Radu, O. Mihalache and A. Savin. "Calculation of the Induced Electromagnetic Field Created by an Arbitrary Current Distribution Located outside a Conductive Cylinder." *Journal of Physics D: Applied Physics*. Vol. 30. Melville, NY: American Institute of Physics (1997): p 2285-2291.
44. Grimberg, R., A. Savin, E. Radu and O. Mihalache. "Nondestructive Evaluation of the Severity of Discontinuities in Flat Conductive Materials by an Eddy-Current Transducer with Orthogonal Coils." *IEEE Transactions on Magnetics*. Vol. 36, No. 1. New York, NY: Institute of Electrical and Electronics Engineers (2000): p 299-307.
45. Hannakam, L. and G. Mrozynski. "Transienter Skineffekt in der Kugel bei Beliebiger Form der Erregender Leiterschleife." *Archiv für Elektrotechnik*. Vol. 55. Berlin, Germany: Springer-Verlag (1973): p 299-309.
46. Theodoulidis, T.P., N.V. Kantartzis, E.E. Tsiboukis and E.E. Kriezis. "Analytical and Numerical Solution of the Eddy-Current Problem in Spherical Coordinates Based on the Second-Order Vector Potential Formulation." *IEEE Transactions on Magnetics*. Vol. 33, No. 4. New York, NY: Institute of Electrical and Electronics Engineers (1997): p 4461-2472.
47. Mrozynski, G. "Analytical Determination of Eddy Currents in a Hollow Sphere Excited by an Arbitrary Dipole." *IEEE Transactions on Magnetics*. Vol. 34, No. 6. New York, NY: Institute of Electrical and Electronics Engineers (1998): p 3822-3829.
48. Auld, B.A., F.G. Muennemann and M. Riaziat. "Quantitative Modeling of Flaw Responses in Eddy Current Testing." *Research Techniques in Nondestructive Testing*. Vol. 7. London, United Kingdom: Academic Press (1984).
49. Burke, S.K. "Impedance of a Horizontal Coil above a Conducting Half-Space." *Journal of Physics D: Applied Physics*. Vol. 19. Melville, NY. American Institute of Physics (1986): p 1159-1173.
50. Burke, S.K. "Eddy-Current Induction in a Uniaxially Anisotropic Plate." *Journal of Applied Physics*. Vol. 68, No. 7. Melville, NY: American Institute of Physics (1990): p 3080-3090.

51. Theodoulidis T.P. and E.E. Kriezis. "Impedance Evaluation of Rectangular Coils for Eddy Current Testing of Planar Media." *NDT&E International*. Vol. 35. Kidlington, United Kingdom: Elsevier Science Limited (2002): p 407-414.
52. Theodoulidis, T.P. "Analytical Modeling of Wobble in Eddy Current Tube Testing with Bobbin Coils." *Research in Nondestructive Evaluation*. Vol. 14, No. 2. Columbus, OH: American Society for Nondestructive Testing (June 2002): p 111-126.
53. Kolyshkin, A.A. and R. Vaillancourt. "Method of Solution of Forward Problems in Eddy-Current Testing." *Journal of Applied Physics*. Vol. 77, No. 10. Melville, NY: American Institute of Physics (1995): p 4903-4913.
54. Satveli, R., J.C. Moulder, B. Wang and J.H. Rose. "Impedance of a Coil near an Imperfectly Layered Metal Structure: The Layer Approximation." *Journal of Applied Physics*. Vol. 79, No. 6. Melville, NY: American Institute of Physics (1996): p 2811-2821.
55. Burke, S.K. "A Perturbation Method for Calculating Coil Impedance in Eddy-Current Testing." *Journal of Physics D: Applied Physics*. Vol. 18. Melville, NY: American Institute of Physics (1985): p 1745-1760.
56. Hannakam, L. and M. Albach. "Induzierte Wirbelstrome in einer Kreisscheibe und in einem Rechteckzylinder Endlicher Hohe." *Archiv für Elektrotechnik*. Vol. 64. Berlin, Germany: Springer-Verlag (1981): p 127-134.
57. Hannakam, L. and A. Kost. "Leitender Rechteckkeil im Felde einer Doppelleitung." *Archiv für Elektrotechnik*. Vol. 65. Berlin, Germany: Springer-Verlag (1982): p 363-368.
58. Nethe, A. "Einschalstrome in Spulen mit Leitendem Permeablem Kern bei Berucksichtigung der Induzierten Wirbelstrome." *Archiv für Elektrotechnik*. Vol. 74. Berlin, Germany: Springer-Verlag (1991): p 389-401.
59. Theodoulidis, T.P. "Model of Ferrite-Cored Probes for Eddy Current Nondestructive Evaluation." *Journal of Applied Physics*. Vol. 93, No. 5. Melville, NY: American Institute of Physics (March 2003): p 3071-3078.
60. Förster, F. "The First Picture: A View of the Initial Steps in the Development of Eight Branches of Nondestructive Material Testing." *Materials Evaluation*. Vol. 41, No. 13. Columbus, OH: American Society for Nondestructive Testing (December 1983): p 1477-1488.
61. British Intelligence Objectives Sub-Committee. *Metallurgical Research and Testing Laboratories in the Stuttgart Area*. London, United Kingdom: Her Majesty's Stationery Office (1947).
62. Sabbagh, H.A. "A Model of Eddy-Current Probes with Ferrite Cores." *IEEE Transactions on Magnetics*. Vol. 1, No. 3. New York, NY: Institute of Electrical and Electronics Engineers (May 1987): p 1888-1904.
63. Theodoulidis, T. "Application of the Eigenvalues Method in Eddy Current NDE: A Model of Eddy Current Ferrite Cored Probes." *Electromagnetic Nondestructive Evaluation (VII)*. Amsterdam, aerlands: IOS Press (2004).
64. Burrows, M.L. *A Theory of Eddy-Current Flaw Detection*. Ph.D. dissertation. Ann Arbor, MI: University of Michigan (1964).
65. Stratton, J.A. *Electromagnetic Theory*. New York, NY: McGraw-Hill (1941).
66. Harrington, R.F. *Time Harmonic Electromagnetic Fields*. New York, NY: McGraw-Hill (1961).
67. Chari, M.V.K. and T.G. Kincaid. "Finite-Element Analysis of Eddy-Current Flaw Detection." *Eddy Current Characterization of Materials and Structures*. Special Technical Publication 722. West Conshohocken, PA: ASTM International (1981): p 59-75.
68. Raiche, A.P. "An Integral Equation Approach to Three Dimensional Modelling." *Geophysical Journal of the Royal Astronomical Society*. Vol. 36. Oxford, United Kingdom: Blackwell Scientific Publications, for the Royal Astronomical Society (1974): p 363-376.
69. Weidelt, P. "Electromagnetic Induction in Three Dimensional Structures." *Journal of Geophysics*. Vol. 41. Oxford, United Kingdom: Blackwell Scientific Publications, for the Royal Astronomical Society (1975): p 85-109.
70. Wannamaker, P.E., G.W. Hohmann and W.A. San Filipo. "Electromagnetic Models of Three Dimensional Bodies in Layered Earths Using Integral Equations." *Geophysics*. Vol. 49, No. 1. Tulsa, OK: Society of Exploration Geophysicists (1984): p 60-74.

71. McKirdy, D.M. "Recent Improvements to the Application of Volume Integral Method of Eddy Current Modelling." *Journal of Nondestructive Evaluation*. Vol. 8. New York, NY: Plenum (1989): p 45-50.
72. Bowler, J.R., S.A. Jenkins, L.D. Sabbagh and H.A. Sabbagh. "Eddy Current Probe Impedance Due to a Volumetric Flaw." *Journal of Applied Physics*. Vol. 70, No. 3. Melville, NY: American Institute of Physics (1991): p 1107-1114.
73. Bowler, J.R., L.D. Sabbagh, H.A. Sabbagh and S.A. Jenkins. "Differential Eddy-Current Probe Response Calculation Using Volume Elements." *COMPEL — The International Journal for Computation and Mathematics in Electrical and Electronic Engineering*. Vol. 9. Bradford, United Kingdom: MCB UP Limited, Emerald (1990): p 143-146.
74. Kahn, A.H., R. Spal and A. Feldman. "Eddy-Current Losses Due to Surface Crack in Conducting Material." *Journal of Applied Physics*. Vol. 48, No. 11. Melville, NY: American Institute of Physics (November 1977): p 4454-4459.
75. Sommerfeld, A. *Optics*. New York, NY: Academic Press (1964).
76. Harrison, D.J., L.D. Jones and S.K. Burke. "Benchmark Problems for Defect Size and Shape Determination in Eddy-Current Nondestructive Evaluation." *Journal of Nondestructive Evaluation*. Vol. 15. New York, NY: Plenum (1996): p 21-34.
77. Tai, C.-T. *Dyadic Green Functions in Electromagnetic Theory*. New York, NY: Institute of Electrical and Electronics Engineers, with Oxford University Press (1996).
78. Cheng, D.K. *Field and Wave Electromagnetics*. Boston, MA: Addison-Wesley (1989).
79. Van Bladel, J. *Electromagnetic Fields*. Berlin, Germany: Springer-Verlag (1985).
80. Tai, C.-T. *Generalized Vector and Dyadic Analysis: Applied Mathematics in Field Theory*. New York, NY: Institute of Electrical and Electronics Engineers, with Oxford University Press (1999).
81. Rumsey, V.H. "Reaction Concept in Electromagnetic Theory." *Physical Review*. Vol. 94, No. 6. Melville, NY: American Institute of Physics, for American Physical Society (June 1954): p 1483-1491.
82. Nair, S.M. and J.H. Rose. "Low-Frequency Asymtotics for Eddy Currents in a Conducting Half-Space in the Absence and Presence of Inhomogeneities." *Journal of Applied Physics*. Vol. 70, No. 4. Melville, NY: American Institute of Physics (1991): p 1924-1937.
83. Harfield, N., Y. Yoshida and J.R. Bowler. "Low Frequency Eddy Current Interaction with a Crack." *Journal of Applied Physics*. Vol. 80, No. 7. Melville, NY: American Institute of Physics (1996): p 4090-4100.
84. Harfield, N. and J.R. Bowler. "Analysis of Eddy-Current Interaction with a Surface-Breaking Crack." *Journal of Applied Physics*. Vol. 76, No. 8. Melville, NY: American Institute of Physics (1994): p 4853-4856.
85. Noble, B. *Methods Based on the Wiener-Hopf Technique for the Solution of Partial Differential Equations*. New York, NY: Pergamon Press (1958).
86. Harfield, N. and J.R. Bowler. "A Geometrical Theory for Eddy-Current Non-Destructive Evaluation." *Proceedings of the Royal Society:* Series A, *Mathematical and Physical Sciences*. Vol. 453. London, United Kingdom: Royal Society (1997): p 1121-1152.
87. Binns, K.J. and P.J. Lawrenson. *Analysis and Computation of Electric and Magnetic Field Problems*. Oxford, United Kingdom: Pergamon Press (1973).
88. Nehari, Z. *Conformal Mapping*. New York, NY: Dover Publications (1952): p 194.
89. Press, W.H., S.A. Teukolsky, W.T. Vetterling and B.P. Flannery. *Numerical Recipes in Fortran: The Art of Scientific Computing*, second edition. Cambridge, United Kingdom: Cambridge University Press (1992).
90. Förster, F. "Nondestructive Inspection by the Method of Magnetic Leakage Fields: Theoretical and Experimental Foundations of the Detection of Surface Cracks of Finite and Infinite Depth." *Defektoskopiya*. Vol. 11. New York, NY: Plenum/Consultants Bureau (1982): p 3-25.
91. Bowler, J.R. and N. Bowler. "Evaluation of the Magnetic Field near a Crack with Application to Magnetic Particle Inspection." *Journal of Physics D: Applied Physics*. Vol. 35. Bristol, United Kingdom: Institute of Physics (2002): p 2237-2242.

92. Wilton, D.T., B.K. Middleton and M.M. Aziz. "Exact Harmonic Coefficients for a Magnetic Ring Head." *IEEE Transactions on Magnetics.* Vol. 35, No 3. New York, NY: Institute of Electrical and Electronics Engineers (May 1999): p 2042-2047.
93. Van Bladel, J. *Singular Electromagnetic Fields and Sources.* Oxford, United Kingdom: Oxford Scientific Publications (1991).
94. Bowler, J.R. "Eddy-Current Interaction with an Ideal Crack I: The Forward Problem." *Journal of Applied Physics.* Vol. 75, No. 12. Melville, NY: American Institute of Physics (1994): p 8128-8137.
95. Burke, S.K. and L.R.F. Rose. "Interaction of Induced Currents with Cracks in Thin Plates." *Proceedings of the Royal Society of London.* Vol. A 418. London, United Kingdom: Royal Society (1988): p 229-246.
96. Bowler, J.R., S.J. Norton and D.J. Harrison. "Eddy-Current Interaction with an Ideal Crack II: The Inverse Problem." *Journal of Applied Physics.* Vol. 75, No. 12. Melville, NY: American Institute of Physics (1994): p 8138-8144.
97. Harrington, R.F. *Field Computation by Moment Methods.* New York, NY: Macmillan (1968).
98. Wang, J.J.H. *Generalized Moment Methods in Electromagnetics: Formulation and Computer Solution of Integral Equations.* New York, NY: Wiley-Interscience (January 1991).
99. Beltrame, P. and N. Burais. "Computing Methods of Hypersingular Integral Applied to Eddy-Current Testing." *IEEE Transactions on Magnetics.* Vol. 38, No. 2. New York, NY: Institute of Electrical and Electronics Engineers (2002): p 1269-1272.
100. Bowler, J.R. "Inversion of Open Cracks Using Eddy-Current Probe Impedance Measurements." *Review of Progress in Quantitative Nondestructive Evaluation* [Montreal, Canada, July 1999] Vol. 19A. New York, NY: Plenum (2000): p 529-533.
101. Michael, D.H., R. Collins and K.B. Ranger. "The AC Fuels around a Plane Semi-Elliptical Crack in a Metal Surface." *Proceedings of the 13th Symposium on Nondestructive Evaluation.* San Antonio, TX: Nondestructive Testing Information and Analysis Center (1981): p 470-479.
102. Michael, D.H., R.T. Waechter and R. Collins. "The Measurement of Surface Cracks in Metals by Using A.C. Electric Fields." *Proceedings of the Royal Society of London:* Series A, *Mathematical and Physical Sciences.* Vol. 381, No. 1780. London, United Kingdom: Royal Society (May 1982): p 139-157.
103. Auld, B.A., S.R. Jefferies and J.C. Moulder. "Eddy-Current Signal Analysis and Inversion for Semielliptical Surface Cracks." *Journal of Nondestructive Evaluation.* Vol. 7. New York, NY: Plenum (1988): p 79-94.
104. Lewis, A.M., D.H. Michael, M.C. Lugg and R. Collins. "Thin-Skin Electromagnetic Fields around Surface-Breaking Cracks in Metals." *Journal of Applied Physics.* Vol. 64, No. 8. Melville, NY: American Institute of Physics (1988): p 3777-3784.
105. Lewis, A.M. "A Theoretical Model of the Response of an Eddy-Current Probe to a Surface-Breaking Metal Fatigue Crack in a Flat Test-Piece." *Journal of Physics D, Applied Physics.* Vol. 25. Bristol, United Kingdom: Institute of Physics (1992): p 319-326.
106. Felsen, L.B. and N. Marcuvitz. *Radiation and Scattering of Waves.* Upper Saddle River, NJ: Prentice-Hall (1973).
107. Harfield, N. and J.R. Bowler. "Theory of Thin-Skin Eddy-Current Interaction with Surface Cracks." *Journal of Applied Physics.* Vol. 82, No. 9. Melville, NY: American Institute of Physics (1997): p 4590-4603.
108. Yoshida, Y. and J.R. Bowler. "Vector Potential Integral Formulation for Eddy-Current Probe Response to Cracks." *IEEE Transactions on Magnetics.* Vol. 36, No. 2. New York, NY: Institute of Electrical and Electronics Engineers (March 2000): p 461-469.
109. Bowler, J.R. and N. Harfield. "Evaluation of Probe Impedance Due to Thin-Skin Eddy-Current Interaction with Surface Cracks." *IEEE Transactions on Magnetics.* Vol. 34. New York, NY: Institute of Electrical and Electronics Engineers (1998): p 515-523.
110. Bowler, J.R. and N. Harfield. "Thin-Skin Eddy-Current Interaction with Semi-Elliptical and Epi-Cyclic Cracks." *IEEE Transactions on Magnetics.* Vol. 36, No. 1. New York, NY: Institute of Electrical and Electronics Engineers (2000): p 281-291.

111. Hochschild, R. "Electromagnetic Methods of Testing Metals." *Progress in Non-Destructive Testing*. Vol. 1. New York, NY: Macmillan (1959): p 59-109.
112. Lord, W. and R. Palanisamy. "Development of Theoretical Models for Nondestructive Testing Eddy Current Phenomena." *Eddy Current Characterization of Materials and Structures*. Special Technical Publication 722. West Conshohocken, PA: ASTM International (1979): p 5-21.
113. McMaster, R.C. "The Origins of Electromagnetic Testing." *Materials Evaluation*. Vol. 43, No. 7. Columbus, OH: American Society for Nondestructive Testing (July 1985): p 946-956.
114. Hughes, D.E. "Induction Balance and Experimental Researches Therewith." *Philosophical Magazine*. Series 5, Vol. 8. Abingdon, United Kingdom: Taylor and Francis (1879): p 50.
115. Förster, F. and K. Stambke. "Theoretische und Experimentelle Grundlagen der Zerstörungsfreien Werkstoffprüfung mit Wirbelstromverfahren, III: Verfahren mit Durchlaufspule zür Quantitativen Zerstörungsfreien Werkstoffprüfung." *Zeitschrift für Metallkunde*. Vol. 45, No. 4. Stuttgart, Germany: Riederer-Verlag (1954): p 166-179.
116. Libby, H.L. *Introduction to Electromagnetic Nondestructive Test Methods*. New York, NY: Wiley-Interscience (1971).
117. Weidelich, D.L. and C.J. Renken. "The Impedance of a Coil near a Conductor." *Proceedings of the National Electronics Conference*. Vol. 12. Chicago, IL: National Engineering Conference (1956): p 188-196.
118. Vine, J. "Impedance of a Coil Placed near to a Conducting Sheet." *Journal of Electronics and Control*. Vol. 16. London, United Kingdom: Taylor and Francis (1964): p 569-577.
119. Cheng, D.H.S. "The Reflected Impedance of a Circular Coil in the Proximity of a Semi-Infinite Medium." *IEEE Transactions on Instrumentation and Measurement*. Vol. 14, No. 3. New York, NY: Institute of Electrical and Electronics Engineers (September 1965): p 107-116.
120. Dodd, C.V. *Solutions to Electromagnetic Induction Problems*. Ph.D. dissertation. Knoxville, TN: University of Tennessee (June 1967).
121. Graneau, P. and S.A. Swann. "Electromagnetic Fault Detection in Non-Ferrous Pipes." *Journal of Electronics and Control*. Vol. 8. London, United Kingdom: Taylor and Francis (1960): p 127-147.
122. Graneau, P. "Frequency Dependence of Induced Currents." *Journal of Electronics and Control*. Vol. 10. London, United Kingdom: Taylor and Francis (1961): p 383-401.
123. Beland, B. "Eddy Currents in Circular, Square and Rectangular Rods." *IEE Proceedings*. Vol. 130, Part A, No. 3. London, United Kingdom: Institution of Electrical Engineers (March 1983): p 112-121.
124. Burrows, M.L. "An Examination of the Coupled-Circuit Theory of Eddy Currents." *Journal of Electronics and Control*. Vol. 16. London, United Kingdom: Taylor and Francis (1964): p 659-668.
125. Vlasov, V.V. and V.A. Komarov. "The Magnetic Field of Eddy Currents above a Surface Crack in Metal with Excitation of Them by an Applied Inductor." *Defektoskopiya*. No. 6. New York, NY: Plenum/Consultants Bureau (1971): p 63-76.
126. Zatsepin, N.N. and V.E. Shcherbinin. "Calculation of the Magnetostatic Field of Surface Defects, I: Field Topography of Defect Models." *Defektoskopiya*. No. 5. New York, NY: Plenum/Consultants Bureau (September-October 1966): p 50-59.
127. Vein, P.R. "Inductance between Two Loops in the Presence of Solid Conducting Bodies." *Journal of Electronics and Control*. Vol. 13. London, United Kingdom: Taylor and Francis (1962): p 471-494.
128. Dodd, C.V., W.E. Deeds, J.W. Luquire and W.G. Spoeri. ORNL-4384, *Some Eddy Current Problems and Their Integral Solution*. Oak Ridge, TN: Oak Ridge National Laboratory (1969).
129. Dodd, C.V., C.C. Cheng, C.W. Nestor, Jr. and R.B. Hofstra. ORNL-TM-4175, *Design of Induction Probes for Measurement of Level of Liquid Metals*. Oak Ridge, TN: Oak Ridge National Laboratory (1973).
130. Luquire, J.W., W.E. Deeds and W.G. Spoeri. ORNL-TM-2501, *Computer Programs for Some Eddy Current Problems*. Oak Ridge, TN: Oak Ridge National Laboratory (August 1969).
131. Brown, M.L. "Calculation of Three-Dimensional Eddy Currents at Power Frequencies." *IEEE Proceedings*. Vol. 129, Part A, No. 1. New York, NY: Institute of Electrical and Electronics Engineers (January 1982): p 46-53.

132. Hammond, P. "Use of Potentials in Calculation of Electromagnetic Fields." *IEEE Proceedings*. Vol. 129, Part A, No. 2. New York, NY: Institute of Electrical and Electronics Engineers (March 1982): p 106-112.
133. Sarma, M.S. "Potential Functions in Electromagnetic Field Problems." *IEEE Transactions on Magnetics*. Vol. MAG-6, No. 3. New York, NY: Institute of Electrical and Electronics Engineers (September 1970): p 513-518.
134. James, M.L., G.M. Smith and J.C. Walford. *Applied Numerical Methods for Digital Computation*. New York, NY: Harper and Row (1977).
135. Erdelyi, E.A. and E.F. Fuchs. "Nonlinear Magnetic Field Analysis of DC Machines. Part I: Theoretical Fundamentals." *IEEE Transactions on Power Apparatus and Systems*. Vol. PAS-89. New York, NY: Institute of Electrical and Electronics Engineers (1970): p 1546-1554.
136. Erdelyi, E.A. and E.F. Fuchs. "Nonlinear Magnetic Field Analysis of DC Machines. Part II: Application of the Improved Treatment." *IEEE Transactions on Power Apparatus and Systems*. Vol. PAS-90. New York, NY: Institute of Electrical and Electronics Engineers (1970): p 1555-1564.
137. Demerdash, N.A., H.B. Hamilton and G.W. Brown. "Simulation for Design Purposes of Magnetic Fields in Turbogenerators with Symmetrical and Asymmetrical Rotors. Part I: Model Development and Solution Technique." *IEEE Transactions on Power Apparatus and Systems*. Vol. PAS-91. New York, NY: Institute of Electrical and Electronics Engineers (1972): p 1985-1992.
138. Demerdash, N.A., H.B. Hamilton and G.W. Brown. "Simulation for Design Purposes of Magnetic Fields in Turbogenerators with Symmetrical and Asymmetrical Rotors. Part II: Model Calibration and Applications." *IEEE Transactions on Power Apparatus and Systems*. Vol. PAS-91. New York, NY: Institute of Electrical and Electronics Engineers (1972): p 1992-1999.
139. Sieminieniuch, J.L. and I. Gladwell. "Analysis of Explicit Difference Methods for a Diffusion-Convection Equation." *International Journal for Numerical Methods in Engineering*. Vol. 12, No. 6. New York, NY: Wiley (1978): p 899-916.
140. Dufort, E.C. and S.P. Frankel. "Stability Conditions in the Numerical Treatment of Parabolic Differential Equations." *Mathematical Tables and Aids to Computation*. Vol. 7. Washington, DC: National Research Council (1953): p 135-152.
141. Richtmyer, R.D. and K.W. Morton. *Differential Methods for Initial Value Problems*. London, United Kingdom: John Wiley (1967).
142. Jennings, A. and G.M. Malik. "The Solution of Sparse Linear Equations by the Conjugate Gradient Method." *International Journal for Numerical Methods in Engineering*. Vol. 12, No. 1. New York, NY: John Wiley (1978): p 141-158.
143. Zienkiewicz, O.C. *The Finite Element Method in Engineering*, third edition. London, United Kingdom: McGraw-Hill (1977).
144. Oden, J.T. and G.F. Carey. *Finite Elements, Special Problems in Solid Mechanics*. Vol. 5. Upper Saddle River, NJ: Prentice-Hall (1984).
145. Melash, R.J. "Numerical Analysis of Automobile Structures." *Advances in Computational Methods in Structural Mechanics and Design*. Huntsville, AL: UAH Press (1972): p 746-766.
146. Sabir, A.B. "Finite Element Analysis of Arch Bridges." *Finite Elements for Thin Shells and Curved Members*. London, United Kingdom: John Wiley and Sons (1974): p 223-242.
147. Bruch, J.C. and G. Zyroloski. "Transient Two-Dimensional Heat Conduction Problems Solved by the Finite Element Method." *International Journal for Numerical Methods in Engineering*. Vol. 8, No. 3. New York, NY: John Wiley (1974): p 481-494.
148. Le Provost, C. and A. Poucet. "Finite Element Method for Spectral Modeling of Tides." *International Journal for Numerical Methods in Engineering*. Vol. 12, No. 5. New York, NY: John Wiley (1978): p 853-872.
149. Arlet, P.L. et al. "Application of Finite Elements to the Solution of Helmholtz' Equation." *Proceedings of the IEEE*. Vol. 115, No. 12. New York, NY: Institute of Electrical and Electronics Engineers (December 1968).
150. McNeice, G.M. and H.C. Amstutz. "Finite Element Studies in Hip Reconstructions." *Biomechanics*. Vol. A. Baltimore, MD: University Park Press (1976): p 394-405.
151. Silvester, P.P. and M.V.K. Chari. "Finite Element Solution of Saturable Magnetic Field Problems." *IEEE Transactions on Power Apparatus and Systems*. Vol. 89. New York, NY: Institute of Electrical and Electronics Engineers (1970): p 1642-1651.

152. Anderson, O.W. "Transformer Leakage Flux Program Based on the Finite Element Method." *IEEE Transactions on Power Apparatus and Systems*. Vol. 92. New York, NY: Institute of Electrical and Electronics Engineers (March-April 1973): p 682-689.
153. Chari, M.V.K. "Finite Element Solution of the Eddy Current Problem in Magnetic Structures." *IEEE Transactions on Power Apparatus and Systems*. Vol. PAS-93. New York, NY: Institute of Electrical and Electronics Engineers (1974): p 62.
154. Aoki, S. "Three Dimensional Magnetic Field Calculation of the Levitation Magnet for HSST by the Finite Element Method." *IEEE Transactions on Magnetics*. Vol. MAG-16. New York, NY: Institute of Electrical and Electronics Engineers (September 1980): p 725-727.
155. Kaminga, W. "Finite Element Solutions for Devices with Permanent Magnets." *Applied Physics*. Vol. 8, No. 7. Berlin, Germany: Springer-Verlag (May 1975): p 841-855.
156. Demerdash, N.A. and T.W. Nehl. "An Evaluation of the Methods of Finite Elements and Finite Differences in the Solution of Nonlinear Electromagnetic Fields in Electrical Machines." *IEEE Transactions on Power Apparatus and Systems*. Vol. PAS-98, No. 1. New York, NY: Institute of Electrical and Electronics Engineers (January-February 1978): p 74-87.
157. Huebner, K.H. *The Finite Element Method for Engineers*. New York, NY: Wiley-Interscience (1975).
158. Palanisamy, R. *Finite Element Modeling of Eddy Current Nondestructive Testing Phenomena*. Ph.D. dissertation. Fort Collins, CO: Colorado State University (1980).
159. Palanisamy, R. and W. Lord. "Finite Element Modeling of Electromagnetic NDT Phenomena." IEEE *Transactions on Magnetics*. Vol. MAG-15, No. 6. New York, NY: Institute of Electrical and Electronics Engineers (November 1979): p 1479-1481.
160. Ida, N. *Three Dimensional Finite Element Modeling of Electromagnetic Nondestructive Testing Phenomena*. Ph.D. dissertation. Fort Collins, CO: Colorado State University (1983).
161. Ida, N. and W. Lord. "A Finite Element Model for Three-Dimensional Eddy Current NDT Calculations." *IEEE Transactions on Magnetics*. Vol. MAG-21, No. 6. New York, NY: Institute of Electrical and Electronics Engineers (November 1985): p 2635-2643.
162. Wood, W.L. "A Further Look at Newmark, Houbolt, etc., Time Stepping Formulae." *International Journal for Numerical Methods in Engineering*. Vol. 20, No. 6. New York, NY: John Wiley (1984): p 1009-1018.
163. Oden, J.T. "A General Theory of Finite Elements I: Topological Considerations." *International Journal for Numerical Methods in Engineering*. Vol. 1, No. 2. New York, NY: John Wiley (1969): p 205-221.
164. Oden, J.T. "A General Theory of Finite Elements II: Applications." *International Journal for Numerical Methods in Engineering*. Vol. 1, No. 3. New York, NY: John Wiley (1969): p 247-259.
165. Demerdash, N.A., T.W. Nehl, F.A. Fouad and O.A. Mohammed. "Three Dimensional Finite Element Vector Potential Formulation of Magnetic Fields in Electrical Apparatus." *IEEE Transactions on Power Apparatus and Systems*. Vol. PAS-100, No. 4. New York, NY: Institute of Electrical and Electronics Engineers (April 1981): p 4104-4111.
166. Cook, W.A. LA-9402-MS, *INGEN: A General-Purpose Mesh Generator for Finite Element Codes*. Los Alamos, NM: Los Alamos Scientific Laboratory (June 1982).
167. Ida, N. "A Mesh Generator with Automatic Bandwidth Reduction for 2-D and 3-D Geometries." *Computational Electromagnetics*. New York, NY: Elsevier Science Publishers (1986): p 13-22.
168. Demerdash, N.A., O.A. Mohammed, T.W. Nehl, F.A. Fouad and R.H. Miller. "Solution of Eddy Current Problems Using Three-Dimensional Finite Element Complex Magnetic Vector Potential." *IEEE PES Winter Meeting*. New York, NY: Institute of Electrical and Electronics Engineers (1982).
169. Ida, N., H. Hoshikawa and W. Lord. "Finite Element Prediction of Differential EC Probe Signals from Fe_3O_4 Deposits in PWR Steam Generators." *NDT International*. Vol. 18, No. 6. Kidlington, United Kingdom: Elsevier Science Limited (December 1985): p 331-338.

170. Ida, N. and W. Lord. "Simulating Electromagnetic NDT Probe Fields." *IEEE Computer Graphics and Applications.*" Vol. 3, No. 3. New York, NY: Institute of Electrical and Electronics Engineers (May-June 1983): p 21-28.
171. Ida, N., R. Palanisamy and W. Lord. "Eddy Current Probe Design Using Finite Element Analysis." *Materials Evaluation.* Vol. 41, No. 12. Columbus, OH: American Society for Nondestructive Testing (November 1983): p 1339-1394.
172. Hoshikawa, H., R.M. Li and N. Ida. "Finite Element Analysis of Eddy Current Surface Probes." *Review of Progress in Quantitative Nondestructive Evaluation.* Vol. 3A. New York, NY: Plenum (1984): p 675-682.
173. Ida, N., K. Betzold and W. Lord. "Finite Element Modeling of Absolute Eddy Current Probe Signals." *Journal of Nondestructive Evaluation.* Vol. 3, No. 3. New York, NY: Plenum (1983): p 147-154.

5

CHAPTER

Probes for Electromagnetic Testing

Satish Udpa, Michigan State University, East Lansing, Michigan

Nathan Ida, University of Akron, Akron, Ohio (Parts 1 and 2)

Gerald L. Fitzpatrick, PRI Research and Development Corporation, Kirkland, Washington (Part 5)

Tatsuo Hiroshima, Marktec Corporation, Taiei, Japan (Part 4)

Michael L. Mester, Lower Burrell, Pennsylvania (Part 4)

William C.L. Shih, PRI Research and Development Corporation, Torrance, California (Part 5)

Roderic K. Stanley, NDE Information Consultants, Houston, Texas (Parts 3 and 4)

Lalita Udpa, Michigan State University, East Lansing, Michigan (Part 1)

PART 1. Introduction to Eddy Current Probes

Basic Operation of Eddy Current Probes

Nondestructive testing involves the application of a suitable form of energy to a test object and measuring the manner in which the energy interacts with the material. An electromagnetic measurement is usually made with a probe, or transducer, that converts the energy into an electrical signal. The probe output is fed to an appropriate instrument, which calculates the measurement variable of interest. The measured variable or signal is then either manually interpreted or analyzed by using signal processing algorithms to determine the state of the test object. The probes typically used in electromagnetic nondestructive testing are described below.

In nondestructive testing, the words *probe* and *transducer* are synonymous. Electromagnetic testing probes come in several forms, types and sizes. The most common types are coils, hall effect detectors and magnetic particles.

The electromagnetic field that is measured in nondestructive testing is usually three-dimensional and varies as a function of space. In most cases, the field varies as a function of time. Full characterization of the state of the test object usually makes it necessary to obtain as much information about the field as possible. Because the field varies as a function of time and space, the test object is usually scanned and measurements are taken at multiple points along the surface of the test object. In the case of magnetic particle testing, the particles are sprayed over the test area. Another approach is to use an array of sensors to cover the area of interest. An alternate technique is to use magnetooptic imaging devices.

When the field varies as a function of time, the probe and the associated instrumentation should have the necessary bandwidth to support the measurement. Inadequate bandwidth can lead to erroneous measurements.

The field, being three-dimensional, is characterized by three independent components. An appropriate coordinate system (cartesian, cylindrical or spherical) is usually chosen and the field values are referenced accordingly. It is customary to use a point in the test object as the origin of the coordinate system and align one of the coordinates along the surface of the test object. Most probes are directional, sensitive to fields along a specific direction. Thus, a flat or pancake eddy current coil is sensitive to fields that are perpendicular (normal) to the plane of the coil. Similarly, a hall element detector output is proportional to the plane of the hall element. Both devices are insensitive to components of the field in the other directions. It is possible to use two or more of these devices to measure components in the field in other directions. The orientation of the probe with respect to the test object is critical.

Consider the situation shown in Fig. 1, where a steel billet is being tested with the magnetic flux leakage technique. The test object is magnetized by passing a current through the billet. Figure 1b shows a cross section of the billet with the flux contours. The leakage flux has only two components if the billet is very long and the rectangular discontinuity runs along the entire length of the test object (Fig. 1c). If a hall element, whose plane is parallel to the surface of the billet, is passed over the surface, the output of the detector will appear as shown in Fig. 1d. If a hall element whose plane is perpendicular to the surface scans the billet in the direction indicated in the figure, then the signal shown in Fig. 1e is observed. The signals depicted in Figs. 1d and 1e are called the *tangential component* and the *normal component,* respectively. In this specific case, the component along the axis of the billet is zero. If the flux density components along the axial, normal and tangential components are denoted by B_X, B_Y and B_Z, respectively, then the magnitude of the total flux density is:

$$(1) \quad B = \sqrt{B_X^2 + B_Y^2 + B_Z^2}$$

Forms of Coil Probes[1]

Coil probes are used extensively in eddy current as well as magnetic flux leakage nondestructive test applications. The popularity of these probes can be attributed to their simple and robust construction, low cost and design flexibility.

Configuration

Design flexibility lets probes be configured in different ways. Three of the most common eddy current configurations are (1) absolute probes, (2) differential probes and (3) absolute and differential array probes.

Absolute eddy current probes consist of a single coil. In this type of probe, the impedance or the induced voltage in the coil is measured directly (the absolute value rather than changes in impedance or induced voltage is considered). In general, absolute eddy current probes are the simplest and perhaps for this reason are widely used.

Differential eddy current probes consist of a pair of coils connected in opposition so that the net measured impedance or induced voltage is cancelled out when both coils experience identical conditions. The coils sense changes in the test material, so differential eddy current probes are used to react to changes in test materials while canceling out noise and other unwanted signals that affect both coils. Their sensitivity to discontinuities in materials is higher than that of absolute probes. Their sensitivity to liftoff variations and probe wobble is reduced because those effects tend to affect both coils equally.

Array probes consist of coils arranged in a circular, rectangular or some other form of an array.

Sensing Technique

A second kind of probe classification is based on the technique used for sensing changes in probe characteristics: either (1) the impedance technique or (2) the transmit-receive technique. Because impedance changes in the coil cause changes in the coil voltage (for a constant current source) or in the coil current (for a constant voltage source), it is possible to monitor the driving coil to sense any material parameters that result in impedance changes. The transmit-receive technique uses a separate driving coil (or

FIGURE 1. Magnetic field in long steel billet: (a) diagram showing rectangular slot on surface that carries current; (b) magnetic flux contours within billet; (c) magnetic flux contours around slot; (d) tangential component of leakage field as region above slot is scanned; (e) normal component of leakage field.

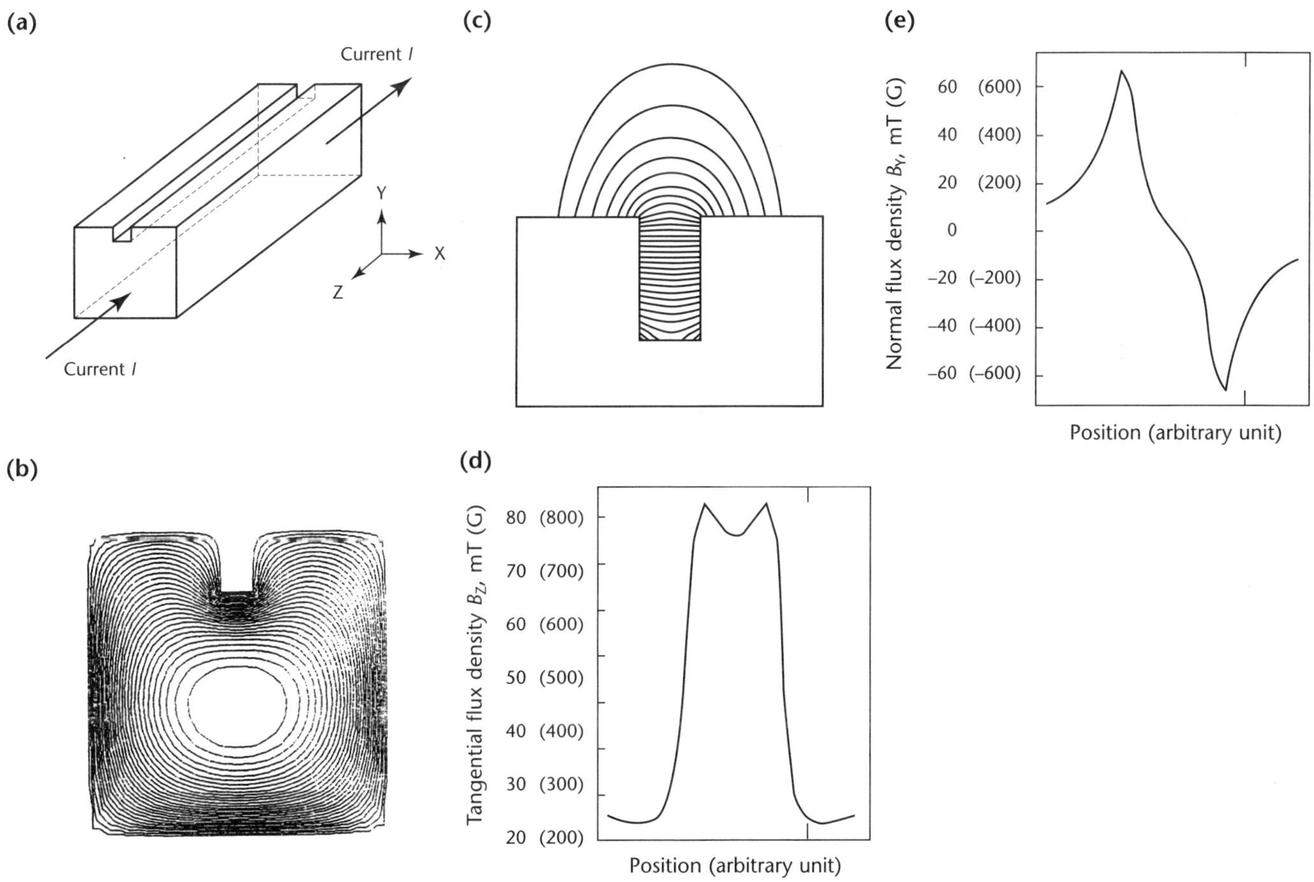

coils) and pickup coil (or coils). In this case, the voltage induced across the pickup coils is measured.

Geometry

A third way to classify probes is according to geometry. Common probe designs include (1) inside diameter probes, (2) encircling coils (outside diameter probes), (3) surface probes such as pancake units and (4) special designs such as plus point probes. The pancake probe has a coil whose axis is normal to the surface of the test material and whose length is not larger than the radius. The plus point probe consists of two coils that lie at a right angle to each other.

Inside diameter probes consist of circular coils inserted in tubes or circular holes. Encircling coils are similar in structure to inside diameter probes except for the fact that the test material is passed inside the coils. They are primarily used to test the outside surface of round materials such as tubes and rods. Surface coils are some of the most widely used eddy current probes. In most cases, they consist of flat coils and are used to test flat surfaces or surfaces with relatively large curvatures relative to their size. Surface probes may be curved to fit contours of the test object.

All of these probes may be used in any of the configurations described above. Thus, for example, an inside surface probe may be absolute or differential and either the impedance or the induced voltage may be measured.

Factors Affecting Eddy Current Probes[1]

Liftoff Curve

An eddy current probe has an initial impedance (quiescent impedance) that depends on the design of the probe itself. This is an intrinsic characteristic of any eddy current probe and is sometimes called *infinite liftoff impedance.* As the probe is moved closer to the test object, the real and imaginary parts of the impedance begin to change until the probe touches the material surface. This is called the *zero liftoff impedance.* The impedance curve described by the probe as it moves between these two points is the liftoff curve and is a very important factor to consider in eddy current testing. Because of the nature of the eddy current probes, the curve is not linear (the change in the field is larger close to the coils). In many cases, especially with small diameter probes for which the field decays rapidly, the range in which measurements may be taken is very small and the effect of liftoff can be pronounced. In other cases, such as with large diameter probes or with forked probes, the effect may be considerably smaller.

Liftoff, because it is troublesome in many cases, is often considered an effect to be minimized. Liftoff effects may be reduced by techniques such as surface riding probes[2] or compensated for by making multifrequency measurements.[3] At the same time, some important eddy current tests depend on the liftoff effect. Measurements of nonconductive coating thicknesses over conducting surfaces and testing for surface evenness are two such tests.

Fill Factor

For encircling coils, the coupling factor, analogous to the liftoff effect, is referred to as *fill factor.* Fill factor is a measure of how well the tested article fills the coil. The largest signal is obtained with the material completely filling the coil — fill factor is almost equal to 1.0. Although it is usually desirable to maximize fill factor, some tests rely on fill factor variations. Fill factor is determined by the intersection of the impedance curve with the vertical or imaginary axis of the impedance plane.

Depth of Penetration

When the eddy current probe is placed on the test object, the eddy currents induced in the test object are not uniformly distributed throughout the material. The eddy current density is high at the surface and decays exponentially with depth in the material; the phenomenon that accounts for this density difference is called the *skin effect.* A measure of the depth to which eddy currents penetrate the material is called the *depth of penetration,* or *skin depth. The standard depth of penetration* can be defined as:

$$\textbf{(2)}\quad S = \sqrt{\frac{1}{\pi f \mu_0 \mu_r \sigma}}$$

where f is frequency (hertz), S is the standard depth of penetration (meter), μ_0 is the magnetic permeability of free space, μ_r is the relative magnetic permeability and σ is the conductivity of the material.

The standard depth of penetration is a convenient figure at which, under precisely controlled conditions, the eddy current density has decayed to $1 \cdot e^{-1}$ (37 percent) of its surface value. It is an important figure for practical purposes because, at about five standard depths of penetration (under precisely defined conditions), the eddy current density is less than 0.7 percent of the surface value.

As Eq. 2 shows, the standard depth of penetration depends on conductivity, permeability and frequency but is relatively small for most metals, about 0.2 mm (0.008 in.) for copper at 100 kHz. The skin effect has two important effects on the design of eddy current probes: (1) the probes are more useful for surface testing and (2) lower frequencies may be necessary for subsurface testing. The standard depth of penetration can be increased in the case of ferromagnetic test objects by magnetically saturating them, thereby reducing their relative magnetic permeability μ_r.

PART 2. Design of Eddy Current Probes

Eddy current probes are based on relatively simple principles and consist of one or more coils. The shape of the coils, their cross section, size, configuration and sources are all parameters that are chosen by the designer to accomplish a particular purpose. Practical eddy current probes may range from tiny coils less than 0.5 mm (0.02 in.) to over 300 mm (12 in.) in diameter, may be long or short and may have square, round or elliptical cross sections, with magnetic or nonmagnetic cores or shields. Parameters of interest in the design may include (1) coil inductance, (2) coil resistance, (3) field distribution in space, (4) coil response to relevant material property changes, (5) liftoff characteristics and (6) response to a notch, drilled hole or other simulated discontinuities. In addition, the design may be influenced by other constraints intrinsic to the test environment, such as weather or access requirements for a specific shape or size. Some of these requirements may in fact be contradictory. The design process is usually iterative, proceeding by trial and error.

There are three basic techniques of probe design. Although these will be considered separately, a combination of the techniques is perhaps the most appropriate approach. The techniques can be classified as follows: (1) experimental or empirical design, (2) analytical design and (3) numerical design.

A practical way to design a probe would be to start with analytical expressions (exact or approximate), design a probe based on some set of initial requirements, construct the probe and then evaluate its performance experimentally. If necessary, the process can be repeated until an acceptable design is obtained. Analytical expressions are not accurate except for the simplest probe geometries and numerical tools are often used in practice. The numerical design of probes has several advantages.

1. The probe, with all its components (coils, core and shield) and the surrounding medium are analyzed. The probe characteristics in the actual test environment can be obtained.
2. A more accurate design is obtained before the probe is actually built by numerically experimenting with the probe parameters.
3. The numerical technique is applicable to situations that cannot be analyzed analytically or simulated experimentally (subsurface discontinuities, layered materials and others).

The following discussion focuses on analytical and numerical approaches involving an iterative approach in which the test results from a specific design lead to improvements to that design.

Some of the avenues available to a designer are outlined below. In particular, a numerical approach to probe design is highlighted. The discussion below uses the finite element technique but the considerations and the treatment of the problem are similar for other numerical techniques.

Experimental Design of Eddy Current Probes

Probe design literature[4-9] reveals the experimental nature of eddy current research. This approach was dominant in the early days of nondestructive testing.[10]

Three questions are associated with any empirical approach. What is the experimental design procedure? How is the outcome of the design evaluated? For the application, is this technique the best approach, merely an acceptable approach or the only feasible approach?

Analytical Design of Eddy Current Probes

The design of an eddy current probe may proceed either (1) by calculating the coil impedances for a given geometry or (2) by determining the appropriate dimensions for a probe with a predetermined impedance. Not all probe parameters may be designed independently. For example, if a certain probe diameter and reactance at a given frequency are required, it may not be possible to design such a probe or the design may not be acceptable for the test at hand.

In the following discussion, the basic relations necessary for probe design are outlined. First, the design of air core coils is presented for single-coil and multiple-coil probes. The discussion on air core coils is followed by remarks on magnetic (ferritic) core probes and a short section on probe shielding.

Calculation of Probe Resistance

The impedance Z of any coil consists of a real part R and imaginary part $j\omega L$:

$$(3) \quad Z = R + j\omega L$$

where j is $\sqrt{(-1)}$, L is inductance (henry) and ω is angular frequency ($\omega = 2\pi f$, where f is frequency in hertz). The real component of impedance Z is the direct current resistance R (ohm). R is calculated from Ohm's law:

$$(4) \quad R = \frac{\rho \ell}{a}$$

where a is the cross sectional area of the wire (square meter), ℓ is the total length (meter) of wire and ρ is the conductor resistivity (ohm meter).

Because the diameter of the coil is important in probe design, the equation may be written in terms of the probe mean diameter:

$$(5) \quad R = \frac{\pi \rho d N}{a}$$

where d is mean coil diameter (meter), N is the number of turns in the coil and πdN is the total wire length (meter). In the more general case of conducting materials in the vicinity of the coil, the real part of the impedance also includes the effects of eddy current losses in the conducting bodies.

The simple calculation of the real part of the probe impedance is applicable only to air core coils at low frequencies. It cannot be used if the cores are magnetic or conducting or if the impedance of an air coil in the vicinity of conducting or magnetic bodies is required. The effective resistance increases due to losses in the core and winding. These losses can be estimated[11] but no general, simple expressions exist for their calculation. In particular, losses in ferrites are difficult to estimate except for particular shapes (such as cup cores) for which empirically derived estimates of the losses may be available.

Calculation of Probe Reactance

In its simplest form, coil reactance can be calculated by assuming (1) that only the inductive reactance (ohm) in Eq. 3 exists and (2) that the mutual inductance is negligible. Then, the coil inductance in air may be calculated using well known formulas.

As a start, consider the inductance L (henry) of a long, circular current sheet:[11,12]

$$(6) \quad L = \frac{4\pi a}{\ell} \times 10^{-7}$$

where a is the cross sectional area of the coil (square meter) and ℓ is the length (meter) of the current sheet. (A current sheet is a conductive surface, such as a flat, energized sheet of copper foil, in which the current density measured in ampere per square meter is uniform at every point.)

This expression's usefulness for design purposes is limited because actual coils are made of individual wires of round cross section and are usually relatively short. Thus, the assumptions of uniform current distribution and long coil in Eq. 6 are seldom met. The equation is useful, however, insofar as the expressions found in the literature for a variety of coils are written as corrections to this simple expression.

If the length ℓ of the solenoid is not large compared with the coil mean radius r (that is, if $r \cdot \ell^{-1}$ is not small compared to unity) the nagaoka end correction must be used in Eq. 6 and inductance L for a short solenoidal current sheet becomes:

$$(7) \quad L = \frac{4\pi a}{\ell} K \times 10^{-7}$$

The K value may be found from the nagaoka formula[12] or from tables[11] where K is customarily expressed in terms of the $r \cdot \ell^{-1}$ value. The value for K tends toward unity as $r \cdot \ell^{-1}$ approaches zero. A further correction is necessary to account for the differences between a current sheet and an equivalent solenoid made of round, insulated wires:[13]

$$(8) \quad L = \frac{4\pi^2 r^2 N^2}{\ell} K \left[1 - \frac{\ell (A + B)}{\pi r N K} \right]$$

where $A = 2.3 \log_{10} 1.73\, d \cdot p^{-1}$; $B = 0.336 \times (1 - 2.5 \cdot N^{-1} + 3.8 \cdot N^{-2})$; K is the nagaoka constant; p is the winding pitch; and r is the coil mean radius (meter). The factor A depends on wire diameter d and pitch p; B depends on the number of turns N. These factors may be found in tables elsewhere.[12]

Eddy current probes are usually made of short, multilayered coils of rectangular cross section. For rectangular cross section coils with any desired proportion between length and thickness of the windings (b and c in Fig. 2), two expressions exist: one for relatively long coils ($b > c$) and one for short, flat coils ($b < c$).[11] In the intermediate range ($b \cong c$), both formulas are useful. The formula for the first case is:

$$(9)\quad L = 0.019\,739\,\frac{2r}{b}\,N^2 r\left(K - k\right)$$

The K value is again the nagaoka constant for a solenoid of length b and may be found tabulated as a function of $2r \cdot b^{-1}$ or $b \cdot (2r)^{-1}$.[11] The quantity k takes into account the decrease in inductance caused by the separation of turns in the radial direction. Again, k may be found tabulated as a function of $c \cdot (2a)^{-1}$, $b \cdot c^{-1}$ or $c \cdot b^{-1}$.[11]

A similar expression exists for pancake coils:[11]

$$(10)\quad L = 0.001\left(N^2 rPF\right)$$

The term P is a function of $c \cdot (2a)^{-1}$ whereas F takes into account the inductance reduction caused by turn separation in the radial direction. It is a function of $c \cdot (2r)^{-1}$ and either $b \cdot c^{-1}$ or $c \cdot b^{-1}$.[11]

These formulas are very accurate but require values to be interpolated from tables or graphs. In many cases, it is more convenient and almost as accurate to use simplified, approximate formulas. Two of the more popular formulas are summarized elsewhere.[12,14]

FIGURE 2. Coil cross section and dimensions used for analytical calculations.

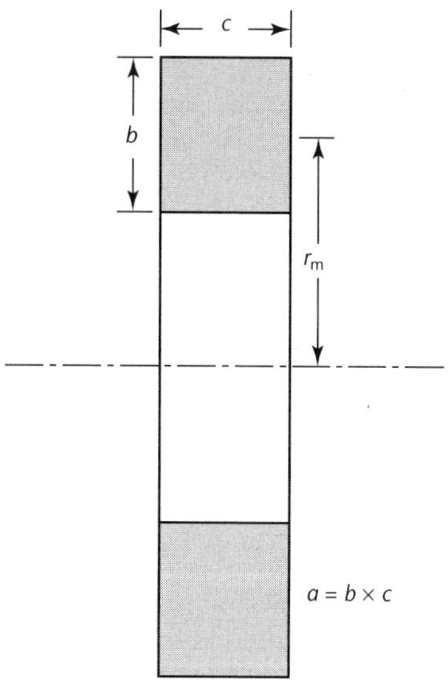

Legend

- a = cross sectional area
- b = radial thickness, or length
- c = axial thickness
- r_m = coil mean radius

For a long, thin coil ($b \cdot c^{-1} > 10$), the inductance may be approximated by Eq. 11:

$$(11)\quad L \cong \frac{r^2 N^2}{9r + 10b} - \frac{crN^2}{10\,\pi b}$$

For short coils, where both b and c are smaller than the radius of the coil, a useful formula is:

$$(12)\quad L \cong \frac{rN^2}{13.5}\,\log_{10}\frac{4}{\sqrt{b^2 + c^2}}$$

Multiple Coil Probes

In the case of multiple coils, the calculation of inductance is somewhat more complicated because the mutual inductance of the coils may have to be taken into account. This is not always the case. For example, in the case of coils spaced relatively far apart, the mutual inductance may be very small. No general solution to this problem exists but, for the most common arrangement of two coils close together, the mutual inductance may be found:[11]

$$(13)\quad M_{1,2} = \left(N_1 N_2\right) f_{tab} \sqrt{r_1 r_2}$$

where f_{tab} is a tabulated value that depends on the dimensions of the two coils, N_1 and N_2 are the number of turns in the coils and r_1 and r_2 are the mean radii (meter) of the two coils.

The total inductance of the two coils is altered by the value of the mutual inductance depending on the way the coils are connected. If the configuration is series aiding, the mutual inductance is added. If their configuration is series opposing, the mutual inductance is subtracted:

$$(14)\quad L = L_1 + L_2 \pm 2M_{1,2}$$

where L_1 and L_2 are the self-inductances of the two coils and $M_{1,2}$ is their mutual inductance.

The same relation may be applied for multiple-coil arrangements by calculating the mutual inductance of each pair of coils as separate values and then summing them.

Analytical Design of Complex Probes

The probes and coils described above represent the simplest of configurations. They consist of air core coils and do not consider the effects of conducting bodies nearby.

The introduction of conducting and magnetic cores within the probe creates two major difficulties as far as the design is concerned: (1) the need to calculate losses within the core and any conducting bodies within the probe's field and (2) the frequency dependence of the probe response.

The second of these difficulties is relatively easy to handle by calculating the coil parameters at a fixed frequency (or at a few frequencies) within the expected range of application.

The calculation of losses within the conducting bodies is far more complicated but is absolutely necessary to estimate the probe impedance. Similarly, calculation of the probe reactance becomes complicated unless the magnetic path is very simple. Analytical tools for designing such probes are discussed elsewhere.

Numerical Design

Numerical tools to aid in the search for better designs and the tighter requirements imposed on systems offer some unique opportunities. The chapter on modeling describes some of the more common numerical approaches that can be used to design eddy current coil probes.

Numerical techniques offer certain advantages over other design techniques. The probe response is calculated from the true physical description of material interactions with the electromagnetic field. The inclusion of discontinuities, material properties and coil parameters are therefore an integral part of the model. Very few assumptions are made. In many cases, however, some assumptions may be useful in reducing the effort and cost involved in the application of the model. Linearity, two dimensionality and axisymmetric formulations are examples of such assumptions.

In addition, the probe response is a complete simulation of the test performed. This is extremely important because it reveals the probe characteristics in a way very similar to the real test and leads to a better design.

Unconventional probe shapes can be modeled whereas analytical techniques can only handle a limited number of coil shapes, such as round coils with rectangular cross sections.

Finally, probes can be optimized to detect specific types of discontinuities. Because numerical techniques can model complex discontinuities (subsurface discontinuities), the probe response can be optimized for any type of discontinuity and test condition.

To illustrate some of the foregoing arguments, the design of a reflection probe specifically intended for measurement of crack depth, coating thickness and corrosion effects is presented below, using a finite element eddy current model. Two types of probes are considered: (1) a simple absolute, surface probe and (2) a reflection probe consisting of a driving coil and a pickup coil. This latter probe will be called a *double-coil probe* to distinguish it from the single-coil probe.

In many applications, single (absolute) coils are used and the coil impedance is monitored by using an alternating current bridge circuit. Homogeneity information is extracted from changes in the coil impedance. Double-coil probes consist of a driving coil and a smaller, concentric pickup coil. In this case, the induced voltage in the pickup coil is measured. The following discussion describes the application of the finite element technique to the analysis and design of both single-coil and double-coil surface probes. Double-coil probes are shown to be superior for the applications mentioned above because of better linearity of the liftoff curve and wider useful range. The rate of change in the induced voltage for a double-coil probe is shown to be larger than the rate of change in the impedance for single-coil probes with given parameter changes. Moreover, in the case of corrosion depth measurements, the noise generated by liftoff variations can be minimized with an appropriately designed double-coil probe. Some of the results presented have been verified experimentally.

Single-Coil versus Double-Coil Probes

The two probes considered here are shown in Fig. 3 as they relate to liftoff measurements. Figure 3a represents a small diameter absolute coil over a conducting surface. Because of the localized nature of the probe fields, the range of liftoff measurement is limited. A larger diameter coil such as the coil in Fig. 3b can extend this range but more sensitivity is obtained by introducing a small diameter pickup at the center of the larger coil (as in Fig. 3c). Because the flux lines through the pickup coil are essentially perpendicular to the conducting surface, the double-coil probe typically offers better linearity and a wider dynamic range. These effects can be seen in Fig. 4, where the effect of a flat bottom hole is shown. In Fig. 4a, there is little disturbance of the coil field from the localized field pattern. In Fig. 4b, the large diameter of the coil tends to mask the hole signal because there is little change caused by the hole. A double coil has the advantage of both of these characteristics

FIGURE 3. Surface probe over conducting surface: (a) small diameter single-coil probe; (b) large diameter single-coil probe; (c) double-coil probe.

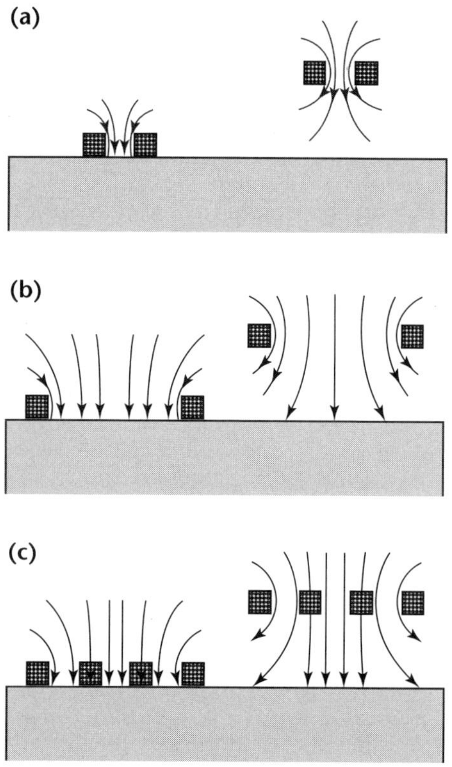

FIGURE 4. Surface probes in presence of flat bottom holes: (a) small diameter single-coil probe; (b) large diameter single-coil probe; (c) double-coil probe.

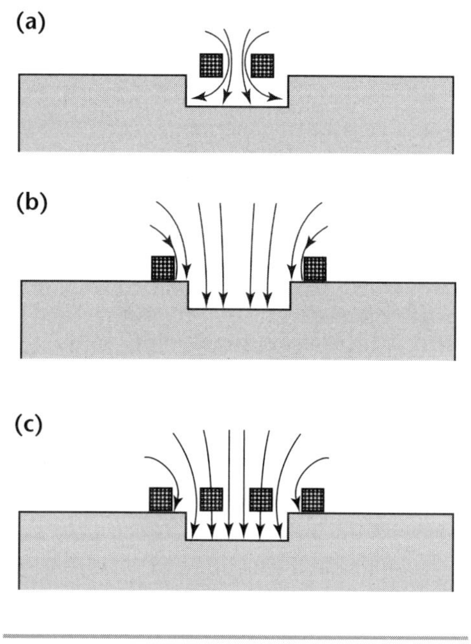

as shown in Fig. 4c. Because the bottom of the hole affects the flux passing through the pickup coil, it should be possible to measure a wide range of hole depths.

Based on these considerations, three different absolute coils with 4, 10 and 30 mm (0.16, 0.4 and 1.2 in.) diameters plus a double coil consisting of an exciting coil 30 mm (1.2 in.) in diameter and a concentric pickup coil 4 mm (0.16 in.) in diameter were evaluated. In all cases, the coil thickness was 3.9 mm (0.15 in.). The performance of these probes was calculated for magnetic and nonmagnetic materials. The magnetic material was carbon steel with conductivity equal to 5 $MS{\cdot}m^{-1}$ and relative permeability of 50. The nonmagnetic material was a solid solution, strengthened, nickel chromium alloy (Unified Numbering System N06600) with a conductivity of 1.1 $MS{\cdot}m^{-1}$. The magnetic material was tested at 1 kHz and the nonmagnetic material at 10 kHz. For each of the materials, the liftoff curves resulting from depth changes in flat bottom holes were obtained and compared. The excitation levels were assumed to be small, so nonlinearities and hysteresis effects could be neglected.

Liftoff Parameter

Figures 5 and 6 summarize the results for various probes with respect to liftoff. The impedance change rate is plotted for the absolute probes and the rate of change of induced voltage is plotted for the double-coil probe in Fig. 5; Z_0 and V_0 represent the probe impedance and induced voltage at zero liftoff. Figure 5a shows the rates of change of impedance and induced voltage caused by liftoff in the nonmagnetic material. The curve for the double-coil probe indicates that it provides a greater output voltage over a greater liftoff range than any of the absolute probes. Figure 5b shows similar calculations for carbon steel test objects. The change in the induced voltage in the double probe also occurs over a wider range of liftoff values than for the absolute probes.

To clarify the differences in range and linearity of liftoff calculations between the various probes, the normalized curves shown in Fig. 6 are useful. The change in impedance defined as $\Delta Z = Z - Z_0$ is normalized with respect to Z_{max}, defined as $Z_{max} = Z_\infty - Z_0$, where Z_∞ is the coil impedance at infinite liftoff. Similarly, the change in induced voltage is normalized with respect to ΔV_{max}. The normalized rates of change are defined as $\Delta Z{\cdot}\Delta Z_{max}^{-1}$ and $\Delta V{\cdot}\Delta V_{max}^{-1}$. The double coil exhibits a more useful range for liftoff measurement and better linearity than the absolute probe exhibits. Range and linearity are shown for nonmagnetic material in Fig. 6a and for magnetic material in Fig. 6b.

Hole Depth Parameter

To simulate a test for evaluating the depth of discontinuities in metal, flat bottom holes of various diameters were used with the probes described above. The probe was located over the hole, flush with the metal surface and centered with the hole. The hole depth was then changed and the signal from the probe was plotted. Figure 7 shows the rate of change of impedance or induced voltage corresponding to the change in depth of a 10 mm (0.4 in.) diameter hole. Similarly, Fig. 8 represents the normalized rates of change for the same test. Again, the sensitivity, useful range and linearity are better in the case of the double-coil probe.

Noise Reduction Effects

When eddy current probes are used for applications such as crack and corrosion detection, the noise caused by liftoff

FIGURE 5. Finite element prediction of change in impedance Z and induced voltage V due to liftoff changes: (a) nonmagnetic material at 10 kHz; (b) magnetic material at 1 kHz.

(a)

(b)

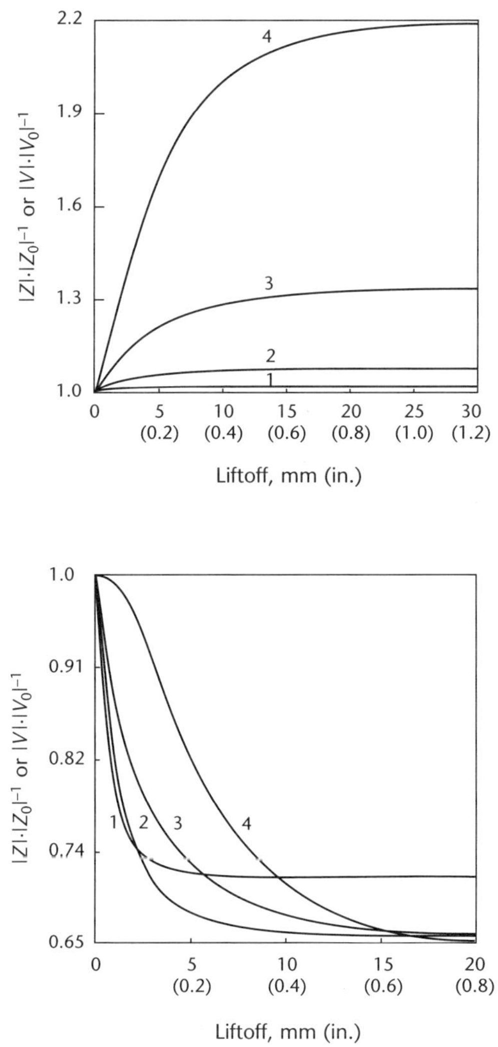

Legend

1. 4 mm (0.16 in.) coil.
2. 10 mm (0.4 in.) coil.
3. 30 mm (1.2 in.) coil.
4. Double coil.

FIGURE 6. Normalized rates of change of impedance Z and induced voltage V due to liftoff changes: (a) nonmagnetic material; (b) magnetic material.

(a)

(b)

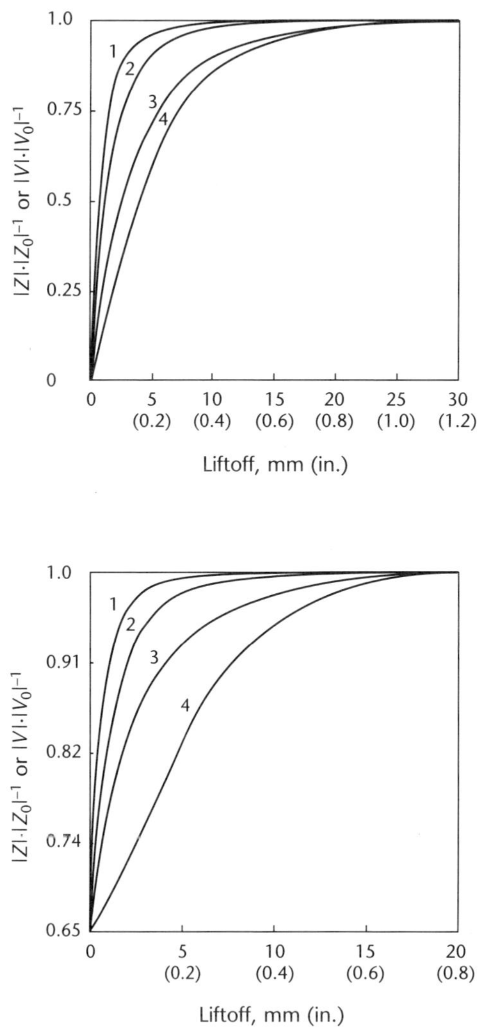

Legend

1. 4 mm (0.16 in.) coil.
2. 10 mm (0.4 in.) coil.
3. 30 mm (1.2 in.) coil.
4. Double coil.

variations needs to be minimized or, if possible, suppressed. The most common technique of liftoff suppression is the phase discrimination technique, which uses the phase difference between the signal and noise. This technique works best when the phase difference is close to 90 degrees but is of little use if phase differences between the signal and the liftoff signal are small. Figure 9a shows normalized impedance of a 4 mm (0.16 in.) diameter, single-coil probe in the case of a magnetic material. The phase between the hole depth impedance change is very small, so noise suppression is difficult. Figure 9b shows the normalized induced potential (real versus imaginary components normalized with respect to magnitude) of a double-coil probe for the same test as in Fig. 9a. The phase difference is significantly larger, so noise can be suppressed.

Experimental Verification of Numerical Model Probe Design

Experimental verification entails actually building the designed probe. A 30 mm (1.2 in.) diameter coil and a 4 mm

FIGURE 7. Predicted change rates of impedance Z and induced voltage V with hole depth changes: (a) nonmagnetic material; (b) magnetic material. Hole diameter is 10 mm (0.4 in.).

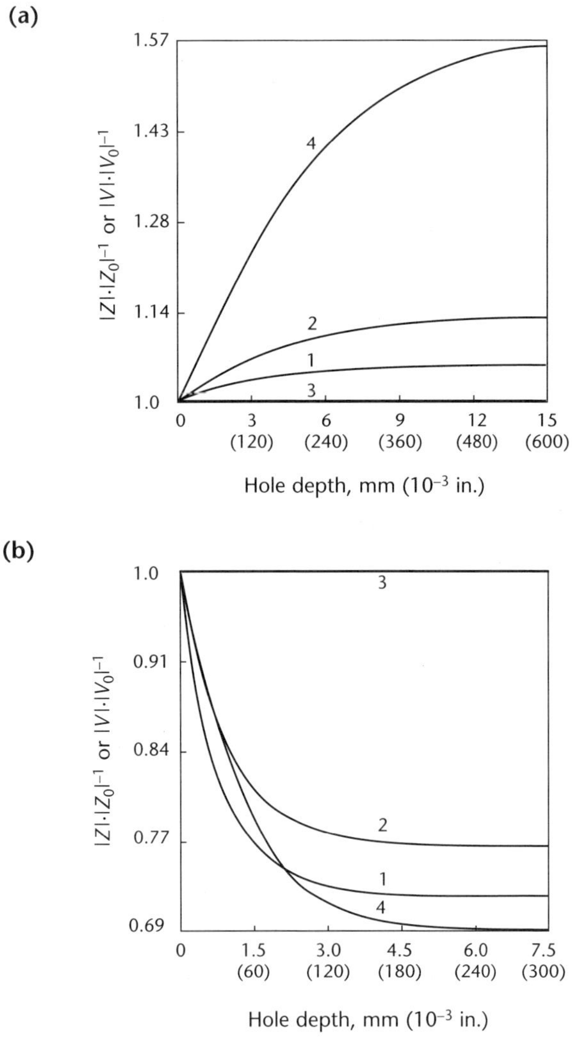

Legend

1. 4 mm (0.16 in.) coil.
2. 10 mm (0.4 in.) coil.
3. 30 mm (1.2 in.) coil.
4. Double coil.

FIGURE 8. Normalized change rates of impedance Z and induced voltage V with hole depth changes: (a) nonmagnetic material; (b) magnetic material. Hole diameter is 10 mm (0.4 in.).

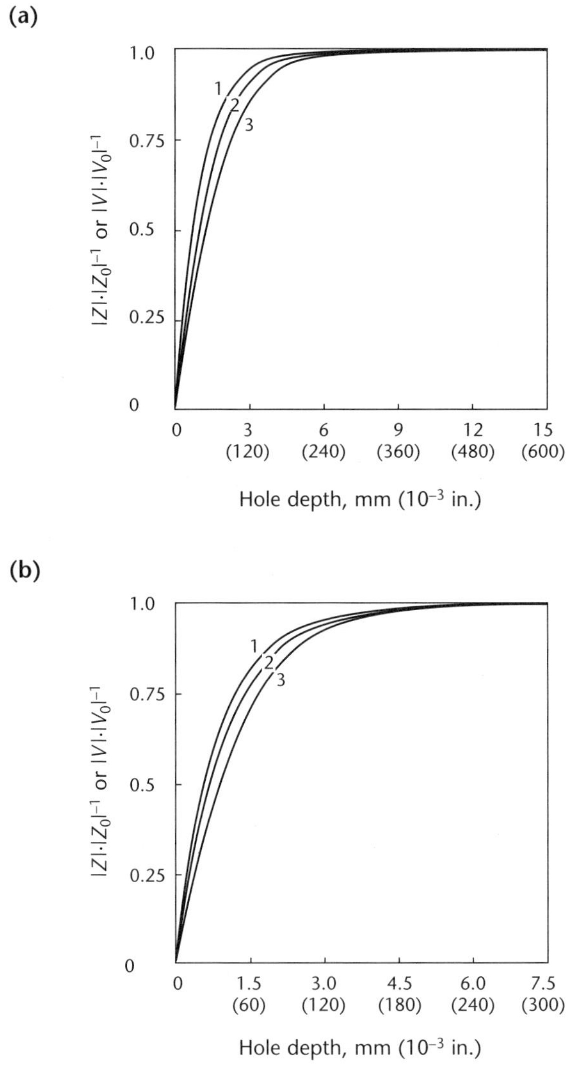

Legend

1. 4 mm (0.16 in.) coil.
2. 10 mm (0.4 in.) coil.
3. Double coil.

(0.16 in.) diameter coil were constructed to form a double-coil probe but each coil could also be used separately as an absolute probe. Figure 10 shows experimental results giving the normalized rates of change for the impedance of single-coil probes and the induced voltage of the double-coil probe. In this figure, V_0 is induced potential (volt) at zero liftoff and V_a is induced potential (volt) when the coils are in air, or far from the material. This shows, as for the finite element results in Fig. 6, that the double-coil probe has a potentially larger useful measurement range.

The results presented indicate the versatility of the finite element model for probe design. In this particular case, they also show a superiority of the double-coil absolute probe over single-coil absolute probes.[15]

The numerical approach to probe design should be viewed as a test bed or a numerical experiment tool. The results presented demonstrate only a fraction of what is possible with a numerical model. Full optimization can be performed, in which parameters such as probe thickness, relations between the two coil sizes, offset between the two coils, frequency response and response to various materials and discontinuities can be studied.

FIGURE 9. Finite element prediction of normalized impedance and induced voltage with liftoff and hole depth changes: (a) 4 mm (0.16 in.) single-coil absolute probe at 1 kHz on magnetic material; (b) double-coil probe at 1 kHz on same material.

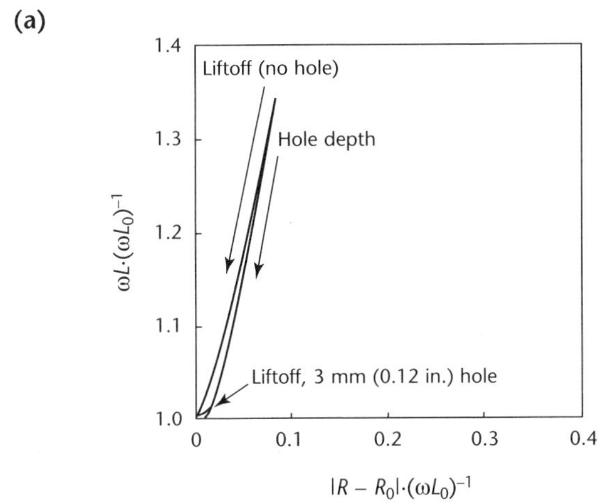

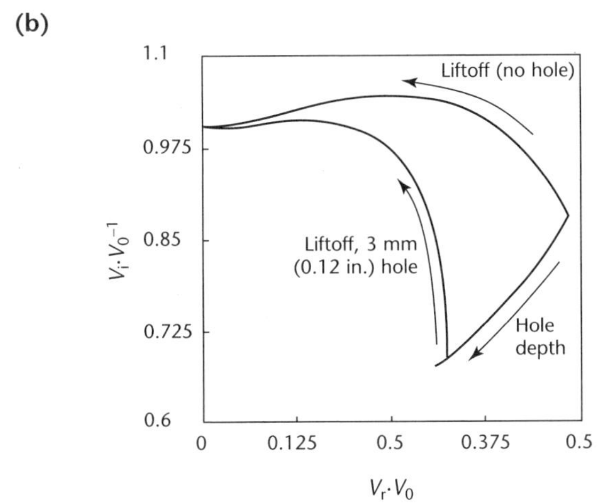

Legend

L = inductance (henry)
R = resistance (ohm)
V_0 = induced potential (volt) at zero liftoff
V_i = imaginary component of induced potential (volt)
V_r = real component of induced potential (volt)
ω = radial frequency (hertz)

Ferrite Core Probes

Ferrite cores can be used to improve the spatial resolution of the probe, to minimize interference caused by other physical structures near the probe, to reduce the overall size of the probe, to provide shielding and to address impedance matching problems. Ferrite cores can also be used to reduce the coil's *footprint* — that is, the normal or tangential component of the fields as they extend in the direction parallel to the surface of the coil.

The simplest way of improving the magnetic path is to wind the coil on (or inside) a core made of high permeability material such as one of the various

FIGURE 10. Experimental impedance and induced voltage versus liftoff.

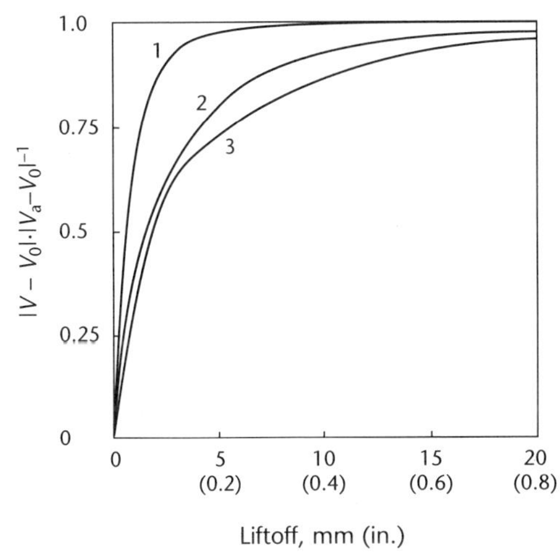

Legend

1. 4 mm (0.16 in.) coil.
2. 10 mm (0.4 in.) coil.
3. Double coil.

ferrites. A coil wound on a magnetic core has a reduced magnetic reluctance while a sleeve outside the coil also acts as a shield and reduces the leakage field from the tested area. Figure 11 shows the field distribution for three types of probes. Figure 11a is a simple single-coil absolute probe in air at 100 kHz. Figure 11b shows the same coil at the same frequency but the coil is wound on a cylindrical ferrite core 8 mm (0.3 in.) in length and 20 mm (0.8 in.) in diameter. Figure 11c again shows the same coil wound inside one half of a cup core. In all three field plots, corresponding lines (starting from the outer line) represent identical flux density values.

Cup Core Probes

The cup core probe has much higher flux densities, as shown by the number and density of the flux lines. The cylindrical core has a field distribution that is only slightly different from that of the air coil because the portion of the magnetic path occupied by the core is small compared to the total magnetic path. In contrast, about half of the main magnetic path is occupied by the core itself in the case of the cup core probe. More significant differences can be seen by testing the impedances of the three probes shown in Fig. 11. The first noticeable difference is in the appearance of the real part in the impedance of the core probes, a change caused by losses in the cores themselves because ferrites have low conductivity. The air core coil has no losses except from the direct current resistance of the coil, which has been removed from consideration here. The actual losses in the windings caused by eddy currents have been neglected as insignificant.

The shielding effect mentioned above is noticeable in the cup core coil because the flux distribution in the back of the core is quite different from that in the front and the line spacings are larger (lower gradients in the field). No such shielding effect can be observed for the ferrite core coil.

To further evaluate the merit of cup core coils, it is useful to examine Fig. 12. Figures 12a to 12c represent the same three probes of Fig. 11 above a nonmagnetic, conducting material (Unified Numbering System N06600) at a liftoff of 4 mm (0.16 in).

Figures 12d to 12f are identical to the previous three except that the test object is ferromagnetic (carbon steel). The differences between the probes become more accentuated with ferromagnetic materials. In particular, the field around the cup core probe has changed drastically and no leakage field is noticeable. The field lines have the same meaning as in Fig. 11 (lines starting from the outer edges represent the same flux values but they are crowded together to create a relatively high field in the vicinity of the probe). In the case of the ferrite core probe and the air core coil, the projection of the field in space has been reduced by the presence of the test material but a significant portion of this field can still be seen. The probe impedances have changed considerably because of the presence of the conducting and magnetic media. Also noticeable is the skin effect: the same flux density at a given depth in the nonmagnetic material is much higher than in the magnetic material.

Figure 13 repeats Fig. 12 but with no liftoff. A 1 mm (0.04 in.) liftoff is actually used for the air core coil so that in all three probes the coil is at the same location relative to the test material. The phenomena mentioned above are present but they are further accentuated by the absence of liftoff.

Figure 11. Flux distribution plots for three different probes using same coil at 100 kHz: (a) air core coil; (b) coil wound on 20 mm (0.8 in.) diameter cylindrical ferrite core, 8 mm (0.3 in.) long; (c) coil inside one half of cup core made of same material.

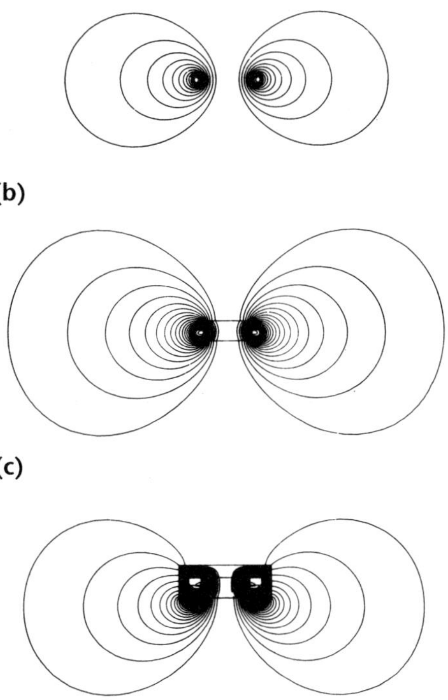

(a)

(b)

(c)

Field Projection Data

The field projection in space is shown in Figs. 14 and 15. Figure 14a is the

FIGURE 12. Flux distribution for three coils at liftoff of 4 mm (0.16 in.) over thick conducting materials: (a) air coil over nonmagnetic material (Unified Numbering System N06600); (b) cylindrical ferrite core probe over same nonmagnetic material; (c) cup core over same nonmagnetic material; (d) air core coil over magnetic material (carbon steel); (e) cylindrical ferrite core probe over same magnetic material; (f) cup core probe over same magnetic material.

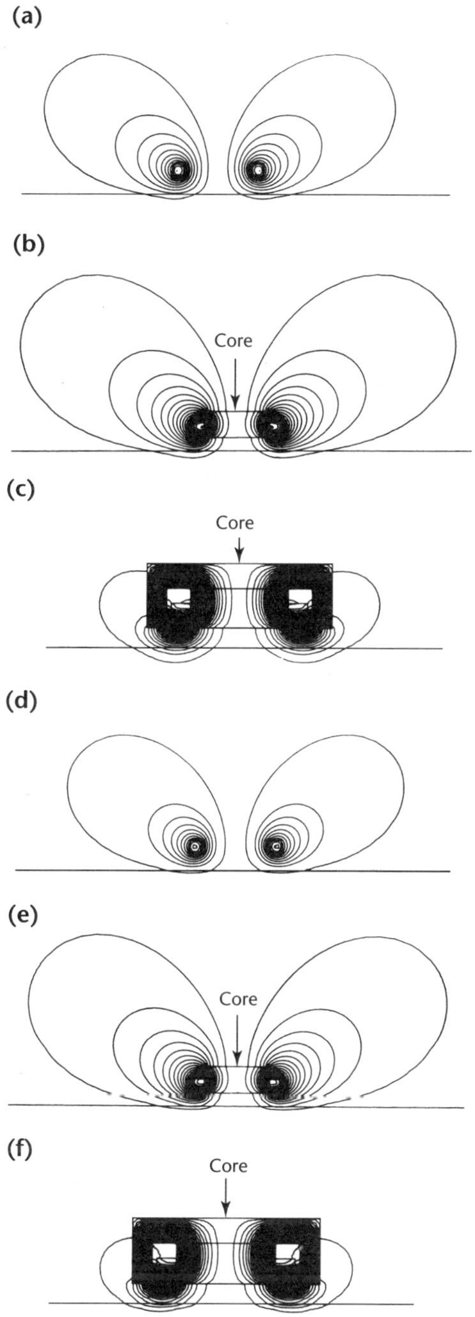

FIGURE 13. Flux distribution for three coils at liftoff of 1 mm (0.04 in.) over thick conducting material: (a) air core coil over nonmagnetic material (Unified Numbering System N06600); (b) cylindrical ferrite core probe over same nonmagnetic material; (c) cup core over same nonmagnetic material; (d) air core coil over magnetic material (carbon steel); (e) cylindrical ferrite core probe over same magnetic material; (f) cup core probe over same magnetic material.

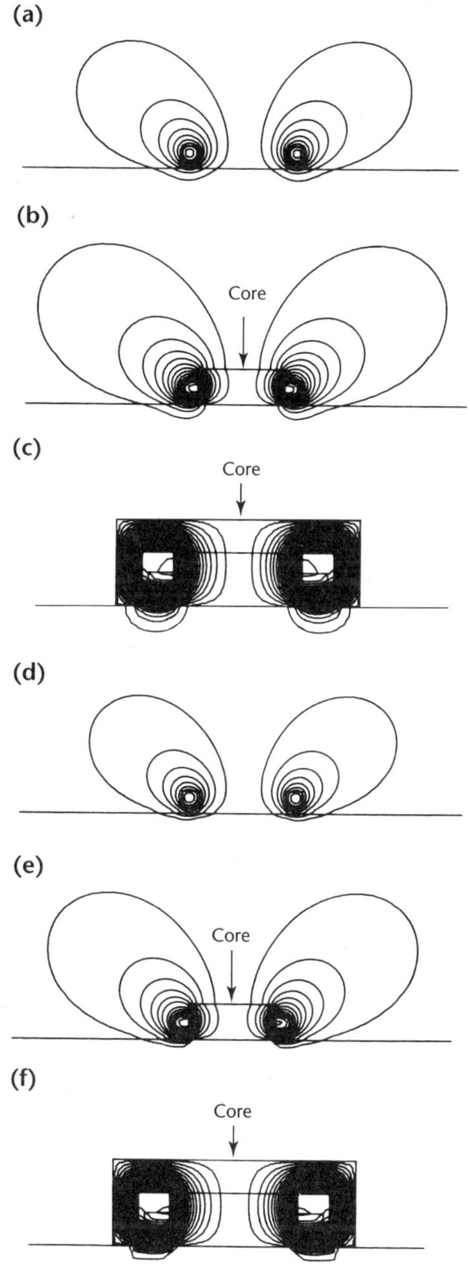

normalized field component for the air core coil, Fig. 14b is the normalized field component for the cylindrical ferrite core probe and Fig. 14c is the normalized component of the field for the cup core probe. For each probe, the field is plotted at seven different locations starting

FIGURE 14. Normal component of magnetic field at seven axial locations below probe surface 0.5 mm to 125 mm (0.02 to 5.0 in.) axial distance: (a) air core coil probe; (b) cylindrical ferrite core probe; (c) cup core probe.

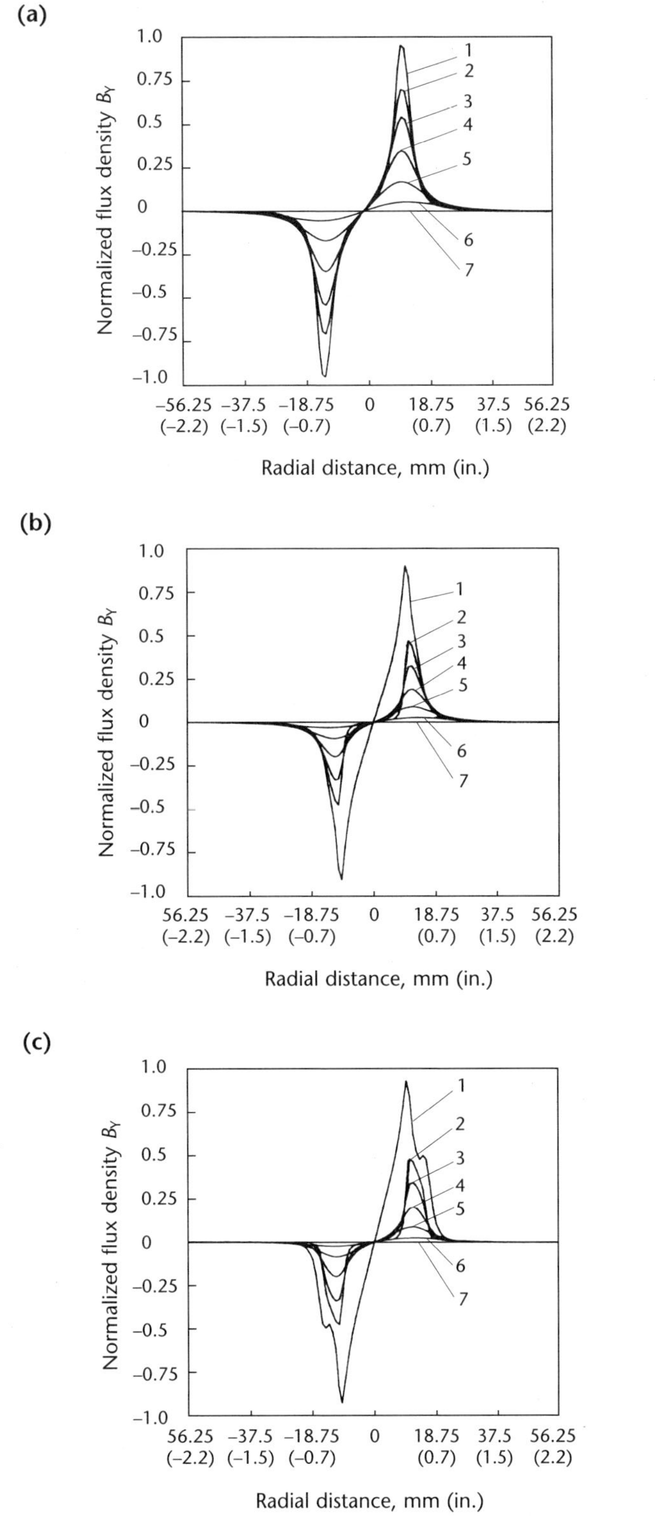

Legend
1. 0.5 mm (0.02 in.).
2. 2 mm (0.08 in.).
3. 5 mm (0.20 in.).
4. 10 mm (0.40 in.).
5. 20 mm (0.80 in.).
6. 50 mm (2.0 in.).
7. 125 mm (5.0 in.).

FIGURE 15. Tangential component of magnetic field at seven axial locations below probe surface: (a) air core coil probe; (b) cylindrical ferrite core probe; (c) cup core probe.

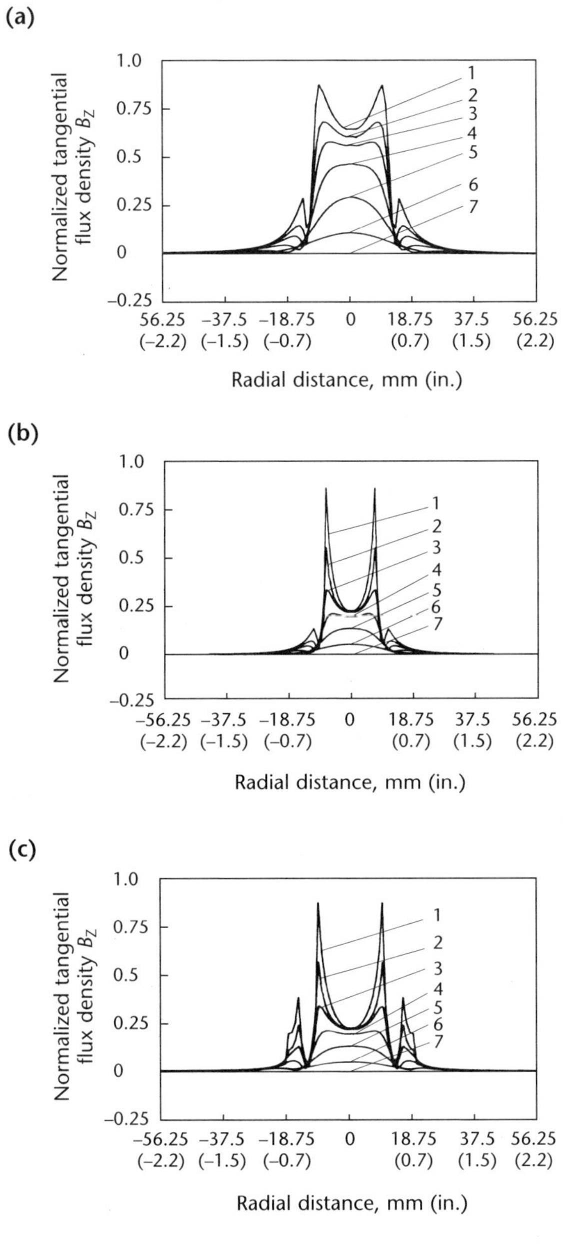

Legend
1. 0.5 mm (0.02 in.).
2. 2 mm (0.08 in.).
3. 5 mm (0.20 in.).
4. 10 mm (0.40 in.).
5. 20 mm (0.80 in.).
6. 50 mm (2.0 in.).
7. 125 mm (5.0 in.).

0.5 mm (0.02 in.) from the lower surface of the ferrite cores for the largest amplitude curve and ending at 125 mm (5.0 in.) away from the coil with the zero curve.

The tangential component (parallel to the coil plane) is shown in Fig. 15. This plot is similar to the plots in Fig. 14 but the amplitudes and the general shape of the curves are different. The curves in Fig. 15 are plotted at the same locations as in Fig. 14. Although the field obtained with the cup core is higher, it is useful to compare these three probes directly.

Figure 16 shows the normalized component of the field at 0.5 mm (0.02 in.) beneath the core for the three probes (these are the three largest curves in Figs. 14a to 14c, plotted on a single scale). The amplitude of the field obtained with the cup core beneath the coil itself is about five times as large as for the air core probe and twice as large as that of the ferrite core probe. From these data and on examination of Fig. 11, it is evident that the cup core field extends further in space and therefore will have deeper penetration into conducting materials. A similar field distribution exists in Fig. 17 where the tangential component of the field is plotted. The differences in amplitudes are roughly the same as those for the normalized component in Fig. 16.

At the same time, the lateral field extension (footprint) in Fig. 11 does not vary as much. This can be seen in Fig. 16 or 17 to the right or left of the peaks. The field extends only about 35 to 40 mm (1.4 to 1.6 in.) in this direction (about two coil diameters) and even at such short distances is very weak.

FIGURE 16. Normal component of magnetic field at 0.5 mm (0.02 in.) below core surface for three probes 1.5 mm (0.06 in.) below coil.

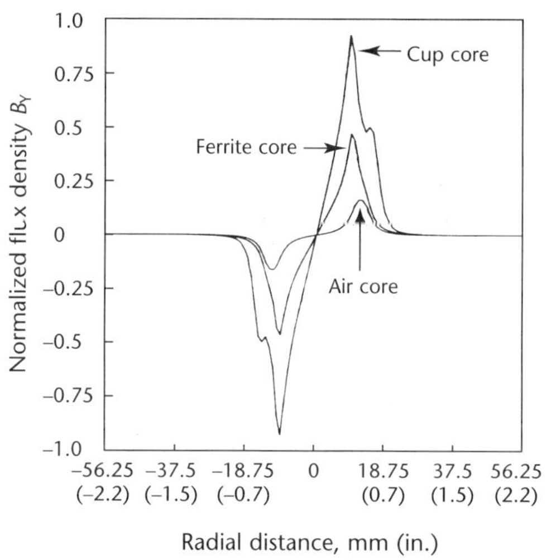

Figure 18 is a plot of the normal component of the field as calculated on a vertical line passing outside the coil at a distance of 0.5 mm (0.02 in.) from the coil's outer diameter. The distortion of the field pattern by the cup core is particularly noticeable in Fig. 18d. The field at the back of the core is low but indicates a very sharp discontinuity at the boundary of the core. The field peaks inside the core and then, as the coil area is encountered, the field reduces rapidly.

A second peak, outside the core and below it, is more significant because it is available for testing. From here on, the field profile is similar to that of the cylindrical ferrite core probe and the air core coil shown in Figs. 18b and 18c respectively. The three profiles are summarized in Fig. 18a, where the relative size and location of the peaks can be seen. Note that the peak under the core for the cup core is shifted down, relative to the other two probes. Although the peak amplitude is smaller than for the cylindrical ferrite core probe because of this downward shift, the field available at the location of the peak is significantly higher than in the ferrite core probe. The extension of the field in the direction where actual testing is performed is only on the order of 20 to 30 mm (0.8 to 1.2 in.) and the field decays very rapidly. Again, the field associated with the cup core probe extends somewhat further but not significantly.

Liftoff Characteristics

The liftoff characteristics of the three probes considered here, as they relate to testing nonmagnetic materials, are shown in Fig. 19. A close inspection of these curves reveals a striking similarity in shape, although the impedance values are quite different. It is also useful to plot all three curves on a single plot as in Fig. 20,

FIGURE 17. Tangential component of magnetic field at 0.5 mm (0.02 in.) below core surface for three probes 1.5 mm (0.06 in.) below coil.

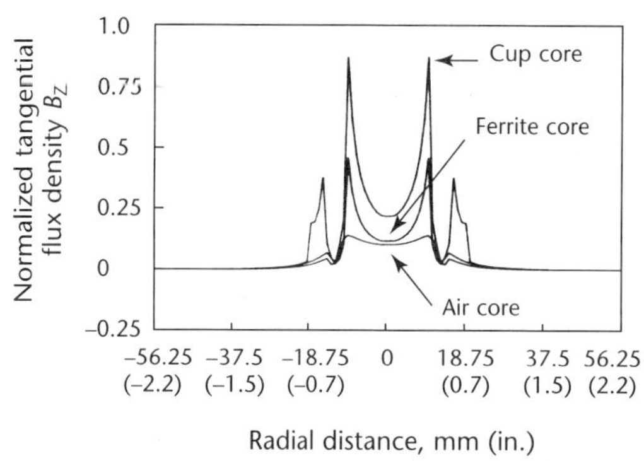

where the impedance values are normalized. Normalization is often used in practice and was done here to make an important point: Fig. 20 indicates (in the absence of additional information) that there is little difference between the three probes. The result above is obtained by normalizing each curve as $(Z - Z_{min})\cdot(Z_{max} - Z_{min})^{-1}$, which produces a curve between zero and one in each case. The danger in normalization is that some information is lost in the process. For example, Fig. 21 plots all three curves without normalization. The differences between the curves are quite large, indicating a greater sensitivity to liftoff variations in the case of the cup core probe. Liftoff sensitivity may be well known but cannot be deduced from the normalized curves in Fig. 20.

A similar set of curves is obtained for magnetic materials. These are shown in Fig. 22 for carbon steel.

Shielded Probes

Shielded eddy current probes may be necessary for some tests to prevent the field generated by the probe from interacting with certain objects near the probe. The primary concern is the interaction with conducting and magnetic bodies that are not part of the test but are near and may produce false indications or mask discontinuity signals nearby. Testing for discontinuities near edges (such as testing fastener holes) is an example. Another effect of shielding is that a larger part of the available flux may be concentrated below the probe. This concentration occurs only when high permeability, low conductivity materials are used for shielding.

Shielding of eddy current probes can be done in three ways: (1) magnetic shielding, (2) active shielding and (3) eddy current shielding.

Magnetic shielding is achieved by creating a low reluctance path for field lines within the required area and away

FIGURE 18. Normal component of the field as calculated on vertical cross sections at 0.5 mm (0.02 in.) beyond the coil's outer diameter 10.5 mm (0.42 in.) radial distance, for three probes: (a) all three fields; (b) air core coil probe field; (c) cylindrical ferrite core probe field; (d) cup core probe field.

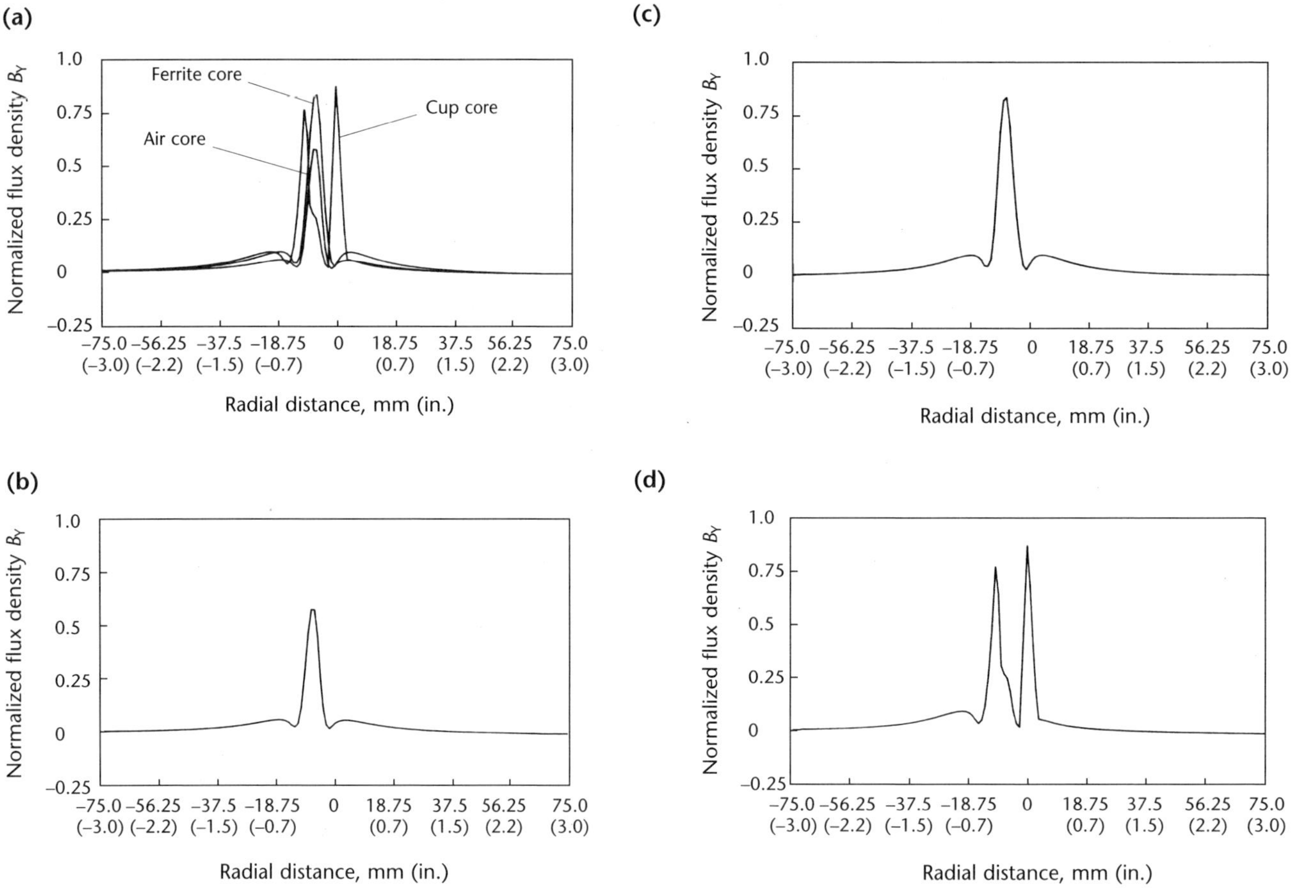

FIGURE 19. Liftoff calculated over nonmagnetic material (Unified Numbering System N06600): (a) simple air core coil; (b) ferrite core probe; (c) cup core probe.

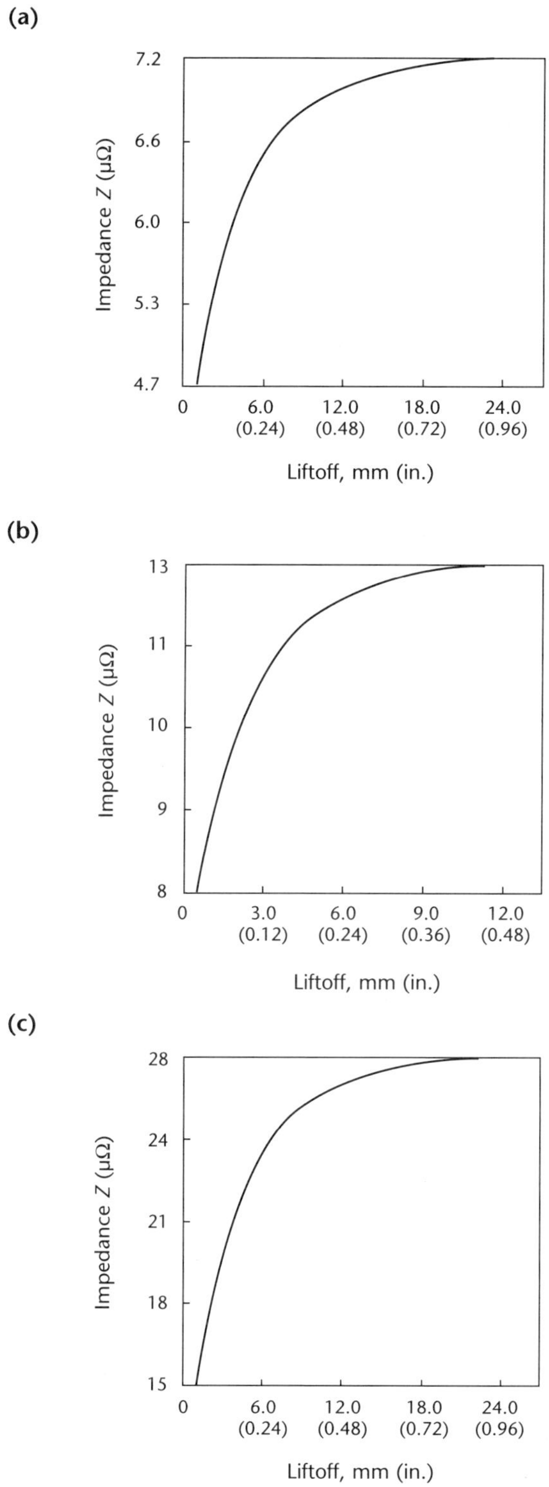

FIGURE 20. Comparison of normalized liftoff curves for three probes over nonmagnetic material.

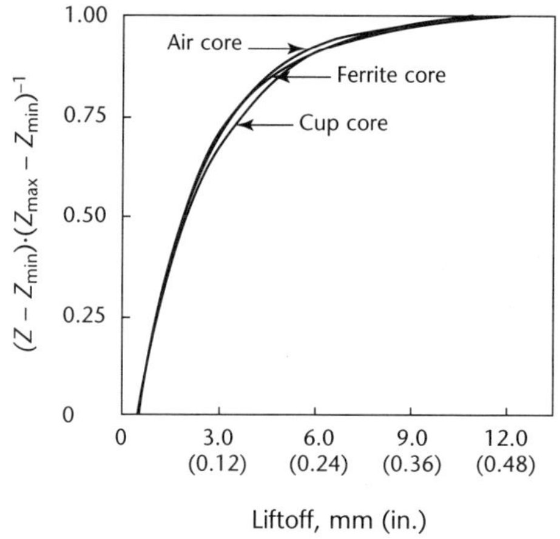

FIGURE 21. Comparison of liftoff curves for three probes over nonmagnetic material.

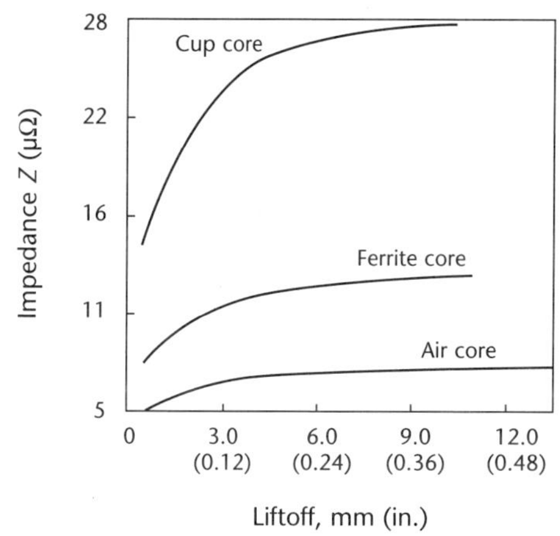

FIGURE 22. Comparison of liftoff curves for three probes over magnetic (carbon steel) material.

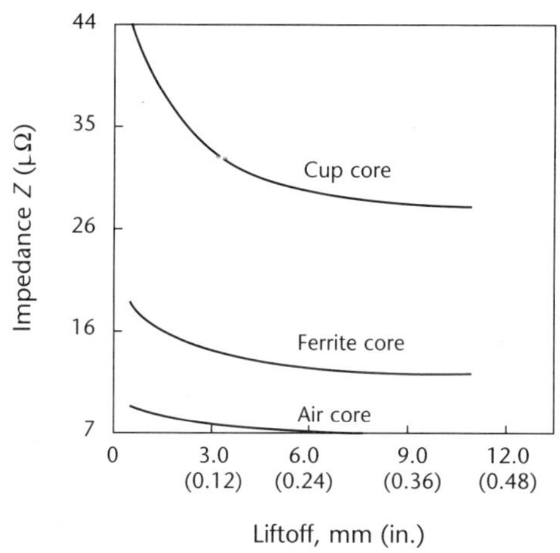

from unwanted regions. Cup core probes, for example, are actually shielded probes. Thus, there is an increased field gradient under the probe and larger sensitivity is obtained. A very simple shielded probe could be built by covering the coil (with or without a ferrite core) with a sleeve of high permeability, low conductivity material such as ferrite.[16]

In active shielding, an active field is generated by means of a coil or system of coils to cancel part of the original field in specific areas. This technique has been used extensively for shielding large structures[17] because magnetic shielding is expensive and bulky for such structures. This technique has not been extensively explored for eddy current tests, perhaps because of the small size of the probes and the ease with which magnetic or eddy current shielding can be achieved.

Eddy current shielding uses the skin effect to prevent the magnetic field from extending to its normal limits. At higher frequencies, all that is needed is to enclose the eddy current probe in a conducting shell (usually copper), leaving open the part of the probe that comes in contact with the test material. This has a drastic effect on the field distribution and on the probe impedance. Because a relatively large portion of the magnetic field energy is absorbed in the shield (depending on the shield material and its proximity to the coil), the field is considerably attenuated. Here lies the main difference between this technique and magnetic shielding. Shielding is obtained by sharply attenuating the field outside a prescribed region rather than by changing the magnetic path.

A simple shielded probe is shown in Fig. 23a. It consists of a coil, 20 mm (0.8 in.) in diameter and 4 × 4 mm (0.16 × 0.16 in.) in cross section, with a copper can over it. The coil is located 1 mm (0.04 in.) from the opening in the shield. The field distribution for this arrangement is shown in Fig. 23b. This, when compared to Fig. 11a (same coil, without shield), immediately reveals the disadvantage of eddy current shielding. The flux lines in Figs. 23b and 11a can be compared directly because, starting with the outermost line, each line represents a flux density value identical to that of the corresponding line in the other figure.

Figures 23c and 23d show the same probe at a liftoff of 4 mm (0.16 in.) over a nonmagnetic slab (Fig. 23c) and over a magnetic slab (Fig. 23d). It should be noted that, because of the weak field, relatively little penetration occurs in either case. Figures 23e and 23f show corresponding results for zero liftoff. Here the penetration is considerably higher

The field distribution for the copper shielded probe described here is summarized in Figs. 24 to 26. Figure 24 describes the normal component of flux density below the probe at seven different locations, 0.5 through 125 mm (0.02 through 5.0 in.). The peaks caused

FIGURE 23. Copper shielded eddy current: (a) cross section; (b) shielded probe in air; (c) shielded probe at liftoff of 4 mm (0.16 in.) over nonmagnetic material; (d) shielded probe at liftoff of 4 mm (0.16 in.) over magnetic material; (e) shielded probe at liftoff of 0 mm over nonmagnetic material; (f) shielded probe at liftoff of 0 mm over magnetic material.

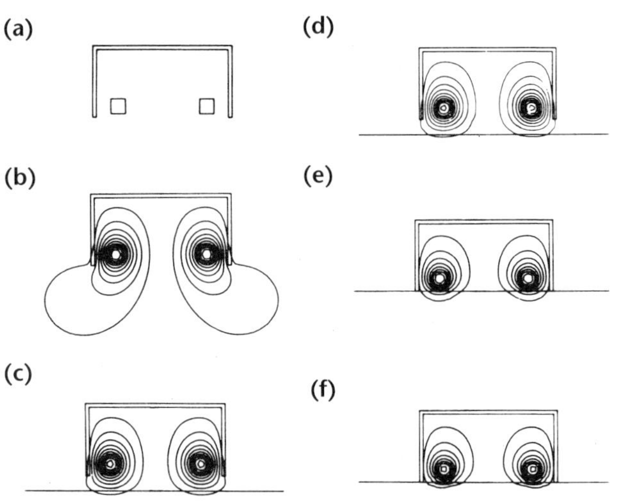

FIGURE 24. Normal component of magnetic field for shielded probe; field distribution shown at seven axial locations 0.5 mm to 125 mm (0.02 to 5.0 in.) below probe's surface.

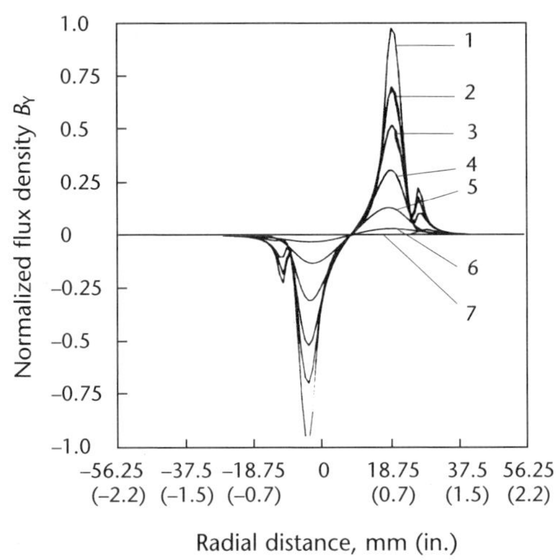

Legend

1. 0.5 mm (0.02 in.).
2. 2 mm (0.08 in.).
3. 5 mm (0.20 in.).
4. 10 mm (0.40 in.).
5. 20 mm (0.80 in.).
6. 50 mm (2.0 in.).
7. 125 mm (5.0 in.).

by the coils are present, together with (1) minima on both sides of the coils caused by the shield and (2) two minor peaks beyond the shield location, indicating that the field extends further in space. This can be seen in Fig. 23b and is a result of the fact that the shielding is only partial. Figure 25 represents the tangential component of the field at the locations mentioned above and is very similar to Fig. 24 in terms of minima and maxima. Figure 26 is the normal component of the field on a vertical line passing through the shield (or outside it) for six locations at radial distances of 8.5 mm (0.33 in.) through 125 mm (5.0 in.) from the center of the coil. The shape of the field inside the shield (three largest curves) is normal but is very low outside the shield (three lowest curves). Also notable is the complete shielding above the shield, as indicated by the abrupt decay to zero on the right side of the curves.

FIGURE 25. Tangential component of the magnetic field for copper shielded probe at same locations as in Fig. 24.

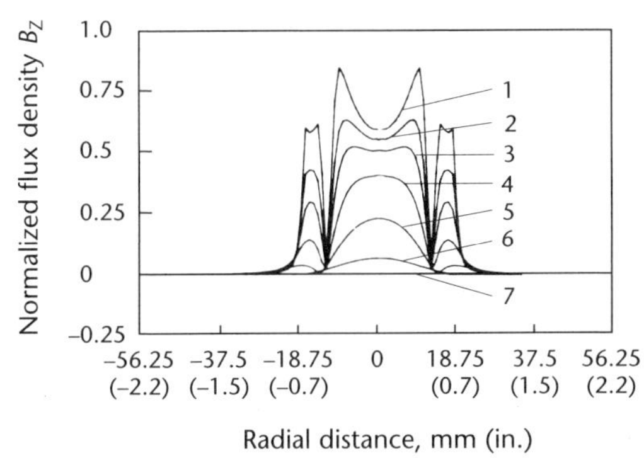

Legend
1. 0.5 mm (0.02 in.).
2. 2 mm (0.08 in.).
3. 5 mm (0.20 in.).
4. 10 mm (0.40 in.).
5. 20 mm (0.80 in.).
6. 50 mm (2.0 in.).
7. 125 mm (5.0 in.).

FIGURE 26. Normal component of magnetic field on vertical cross sections at six radial locations.

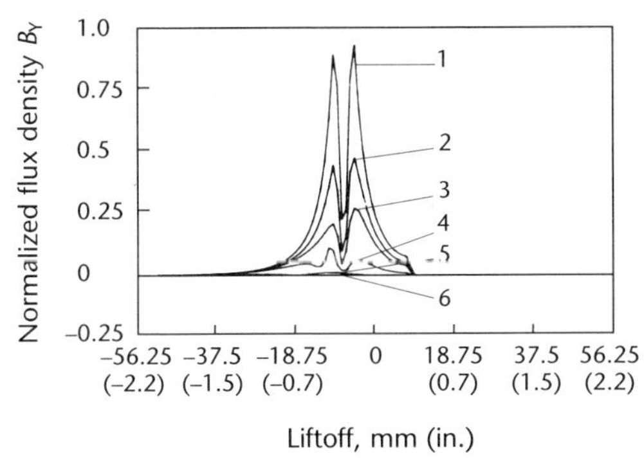

Legend
1. 8.5 mm (0.33 in.).
2. 9.5 mm (0.37 in.).
3. 11.0 mm (0.43 in.).
4. 15.0 mm (0.59 in.).
5. 50.0 mm (1.97 in.).
6. 120.0 mm (4.72 in.).

Liftoff Characteristics

The liftoff characteristics of shielded probes are shown in Fig. 27 and are compared to those of a simple absolute air core probe of identical dimensions in Fig. 28. In both cases, the frequency used is 100 kHz.

The copper shield used here is a simple example of shielding. A similar shield could be used at a larger distance from the coil to reduce losses and reduce the decrease in probe field. As pointed out

FIGURE 27. Liftoff curves for shielded probes: (a) over nonmagnetic material; (b) over magnetic material.

(a)

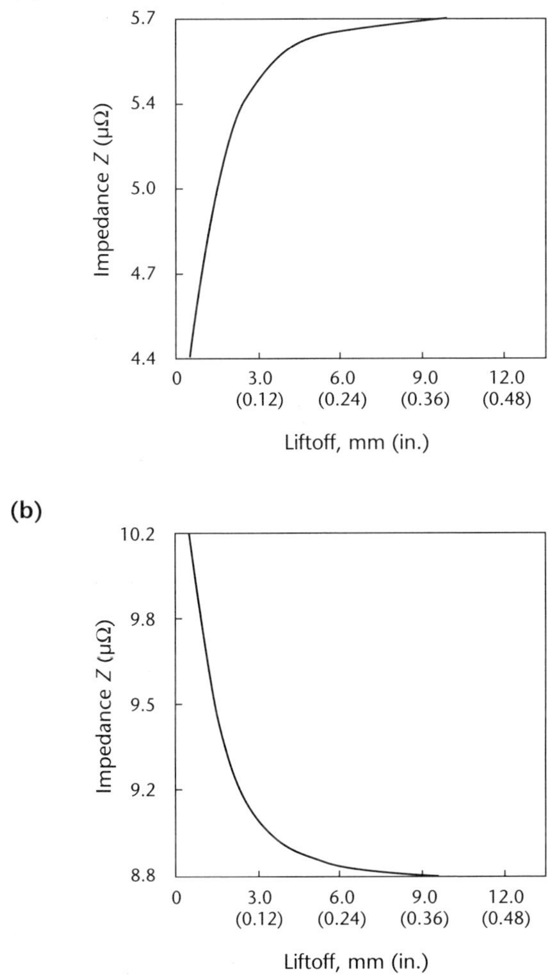

above, the field is greatly diminished by this type of shielding. In some applications, this is acceptable or even desirable (as in the case of testing nonconductive coating thickness). Otherwise, magnetic shielding should be used.

Inside Diameter Probes

In addition to surface probes, inside diameter or feed-through probes are widely used. There are two distinct types of inside diameter probes, absolute and differential.

Absolute probes consisting of a single coil are very common but differential probes usually offer superior noise rejection characteristics and high sensitivity to sudden changes in geometry or material properties.

Unlike surface probes, where the probe axis is frequently perpendicular to the test material, inside diameter probes usually lie parallel to the test object axis and therefore the dominant field distribution is called the *lateral field*.

When inside diameter probes are used to test relatively thin walled tubes made of nonmagnetic materials, it can be expected that through-wall penetration will occur and both tube surfaces can be tested. With thicker walls or with magnetic materials, where the standard depth of penetration is significantly smaller than the wall thickness, the test becomes a test more of the surface than of the tube wall.

Both absolute and differential inside diameter eddy current probes are treated extensively in the literature.[18-22] The treatment here will therefore be limited to basic examples of both as they relate to testing of tubes and circular cavities.

In many inside diameter probe applications, the probe impedance is monitored. For other inside diameter probes, the induced voltage is measured.[5,23] The treatment in this discussion precludes neither of these techniques but uses some simple examples to represent typical aspects of testing with inside diameter probes.

The field distribution of a simple absolute coil in air is presented in Fig. 11a. This coil, when inserted in a nonmagnetic tube (Unified Numbering System N06600 nickel chromium alloy) with a wall thickness of 3 mm (0.12 in.) produces the field distribution shown in Fig. 29a. The field distribution and

FIGURE 28. Comparison of liftoff characteristics of shielded and unshielded probes: (a) liftoff curves over nonmagnetic materials; (b) liftoff curves over magnetic materials.

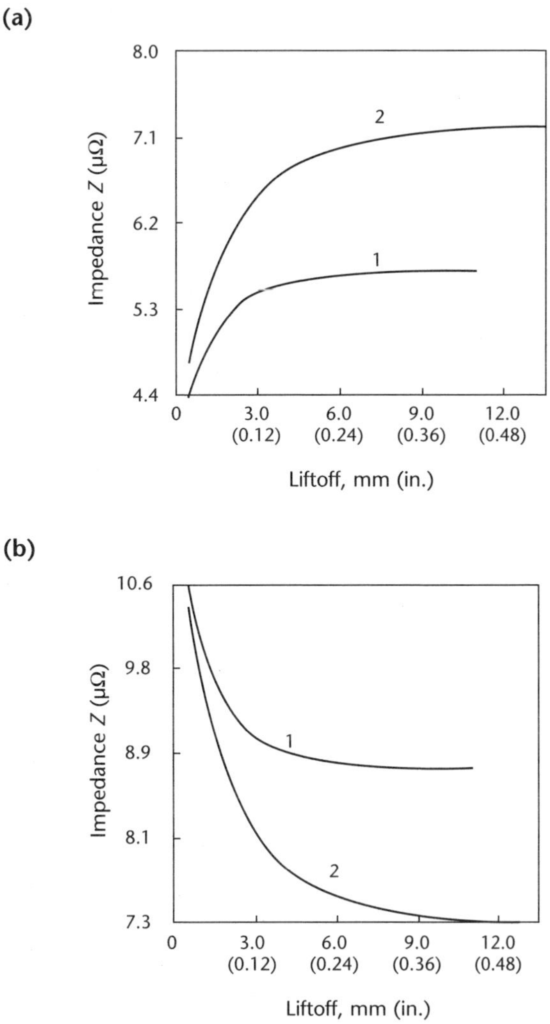

Legend
1. Shielded probe.
2. Unshielded probe.

FIGURE 29. Absolute inside diameter probe inside tube with 3 mm (0.12 in.) wall: (a) nonmagnetic tube material (nickel chromium alloy); (b) magnetic tube material (carbon steel).

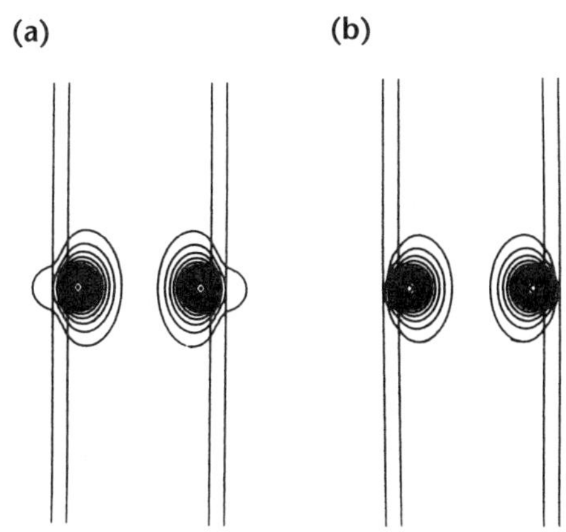

impedance have changed. The flux has the same general form but a smaller spatial distribution, penetrating through the wall as would be expected with a relatively thin, nonmagnetic material. The same coil, when inserted in a magnetic (carbon steel) tube has the distribution shown in Fig. 29b. The flux penetrates through only part of the tube because of the high permeability and smaller standard depth of penetration.

The patterns in Fig. 3 show the magnetic fields when the coil currents flow parallel to each other, that is, in the same direction relative to each other.[24]

As a striking example of the flexibility of inside diameter absolute probes, the results obtained from simulation of impurities in the crevice gap region of a nuclear power plant steam generator[18] are presented here.

The crevice gap tends to fill with foreign materials that are either magnetic (magnetite) or nonmagnetic (copper). Figure 31 helps classify the test results. Figure 31b is the signal (impedance plane trajectory) obtained from a clean gap with an absolute probe. Figures 31c and 31f show the signals obtained from the crevice gap filled with magnetite and copper, respectively. Figure 31d and 31g represent the signals obtained by simulating a 2 mm (0.08 in.) ring (region 2) of magnetite (Fig. 31d) or copper (Fig. 31g) to represent the common ridge of foreign material found in steam generators. The crevice gap itself is clean in this simulation. Figures 31e and 31h are the impedance plane trajectories when both the crevice gap and the ring are present.

The signals in all cases are distinct and it is possible to distinguish between the various accumulations. This point is important and is further explained in Fig. 32, where three representative signals are plotted together. All three signals start at the same point but describe significantly different paths. Figure 33 shows the field distribution around the absolute coil in the vicinity of the crevice gap. The field penetrates the stainless steel tube but penetrates very little into the carbon steel tube sheet. This behavior is useful because it tends to concentrate the field in the crevice gap and creates an enhanced sensitivity to materials in the crevice gap.

The field projection of absolute probes has been discussed earlier. It may be useful however to examine the projection of differential eddy current probe fields as well. Figure 34 is the normal component of the field below the probe (at the seven locations mentioned above). Figure 35 shows the tangential component of flux density for six locations at radial distances from 8.5 mm (0.33 in.) to 125 mm (5 in.). The three peaks in the field (one between the probes and two on the outer side of each coil) are clearly visible.

FIGURE 30. Differential probe testing of tubes.

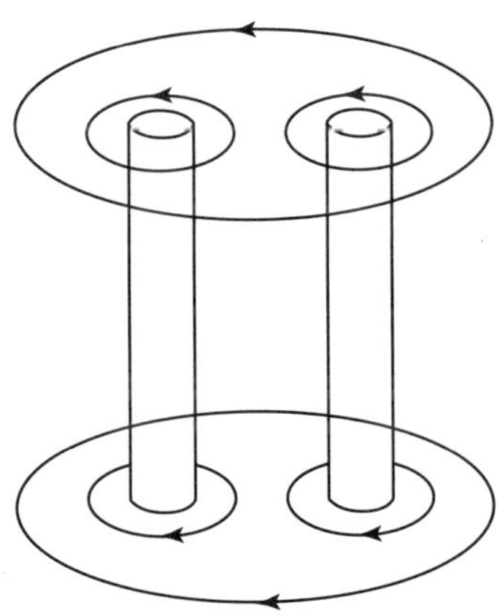

FIGURE 31. Impedance plane trajectories for different conditions in gap and ring of steam generator: (a) schematic diagram; (b) clean gap; (c) gap full of magnetite; (d) gap full of copper deposits; (e) ring of magnetite around tube; (f) ring of copper around tube; (g) gap and ring full of magnetite; (h) gap and ring full of copper deposits.

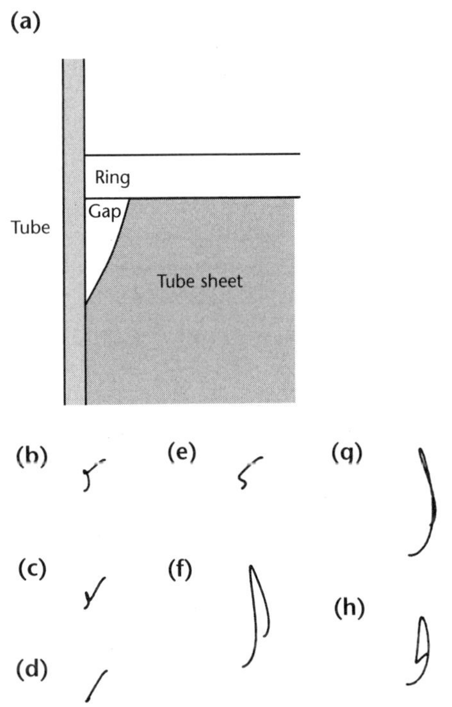

FIGURE 32. Comparison of impedance signals obtained from simulation of crevice gap conditions.

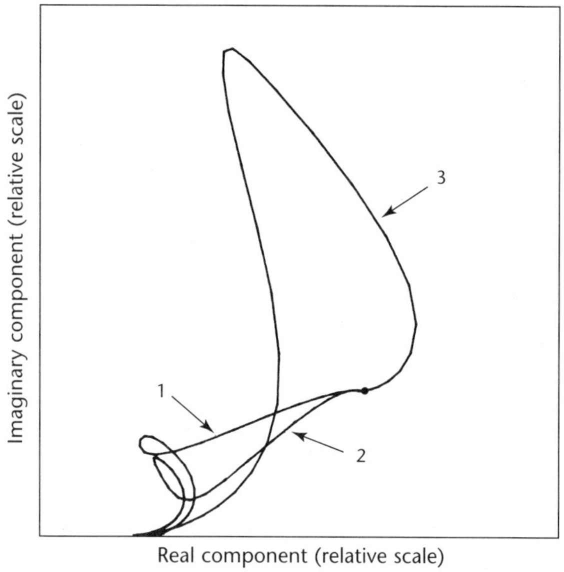

Legend

1. Clean gap.
2. Gap full of magnetite.
3. Gap full of copper deposits.

FIGURE 33. Flux distribution around coil in gap region.

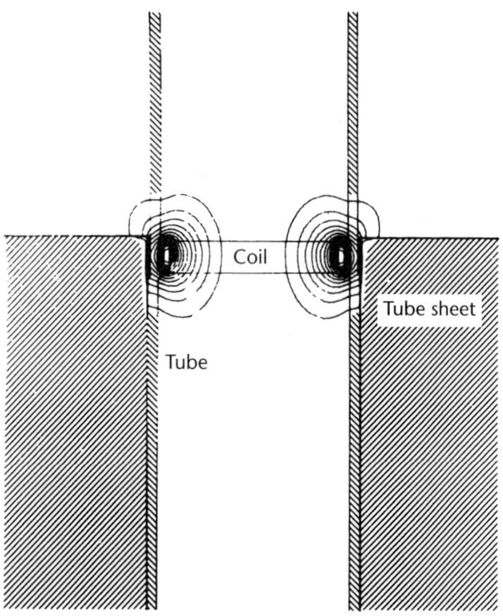

FIGURE 34. Normal component of magnetic field for differential probe. Field is calculated at seven axial locations below lower coil.

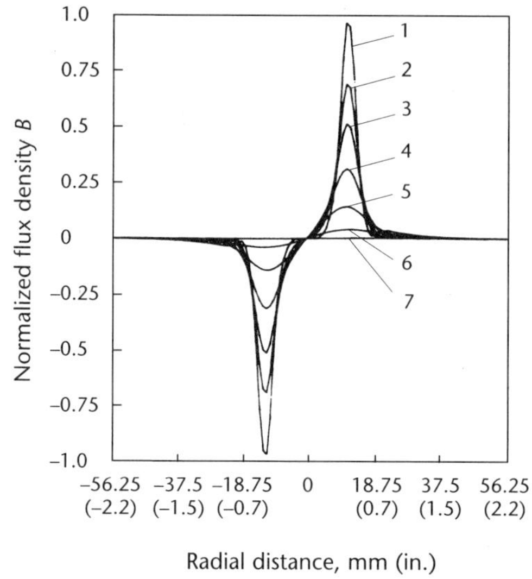

Legend

1. 0.5 mm (0.02 in.).
2. 2 mm (0.08 in.).
3. 5 mm (0.20 in.).
4. 10 mm (0.40 in.).
5. 20 mm (0.80 in.).
6. 50 mm (2.0 in.).
7. 125 mm (5.0 in.).

FIGURE 35. Tangential component of magnetic field in same locations as in Fig. 34.

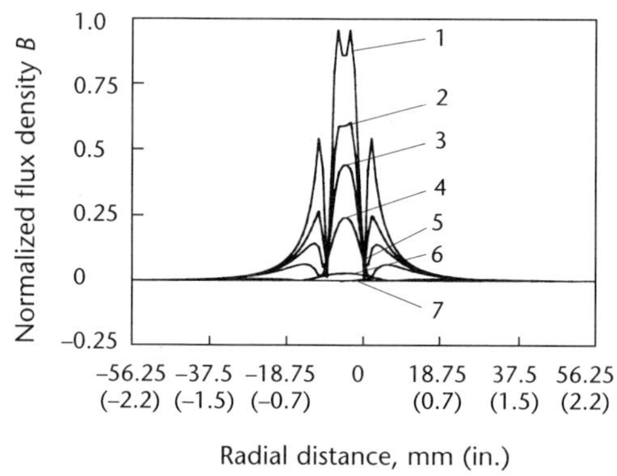

Legend

1. 0.5 mm (0.02 in.).
2. 2 mm (0.08 in.).
3. 5 mm (0.20 in.).
4. 10 mm (0.40 in.).
5. 20 mm (0.80 in.).
6. 50 mm (2.0 in.).
7. 125 mm (5.0 in.).

To illustrate the dependence of the signal on permeability, Fig. 36 shows the flux distribution of the absolute probe used in Fig. 29.

Encircling Coil Probes

Encircling coils are by design identical to inside diameter coils but they differ in function. Encircling coils are used primarily to test the outside diameter surfaces of objects that pass through the coils. Although the coil is the same in encircling and inside diameter probes, the field distribution is somewhat different. The flux density gradient tends to be more uniform inside the coil and decreases to zero at the center of the tested material. Thus, even with low frequency measurements, it is only feasible to test part of the material's volume. All other aspects relating to eddy current testing (such as skin effect) are present and must be considered.

Figure 37a represents the coil mentioned above, around a 16 mm (0.63 in.) diameter nonmagnetic round bar. Figure 37b shows the field distribution when the bar is magnetic. The field penetration into the bar is clearly visible in both cases but is greater in Fig. 37a.

FIGURE 36. Absolute probe inside 28 mm (1.12 in.) diameter hole: (a) nonmagnetic material; (b) magnetic material.

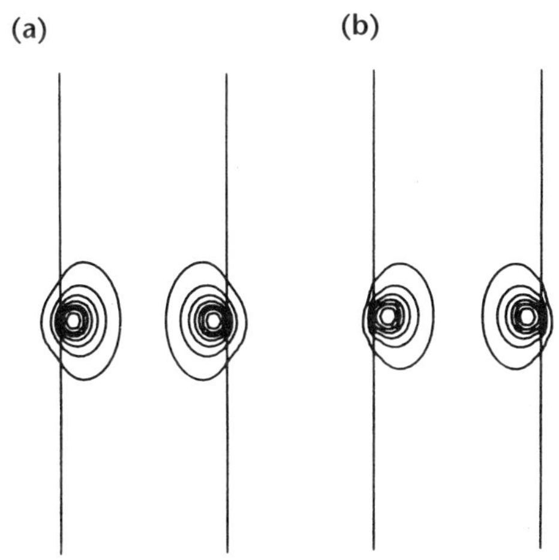

FIGURE 37. Absolute probe over round bars: (a) nonmagnetic bar; (b) magnetic bar.

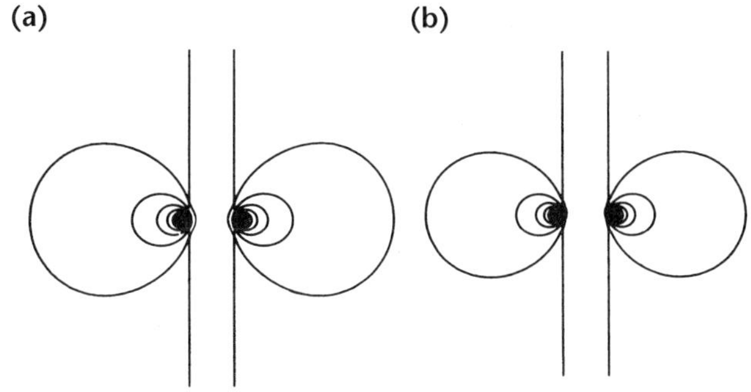

PART 3. Hall Effect Detectors[25]

Principles of Hall Effect Detectors

The hall effect creates potential differences at right angles to the direction of current flow in a conductor when a magnetic field is present. Detectors use the hall effect extensively in magnetic flux leakage test applications and to a lesser extent in eddy current probes. The following discussion briefly describes these detectors.

An external magnetic field reacts with electrical charge carriers within solid state conductors in much the same way that it reacts with electric currents flowing in a wire. The component of magnetic flux density normal (perpendicular) to the direction of electric current flow results in a force on the charge carriers that tends to move them at right angles both to the direction of current flow and to the direction of the magnetic field. In the case of a metallic wire, where conduction is provided by the motion of electrons (negative charges), the magnitude of this transverse force on a charge carrier electron e is given by:

$$F_e = evB_n \tag{15}$$

where B_n represents the normal component of magnetic induction (in tesla, where 1 T = 1 Wb·m^{-2}), e is the electron charge (-1.6×10^{-19} C), F_e is the transverse force (newton) acting on the electrons and v is the velocity (meter per second) of electrons along the wire.

Charge Carriers in Hall Effect Detectors

The charge carriers in semiconductors can consist of: (1) negatively charged electrons only, in the case of n type semiconductors; (2) positively charged holes or vacancies at locations where an electron is missing from the normal equilibrium distribution of charge (such holes can move in response to electrical fields like a positive charge whose magnitude equals that of an electron, for the case of p type semiconductors); and (3) combined movements of negative electrons and positive holes (in opposite directions) under the action of an electric field. The total current is the sum of the contributions of the motions of positive and negative charge carriers.

If the crystal lattice provides electron carriers, then each displaced electron leaves a positive hole at its former site. Such holes can appear to move or be diffused when a nearby electron falls into the first vacant hole and this electron in turn leaves its own place empty. Small proportions of impurities or elements added to an otherwise pure element in a semiconductor can influence the number of charge carriers and their lifetimes. Special combinations of elements (as from Groups III and V of the periodic table) can be selected to optimize selected response characteristics, such as the response of semiconductors to magnetic fields.

Action of Magnetic Fields on Semiconductor Charge Carriers

In semiconducting devices, when an external magnetic field acts on (1) the electrons, (2) the positive holes or (3) both types of charge carriers simultaneously, the external magnetic field creates transverse forces, in accordance with Eq. 15, on both the moving electrons and the oppositely moving positive holes. Because the signs (+ or –) of the electric charge e and the velocity v are both reversed, the magnetic forces tend to deflect both types of charge carriers in the same transverse direction. The net result is transverse deflection of the normal flow lines of the electric current within the semiconducting material.

This magnetic disturbance of the current flow lines produces two detectable effects within hall effect detectors. The first effect is to develop potential differences at right angles to the current flow lines, through the phenomenon known as the *hall effect.* The second effect is to change the resistance along the direction of the current flow paths, through the phenomenon of the magnetoresistive effect. The hall effect is readily applied to detection of weak magnetic field intensities and directions, whereas the magnetoresistive effect is most evident with very strong magnetic fields.

Hall Element

The hall effect results from the action of externally applied magnetic fields on charge carriers in metals or semiconductors. Figure 38 shows a sketch of a simple hall element, consisting of a thin layer of semiconducting material in the form of a rectangle. A typical hall device might consist of indium arsenide, indium antimony, germanium or other semiconducting materials selected for large hall effect and for minimum response to temperature variations. Typical dimensions of the rectangular layer might be 0.4 mm (0.015 in.) wide by 0.8 mm (0.03 in.) long by 0.05 mm (0.002 in.) thick. However, for such an element, the effective sensing area might be about 0.4 mm (0.015 in.) in diameter. Smaller hall effect detectors are feasible, with maximum dimensions of only 0.1 mm (0.005 in.). Precise placing of electrical connections on the hall element becomes more difficult as size is reduced.

Electrical Circuit of Hall Element

As shown in Fig. 38a, an external source supplies a steady direct current control current (typically a few milliampere) to electrodes at either end of the hall device.[25] This control current I_c passes longitudinally along the semiconductor, providing a uniform current density $\boldsymbol{J}$ across the width of the layer in the absence of external magnetic fields. Signal electrodes are located at the midpoints of the long sides of the rectangular layer.

These connections are placed carefully to minimize the signal voltage when control current flows along the layer longitudinally in the absence of external magnetic fields. Ideally, the hall voltage V_H appearing across the signal electrodes in the absence of a magnetic field should be zero but usually it is not. External correction circuits must be used to reduce this signal to zero and to provide temperature compensation.

When an external magnetic field with a normal induction B_n acts perpendicularly on the face of the rectangular layer, the charge carriers are deflected by the force acting transversely (across the width of the layer, between the two signal electrodes). The control current paths are no longer longitudinal but instead are bent to one side by the magnetic field (see Fig. 38b). The two signal electrodes are no longer at the *same voltage potential*. Instead, a voltage difference V_H, known as the *hall voltage*, appears across the signal electrodes. The hall voltage is proportional to the product of the control current magnitude I_c and the magnitude of the normal component of the external magnetic field B_n.

$$V_H = K_H I_c B_n \tag{16}$$

where K_H is the hall coefficient, or hall constant.

FIGURE 38. Hall element circuit and operation: (a) basic electric circuit of hall effect detector (uniform control current flowing along length of hall element in absence of magnetic field); (b) bending of current paths when magnetic field acts vertically through face of hall element.

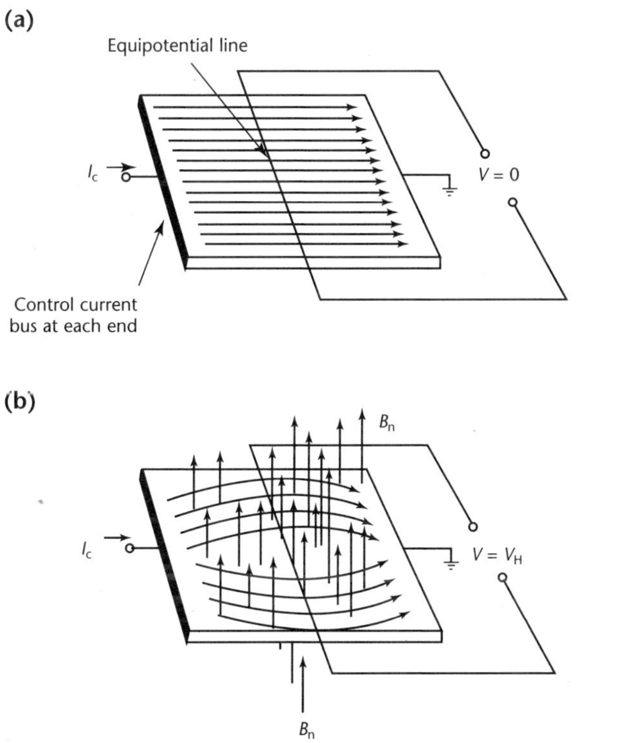

Operating Characteristics of Hall Effect Detectors

With constant magnitude direct control current, the output signal voltage from the hall effect detector is directly proportional to the instantaneous magnitude of the normal component B_n of the external magnetic field. Magnetic field components parallel to the face of the rectangular element have no effect on the hall voltage between the signal electrodes. However, it is essential to minimize the cross sectional area of the loop created by the external signal leads, because varying magnetic flux passing through this loop can produce induced voltages (90 degrees out of phase with the hall voltage and proportional to the frequency of the varying magnetic field). Special lead wire arrangements have been developed to limit the spurious induced voltage signals to negligible magnitudes for test frequencies up to 100 kHz.

Hall effect detectors developed for electromagnetic test systems can be used for very low test frequencies including direct current magnetization and can provide constant sensitivity to magnetic field magnitudes over the operating range from direct current to 100 kHz. Consequently, they can also be used as detectors with pulse, square wave, multiple frequencies and other complex waveforms within their frequency limitations.

Also, because the hall effect detector is essentially an instantaneous multiplier of the control current I_c and the normal component of the external magnetic field B_n, other forms of control current can be used for special purposes. If an alternating control current of a single frequency is used, a hall signal voltage related only to the identical frequency component of a complex magnetizing field waveform can be selected. If the control current I_c is used as a gate (by using rectangular pulses of limited duration during each cycle of the test frequency), signals indicative of a small segment of each cycle of the magnetizing coil frequency can be selected.

FIGURE 39. Directional response characteristics of hall effect detectors: (a) magnetic field directed at angle θ to normal (vertical dashed line perpendicular to horizontal face of semiconductor): (b) directional effect upon hall effect detector signal voltage, showing $B_n = \boldsymbol{B} \cos \theta$. Only normalized component B_n contributes to hall effect detector output signal voltage V_H.

(a)

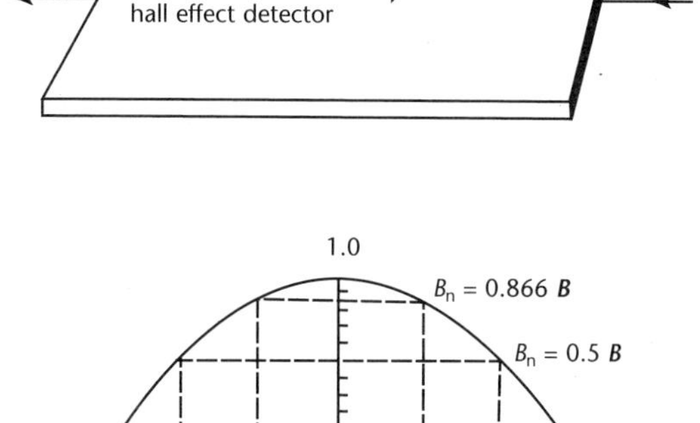

(b)

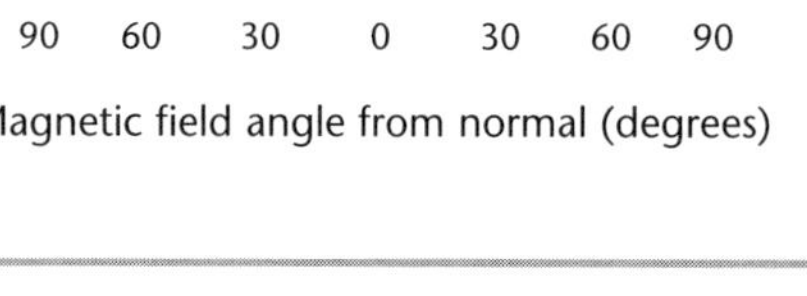

Two or more magnetizing coils, of different shapes or sizes and operating at different frequencies, can be used to explore the test material simultaneously. The hall effect detector signal will then contain both test frequency signals; separation into discrete frequency bands can be done in external circuitry. In addition, multiple hall effect detectors can be placed in any desired position and orientation in the magnetic field of the test system to provide local measurements of the magnetic field intensity.

Directional Response of Hall Effect Detectors

With detectors such as the one shown in Fig. 38, the output hall voltage signal is responsive to the angle that the local external magnetic field makes with the normal direction. The normal axis is a line perpendicular to the face of the rectangular hall element. If, as sketched in Fig. 39a, the angle between the magnetic field direction and the normal is θ, the output signal is given by:

$$V_H = K_H I_c \boldsymbol{B} \cos \theta \tag{17}$$

where $\boldsymbol{B}$ is the total magnetic flux density (tesla) and the normal component of this magnetic induction $B_n = \boldsymbol{B} \cos \theta$.

The maximum output signal is obtained when θ = 0. As θ increases, the test signal is reduced along the cosine curve of Fig. 39b. When the external magnetic field direction is parallel to the face of the hall device, the output hall voltage signal is zero. Thus, hall effect detectors can be arranged to select any desired directional component of the magnetic field intensity $\boldsymbol{B}$ in three-dimensional space. The face of the hall device is simply placed perpendicular to the direction of the magnetic flux lines to be measured.

Hall Effect Detector Configurations

Multidimensional Arrays

Because of their directional response characteristics, two or three hall effect detectors can be mounted in mutually perpendicular planes, in a compact probe that simultaneously measures magnetic field intensity components in two or three spatial directions. From such multiple measurements, the intensity and direction of the magnetic field can be determined precisely, regardless of its orientation. In many cases, the direction of the magnetic field distortion in electromagnetic tests contains useful information concerning discontinuities and local property variations.

Such local field distortions, when small in area compared to conventional coil probes, usually cannot be detected by the larger coils because they integrate the effect of all flux lines enclosed within the circumference of the coils, regardless of their spatial direction. Because of the small dimensions of hall effect detectors, they can often resolve such effects clearly. Figure 40 illustrates two-dimensional and three-dimensional arrays of hall effect detectors. Other directionally sensitive detector arrays are specially designed for unique application problems.

Linear Multichannel Arrays

Because of their relatively small dimensions, numerous hall effect detectors can be arranged adjacent to one another in the same plane, as shown in Fig. 41a. The density of these detectors can be very high, depending on the width of each hall element. If the multiple hall effect detectors are arranged along a straight line, within a suitable magnetizing coil field, adjacent areas of the test object can be scanned through separate signal channels. If a test object, such as a fusion weld, is scanned longitudinally, each detector can sense a different area across the width of the weld. This permits separation of cracks and property variations associated with (1) the fusion zone, (2) the heat affected zones and (3) the adjacent areas of the unaffected base material, simultaneously in one test operation.

In another arrangement, linear arrays of hall effect detectors can be positioned around the circumference of an encircling coil (through which circular bars or tubes are passed longitudinally) and can be used to provide a sensitive electromagnetic test of the entire 360 degree surface in one pass. One magnetizing coil provides the exciting field for all the hall effect detectors (Fig. 41b). Concentricity of conductors within insulating coatings, or uniformity of conducting coatings on

FIGURE 40. Multidimensional arrays of hall effect detectors used to measure directional components of magnetic field intensity: (a) two-dimensional array of hall effect detectors, in which components of magnetic field in X,Y plane are sensed individually; (b) three-dimensional array of hall effect detectors, in which each detector senses magnetic field component perpendicular to face of semiconductor.

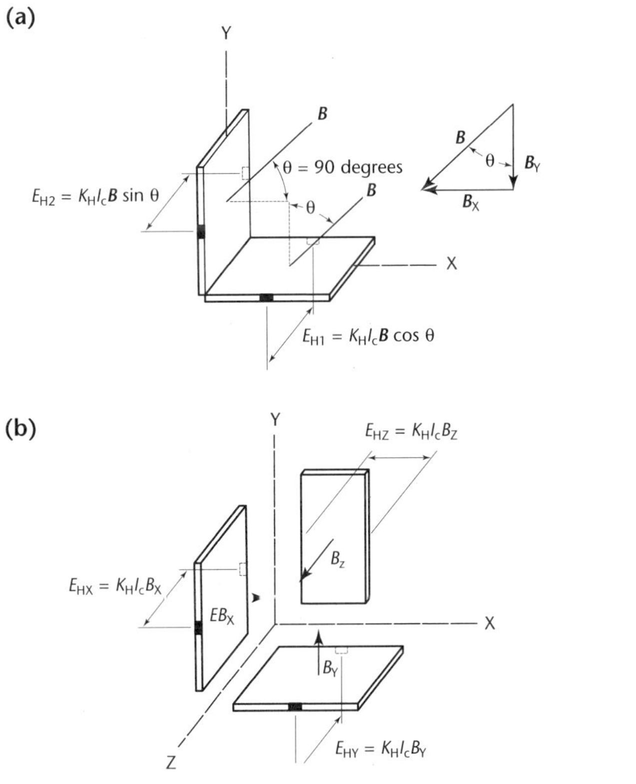

Legend

B = magnetic flux density
E_H = electric field detected in X, Y or Z direction
I_C = control current
K = hall coefficient

FIGURE 41. Linear arrays of hall effect detectors in eddy current magnetizing coils: (a) linear array of hall elements within single magnetizing coil, for multichannel longitudinal scanning of fusion weld zones; (b) linear array of hall effect detectors within circumference of encircling magnetizing coil, for multichannel longitudinal scanning of bars or tubes. For a 25 mm (1 in.) diameter coil, there must be about 100 hall elements each about 0.75 mm (0.03 in.) long for 100 percent coverage.

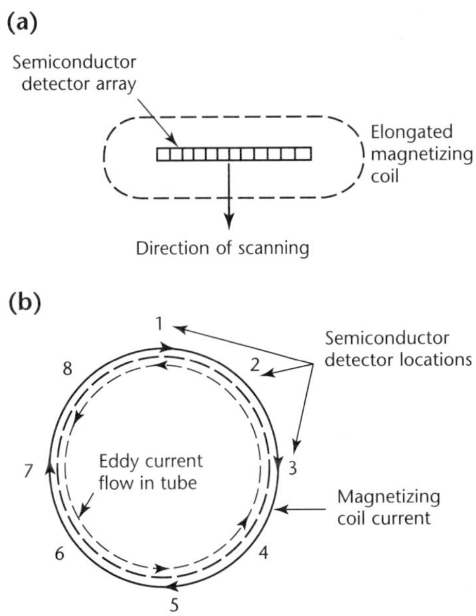

concentric cores, may be evaluated with this type of probe. Longitudinal discontinuities, such as seams and laps in hot rolled bars or tubes, can also be detected continuously, rather than intermittently by, for example, orbiting probe coil test systems.

Bridge Arrays

An alternative arrangement of small area hall effect detectors within a single test coil corresponds to the wheatstone bridge circuit used in many electrical measurements. As sketched in Fig. 42a, the detectors are in the corners of a small square in the test surface and are excited by a single large magnetizing coil. Four local measurements are made simultaneously. Signals from the four discrete channels can be compared in various combinations in the external circuitry of the electromagnetic test

FIGURE 42. Differential and bridge arrays of hall effect detectors within field of circular magnetizing coil: (a) simple bridge array of hall effect detectors within circular magnetizing coil (signals are compared in external circuitry to determine parameters such as anisotropy, crack direction, probe alignment or probe position relative to weldments); (b) differential arrangement of two hall effect detectors in symmetric locations of magnetizing coil field (subtraction of signals in external circuitry permits local property variations or discontinuities to be detected while average test object properties are nulled out).

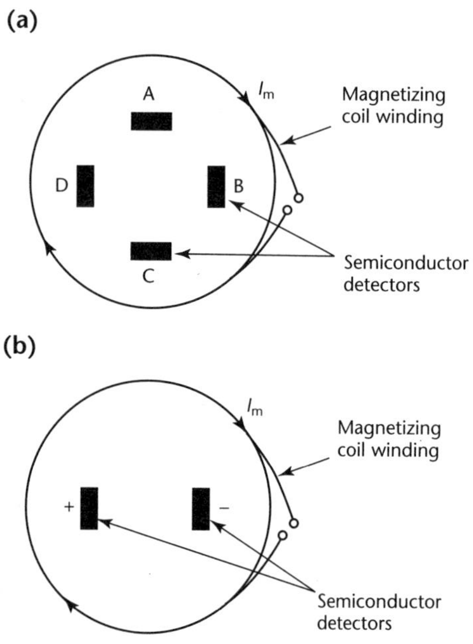

Legend
A, B, C, D = hall effect detectors
I_m = magnetizing current

instrumentation. Such an arrangement is responsive to the degree and orientation of anisotropic properties in test materials or to the directions and severity of discontinuities such as cracks. An alternative is to align coils precisely with the flat surfaces of test materials to measure tilt or surface waviness. Other arrangements of multiple hall effect detectors over an area of the test object surface can be applied for special measurement problems.

Differential Systems

Two hall effect detectors can be located at various spacings and their signals subtracted from one another in the external circuitry, to provide differential measurements. A single coil can provide excitation for both detectors (Fig. 42b).

With a differential array (as with differentially connected electromagnetic coil transducers), average or slowly varying properties of the test material and conditions of the eddy current test can be eliminated from the test indication. For example, with flat, vertical dipole probe coil excitation in electromagnetic tests of sheets, the effects of probe liftoff and the effects of slow variations in test material properties and sheet thicknesses are minimized by differential measurements. On the other hand, local discontinuities and property variations can be easily indicated, because any variation from a null output signal can be greatly amplified.

Systems using differential hall effect detectors within the field of a single exciting coil have been applied effectively to locate and follow weldments in tubes and pipe or to guide welding equipment along the groove for butt weld preparations with high precision. For this purpose, two hall effect detectors are put on a line transverse to the length of the weld to sense material property variations in the weld fusion zone or its heat affected zone. When the probe coil and detectors are centered over the weld line, the differential test signal is zero. If the probe is displaced (even slightly) from exact positioning over the weld line, the differential hall effect detector output is used to provide appropriate phase and amplitude data to servo mechanisms that move the probe to the desired center of the weld zone or weld preparation groove.

In weld testing, a second differential probe system is used with the hall effect detectors displaced along the weld line, to compare adjacent areas of the weld for local discontinuities. Here, the lack of response to liftoff and to slowly varying properties such as wall thickness are helpful in permitting detection of small discontinuities.

PART 4. Probes for Magnetic Flux Leakage[26]

Description of Method[27]

Magnetic flux leakage testing is an electromagnetic technique that can provide a quick assessment of the integrity of ferromagnetic materials. The technique is often used instead of eddy current testing when the test object is ferromagnetic. The test object is magnetized during testing and the sensor detects the magnetic flux that is said to *leak* from the magnetic field at a surface discontinuity.

Flux leakage tests are used in many industries to detect a wide variety of discontinuities. Typical applications of magnetic flux leakage testing are by producers of steel products — blooms, billets, rods, bars, tubes and ropes. Examples of inservice applications are the testing of used wire rope, installed tubing and retrieved oilfield tubular goods.

Coil Probes

Coil probes are sometimes used in magnetic flux leakage applications. Consider a pancake coil whose dimensions are extremely small in the context of the spatial distribution of the magnetic field to be measured. The voltage V is induced in the small coil:

$$(18) \quad V = -N \frac{d\phi_n}{dt}$$

where d indicates a derivative, N is the number of turns in the coil, t is time (second) and ϕ_n represents the flux normal to the plane of the coil.

If the variations in the flux density over the area of the coil are negligible, then the component B_n of the flux density normal to the plane of the coil is given by:

$$(19) \quad B_n = \frac{\phi_n}{A} = \frac{-1}{NA} \int_0^t V\,dt$$

where A is coil area (square meter) and V is potential (volt). Consequently, to measure B_n, it is necessary to integrate the output of the coil as shown in Fig. 43.

The coil cannot be used to measure the magnetic leakage field under static conditions, for $d\phi_n \cdot (dt)^{-1}$ and hence V would then be zero. Coils can be used to measure the magnetic field either when the field is time varying or when the test object is moving relative to the coil probe.

FIGURE 43. Integration used in combination with coil to measure magnetic flux leakage field.

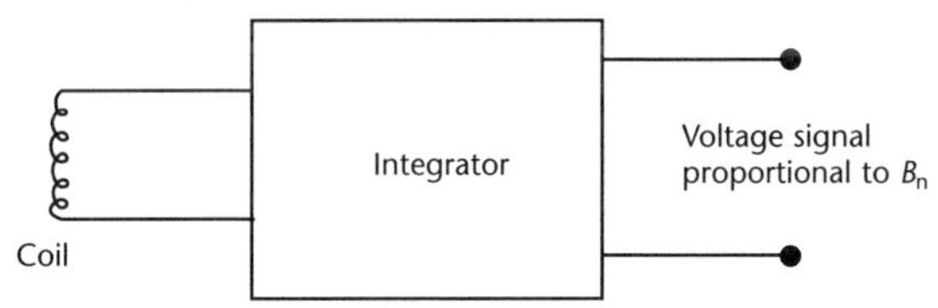

Other Indicating Means for Magnetic Flux Leakage

It is possible to use transducers other than inductive coils for magnetic flux leakage testing.

Magnetodiode

The magnetodiode is a solid state device, the resistance of which changes with magnetic field intensity. It consists of p zones and n zones of a semiconductor, separated by a region of material that has been modified to create a recombination zone (Fig. 44). Active areas typically

FIGURE 44. Schematic of magnetodiode.

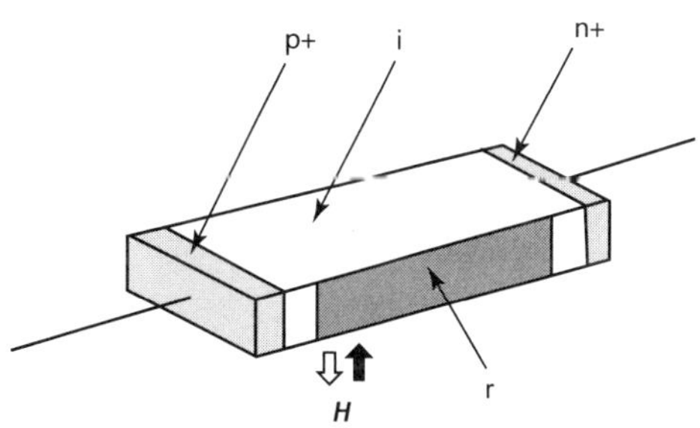

Legend
H = magnetic field
i = intrinsic zone
n = n zone
p = p zone
r = recombination zone

measure 3.0 × 0.6 × 0.4 mm (0.12 × 0.024 × 0.016 in.) and output signals are generally larger than for hall elements, although the response to field intensity is not so linear for higher fields as shown in Fig. 45.

Figure 46 shows that frequency response is flat from direct current field to 3 kHz and Fig. 47 shows that sensitivity is stable without temperature dependence in the range of –10 to 50 °C (14 to 122 °F).

Magnetic Recording Tape

For the testing of flat plates and billets, it is possible to scan the surface with wide strips of magnetic recording tape. Discontinuity signals are taken from the tape by an array of tape recorder heads. Elongated magnetic balloons also exist for the testing of the inside surface of tubes. Scale, dirt or oil on the test surface can contaminate the tape. Surface roughness can tear the tape.

Magnetic Particles

Magnetic particles are finely ground high permeability magnetic material, sometimes dyed for visible contrast with the test surface. Ideal test conditions occur when a fine spray of such particles is intercepted by a magnetic flux leakage field and some of them stick to the field. An advantage over other forms of magnetic indicators is that the particles have zero liftoff from the discontinuity field. In a simple approximation, the force F_{mag} holds the particles in the leakage field:

$$F_{mag} = \alpha \mu_0 V_{vol} \left(\boldsymbol{H} \cdot \nabla \right) \boldsymbol{H} \tag{20}$$

where ∇ is the vector differential operator (gradient operator), $\boldsymbol{H}$ is the local flux leakage field intensity (ampere per meter), V_{vol} is the volume (cubic meter) of the particle, α is a factor related to the demagnetization factor of the particle and μ_0 is the permeability of free space ($\mu_0 = 4\pi \times 10^{-7}$ H·m^{-1}).

The force that holds the particle to the discontinuity leakage field is proportional to the result of a vector calculus operation on the leakage field. This force can be computed for simple leakage fields by using approximations such as those by Förster[28] and by Zatsepin and Shcherbinin[29] or more accurately by using finite element techniques such as those described elsewhere, in this volume's chapter on modeling. The force is also

FIGURE 45. Response of magnetodiode is linear up to about 40 kA·m^{-1} (500 Oe) at ambient temperature of 25 °C (77 °F) and potential of 6 V.

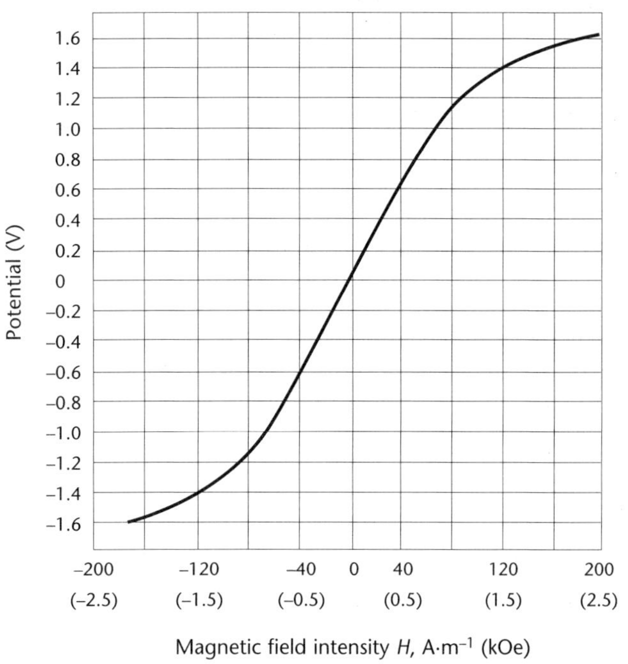

FIGURE 46. Frequency response of magnetodiode at ambient temperature of 25 °C (77 °F).

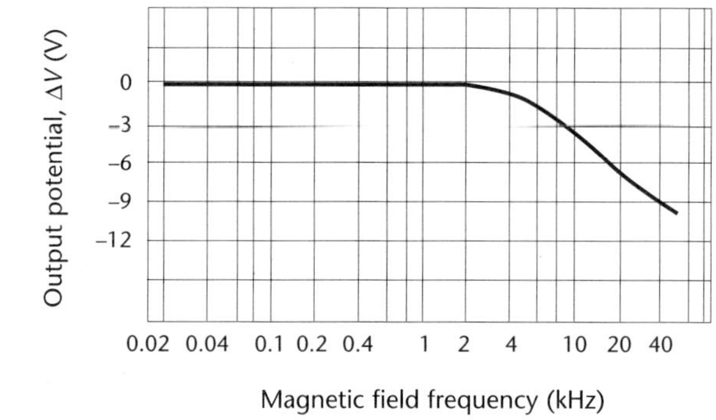

FIGURE 47. Temperature dependence of magnetodiode.

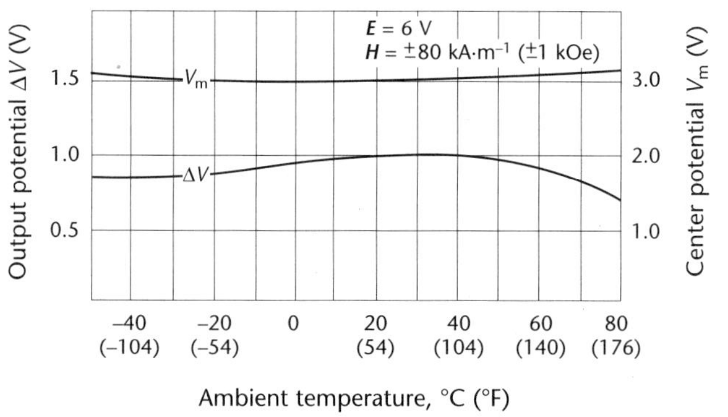

related to the shape of the particle, which is implied by the term α. For spherical particles, $\alpha = 3$. For particles with a prolate ellipsoidal shape and an axis ratio of 2:1, $\alpha = 5.8$. As particles become more elongated, α increases and the magnetic force on them in a given leakage field also increases. Elongation of the particle decreases the demagnetization factor for the particle. Additional decreases in demagnetization factor (or increases in α) are achieved when particles stick together, end to end.

Other forces acting on the particles are (1) gravity, (2) friction and (3) spray force.

1. The pull of gravity is reduced if particles are very small. Large particles are used to detect only deep cracks that scale off during reheating.
2. Friction is reduced if the particles are suspended in water or light oil.
3. Spray application should be gentle to avoid washing particles off cracks.

Optimum visual conditions occur when the particles emit light in that portion of the visible spectrum where the eye is most sensitive (yellow green). This is particularly effective when all other visible light has been removed. This situation is achieved when low energy ultraviolet radiation is used to irradiate the particles and their dye absorbs this energy and reemits it as visible light.

Magnetic particle testing is performed, either in active or residual field, for a wide variety of parts. It is performed as the primary test, or as a followup test when discontinuities have been found by other methods. The method is described in detail elsewhere.[30]

PART 5. Eddy Current Imaging with Magnetooptic Sensors

Magnetooptic Imaging Principles

Magnetooptic imaging is a real time eddy current imaging technology that relies on the faraday magnetooptic effect. This technology has been used to image cracks and other discontinuities in electrical conductors such as aging aluminum airframes. The following discussion briefly describes such imaging devices and gives examples of both surface and subsurface indications obtained with this approach.

Both conventional and unconventional eddy current techniques rely on Faraday's law of electromagnetic induction:

$$(21) \quad \nabla \times \boldsymbol{E} = \frac{-1}{c} \frac{\partial \boldsymbol{B}}{\partial t}$$

In differential form this law describes the connection between magnetic field vector ***B*** and electric field vector ***E*** at points of three-dimensional space at time *t* (second) — including points located inside electrical conductors.[31,32] In particular, this law shows that a time varying magnetic field ***B*** (produced by a moving permanent magnet, the field from a coil of wire carrying a changing current or some other source) in the vicinity of any electrical conductor having conductivity σ, will induce a time varying electric field ***E*** and thus a time varying eddy current density ***J*** at an arbitrary point near the surface of the conductor:

$$(22) \quad \boldsymbol{J} = \sigma \boldsymbol{E}$$

To gain a general understanding of the relevant relationships and concepts, consider the familiar case of a time varying magnetic field ***B*** produced by an external alternating current in a coil of wire.

By Lenz's law,[31] the direction of the vector ***J*** at any point in the conductor and hence the direction of the eddy currents is always opposed to the change in the direction of the external electric currents that produced ***B*** in the first place. This opposition is illustrated in Fig. 48, where a standoff or noncontact coil carrying an alternating current near a conducting plate has induced eddy currents in the plate. The magnetic fields in Fig. 48 tend to be excluded from the conducting plate, especially at high frequencies. The magnitude of the eddy currents diminish as the depth increases.

Sheet Eddy Current Generation

Coils similar to those in Fig. 48 form the basis for conventional coil based eddy current techniques. Note that the induced eddy currents form a kind of image of the coil currents in the conducting plate, meaning that the magnitude of the eddy currents is greatest just under the circular footprint of the coil. Consider an unconventional eddy current induction technique called *sheet current induction* as illustrated in Fig. 49.[33,34]

In Fig. 49, just as in the case of the coil of Fig. 48, Lenz's law ensures that the

FIGURE 48. Eddy currents being induced in electrically conducting plate by magnetic field ***B*** produced by coil of wire carrying alternating current: (a) view of eddy current distribution in plate; (b) view of eddy current penetration into plate. By Lenz's law, direction of induced eddy currents (current density $J = \sigma E$ at one point) is opposed to a change in direction of currents in coil (solid and dotted lines represent currents and fields roughly 180 degrees out of phase).

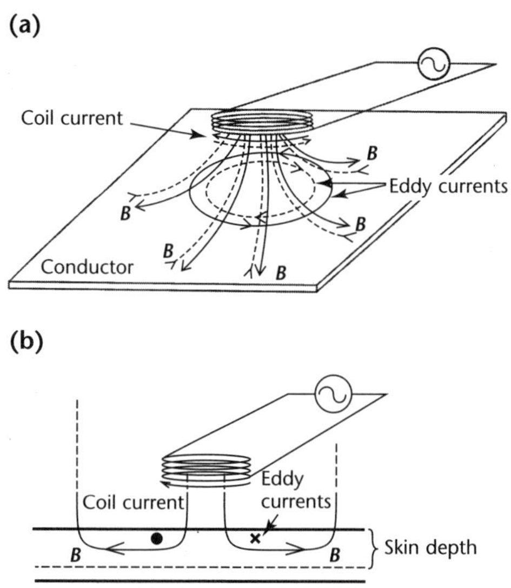

direction of the eddy currents is opposed to the change in the direction of the currents in the foil (solid and dotted lines represent currents roughly 180 degrees out of phase). As in the case of the coil of Fig. 48, eddy currents diminish as the depth increases. Note that the eddy currents under the footprint of the foil — unlike eddy currents concentrated near the coil (Fig. 48) — tend to be very uniform. Moreover, these eddy currents have return paths outside the footprint of the foil, unlike a coil where the return paths are located in a circular region falling mostly within the footprint of the coil. This technique of inducing eddy currents is crucial for the operation of magnetooptic eddy current imaging devices.

Magnetooptic Imaging and Eddy Current Principles Combined

To understand how magnetooptic eddy current devices produce eddy current images, it is useful to consider the magnetooptic part of the technology and how it can be combined with the sheet current induction technique (Fig. 49) to make real time eddy current images of cracks and other conditions, such as corrosion, in such objects as aluminum lap joints in aging aircraft.

Magnetooptic Image Displays

Figure 50 illustrates a reflection optical arrangement for producing images of the magnetic fields ***H*** (static or dynamic) associated with discontinuities such as cracks. These images are produced by observing the effect on the magnetization ***M*** of the magnetooptic sensor using a source of polarized light and an analyzer, which is another polarizer.[33,34] It happens that the magnetooptic sensors can have only two states of magnetization (that is, +***M*** or –***M***) as shown by the arrows representing ***M*** in the two drawings in Figs. 50b and 50c. When viewed in polarized light, one of these two states of magnetization (+***M*** in this case) can be made to appear dark by rotating the analyzer; –***M*** will then appear light by comparison. Hence, a magnetooptic sensor is essentially a two-state (dark and light) imaging device. Any magnetic field ***H***, including a magnetic field associated with eddy current distortions caused by

FIGURE 49. Eddy currents being induced in conducting plate by magnetic field (not shown) from flat, electrically conducting foil carrying alternating current produced by transformer arrangement shown above conducting plate.

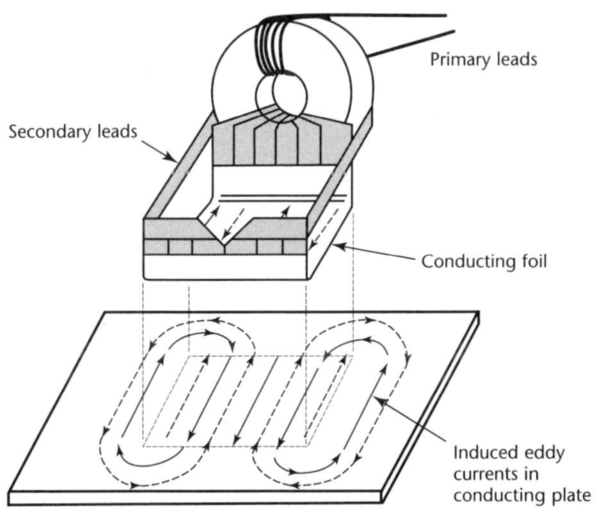

FIGURE 50. Arrangements for production of magnetooptic images: (a) reflection optical arrangement; (b) sensor magnetization and resultant image at different time than for Fig. 50c; (c) sensor magnetization and resultant image but at different time than for Fig. 50b.

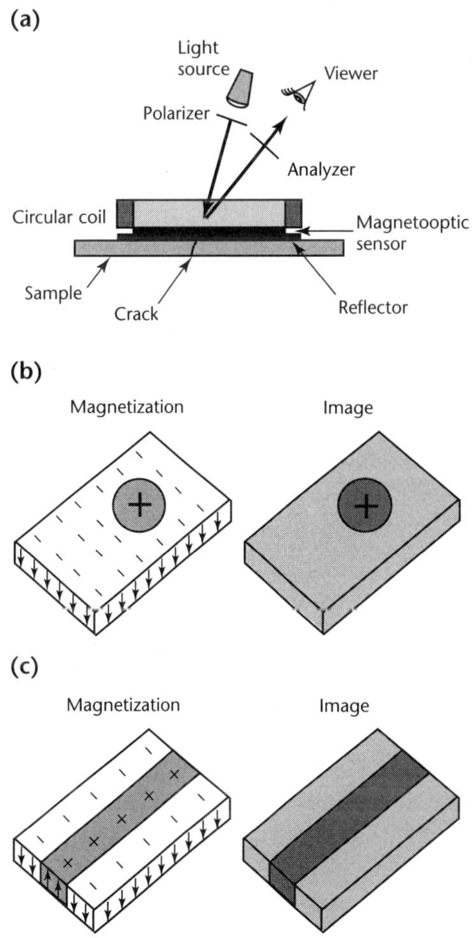

discontinuities, can with such an optical arrangement produce magnetooptic images of discontinuities that closely resemble the actual discontinuities.

Although the present discussion emphasizes nonmagnetic electrical conductors similar to aluminum, it should be clear from Fig. 50 that magnetooptic displays could, in principle, be used to form images of surface breaking cracks in magnetic materials such as steel. In this case, either static or slowly varying magnetic fields are used to magnetize the material and the magnetooptic display is used to form an image of the magnetic flux leakage from the crack.[35]

Magnetooptic Imaging of Sheet Current Induction

By combining the two technologies illustrated in Figs. 49 and 50 and adding a bias coil,[33,34] it is possible to make real time eddy current images as illustrated in Fig. 51. Figure 51 illustrates how *linear* sheet current excitation — *rotating* sheet current excitation is discussed below — is combined with the reflection magnetooptic imaging technique of Fig. 50 to produce real time magnetooptic eddy current images.

Figure 51 illustrates the current paths in both the conducting foil and the work piece. Currents in the first half cycle of a full current cycle are indicated by solid lines whereas currents in the second half cycle are indicated by dotted lines.

FIGURE 51. Magnetooptic sensor must be modulated by bias magnetic field to provide magnetooptic image of discontinuity (open hole) by using linear sheet eddy current induction in reflection optical arrangement. Light source and crossed polarizers are not shown. Erase pulse precedes image formation.[33,34]

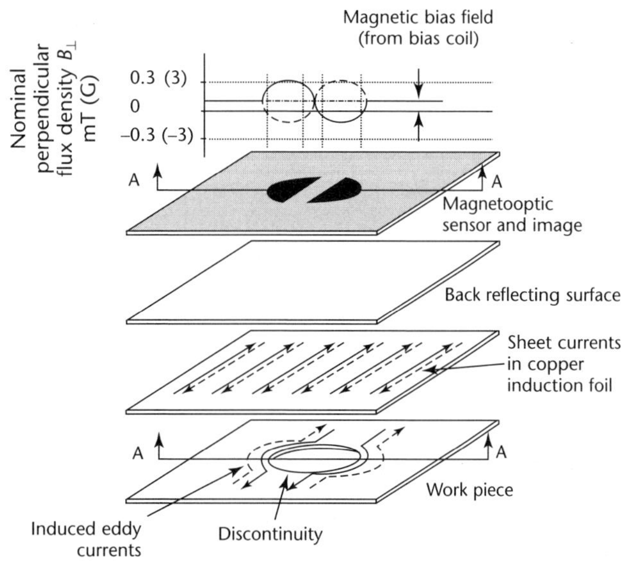

Distortions in the induced eddy currents caused by electrical discontinuities in the test object such as rivets or cracks, produce magnetic fields perpendicular (normal) to the surface of the work piece. These are the only magnetic fields that can be detected by the magnetooptic probe because only these magnetic fields lie parallel to the sensor's easy axis of magnetization.

Solid and dotted lines at the top of Fig. 51 also illustrate the normal (perpendicular) component magnetic fields corresponding, respectively, to the eddy currents induced in the test object during the first and second half cycles of a given current cycle. These time varying magnetic fields $\boldsymbol{H}$, associated in this case with a hole, are able to alter and then temporarily maintain the probe's magnetization $\boldsymbol{M}$ because of the presence of a magnetic field with a nearly static bias. This magnetic field points in one direction only and adds to the field from the discontinuity in such a way that only those fields directed along the bias field can alter and then temporarily maintain the magnetization $\boldsymbol{M}$. Only these regions of the sensor where $\boldsymbol{M}$ has been so altered and maintained can form a high contrast visible image in crossed polarizers (dark for $+\boldsymbol{M}$ and light for $-\boldsymbol{M}$). In the absence of a bias magnetic field, the time averaged magnetooptic image would be washed out because the magnetization $\boldsymbol{M}$ (Fig. 51) would be rapidly switching from $+\boldsymbol{M}$ to $-\boldsymbol{M}$ at the peak of each current half cycle.

However, with the bias present only half of the image is produced during the first half cycle (see one of the two half moon shaped image segments illustrated in Fig. 51). Because the magnetooptic probe also has an effective memory, this image is remembered by the probe until the next half cycle, which completes the other half of the image to form a

FIGURE 52. Schematic illustration of magnetooptic eddy current imager. This reflection geometry has reflective surface behind sensor. Bias coil supplies bias magnetic field needed in image formation (see Fig. 51).

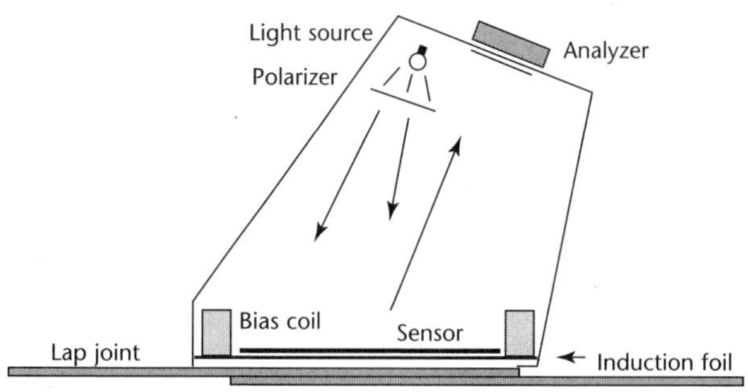

composite image that looks like the head of a slotted screw (see Fig. 51).

The magnetooptic eddy current imager (Fig. 52) combines the concepts discussed above. The vertical extent and position of the bias coil (not shown to scale or actual position) can be chosen in such a way that the component of the bias magnetic field perpendicular to the probe is essentially constant across the entire sensor. Consequently, every rivet or other area imaged in the field of view experiences essentially the same (normal component) bias field. There is no need to compensate for variations across the field of view.

Magnetooptic Indications

Figure 53 illustrates idealized and actual magnetooptic eddy current images for two different techniques of eddy current excitation. Magnetooptic eddy current images are always accompanied by a background of serpentine magnetic domains that are required to form the images and cannot be completely eliminated. Adjusting the bias magnetic field can minimize background domains. The eddy current frequency used in the images of Fig. 53 is 50 kHz.

Linear versus Rotating Induction

Figure 51 shows how images are formed when linear sheet current induction — produced by the transformer arrangement in Fig. 49 — is used. However, by modifying the transformer of Fig. 49 as described in greater detail elsewhere,[33,34] it is possible to obtain rotating sheet current induction.

The practical significance of this is that the orientation of the magnetooptic eddy current imager, on the surface of the test object, is irrelevant when rotating eddy current excitation is used but orientation can be important when linear eddy current excitation is used. As shown in Fig. 53a, the linear eddy current excitation direction must be perpendicular to the long axis of the crack to detect the crack. That is, if the eddy current direction is parallel to the crack, the crack will not be detected because there is no effect on the paths of the eddy currents in this case. Clearly, this is not a problem when rotating eddy current excitation is used (see Fig. 53b).

Although there are advantages to using linear instead of rotating eddy current excitation in some applications, rotating eddy current excitation generally produces better defined images because they are not split as in the case of linear excitation. This advantage is especially important for irregularly shaped discontinuities such as areas of subsurface corrosion that may be difficult to detect when the corresponding magnetooptic eddy current image is split in half.

FIGURE 53. Magnetooptic eddy current images of rivet with crack extending to right: (a) linear eddy current excitation mode; (b) rotating eddy current excitation mode.

(a)

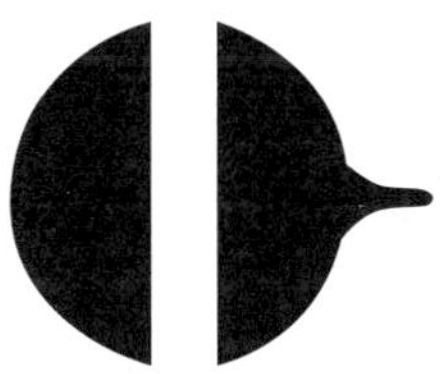

Idealized image

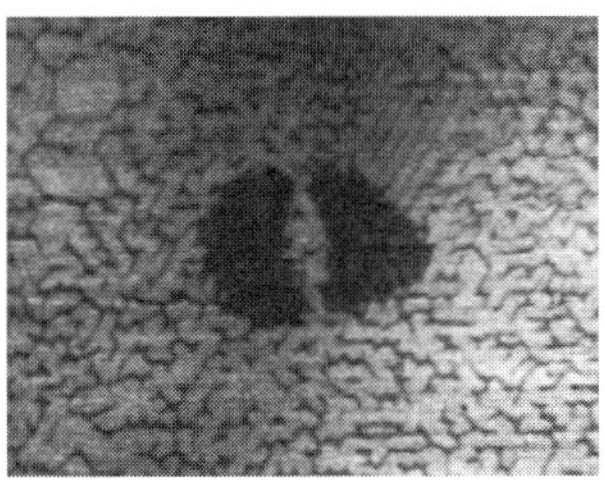

Actual image

(b)

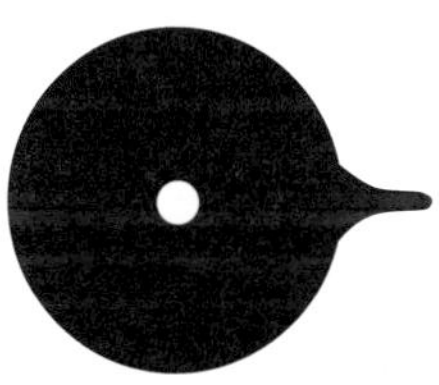

Idealized image

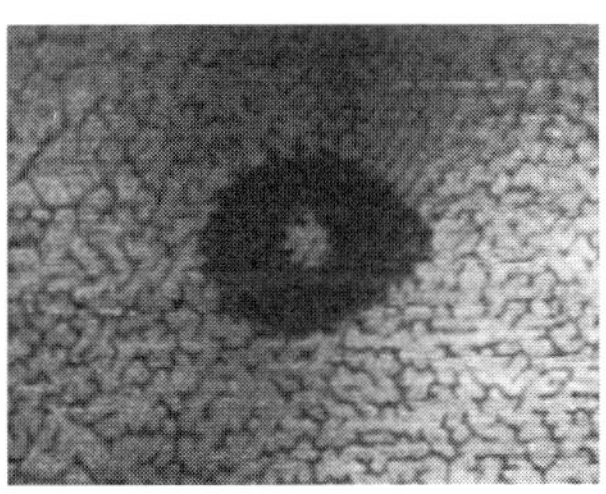

Actual image

Magnetooptic Image Interpretation

Figures 53 to 55 demonstrate that, although the length of surface breaking cracks can be estimated roughly from magnetooptic eddy current images, these same images are not at all appropriate for determining crack width. For example, the apparent crack width in Figs. 53 and 55 is considerably wider than the actual crack width because these are images of the magnetic fields near the crack that are produced by the flow of eddy currents around the crack. These eddy currents are significant only within about one standard depth of penetration from the crack as measured parallel to the surface of the test object.

A rough rule of thumb for surface breaking cracks is that the apparent width of the crack is roughly twice the standard depth of penetration δ, so there is no way to estimate actual crack width. The actual crack length is roughly given by the apparent crack length, as seen on the image, minus the standard depth of penetration δ. This approximation works best at high frequencies and low power where the rivet edge is discernable and readily located.

Using the foregoing rule of thumb, the higher the eddy current frequency, the smaller the standard depth of penetration δ and the narrower the apparent crack width $w = 2\delta$. The magnetooptic eddy current images in Fig. 53 were made at an eddy current frequency of 50 kHz with a standard depth of penetration $\delta = 544$ µm (0.021 in.) in a wrought aluminum alloy (Unified Numbering System A97075, temper 6, or 7075-T6) but current technology permits images at frequencies up to 200 kHz with a standard depth of penetration $\delta = 305$ µm (0.012 in.) in the same material. Hence, the apparent crack width in aluminum at 200 kHz will be $w = 533$ µm (0.021 in.) whereas at 50 kHz the apparent crack width will be twice as large, namely, $w = 109$ µm (0.043 in.). At still lower frequencies in aluminum, the apparent crack width is wider still. At 10 kHz in the wrought aluminum alloy, for example, the standard depth of penetration $\delta = 117$ µm (0.046 in.) and therefore the apparent crack width $w = 2.34$ mm (0.092 in.).

To make magnetooptic eddy current images that resemble actual discontinuities as closely as possible, the highest possible eddy current frequency that still permits detection of the discontinuity should always be used. This is true for both surface and subsurface discontinuities.

Discontinuities that are more than one standard depth of penetration below the surface of a material can sometimes be difficult to detect unless the eddy current magnitude is sufficient. Accordingly, high power settings are invariably used when attempting to detect subsurface discontinuities or corrosion at depths of two or more standard depths of penetration from the surface of materials such as aging aluminum airframes. Typically, the highest possible power level

FIGURE 54. Two-layer setup standard in which each layer is 1 mm (0.04 in.) thick aluminum (Unified Numbering System A82024, temper 3) containing electric discharge machined notches and second layer hole.

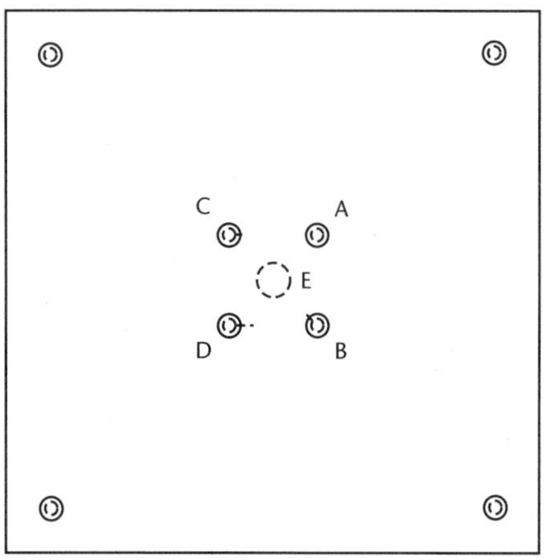

Legend

A. Rivet without anomaly.
B. First layer electric discharge machined notch of 45 degrees, 2.5 mm (0.10 in.) long.
C. Horizontal electric discharge machined notch, 1.8 mm (0.07 in.) long.
D. Second layer electric discharge machined notch, 5.0 mm (0.20 in.) long.
E. Hole in second layer, 9.55 mm (0.376 in.) in diameter, mimicking corrosion.

FIGURE 55. Magnetooptic eddy current images made with rotating eddy current excitation and corresponding to artificial discontinuities in Fig. 54: (a) image of discontinuity free rivet, corresponding to notch A, made at 100 kHz; (b) image of notch B, made at 100 kHz; (c) image of notch C, made at 100 kHz; (d) image of notch D, made at 10 kHz; (e) image of hole E, made at 10 kHz.

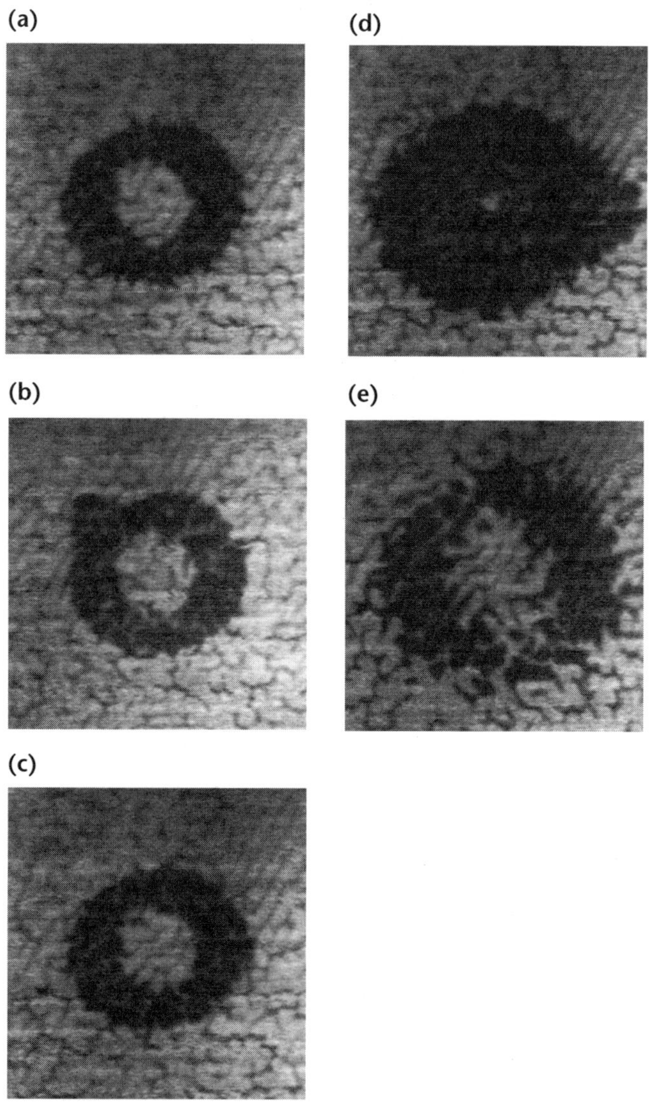

is selected and the highest workable eddy current frequency is one that achieves the best possible discontinuity resolution. Some designs of magnetooptic eddy current imaging devices permit eddy current magnitude to be much greater than normal. This improvement permits discontinuities at three or four standard depths of penetration below the surface of a material to be detected, depending on the size and nature of the discontinuity.

Reference Standards and Selected Magnetooptic Images

Figure 55 shows magnetooptic images of a small set of surface breaking electric discharge machined notches shown in Fig. 54. Note that the small notch 1.8 mm (0.07 in.) from rivet shank C in Fig. 54 produces a small bump or elongation on the right side of the rivet image in Fig. 55c. The shank diameter of rivets shown is 4 mm (0.16 in.) and exposed rivet heads are 6 mm (0.24 in.) in diameter and flush with the outer surface. Notch lengths are measured from the rivet shank.

In a still picture such as this, it can sometimes be difficult to detect small discontinuities. However, in practice, when the magnetooptic eddy current imager is moved, this small notch indication appears to move along the rivet circumference as the imager is rotated slightly in the plane of the sample. In general, moving the magnetooptic eddy current imager makes images appear to move in various ways. This fact, combined with the capabilities of the human eye to process moving images, has been found to improve the ability to detect discontinuities, especially discontinuities such as corrosion that are small or whose boundaries are difficult to define.

Figure 56 shows a simulated aircraft lap joint and fatigue cracks. Finally, Fig. 57 shows a mosaic of actual magnetooptic eddy current images that resulted from an examination of the simulated aluminum aircraft lap joint illustrated in Fig. 56. The mosaic shows rivets and in some cases fatigue cracks radiating from the rivets. Acceptable rivets show perfectly circular images whereas rejectable rivets have easily detected crack indications. Figure 57 shows the visual impression when such a lap joint is scanned in real time. A section of rivets this long can be scanned in about 12 s (one rivet per second) with a magnetooptic eddy current imager.

Figure 56. Simulated aluminum aircraft lap joint consisting of three rows of rivets spaced 25 mm (1.0 in.) apart: (a) plane view; (b) cross section. Numerous radial surface breaking fatigue cracks indicated on upper row were induced by thousands of cycles under tensile load directed vertically in illustration. Thickness of aluminum sheets was 1.0 mm (0.04 in.).

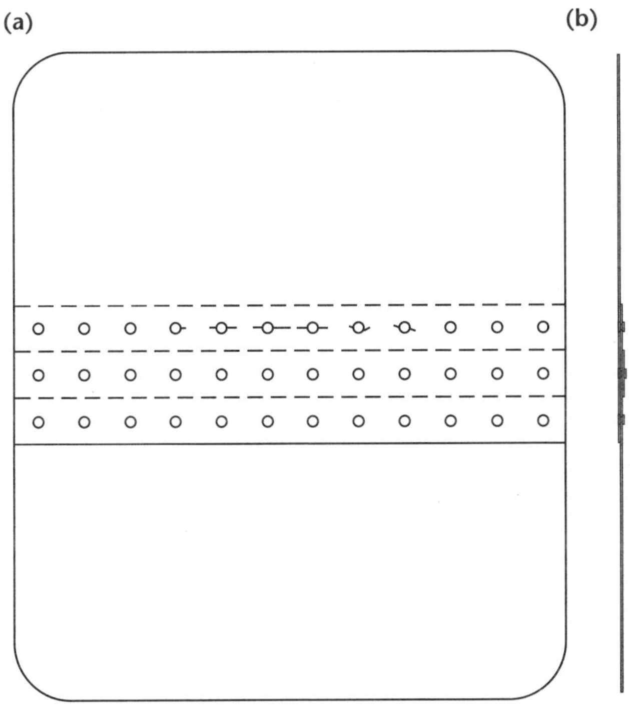

Summary

Magnetooptic imaging displays are combined with unconventional sheet eddy current induction techniques to form real time magnetooptic eddy current images of cracks and discontinuities in electrical conductors such as aging aluminum airframes. Magnetooptic eddy current images of simple test objects reveal both surface and subsurface discontinuities, such as cracks in a simulated aircraft lap joint.

Figure 57. Mosaic image of upper row of rivets spaced 25 mm (1.0 in.) apart and associated with aluminum aircraft lap joint in Fig. 56. Numerous radial surface breaking fatigue cracks are visible. Eddy current frequency was 200 kHz.

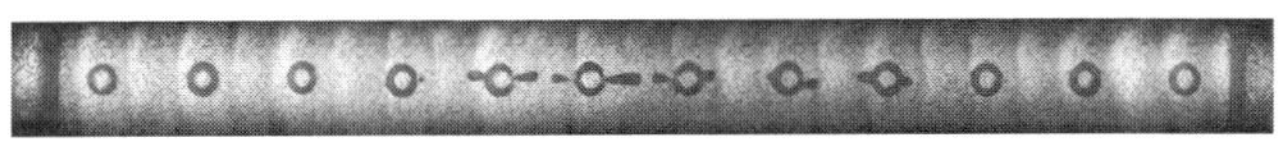

References

1. Ida, N. Section 3, "Eddy Current Transducers." *Nondestructive Testing Handbook,* second edition: Vol. 4, *Electromagnetic Testing.* Columbus, OH: American Society for Nondestructive Testing (1986): p 53-88.
2. Rogel, A.P. and J.J. Scalese. "Automatic Eddy Current Bolt-Hole Scanning System." *Materials Evaluation.* Vol. 41, No. 7. Columbus, OH: American Society for Nondestructive Testing (June 1983): p 839-843.
3. Davis, T.J. "Multifrequency Eddy Current Inspection with Continuous Wave Methods." *Materials Evaluation.* Vol. 38, No. 1. Columbus, OH: American Society for Nondestructive Testing (January 1980): p 62-68.
4. Cecco, V.S. and C.R. Box. "Eddy Current In-Situ Inspection of Ferromagnetic Monel Tubes." *Materials Evaluation.* Vol. 33, No. 1. Columbus, OH: American Society for Nondestructive Testing (January 1975): p 1-4.
5. Lassahm, G.D. "A Comparison of Three Types of Eddy Current Systems." *Materials Evaluation.* Vol. 32, No. 9. Columbus, OH: American Society for Nondestructive Testing (September 1974): p 187-192.
6. Baranger, J.P. "Eddy Current Jet Engine Disk-Crack Monitor." *Materials Evaluation.* Vol. 42, No. 11. Columbus, OH: American Society for Nondestructive Testing (October 1984): p 1374-1378.
7. Thompson, A. "Evolution of an Automated Eddy Current Inspection System." *Materials Evaluation.* Vol. 42, No. 12. Columbus, OH: American Society for Nondestructive Testing (November 1984): p 1511-1514.
8. Dilbeck, R.A. and T.J. Davis. "An Eddy Current System for High Speed Inspection of M-16 Cartridge Cases." *Materials Evaluation.* Vol. 39, No. 4. Columbus, OH: American Society for Nondestructive Testing (March 1981): p 396-400.
9. Ammirato, F.V. "Automated In-Service Eddy Current Inspection of Gas Turbine Rotor Bores." *Materials Evaluation.* Vol. 39, No. 9. Columbus, OH: American Society for Nondestructive Testing (August 1981): p 849-853.
10. Hughes, D.E. "Induction-Balance and Experimental Researches Therewith." *Philosophical Magazine.* Series 5, Vol. 8. London, United Kingdom: Taylor and Francis (1879): p 50-57.
11. Grover, F.W. *Inductance Calculations, Working Formulas and Tables.* New York, NY: Van Nostrand Company (1946).
12. Welsby, V.G. *The Theory and Design of Inductance Coils.* London, United Kingdom: MacDonald Press (1960).
13. Ida, N. and W. Lord. "A Finite Element Model for 3-D Eddy Current NDT Calculations." *IEEE Transactions on Magnetics.* Vol. MAG-21, No. 6. New York, NY: Institute of Electrical and Electronics Engineers (November 1985): p 2635-2643.
14. *Radiotron Designer's Handbook.* Sydney, Australia: Wireless Press (1952).
15. Hoshikawa, H., R.M. Li and N. Ida. "Finite Element Analysis of Eddy Current Surface Probes." *Review of Progress in Quantitative Nondestructive Evaluation.* Vol. 3A. New York, NY: Plenum (1984): p 675-682.
16. Bailey, D.M. "Shielded Eddy Current Probes." *Materials Evaluation.* Vol. 41, No. 7. Columbus, OH: American Society for Nondestructive Testing (June 1983): p 776-778.
17. Kelha, V., R. Peltonen, J. Puki, J. Heino, R. Ilomoniemi and A. Penttinen. "Design, Construction and Performance of a Large Volume Magnetic Shield." *IEEE Transactions on Magnetics.* Vol. MAG-18. New York, NY: Institute of Electrical and Electronics Engineers (1982): p 260-270.
18. Palanisamy, R. and W. Lord. "Prediction of Eddy Current Probe Signal Trajectories." *IEEE Transactions on Magnetics.* Vol. MAG-16, No. 5. New York, NY: Institute of Electrical and Electronics Engineers (September 1980): p 1083-1085.
19. Palanisamy, R. and W. Lord. "Finite Element Modeling of Electromagnetic NDT Phenomena." *IEEE Transactions on Magnetics.* Vol. MAG-15, No. 6. New York, NY: Institute of Electrical and Electronics Engineers (November 1979): p 1479-1481.

20. Ida, N. and W. Lord. "Simulating Electromagnetic NDT Probe Fields." *IEEE Computer Graphics and Applications.*" Vol. 3, No. 3. New York, NY: Institute of Electrical and Electronics Engineers (May-June 1983): p 21-28.
21. Ida, N., H. Hoshikawa and W. Lord. "Finite Element Prediction of Differential EC Probe Signals from Fe_3O_4 Deposits in PWR Steam Generators." Vol. 18, No. 6. *NDT International.* Kidlington, United Kingdom: Elsevier Science Limited (December 1985): p 331-338.
22. Ida, N., R. Palanisamy and W. Lord. "Eddy Current Probe Design Using Finite Element Analysis." *Materials Evaluation.* Vol. 42, No. 12. Columbus, OH: American Society for Nondestructive Testing (November 1984): p 1389-1394.
23. Dodd, C.V. "The Use of Computer-Modeling for Eddy-Current Testing." *Research Techniques in Nondestructive Testing.* Vol. 3. New York, NY: Academic Press (1977): p 429-479.
24. Sullivan, S.P., S.P. Smith and F.L. Sharp. *Simultaneous Absolute and Differential Operation of Eddy Current Bobbin Probes for Heat Exchanger Tube Inspection.* Report COG-96-045-I. Chalk River, Canada: Chalk River Laboratories (1996).
25. McMaster, R.C. Section 12, "Electromagnetic Tests with Hall Effect Devices." *Nondestructive Testing Handbook,* second edition: Vol. 4, *Electromagnetic Testing.* Columbus, OH: American Society for Nondestructive Testing (1986): p 315-334.
26. Stanley, R.K., T. Hiroshima and M. Mester. Section 22, "Diverted Flux Applications." *Nondestructive Testing Handbook,* second edition: Vol. 4, *Electromagnetic Testing.* Columbus, OH: American Society for Nondestructive Testing (1986): p 631-651.
27. Stanley, R.K. Section 21, "Diverted Flux Theory." *Nondestructive Testing Handbook,* second edition: Vol. 4, *Electromagnetic Testing.* Columbus, OH: American Society for Nondestructive Testing (1986): p 607-630.
28. Förster, F. "Nondestructive Inspection by the Method of Magnetic Leakage Fields: Theoretical and Experimental Foundations of the Detection of Surface Cracks of Finite and Infinite Depth." *Defektoskopiya.* Vol. 11. New York, NY: Consultants Bureau (1982): p 3-25.
29. Zatsepin, N.N. and V.E. Shcherbinin. "Calculation of the Magnetostatic Field of Surface Defects: I, Field Topography of Defect Models; II, Experimental Verification of the Principal Theoretical Relations." *Defektoskopiya.* Vol. 2. New York, NY: Consultants Bureau (1966): p 50-65.
30. *Nondestructive Testing Handbook,* second edition: Vol. 6, *Magnetic Particle Testing.* Columbus, OH: American Society for Nondestructive Testing (1989).
31. Jackson, J.D. *Classical Electrodynamics.* New York, NY: John Wiley and Sons (1962).
32. Feynman, R.P., R.B. Leighton and M. Sands. *The Feynman Lectures on Physics,* Vol. 2. Reading, MA: Addison-Wesley (1964).
33. Fitzpatrick, G.L., D.K. Thome, R.J. Skaugset, E.Y.C. Shih and W.C.L. Shih. "Novel Eddy Current Field Modulation of Magneto-Optic Garnet Films for Real-Time Imaging of Fatigue Cracks and Hidden Corrosion." *Nondestructive Inspection of Aging Aircraft.* SPIE Proceedings, Vol. 2001. Bellingham, WA: International Society for Optical Engineering (1993): p 210-222.
34. Fitzpatrick, G.L., D.K. Thome, R.J. Skaugset, E.Y.C. Shih and W.C.L. Shih. "Magneto-Optic/Eddy-Current Imaging of Aging Aircraft: A New NDE Technique." *Materials Evaluation.* Vol. 51, No. 12. Columbus, OH: American Society for Nondestructive Testing (December 1993): p 1402-1407.
35. Fitzpatrick, G.L., R.L. Skaugset, D.K. Thome and W.C.L. Shih. "New Methods for Inspecting Steel Components Using Real-Time Magneto-Optic Imaging." *Topics on Nondestructive Evaluation Series:* Vol. 2, *Nondestructive Testing and Evaluation of Infrastructure.* Columbus, OH: American Society for Nondestructive Testing (1998): p 261-277.

Bibliography

Amin, K.E. and K. Peck. "Eddy Current Sensors for Measuring Fiber Coating Thickness." *Materials Evaluation.* Vol. 56, No. 1. Columbus, OH: American Society for Nondestructive Testing (January 1998): p 53.

Cecco, V.S. "Design and Specification of a High Saturation Absolute Eddy-Current Probe with Internal Reference." *Materials Evaluation.* Vol. 37, No. 13. Columbus, OH: American Society for Nondestructive Testing (December 1979): p 51-58.

Dodd, C.V. *Solutions to Electromagnetic Induction Problems*. Ph.D. dissertation. Knoxville, TN: University of Tennessee (June 1967).

Dodd, C.V., W.E. Deeds and J.W. Luquire. "Integral Solutions to Some Eddy Current Problems." *International Journal of NDT*. Vol. 1. Kidlington, United Kingdom: Elsevier Science Limited (1969-1970): p 29-90.

Dodd, C.V., C.C. Cheng, W.A. Simpson, D.A. Deeds and J.H. Smith. *The Analysis of Reflection Type Coils for Eddy-Current Testing*. ORNL-TM-4107. Oak Ridge, TN: Oak Ridge National Laboratory (1973).

Dodd, C.V. and W.A. Simpson. "Thickness Measurements Using Eddy Current Techniques." *Materials Evaluation.* Vol. 31, No. 5. Columbus, OH: American Society for Nondestructive Testing (May 1973): p 73-84.

Dodd, C.V., J.H. Smith and W.A. Simpson. "Eddy Current Evaluation of Nuclear Control Rods." *Materials Evaluation.* Vol. 32, No. 5. Columbus, OH: American Society for Nondestructive Testing (May 1974): p 93-99.

EPCOS Data Book. Munich, Germany: EPCOS AG (2001).

Grimberg, R., A. Savin, E. Radu and O. Mihalache. "Nondestructive Evaluation of the Severity of Discontinuities in Flat Conductive Materials by an Eddy-Current Transducer with Orthogonal Coils." *IEEE Transactions on Magnetics*. Vol. 36, No. 1. New York, NY: Institute of Electrical and Electronics Engineers (2000): p 299-307.

Grubinskas, R.C. "An Evaluation of Probe Coils with Ferrite Cores for Use in Electromagnetic Testing." *Materials Evaluation*. Vol. 24, No. 10. Columbus, OH: American Society for Nondestructive Testing (October 1966): p 557-563.

Hagemaier, D.J., A.H. Wendelbo and Y. Bar-Cohen. "Aircraft Corrosion and Detection Methods." *Materials Evaluation*. Vol. 43, No. 4. Columbus, OH: American Society for Nondestructive Testing (March 1985): p 426-437.

Hall Effect Transducers. Freeport, IL: MicroSwitch (1982).

Hallet, J.B., G. Van Drunen and V.S. Cecco. "An Eddy Current Probe for Separating Defects from Resistivity Variations in Zirconium Alloy Tubes." *Materials Evaluation*. Vol. 42, No. 10. Columbus, OH: American Society for Nondestructive Testing (September 1984): p 1276-1280.

Hochschild, R. "Electromagnetic Methods of Testing Metals." *Progress in Non-Destructive Testing*. Vol. 1. New York, NY: MacMillan Company (1959): p 59-109.

Hoshikawa, H. and K. Koyama. "A New Eddy Current Probe Using Uniform Rotating Eddy Currents." *Materials Evaluation*. Vol. 56, No. 1. Columbus, OH: American Society for Nondestructive Testing (January 1998): p 85-89.

Ida, N. "Development of a 3-D Eddy Current Model for Nondestructive Testing Phenomena." *Review of Progress in Quantitative Nondestructive Evaluation* [Santa Cruz, CA, August 1983]. Vol. 3A. New York, NY: Plenum (1984): p 547-555.

Ida, N. *Three Dimensional Finite Element Modeling of Electromagnetic Nondestructive Testing Phenomena*. Ph.D. dissertation. Fort Collins, CO: Colorado State University (1983).

Lang, D.J. "Inspection of Aircraft Surfaces Using Two Single Coil Eddy Current Techniques." *Materials Evaluation.* Vol. 32, No. 4. Columbus, OH: American Society for Nondestructive Testing (April 1974): p 87-92.

Lewis, A.M. "A Theoretical Model of the Response of an Eddy-Current Probe to a Surface-Breaking Metal Fatigue Crack in a Flat Test-Piece." *Journal of Physics D, Applied Physics*. Vol. 25. Bristol, United Kingdom: Institute of Physics (1992): p 319-326.

Lord, W. and R. Palanisamy. "Development of Theoretical Models for Nondestructive Testing Eddy Current Phenomena." *Eddy-Current Characterization of Materials and Structures*. West Conshohocken, PA: ASTM International (1979): p 5-21.

Lorenzi, D.E., H. Migel and D.T. O'Connor. *Eddy Current Crack Detector System Using Crossed Coils*. United States Patent 3 495 166 (1970).

Maxwell, J.C. *A Treatise on Electricity and Magnetism*, third edition. New York, NY: Dover Press (1954).

Mottl, Z. "The Quantitative Relations between True and Standard Depth of Penetration for Air-Cored Probe Coils in Eddy Current Testing." *NDT International*. Vol. 23, No. 1. Kidlington, United Kingdom: Elsevier Science Limited (1990): p 11-18.

Nikitin, A.I. and L.V. Babushkina. "Solution of the Problem of Eddy Currents in a Conducting Sphere Situated in the Field of a Superposed Transducer." *Russian Journal of Nondestructive Testing.* New York, NY: Plenum/Consultants Bureau (1989): p 863-869.

Palanisamy, R. *Finite Element Modeling of Eddy Current Nondestructive Testing Phenomena.* Ph.D. dissertation. Fort Collins, CO: Colorado State University (1980).

Prestidge, F.L. and M.L. Stanley. "In-Core Eddy Current Measurements of ATR Fuel Element Channel Spacing." *Materials Evaluation.* Vol. 30, No. 10. Columbus, OH: American Society for Nondestructive Testing (October 1972): p 214-216.

Sabbagh, H.A. "A Model of Eddy-Current Probes with Ferrite Cores." *IEEE Transactions on Magnetics.* Vol. 1, No. 3. New York, NY: Institute of Electrical and Electronics Engineers (May 1987): p 1888-1904.

Sadeghi, S.H.H. and A.H. Salemi. "Electromagnetic Field Distributions around Conducting Slabs, Produced by Eddy-Current Probes with Arbitrary Shape Current-Carrying Excitation Loops." *IEE Proceedings: Science, Measurement and Technology.* Vol. 148, No. 4. London, United Kingdom: Institution of Electrical Engineers (2001): p 187-192.

Sapunov, V.M. and P.I. Beda. "Eddy-Current Inspection of Sheet of Nonmagnetic Material by Superposed Transducer Excited by Pulsed Current with Nonideal Shape." *Russian Journal of Nondestructive Testing.* Vol. 27, No. 10. New York, NY: Plenum/Consultants Bureau (1991): p 743-750.

Sun, Y.S., S. Udpa, W. Lord and D. Cooley. "A Remote Field Eddy Current NDT Probe for the Inspection of Metallic Plates." *Materials Evaluation.* Vol. 54, No. 4. Columbus, OH: American Society for Nondestructive Testing (April 1996): p 510-512.

Theodoulidis, T.P. "A Model of Ferrite-Cored Probes for Eddy Current Nondestructive Evaluation." *Journal of Applied Physics.* Vol. 93, No. 5. Melville, NY: American Institute of Physics (March 2003): p 3071-3078.

6

CHAPTER

Eddy Current Instrumentation

James E. Cox, Zetec, Incorporated, Issaquah, Washington

David J. Brown, Zetec, Incorporated, Issaquah, Washington

Eric J. Strauts, TEEM Electronics, Park Ridge, Illinois

PART 1. Introduction to Eddy Current Instrumentation

Four techniques of electromagnetic testing are well developed and used for commercial applications: alternating current field measurement, eddy current testing, magnetic flux leakage testing and remote field testing. Each of these four electromagnetic techniques has specific application areas where it has shown some advantage over the other three as well as other nondestructive test methods. All four share core instrumentation concepts and many of the following details are applicable to all four techniques. A general purpose eddy current instrument can be adapted to perform any of these tests.

In addition to these four techniques, several additional electromagnetic test processes provide some test capabilities but are less developed. This chapter focuses specifically on instrumentation for the eddy current technique.

Functions

Purpose

The purpose of an eddy current instrument is to drive an eddy current transducer, or probe, with an excitation signal and to analyze the signal modulated by that transducer for information pertinent to the application.

Applications for eddy current instrumentation vary from use on a manufacturing floor for verification of whether a part has been heat treated to evaluation of nuclear steam generator tubing.

Test frequencies may range from a few hertz for testing ferromagnetic parts to megahertz for testing thin titanium. The instrument may use a fixed frequency for testing a single property, as does a simple conductivity meter, or it may use multiple frequencies for discrimination of multiple conditions when analyzing heat treated materials or heat exchanger tubing.

A wide variety of instrumentation exists today. Single-frequency units with an analog meter allow an operator to distinguish discontinuities by needle position. Other single-frequency units use a bar graph to display a single parameter and trigger an alarm to allow a controller to properly dispose of a bad part. Complex, general purpose instruments may be used in multifrequency tests of up to 32 frequencies over six orders of magnitude, from 10 Hz to 10 MHz. These instruments range from handheld and desktop units to remote data acquisition systems on networks of dedicated workstations that also control robotics and run sophisticated data analysis programs.

Critical Specifications

Several specifications critical to an eddy current test must be carefully considered by the instrumentation designer.

Frequency Accuracy. The frequency is absolutely critical because it determines eddy current depth of penetration and the amplitude and phase of a discontinuity response. Distortion of the applied waveform should be minimized and quantified because distortion is caused by other frequency components in the signal.

Drive Accuracy. The amplitude of the drive determines the amplitude of the response. It should be well controlled and its frequency response should be specified.

Gain Linearity and Accuracy. The amplitude and phase characteristics of all gain stages must be qualified for adherence to a standard or specification appropriate to the application, especially so that adjustment of variable gain stages does not distort a signal.

Horizontal and Vertical Deviation. Gain of the in-phase and out-of-phase components of a signal must be controlled to prevent unwanted distortion of a signal. A flattened appearance caused by unequal gains is a liability for some applications but is very useful in others. In a typical heat exchanger test, a flattened appearance would cause misrepresentation of the data whereas, in the rotating test of a rivet hole for surface breaking cracks, it can be used to minimize liftoff noise and accentuate a crack signal.

Quadrature Accuracy. The phase of reference signals must be well controlled and the response of the display must be designed to ensure that the in-phase and out-of-phase components are truly at 90 degrees electrically and that they are displayed orthogonally.

Digitization Rate. The digitization rate for a digital system is the number of samples per unit of transducer travel. This rate is critical for determining the response to a discontinuity. Too few samples over the length of the test object may cause a discontinuity to be missed altogether or at least to be highly distorted. Digitization rate must be determined from the application criteria. If detection is the only requirement, a lower digitization rate may be sufficient to produce a signal that breaks the desired alarm level. If analysis of the signal is required, the digitization rate must be high enough that an accurate picture of the discontinuity signal is presented without distortion.

Sample Rate. The sample rate for a digital system is the number of interrogations per unit of time, often given as *samples per second*. To calculate the required sample rate for a test, the transducer speed is multiplied by the desired digitization rate. Because the sample rate is the determining factor for the digitization rate, it must be high enough to permit the required transducer speed and must be accurate so that the digitization of signals is evenly spaced along the surface of the test object(s).

Some instruments have an external sample trigger that may be activated by an encoder system measuring actual distance traveled. This trigger ensures an accurate digitization rate for varying transducer speed as long as the maximum sample rate of the instrument is not exceeded.

Bandwidth. Bandwidth is not to be confused with sample rate. Rather than the number of samples taken, bandwidth governs the response of the system as a function of frequency and is measured in hertz. It is usually determined by demodulator and filter characteristics. Exceeding the bandwidth of the instrument will cause signals to be attenuated and distorted because some of their frequency components will be filtered out. Although distortion may be acceptable in situations where detectability can be demonstrated, as with known types of discontinuities on a high speed manufacturing line, distortion may prevent detection of discontinuities or adversely affect the ability to characterize them. A variable frequency, external modulator is often used to measure and characterize bandwidth through all stages of an instrument.

Stability. System stability and response to temperature and aging variations must be demonstrated to ensure that excessive drift does not occur between calibration intervals. This is especially important in the manufacturing world where a very small alarm may be placed around a baseline signal and the equipment run continuously for months or even years at a time.

Two factors govern the design of any test instrumentation: (1) signal-to-noise ratio and (2) dynamic range. These factors are linked to one another and, in the design of an instrument, care must be taken to maximize the signal-to-noise ratio and to allow adequate dynamic range for signals of interest.

Ultimately, the goal of a nondestructive test is to produce a discontinuity signal distinguishable from surrounding noise and to view a desired range of signals without distortion.

Dynamic range can be defined as the ratio of the total signal range of an instrument divided by the amplitude of the noise signal. This range can be expressed either as a ratio or as decibels or bits in a digital system. For example, a 16-bit analog-to-digital converter with a ±10 V input may have one bit (305 µV) of noise. This noise level yields a dynamic range of 15 bits (range of 16:1 noise bit), 65 574:1 (20 V to 305 µV) or 96 dB.

If the minimally acceptable signal-to-noise ratio for a test is 3:1, then the dynamic range available for signals is the total range divided by the minimum signal or is the total range divided by three times the noise.

Noise is a very general term and can be defined as any unwanted signal affecting the test. Common sources of noise in electromagnetic testing are the following: (1) electronic noise inherent to the instrumentation; (2) external electromagnetic interference, radiated or conducted; (3) transducer imbalance; (4) triboelectric, or microphonic, noise in the transducer cabling; and (5) test object irregularities, such as wall thickness variations in a rotating pancake coil tube test. These all contribute to a reduction in both signal-to-noise ratio and dynamic range.

Each stage of the instrumentation acts as a dynamic range window, or bottleneck, and adds some amount of noise to a test.

Presentation of Eddy Current Data

Any eddy current or remote field system is measuring an alternating magnetic field as modified by a test object. The voltage signal from the transducer may be represented as a phasor in the complex plane with an amplitude and phase angle (Fig. 1). The technique used to display the phasor depends on the application of the test being performed.

In the simplest instruments, the amplitude of this phasor alone may contain the desired information. A voltage readout in the form of a meter or bar graph may constitute the only display, usually supplemented by an alarm output, triggered when the voltage exceeds a set level.

In most instruments, however, the incoming signal is processed to obtain either amplitude and phase or real and imaginary components of the phasor. This provides two known components per phasor, which allows the separation of two variables. Multiple frequencies add additional known values for discriminating additional variables. These components are most commonly displayed on an oscilloscope display, where the resultant dot moves around the display as the signal changes. Different mechanisms produce characteristic patterns of change of the signal voltage and these patterns may be scrutinized in detail to evaluate the object being tested.

To maximize the dynamic range and signal-to-noise characteristics of an instrument, various nulling (balancing) techniques are used. These basically subtract out the phasor of a null signal from a transducer on a good area of a test object so that the resultant dot is centered at the zero point of the display. The signal patterns are then rotated around the zero to show the discontinuity patterns as desired. It is common for the operator to choose a frequency at which the liftoff of the transducer from the test object lies on the horizontal axis of the display and the desired discontinuity signal displays vertically. Operators can then watch for vertical indications and ignore the liftoff signal. A number of simple eddy current instruments use this technique and display only one component on a meter or bar graph with capabilities of triggering an alarm on signals that exceed a set level.

FIGURE 1. Phasor in complex plane.

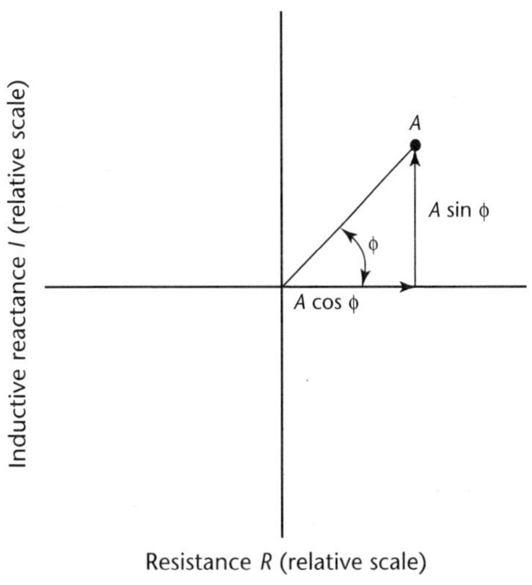

Legend

A = amplitude
ϕ = phase

A number of specialized instruments process the phasor signal for a particular quantity. Conductivity meters are a typical example. The signal is processed by algorithms that calculate the conductivity of a specimen and display it as a number on a digital readout. Some conductivity meters also calculate and provide a number for the distance of the transducer from the specimen for measurement of coating thickness. There are dedicated coating thickness gages available too.

Hugo Libby laid his groundwork measuring the impedance of eddy current coils with which he performed his test.[1] He called the complex plane on which he displayed data the *impedance plane.* The impedance plane has become the most common display for electromagnetic test data, whether used with impedance probes, reflection (driver pickup or send/receive) probes or even display of remote field data.

When Libby built his first eddy current instrument, he attached the cables to the oscilloscope to display the in-phase component (which he called *imaginary*) on the horizontal axis and the out-of-phase component (called *real*) on the vertical axis. He then rotated the field to get the real component back onto the horizontal axis. This rotation leaves a phase angle of zero on the left horizontal axis and phase angle increasing in a clockwise direction. This is in contrast to the true complex plane, which has a phase angle of zero on the right horizontal axis and phase angle increasing in the counterclockwise direction. This interesting artifact is often confusing for engineers new to nondestructive testing. In practice, most testing is done relative to a reference standard with the field rotated such that *zero phase* is defined to be a particular indication, most often the transducer liftoff signal.

Part 2. Eddy Current System Functions

For an eddy current system to provide information to an inspector, five functional steps have to be performed (Fig. 2): excitation, modulation, signal preparation, signal demodulation and signal display. An optional sixth step would be test object handling equipment.

Excitation

The excitation portion of an eddy current instrument consists of signal generation and amplifiers to drive the transducers.

The signal generator (or oscillator) provides sine wave excitation for the test coil. Single-frequency systems have one fixed frequency whereas multifrequency systems can apply several frequencies to provide multiple-parameter options. The application determines the required frequency and the number of frequencies to be used. Additional frequencies are selected to provide additional test parameters for specimens with multiple variables. In this way, the number of measured parameters will equal or exceed the number of variables to allow discrimination of all desired discontinuity types. More complex test specimen problems require more sophisticated instrumentation to test the specimens adequately.

Early systems had what were called *frequency modules,* designed to generate one frequency only. If a different frequency was wanted, then the physical module was replaced with another that would operate at a different frequency. Most modern systems have a frequency generator that will operate over a very wide frequency range. This means that the operator can make the best choice of frequencies to apply to a given application. On analog systems, this would be controlled through a switch. On digital systems, it can be accomplished through defining the digital parameters on a menu of available options.

The signal generators themselves vary from simple, fixed frequency, schmitt trigger oscillators, to phase locked loops, sine lookup tables and digital data synthesizers. Some digital systems filter a square wave clock to provide an adequately clean sine wave to apply to the transducer. The design emphasis in this stage is to produce a signal of adequate frequency accuracy, frequency stability and low distortion for a reliable test. If frequency is inaccurate, depth of penetration is not as expected and analysis of any resulting signal is

Figure 2. Internal functions of eddy current test instrument.

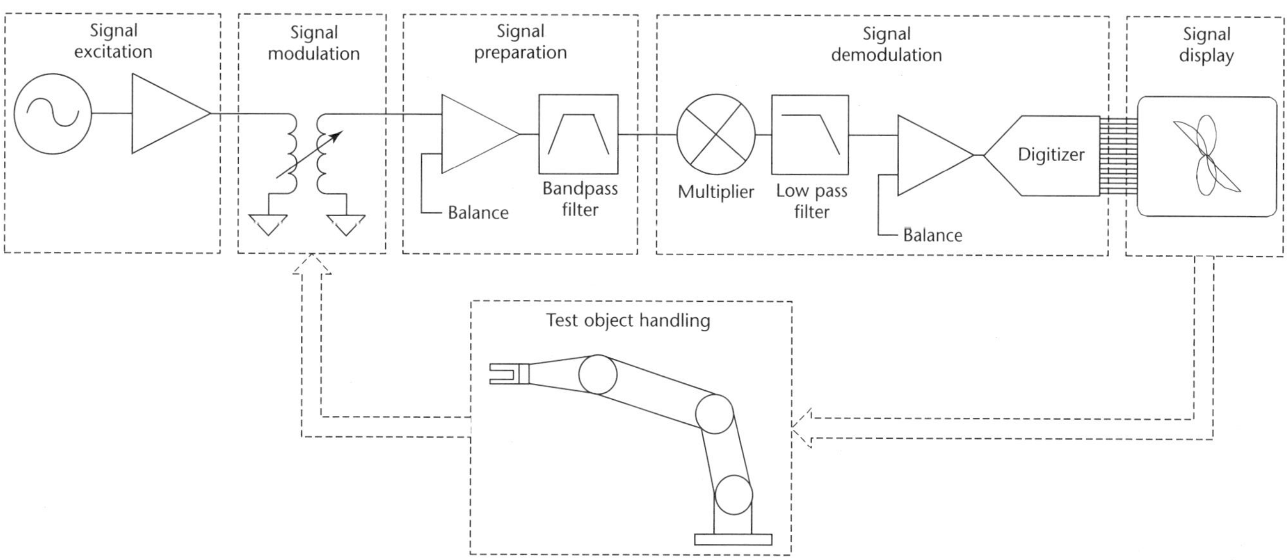

correspondingly inaccurate. If the frequency is allowed to drift, results will vary accordingly throughout the test, with the result that good parts may be rejected or that critical discontinuities may be ignored — or both. Excessive distortion on the excitation signal introduces into the test additional frequencies that may cause undesired signal results or may simply reduce the signal-to-noise ratio of the test. A certain amount of broadband electronic noise is present on any signal. It is important that it be minimized here because excitation is the beginning of the signal train through the instrument.

Many instruments also derive the timing functions for the demodulation stage from the oscillator. The amplitude and frequency characteristics must be well controlled because any inaccuracy, frequency jitter or voltage noise on these reference signals will be added into the test signal in the demodulator stage. The quadrature, the relative timing, of these signals must be accurately controlled so that the in-phase and out-of-phase components of the test signals are accurately demodulated at 90 degrees from each other.

Modulation

Multiplexing

Single-frequency testers apply one selected frequency to the transducer. This frequency may be permanently set as in a 60 kHz conductivity meter, which uses a dedicated coil to discriminate liftoff and conductivity.

Single-frequency instruments intended for general purposes allow selection of the frequency to fit the application and are capable of driving a wide range of transducers.

Multifrequency testers are normally set up to generate from two to 32 different frequencies selectable by the operator. Multifrequency testing is accomplished through three different techniques: (1) time domain multiplexing, (2) frequency domain multiplexing and (3) pulsed frequency testing.

Multiplexing means sending multiple frequencies over a single channel to the test coil. Instruments using time domain and frequency domain multiplexing generate and switch among a number of discrete frequencies. Each technique has advantages and disadvantages. Some instruments can multiplex in the frequency domain, in the time domain or in both concurrently.

With time domain multiplexing, multiple frequencies are generated sequentially (Fig. 3). In digital systems, a number of frequency time slots are selected. Analog systems usually have a fixed number of time slots. The coil is energized at one frequency for a predetermined period of time. The frequency generator is then switched to a different frequency in the second time slot for another specific period of time. This process continues until all selected frequencies have been applied to the coil. The process then starts again with frequency number one. Timing of each specific frequency is critical to accurately process and display the information from each individual time slot on the output device.

With frequency domain multiplexing, multiple frequencies are applied to a coil at the same time, so this means of multiplexing is commonly called *simultaneous injection* (Fig. 4). In this scenario, all of the selected frequencies are applied to the coil continuously. The recurring pattern in Fig. 4 is not caused by sequential input of frequencies as in the time domain multiplexing of Fig. 3.

Pulsed frequency instruments apply a unit pulse to a transducer. A unit pulse is a short, rapidly rising and falling pulse containing an infinite series of harmonics. This pulse can be mathematically shown to contain these harmonics. (Harmonics are an infinite series of frequencies that are multiples of the base frequency.) These instruments usually display a time based result and allow gating of the signal at selectable time intervals to analyze the data for a specific frequency. This gating is very familiar to operators of ultrasonic testing equipment and provides a

FIGURE 3. Time domain multiplexed waveform.

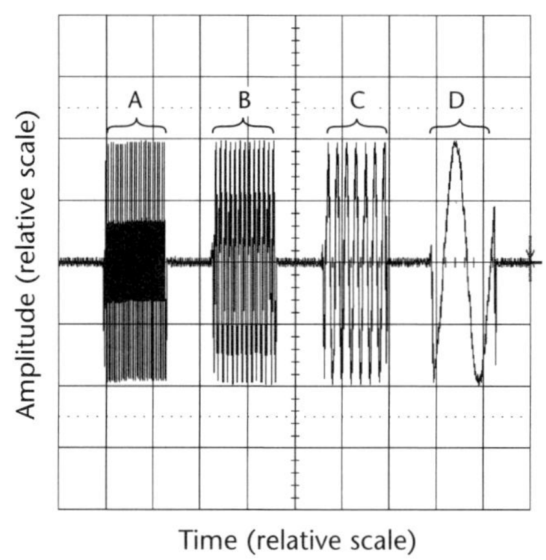

Legend

A. 400 kHz.
B. 200 kHz.
C. 100 kHz.
D. 20 kHz.

corollary to the time of flight of a sound wave.

The advantages of time domain multiplexing are that it permits maximum power to be applied to the transducer and that it permits maximum use of the dynamic range of an instrument at each discrete frequency. Drive and gain stages can be optimized for each time slot. Optimizing can be very helpful for analyzing a wide range of frequencies on the same transducer, where the response at certain frequencies may be much greater than at others. A large number of frequencies can be used for a single test with instruments capable of 32 time slots, each addressable to a different frequency. The test frequency applied during an individual time slot in a well designed system is monotonic and free of spurious signals. Although hardware is minimized, timing processes become very critical and adequate settling time must be provided when switching time slots.

Disadvantages of time domain multiplexing include ringing on high inductance coils. Ringing is feedback that occurs when a coil is switched on or off. Ringing may be suppressed in some applications with passive networks at the transducer interface. This technique does not lend itself to remote field testing, because the high inductance coils respond to the multiplexing frequency more than to the multiplexed waveforms.

The advantage of frequency domain multiplexing is the application of a continuous waveform to the transducer. There are no switching transients between time slots, so there is no ringing of high inductance coils. This lack of ringing is advantageous particularly in low frequency situations such as remote field testing. Some fixed frequency instruments with wide frequency separation may have very high bandwidths and, if they have the same demodulation filters, may be capable of testing at higher speeds than time domain multiplexed instruments. In practice, however, the additional filtering required to separate the multiple frequencies narrows the bandwidth, so the speed capabilities are somewhat similar. The tradeoff is between frequency separation and bandwidth. If selected test frequencies are separated by less than the pass band of the demodulators, then there is interference between the test frequencies. Some instruments use bandpass filters to separate the individual frequencies before the demodulator stage in order to increase the dynamic range of each demodulator and these then limit the test frequency separation.

The major disadvantage of frequency domain multiplexing is that all frequencies share the dynamic range of the instrument. Total drive to the transducer is divided among the individual frequencies. Gain in the amplifier stages and dynamic ranges of the demodulators in some instruments are limited by the response of the frequency that yields the greatest signal. With an increasing number of frequencies applied, the drive and dynamic range of each frequency is reduced.

Pulse frequency instruments have the advantage of showing a response over a wide sweep of frequencies and allowing analysis of the frequency response itself. The dynamic range is limited by the maximum response and the signal-to-noise ratio diminishes with separation from the frequency at which this occurs. A disadvantage is that the drive power is spread over the entire spectrum. The advantages of these instruments are in data display and data analysis. The circuitry is very simple, requiring little hardware, and most of the processing is performed digitally.

FIGURE 4. Frequency domain multiplexed waveform, for simultaneous application of frequencies at 400, 200, 100 and 20 kHz.

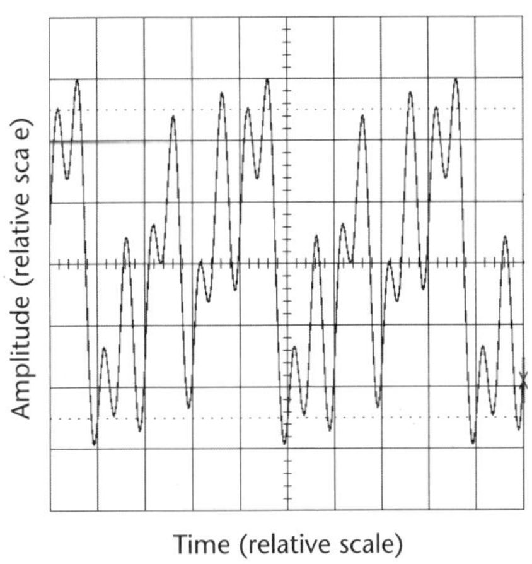

Drive Configuration

The eddy current transducer is a coil that presents an inductive load, so the total power is given with the compound unit of volt ampere (V·A), accounting for the phase angle between the current and the voltage being driven. This power appears greater than the true power (in watts) dissipated by the current over the resistance of the load.

The other part of the instrument required for excitation is the drive amplifier, which has many variants tailored to different applications. Handheld testers usually have a drive capability limited to a few volts to conserve battery power whereas remote field testers may apply up to 40 V·A to a coil to project a magnetic field through thick walled ferromagnetic tubing. The basic requirement of the driver is to have the current capability to drive the specified voltage to the transducer over the specified frequency range. There are three basic drive types used in eddy current instruments: absolute, differential bridge and driver pickup (Fig. 5).

An absolute drive consists of an amplifier powering a single-coil transducer through a drive resistor. The voltage change at the junction of drive resistor and coil is monitored with a single-ended amplifier. Impedance changes in the coil modulate this voltage. This technique has an inherently limited dynamic range because there is always a background noise level of the applied waveform using part of the dynamic range of the input amplifier, limiting its sensitivity.

FIGURE 5. Drive coil arrangements: (a) absolute; (b) differential bridge; (c) driver pickup, shown with differential driver pickup.

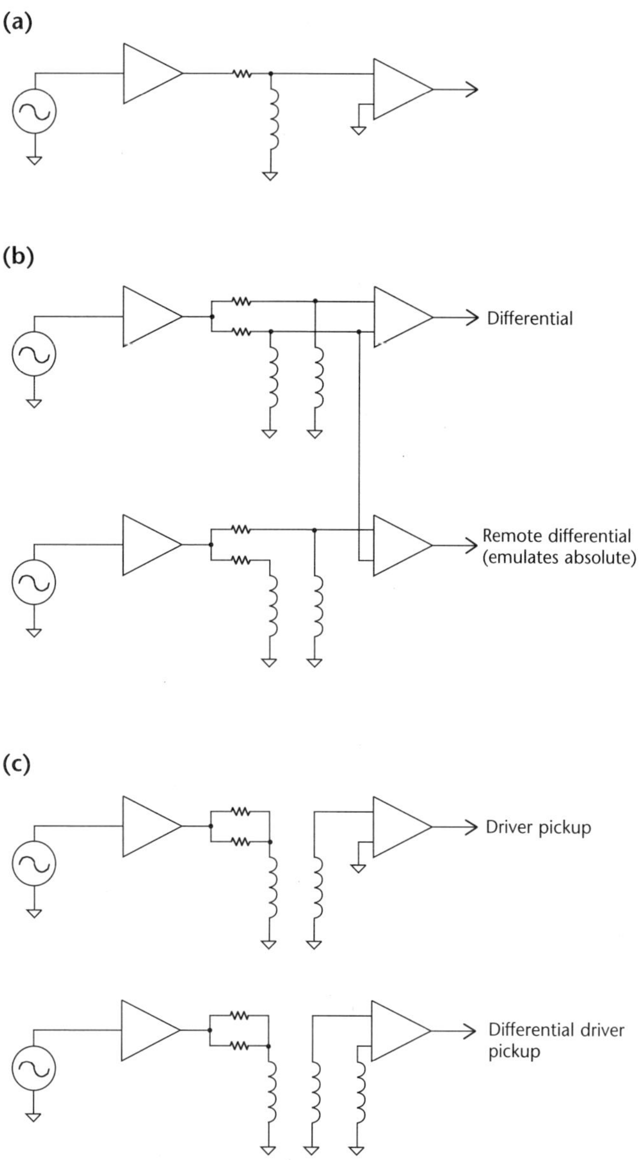

The differential bridge system reduces the handicaps of the absolute system by driving two coils through separate resistors from the same driver. The voltage across one coil is then subtracted from the voltage across the other coil and the difference becomes the test signal. In a well designed system, with balanced transducers at equal temperature and on identical test specimens, the residual carrier signal will be zero and the full dynamic range of the instrument will be available for the test. All of the inherent advantages of a differential bridge minimizing noise common to both transducers allow for a highly sensitive test with great noise rejection.

It should be noted that the term *absolute* is commonly applied to a remote differential test. In a remote test, a reference transducer is applied to a known test object or reference standard and a separate test transducer performs the actual test. The resultant signals form identically to those from a true absolute probe but have the advantages of wide dynamic range and noise rejection of the differential bridge. One limitation of this test is that the two transducers are in physically separate locations — noise may appear on one transducer but not the other and hence not be cancelled out by the bridge configuration. Like any absolute test, this test technique is susceptible to electromagnetic interference, as well as to temperature differences between the two transducers and between test and reference parts.

Drive impedance is an important specification to consider. Most eddy current instruments operate at either 50 or 100 Ω. Because most eddy current probes consist of a coil on the end of a cable, they appear electrically as a resonant tank circuit and have the highest sensitivity near the resonant frequency. The tradeoff is that higher impedance instruments have greater sensitivity whereas lower impedance instruments have greater bandwidth. Correspondingly, a 50 Ω probe on a 100 Ω instrument will have a peak response at twice the frequency as on a 50 Ω instrument.

The driver pickup technique has as many names as variations. It is also known as the *exciter pickup, send/receive* and *reflection* technique. The basic concept is that the drive coil or coils in the transducer are coupled only by magnetic field to the sensor or sensors in the transducer. Sensors are usually coils but may also be hall effect devices or magnetoresistors.

The drive amplifiers for driver pickup transducers are intended to create the magnetic field. In some instruments, a current driver circuit is used to maintain a constant current and therefore a constant magnetic field if the impedance of the driver coil changes. This constant field is especially useful for eliminating the secondary effects of the changing driver impedance in an absolute measurement device such as a conductivity meter. Some instruments use a bridge drive, driving either side of a drive coil at 180 degrees phase difference to apply greater current to high impedance coils.

The remote field technique is inherently a driver pickup technique. It uses a characteristically high impedance drive coil driven by an amplifier capable of supplying tens of volt amperes to create a magnetic field that can penetrate ferromagnetic tubing at the frequencies required.

PART 3. Demodulation

Signal modulation occurs in the electromagnetic field of the transducer assembly. It is the primary magnetic field created by the transducer that provides the energy transfer into the test specimen. This magnetic energy is modulated by the test specimen and the resultant signal is returned to the instrument for processing.

Preparation

After modulation, the signal is processed for demodulation and analysis. The purpose of this step is to amplify the probe signal and reject extraneous noise. This part of the instrument may consist of a single-ended amplifier in an absolute or simple driver pickup system or differential amplifiers for a differential bridge or a more sophisticated system of driver and pickup. Eddy current signal levels at the transducer are often in the tens of microvolts. Remote field systems have very high gain input amplifiers capable of resolving signals down to the nanovolt level.

FIGURE 6. Residual carrier signal and dynamic range of instrument.

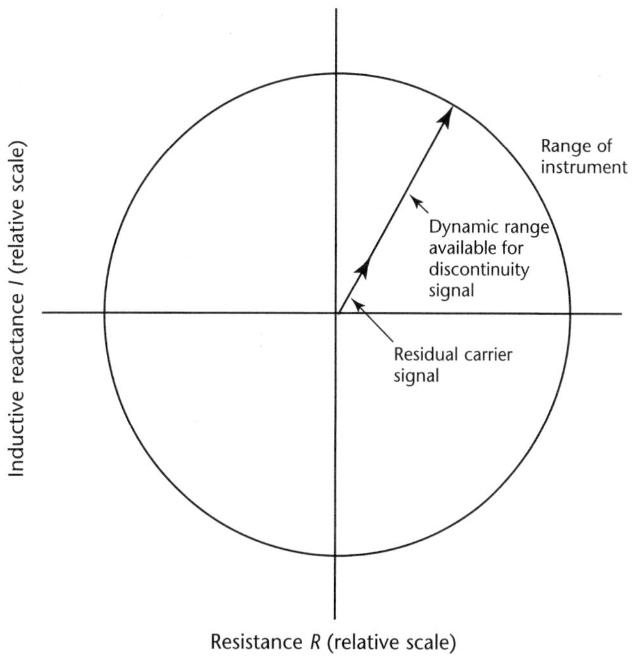

The noise to be rejected at this point consists of external electromagnetic interference, thermal variations in the transducer and a residual carrier signal due to transducer imbalance (Fig. 6). Electromagnetic interference and thermal variations can be minimized with a differential input that rejects these as common mode noise. This stage may also include a balancing network to minimize the residual carrier signal and maximize the dynamic range of the signal fed to the demodulator (Fig. 7).

Ideally, the balanced network should be the transducer. With a minimized carrier signal and balanced impedance into differential inputs, an amplifier can be configured for maximum common mode noise rejection. For a single-frequency instrument, a balancing network can be incorporated as part of the instrument input. The network may consist of a balance coil (for an absolute probe), an adjustable balance impedance or an adjustable bridge balance (Fig. 8).

Carrier suppression by signal injection is also used in some instruments to null out the residual signal (Fig. 9). In this technique, the unbalanced transducer signal is summed with the carrier signal and a carrier signal that is phase shifted, usually by 90 degrees. By adjusting the amplitude of these summing signals, the residual carrier can be nulled out. This technique can be applied to both time domain multiplexed and frequency domain multiplexed, multifrequency

FIGURE 7. Inductive carrier suppression controls.

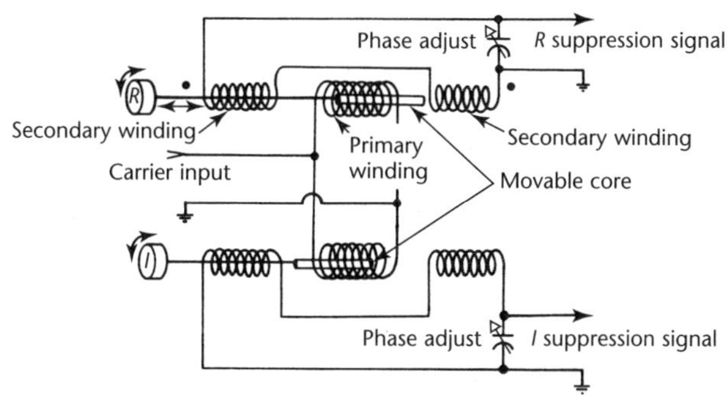

Legend
I = inductive reactance
R = resistance

systems. It should be noted that whenever a signal is added into the system, some amount of noise is also added, reducing the signal-to-noise ratio of the entire system. It is therefore very important to ensure the minimum noise possible on any injected waveforms. In addition to this, the input gain in these systems has to be lower to accommodate the transducer imbalance, reducing the signal size and therefore reducing the signal-to-noise ratio at the input stage. The cleanest signals possible will always come from a well balanced transducer.

Some frequency domain multiplexed instruments also contain bandpass filters in the signal preparation stage to separate the individual frequencies and maximize the dynamic range of the demodulator for each frequency. Bandpass filtering works the best for fixed frequency instruments because adjustable frequency bandpass filters are difficult and expensive to construct.

FIGURE 8. Bridge balance with *I* and *R* controls.

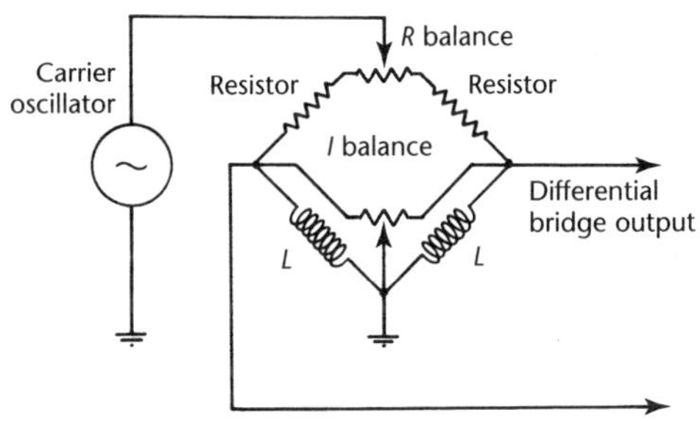

Legend
I = inductive reactance
L = inductance
R = resistance

FIGURE 9. Carrier suppression by signal injection.

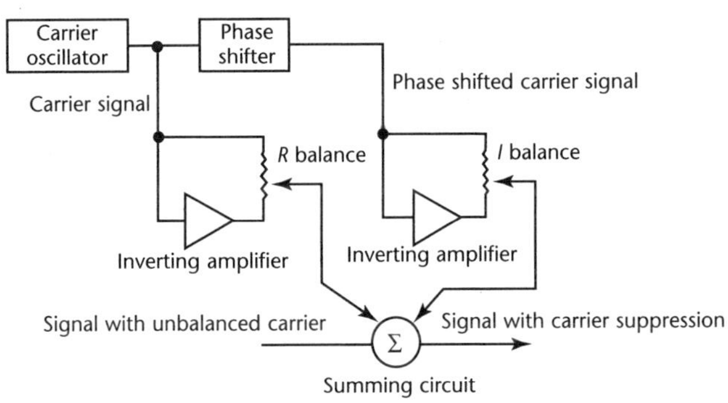

Legend
I = inductive reactance
R = resistance

Demodulation and Analysis

In the next functional step of instrumentation, data about the test object are extracted from the carrier signal.

If only a single parameter such as hardness or presence of a crack is being measured with no other signals present, then only the amplitude of the signal is required. An amplitude detector can be as simple as a single diode detector followed by a resistor and capacitor integrator (Fig. 10). This design will yield a signal following the envelope of the carrier as it is modulated by the transducer. The bandwidth of the system is determined by the integrator. A precision rectifier may be constructed using operational amplifiers followed by a low pass filter, which eliminates diode drops and distortion to provide a signal that follows the envelope of the carrier with high accuracy. The bandwidth of the system is determined by the low pass filter. Very high linearity detectors can also be constructed using phase locked loops.

A phase sensitive detector is required to reduce the signal for both phase and amplitude information. The classic lock-in amplifier or synchronous demodulator technique is widely used to perform this function. The signal from the transducer is demodulated by a reference signal in phase with the drive waveform and by a reference signal 90 degrees out of phase from the drive waveform. These in-phase and out-of-phase components of the signal are then filtered and presented as the horizontal and vertical components of the phasor representing the signal waveform. This synchronous demodulation essentially subtracts the carrier frequency and the postdemodulator filter appears as a bandpass filter centered at the test frequency.

The simplest circuit is the half wave averaging phase sensitive detector, in which a simple field effect transistor switch is driven by the reference waveform and switches the signal into a resistor capacitor filter (Fig. 11). An

FIGURE 10. Diode amplitude detector with capacitor integrator.

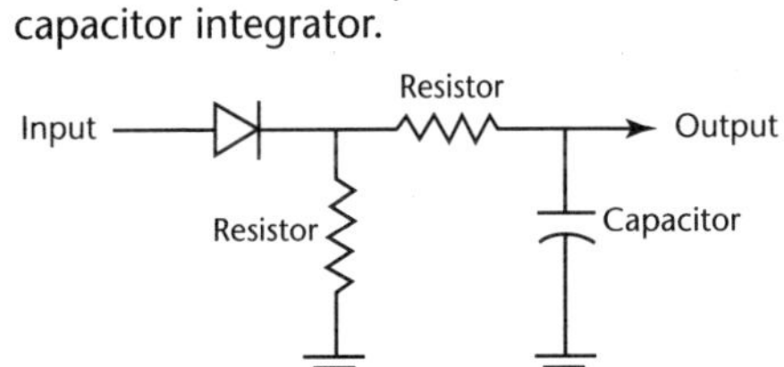

inverting amplifier may be added to produce a full wave averaging, phase sensitive detector. The phase sensitive detector provides bipolar switching and yields lower ripple, allowing less filtering and therefore greater bandwidth (Fig. 12). A single-diode phase sensitive detector can be used for sine wave demodulation as well (Fig. 13).

A sampling phase sensitive detector takes samples of the test waveform at two points separated by 90 degrees and usually digitizes these points directly (Fig. 14). This approach allows a fast system response but requires a very clean signal because the detector is very sensitive to whatever noise is present at the instant of sampling and has no postdemodulator filtering. Pulsed eddy current instruments may use this technique to sample at a specific time delay to obtain response relating to a particular frequency.

The four-quadrant multiplier is the most accurate demodulator (Fig. 15). It has no diode drops or switch artifacts to cause nonlinearity. In a four-quadrant multiplier, both the reference and the signal inputs may swing in both positive and negative directions. This flexibility is in contrast to the operation of a two-quadrant multiplier (Fig. 11).

Integrated circuit multipliers with low noise and high linearity are available off the shelf from manufacturers. Even so, the multiplier is usually the noise bottleneck of the system. It has a characteristic noise level and a limited input range, so it is desirable to maximize the use of that input range for optimal signal-to-noise performance.

FIGURE 11. Half wave averaging, two-quadrant, phase sensitive detector: (a) circuit diagram (V_{ref} = 1 when switch closed; V_{ref} = 0 when switch open); (b) V_{sig} in phase with V_{ref}; (c) V_{sig} 90 degrees out of phase with V_{ref}; (d) V_{sig} 180 degrees out of phase with V_{ref}. Shading indicates part of signal that is being averaged.

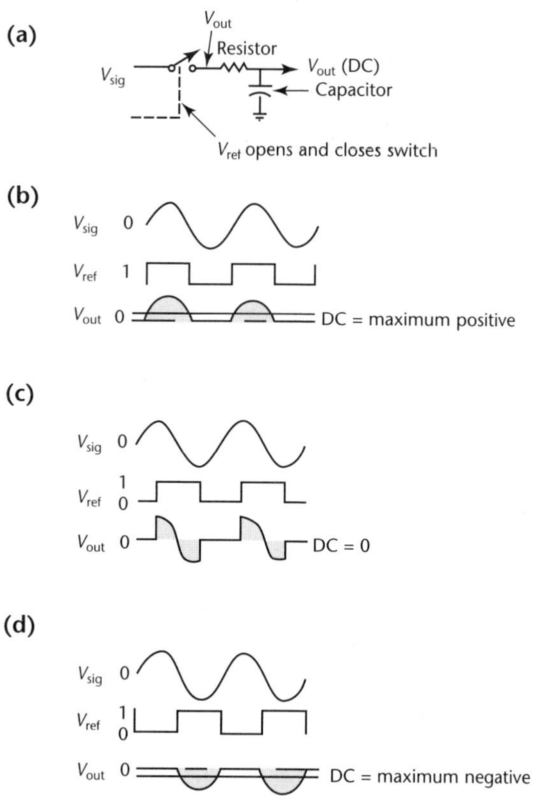

Legend

DC = averaged voltage of direct current component of waveform
V_{out} = output signal amplitude
V_{ref} = reference signal amplitude
V_{sig} = test signal amplitude

FIGURE 12. Full wave averaging phase sensitive detector: (a) circuit diagram; (b) V_{sig} in phase with V_{ref}; (c) V_{sig} 90 degrees out of phase with V_{ref}; (d) V_{sig} 180 degrees out of phase with V_{ref}.

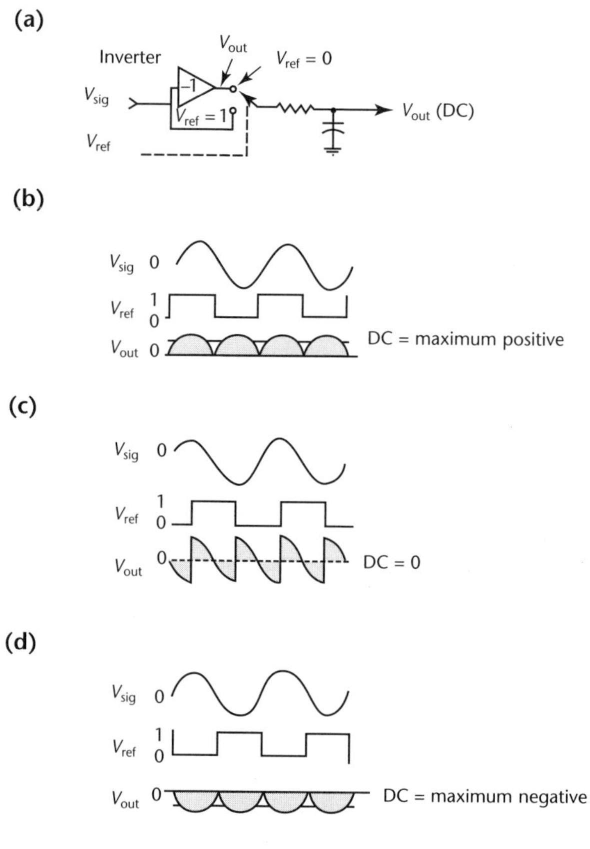

Legend

DC = averaged voltage of direct current component of waveform
V_{out} = output signal amplitude
V_{ref} = reference signal amplitude
V_{sig} = test signal amplitude

The reference signals used for demodulation may be either sine waves or square waves. The square waves are easily generated from logical timing signals and require little hardware to produce a clean and accurate reference. The signal can be expanded in a series of odd harmonic terms:

$$(1) \quad V = \frac{4V_m}{\pi} \sum \sin(\omega_0 t) + \frac{1}{3}\sin(3\omega_0 t) + \frac{1}{5}\sin(5\omega_0 t) + \ldots + \frac{1}{n}\sin(n\omega_0 t)$$

where n is the nth frequency term, V is the amplitude (volt) of the signal at any time t (second), V_m is the maximum or peak amplitude (volt) and ω_0 is 2π times the fundamental frequency (hertz).

This results in demodulation of the odd multiples of the carrier frequency as well. In a frequency domain multiplexed system, this would prevent the instrument from running frequencies that are odd multiples of one another. In any system, this restriction makes the test susceptible to noise occurring at those frequencies. For example, a 300 kHz test being performed on a steam generator may be contaminated by 900 kHz noise from a nearby motor drive that is actually switching at 100 kHz but generating an odd harmonic that coincides with one being demodulated in the eddy current instrument.

The purest means of demodulation then is to put a sine wave reference signal into a high quality multiplier. This demodulation requires extra hardware to generate a second reference sine wave at

FIGURE 13. Single-diode phase sensitive detector: (a) circuit diagram; (b) V_{sig} in phase with V_{ref}; (c) V_{sig} 90 degrees out of phase with V_{ref}; (d) V_{sig} 180 degrees out of phase with V_{ref}.

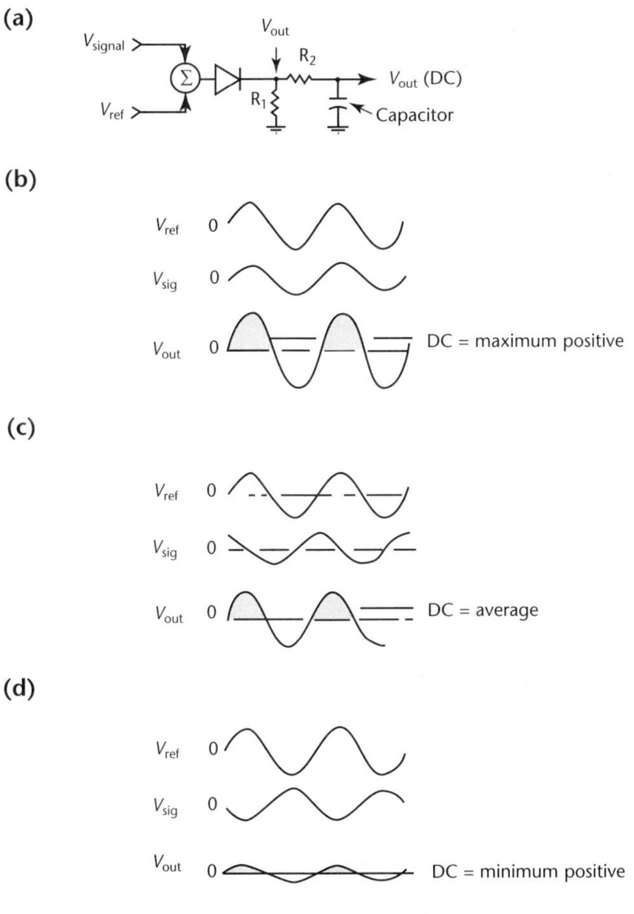

Legend

DC = averaged voltage of direct current component of waveform
R_1, R_2 = resistors
V_{out} = output signal amplitude
V_{ref} = reference signal amplitude
V_{sig} = test signal amplitude

FIGURE 14. Sampling phase sensitive detector: (a) circuit diagram; (b) V_{sig} in phase with V_{ref}; (c) V_{sig} 90 degrees out of phase with V_{ref}.

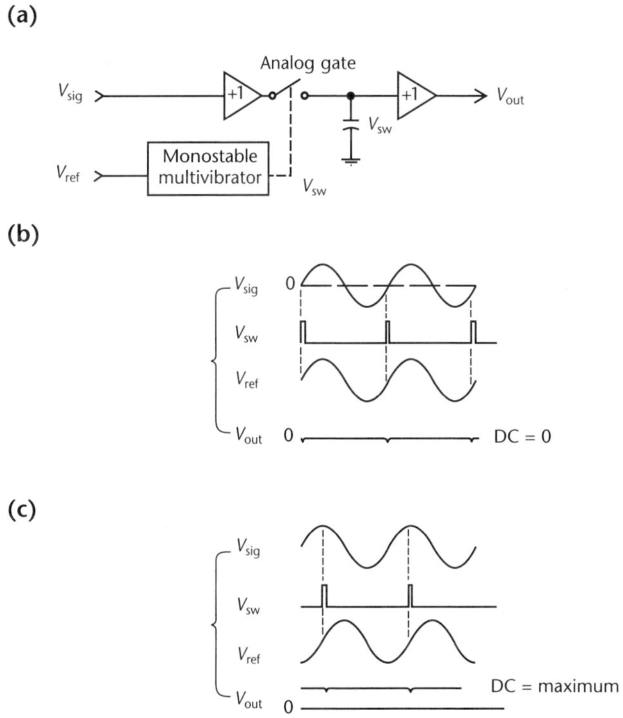

Legend

DC = averaged voltage of direct current component of waveform
V_{out} = output signal amplitude
V_{ref} = reference signal amplitude
V_{sig} = test signal amplitude
V_{sw} = sampled waveform amplitude

90 degrees out of phase from the driving waveform. When these two reference signals are clean and free of harmonic distortion, the in-phase and out-of-phase points will be the truest representation of the end points of the phasor that represents the test signal.

Note that an eddy current transducer acts as a bandpass filter, filtering out noise that may be present on the driving signal and yielding a signal actually cleaner than the reference signal. If this reference signal is used for demodulation, the noise is put right back into the signal. Therefore, attention must be taken to properly filter the sine wave references in order to minimize noise. It is easier and requires much less hardware to generate a very low noise square wave for demodulation than it is to produce a sine wave of similar noise level in a digital system, especially when that system is required to operate over a broad frequency range.

FIGURE 15. Analog multiplier, four-quadrant, phase sensitive detector: (a) circuit diagram; (b) V_{sig} in phase with V_{ref}; (c) V_{sig} 90 degrees out of phase with V_{ref}; (d) V_{sig} 180 degrees out of phase with V_{ref}.

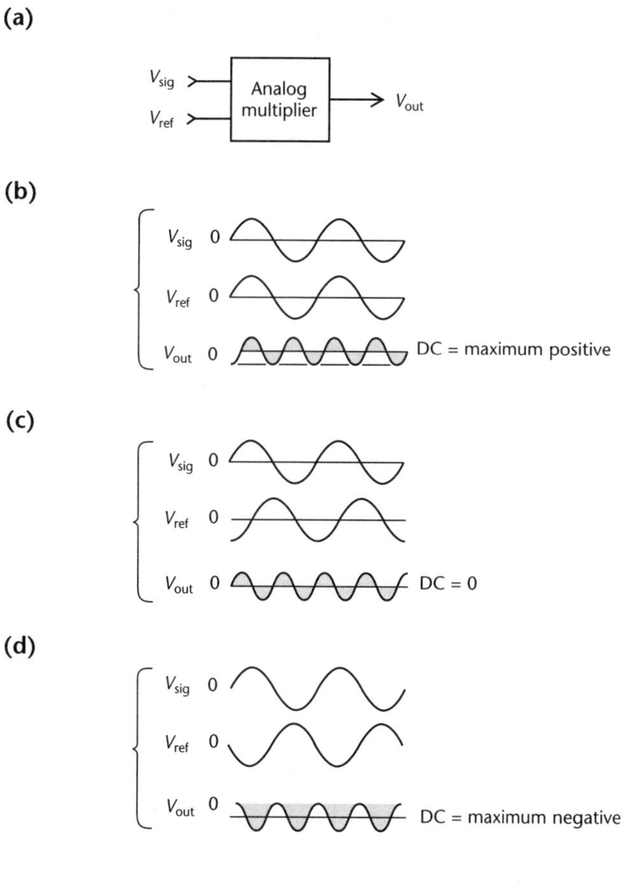

Legend
DC = averaged voltage of direct current component of waveform
V_{out} = output signal amplitude
V_{ref} = reference signal amplitude
V_{sig} = test signal amplitude

Time domain multiplexed instruments require a single set of demodulators for each transducer input. The reference waveforms are multiplexed synchronously with the test signal and the individual frequencies are demodulated in a serial fashion. Frequency domain multiplexed instruments require a set of demodulators operating at each frequency for each transducer input. The synchronous demodulator acts as a bandpass filter centered on the carrier frequency. The postdemodulator filter on each channel must be narrow enough to reject the other test frequencies. Time domain multiplexed instruments typically have much greater flexibility with frequency selection because the filters are not required to reject the other frequencies.

After the signal has been demodulated, the system presents a useful representation of the condition of the test object. At this point, the results can be analyzed by many means. The signal may be directed to an analog meter or bar graph for display or it may be digitized for further manipulation and analysis.

Simple instruments used for accept/reject testing of easily distinguishable components on a manufacturing line need only a rudimentary display and will have a comparator that analyzes the signal to determine if it exceeds a set limit. The only controls on such an instrument are frequency, gain, phase and alarm level. The frequency is determined by the application. The gain is set to get an adequate signal, the phase is rotated so that the parameter of interest is maximized on the display, the alarm level is set and the instrument is ready. The alarm may be audible or visible for indication to a technician or it could activate a sorting gate or a marking or cutting device.

Analog instruments requiring two or more parameters in analysis may use the in-phase and out-of-phase signals as horizontal and vertical components for display on a cathode ray tube.

These analog instruments are limited in bandwidth only by the operating frequency, the number of frequencies used and the bandwidth of the filters. They can be constructed to perform extremely fast testing. For example, there are instruments with up to 60 kHz bandwidth for use in bar, tube and wire applications at part speeds up to 150 m·s^{-1} (29 500 ft·min^{-1}).

Digital instruments use an analog-to-digital converter to digitize the in-phase and out-of-phase components for analysis and display. Time domain multiplexed instruments use a pair of analog-to-digital converters for each input and digitize the signal from each frequency sequentially. Frequency domain multiplexed instruments may have a pair of analog-to-digital converters sampling simultaneously for each input or they may have simultaneous sample-and-hold circuits that are then sequentially switched into a pair of analog-to-digital converters.

The resolution and range of the analog-to-digital converters is critical. Generally, converters with a larger number of bits have better signal-to-noise performance. The greater the number of bits, the greater the resolution, so a converter must be selected for the resolution desired for the application. With a ±10 V input, an eight-bit analog-to-digital converter has 78 mV per bit resolution, a 12-bit analog-to-digital converter has about 5 mV resolution and a 16-bit analog-to-digital converter has 305 µV resolution. To take advantage of 16 bits of resolution, the instrument noise must be less than 1 mV.

Analog-to-digital converters have an inherent noise floor and a limited input range, so null and gain circuitry is often used before the analog-to-digital conversion to maximize the dynamic range of the signals being digitized. Generally a digital-to-analog converter is used to generate a signal subtracted from the test signal to obtain a null value. The transducer is normally placed on a known good part or on a clean reference standard and a software routine calculates the null values required and writes them to the digital-to-analog converter. A digitally adjustable gain may then be applied to the signal to maximize use of the dynamic range of the analog-to-digital converter.

Part 4. Output

Display

The displays of analog instruments are relatively simple, enabling the user to view and perform simple manipulations of raw data. Generally, controls for phase, gain, alarm levels and some filtering are available. Analog recording of data via magnetic tape and strip chart recorders was common in the twentieth century but has largely been replaced by digital data storage. Analog instruments are used in a few niche applications.

Digital data are generally displayed in a complex plane presentation with supporting strip chart and C-scan displays as required by the application. The point described by the in-phase and out-of-phase components of the signal is displayed as a flying dot and the digital capabilities of the instrument allow variable persistence, centering of the dot, rotation of the signal and scaling of the display. Digital systems allow setup of calibration curves constructed from stored data and automated analysis of signals as compared to these curves.

Digital conductivity meters, calibrated from conductivity reference standards, feed subsequently acquired data into algorithms that calculate conductivity and distance of the conductivity probe from the material surface for display in a numerical format.

Filtering[2]

Digital systems allow considerable filtering capability. The primary use for filtering of demodulated signals is to separate desired from undesired signals generated by the eddy current test. For instance, if the signal contains a low frequency component from probe motion and a high frequency component from a crack to be detected, then a high pass filter could be used to attenuate the low frequency component from the probe motion while still passing the high frequency component from the crack.

Time differentiation or high pass filtering of test signals can be used if the change in the state of the eddy current signal conveys useful information but the steady state output has no useful information or masks the effect of the useful information. An example of this is the detection of cracks in wire or tubing with a through-coil transducer. The steady state signal does not convey useful information. The signal will be generated by motion of the wire laterally within the coil or by the drift of the test system. These effects would interfere with the system's ability to detect discontinuities of a dynamic, time changing nature due to the constant motion of the wire through the coil. By differentiating the signal from the demodulator, the effects of drift and lateral motion will be rejected whereas the discontinuity signals will still be seen.

Time integration of test signals can also be used to minimize undesired information. When unwanted noise is present, it will often have a random and widely spread frequency distribution whereas the desired signal will not. If a proper integration time constant is selected, random noise can often be reduced whereas the desired signal is left relatively unchanged.

Time domain differentiation and integration can also be interpreted in the frequency domain. Differentiation removes direct current components and enhances high frequency components. It acts as a high pass or low cut filter. Conversely, a time integrator accentuates direct current components and tends to suppress high frequency components. It acts as a low pass or high cut filter.

These filters can be combined with the proper selection of cutoff frequencies (time constants) both to remove low frequency components from drift and to suppress higher frequency random noise. This composite filter is essentially a bandpass filter.

Signal Recognition

Digital mixing, the combination of components from different test frequencies, allows the suppression of unwanted parameters or signals from structures such as support plates in tubing applications while retaining the signature of discontinuities beneath those structures.

The combination of filtering, mixing and signal recognition can be combined with decision making algorithms to produce powerful automated analysis systems used in some applications.

Alarms can be constructed digitally as amplitude levels, boxes, ellipses or whatever shape can be digitally described for rejection or acceptance of parts. Alarms from various frequencies or transducers can be tagged *and* or *or* to allow discrimination of different test object conditions.

Control

Eddy current instruments are often used as a component in a test system or material handling system and so require input and output capability to interface with the controllers.

In the nuclear industry, robotic manipulators are used to position and manipulate the probes. The instruments are part of an ethernet network, which consists of data acquisition stations, data analysis stations, data management stations and control systems, some of which may be thousands of kilometers away. The instrument itself essentially feeds data into this system for analysis and control of the job.

In the manufacturing industry, an eddy current instrument may communicate with a programmable logic controller that runs a material handling station or the instrument may be required to perform material control functions itself. These instruments must have application specific input and output capability to provide the communications that can perform the required control in a demanding industrial environment. Instruments are basically required to trigger an alarm on certain conditions and provide an output to mark, cut or reject a part. Accept and reject information may be digitally stored by an instrument or accessed over a network for process control.

References

1. Libby, H.L. *Introduction to Electromagnetic Nondestructive Test Methods.* New York, NY: John Wiley and Sons (1971).
2. Strauts, E.J. Section 11, "Electronic Analysis Circuits for Eddy Current Tests." *Nondestructive Testing Handbook,* second edition: Vol. 4, *Electromagnetic Testing.* Columbus, OH: American Society for Nondestructive Testing (1986): p 265-314.

7

CHAPTER

Signal and Image Processing for Electromagnetic Testing

Lalita Udpa, Michigan State University, East Lansing, Michigan

Part 1. Signal Enhancement

Signal and image processing techniques are valuable for the accurate and consistent interpretation of signals in nondestructive testing. Signal processing performs important functions in data analysis — ranging from simple noise filtering for enhancing the ratio of signal to noise to automated signal classification for improving discontinuity detectability. This chapter focuses on some of the more advanced signal processing techniques. Classical texts can provide a fundamental understanding of the subject.[1,2] A schematic diagram for the overall approach used in nondestructive test signal analysis is shown in Fig. 1.

Techniques of signal processing can be broadly classified into procedures for (1) enhancement, (2) restoration, (3) classification and (4) characterization. These techniques are described below.

Signal enhancement techniques are used to minimize high frequency noise and artifacts in a signal. These techniques generally do not require a precise understanding of the factors that contribute to the distortion. Techniques for enhancing the ratio of signal to noise can range from simple averaging and low pass filtering[1] to more sophisticated techniques such as wavelet shrinkage denoising. In general, the noise contained in a signal can be attributed to several sources, including instrumentation, probe wobble and variations in liftoff and surface roughness. Signals can be enhanced using simple standard linear low pass filters, band stop filters and band pass filters.[2] These filters are often implemented in either hardware or software and available as special features in the instrument. However, these filters are effective only when the signal is stationary. A signal is considered stationary when its statistical properties such as mean or variance do not vary with time. Nondestructive test signals that contain time localized discontinuity indications are, as a rule, nonstationary. Such problems are addressed using techniques such as wavelet shrinkage denoising,[3] described next.

Figure 1. Overall approach for signal analysis in nondestructive testing.

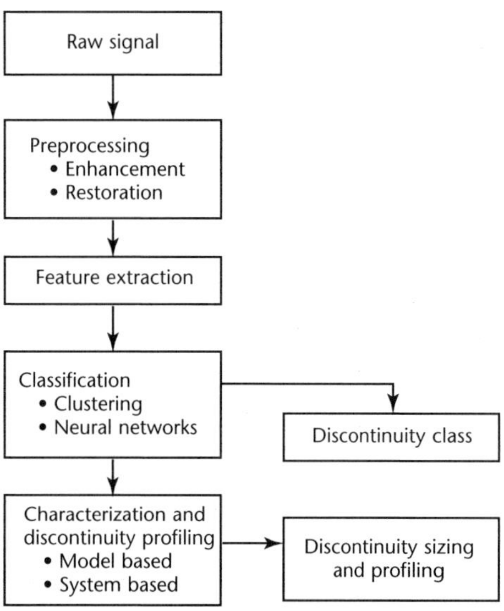

Wavelet Shrinkage Denoising Filter

Consider a noisy signal y_i represented by the discrete time sequence:

$$(1)\quad y_i = x_i + \sigma z_i$$

where $i = 0, 1, \ldots, n-1$; subscript i is the time index of the signal; n is the length of the time sequence; x_i is the desired signal; and z_i represents conventional white noise (indicating that the noise is uncorrelated[1,2]) with standard deviation σ. The discrete wavelet transform decomposes a signal y into a weighted sum of basis functions $\psi_{v,k}$:

$$(2)\quad y(n) = \sum_v \sum_k c_{v,k}\, \psi_{v,k}(n)$$

where v and k are integer values. The basis functions $\psi_{v,k}$ are derived by using dilations and translation operations from a single function ψ, referred to as the *mother wavelet:*

$$(3)\quad \psi_{v,k}(n) = 2^{-\frac{v}{2}}\, \psi\left[2^{-\frac{v}{2}}(n-k)\right]$$

where v and k are the dilation and translation parameters respectively. (Here, the terms *dilation* and *translation* should not be confused with the morphological operations called *erosion* and *dilation*.) Dilation is a scaling operation that compresses the time axis (that is, the X axis and not the Y axis) of a signal. Translation involves shifting a signal in time. As in the case of fourier series coefficients,[1,2] the wavelet transform coefficients $c_{v,k}$ are determined by projecting the signal onto the wavelet basis set $\psi_{v,k}$. The wavelet transform W of Eq. 1 can be expressed as:

$$(4)\quad W(y_i) = W(x_i) + \sigma W(z_i) = W(x_i) + \sigma w_i$$

for orthonormal basis functions (such as fourier bases), where w_i represents $W(z_i)$. Solving for x_i yields:

$$(5)\quad x_i = W^{-1}\left[W(y_i) - \sigma w_i\right]$$

Because σw_i is in general unknown, denoising can be accomplished by removing the noise contribution from each wavelet coefficient by applying a data dependent soft threshold.[3]

Wavelet shrinkage has found increasing use in nondestructive test applications. Programs for fast implementation of the discrete wavelet transform are available commercially. The result of implementing the wavelet shrinkage denoising filter on a typical eddy current signal in Fig. 2a is shown in Fig. 2b. Although the resulting signal is relatively noise free, it contains low frequency trends, which are undesirable.

In many inspection processes, the statistical properties of noise can vary spatially, making conventional invariant position filters not very useful. In such situations, adaptive filtering procedures are needed for noise removal.

FIGURE 2. Eddy current signals: (a) raw signal; (b) wavelet denoised signal.

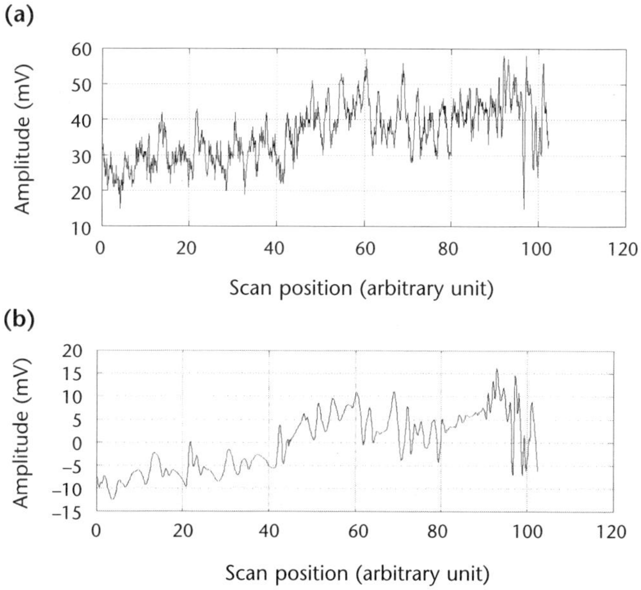

Adaptive Filtering

Adaptive filtering approaches make use of statistical correlation properties of noise and target signals. An adaptive noise cancellation scheme is shown in Fig. 3.[4,5] The reference signal d_i and the primary input signal u_i, obtained from adjacent positions of the transducer, are applied to the filter $H(z)$ and the filter output is represented by y_i. Assuming that the noise is uncorrelated with the input signal, an adaptive filter can be designed to cancel the noise by minimizing the square error. The mean square error is given by the expectation E of the squared error:

$$(6)\quad E\left(\varepsilon_i^2\right) = E\left(s_i - y_i + n_i\right)^2$$

If it is assumed that the reference signal d_i contains both signal and noise (that is, $d_i = s_i + n_i$ where s_i and n_i are the discontinuity signal and grain noise components respectively), then:

$$(7)\quad E\left(\varepsilon_i^2\right) = E\left(s_i - y_i\right)^2 + E\left(n_i\right)^2 + 2E\left(n_i\right)\left(s_i - y_i\right)$$

If n_i is uncorrelated with s_i and y_i the last term reduces to zero:

$$(8)\quad E\left(n_i\right)\left(s_i - y_i\right) = 0$$

and:

FIGURE 3. Overall schematic diagram of adaptive noise cancellation.

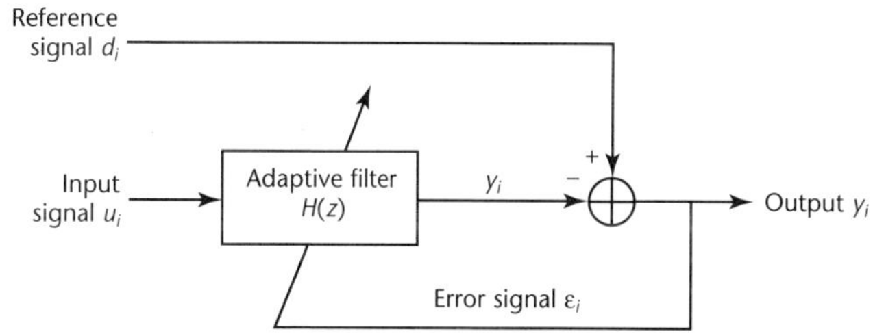

$$E\left[\varepsilon_i^2\right] = E\left[\left(s_i - y_i\right)^2\right] + p_n \quad (9)$$

where p_n is the noise power.

When the filter is optimized, the error is a minimum p_n and the filter output $y_i = s_i$ is completely noise free.

Figure 4a shows a magnetic flux leakage signal obtained during tests of gas transmission pipelines. The measurements are corrupted by a periodic noise caused by helical variations introduced by the manufacturing process. The result of applying adaptive filtering (Fig. 3) is shown in Fig. 4b.[5]

Signal Restoration

Signal restoration procedures are used when the distortion processes that introduce specific artifacts in the signal are known and can be expressed in the form of a mathematical function. Two classes of signal restoration procedures are discussed below: (1) low frequency trends and (2) the effect of the transducer footprint. These distortions can be eliminated by using restoration procedures such as detrending and deconvolution, which are described next.

Detrending

Noise and trends are common forms of distortion that are often present in eddy current and magnetic flux leakage signals. Trends are low frequency changes in the signal levels caused by several factors including instrument drift and gradual variations in probe orientation. In the case of eddy current nondestructive testing, low frequency trends are introduced in the signal because of gradual variations in probe liftoff. The raw eddy current signal in Fig. 2a shows typical distortion introduced by slowly varying trends. A commonly used technique for eliminating such artifacts is based on zero phase high pass filters, which can be implemented by using the discrete cosine transform.

Discrete Cosine Transform

The discrete cosine transform is a special case of the discrete fourier transform[1] where the basis functions consist of cosines instead of complex exponentials. The discrete cosine transform of an N-dimensional discrete time signal x is given by Eq. 10:

FIGURE 4. Results obtained from application of adaptive noise cancellation algorithm: (a) raw magnetic flux leakage data; (b) output after noise cancellation.

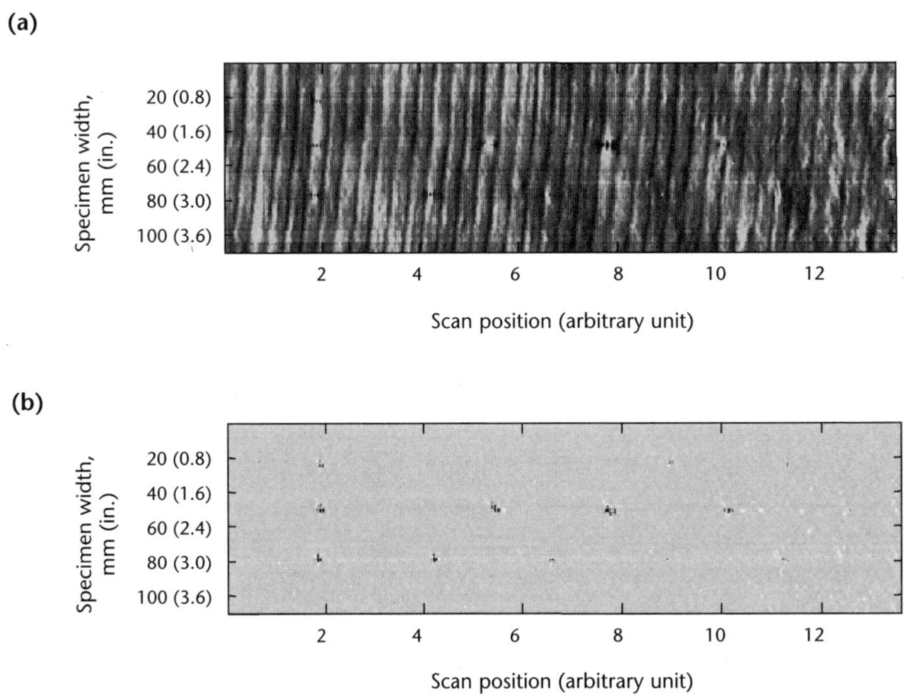

$$B(k) = c(k)\sum_{i=0}^{N-1} x(i)\cos\left[\frac{\pi k}{2N}(2i+1)\right] \quad (10)$$

where $B(k)$ denotes the N transformed values and where the coefficients for $c(k)$ are given by:

$$c(0) = \sqrt{\frac{1}{N}} \quad (11)$$

and:

$$c(k) = \sqrt{\frac{2}{N}} \quad (12)$$

for $1 \le k \le N - 1$.

To get rid of the low frequency trends, the low frequency coefficients in the transformed signal $B(k)$ are set to zero and the signal is reconstructed. This technique results in removing the low frequency trends and the mean without affecting the phase information, crucial in eddy current signals. Figure 5 shows the results of discrete cosine transform based detrending of the eddy current signal in Fig. 2a.

Deconvolution

A second source of distortion is the blurring of a signal because of the point spread function or impulse response of the transducer. Deconvolution techniques can be used to eliminate transducer responses from signals and thereby estimate the true response. One of the most popular restoration techniques uses the wiener filter.

Wiener Filtering

The wiener filter models the measured signal $y(t)$ as the output of a linear time invariant system corrupted by noise $n(t)$:[6]

FIGURE 5. Eddy current signal after detrending of signal in Fig. 2a.

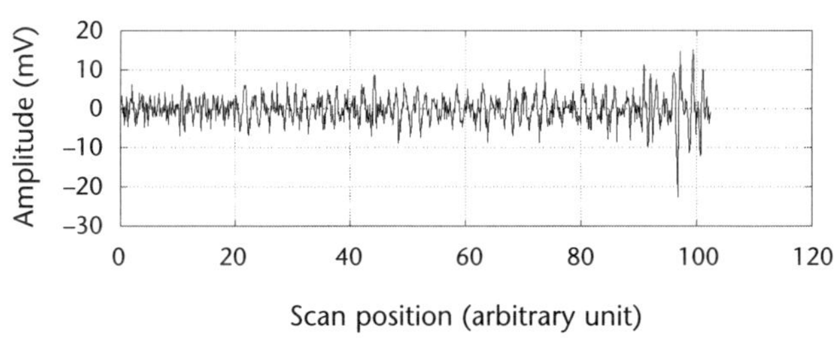

$$y(t) = x(t) * h(t) + n(t) \quad (13)$$

where $h(t)$ represents the impulse response of the measurement system or the point spread function, $x(t)$ represents the true signal and the symbol $*$ is the convolution operator. The fourier transform of Eq. 13 yields:

$$Y(f) = X(f)H(f) + N(f) \quad (14)$$

The estimation of the true function $X(f)$ from $Y(f)$ is performed using the wiener filter $\hat{H}(f)$, characterized in the frequency domain by:[6]

$$\hat{H}(f) = \frac{H*(f)}{\left|H(f)\right|^2 + Q} \quad (15)$$

where $H*(f)$ is the complex conjugate of $H(f)$ and Q is related to the ratio of signal to noise. Wiener filters[6] use a constrained least squares minimization procedure to estimate $x(t)$ from the measurement $y(t)$. Although this technique is well established, it requires knowledge of $H(f)$. In cases where the distortion kernel $H(f)$ is not available, blind deconvolution algorithms can be used as described below.

Blind Deconvolution

Blind deconvolution techniques are particularly attractive because they do not require specification of the distortion kernel. These algorithms iteratively estimate the kernel (in this case the point spread function of the probe) from the available data. Another advantage associated with blind deconvolution algorithms is the ease with which constraints can be added. This property can be used to constrain the size of the kernel based on the size of the probe used. The blind deconvolution technique based on the richardson-lucy algorithm[7] uses maximum likelihood principles and obtains high quality reconstructed images even in the presence of noise. Consider the experimental (convolved) eddy current image data $c(x,y)$ obtained using a raster scan of a test object. Let $f(x,y)$ represent the true image that would be obtained with a point sensor.

The richardson-lucy algorithm is based on Bayes' theorem:

$$P(x|y) = \frac{P(y|x)\,P(x)}{\int P(y|x)\,P(x)\,dx} \quad (16)$$

where $P(y|x)$ is the conditional probability of an event y, and the probability $P(x)$ of an event X. In the context of deconvolution, $P(x)$ and $P(y)$ are identified with the unknown $f(x,y)$ to be estimated and the convolved (or measured) image $c(x,y)$, respectively. Also, the conditional probability $P(y|x)$ is identified as the kernel function or the point spread function centered at (x,y), that is, $g(x,y)$. Equation 16 can be used to derive the iterative form of the deconvolution algorithm[7] in the discrete domain as:

$$(17) \quad f_{i+1}(m,n) = \left\{ \left[\frac{c(m,n)}{f_i(m,n) * g(m,n)} \right] * g(-m,-n) \right\} f_i(m,n)$$

where i is the iteration number. All quantities in Eq. 17 are two-dimensional and depend on two spatial variables m and n. Given the point spread function $g(m,n)$ and an initial guess of the original image $f(m,n)$, the reconstructed image can be obtained by iteratively applying Eq. 17 until convergence.

The inverse iterative equation, derived by reversing the role of the reconstructed image and the point spread function in Eq. 17, is given by:

$$(18) \quad g_{i+1}(m,n) = \left\{ \left[\frac{c(m,n)}{g_i(m,n) * f^{k-1}(m,n)} \right] * f(-m,-n) \right\} g_i(m,n)$$

The inverse iterative equation is also referred to as a *richardson-lucy operation.*

The blind deconvolution algorithm consists of a two-step procedure. At the kth iteration, the point spread function $g^k(y,x)$ is calculated from Eq. 19 by performing a specific number of richardson-lucy operations, given the knowledge of the reconstructed image $f^{k-1}(m,n)$ obtained from Eq. 20 after the $(k-1)$th iteration:

$$(19) \quad g_{i+1}^k(m,n) = \left\{ \left[\frac{c(m,n)}{g_i^k(m,n) * f^{k-1}(m,n)} \right] * f^{k-1}(-m,-n) \right\} g_i^k(m,n)$$

Here, k denotes the iteration number and i denotes the number of richardson-lucy operations during each iteration. The new estimate of the deconvolved image $f^k(m,n)$ is obtained by performing the same number of richardson-lucy operations (Eq. 20), given the point spread function $g^k(m,n)$ obtained from Eq. 19 above:

$$(20) \quad f_{i+1}^k(m,n) = \left\{ \left[\frac{c(m,n)}{f_i^k(m,n) * g^k(m,n)} \right] * g^k(-m,-n) \right\} f_i^k(m,n)$$

These two steps are repeated until convergence is achieved.

Some typical results of applying this algorithm to data from an eddy current pancake coil probe are presented in Fig. 6. Figure 7 shows the results of blind deconvolution on experimental eddy current measurements, illustrating how the process of deconvolution can help in resolving two closely spaced discontinuities.

FIGURE 6. Images from pancake coil probes: (a) true image; (b) gaussian point spread function kernel; (c) measured, convolved image; (d) result of blind deconvolution.

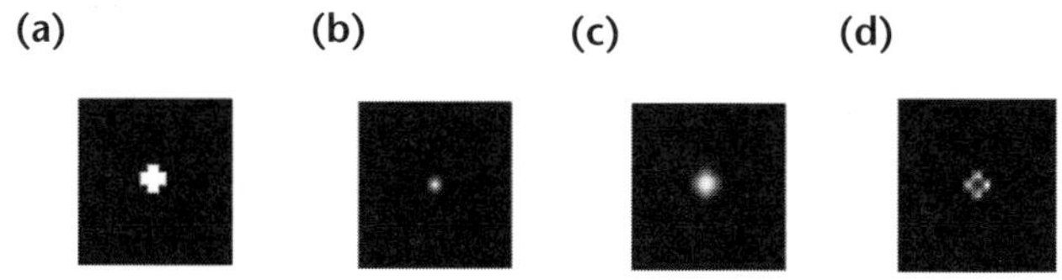

FIGURE 7. Eddy current test with two closely spaced discontinuities: (a) raw image; (b) result of blind deconvolution algorithm.

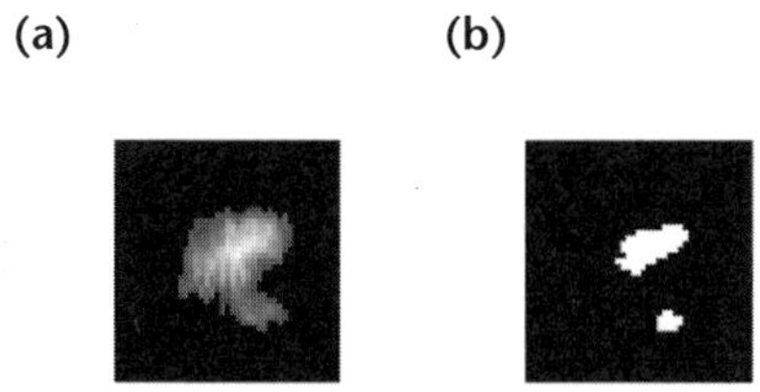

Part 2. Signal Classification

Signal classification techniques often rely on pattern recognition for interpreting nondestructive test data. Pattern recognition techniques help to classify signals into one of a known set of classes. Such techniques may be used, for example, to discriminate between multiple types of discontinuities or between discontinuities and benign sources. In the case of steam generator tube tests, for example, such techniques could be used to distinguish eddy current signals from those caused by cracks, tube supports and antivibration bars. The parameters of the classifier are generally determined by using a data bank of signals from expected discontinuity types. The collection of signals, referred to as the *training database,* is used for training the classification algorithm. Most classification techniques use a two-step procedure.

1. In feature extraction, characteristic features in the signal that carry discriminatory information are identified and extracted. These features serve as a compact signature of the signal.
2. The feature vector is classified by using a standard pattern classification technique such as a clustering algorithm or a neural network.

Feature Extraction

Feature extraction serves two major functions, namely data compression and invariance with respect to parameters such as frequency and gain. Features are data attributes that capture similarities in signals from the same class as well as dissimilarities in signals from different classes. In addition to containing discriminatory information, the feature vector is typically of lower dimension than the signal, resulting in data compression. Features can be physical or structural (peak value, rise time, peak-to-peak separation and others) or transform based (fast fourier transform, discrete cosine transform and others). Transform based features are more easily implemented by using numerical algorithms. Examples of commonly used transform based features are discrete fourier transform coefficients,[1] discrete cosine transform coefficients[8] and scale based features such as discrete wavelet transform coefficients,[9] principal components[10] and linear predictive coding coefficients.[11] These feature extraction schemes are described below.

Discrete Fourier Transform

One of the earliest and simplest techniques used for feature extraction is the discrete fourier transform. The discrete fourier transform of a signal $x(n)$ can be expressed as the weighted sum of complex exponential basis functions. For a series $x(n)$ with N samples, the discrete fourier transform is expressed as:

$$X(k) = \frac{1}{N}\sum_{n} x(n)\, e^{-\frac{jn2\pi}{N}} \tag{21}$$

where $k = 0, 1, \ldots, N-1$.

For smooth signals, the magnitude of the coefficients can be shown to decay at the rate of n^{-2} with the result that the energy can be compacted in very few discrete fourier transform coefficients. Magnitudes of the discrete fourier transform coefficients are the simplest and most commonly used feature vector for representing signals for classification.

Discrete Cosine Transform

The discrete cosine transform is a special case of the discrete fourier transform where the basis functions consist of cosines instead of complex exponentials. Repeating Eq. 10, the discrete cosine transform of an N-dimensional discrete time signal x is given by:

$$B(k) = c(k)\sum_{i=0}^{N-1} x(i)\cos\left[\frac{\pi k}{2N}(2i+1)\right] \tag{22}$$

where B_k denotes the N transformed values and where the coefficients for $c(k)$ are given by:

$$c(0) = \sqrt{\frac{1}{N}} \tag{23}$$

and:

$$(24)\quad c(k) = \sqrt{\frac{2}{N}}$$

for $1 \le k < N - 1$.

This technique results in a feature vector composed of a smaller number of coefficients B_k that can represent the signal.

Discrete Wavelet Transform

The discrete wavelet transform[9] of a signal $x(n)$ is a joint time scale transform that provides both time and frequency localization of a signal. The discrete wavelet transform can be expressed as the weighted sum of basis functions:

$$(25)\quad x(n) = \sum_v \sum_k C_{v,k}\, \psi_{v,k}(n)$$

where realizations of the wavelet basis function $\psi_{v,k}(n)$ are derived from a single function $\psi(n)$, referred to as the *mother wavelet,* by dilations v and translations k according to:

$$(26)\quad \psi_{v,k}(n) = 2^{-\frac{k}{2}}\, \psi\left[2^{-k}(n-k)\right]$$

The discrete wavelet transform coefficients $C_{v,k}$ are determined by projecting the signal $x(n)$ onto the wavelet basis set $\psi_{v,k}(n)$. It is usually implemented as a series of subband filters. The most common version is the two-band discrete wavelet transform, which uses two finite impulse response filters — a low pass filter and a high pass filter.

The computation of the discrete wavelet transform coefficients for a data vector $\boldsymbol{x}$ of length n (where n indicates the number of points in the signal) is indicated in Fig. 8, which presents a fast implementation of the discrete wavelet transform using a filter bank approach. The output of each filter is downsampled by a factor of 2 by discarding every other sample. The output of the high pass filter represents the discrete wavelet transform coefficients at the first resolution level. The output of the low pass filter is then applied to the same set of filters and sampled again. The output of the high pass filter is retained as the discrete wavelet transform coefficients at the

FIGURE 8. Filter bank approach for discrete wavelet transform computation.[9]

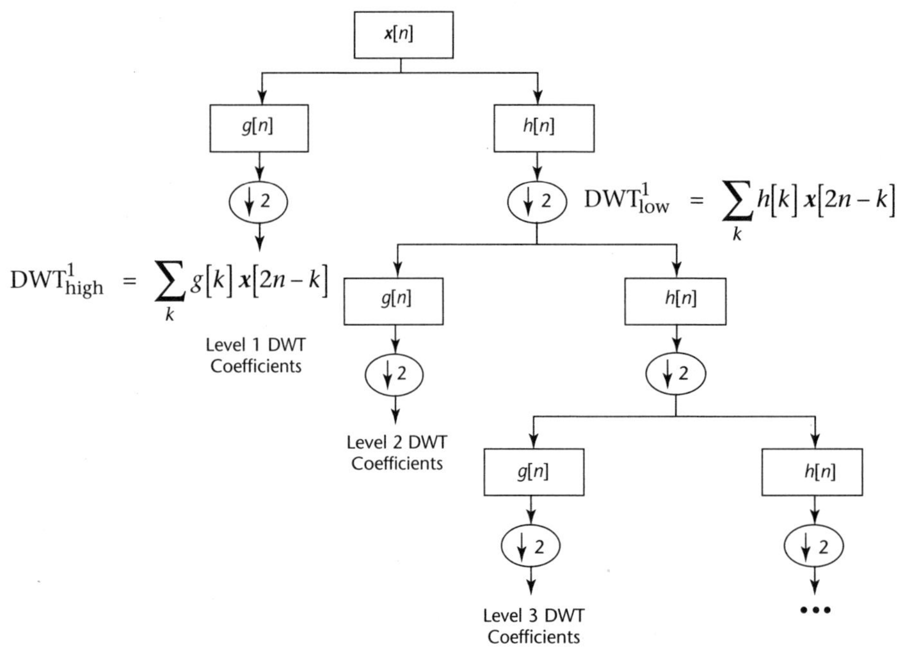

Legend

DWT = discrete wavelet transform
g = high pass function
h = low pass function
k = iteration number
n = number of points in signal
$\boldsymbol{x}$ = data vector

second resolution level. This process is repeated until the number of samples is reduced to 1. The number of possible resolution levels is given by α:

$$\alpha = \log_2 n \tag{27}$$

Because the discontinuity related information is typically present in the discontinuity scale subspace, an appropriate set of coefficients in the discontinuity subspace can be used as features.

Principal Component Analysis

Principal component analysis is a statistical technique that linearly transforms a time series sequence into a substantially smaller set of uncorrelated variables that contains most of the information in the original data set.[10] The overall goal of principal component analysis is to reduce the dimensionality of the original data set. Principal component analysis allows the reconstruction of the original pattern from linear projections required to have sequentially maximal variances. The basis vectors of the representation are constrained to be mutually orthonormal. If $\mathbf{X}$ is an $n \times n$ data matrix of measurement vectors with mean M_x and covariance matrix Σ_x (where subscript x represents a datum) an orthogonal set of eigenvectors may be found that diagonalizes the covariance matrix. By arranging the eigenvectors in a matrix in accordance with decreasing eigenvalues (largest first), an ordered orthogonal basis may be created that has the greatest degree of variability of the data along the first eigenvector. Retaining only p largest eigenvalues provides a feature extraction operator $\mathbf{\Phi}$, a $p \times p$ matrix of p eigenvectors. Using this transformation matrix, a data set $\mathbf{X}$ may be transformed to matrix $\mathbf{Y}$:

$$\mathbf{Y} = \Phi\left(\mathbf{X} - \mathbf{M_X}\right) \tag{28}$$

making $\mathbf{Y}$ an orthonormal projection of $\mathbf{X}$ onto the columns of the transformation matrix. The inverse transformation may be used to reconstruct the original data set $\mathbf{X}$ by:

$$\mathbf{X} = \Phi^{\mathrm{T}}\mathbf{Y} + \mathbf{M_X} \tag{29}$$

where $\mathbf{\Phi}^{\mathrm{T}}$ represents the transpose of matrix $\mathbf{\Phi}$. The matrix $\mathbf{Y}$ represents $\mathbf{X}$ in the domain spanned by the vectors $\phi_1, \ldots, \phi_p$. These columns of $\mathbf{Y}$ are referred to as the *principal components* of the data set $\mathbf{X}$ and are of a lower dimension than the original data vectors.

Linear Predictive Coding Coefficients

Linear predictive modeling[11] is commonly used in the processing of speech signals. Linear predictive coding coefficients are known to accurately represent speech signals with a small set of parameters. The approach can be used also for extracting features from test signals.

In linear predictive coding analysis, it is assumed that the present value of the sample $s(n)$ can be represented as a weighted sum of the past samples. The linear predictive coding coefficients are estimated by minimizing the mean squared error between the predicted value and true value. The error $\varepsilon(n)$ is given by:

$$\varepsilon(n) = s(n) - \sum_{j=1}^{p} \alpha_j \, s(n-j) \tag{30}$$

where α_j represents the estimates of the linear predictive coding coefficients. Setting the partial derivatives of the mean squared error with respect to α_j to zero for $j = 1, 2, \ldots, p$ gives:

$$E\left\{\left[s(n) - \sum_{j=1}^{p} \alpha_j \, s(n-j)\right] s(n-i)\right\} = 0 \tag{31}$$

for $i = 1, 2, \ldots, p$. Equation 31 can be rearranged:

$$\sum_{j=1}^{p} \alpha_j \, \phi_n(i,j) = \phi_n(i,0) \tag{32}$$

for $i = 1, 2, \ldots, p$, where ϕ_n is the autocorrelation function:

$$\phi_n(i,j) = E\left\{s(n-i)\, s(n-j)\right\} \tag{33}$$

The linear predictive coding coefficients in Eq. 33 can be solved recursively by using Durbin's algorithm.[11] Nondestructive test signals can be represented by a small set of linear predictive coding coefficients, thereby achieving data reduction and compaction. The coefficients represent the signal and serve as a reduced dimensional feature vector.

Feature Evaluation

Once the features are computed, a feature evaluation and selection step may be used to eliminate redundancy in the representation and to evaluate the features on the basis of the discriminatory information. More importantly, the selection step offers an opportunity to choose features invariant either to changes in test conditions or to some selected aspect of the test object properties. Because nondestructive test signals are acquired under varying test conditions, the results are sensitive to instrument drift and to variations in probe characteristics, scanning speeds, gain settings, operating frequencies and test object conductivity and permeability. A major challenge lies in the development of signal processing schemes to compensate the signal for variations in experimental test parameters. Such schemes are crucial for rendering the overall signal classification performance insensitive to the environment in which the signal was acquired.

A number of procedures capable of selecting features on the basis of discriminatory information in them have been proposed. The process begins with the selection of candidate features.[12,13] Each candidate is then evaluated and either accepted or rejected on the basis of the amount of discriminatory information contained in it. In the second step, the goal is to identify features that contain the greatest amount of discriminatory information. One popular technique for feature reduction is called *fisher linear discrimination*.[14] The technique uses a statistical weight function for each feature to determine the optimum feature set with the greatest amount of discriminatory information to be selected for signal classification. Fisher linear discrimination quantifies the discriminatory content of the different features.

A typical fisher linear discriminant implementation is carried out by using scatter matrix analysis. The within-class and between-class scatter matrices are computed as follows:

Let $\mathbf{\Sigma}_i$ be the scatter matrix of the sample vector $\boldsymbol{x}$ in class C_i around their respective mean $\boldsymbol{m}_i$:

$$(34) \quad \Sigma_i = E\left[\left(\boldsymbol{x}-\boldsymbol{m}_i\right)\left(\boldsymbol{x}-\boldsymbol{m}_i\right)^{\mathrm{T}} \mid C = C_i\right]$$

Then $\boldsymbol{S}_{\mathrm{w}}$ can be defined:

$$(35) \quad \boldsymbol{S}_{\mathrm{w}} = \frac{1}{M}\sum_{i=1}^{M} P\left(C_i\right)\Sigma_i$$

and:

$$(36) \quad S_{\mathrm{b}} = \frac{1}{M}\sum_{i=1}^{M} P\left(C_i\right)\left(\boldsymbol{m}_i-\boldsymbol{m}\right)\left(\boldsymbol{m}_i-\boldsymbol{m}\right)^{\mathrm{T}}$$

where $\boldsymbol{S}_{\mathrm{w}}$ is the within-class scatter matrix showing the average scatter $\mathbf{\Sigma}_i$ of the sample vector $\boldsymbol{x}$ in class C_i around their respective mean $\boldsymbol{m}_i$, where $P(C_i)$ is the prior probability of class C_i and where $\boldsymbol{S}_{\mathrm{b}}$ is the between-class scatter matrix, representing the scatter of the conditional mean vectors $\boldsymbol{m}_i$ around the overall mean vector $\boldsymbol{m}$.

Various measures are available for quantifying the discriminatory power, the commonly used one being:

$$(37) \quad J\left(\boldsymbol{W}\right) = \frac{\left\|\boldsymbol{W}^{\mathrm{T}} S_{\mathrm{w}} \boldsymbol{W}\right\|}{\left\|\boldsymbol{W}^{\mathrm{T}} S_{\mathrm{b}} \boldsymbol{W}\right\|}$$

Here $\boldsymbol{W}$ is the optimal discrimination projection vector, which can be obtained by solving the generalized eigenvalue problem:

$$(38) \quad S_{\mathrm{b}}\boldsymbol{W} = \lambda \boldsymbol{S}_{\mathrm{w}} \boldsymbol{W}$$

where λ is an eigenvalue.

An example of a feature extraction procedure that offers dimensionality reduction as well as invariance properties involves fourier descriptors.[15] The technique has been used for representing eddy current impedance plane trajectories. The model not only represents the signal by a few coefficients, which are invariant under rotation, translation and scaling of the eddy current impedance plane trajectory, but also allows the resynthesis of the original signal from the stored coefficients.

Classification Algorithms

The features computed in the previous step are applied as input to a classification algorithm for data interpretation. Two of the most widely used pattern classification techniques are (1) clustering algorithms and (2) neural networks. These techniques are described next.

K Means Clustering

Clustering algorithms treat a feature vector as a point in the *N*-dimensional feature space.[16] Feature vectors from a similar class of signals then form a cluster in the feature space. The most popular of the clustering algorithms is the *K* means clustering algorithm, which uses an iterative procedure that classifies each input signal into one of *K* classes.

K Means Algorithm

The objective of the *K* means clustering algorithm is to partition the feature space into *K* mutually exclusive regions. The partitioning is performed in a way that minimizes a performance index or cost function F equal to the sum of the square of distance between the cluster center and all points within the cluster.

Let the number of patterns be N_c.

1. Assign any K (first K, randomly selected K or user assigned K) patterns as the K cluster centers z_i, where $i = 1, 2, \ldots, K$.
2. Assign each of the remaining $N_c – K$ patterns at the jth iteration to one of the K clusters whose center is closest (using the euclidian norm):

$$x \in w_m^j \tag{39}$$

where w_m^j is the mth cluster in the jth iteration and:

$$\left\| x - z_m^j \right\| < \left\| x - z_n^j \right\| \tag{40}$$

for $1 \le m, n \le K$.

3. Update the cluster centers z_i^{j+1}, $i = 1, 2, \ldots K$, in a manner that minimizes the performance index:

$$F_i^j = \sum_{x \in N_c^j} \left\| x - z_i^{j+1} \right\|^2 \tag{41}$$

where F_i^j is the cost function corresponding to the ith cluster in the jth iteration and N_c^j is the number of patterns in the cth cluster in the jth iteration. It can be shown that the centers z_i^{j+1}, $i = 1, 2, \ldots, K$), which minimize the above performance index, are the sample mean of all points within the cluster:

$$z_i^{j+1} = \frac{1}{N_c^j} \sum_{x \in w_i^j} x \tag{42}$$

4. If $z_i^{j+1} = z_i^j$ for all $i = 1, 2, \ldots, K$, the algorithm has converged and the process can be terminated. Otherwise go to step 2.

The K means algorithm converges if the classes are linearly separable and the performance generally is better if the initial cluster centers are chosen from the K classes.

Neural Networks

Neural networks provide an alternate approach for classification. Interest in this approach arose from a desire to mimic biological nervous systems with respect to architecture as well as information processing strategies.[17] The network consists of simple processing elements interconnected by weights. The network is first trained using an appropriate learning algorithm for the estimation of interconnection weights. Once the network is trained, unknown test signals can be classified. The class of neural networks used most often for classification tasks is the multilayer perceptron network.

The multilayer perceptron network (Fig. 9) generally consists of an input layer of nodes, one or more hidden layers of nodes and an output layer of nodes. Nodes within the same layer are not connected. However, each layer of nodes is fully interconnected to the nodes in the next layer. All units within a layer process data in parallel but the outputs of different layers are calculated sequentially

FIGURE 9. Architecture of multiplayer perceptron neural network.

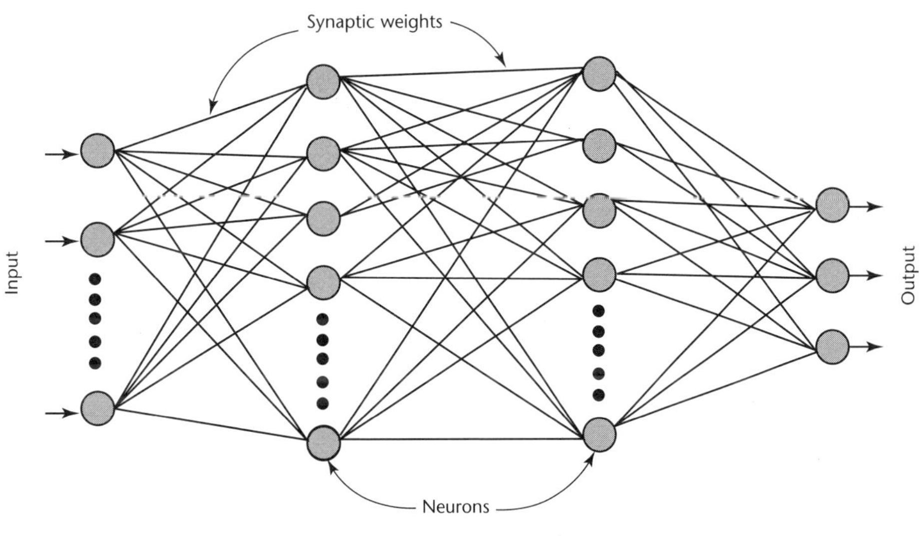

starting from the input layer and moving toward the output layer. Each node generates an output that is a nonlinear function of the weighted sum of all its input signals. This nonlinear function is primarily used to limit the output of a node between the values of 0 and 1.

The network is trained using the backward error propagation algorithm[18] where training patterns are sequentially applied to the network. The overall algorithm is summarized in Fig. 10. The algorithm uses a gradient search technique for minimizing the squared error between the actual output and the desired output by adapting the interconnection weights iteratively. The algorithm cycles through the training data repeatedly until the error drops below a specified threshold value. Neural networks have been used with success for the classification of eddy current and ultrasonic signals.[19]

FIGURE 10. Flow chart of backpropagation training algorithm for multilayer perceptron networks.

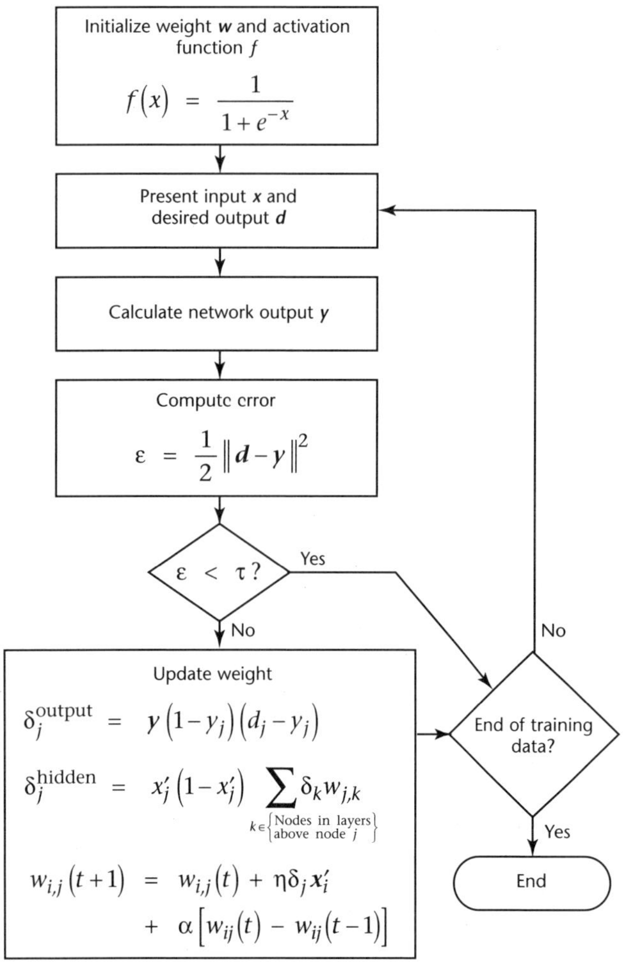

Legend

d = output at node *j*
t = time
w = weighting factor
x = input signal
y = network output at node *j*
α = momentum parameter
δ = variable defined by equation
η = learning parameter
τ = preset threshold value for error

PART 3. Signal Characterization

Signal characterization involves a more complete solution to the inverse problem. In material science, the inverse problem involves reasoning from effects (that is, indications) in order to draw inferences about test objects. Characterization techniques use information contained in the signal to estimate the size, shape and location of discontinuities. In other words, characterization procedures involve the full two-dimensional or three-dimensional reconstruction of discontinuity profiles in terms of the spatial distribution of the material properties of the test object. In general, the objective of the signal or discontinuity characterization procedure can be described as the identification of a mapping f such that:

$$D = f(\mathbf{S}) \tag{43}$$

where $\mathbf{S}$ represents the measurement vector from a scan in two dimensions M and Q:

$$\mathbf{S} = \{s_{ij}\}_{M \times Q} \tag{44}$$

and D represents the discontinuity profile:

$$D = \{d_{ij}\}_{R \times P} \tag{45}$$

The value of d_{ij} represents the depth of the discontinuity at a location (i,j).

Several approaches have been developed for solving the inverse problem in nondestructive testing. These solutions can be categorized as either phenomenological or nonphenomenological. Phenomenological techniques are based on the underlying physical process of the nondestructive test technique. Examples of the phenomenological approach for inversion are based on analytical solutions of the underlying governing equation, which is in general a difficult problem. Nonphenomenological approaches do not depend on the physics of the inspection technique. These approaches model the nondestructive test system as a black box or as a linear system and use signal processing techniques to invert the measured signal. Typical signal processing approaches for inversion use neural networks for solving the discontinuity characterization problem. An approach using a radial basis function neural network for the inversion of magnetic flux leakage signals is described next.[20]

Radial Basis Function Networks

Radial basis function networks can be viewed as tools for multivariate interpolation.[20] Such networks can be used for estimating a hypersurface that provides what can be called the *best fit* to the training data. The architecture of the radial basis function network is in many respects similar to that of a multilayer perceptron, defined above. A nonlinear transformation of the signal is performed between the input and hidden nodes followed by a linear transformation between the hidden and output nodes. Mathematically, the radial basis function network computes a multidimensional function:

$$f(\mathbf{x}) = \sum_{i=1}^{N} w_i \phi_i \left(\left\| \mathbf{x} - c_i \right\| \right) \tag{46}$$

where ϕ_i is a set of basis functions, c_i are the basis centers and w_i are the weights. Substituting the values in the training data $\{x_i, f(x_i), i = 1, \ldots, N\}$ in Eq. 46 makes it possible to derive the matrix equation:

$$f = \mathbf{\Phi w} \tag{47}$$

The training of the radial basis function network consists of estimating the expansion coefficients, which can be done by inverting Eq. 47:

$$w = \Phi \Phi^{-1} \cdot f \tag{48}$$

Once the weights are estimated by using the training data, the radial basis function network can be used to invert a test signal $\mathbf{x}$ according to Eq. 46.

Reconstruction results can be further improved by using a variation of the radial basis function network, a multiresolution approach that uses neural networks with wavelet basis functions.[21]

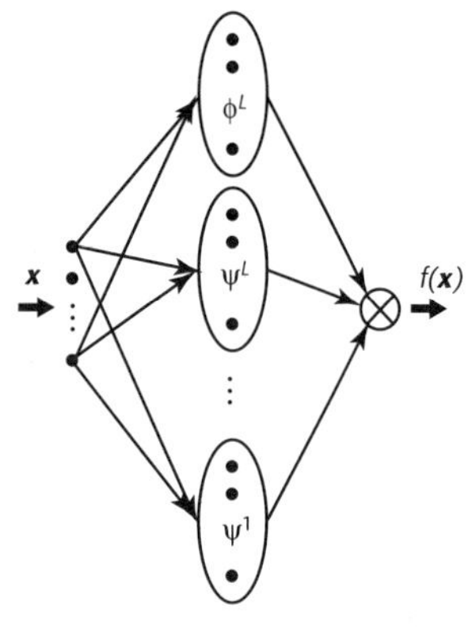

FIGURE 11. Architecture of wavelet basis function network.

Legend

- L = scale index (superscript)
- x = input to network
- $f(x)$ = function output
- ϕ = scaling function
- ψ = wavelet basis function

These networks, called *wave nets,* use a hierarchical architecture associated with multiple levels of resolution or scale for both global and local interpolation as shown in Fig. 11. The network is trained hierarchically, first to learn the mapping between inputs and outputs at the coarsest resolution and later to augment the mapping with details at higher resolutions.

Mathematically, the approximation of a scalar function $f(t)$ can be expressed as:

$$(49)\quad f(x) = \sum_{j=1}^{N_L} s_j^L \phi_j^L(x) + \sum_{k=1}^{L} \sum_{j=1}^{N_k} d_j^k \psi_j^k(x)$$

where d_j^k denotes the corresponding expansion coefficients; $f(x)$, the mapping function to be estimated; s_j^L, the scaling coefficients; $\psi_j^k(x)$, the wavelet basis functions generated via translations and

FIGURE 12. Neural network characterization of magnetic flux leakage signal from 75 mm (3 in.) long and 75 mm (3 in.) wide discontinuity: (a) magnetic flux leakage signal; (b) discontinuity profile; (c) prediction using radial basis function network; (d) prediction using wavelet basis function network.[21]

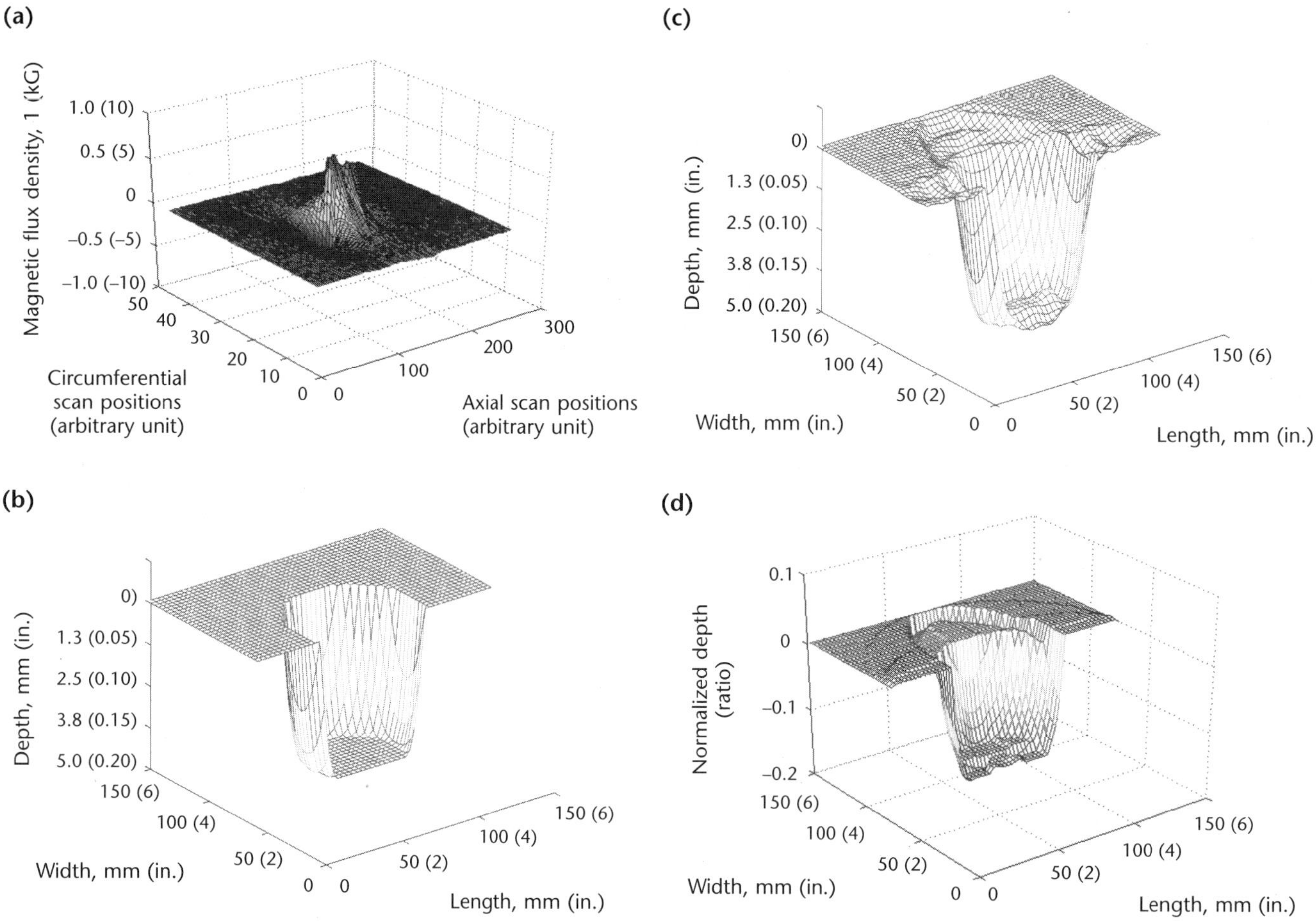

dilations of a mother wavelet $\psi(x)$; and ϕ_j^L, the scaling basis functions at resolution L.

The first term represents the approximation of the function at the Lth resolution and the second term represents additional information pertaining to the details at the corresponding resolution. The accuracy of the discontinuity reconstruction can then be controlled by selecting the number of resolution levels in the network architecture. The training algorithm for wave nets is similar to that used for training a radial basis function network.

Initial results obtained using both radial basis function networks and wavelet basis function networks using a two-dimensional magnetic flux leakage signal as input are shown in Fig. 12. The magnetic flux leakage signal is obtained from a rectangular notch machined on the pipe wall. The results obtained using the trained radial basis function network and the wavelet basis function network are compared with the true profile. These results show that such nonphenomenological techniques for inversion can be trained to perform well for measurements as long as the measurements are similar to those used for training the network.

Summary

Advances in digital processing have made sophisticated signal and image processing techniques available for practical applications in nondestructive testing. When integrated in software programs for discontinuity classification, signal processing algorithms make possible the automation of diagnostic procedures and quality assurance protocols.

References

1. Oppenheim, A.O. and R.W. Schafer. *Discrete Time Signal Processing.* Upper Saddle River, NJ: Prentice Hall (1989).
2. Mitra, S.K. *Digital Signal Processing: A Computer-Based Approach.* Boston, MA: McGraw-Hill/Irwin (2001).
3. Donoho, D.L. "Nonlinear Wavelet Methods for Recovery of Signal Densities and Spectra from Indirect and Noisy Data." *Proceedings of Symposia in Applied Mathematics.* Vol. 47. Providence, RI: American Mathematical Society (1993): p 173-205.
4. Kim, J., L. Udpa and S.S. Udpa. "Multistage Adaptive Noise Cancellation for Ultrasonic Nondestructive Evaluation." *Review of Progress in Quantitative Nondestructive Evaluation* [Snowbird, Utah, July 1998]. Vol. 18A. New York, NY: Plenum Press (1999): p 781-787.
5. Afzal, M., S.S. Udpa, L. Udpa and W. Lord. "Rejection of Seamless Pipe Noise in Magnetic Flux Leakage Data Obtained from Gas Pipeline Inspection." *Review of Progress in Quantitative Nondestructive Evaluation* [Montreal, Canada, July 1999]. Vol. 19B. New York, NY: Plenum Press (2000): p 1589-1596.
6. Neal, S. and D.O. Thompson. "An Examination of the Application of Wiener Filtering to Ultrasonic Scattering Amplitude Estimation." *Review of Progress in Quantitative Nondestructive Evaluation* [Williamsburg, VA, June 1985]. Vol. 5A. New York, NY: Plenum Press (1986): p 737-746.
7. Richardson, W.H. "Bayesian-Based Iterative Method of Image Restoration." *Journal of the Optical Society of America.* Vol. 62, No. 1. Washington, DC: Optical Society of America (January 1972): p 55-59.
8. Gonzalez, R.C. and R.E. Woods. *Digital Image Processing,* second edition. Upper Saddle River, NJ: Prentice Hall (2002).
9. Strang, G. and T. Nguyen. *Wavelets and Filter Banks.* Wellesley, MA: Wellesley-Cambridge Press (1996).
10. Haykin, S.S. *Neural Networks: A Comprehensive Foundation,* second edition. Upper Saddle River, NJ: Prentice Hall (1999).
11. Makhoul, J. "Linear Prediction: A Tutorial Review." *Proceedings of the IEEE.* Vol. 62. New York, NY: Institute of Electrical and Electronics Engineers (April 1975): p 561-580.
12. Doctor, P.G., T.P. Harrington, T.J. Davis, C.J. Morris and D.W. Fraley. "Pattern Recognition Methods for Classifying and Sizing Flaws Using Eddy-Current Data." *Eddy Current Characterization of Materials and Structures.* Special Technical Publication 722. West Conshohocken, PA: ASTM International (1980): p 464-483.
13. Burch, S.F., A.R. Lomas and A.T. Ramsey. "Practical Automated Ultrasonic Signal Characterization of Welding Defects." *British Journal of Non-Destructive Testing.* Vol. 32, No. 7. Northampton, United Kingdom: British Institute of Non-Destructive Testing (July 1990): p 347-350.
14. Duda, R.O. and P. Hart. *Pattern Classification and Scene Analysis.* New York, NY: John Wiley and Sons (1973).
15. Udpa, S.S. and W. Lord. "A Fourier Descriptor Classification Scheme for Differential Probe Signals." *Materials Evaluation.* Vol. 42, No. 9. Columbus, OH: American Society for Nondestructive Testing (August 1984): p 1136-1141.
16. Tou, J.T. and R.C. Gonzalez. *Pattern Recognition Principles.* Reading, MA: Addison-Wesley (1974).
17. Lippmann, R.P. "An Introduction to Computing with Neural Nets." *IEEE ASSP Magazine.* Vol. 4. New York, NY: Institute of Electrical and Electronics Engineers (April 1987): p 4-22.
18. Rumelhart, D.E., G.E. Hinton and R.J. Williams. "Learning Internal Representations by Error Propagation." *Parallel Distributed Processing: Exploration in the Microstructure of Cognition:* Vol. 1, *Foundations.* Cambridge, MA: MIT Press (1986): p 318-362.
19. Udpa, L. and S.S. Udpa. "Neural Networks for Classification of NDE Signals." *IEE Proceedings F: Communications, Radar, and Signal Processing.* Vol. 138. Stevenage, Hertfordshire: Institution of Electrical Engineers (1981): p 41-45.

20. Broomhead, D.S. and D. Lowe. "Multivariate Functional Interpolation and Adaptive Networks." *Complex Systems*. Vol. 2. Champaign, IL: Complex Systems Publications (1988).
21. Hwang, K., S. Mandayam, S.S. Udpa and W. Lord. "A Multiresolution Approach for Characterizing MFL Signatures from Gas Pipeline Inspections." *Review of Progress in Quantitative Nondestructive Evaluation* [Brunswick, ME, July-August 1996]. Vol. 16A. New York, NY: Plenum Press (1997): p 733-739.

Bibliography

McClelland, J.L. and D.E. Rumelhart. *Explorations in Parallel Distributed Processing*. Cambridge, MA: MIT Press (1988).

Stanley, W.D. *Digital Signal Processing*. Reston, VA: Reston Publishing (1975).

Strauts, E.J. Section 11, "Electronic Analysis Circuits for Eddy Current Tests." *Nondestructive Testing Handbook,* second edition: Vol. 4, *Electromagnetic Testing*. Columbus, OH: American Society for Nondestructive Testing (1986): p 265-314.

8

CHAPTER

Remote Field Testing

David E. Russell, Russell NDE Systems, Edmonton, Alberta, Canada

David D. Mackintosh, Queens University, Kingston, Ontario, Canada

Ad A. Shatat, Russell NDE Systems, Incorporated, Edmonton, Alberta, Canada

PART 1. Background

Introduction

Remote field testing is popular because of its ability to inspect regions not only near the probe but also throughout the thickness of the material. This characteristic is especially valuable for the testing of thick walled ferromagnetic tubing, where the technique shows high sensitivity to inside and outside surface pipe wall heterogeneities. Remote field testing often does not require cleaning of the pipe and is not sensitive to internal coatings. Since 1970, remote field testing has grown into a mature and recognized nondestructive testing technology. This chapter introduces remote field testing — its history, applications, strengths and limitations.

History

The remote field effect was first noted in the 1940s and was patented by W.R. MacLean in 1951.[1] In the late 1950s, Thomas R. Schmidt independently rediscovered the technique while developing a tool for the inspection of oil well casings. Shell Development purchased the patent rights from MacLean and had great success with the tool. At that time, no electromagnetic techniques for examining the casings nondestructively were available.[2] Schmidt spearheaded the development of the technique and named it *remote field eddy current testing* to distinguish it from conventional eddy current testing. The technique as used in industry is now referred to as *remote field testing*. The term minimizes confusion with conventional eddy current testing and emphasizes the magnetic field interactions exhibited by remote field testing.[3] Shell encouraged the commercialization of the technique by licensing it to interested parties in the 1980s. Several manufacturers immediately recognized the value of remote field testing for the examination of ferrous heat exchanger tubes and began manufacturing remote field test equipment.

Over the last 20 years remote field testing has attracted the interest of researchers around the world.[4] The research was triggered by Schmidt's 1984 publication on the usage of a circumferential array of detectors.[5] W. Lord and others gave remote field testing a firm theoretical basis by publishing the first in-depth finite element study.[6] Later Mackintosh and Atherton developed powerful analysis tools by recognizing remote field testing's through-transmission character.[7] The improved understanding of how remote field testing worked increased its accuracy and acceptance. This resulted in tremendous growth in the late 1980s and early 1990s. Systems were developed using internal probes to examine gas distribution pipelines, oil and gas well casings, cast iron and steel water mains, heat exchangers and boilers. Developments since 1990 include the testing of flat plates (for example, storage tank floors)[8] and steel pipes using external probes that use a technique similar to remote field testing.

Early remote field testing systems were much like the eddy current test systems of the 1980s: both used analog circuits, cathode ray tube displays and paper chart recorders for data storage. Modern instruments use computers to display and store data and more advanced systems also have automated signal analysis routines.

Part 2. System Components

Probe Configuration

In the basic remote field testing probe (Fig. 1), there is one exciter coil and one detector coil. Both coils are wound coaxially with respect to the tested tube and are separated by a distance greater than twice the tube diameter. This axial distance is very characteristic of remote field testing. If the exciter and detector were to be placed close together the detector would measure only the field generated by the exciter in its vicinity. In that case the setup would basically be a standard eddy current test setup in send and receive mode.

To observe remote field testing's unique through-wall transmission effect the detector needs to be moved away from the exciter. The actual separation depends on the application and the probe manufacturer but will always be a minimum of two pipe diameters. It is this separation that gives remote field testing its name — the detector measures the electromagnetic field *remote* from the exciter. Although the fields have become very small at this distance from the exciter they contain information on the full thickness of the pipe wall.

The dimensions of the coils will vary from manufacturer to manufacturer. The *fill factor* is the ratio of the effective cross sectional area of the primary internal probe coil to the cross sectional area of the tube interior. Although the fill factor of the coils can be as low as 70 percent it will usually be similar to the fill factor for eddy current probes: 85 percent or more. A lower fill factor reduces sensitivity to small discontinuities but does not otherwise affect the quality of remote field testing data. The ability to function with a low fill factor makes remote field testing especially attractive for pipes with internal coatings and tight bends.

Remote field testing probes often contain arrays of receiver coils. The coils are connected to a remote field testing instrument by coaxial cable, where the outer conductor is used to shield the inner conductors from ambient noise. The coaxial cable is usually housed in a stiff plastic tube that lets the probe be pushed into a heat exchanger tube or pipe for distances up to 30 m (100 ft).

Figure 1. Simple probe for remote field testing.

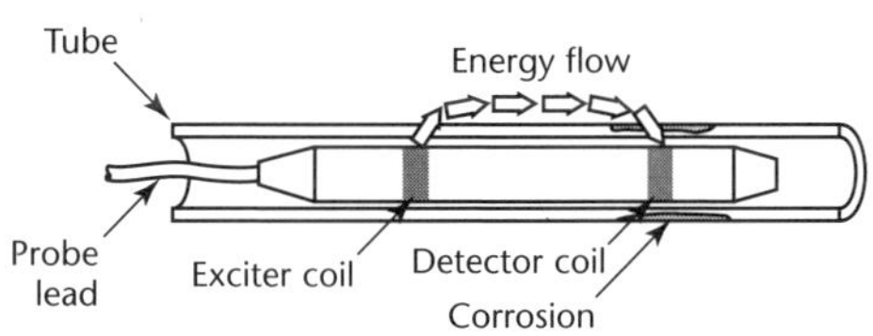

Instrumentation

Besides the coils and coaxial cable a remote field testing system contains four other major components.[9]

1. An oscillator is used as the signal source for the exciter coil and as a reference for the detector signal.
2. A power amplifier increases the power level from the oscillator signal so that it can be used to drive the exciter coil.
3. The phase and amplitude detector measures the detector coil signal.
4. A microcomputer based storage device processes and stores the data.

Figure 2 shows how the different electronic components interact.

Driving of Remote Field Test Probe

The exciter coil is energized with alternating current at frequencies ranging from 50 Hz to 1 kHz for ferrous materials. Higher frequencies are used for nonferrous tubes. The exciter coil typically carries a current of 0.1 to 1.0 A,

Figure 2. Electronic components of remote field test system.

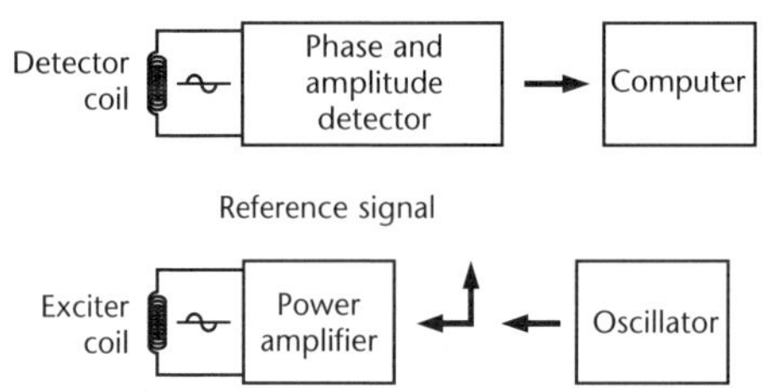

the limitation being probe temperature and, for some probes, magnetic saturation of the probe core. The detector signal depends directly on the exciter current and frequency. Thicker wall penetration can be obtained by lowering the frequency and increasing the exciter current. The actual test parameters (such as test frequency, drive voltage and sample rate) are chosen by taking into consideration factors such as probe pull speed, discontinuity sensitivity requirements, magnetic permeability μ, tube electrical conductivity σ and tube wall thickness τ.

In general, a lower frequency (up to 250 Hz) is used for thick walled and high permeability pipe. For superior sensitivity or high test speeds, a higher frequency is preferred. The frequency is chosen as high as possible while minimizing noise and remaining in the remote field zone. Because remote field testing is used in very diverse applications, it is important to check for the presence of any electromagnetic interference noise sources, such as welders, electric motors, power inverters and pumps. Such devices tend to generate noise in the frequency range of remote field testing. Modern day remote field testing equipment manufacturers sometimes provide the user with a noise spectrum, showing the environmental noise for a range of frequencies. Some technicians prefer to place the probe in a thick walled block to determine the baseline noise at a given frequency.

Because remote field testing signals are often quite small (1 to 10 µV) it is advisable to avoid using line voltages and their harmonics (60 Hz or, in Europe, 50 Hz), which can cause interference.

The inspector needs to minimize the background noise while keeping in mind the test factors mentioned above.

PART 3. Detector Signal

Remote Field Energy Zones

For a remote field probe, there are two distinct sensing zones with a transition zone between them (Fig. 3).[10] In order, the zones are the direct field zone, the transition zone and the remote field zone.

As the detector coil distance from the exciter coil is increased, the dominant field energy changes from direct coupled energy (between the exciter and detector coils, inside the tube) to energy that is coupled to the detector coil primarily by transmission through the tube wall. Between these two distinct zones, there is a transition zone where the direct coupled energy and the indirect coupled energy are comparable in magnitude. The location of the transition zone changes with frequency, wall thickness, permeability and conductivity.

In an idealized situation (infinite alternating current sheet over a conducting half space), the eddy current density in the case of conventional eddy current techniques decays exponentially with depth. This phenomenon, called *skin effect,* in general limits the application of conventional eddy current techniques to the detection of surface or shallow heterogeneities. The remote field eddy current technique seemingly violates the skin effect limitation in that it is equally sensitive to inside surface and outside surface discontinuities. Lord and others simulated the underlying physical process and examined the field distribution to look for clues to explain this seeming contradiction.[6] A brief summary of their findings are reported below.

Finite Element Simulation

Electromagnetic induction phenomena associated with conventional and remote field eddy current nondestructive techniques are both described by the complex form of the vector poisson equation:

FIGURE 3. Zones in remote field testing. Profiles of *B* field just inside and outside pipe wall are used to indicate direct field region, transition and remote field zones.[10]

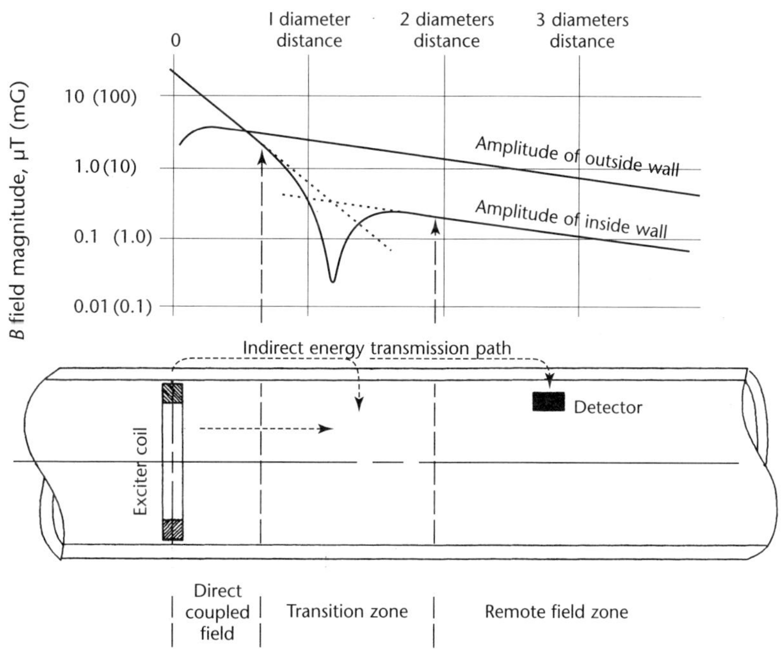

$$(1)\quad \frac{1}{\mu}\nabla^2 A \;=\; -J_s \;+\; j\omega\sigma A$$

where A is magnetic vector potential (volt), J is applied current density (ampere per square meter), μ is magnetic permeability (henry per meter), σ is conductivity (siemens per meter) and ω is angular frequency (radian per second). Lord and others used the finite element technique to study a test geometry consisting of an axisymmetric excitation and detector coils in a ferromagnetic pipe.[6]

Figure 4 shows experimental and model predictions of the detector coil signal magnitude and phase as a function of the distance (expressed in multiples of the pipe diameter) between the excitation and detector coils. The phase measurements are made with reference to the excitation signal. The magnitude of the detector signal decreases rapidly with distance until a transition region is reached (after about two pipe diameters in length), after which the rate of decrease is much slower. The phase angle, in contrast, changes rapidly in the transition region. Some insight into the underlying physical process can be gained by examining the magnetic flux plots.

Figure 5 shows the variation in the flux distribution at various points in the sinusoidal excitation cycle. The excitation frequency is 40 Hz and the plots show the evolution of the magnetic flux pattern as a function of time in steps of 4.1667 ms, which correspond to 0.524 rad (30 deg) increments in ωt. The flux lines are plotted on a logarithmic basis to increase the dynamic range of the plot. The plots show that the flux density decays rapidly close to the exciter (near field). The decay

FIGURE 4. Experimental and finite element predictions of signal magnitude and phase for detector coil for 38.1 mm (1.5 in.) inside diameter ferromagnetic pipe of 5.1 mm (0.2 in.) wall thickness. Test frequency f = 40 Hz; relative permeability μ_r = 250; conductivity σ = 143 S·m^{-1}.

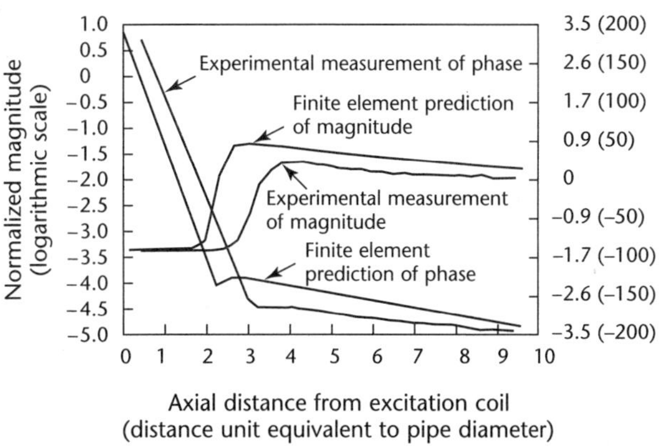

FIGURE 5. Flux distribution in pipe as function of time in steps of 4.1667 ms, which corresponds to 0.5 rad (30 deg) increments of ωt (product of angular frequency ω and time t) when frequency f = 40 Hz: (a) 0 rad (0 deg); (b) 0.5 rad (30 deg); (c) 1.0 rad (60 deg); (d) 1.5 rad (90 deg); (e) 2.0 rad (120 deg); (f) 2.5 rad (150 deg). Only top half of pipe longitudinal section is shown. Flux lines are plotted logarithmically to increase dynamic range of plot. Test conditions and specimen are similar to those in Fig. 4.

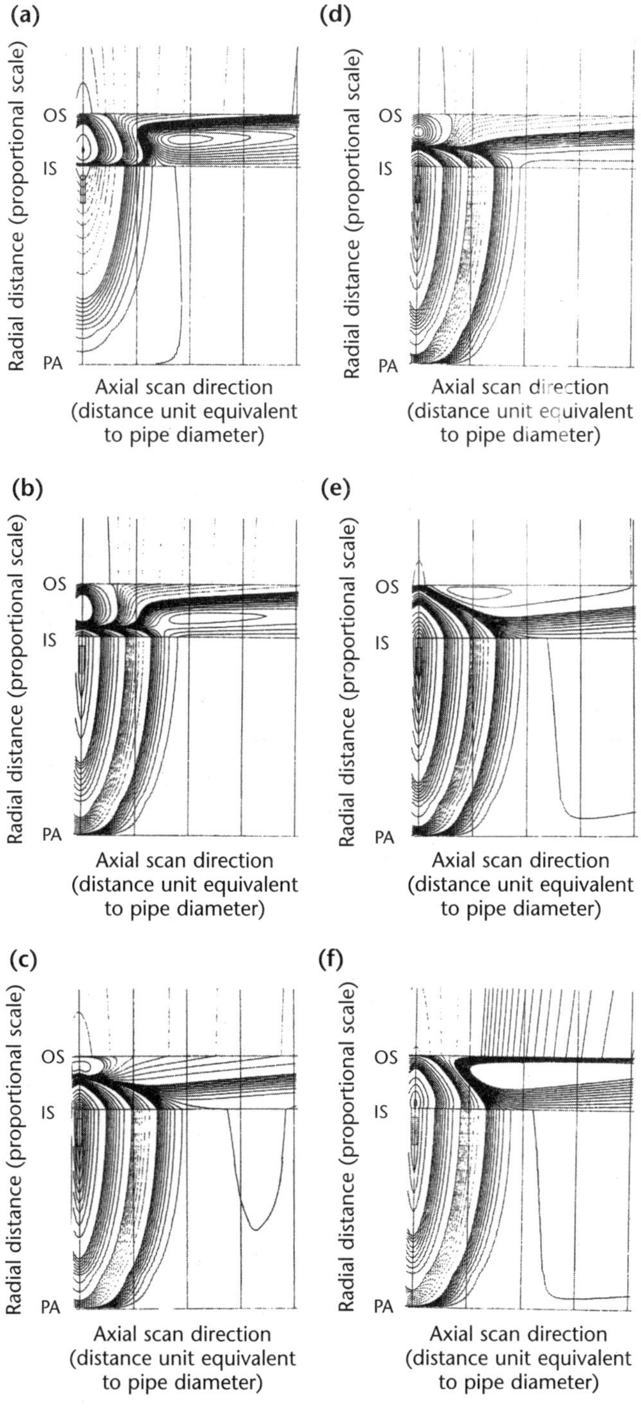

Legend

IS = inside surface
OS = outside surface
PA = pipe axis

rate is much lower in the remote field zone. As an example, at $\omega t = 1.57$ rad (90 deg), the four bands of flux lines contain 90 percent, 9 percent, 0.9 percent and 0.09 percent of the total flux.

Additional insight can be gained by reviewing the equivector magnetic potential contour and equiphase contour plots shown in Figs. 6 and 7. The equivector magnetic potential contour plot shows the existence of a potential valley (that is, a point where the magnetic vector potential reaches a minimum) in the transition region where the root mean square of the vector magnetic potential is zero. Similarly, Fig. 7 shows the existence of a phase knot (that is, a point where the phase is undefined) in the transition region. The presence of these unusual artifacts indicates that the energy flows in an unusual pattern.

Figure 8 shows model predictions of the real component of the poynting vector, which shows the magnitude and direction of the energy flow at each point in space. The plot shows that the field pattern in the transition region arises as a consequence of the interaction between two energy streams. The outward bound energy stream interacts with the inwardly directed energy stream in the transition region, giving rise to the potential valley and phase knot. The presence of potential valley leads directly to the magnitude plot shown in Fig. 4. Because the energy flow loop includes the regions near the outer and inner walls of the pipe and because the detector coil output is a function of the field that it is immersed in, the technique is sensitive to both inside surface and outside surface discontinuities. In other words, the field in the vicinity of the detector coil is a consequence of the superposition of the fields generated by the energy flows near the inner and outer walls of the pipe. Consequently, the method is equally sensitive to inside surface and outside surface discontinuities.

A detector placed in the remote field zone will provide information from two pipe wall transitions: one at the exciter and one at the detector. The total attenuation (in decibel) and phase change

FIGURE 6. Equivector magnetic contour plots showing existence of potential valley (point where magnetic vector potential reaches minimum) in transition region. Test conditions and specimen are similar to those in Fig. 4.

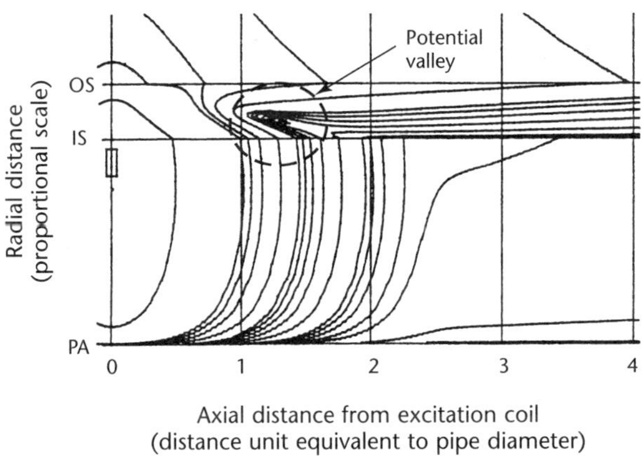

Legend
IS = inside surface
OS = outside surface
PA = pipe axis

FIGURE 7. Equiphase contour plots corresponding to Fig. 8, indicating existence of phase knot (point where phase is undefined) in transition region. Test conditions and specimen are similar to those in Fig. 4.

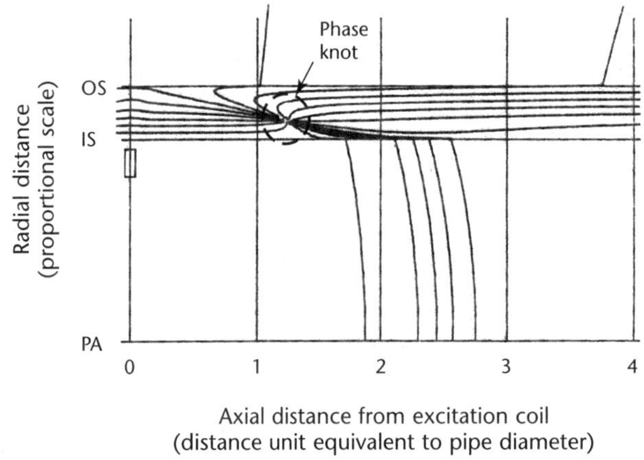

Legend
IS = inside surface
OS = outside surface
PA = pipe axis

FIGURE 8. Finite element model predictions of real component of poynting vector, showing magnitude and direction of energy flow pattern in region.

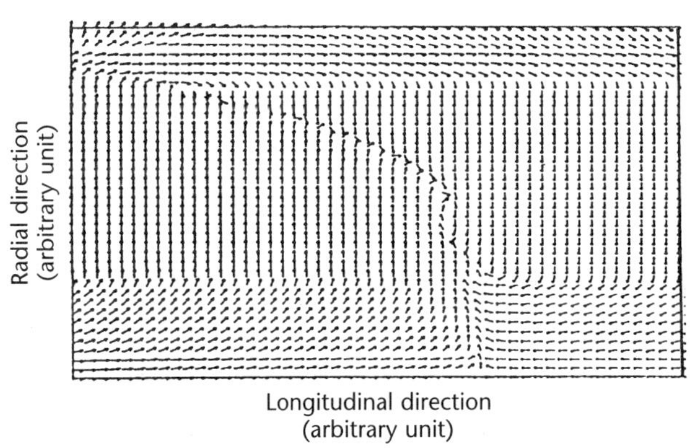

(in degrees) is the sum of the attenuation and phase change at both locations. Schmidt[5] discovered that the attenuation and phase change could be derived approximately by using the skin effect equation for depth of penetration:

$$\text{(2)} \quad \delta = \sqrt{\frac{1}{\pi f \mu_0 \mu_r \sigma}}$$

and:

$$\text{(3)} \quad \text{Phase lag} = \tau \sqrt{\pi f \mu_0 \mu_r \sigma}$$

where f is frequency (hertz), δ is standard depth of penetration (meter), μ_0 is magnetic permeability of free space (ratio), μ_r is relative magnetic permeability (ratio), σ is tube electrical conductivity (siemens per meter) and τ is tube wall thickness (meter).

When the probe is in the nominal tube wall setting, the signal from the remote field testing detector coil is used as the reference voltage against which changes caused by wall thinning and other discontinuities are measured. The attenuation and phase rotation angles of remote field test signals are governed by rules like those for eddy current signals. Equation 1 shows that, if the frequency, conductivity or relative permeability goes up, then the standard depth of penetration into the material goes down — and vice versa. Therefore, a relatively high test frequency (500 Hz to 1 kHz) would be chosen to test thin materials — for example, a wall less than 1 mm (0.04 in.) — or to test materials that have a low permeability such as magnetic stainless steel or low electrical conductivity such as nickel. Conversely, if the tube wall is thick — for example, greater than 3 mm (0.12 in.) — or has a high relative permeability (for example, seamless carbon steel), then a lower frequency would be more appropriate (for example, 50 to 200 Hz). The quantity that sets the standard depth of penetration is the product of the frequency, permeability and conductivity. Instead of resolving these factors separately, it is often more convenient to determine their product.

To analyze remote field discontinuity signals using the skin effect theory, the wall thickness must be expressed in standard depths of penetration, where a standard depth of penetration is equivalent to a radian (rad) and 1 rad = $180 \cdot \pi^{-1} \cong 57.3$ degrees. Consider a case where the nominal wall thickness of the pipe is five standard depths of penetration and wall loss at one point is fully circumferential and 1.5 standard depths of penetration deep. When diffusing through the pipe wall, the signal will be delayed by five standard depths of penetration or 5 rad (286 degrees) through nominal thickness and by 3.5 rad (201 degrees) at the circumferential discontinuity. As the detector coil moves from nominal pipe into the area of metal loss, its phase lag will decrease by 1.5 rad (86 degrees), which makes the phase increase by 1.5 rad (86 degrees), a positive change at metal loss. A similar analysis can be applied to the amplitude. In nominal pipe, the signal is attenuated by a factor of exp (5) = 148 whereas at the discontinuity location the amplitude is reduced by a factor of exp (3.5) = 33. The detector coil output will therefore *increase* in amplitude by a factor of exp (1.5) = 4.5.

In general, the relatively simple skin effect equation describes remote field testing behavior very well. However, when a discontinuity's depth is such that the remaining wall thickness is less than one standard depth of penetration, the remote field response will actually start to deviate from the skin effect equation.[7] The deviations are especially noticeable in amplitude. In those cases, more sophisticated techniques such as those described elsewhere, in this volume's chapter on modeling, are required.

PART 4. Selection of Remote Field Testing

Features of Remote Field Testing[11]

Remote field testing can be used for all conventional carbon steel material specifications, diameters and wall thicknesses. It is therefore used in many different types of heat exchangers, including fossil fuel boilers (especially in water wall and generator bank tubes), black liquor recovery boilers, shell and tube exchangers and air fin coolers.

Remote field testing operates at relatively low frequencies. Typical frequencies are in the range of 40 to 500 Hz.

The test speed for carbon steel tubes is about 150 $mm \cdot s^{-1}$ (30 $ft \cdot min^{-1}$). A two-person crew can examine from 200 to 500 tubes measuring 9 m (30 ft) long in an 8 h shift, depending on access, setup time, number of discontinuities encountered and other factors.

Remote field testing is a noncontact technique, so the probes have minimal friction with the pipe wall and require no couplant.

Sensitivity

The accuracy for remote field testing in the straight part of the tubes is about 10 percent of wall thickness for general wall loss. The accuracy is generally less (20 percent of wall) for highly localized discontinuities and in bends or near external conducting objects because of the changes in magnetic properties of the tube in the bend area and because of shielding effects of external objects.

Remote field testing is also equally sensitive to inside and outside surface discontinuities but usually cannot discriminate between them without the help of near field coils. Remote field testing is relatively insensitive to scale and magnetic debris.

A large fill factor is not required for remote field testing and centralization is not critical (as it is with ultrasonic, eddy current and flux leakage testing). However, a small fill factor will result in decreased sensitivity to small discontinuities.

Other Test Techniques

To fully appreciate the strengths and weaknesses of remote field testing, it is useful to compare it to other techniques used in industry. Included here are eddy current testing, saturation eddy current testing and magnetic flux leakage testing.

Although eddy current testing is predominantly used for the testing of nonferromagnetic tubing, it can be used to test slightly ferromagnetic materials such as nickel copper alloy by using it with a direct current biasing field strong enough to magnetically saturate the pipe wall. This *saturation eddy current* technique is unsuited for thick walled, highly magnetic material, such as carbon steel; however, various methods have been found useful for carbon steel tubes and pipes. These include the magnetic flux leakage technique and ultrasonic testing. Table 1 provides a quick comparison of different nondestructive test methods for heat exchanger testing.

TABLE 1. Techniques used for nondestructive testing of heat exchanger tubes.

	Electromagnetic Techniques				
	Eddy Current Techniques			Magnetic	Ultrasonic
Characteristic	Conventional	Saturation	Remote Field	Flux Leakage	Testing
Skilled technicians required	yes	yes	yes	yes	yes
Speed ($m \cdot s^{-1}$)	0.9 to 2.0	0.61	0.61	0.61	0.04
Wall loss identified[a]	yes	yes	no	yes	yes
Minimum probe fill factor	0.8	0.7	0.7	0.8	note[b]
Ferromagnetic tube test	no	slight	yes	yes	yes

a. Inside diameter versus outside diameter.
b. Ultrasonic testing requires seals to center the probe and retain fluid coupling.

Eddy Current Testing versus Remote Field Testing

In typical eddy current testing instruments, the impedance of the inspection coil is measured. Usually the coil is part of a bridge circuit that becomes unbalanced as the coil passes over a change in material thickness, permeability or conductivity. Discontinuities are characterized and sized by the phase rotation and attenuation of the signal as compared to a reference standard. The test coil in eddy current testing can be an energized coil or it can be a passive coil that receives its energy from a separate energized coil in close proximity (send and receive configuration). Common coil configurations are absolute or differential coils; axial or radial coils; and bobbin or pancake coils.

Remote field testing has many similarities to eddy current testing but there are also major differences.

1. In remote field testing, the exciter coil is always separated from the receiver coil or coils by at least two tube diameters. As such, remote field test coils are always in a send and receive configuration.
2. In remote field testing, the energy from the exciter coil passes through the tube wall twice, once when leaving the exciter and again when passing back through the wall at the detector.
3. The sensitivity to discontinuities on the outside of a tube is reduced in the eddy current technique whereas remote field testing maintains almost equal sensitivity to discontinuities either inside or outside the tube.
4. Remote field test systems measure the phase and amplitude of a signal. Eddy current test systems may measure the same quantities in send and receive configurations or may measure the impedance of the test coil.
5. Eddy current technique probes are sensitive to changes in the proximity of the test coil to the tube surface. This change is known as *probe wobble* — as the probe passes through the tube, it can be pushed to one side of center by internal scale or dents. Even if the tube is clean, the eddy current testing probe can wobble unless it is centered with mechanical guides.
6. Remote field probes are relatively insensitive to probe wobble and are forgiving if the probe is undersized or pushed to one side of the tube.
7. Because of the much lower test frequencies used for remote field testing in steel (and because the measurement of phase usually requires at least one time period of the excitation signal), remote field probes must be moved more slowly than eddy current probes. *The remote field probe must be near the smallest discontinuity required to be detected for at least one cycle in order to detect the discontinuity.*
8. Absolute coils for both remote field and conventional eddy current techniques are both sensitive to temperature variations over the length of the tube.
9. In steel tubes, remote field testing is more sensitive to circumferential cracks that interrupt the lines of magnetic flux. Eddy current bobbin probes are more sensitive to axial cracks in tubes, which interrupt the eddy currents.
10. Because of its so called *through-transmission* nature, remote field testing can examine thicker materials than eddy current techniques can.
11. Because it is commonly used for steel and cast iron, remote field testing is generally carried out at a lower frequency than eddy current techniques are.

Magnetic Flux Leakage versus Remote Field Testing

With a magnetic flux leakage probe, the energy source is an axially aligned magnetic field (as with a remote field testing probe). Discontinuities are detected by the magnetic flux leakage probe as some of the magnetic flux lines *leak* out of the pipe and are detected by passive coils or sensors passing though the leakage field. This arrangement is similar to that for remote field testing except that the remote field test probe generates an alternating current field whereas magnetic flux leakage testing uses a direct current field. When pickup coils are used, the remote field testing probe itself does not need to be moving to measure the wall thickness, because the excitation is alternating.

The remote field eddy current signal is likely to offer additional information because both phase and amplitude of the signal can be analyzed. In contrast, only the amplitude of the magnetic flux leakage signal is available. Consequently, remote field testing provides two pieces of information, usually enough to permit calculation of tube wall thickness.

Part 5. Signal Analysis

Data Presentation

Strip Chart

Remote field test data are recorded in computer memory or hard drive and phase amplitude diagrams (voltage planes) are displayed on instrument monitors in near real time as the test progresses. The raw data from the detector are stored either in phase amplitude format or as in-phase and quadrature components. The data can be recalled for display, analysis and reporting purposes after the test process is completed.

A strip chart displays coordinates from the phase amplitude diagram (for example, an x,y display, a phase display or a log amplitude display) as a function either of time or of the axial distance along the length of the tube. In the strip chart of Fig. 9, which shows phase and log amplitude for an absolute coil, deflections to the left represent metal loss and, on the right, wall thickening as in the case of tight fitting support plates or baffle plates. Phase and log amplitude are the preferred quantities for the absolute coil strip chart display because they are both linear indicators of overall wall thickness (as opposed to the in-phase and quadrature components, which make a more suitable display for differential coils). To display phase and amplitude on the same strip chart, the amplitude is multiplied by the factor for conversion from radian to degree ($180 \cdot \pi^{-1}$).

FIGURE 9. Strip chart recordings: (a) phase and log amplitude signals for absolute probe; (b) x,y voltage signals; (c) differential signals; (d) mixed frequency signals.

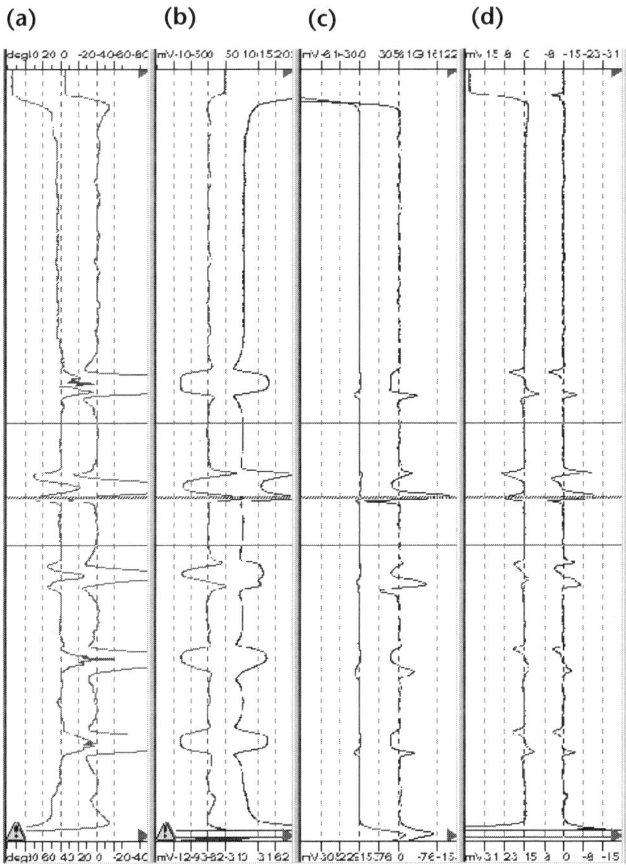

Complex Plane Displays

The voltage plane and x,y displays provide maps of the detector coil output in polar coordinates (Figs. 10 and 11). On polar

FIGURE 10. Phase amplitude diagram: voltage plane recording.

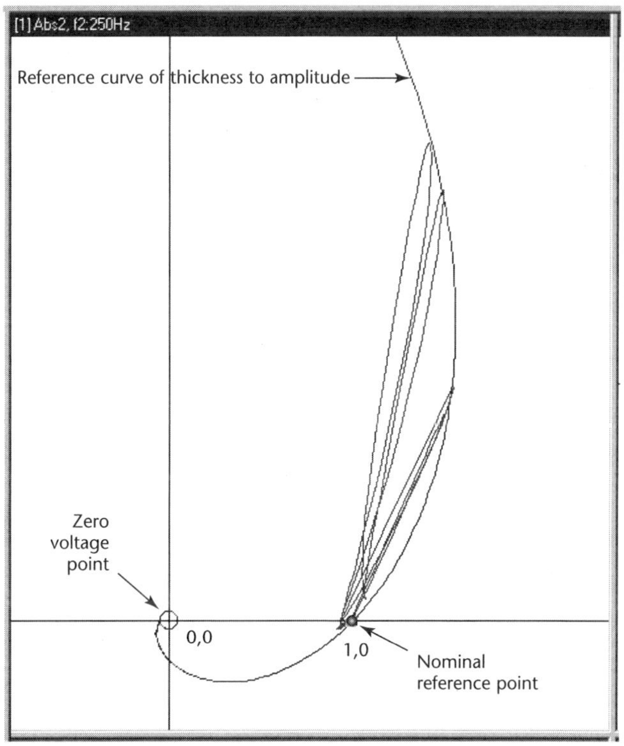

displays, signals are drawn as vector points with the angle representing the phase and the radius representing the amplitude. Remote field testing signals on the voltage plane or *x,y* display are scaled and rotated to a convenient position for viewing. This manipulation makes it easier to measure and recognize deviations from the nominal position. With absolute detectors the signals can be scaled and rotated about the origin to place the nominal signal at (1,0) in rectangular coordinates. In other words the signal for nominal tube is placed at 1 V normalized at zero degrees. With differential detectors the signal is often shifted to place it at the origin (0,0) even though the signal is usually not actually zero.

Besides the detector trace, the voltage plane has a number of static components: the origin, the exponential skin depth reference curve and the X and Y axes. The reference curve is a feature unique to remote field testing and is very helpful when identifying and sizing anomalies. The curve starts at 0,0 (that is, zero voltage at origin) and follows an exponential spiral path as the overall wall thickness of a tube is decreased. It is theoretically possible to place thickness values for full circumferential discontinuities directly on this curve; however, the values would be different for short discontinuities versus long discontinuities, as explained below. Although axial distance information is not displayed on a voltage plane, it remains a very powerful tool for sizing discontinuities. Strip charts are useful because they usually show distance information.

As mentioned before, the strip chart records what the detector coils senses as the probe is pulled through the tube being tested whereas the voltage plane shows selected discontinuity signatures at any point along the strip chart. As a result, the strip chart displays and *x,y* voltage plane displays are often put side by side. This presentation lets the technician use the complex planes to identify and size discontinuities while simultaneously using the strip chart to find the axial location of the discontinuity. Combining both displays accelerates discontinuity reporting.

Signatures of Short and Long Discontinuities

In the remote field technique, discontinuity indications are recognized by their shape and size. As each of the exciter and detector coils in the basic probe passes a discontinuity, there is a change in the voltage of the receiver coil. Therefore, if the discontinuity is shorter than the distance from exciter to detector, there will be two distinct signals — one as the exciter passes the discontinuity and a second one as the detector passes it (Fig. 12). When the physical sizes of the exciter and detector coils are about equal, the two strip chart responses will be

FIGURE 11. Phase amplitude diagram: *x,y* plane recording.

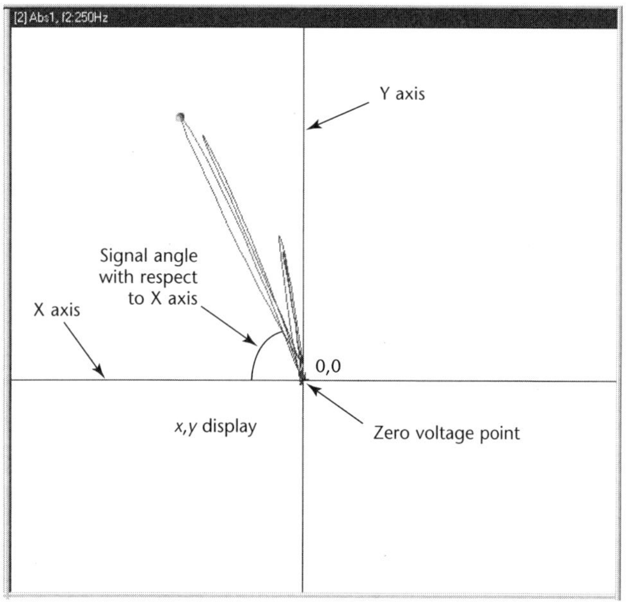

FIGURE 12. Strip charts and phase amplitude diagrams show indications from short discontinuity.

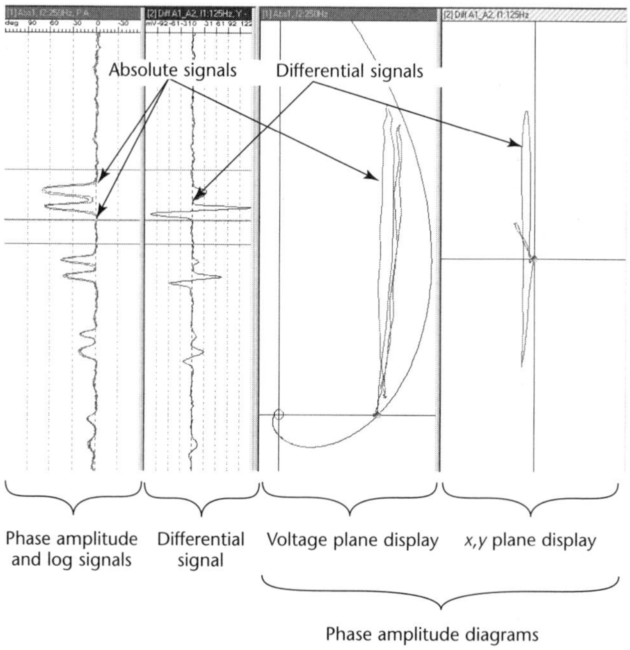

comparable. However, in some cases, the exciter coil design is optimized for generating the electromagnetic field whereas the detector dimensions are optimized for resolution and sensitivity. As a result, the exciter coil may be longer than the detector.

The voltage plane trace in Fig. 12 corresponds to the gated area in the strip charts. The double response is also exhibited on the voltage plane by two distinct traces moving from the normal wall point (0,0 for differential and 1,0 for absolute coils) vertically to the exponential skin depth spiral. The fact that the voltage plane traces terminate on the reference spiral shows that the discontinuity must extend fully around the entire circumference of the pipe. The angle of the traces with respect to the horizontal axis permits calculation of the depth of the discontinuity.

The X,Y display in Fig. 12 is characteristic of a differential detector setup. A differential detector coil consists of two identical coils displaced axially from each other and wound in opposite directions. Differential coils are excellent edge detectors and can neutralize to some extent the double response characteristic of remote field testing. As such, differential detectors improve remote field testing's sensitivity to small discontinuities. Differential detectors are less sensitive to tapered or smooth wall loss. The origin for the differential coil trace is chosen at 0,0 because under nominal pipe wall the two halves of the differential detector cancel each other out.

As soon as one half moves underneath a discontinuity, the detector circuit will become unbalanced and a resultant signal is measurable. When the second half of the detector moves underneath the same discontinuity, it will produce a similar trace in the opposite direction. The two opposite responses from a differential detector as it moves underneath a pipe anomaly is very characteristic and is often referred to as a *differential kick*. The differential kicks are clearly visible on the second strip chart in Fig. 12. The angle of the differential kick on the phase amplitude diagram can also be used to estimate the depth of the discontinuity.

Eddy current test technicians will often prefer to rotate the differential signal from a through hole to 40 degrees so that it looks like a familiar signal on an eddy current impedance plane. However, this is where the similarity ends. In remote field testing, signals from increasing wall loss (outside diameter or inside diameter) always rotate counter clockwise. Eddy current test signals from inside diameter discontinuities rotate clockwise and outside diameter discontinuities rotate counter clockwise as they increase in depth.

With the absolute coil, if the discontinuity is longer than the spacing from exciter to detector, the signal phase and the log of the signal amplitude will both double in size, resulting in a *head and shoulders* signal (Fig. 13). This signal is the result of the overlap of the exciter and detector responses. The corresponding trace on the voltage plane is called a *dog leg* because of the bend in the indication with machined discontinuities. Notice that the discontinuity is much harder to characterize with differential detectors because of its long and gradual nature.

Voltage Plane Polar Plot Display

In remote field testing, the material volumes at the exciter and again at the detector are both interrogated simultaneously. As a result, the sensitivity to discontinuities located on the inside of the tube is about the same as the sensitivity to external discontinuities; however, there is no way to tell them apart unless a separate, high frequency coil is added in the direct field (which would be sensitive mainly to internal discontinuities). On the voltage plane polar plot display, thickness decreases (due to metal loss) rotate the signal counter clockwise either along or inside the reference curve (see Figs. 14 and 15).

The shape and orientation of a discontinuity will affect the remote field testing indication. Two small discontinuities with the same amount of

FIGURE 13. Voltage plane signals from long discontinuity show direction of wall thickness changes.

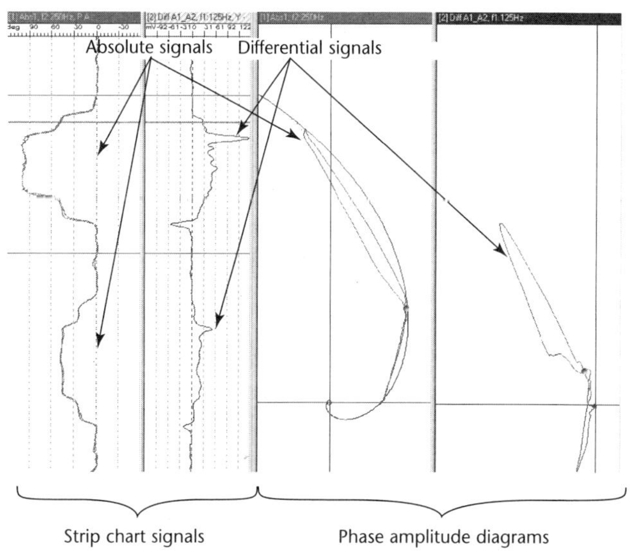

volume loss but different depths will show different indications. In general, deeper discontinuities will show up as larger, more pronounced signals unless they are extremely small in volume.

The orientation of the discontinuity is also of importance. From eddy current testing, it is intuitively expected that discontinuities aligned in the axial direction show larger responses than discontinuities oriented circumferentially: an axially aligned discontinuity interferes more with the circumferentially flowing eddy currents. In remote field testing, the opposite is true: discontinuities oriented in the circumferential direction are more pronounced.

This difference exists because the magnetic field interaction exhibited by remote field testing makes it behave somewhat similar to magnetic flux leakage testing. A circumferential discontinuity forms a large interruption for the axially aligned magnetic flux whereas a thin axial discontinuity barely influences the flux. The disturbance in the magnetic field caused by the circumferential discontinuity results in a larger remote field test signal response, which can then be used to estimate the circumferential extent of the discontinuity.

If the tube wall is locally thinned on one side of the tube, a line can be drawn from the nominal point through the signal tip; the line will point toward the spot on the reference curve that represents the same reduced wall thickness if the thinning were all around the circumference. By measuring the phase angle of the signal the remaining wall thickness can be estimated. By calculating the ratio of the signal height from a one-sided discontinuity to a signal from a circumferentially thinned tube wall, the circumferential extent can also be calculated. Phase angle, wall thickness and circumferential extent can be critically important when characterizing discontinuities.

By observing the signal shape, phase angle and relative size on the voltage plane many discontinuities can be characterized and sized for depth and circumferential extent. The axial length of a discontinuity can also be measured by recording the data on a strip chart as the probe is pulled through the tube. Virtually all remote field test instruments display the data as strip charts and voltage planes. Many instruments have automatic depth sizing and reporting software.

Figure 14. Voltage plane indication of long discontinuity, made with absolute and differential coils. Overall thickness of tube decreases evenly from point A to point B.

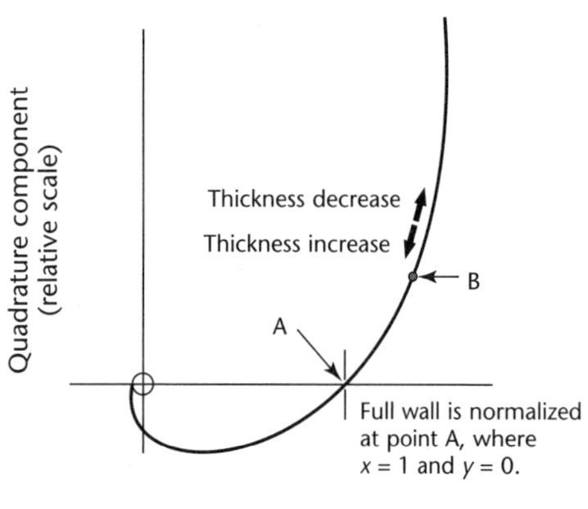

Common Signatures in Remote Field Testing

Signals from tubes can be categorized as follows: (1) general wall loss (with long length and extensive circumferential extent), (2) long one-sided discontinuities, (3) short circumferential discontinuities and (4) small volume, local discontinuities.

General Wall Loss

Discontinuities that are longer than the probe are classified as long discontinuities. These discontinuities, if

Figure 15. Voltage plane signals indicate discontinuity depth and discontinuity volume.

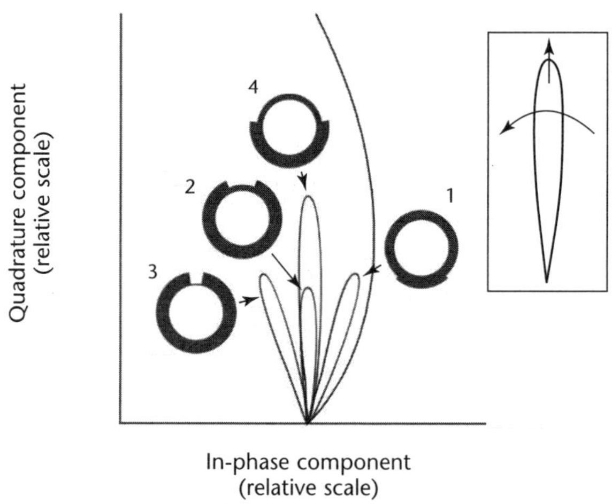

Legend
1. Increasing depth.
2. Increasing depth and increasing volume.
3. Increasing depth.
4. Increasing volume.

they affect the entire circumference of the tube evenly, will produce signals on the voltage plane that closely follow the reference curve. Technicians identify discontinuity length from the strip chart recording and decide if the discontinuity is long or short. On the strip chart, the *phase* and *log amplitude* signals will track in synchronization with each other if the discontinuity is perfectly circumferential.

The phase and log amplitude signals of long discontinuities are roughly double those from short discontinuities of similar depth, so it is important that the technician recognize the characteristics of each signal type. Figure 16 shows general wall loss as well as signals from short discontinuities.

Long One-Sided Discontinuities

Discontinuities that occur on one side of the tube and are longer than the distance from the exciter coil to the detector coil are classified as long one-sided discontinuities. Such discontinuities will produce signals inside the reference curve (that is, left of the curve) that may follow a track that is parallel to the curve and that may have an included loop that does not return all the way to the nominal point at 1,0. These discontinuities often taper at each end back to nominal thickness. On the strip chart recording, the phase and log amplitude signals of one-sided discontinuities do not track together. The phase trace will generally deviate further from the baseline position than the log amplitude trace. Examples are steam impingement erosion and tube to tube fretting (midspan erosion). Carbon dioxide corrosion can also be long and one-sided.

Short Circumferential Discontinuities

Discontinuities that are shorter than the probe (that is, shorter than the distance from exciter to detector coil) and are all of the way around the tube are classified as short circumferential discontinuities (Fig. 17). These discontinuities produce a typical double signal on both voltage plane and strip chart. The phase and log amplitude strip chart traces will track synchronously if the discontinuity is perfectly circumferential. Examples are baffle wear and condensate grooving.

Small Volume, Local Discontinuities

Discontinuities that are primarily on one side of the tube and have limited axial and circumferential length are classified as small volume, local discontinuities. Examples are pitting and cracking. These discontinuities produce signals that are very small in amplitude and are left of the reference curve on the voltage plane. The signals are sharp and will repeat as each

FIGURE 16. Examples of general wall loss plus local pitting (chelant corrosion, inlet erosion and general pitting corrosion).

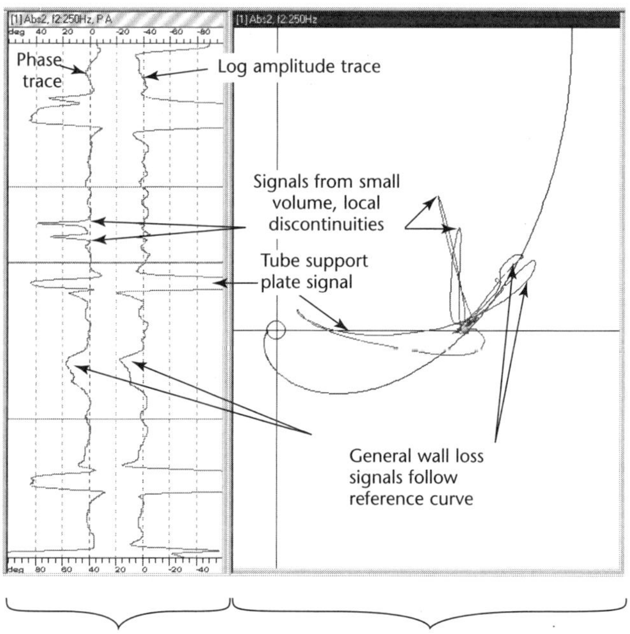

FIGURE 17. Short, circular discontinuity caused by condensate grooving next to support plate.

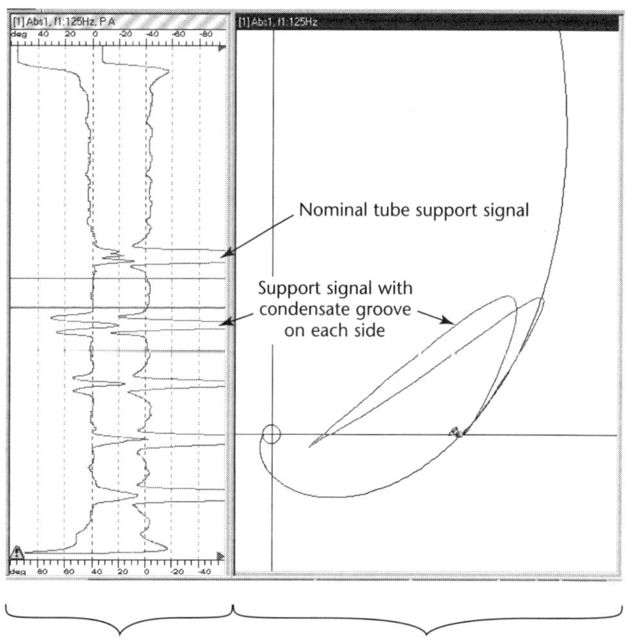

coil passes the discontinuity. Differential detector coils are preferred for detection of small volume discontinuities because they produce a signal twice as large as that from an absolute probe.

Reference Standards

All nondestructive test methods use reference standards to compare discontinuity signals with those from known machined discontinuities. Remote field testing is no different: its reference standards are tubes with artificial discontinuities for calibration. However, remote field testing requires reference standards for each variation in tube diameter, wall thickness (tube gage), conductivity and permeability. In each reference standard, reference discontinuities must be machined to closely simulate the discontinuities expected in the tube or pipe being examined. ASTM E 2096-00 mentions two possible reference tube styles.[11] Customers may specify different discontinuity types if they expect to encounter discontinuities other than the suggested types.

When ordering the manufacture of a reference tube it is important to provide the following specifications.

1. Specify the tube material.
2. Specify the tube diameter and gage.
3. Specify the tube manufacturing technique, whether seamless or electric resistance welded.
4. Describe any heat treating that the tubes to be tested have undergone.
5. Space the discontinuities at least four tube diameters apart and at least four tube diameters from each end.
6. Require the reference discontinuities to be machined with a series of small cuts using sharp machine tools and copious coolant. This machining prevents local heating that can change the permeability and conductivity of the tube material.
7. Specify the tolerances on the discontinuity depths and ask for them to be machined according to the actual, measured, tube wall — not the nominal wall thickness that the tube is specified to be. Tubes are manufactured to tolerances of, for example, +12 percent and –10 percent.
8. Specify that all edges of discontinuities are to be radiused to avoid edge signals.
9. Specify how the tube is to be identified with a permanent label, listing the discontinuity depths, reference serial number, material and wall thickness.
10. Specify whether the tube is to be protected with a coat of paint.

In an ideal world the reference tube would be made from the same material as those in the tube bundle to be tested; however, this is usually not possible. Differences between the reference tube and the tubes tested will introduce inaccuracies in data analysis. Consider the fact that the tube wall thickness can vary by +12 percent to –10 percent (seamless tubes can vary by this much from one side of the tube to the other). The chemical makeup of the tube can have similar tolerances. Add to this the machining tolerances; differences in heat treatment and the magnetic history of the reference tube and the tube to be examined and these combined parameters can lead to substantial differences when comparing the signals from the two tubes. An informed inspector can compensate for most of these differences if that inspector can identify them.

Effects of Probe Speed

The ability of remote field testing to detect and quantify discontinuities relies on the quality of the signal received. The probe must be in the vicinity of the discontinuity for at least one excitation cycle in order to detect it. Ideally the probe senses a discontinuity for several cycles so that any noise signals can be averaged out of the data.

If a test frequency of 100 Hz is being used, there are 100 opportunities per second to measure the signal. If the smallest discontinuity that is required to be detected has a length of 3 mm (0.12 in.), then the fastest speed possible is 300 $mm \cdot s^{-1}$ (12 $in. \cdot s^{-1}$). To build in a safety margin, the test speed should be no faster than half this speed. The probe pull speed should be slow enough so that the digital sample rate allows the field profile near the probe to be accurately recorded.

It is equally important that the speed of testing be as constant as possible. Sudden changes in speed can result in anomalous signals.

Tube Support Plates

Tube support plates are normally made of steel and are common in heat exchangers. These plates have the effect of absorbing magnetic energy from the probe. In the vicinity of a tube support plate, the remote field test signal tends to concentrate in the plate, creating a momentary increase in signal amplitude followed by a substantial decrease when the exciter and detector coils are on opposite sides of the plate. Luckily, if there is any wall loss caused by corrosion or erosion at, or under, the tube support plate, there is usually a residual remote field test signal that can be used to detect the loss.

Figure 18 shows a remote field test support plate response on the voltage plane. On the voltage plane, the support plate causes a characteristic *whale* shaped signal. The *nose* of the whale corresponds to the increase in amplitude just before the detector or exciter goes beneath the support plate. This increased amplitude is caused by the electromagnetic wave taking the path of least reluctance (that is, through the support plate), thus causing a local increase in the field intensity.[12] After the energy has diffused through the air gap, it suffers heavy attenuation in the metal. If the tube support plate is thick and in tight contact with the tube, there may be no measurable signal left until the probe has completely passed the plate. The *tail* of the whale corresponds to the situation when transmitter and receiver are located on opposite sides of the support plate. The tail is shorter when the gap between pipe and plate is larger or when the plate is relatively thin. The nose of the whale usually occurs at around 1 to 5 mm (0.04 to 0.2 in.) from the edge of the support plate.

Support plates signatures are sometimes modeled for calibration purposes by metal rings with the same thickness as the desired support plate. The rings can be too small, causing the electromagnetic wave to wrap itself around the outside of the ring rather than squeezing through the gap between tube and ring. Proper calibration rings should have diameters of at least 140 mm (5.5 in.).

FIGURE 18. Typical tube support plate signal on remote field testing voltage plane display.

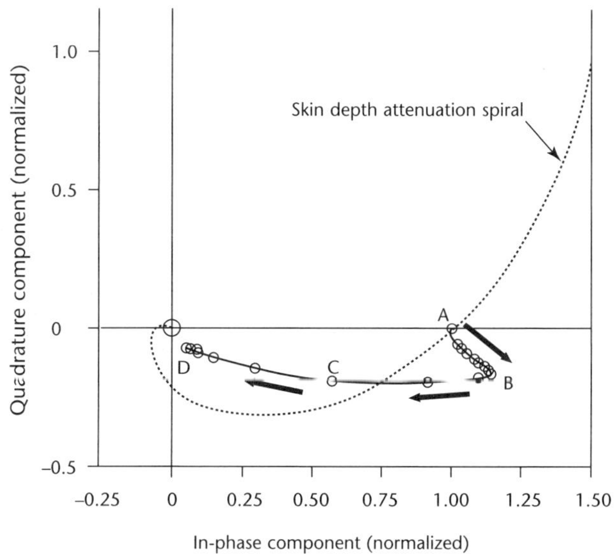

Legend

A. Both transmitter and receiver are located under nominal pipe wall.
B. Amplitude increases as receiver approaches plate.
C. Amplitude decreases rapidly as receiver moves beneath plate.
D. Receiver and transmitter are on different sides of plate.

FIGURE 19. Remote field testing signals at tube support plate: before and after frequency mixing.

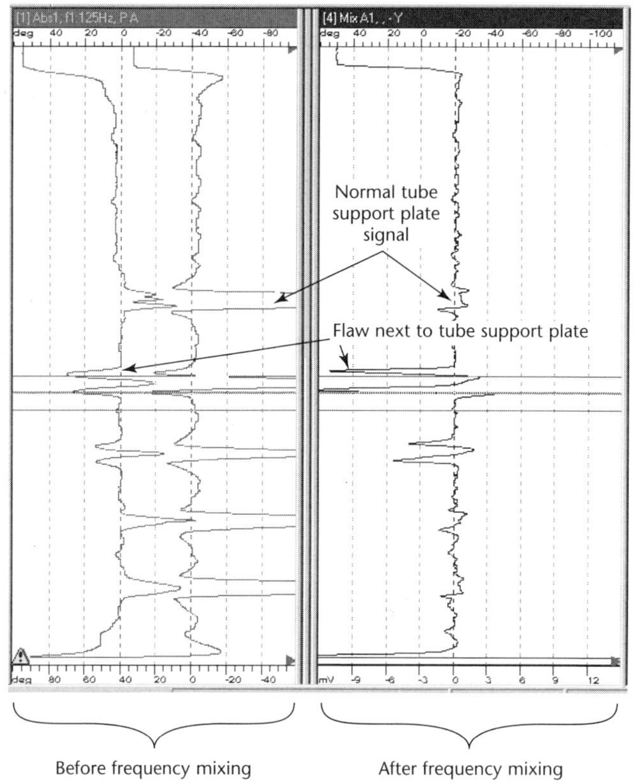

Mixed Signals

An effective way to cancel the tube support plate signal is to mix the signals from two frequencies in order to enhance detection of discontinuities under or near the plate. Figure 19 shows signals from discontinuities near a tube support plate both before and after mixing. In remote field testing, the tube support plate indication is only suppressed and not removed. There is still less accuracy and sensitivity near a tube support plate.

Part 6. Field Operation

Test Considerations[12-14]

The purpose of a typical remote field test is to determine the depth and type of metal loss in each tube. A typical boiler inspection will have the following considerations and steps.

1. A technician may decide that cleaning is required to allow passage of the probe. In most cases, boiler tubes do not require cleaning before a remote field test.
2. The boiler operator generally provides drawings of the boiler showing the locations of soot blowers, access openings, test ports and other areas of interest. A tube numbering system is established.
3. If the tubes are 6 m (20 ft) or less in length, the examination can be conducted from the mud drum. If the tubes are longer, it is generally advisable to remove the steam drum internals and do the examination from there.
4. It may be decided to test all the tubes or only selected tubes in suspected trouble areas. If damage on the bends is suspected, it is necessary to examine many tubes in the same row so that bend signatures can be compared.
5. During data acquisition, an assistant handles the probe while the technician operates the instrument. All data are stored.
6. When the data are being analyzed, the technician pays special attention to the location of soot blowers, to the direction of gas flow on the hot side, to types of water treatment and to the history of previous failures. Knowledge of these factors helps analysis by giving insight into types of discontinuities that might be found.
7. A field report is written on site at the completion of the job. In this way, the boiler operator can make immediate decisions concerning tube plugging, repairs or replacement. A computer report can be created that may include a color coded tube sheet map of the boiler.

Special Probes

To improve sensitivity of the remote field technique to discontinuities close to a tube support plate or tube sheet, probes can contain detector coils on each side of an exciter coil. Alternatively, two exciters can be used with one or two detectors placed between them. The disadvantage of these probes is their increased length and more complex data.

Pipes with diameters larger than 25 mm (1 in.) often require more than one detector coil to improve detection of local discontinuities. Array coils are used to segment the detector section to improve small volume discontinuity sizing. The probe in Fig. 20 is being used to examine a 75 mm (3 in.) diameter pipeline used to transport gas from the wellhead. Figure 21 shows the display narrowed to array probe signals of interest.

External probes are similar in design to remote field testing probes; however, because external probes are sensitive to liftoff, probe travel must be carefully controlled to eliminate false signals.

Array probes require special software to display and analyze the large quantities of multichannel data. It is possible for software to analyze remote field test data automatically. In all cases of automatic data analysis the success of the analysis

Figure 20. Array probe in use on 75 mm (3 in.) pipeline.

FIGURE 21. Array probe signals.

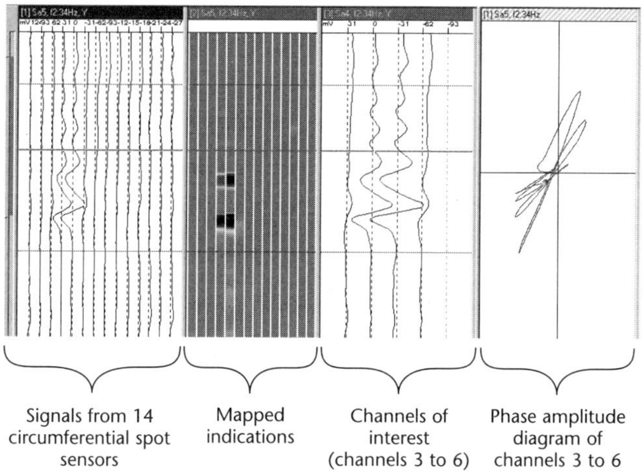

depends on the criteria that the operator uses to train the software to recognize signals of interest. It is also very important to have a skilled technician review a percentage of the calls made by automatic analysis software to verify accuracy. Only one selected channel at a time can be displayed on a voltage plane display.

Summary

Remote field testing is a versatile test technique that can be used effectively to test steel tubes, pipes and plates. It can also be used to test thick walled, nonferrous tubes with equal sensitivity to internal and external discontinuities.

References

1. MacLean, W.R. *Apparatus for Magnetically Measuring Thickness of Ferrous Pipe*. United States Patent 2 573 799 (1951).
2. Schmidt, T.R. "History of the Remote-Field Eddy Current Inspection Technique." *Materials Evaluation*. Vol. 47, No. 1. Columbus, OH: American Society for Nondestructive Testing (January 1989): p 14, 17-18, 20-22.
3. Atherton, D.L. and W.M. Czura. "Finite Element Calculations for Eddy Current Interactions with Collinear Slots." *Materials Evaluation*. Vol. 52, No. 1. Columbus, OH: American Society for Nondestructive Testing (January 1994): p 96-100.
4. Hoshikawa, H., K. Koyama, J. Koido and Y. Ishibashi. "Characteristics of Remote-Field Eddy Current Technique." *Materials Evaluation*. Vol. 47, No. 1. Columbus, OH: American Society for Nondestructive Testing (January 1989): p 93-97.
5. Schmidt, T.R. "The Remote Field Eddy Current Inspection Technique." *Materials Evaluation*. Vol. 42, No. 2. Columbus, OH: American Society for Nondestructive Testing (February 1984): p 225-230.
6. Lord, W., Y.-S. Sun, S.S. Udpa and S. Nath. "A Finite Element Study of the Remote-Field Eddy Current Phenomenon." *IEEE Transactions on Magnetics*. Vol. 24. New York, NY: Institute of Electrical and Electronics Engineers (January 1988): p 435-438.
7. Mackintosh, D.D., D.L. Atherton and P.A. Puhach. "Through-Transmission Equations for Remote-Field Eddy Current Inspection of Small-Bore Ferromagnetic Tubes." *Materials Evaluation*. Vol. 51, No. 6. Columbus, OH: American Society for Nondestructive Testing (June 1993): p 744-748.
8. Sun, Y.-S., L. Udpa, S. Udpa, W. Lord, S. Nath, S.K. Lua and K.H. Ng. "A Novel Remote-Field Eddy Current Technique for Inspection of Thick Walled Aluminum Plates." *Materials Evaluation*. Vol. 56, No. 1. Columbus, OH: American Society for Nondestructive Testing (January 1998): p 94-97.
9. Kilgore, R.J. and S. Ramachandran. "Remote Field Eddy Current Testing of Small-Diameter Carbon Steel Tubes." *Materials Evaluation*. Vol. 47, No. 1. Columbus, OH: American Society for Nondestructive Testing (January 1989): p 32-36.
10. Atherton, D.L., D.D. Macintosh, S.P. Sullivan, J.M.S. Dubois and T.R. Schmidt. "Remote Field Eddy Current Signal Representation." *Materials Evaluation*. Vol. 51, No. 7. Columbus, OH: American Society for Nondestructive Testing (July 1993): p 782-789.
11. ASTM E 2096-00, *Standard Practice for In Situ Examination of Ferromagnetic Heat-Exchanger Tubes Using Remote Field Testing*. West Conshohocken, PA: ASTM International (2000).
12. Shatat, A. and D.L. Atherton. "Remote Field Eddy Current Inspection of Support Plate Fretting Wear." *Materials Evaluation*. Vol. 55, No. 3. Columbus, OH: American Society for Nondestructive Testing (March 1997): p 361-366.
13. Smith, H. and D.D. Mackintosh. "Remote Field Eddy Current Examination of Boiler Tubes." *Proceedings of EPRI Workshop: Electromagnetic NDE Applications in the Electric Power Industry* [Charlotte, NC, August 1995]. Palo Alto, CA: Electric Power Research Institute (1995).
14. Mackintosh, D.D., D.L. Atherton, T.R. Schmidt and D.E. Russell. "Remote Field Eddy Current for Examination of Ferromagnetic Tubes." *Materials Evaluation*. Vol. 54, No. 6. Columbus, OH: American Society for Nondestructive Testing (March 1996): p 652-657.

9

CHAPTER

Magnetic Flux Leakage Testing

Satish S. Udpa, Michigan State University, East Lansing, Michigan

Roderic K. Stanley, NDE Information Consultants, Houston, Texas

PART 1. Introduction to Magnetic Flux Leakage Testing

Magnetic flux leakage testing is part of the widely used family of electromagnetic nondestructive techniques. Magnetic particle testing is a variation of flux leakage testing that uses particles to show indications. When used with other methods, magnetic tests can provide a quick and relatively inexpensive assessment of the integrity of ferromagnetic materials.

The theory and practice of electromagnetic techniques are discussed elsewhere in this volume. The origins of magnetic particle testing are described in the literature[1] and information that the practicing magnetic test engineer might require is available from a variety of manuals and journal articles.

The magnetic circuit and the means for producing the magnetizing force that causes magnetic flux leakage are described below. Theories developed for surface and subsurface discontinuities are outlined along with some results that can be expected.

Industrial Uses[2]

Magnetic flux leakage testing is used in many industries to find a wide variety of discontinuities. Much of the world's production of ferromagnetic steel is tested by magnetic or electromagnetic techniques. Steel is tested many times before it is used and some steel products are tested during use for safety and reliability and to maximize their length of service.

Production Testing

Typical applications of magnetic flux leakage testing are by the steel producer, where blooms, billets, rods, bars, tubes and ropes are tested to establish the integrity of the final product. In many instances, the end user will not accept delivery of steel product without testing by the mill and independent agencies.

Receiving Testing

The end user often uses magnetic flux leakage tests before fabrication. This test ensures the manufacturer's claim that the product is within agreed specifications. Such tests are frequently performed by independent testing companies or the end user's quality assurance department. Oil field tubular goods are often tested at this stage.

Inservice Testing

Good examples of inservice applications are the testing of used wire rope, installed tubing, or retrieved oil field tubular goods by independent facilities. Many laboratories also use magnetic techniques (along with metallurgical sectioning and other techniques) for the assessment of steel products and prediction of failure modes.

Discontinuities[2]

Discontinuities can be divided into two general categories: those caused during manufacture in new materials and those caused after manufacture in used materials.

Discontinuities caused during manufacture include cracks, seams, forging laps, laminations and inclusions.

1. Cracking occurs when quenched steel cools too rapidly.
2. Seams occur in several ways, depending on when they originate during fabrication.
3. Discontinuities such as piping or inclusions within a bloom or billet can be elongated until they emerge as long tight seams or gouges during initial forming processes. They may later be closed with additional forming.
4. Their metallurgical structures are often different but the origin of manufactured discontinuities is not usually taken into account when rejecting a part.
5. Forging laps occur when gouges or fins created in one metal working process are rolled over at an angle to the surface in subsequent processes.
6. Inclusions are pieces of nonmagnetic or nonmetallic materials embedded inside the metal during cooling. Inclusions are not necessarily detrimental to the use of the material.
7. The pouring and cooling processes can also result in lack of fusion within the steel. Such regions may be worked into internal laminations.

Discontinuities in used materials include fatigue cracks, pitting corrosion, erosion and abrasive wear.

Much steel is acceptable to the producer's quality assurance department if no discontinuities are found or if discontinuities are considered to be of a depth or size less than some prescribed maximum. Specifications exist for the acceptance or rejection of such materials and such specifications sometimes lead to debate between the producer and the end user. Discontinuities can either remain benign or can grow and cause premature failure of the part. Abrasive wear can turn benign subsurface discontinuities into detrimental surface breaking discontinuities.

For used materials, fatigue cracking commonly occurs as the material is cyclically stressed. Fatigue cracks grow rapidly under stress or in the presence of corrosive materials such as hydrogen sulfide, chlorides, carbon dioxide and water. For example, drill pipe failure from fatigue often initiates at the bases of pits, at tong marks or in regions where the tube has been worn by abrasion. Pitting is caused by corrosion and erosion between the steel and a surrounding or containing fluid. Abrasive wear occurs in many steel structures. Good examples are (1) the wear on drill pipe caused by hard formations when drilling crooked holes or (2) the wear on both the sucker rod and the producing tubing in rod pumping oil wells. Specifications exist for the maximum permitted wear under these and other circumstances. In many instances, such induced damage is first found by automated magnetic techniques.

Steps in Magnetic Flux Leakage Testing

There are four steps in magnetic flux leakage testing: (1) magnetize the test object so that discontinuities perturb the flux, (2) scan the surface of the test object with a magnetic flux sensitive detector, (3) process the raw data from these detectors in a manner that best accentuates discontinuity signals and (4) present the test results clearly for interpretation. The following discussion deals with the first step, producing the magnetizing force.

Part 2. Magnetization Techniques

Successful testing requires the test object to be magnetized properly. The magnetization can be accomplished using one of several approaches: (1) permanent magnets, (2) electromagnets and (3) electric currents used to induce the required magnetic field.

Excitation systems that use permanent magnets offer the least flexibility. Such systems use high energy product permanent magnet materials such as neodymium iron boron, samarium cobalt and aluminum nickel. The major disadvantage with such systems lies in the fact that the excitation cannot be switched off. Because the magnetization is always turned on, it is difficult to insert and remove the test object from the test rig. Although the magnetization level can be adjusted using appropriate magnetic shunts, it is awkward to do so. Consequently, permanent magnets are very rarely used for magnetization.

Electromagnets, as well as electric currents, are used extensively to magnetize the test object. Figure 1 shows an excitation system where the test object is part of a magnetic circuit energized by current passing through an excitation coil. The magnetic circuit passes through a yoke made of a soft magnetic material and through a test object placed between the poles of the yoke. When the coil wound on the yoke carries current, the resulting magnetomotive force drives magnetic flux through the yoke and the test object. The total magnetic flux ϕ (weber) is given by:

$$(1) \quad \phi = \frac{NI}{S}$$

where I is the current (ampere) in the coil, N is the number of turns in the coil and S is the reluctance (ampere per weber) of the magnetic circuit.

Reluctance S is the sum of the reluctance S_g of air gaps (between the test object and the yoke), test object reluctance S_s and yoke reluctance S_y. The reluctance values of the air gaps, test object and yoke are given by Eqs. 2 to 4:

$$(2) \quad S_g = \frac{L_g}{a_g \mu_0}$$

$$(3) \quad S_s = \frac{L_s}{a_s \mu_0 (\mu_r)_s}$$

$$(4) \quad S_y = \frac{L_y}{a_y \mu_0 (\mu_r)_y}$$

where a_x is the cross sectional area (square meter) of the air gaps, test object or yoke; L_x is the length (meter) of the air gaps, test object or yoke; μ_0 is the permeability of free space ($\mu_0 = 4\pi \times 10^{-7}$ H·m^{-1}); μ_r is relative permeability; and subscripts g, s and y denote the air gaps, test object and yoke, respectively. Note that the magnetic circuit consists of two air gaps, one at each end of the test object. Both air gaps need to be taken into account in calculating the total reluctance of the magnetic circuit.

To obtain maximum sensitivity, it is necessary to ensure that the magnetic flux is perpendicular to the discontinuity. This direction is in contrast to the orientation in techniques that use an electric current for inspection of a test object, where it may be more advantageous to orient the direction of current so that a discontinuity would impede the current as much as possible.

Because the orientation of the discontinuity is unknown, it is necessary to test twice with the yoke, in two directions perpendicular to each other. A

Figure 1. Electromagnetic yoke for magnetizing of test object.

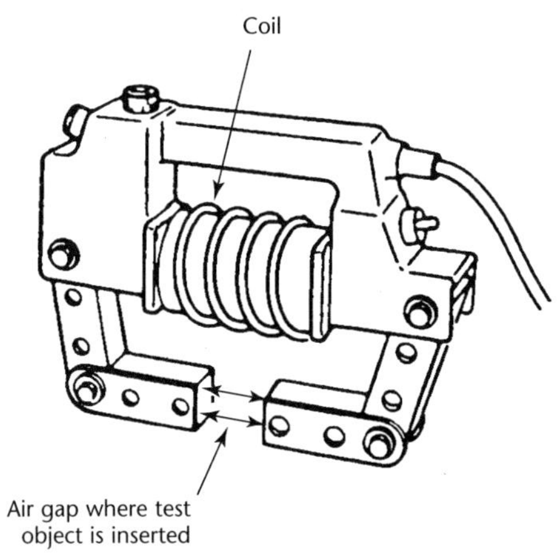

grid is usually drawn on the test object to facilitate the tests.

Magnetizing Coil

A commonly used encircling coil is shown in Fig. 2. The field direction follows the right hand rule. (The right hand rule states that, if someone grips a rod, holds it out and imagines an electric current flowing down the thumb, the induced circular field in the rod would flow in the direction that the fingers point.) With no test object present, the field lines form closed loops that encircle the current carrying conductors. The value of the field at any point has been established for a great many coil configurations. The value depends on the current in the coils, the number of turns *N* and a geometrical factor. Calculation of the field from first principles is generally unnecessary for nondestructive testing; a hall element tesla meter will measure this field.

Introduction of the test object into the field of the coil changes the field. The metal becomes part of the magnetic circuit, with the result that, close to the surface of the test object, magnetic field intensity ***H*** is lower than it would be if the test object were removed. Again, a hall element tesla meter will show the field intensity at the test object. This reduces the need for semiempirical formulas. With the test object inserted, the flux density changes and the flux lines get concentrated within the test object. Thus, the fields inside and outside the test object are not the same. However, two boundary conditions allow assessment of the magnetic state of the test object. The fact that the tangential field is continuous across the air-to-metal interface allows measurement of ***H*** at the point R to yield the value of the tangential field at the test surface. In addition, because the normal component of magnetic flux density ***B*** is continuous, a tesla meter at point S will yield ***B*** inside the test object at that point.

Two totally different situations, common in magnetic flux leakage testing, are described below.

FIGURE 2. Encircling coil using direct current to produce magnetizing force.

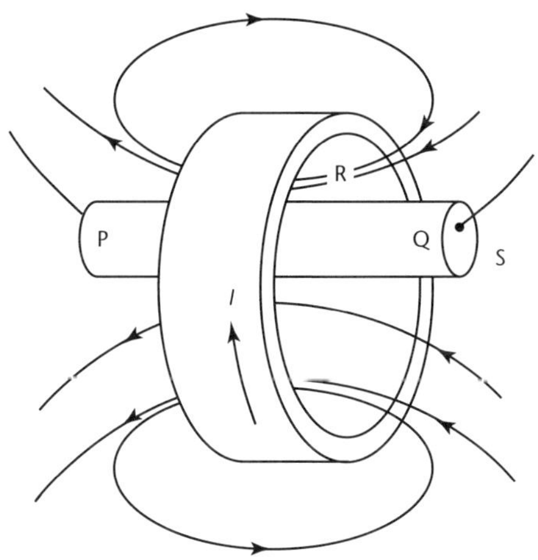

Legend

I = electric current
P, Q = points of discontinuities in example
R = point at which magnetic field intensity ***H*** is measured
S = point at which magnetic flux density ***B*** is measured

Testing in Active Field

In this technique, the test object is scanned by probes near position R in Fig. 2, in the presence of an active field. Air fields of 16 to 24 kA·m^{-1} (200 to 300 Oe) are commonly used. In this situation, application of small fields is sufficient to cause magnetic flux leakage from transversely oriented surface breaking discontinuities. For subsurface discontinuities or those on the inside surface of tubes, larger fields are required. The inspector must experiment to optimize the applied field for the particular discontinuity.

Testing in Residual Field

Test objects are first passed through the coil field and then tested in the resulting residual field. Elongating the coil and placing the test object next to the inside surface of the coil will expose the test object to the largest field that the coil can produce.

This technique is often used in magnetic particle testing. The main problem to avoid is the induction of so much magnetic flux in the test object that the magnetic particles stand out like fur along the field lines that enter and leave the test object, especially close to its ends. Optimum conditions require that the test object be somewhat less than saturated. The inspector should experiment to optimize the coil field requirements for the test object because this field depends on test object geometry.

Applied Direct Current

If an electric current is used to magnetize the test object, it may be more advantageous to orient the direction of current in a manner where the presence of a discontinuity impedes the current flow as much as possible. Bars, billets and tubes are often magnetized by application of a direct current I to their ends (Fig. 3).

Figure 4 shows a system where the current I is passed directly through a tubular test object to magnetize the test object circularly. Figure 5 shows a central conductor energized by a current source I, again, to establish a circular magnetic field intensity $\boldsymbol{H}$ (ampere per square meter) in a tubular test object:

$$(5) \quad H = \frac{I}{2\pi a}$$

where a is area (square meter).

Capacitor Discharge Devices

For the circular magnetization of tubes or the longitudinal magnetization of the ends of elongated test objects, a capacitor discharge device is sometimes used.[3,4] The capacitor discharge unit represents a practical advance over battery packs and consists of a capacitor bank charged to a voltage V and then discharged through a rod, a cable and a silicon controlled rectifier of total resistance R.

The full system, considered mathematically, also contains a variable amount of inductance, so that if the current I_c were allowed to oscillate, it would do so according to the theory of LCR circuits (that is, circuits described by inductance L, capacitance C and resistance R). The theory is complicated by the time required to magnetize the material and to induce an eddy current in the test object. Typical configurations shown in Fig. 6 illustrate the complexity of the situation. In the case of the magnetization of a tube, the current I_c first rises rapidly, inducing magnetic flux in the tube. This time varying flux changes rapidly and induces an electromotive force in the tube, as dictated by Faraday's law, the result being that an eddy current I_e flows around the tube as shown in Fig. 6a, where the dashed line is the inner surface eddy current and the solid line is the outer surface current.

The net result is a lack of penetration of the field caused by the capacitor discharge current I_c. For a centered rod, in effect, the magnetic field intensity in the test object at radius r is given not by $\boldsymbol{H} = I_c \cdot (2\pi r)^{-1}$ but rather by Eq. 6:

$$(6) \quad H(r) = \frac{I_c - I_e}{2\pi r}$$

Here I_e is the amount of eddy current (ampere) contained within the cylinder of radius r (meter). Investigation of the effect of the eddy current is theoretically quite complicated because of its effect on the

FIGURE 3. Circumferential magnetization by application of direct current: (a) rectilinear bar; (b) round bar; (c) tube.

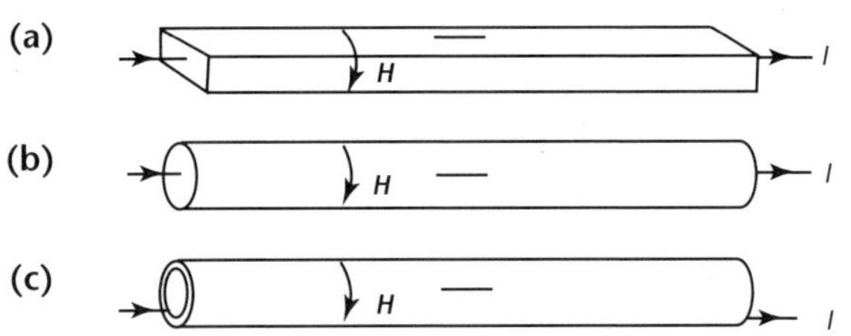

Legend
H = magnetic field intensity
I = electric current

FIGURE 4. Current carrying clamp electrodes used for testing ferromagnetic tubular objects with small diameters.

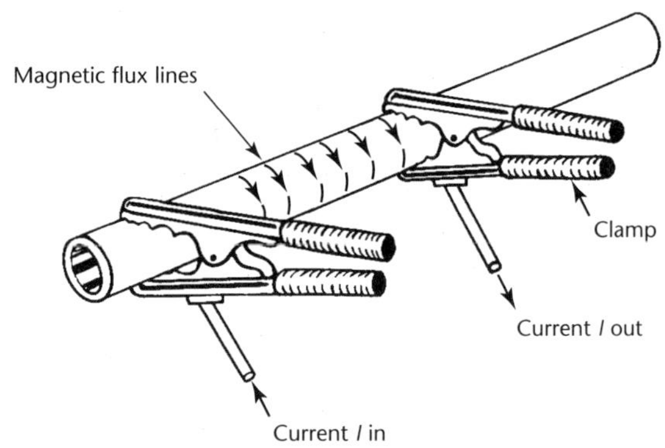

FIGURE 5. Simple technique for circumferential magnetization of ferromagnetic tube.

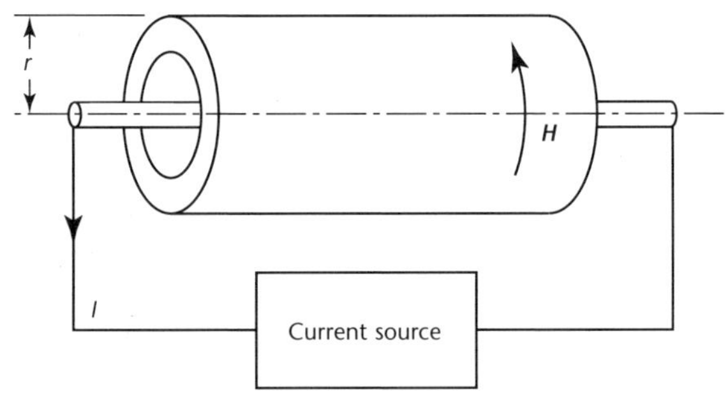

Legend
H = magnetic field intensity
I = electric current
r = tube radius

inductance, which in turn affects I_c. In practice, however, measurement of the magnetic flux density $\boldsymbol{B}$ in the material will yield the final degree of magnetization of that material.

A good rule is that, if $\boldsymbol{H}(r)$ in Eq. 6 can be maintained at about 3.2 kA·m^{-1} (40 Oe), the material will be magnetized almost to saturation and can be tested for both surface and subsurface discontinuities. Several other practical conclusions can be drawn from the above discussion.

1. Pulse duration plays a greater role than pulse amplitude $I_{c(max)}$ in determining the amount of flux induced in a test object. This is intuitively seen in direct current tests.
2. It is not possible to give simple rules that relate $I_{c(max)}$ to magnetization requirements. This relationship can be shown with a magnetic flux meter.
3. The eddy currents induced during pulse magnetization play an important role in the result. They can shield midwall regions from magnetization.
4. Larger capacitances at lower voltages provide better magnetization than smaller capacitances at higher voltages because larger capacitances at lower voltages lead to longer duration pulses and therefore to lower eddy currents. The lower voltage is an essential safety feature for outdoor use. A maximum of 50 V is recommended.[5]

FIGURE 6. Capacitor discharge configurations causing magnetization perpendicular to current direction: (a) conductor internal to test object creates circular field; (b) flexible cable around test object creates longitudinal field.

(a)

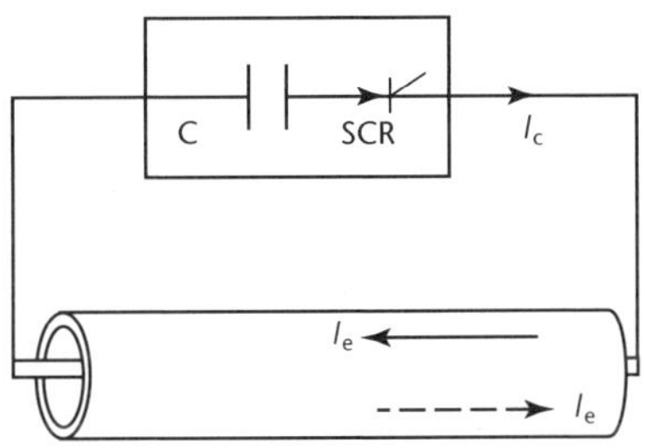

(b)

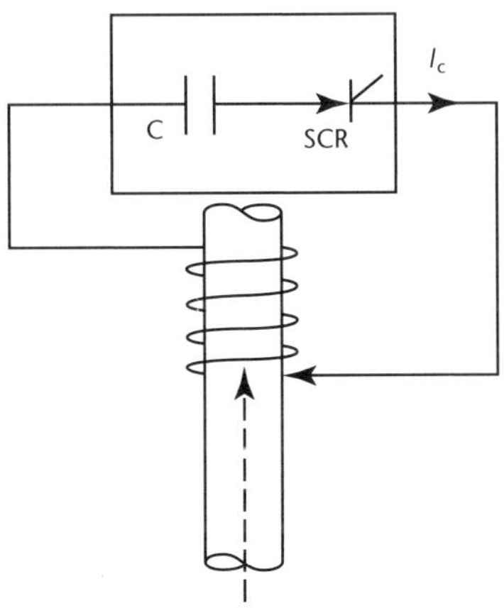

Legend

C = capacitor
I_c = capacitor discharge current
I_e = eddy current
SCR = silicon controlled rectifier

Magnitudes of Magnetic Flux Leakage Fields

The magnitude of the magnetic flux leakage field under active direct current excitation naturally depends on the applied field. An applied field of 3.2 to 4.0 kA·m^{-1} (40 to 50 Oe) inside the material can cause leakage fields with peak values of tens of millitesla (hundreds of gauss). However, in the case of residual induction, the magnetic flux leakage fields may be only a few hundred microtesla (a few gauss). Furthermore, with residual field excitation, an interesting field reversal may occur, depending on the value of the initial active field excitation and the dimensions of the discontinuity.

Optimal Operating Point

Consider raising the magnetization level in a block of steel containing a discontinuity (Fig. 7). At low flux density levels, the field lines tend to crowd together in the steel around the discontinuity rather than go through the nonmagnetic region of the discontinuity. The field lines are therefore more crowded above and below the discontinuity than they are on the left or right. The material can hold more flux as the permeability rises, so there is no significant leakage flux at the surfaces (Fig. 7a).

However, an increase in the number of lines causes $\Delta\boldsymbol{B}\cdot(\Delta\boldsymbol{H})^{-1}$ to fall — the material is becoming less permeable. At about this point, magnetic flux leakage is first noticed at the surfaces. Although the lines are now closer together, representing a higher magnetic flux density, they do not have the ability to crowd closer together around the discontinuity where the permeability is low.

At higher and higher values of applied field, the permeability falls. It is, however,

FIGURE 7. Effects of induction on magnetic flux lines at discontinuity: (a) no surface flux leakage occurs where magnetic flux lines are compressed at low levels of induction around discontinuity; (b) lack of compression at high magnetization results in surface magnetic flux leakage.

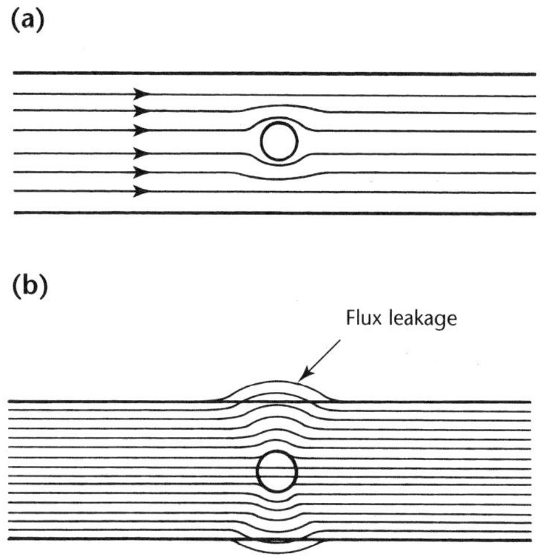

still large compared to the permeability of air, so the reluctance of the path through the discontinuity is still larger than through the metal. As a result, magnetic flux leakage at the outside surface helps provide a sufficiently high flux density in the material for the leakage of magnetic flux from discontinuities (Fig. 7b) while partially suppressing long range surface noise.

For residual field testing, it is best to ensure that the material is saturated. The magnetic field starts to decay as soon as the energizing current is removed.

PART 3. Magnetic Flux Leakage Test Results

Magnetic flux leakage testing continues to be one of the most popular nondestructive test techniques in industry. A number of factors, including low cost and simplicity of the data interpretation process, contribute to this popularity. The underlying principles and modeling techniques are described elsewhere in this volume. The discussion below focuses on probes and excitation schemes to detect and measure magnetic leakage fields.

Magnetic Flux Leakage Probes

The purpose of probes for magnetic testing is to detect and possibly quantify the magnetic flux leakage field generated by heterogeneities in the test object. The leakage fields tend to be local and concentrated near the discontinuities. The leakage field can be divided into three orthogonal components: normal (vertical), tangential (horizontal) and axial directions. Probes are usually either designed or oriented to measure one of these components. Typical plots of these components near discontinuities are shown in this volume's chapter on probes.

A variety of probes (or transducers) are used in industry for detecting and measuring leakage fields.

Pickup Coils

One of the simplest and most popular means for detecting leakage fields is to use a pickup coil.[6] Pickup coils consist of very small coils that are either air cored or use a small ferrite core. The voltage induced in the coil is given by the rate of change of flux linkages associated with the pickup coil.

$$(7)\quad V = \frac{-N\,d\phi}{dt}$$

where N is the number of turns in the coil, V is the voltage induced in the coil and ϕ is the magnetic flux (weber) linking the coil. It must be mentioned that only the component of the flux parallel to the axis of the coil (or alternately perpendicular to the plane of the coil) is instrumental in inducing the voltage. This induction direction makes it possible to orient the pickup coil so as to measure any of the three leakage field components selectively. Thus, a coil A whose axis is perpendicular to the surface of the test object (Fig. 8a), is sensitive only to the normal component. In contrast, the coil in Fig. 8b is sensitive only to the tangential component. Consider the case where the pickup coil is moving over the test object in the X direction. Making use of the fact that $\phi = \boldsymbol{B} \cdot A$, where $\boldsymbol{B}$ is the magnetic flux density (tesla) and A is the cross sectional area (square meter) of the pickup coil, Eq. 7 can be rewritten:

$$(8)\quad V = \frac{-N\,d\phi}{dt} = -NA\frac{d\boldsymbol{B}}{dt} = -NA\frac{d\boldsymbol{B}}{dx}\cdot\frac{dx}{dt}$$

This equation indicates that the output of the pickup coil is proportional to the spatial gradient of the flux along the direction of the coil movement as well as the velocity of the coil. Two issues arise as a result.

1. It is essential that the probe scan velocity (relative to the test object) should be constant to avoid introducing artifacts into the signal through probe velocity variations.
2. The output is proportional to the spatial gradient of the flux in the direction of the coil.

The output of the pickup coil can be integrated for measurement of the leakage flux density rather than of its gradient.

FIGURE 8. Effect of pickup coil orientation on sensitivity to components of magnetic flux density: (a) coil sensitive to normal component; (b) coil sensitive to tangential component.

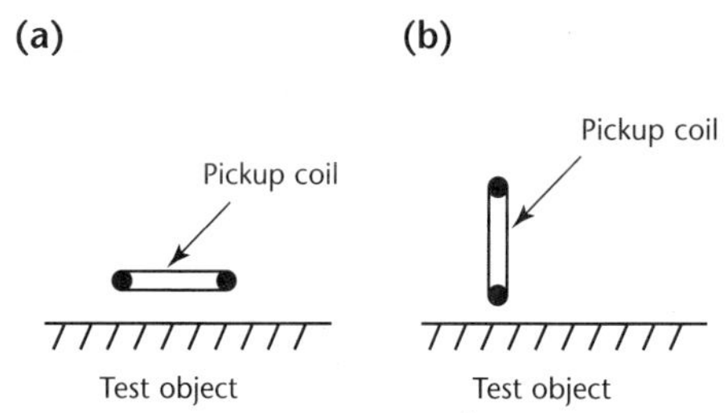

Figure 9 shows the output of a pickup coil and the signal obtained after integrating the output.[7] The coil is used to measure, in units of tesla (or gauss), the magnetic flux density ***B*** leaking from a rectangular slot.

The sensitivity of the pickup coil can be improved by using a ferrite core. Tools for designing pickup coils, as well as predicting their performance, are described elsewhere in this volume.

Magnetodiodes

The magnetodiode is suitable for sensing leakage fields from discontinuities because of its small size and its high sensitivity. Because the coil probe is usually larger than the magnetodiode, it is less sensitive to longitudinally angled discontinuities than the magnetodiode is. However, the coil probe is better than the magnetodiode for large discontinuities, such as cavities.

Hall Effect Detectors

Hall effect detector probes are used extensively in industry for measuring magnetic flux leakage fields in units of tesla (or gauss). Hall effect detector probes are described in this volume's chapter on probes for electromagnetic testing.

FIGURE 9. Pickup coil and signal integrator (magnetic flux leakage) output for rectangular discontinuity.[7]

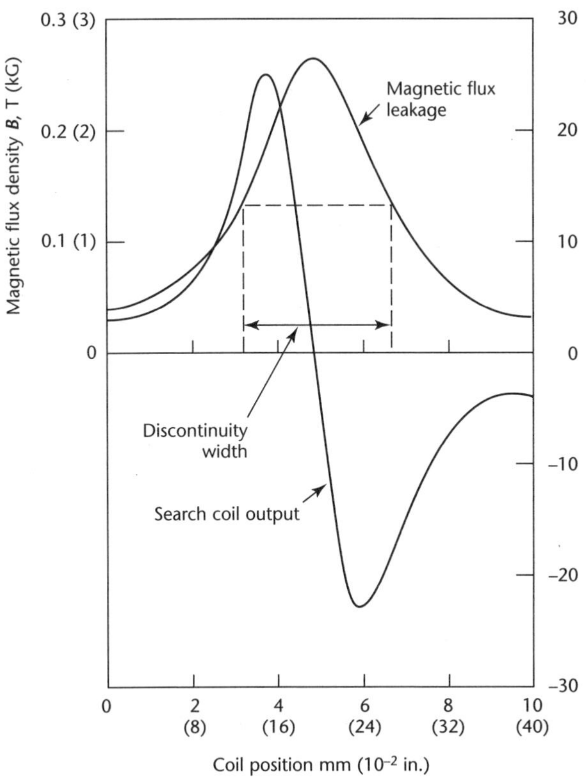

Giant Magnetoresistive Probes

Magnetic field sensitive devices called *giant magnetoresistive probes*,[8,9] at the most basic level, consist of a nonmagnetic layer sandwiched between two magnetic layers. The apparent resistivity of the structure varies depending on whether the direction of the electron spin is parallel or antiparallel to the moments of the magnetic layers. When the moments associated with the magnetic layers are aligned antiparallel, the electrons with spin in one direction (up) that are not scattered in one layer will be scattered in the other layer. This increases the resistance of the device. This is in contrast to the situation when the magnetic moments associated with the layers are parallel where the electrons that are not scattered in one layer are not scattered in the other layer, either.

Giant magnetoresistive probes use a biasing current to push the magnetic layers into an antiparallel moment state and the external field is used to overcome the effect of the bias. The resistance of the device, therefore, decreases with increasing field intensity values. Figure 10 shows a typical response of a giant magnetoresistive probe.

Magnetic Tape

For the testing of flat surfaces, magnetic tape can be used. The tape is pressed to the surface of the magnetized billet and then scanned by small probes before being erased. This technique is sometimes called *magnetography*.

FIGURE 10. Resistance versus applied field for 2 µm (8×10^{-5} in.) wide strip of antiferromagnetically coupled, multilayer test object composed of 14 percent giant magnetoresistive material.[9]

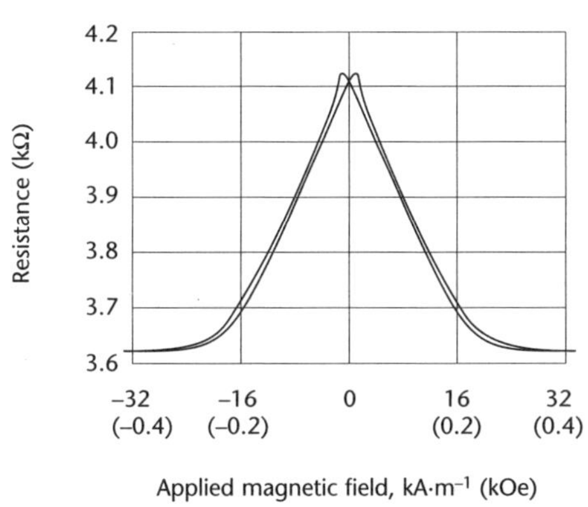

In automated systems, magnetic tape can be fed from a spool. The signals can be read and the tape can be erased and reused.

Unfortunately, the tangential leakage field intensity at the surface of the material is not constant. To optimize the response, the amplification of the signals can be varied.

Scabs or slivers projecting from the test surface can easily tear the tape

Magnetic Particles

Magnetic particles are one of the most popular means used in industry for detecting magnetic fields. Indeed, magnetic particle testing is so popular that an entire volume of the *Nondestructive Testing Handbook* is devoted to the subject.[1] The descriptions below are therefore cursory.

Magnetic particle testing involves the application of magnetic particles to the test object after it is magnetized by using an appropriate technique. The ferromagnetic particles preferentially adhere to the surface of the test object in areas where the flux is *diverted,* or *leaks out.* The magnetic flux leakage near discontinuities causes the magnetic particles to accumulate in the region and in some cases form an outline of the discontinuity. Heterogeneities can therefore be detected by looking for indications of magnetic particle accumulations on the surface of the test object either with the naked eye or through a camera. The indications are easier to see if the particles are bright and reflective. Alternately, particles that fluoresce under ultraviolet or visible radiation may be used. The test object has to be viewed under appropriate levels of illumination with radiation of appropriate wavelength (visible, ultraviolet or other).

Application Techniques

Magnetic particles are applied to the surface by two different techniques in industry.

Dry Testing. Dry techniques use particles applied in the form of a fine stream or an aerosol. They consist of high permeability ferromagnetic particles coated with either reflective or fluorescent pigments. The particle size is chosen according to the dimensions of the discontinuity sought. Particle diameters range from ≤50 to 180 µm (≤0.002 to 0.007 in.). Finer particles are used for detecting smaller discontinuities where the leakage intensity is low. Dry techniques are used extensively for testing welds and castings where heterogeneities of interest are relatively large.

Wet Testing. Wet techniques are used for detecting relatively fine cracks. The magnetic particles are suspended in a liquid (usually oil or water) usually sprayed on the test object. Particle sizes are significantly smaller than those used with dry techniques and vary in size within a normal distribution, with most particles measuring from 5 to 20 µm (2×10^{-4} to 8×10^{-4} in.). As in the case of dry powders, the ferromagnetic particles are coated with either reflective or fluorescent pigments. More information on this subject is available elsewhere.[1]

Imaging of Magnetic Particle Indications

The magnetic particle distribution can be examined visually after illuminating the surface or the surface can be scanned with a flying spot system or imaged with a charge coupled device camera.

Flying Spot Scanners. To illuminate the test object (Fig. 11), flying spot scanners[10,11] use a narrow beam of radiation — visible light for nonfluorescent particles and ultraviolet radiation for fluorescent ones. The source of the beam is usually a laser. The wavelength of the beam is chosen carefully to excite the pigment of the magnetic particles. The incidence of the radiation beam on the test object can be varied by moving the scanning mirror.

The photocell does not sense any light when the test object is scanned by the narrow radiation beam until the beam is directly incident on the magnetic particles adhering to the test object near a discontinuity. When this occurs, a large amount of light is emitted, called *fluorescence* if excited by ultraviolet radiation. The fluorescence is detected by a single phototube equipped with a filter that renders the system blind to the

FIGURE 11. Flying spot scanner for automated magnetic particle testing.[10,11]

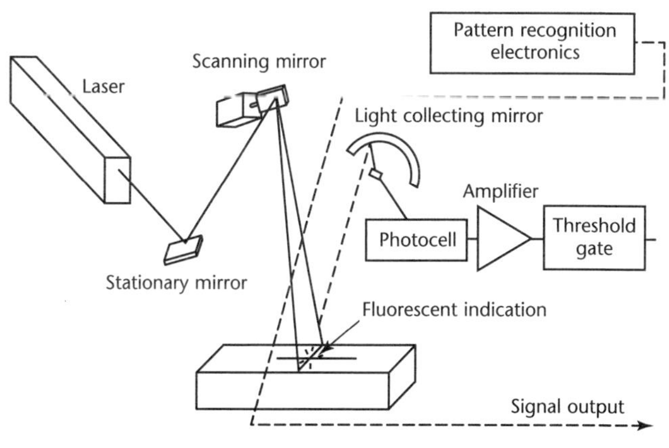

radiation from the irradiating source. The output of the photocell is suitably amplified, digitized and processed by a computer.

Charge Coupled Devices. An alternative approach is to flood the test object with radiation whose wavelength is carefully chosen to excite the pigment of the magnetic particles. Charge coupled device cameras,[12,13] equipped with optical filters that render the camera blind to radiation from the source but are transparent to light emitted by the magnetic particles, can be used to image the surface very rapidly.

In very simple terms, charge coupled devices each consist of a two-dimensional array of tiny pixels that each accumulates a charge corresponding to the number of photons incident on it. When a readout pulse is applied to the device, the accumulated charge is transferred from the pixel to a holding or charge transfer cell. The charge transfer cells are connected in a manner that allows them to function as a bucket brigade or shift register. The charges can, therefore, be serially clocked out through a charge-to-voltage amplifier that produces a video signal.

In practice, charge coupled device cameras can be interfaced to a personal computer through frame grabbers, which are commercially available. Vendors of frame grabbers usually provide software that can be executed on the personal computer to process the image. Image processing software can be used for example to improve contrast, highlight the edges of discontinuity or to minimize noise in the image.

Test Calculations

In determining the magnetic flux leakage from a discontinuity, certain conditions must be known: (1) the discontinuity's location with respect to the surfaces from which measurements are made, (2) the relative permeability of the material containing the discontinuity and (3) the levels of magnetic field intensity $\boldsymbol{H}$ and magnetic flux density $\boldsymbol{B}$ in the vicinity of the discontinuity. Even with this knowledge, the solution of the applicable field equations (derived from Maxwell's equations of electromagnetism) is difficult and is generally impossible in closed algebraic form. Under certain circumstances, such as those of discontinuity shapes that are easy to handle mathematically, relatively simple equations can be derived for the magnetic flux leakage if simplifying assumptions are made. This simplification does not apply to subsurface inclusions.

Finite Element Techniques

An advance in magnetic theory since 1980 has been the introduction of finite element computer codes to the solution of magnetostatic problems. Such codes came originally from a desire to minimize electrical losses from electromagnetic machinery but soon found application in magnetic flux leakage theory. The advantage of such codes is that, once set up, discontinuity leakage fields can be calculated by computer for any size and shape of discontinuity, under any magnetization condition, so long as the $\boldsymbol{B},\boldsymbol{H}$ curve for the material is known.

In the models of magnetic flux leakage discussed so far, the implicit assumptions are (1) that the field within a discontinuity is uniform and (2) that the nonlinear magnetization characteristic ($\boldsymbol{B},\boldsymbol{H}$ curve) of the tested material can be ignored. Much of the early pioneering work in magnetic flux leakage modeling used these assumptions to obtain closed form solutions for leakage fields.

The solutions of classical problems in electrostatics have been well known to physicists for almost a century and their magnetostatic analogs were used to approximate discontinuity leakage fields. Such techniques work reasonably well when the permeability around a discontinuity is constant or when nonlinear permeability effects can be ignored. The major problem that remains is how to deal with real discontinuity shapes often impossible to handle by classical techniques.

Such deficiencies are overcome by the use of computer programs written to allow for nonlinear permeability effects around oddly shaped discontinuities. Specifically, computerized finite element techniques, originally developed for studying magnetic flux distributions in electromagnetic machinery, have also been developed for nondestructive testing. Both active and residual excitation are discussed above. The extension of the technique to include eddy currents is detailed elsewhere in this volume.

PART 4. Applications of Magnetic Flux Leakage Testing[14]

Magnetic flux leakage testing is a commonly used technique. Signals from probes are processed electronically and presented in a manner that indicates the presence of discontinuities. Although some techniques of magnetic flux leakage testing may not be as sophisticated as others, it is probable that more ferromagnetic material is tested with magnetic flux leakage than with any other technique.

Magnetizing techniques have evolved to suit the geometry of the test objects. The techniques include yokes, coils, the application of current to the test object and conductors that carry current through hollow test objects. Many situations exist in which current cannot be applied directly to the test object because of the possibility of arc burns. Design considerations for magnetization of test objects often require minimizing the reluctance of the magnetic circuit, consisting of (1) the test object, (2) the magnetizing system and (3) any air gaps that might be present.

Test Object Configurations

Short Asymmetrical Objects

A short test object with little or no symmetry may be magnetized to saturation by passing current through it or by placing it in an encircling coil. If hollow, a conductor can be passed through the test object and magnetization achieved by any of the standard techniques (these include half-wave and full-wave rectified alternating current, pure direct current from battery packs or pulses from capacitor discharge systems). For irregularly shaped test objects, testing by wet or dry magnetic particles is often performed, especially if specifications require that only surface breaking discontinuities be found.

Elongated Objects

The cylindrical symmetry of elongated test objects such as wire rope permits the use of a relatively simple flux loop to magnetize a relatively short section of the rope. Encircling probes are placed at some distance from the rope to permit the passage of splices. Such systems are also suited for pumping well sucker rods and other elongated oil field test objects.

After a well is drilled, the sides of the well are lined with a relatively thin steel casing material, which is then cemented in. This casing can be tested only from the inside surface. The cylindrical geometry of the casing permits the flux loop to be easily calculated so that magnetic saturation of the well casing is achieved.

As with inservice well casing, buried pipelines are accessible only from the inside surface. The magnetic flux loop is the same as for the well casing test system. In this case, a drive mechanism must be provided to propel the test system through the pipeline.

Threaded Regions of Pipe

An area that requires special attention during the inservice testing of drill pipe is the threaded region of the pin and box connections. Common problems that occur in these regions include fatigue cracking at the roots of the threads and stretching of the thread metal. Automated systems that use both active and residual magnetic flux techniques can be used for detecting such discontinuities.

Ball Bearings and Races

Systems have been built for the magnetization of both steel ball bearings and their races. One such system uses specially fabricated hall elements as detectors.

Relatively Flat Surfaces

The testing of welded regions between flat or curved plates is often performed using a magnetizing yoke. Probe systems include coils, hall effect detectors, magnetic particles and magnetic tape.

Discontinuity Mechanisms

In the metal forming industry, discontinuities commonly found by magnetic flux leakage techniques include overlaps, seams, quench cracks, gouges, rolled-in slugs and subsurface inclusions. In the case of tubular goods, internal mandrel marks (plug scores) can also be identified when they result in remaining wall thicknesses below some specified minimum. Small marks of the same type can also act as stress raisers and cracking can originate from them during quench and temper procedures. Depending on the use to which the material is put, subsurface discontinuities such as porosity and laminations may also be considered detrimental. These types of discontinuities may be acceptable in welds where there are no cyclic stresses but may cause injurious cracking when such stresses are present.

In the metal processing industries, grinding especially can lead to surface cracking and to some changes in surface metallurgy. Such discontinuities as cracking have traditionally been found by magnetic flux leakage techniques, especially wet magnetic particle testing.

Service induced discontinuities include cracks, corrosion pitting, stress induced metallurgy changes and erosion from turbulent fluid flow or metal-to-metal contact. In those materials placed in tension and under torque, fatigue cracking is likely to occur. A discontinuity that arises from metal-to-metal wear is sucker rod wear in tubing from producing oil wells. Here, the pumping rod can rub against the inner surface of the tube and both the rod and tube wear thin. In wire rope, the outer strands will break after wearing thin and inner strands sometimes break at discontinuities present when the rope was made. Railroad rails are subject to cyclic stresses that can cause cracking to originate from otherwise benign internal discontinuities.

Loss of metal caused by a conducting fluid near two slightly dissimilar metals is a very common form of corrosion. The dissimilarity can be quite small, as for example, at the heat treated end of a rod or tube. The result is preferential corrosion by electrolytic processes, compounded by erosion from a contained flowing fluid. Such loss mechanisms are common in subterranean pipelines, installed petroleum well casing and in refinery and chemical plant tubing.

The stretching and cracking of threads is a common problem. For example, when tubing, casing and drill pipe are overtorqued at the coupling, the threads exist in their plastic region. This causes metallurgical changes in the metal and can create regions where stress corrosion cracking takes place in highly stressed areas at a faster rate than in areas of less stress. Couplings between tubes are a good example of places where material may be highly stressed. Drill pipe threads are a good example of places where such stress causes plastic deformation and thread root cracking.

Typical Magnetic Flux Leakage Techniques

Short Parts

For many short test objects, the most convenient probe to use is the magnetic particle. The test object can be inspected for surface breaking discontinuities during or after it has been magnetized to saturation. For active field testing, the test object can be placed in a coil carrying alternating current and sprayed with magnetic particles. Or it can be magnetized to saturation by a direct current coil and the resulting residual induction can be shown with magnetic particles. In the latter case, the induction in the test object can be measured with a flux meter. Wet particles perform better than dry ones because there is less tendency for the wet particles to *fur* (that is, to stand up like short hairs) along the field lines that leave the test object. These techniques will detect transversely oriented, tight discontinuities.

The magnetic flux leakage field intensity from a tight crack is roughly proportional to the magnetic field intensity H_g across the crack, multiplied by crack width L_g. If the test is performed in residual induction, the value of H_g (which depends on the local value of the demagnetization field in the test object) will vary along the test object. Thus, the sensitivity of the technique to discontinuities of the same geometry varies along the length of the test object.

For longitudinally oriented discontinuities, the test object must be magnetized circumferentially. If the test object is solid, then current can be passed through the test object, the surface field intensity being given by Ampere's law:

$$\text{(9)} \quad \oint \boldsymbol{H} \, d\ell = I$$

where $d\ell$ is an element of length (meter), $\boldsymbol{H}$ is the magnetic field intensity (ampere per meter) and I is the current (ampere) in the test object.

If the test object is a cylindrical bar, the symmetry of the situation allows $\boldsymbol{H}$ to be constant around the circumference, so the closed integral reduces:

$$\text{(10)} \quad 2\pi R H = I$$

or:

$$\text{(11)} \quad H = \frac{I}{2\pi R}$$

where R is the radius (meter) of the cylindrical test object. A surface field intensity that creates an acceptable magnetic flux leakage field from the minimum sized discontinuity must be used. Such fields are often created by specifying the amperage per meter of the test object's outside diameter.

Transverse Discontinuities

Because of the demagnetizing effect at the end of a tube, automated magnetic flux leakage test systems do not generally perform well when scanning for transverse discontinuities at the ends of tubes. The normal component H_y of the field outside the tube is large and can obscure discontinuity signals. Test specifications for such regions often include the requirement of additional longitudinal magnetization at the tube ends and subsequent magnetic particle tests during residual induction. This situation is equivalent to the magnetization and testing of short test objects as outlined above.

The flux lines must be continuous and must therefore have a relatively short path in the metal. Large values of the magnetizing force at the center of the coil are usually specified. Such values depend on the weight per unit length of the test object because this quantity affects the ratio of length L to diameter D. Where the test object is a tube, the $L{\cdot}D^{-1}$ ratio is given by the length between the poles divided by twice the wall thickness of the tube. (The distance L from pole to pole can be longer or shorter than the actual length of the test object and must be estimated by the operator.) As a rough example, with L = 460 mm (18 in.) and D = 19 mm (0.75 in.), the $L{\cdot}D^{-1}$ ratio is 24.

The effective permeability of the metal under test is small because of the large demagnetization field created in the test object by the physical end of the test object. An empirical formula is often used to calculate approximately the effective permeability μ:

$$\mu = 6 \frac{L}{D} - 5 \quad (12)$$

so effective permeability μ = 139 in the above example.

For wet magnetic particle testing, the surface tension of the fluids that carry the particles is large enough to confine the particles to the surface of the test object. This is not the case with dry particles, which have the tendency to stand up like fur along lines of magnetizing force.

In many instances, it may be better to use some other test technique for transverse discontinuities, such as ultrasonic or eddy current techniques.

Alternating Current versus Direct Current Magnetization

Alternating current magnetization is more suitable for detection of outer surface discontinuities because it concentrates the magnetic flux at the surface. For equal magnetizing forces, an alternating current field is better for detecting outside surface imperfections but a direct current field is better for detecting imperfections below the surface.

In practice, the ends of tubes are tested for transverse discontinuities by the following magnetic flux leakage techniques.

1. Where there is a direct current active field from an encircling coil, magnetic particles are thrown at the tested material while it is maintained at a high level of magnetic induction by a direct current field in the coil. This technique is particularly effective for internal cracks. Fatigue cracks in drill pipe are often found by this technique.
2. Where there is an alternating current active field from an encircling coil, magnetic particles are thrown at the tested material while it lies inside a coil carrying alternating current. Using 50 or 60 Hz alternating current, the penetration of the magnetic field into the material is small and the technique is good only for the detection of outside surface discontinuities.

When tests for both outer surface and inner surface discontinuities are necessary, it may be best to test first for outer surface discontinuities with an alternating current field, then for inner surface discontinuities with a direct current field.

Liftoff Control of Scanning Head

To obtain a stable detection of discontinuities, liftoff between the probe and the surface of the material must be kept constant. Usually liftoff is kept constant by contact of the probe with the surface but the probe tends to wear with this technique. A magnetic floating technique has been used for noncontact scanning. In this technique, liftoff is measured by a gap probe and the probe holder is moved by a voice coil motor, controlled by the gap signal.[15] This system and related technology are described in this volume's chapter on primary metals applications.

Particular Applications

Wire Ropes

An interesting example of an elongated steel product inspected by magnetic flux leakage testing is wire rope. Such ropes are used in the construction, marine and oil production industries, in mining applications and elevators for personnel and raw material transportation. Testing is performed to determine cross sectional loss caused by corrosion and wear and to detect internal and external broken wires. The type of flux loop used (electromagnet or permanent magnet) can depend on the accessibility of the rope. Permanent magnets might be used where taking power to an electromagnet might cause logistic or safety problems.

By making suitable estimates of the parameters involved, a reasonably good estimate of the flux in the rope can be made. Because discontinuities can occur deep inside the rope material, it is essential to maintain the rope at a high value of magnetic flux density, 1.6 to 1.8 T (16 to 18 kG). Under these conditions, breaks in the inner regions of the rope will produce magnetic flux leakage at the surface of the rope.

The problem of detecting magnetic flux leakage from inner discontinuities is compounded by the need to maintain the magnetic probes far enough from the rope for splices in the rope to pass through the test head. Common probes include hall effect detectors and encircling coils.

The cross sectional area of the rope can be measured by sensing changes in the magnetic flux loop that occur when the rope gets thinner. The air gap becomes larger and so the value of the field intensity falls. This change can easily be sensed by placing hall effect probes anywhere within the magnetic circuit.

Internal Casing or Pipelines

The testing of inservice well casing or buried pipelines is often performed by magnetic flux leakage techniques. Various types of wall loss mechanisms occur, including internal and external pitting, erosion and corrosion caused by the proximity of dissimilar metals.

From the point of view of magnetizing the pipe metal in the longitudinal direction, the two applications are identical. The internal diameters and metal masses involved in the magnetic flux loop indicate that some form of active field excitation must be used. Internal diameters of typical production or transportation tubes range from about 100 mm (4 in.) to about 1.2 m (4 ft).

If the material is generally horizontal, some form of drive mechanism is required. Because the test device (a robotic crawler) may move at differing speeds, the magnetic flux leakage probe should have a signal response independent of velocity. For devices that operate vertically, such as petroleum well casing test systems, coil probes can be used if the tool is pulled from the bottom of the well at a constant speed. In both types of instrument, the probes are mounted in pads pressed against the inner wall of the pipe.

Because both line pipe and casing are manufactured to outside diameter size, there is a range of inside diameters for each pipe size. Such ranges may be found in specifications. To make the air gap as small as possible, soft iron attachments can be screwed to the pole pieces.

For the pipeline crawler, a recorder package is added and the signals from discontinuities are tape recorded. When the tapes are retrieved and played back, the areas of damage are located. Pipe welds provide convenient magnetic markers. With the downhole tool, the magnetic flux leakage signals are sent up the wire line and processed in the logging truck at the wellhead.

A common problem with this and other magnetic flux leakage equipment is the need to determine whether the signals originate from discontinuities on the inside or the outside surface of the pipe. Production and transmission companies require this information because it lets them determine which form of corrosion control to use. The test shoes sometimes contain a high frequency eddy current probe system that responds only to inside surface discontinuities. Thus, the occurrence of both magnetic flux leakage and eddy current signals indicates an inside surface discontinuity whereas the occurrence of a magnetic flux leakage signal indicates only an outside surface discontinuity.

Problems with this form of testing include the following.

1. The magnetic flux leakage system cannot measure elongated changes in wall thickness, such as might occur with general erosion.
2. If there is a second string around the tested string, the additional metal contributes to the flux loop, especially in areas where the two strings touch.
3. A relatively large current must be sent down the wire line to raise the pipe wall to saturation. Temperatures in deep wells can exceed 200 °C (325 °F).
4. The tool may stick downhole or underground if external pressures cause the pipe to buckle.

Cannon Tubes

In elongated tubing, the presence of rifling affects the ability to perform a good test, especially for discontinuities that occur in the roots of the rifling. Despite the presence of extraneous signals from internal rifling, however, rifling causes a regular magnetic flux leakage signal that can be distinguished from discontinuity signals. As a simulated discontinuity is made narrower and shallower, the signal will eventually be indistinguishable from the rifle bore noise. In magnetic flux leakage testing, cannon tubes can be magnetized to saturation and scanned with hall elements to measure residual induction.

Round Bars and Tubes

In some test systems, round bars and tubes have been magnetized by an alternating current magnet and rotated under the magnet poles. Because the leakage flux from surface discontinuities is very weak and confined to a small area, the probes must be very sensitive and extremely small. The system uses a differential pair of magnetodiodes to sense leakage flux from the discontinuity. The differential output of these twin probes is amplified to separate the leakage flux from the background flux. In this system, pipes are fed spirally under the scanning station, which has an alternating current magnet and an array of probe pairs. The system usually has three scanning stations to increase the test rate.

In one similar system, round billets are rotated by a set of rollers while the billet surface is scanned by a transducer array moving straight along the billet axis. Seamless pipes and tubes are made from the round billets.

In another tube test system, the transducers rotate around the pipe as the pipe is conveyed longitudinally. Overlapping elliptical printed circuit coils are used instead of magnetodiodes and are coupled to electronic circuits by slip rings. The system can separate seams into categories according to crack depth.

Billets

A relatively common problem with square billets is elongated surface breaking cracks. By magnetizing the billet circumferentially, magnetic flux leakage can be induced in the resulting residual magnetic field.

Magnetic flux leakage systems for testing tubes exhibit the same general ability to classify seam depth. It is generally accepted that even with the lack of correlation between some of the instrument readings and the actual discontinuity depths, the automatic readout of these two systems still represents an improvement over visual or magnetic particle testing.

One technique, often called *magnetography,* for the detection of discontinuities uses a belt of flux sensitive material, magnetic tape, to record indications. Discontinuity fields magnetize the tape, which is then scanned with an array of microprobes or hall effect detectors. Finally, the tape passes through an erase head before contacting the billet again. Because the field intensity at the corners is less than at the center of the flat billet face, a compensation circuit is required for equal sensitivity across the entire surface.

Damage Assessment

In most forms of magnetic flux leakage testing, discontinuity dimensions cannot be accurately measured by using the signals they produce. The final signal results from more than one dimension and perhaps from changes in the magnetic properties of the metal surrounding the discontinuity.

Signal shapes differ widely, depending on location, dimensions and magnetization level. It is therefore impossible to accurately assess the damage in the test object with existing equipment. Under special circumstances (for example, when surface breaking cracks can be assumed to share the same width and run normal to the material surface), it may be possible to correlate magnetic flux leakage signals and discontinuity depths. This correlation is normally impossible.

Commercially available equipment does not reconstruct all the desired discontinuity parameters from magnetic flux leakage signals. For example, the signal shape caused by a surface breaking forging lap is different from that caused by a perpendicular crack but no automated equipment uses this difference to distinguish between these discontinuities.

As with many forms of nondestructive testing, the detection of a discontinuity and subsequent followup by either nondestructive or destructive methods pose no serious problems for the inspector. Ultrasonic techniques, especially a combination of shear wave and compression wave techniques, work well for discontinuity assessment after magnetic flux leakage has detected them. In some cases, however, the discontinuity is forever hidden. Such is very often the case for corrosion in downhole and subterranean pipes.

References

1. *Nondestructive Testing Handbook,* second edition: Vol. 6, *Magnetic Particle Testing.* Columbus, OH: American Society for Nondestructive Testing (1989).
2. Stanley, R.K. Section 21, "Diverted Flux Theory." *Nondestructive Testing Handbook,* second edition: Vol. 4, *Electromagnetic Testing.* Columbus, OH: American Society for Nondestructive Testing (1986): p 607-630.
3. Stanley, R.K. and G.L. Moake. "Inspecting Oil Country Tubular Goods Using Capacitor Discharge Systems." *Materials Evaluation.* Vol. 41, No. 7. Columbus, OH: American Society for Nondestructive Testing (1983): p 779-782.
4. Moake, G.L. and R.K. Stanley. "Capacitor Discharge Magnetization of Oil Country Tubular Goods." *Electromagnetic Methods of Nondestructive Testing.* New York, NY: Gordon and Breach (1985): p 151-160.
5. NFPA 70, *National Electric Code,* 2002 edition. Quincy, MA: National Fire Prevention Association (2001).
6. Beissner, R.E., G.A. Matzkanin and C.M. Teller. *NDE Applications of Magnetic Leakage Field Methods: A State of the Art Survey.* San Antonio, TX: Southwest Research Institute (1980).
7. Owston, C.N. "The Magnetic Leakage Field Technique of NDT." *British Journal of Non-Destructive Testing.* Vol. 16, No. 6. Northampton, United Kingdom: British Institute of Non-Destructive Testing (November 1974): p 162-168.
8. Daughton, J., J. Brown, R. Beech, A. Pohm and W. Kude. "Magnetic Field Sensors Using GMR Multilayer." *IEEE Transactions on Magnetics.* Vol. 30, No. 6. New York, NY: Institute of Electrical and Electronics Engineers (1994): p 4608-4610.
9. Smith, C. and J. Daughton. "Low-Field Magnetic Sensing with GMR Sensors." *Proceedings of the Sensors Expo* [Baltimore, MD, May 1999]. Cleveland, OH: Advanstar Communications (1999).
10. Flaherty, J. and E.J. Strauts. United States Patent 3 774 162, *Laser Scan Testing System Having Pattern Recognition Means* (1973).
11. Flaherty, J. and E.J. Strauts. "Automatic Scanning of Fluorescent Indications." *Proceedings of the ASM Metals Show.* Materials Park, OH: ASM International (October 1971).
12. Bloke, M.M., N. Sampat and J. Canosa. *Sensors and Camera Systems for Systems for Scientific Industrial and Digital Photography Applications.* Bellingham, WA: SPIE Press (2001).
13. Donati, S. *Photodetectors: Devices, Circuits and Applications.* Upper Saddle River, NJ: Prentice Hall (2000).
14. Stanley, R.K., T. Hiroshima and M.L. Mester. Section 22, "Diverted Flux Applications." *Nondestructive Testing Handbook,* second edition: Vol. 4, *Electromagnetic Testing.* Columbus, OH: American Society for Nondestructive Testing (1986): p 631-651.
15. Sakamoto, T. and T. Hiroshima. "Rotating Eddy Current Machine for Hot Steel Rods and Wires." Paper 5. *QualTest-2 Conference Proceedings* [Dallas, TX, October 1983]. Columbus, OH: American Society for Nondestructive Testing (1983): p 1-11.

Bibliography

ASTM E 570, *Standard Practice for Flux Leakage Examination of Ferromagnetic Steel Tubular Products.* West Conshohocken, PA: ASTM International (1997).

ASTM E 1571, *Standard Practice for Electromagnetic Examination of Ferromagnetic Steel Wire Rope.* West Conshohocken, PA: ASTM International (2001).

Hwang, J.H. *Defect Characterization by Magnetic Leakage Fields.* Ph.D. dissertation. Fort Collins, CO: Colorado State University (1975).

Lord, W., J.M. Bridges, W. Yen and R. [P. Samy]. "Residual and Active Leakage Fields around Defects in Ferromagnetic Materials." *Materials Evaluation.* Vol. 36, No. 7. Columbus, OH: American Society for Nondestructive Testing (July 1978): p 47-54.

Mackintosh, D.D., D.L. Atherton, P.C. Porter and A. Teitsma. "Test Rigs for Magnetic Flux Leakage Inspection Tools for Pipelines." *Materials Evaluation*. Vol. 50, No. 1. Columbus, OH: American Society for Nondestructive Testing (January 1992): p 13-17.

McClurg, G.O. "Theory and Application of Coil Magnetization." *Nondestructive Testing*. Vol. 13, No. 1. Columbus, OH: American Society for Nondestructive Testing (January-February 1955): p 23-25.

Shannon, R.W.E. and L. Jackson. "Flux Leakage Testing Applied to Operational Pipelines." *Materials Evaluation*. Vol. 46, No. 12. Columbus, OH: American Society for Nondestructive Testing (November 1988): p 1516-1518, 1520-1522, 1524.

Stanley, R.K. "Circumferential Magnetization of Tubes and the Measurement of Flux Density in Such Materials." *Materials Evaluation*. Vol. 44, No. 8. Columbus, OH: American Society for Nondestructive Testing (July 1986): p 966-970.

Stanley, R.K. "Simple Explanation of the Theory of the Total Magnetic Flux Method for the Measurement of Ferromagnetic Cross Sections." *Materials Evaluation*. Vol. 53, No. 1. Columbus, OH: American Society for Nondestructive Testing (January 1995): p 72-75.

10

CHAPTER

Alternating Current Field Measurement

Martin C. Lugg, TSC Inspection Systems, Milton Keynes, United Kingdom

Part 1. Introduction to Alternating Current Field Measurement

Background

Alternating current field measurement is an electromagnetic technique that uses induced uniform currents and magnetic flux density sensors to detect and size surface breaking discontinuities without calibration. The term *uniform* means that, at least in the area under the probe, current lines (components of electric field intensity $\boldsymbol{E}$) in the absence of a discontinuity are parallel, unidirectional and equally spaced ($E_x = E_z = 0$ and E_y = constant $\neq 0$).

In the 1980s, there were moves to develop nondestructive test techniques for detecting and sizing fatigue cracks underwater in welded offshore structures. The development focused on two existing techniques — eddy current testing for detection and alternating current potential drop testing for sizing.

Conventional eddy current techniques were not particularly effective because of the scanning patterns needed across the weld and because of the signals received from the weld itself. The developments therefore concentrated on producing eddy current probes that could test the whole weld while being scanned only parallel to the weld. This was achieved by using arrays of coils covering the weld. To assist detection, because the probe was no longer being scanned across the weld toe (and hence the discontinuity), coils of more than one orientation were used to obtain signals from the ends of a discontinuity as well as the center. Both the induction coils and sensing coils were relatively large to give wide coverage rather than high sensitivity because there was no need to detect discontinuities less than 1 mm (0.04 in.) deep in the types of welds found underwater. The development emphasized detection of discontinuities and suppression of liftoff effects. Crack depth had to be sized with calibration slots.[1]

Alternating Current Potential Drop Technique

The alternating current field measurement technique was developed out of work on the potential drop techniques. Potential drop test applications in the 1980s tended to use direct current rather than alternating current.[2]

Conventional potential drop testing injects an alternating current at points on either side of the crack (so that the test object becomes part of the energizing circuit) by using spot welded pins or magnetically attached sprung pins and then measures differences of electric potential at the surface adjacent to and across the crack. The current can, however, be induced into the test site instead of being injected. Current has been induced where direct injection is difficult (such as on threaded connections or on materials where magnetic attachment does not work) or where the electric field intensity would otherwise be too low — for example, on low conductivity metals such as aluminum.

An induced field can be achieved either by laying current carrying wires on the test object surface across the line of the crack or by building an inducing coil into the voltage probe. This means of induction simplifies deployment underwater but in many cases requires the area around the crack to be cleaned (and kept clean) to bright metal in order to allow voltage measurements to be made. When used with an induced field, alternating current potential drop becomes an eddy current technique although it is not usually recognized as such.

The alternating current potential drop technique has the advantage that, being initially uniform, the current flow past a wide range of discontinuities can easily be mathematically modeled. Comparisons can then be made between measured and predicted voltages, allowing estimation of crack depth and shape without calibration.[3-8]

The alternating current potential drop technique struggled when deployed underwater because of the need to make good electrical contact with the metal surface. For the potential drop technique to be usable underwater, means were developed to obviate electrical contact with the test object. Instead of measuring the surface electric field with contacting pins, the magnetic flux density just above the surface was measured with noncontacting coils while retaining the ability of alternating current potential drop to calculate crack depth without calibration.

Magnetic Field Measurement

Work in potential drop testing in the early 1990s studied the surface electric fields to describe the associated magnetic fields. As with alternating current potential drop testing, field measurement made it possible to estimate crack depths without calibration.[9]

As it turned out, making a noncontacting technique from alternating current potential drop meant that the probe could be scanned along a weld very easily. Thus, although alternating current field measurement was developed for noncontact sizing, it was useful also for discontinuity detection.

The modeling of alternating current field measurement does not require the input field to be induced. Indeed, early experiments with the technique often used injected currents to give better large scale uniformity. In this situation, therefore, alternating current field measurement does not use eddy currents, so the boundaries between alternating current potential drop, alternating current field measurement and conventional eddy current test techniques are blurred. There are other electromagnetic techniques of the same general type: uniform field eddy current, current perturbation, electromagnetic array and surface magnetic flux density measurement techniques. Applications include offshore oil platforms,[10] petrochemical equipment,[11-13] threaded connections,[14] cranes,[15] bridges and rails.[16]

Part 2. Alternating Current Field Measurement Technique

Principle of Operation

The alternating current field measurement technique involves inducing a locally uniform current into a test object and measuring the magnetic flux density above the test object surface. The presence of a surface breaking discontinuity perturbs the induced current and the magnetic flux density. Relative, rather than absolute, amplitudes of components of the magnetic flux density are used to minimize variations caused by material properties, instrument calibration and other circumstances. These relative amplitudes are compared with values in sizing tables produced from a mathematical model to estimate discontinuity sizes without the need for calibration using artificial discontinuities such as slots. This feature of alternating current field measurement is useful because calibration on slots is prone to error for several reasons.

1. Calibration adds opportunities for operator error (and one mistake on a calibration setting will affect all subsequent sizing).
2. Slots behave differently, electrically, from real cracks. In particular, the magnetic fields inside the slot width produce extra induction effects.[17]
3. Slots in calibration blocks are often in materials with different properties (for example, parent plate rather than heat affected zone or weld material).
4. Slots often have geometry different from that of a real crack (for example, a rectangular shape rather than the semielliptical shape more typical of fatigue cracks).
5. The range of slots available in a calibration block is limited. In particular, they tend to be of the same length, whereas the signal intensity can be affected by crack length as well as depth, particularly for short cracks.

The sizing tables have been produced by repeated running of the model for semielliptical cracks in a wide range of different lengths and depths. The model is called the *forward problem* for which the discontinuity size is known and the signals are then predicted. For the inverse problem, sizing an unknown discontinuity when the signal variations are known, software is used to interpolate between and within these tables. The mathematical model used in early development work assumes that the incident current is uniform on a scale comparable to the discontinuity. That is, the electric field lines were parallel and equally spaced. The model also assumes that the standard depth of penetration is small compared to the depth of the discontinuity.

The method in its simplest form uses an instrument and a hand held probe containing a uniform field induction

Figure 1. Uniform magnetic field generated by horizontal solenoid: (a) induction of magnetic field by alternating current in coil; (b) uniform field induced by alternating current in metal surface.

(a)

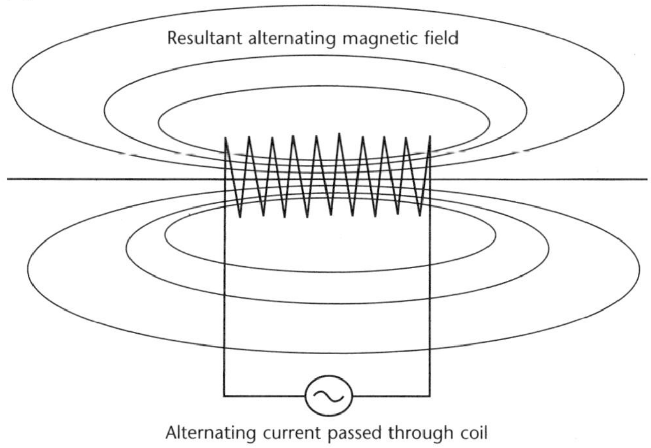

(b)

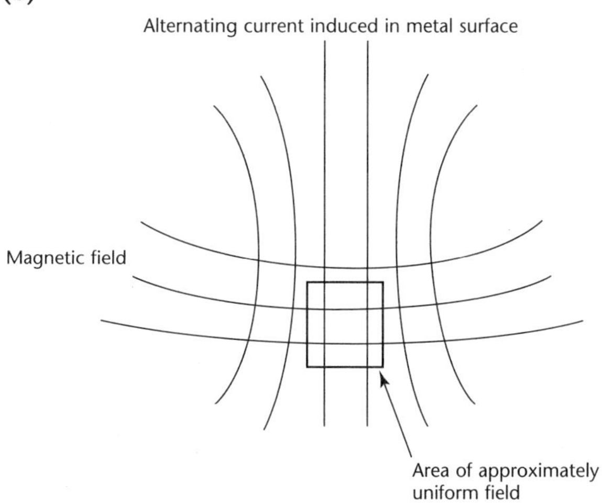

system and two magnetic field sensors. Software on an external personal computer is used to control the instrument and to display and analyze the data.

The required locally uniform magnetic field is induced using one or more horizontal axis solenoids, with or without a yoke (see Fig. 1). By convention, the direction of this electric field ***E*** is designated as the Y axis and the direction of the associated uniform magnetic flux density ***B*** (at right angles to the electric field and parallel to the test surface) is designated as the X axis. The Z axis is then the direction normal to the surface (Fig. 2).

With no discontinuity present and a uniform current flowing in the Y direction, the magnetic field is uniform in the X direction perpendicular to the current flow. Thus, B_x, B_y and B_z are the three orthogonal components (in tesla) of magnetic flux density ***B***. B_x will have a constant positive value whereas B_y and B_z will both be zero.

Figure 3 shows the effect of a surface breaking discontinuity on the magnetic field. The presence of a discontinuity diverts current away from the deepest parts and concentrates it near the ends of a crack. The current distribution produces a broad dip in B_x along the discontinuity with the minimum value coinciding with the deepest point of the discontinuity. The amplitude of this dip is larger for a deeper discontinuity of a given length. At the same time, concentration of current lines where it flows around the discontinuity ends produces small peaks in B_x. The same circulation around the discontinuity ends also produces a nonzero B_z component. The flow is clockwise around one end, producing a negative value of B_z (pointing into the surface) and counterclockwise around the other end producing a positive value of B_z (out of the surface). The locations of the maximum (positive and negative) values of B_z are close to, but not coincident with, the ends of the physical discontinuity.

The B_y component also becomes nonzero in the presence of a discontinuity, producing a peak and a trough at both ends of the discontinuity but these are antisymmetric across the line of the discontinuity. Because a B_y sensor scanning exactly along the line of a discontinuity would see no response, the B_y component is not usually measured in alternating current field measurement.

Measurements of B_x and B_z from sensors in the probe are used with software algorithms to determine the length and depth of the discontinuity. To aid interpretation, the B_x and B_z components are often plotted against each other to produce a closed loop indication. Because of its shape, the display is often called a *butterfly plot* (Fig. 4). This loop's size is insensitive to probe speed, so this display can help to interpret data and evaluate indications.

The actual parameters used by the software can vary but must include the following.

FIGURE 2. Coordinates conventionally used in alternating current field measurement.

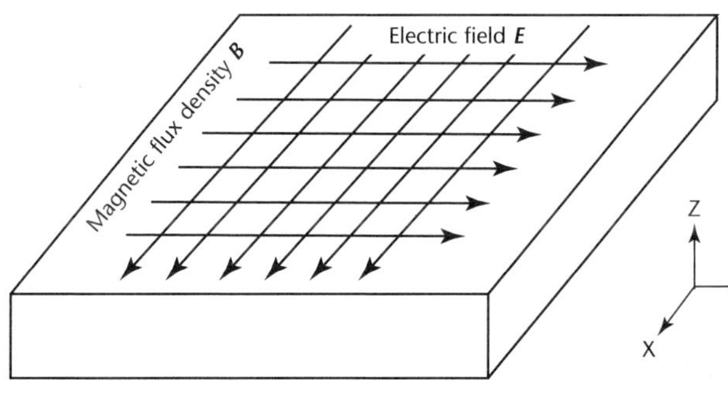

FIGURE 3. Effect of surface breaking discontinuity on magnetic field.

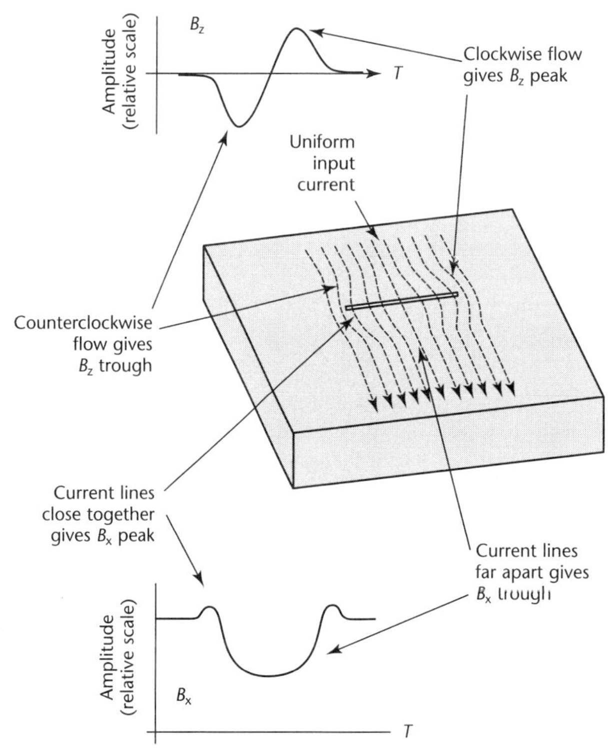

Legend

B_x = magnetic flux component normal to electric field and parallel to test surface
B_z = magnetic flux component normal to test surface
T = time or scan distance (relative scale)

1. The perturbation amplitude is needed for one component of the magnetic flux density produced by the discontinuity (usually B_x but B_z can also be used).
2. The intensity of the input magnetic flux density B_x is used to normalize the perturbation. This background B_x value must therefore be measured in an area of properties similar to the perturbation value. This area is normally next to the discontinuity but outside its influence.
3. A measurement unit is needed to quantify the signal. This unit is usually the distance between the peak and trough in the B_z signal because these signals are sharply defined but the distance between the peaks in B_x can also be used.

Typical Probe Designs

Figure 5 shows components arranged in a typical alternating current field measurement test. The exact parameters used in a probe vary according to the application. The larger dimensions are used where possible because they give the most uniform field and allow the two sensors to be wound concentrically, which gives clear symmetric loops in the butterfly plot. In probes designed for tight access applications or for higher sensitivity, the smaller dimensions are used.

FIGURE 4. Data from longitudinal discontinuity: (a) chart recorder plot; (b) butterfly shaped plot.

(a)

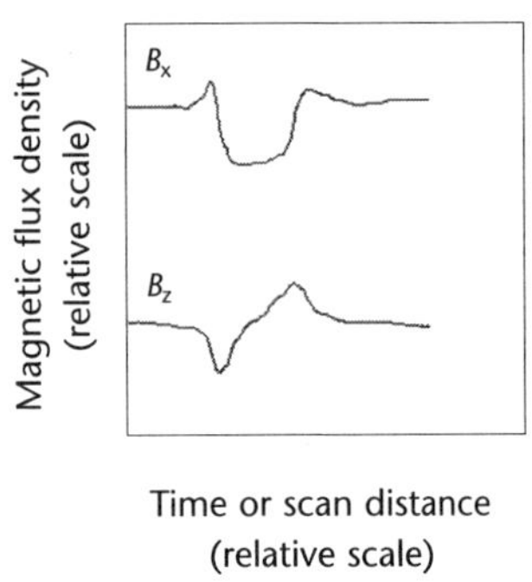

(b)

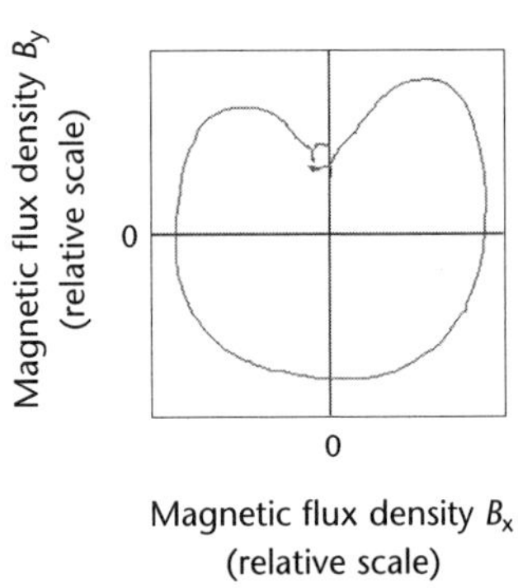

Uniform Field

The alternating current field measurement technique uses a uniform input field to allow comparison of signal intensities with theoretical predictions. A uniform field has advantages and disadvantages compared with conventional eddy currents. The main advantages are (1) the ability to test through coatings several millimeters (one or two tenths of an inch) thick, (2) the ability to obtain depth information on cracks up to 25 mm (1 in.) deep and (3) easier testing at material boundaries such as welds. The main disadvantages are (1) lower sensitivity to small discontinuities, (2) signals obtained from nearby geometry changes (such as plate edges) and (3) dependence of signals on discontinuity orientation relative to probe. These advantages and disadvantages are discussed below.

Advantages

Testing through Coatings. The primary advantage of using a uniform field is that the intensity of the input field decays gradually with distance from the inducing coil; the intensity of the field perturbed by a discontinuity also decays gradually with distance above the surface. The intensity of a uniform field performance does not drop off very rapidly with probe liftoff, so alternating current field measurement can be used to test through thick nonconductive coatings. The technique can be used on painted or rusty surfaces or on structures covered with

FIGURE 5. Typical alternating current field measurement probe layout.

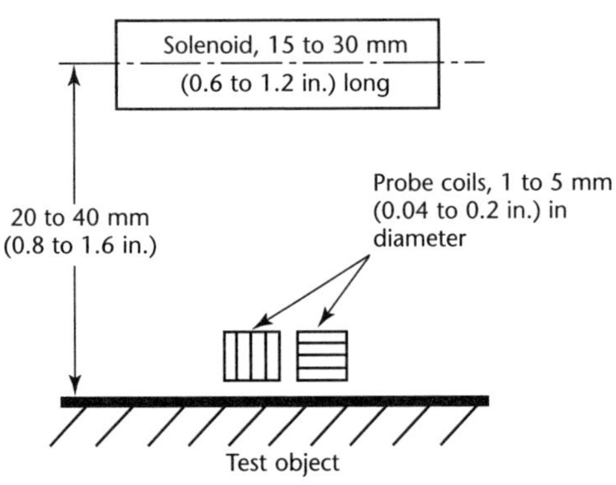

protective or fire resistant coatings several millimeters (one or two tenths of an inch) thick.

Depth Information. The second advantage is that the larger inducing coil forces currents to flow farther down the face of a deep crack. Currents from conventional eddy current probes flow in circles a few millimeters (about an eighth of an inch) across. When a probe lies over a deep crack, the current splits into two separate circles, one on each side of the crack and confined to the top few millimeters (about an eighth of an inch) of the crack face. Because essentially no current then flows to the bottom of the crack, no information can be obtained about where the bottom is, so the depth of the crack cannot be measured.

The same feature occurs with an alternating current field measurement probe but, because the depth of penetration down the crack face is related to the size of the magnetic field inducing coil, an alternating current field measurement probe can measure more deeply, typically 15 to 30 mm (0.6 to 1.2 in.), depending on probe type.

Greater depths could be achieved if a directly injected current were used instead of an induced one but direct injection is infrequently used because it requires a clean metal surface and the current density achieved (and hence the signal intensity) would be much less than with an induced field. In these circumstances, alternating current potential drop testing would be more suitable.

Material Boundaries. A third advantage of a uniform field arises when testing at a weld or other boundary between two metals of different permeability or conductivity. In this case, if the probe is scanning for discontinuities parallel to the boundary, no probe motion is required across the boundary and no signals are caused by the change in material property. Also, the currents are flowing perpendicularly across the boundary, so the effect of this material change is reduced even when scanning up to it.

Disadvantages

Reduced Sensitivity. The main disadvantage of using a uniform field is that sensitivity is reduced. This reduction is of little consequence on welded or rough surfaces, where sensitivity would be reduced anyway. On smooth, clean surfaces, however, alternating current field measurement is less sensitive to short or shallow discontinuities than conventional eddy current techniques. The smallest detectable discontinuity on a good surface with alternating current field measurement is around 2 mm (0.08 in.) long or 0.25 mm (0.01 in.) deep.

Geometry Changes. A second disadvantage of a uniform field is that, because the currents spread out farther, signals are obtained from local geometry changes, such as plate edges and corners. Although these signals do not usually have the same form as a signal from a discontinuity, they can confuse the operator. If many similar geometries are being tested, the operator can learn what signals are caused by the geometry alone and then ignore these. Alternatively, scans from discontinuity free sites with the same geometry can be stored and displayed for comparison or probes with differential sensors can be used to eliminate the large scale signals.

Discontinuity Orientation. A third disadvantage is that the signals obtained from a discontinuity depend on the orientation of the discontinuity. The uniform field theoretical model would suggest that no signal be produced when a probe scans across a transverse discontinuity, because the current flow is then parallel to the discontinuity and would not be perturbed. In fact, in practice, there is a signal produced in this situation (caused by magnetic flux lines jumping the discontinuity) but these do not conform to the signal expected from a discontinuity. The operator is trained to look for the signals caused by a transverse discontinuity in order to detect them. Additional scans must be made along the line of the discontinuity to size it.

Effect of Coating Thickness

One of the main advantages of the uniform field used in alternating current field measurement is that it results in a relatively small reduction in signal intensity with probe liftoff. Consequently, alternating current field measurement can detect cracks through several millimeters (one or two tenths of an inch) of nonconductive coating. Typical coatings include paint, epoxy coatings, oxide layers, fire protection layers and marine growth.

The magnetic field inducer is typically a solenoid, either cylindrical or flat, with or without a steel core, with axis parallel to the surface being tested. The length of the solenoid is typically of the same order as the distance above the metal surface. At such distances, the magnetic flux density decays much slower than the $1 \cdot r^{-3}$ (where r is coil radius) decay that occurs far from the solenoid on the axis of a circular coil.

The maximum coating thickness through which a discontinuity can be detected depends on the discontinuity size, the probe type and the signal noise. Figure 6 shows rates measured at which the magnetic flux density B_x signal amplitude drops with coating thickness for a probe with a flat, 30 mm (1.2 in.) long solenoid 40 mm (1.6 in.) above the base of the probe. The signal variation caused by conditions such as surface roughness and material property variations is usually less than 1 percent.

FIGURE 6. Effect of coating thickness on magnetic flux density B_x for 5 kHz, 30 mm (1.2 in.) long solenoid probe.

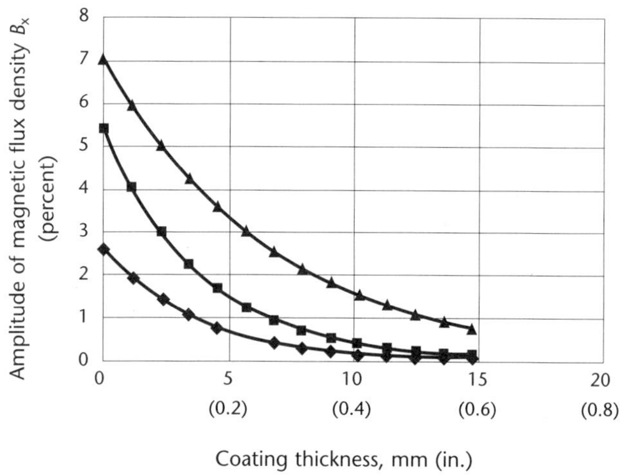

Legend
◆ = 20 × 1 mm (0.8 × 0.04 in.) slot
■ = 20 × 2 mm (0.8 × 0.08 in.) slot
▲ = 50 × 5 mm (2 × 0.2 in.) slot

FIGURE 7. Coating thickness at which magnetic flux density B_x amplitude drops to 1 percent for solenoid probes of three sizes.

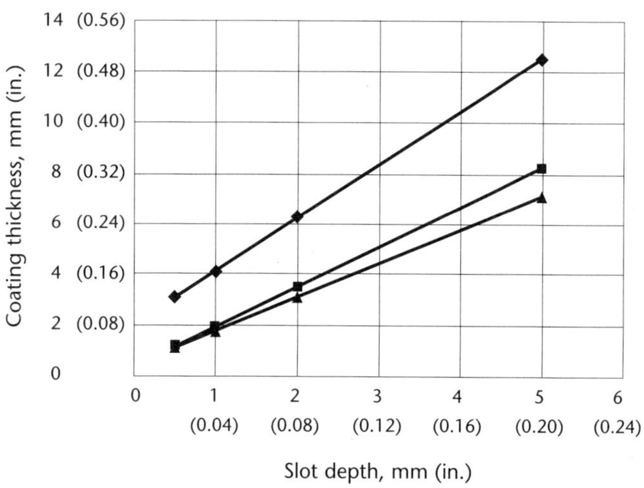

Legend
◆ = probe with 40 mm (1.6 in.) long solenoid
■ = straight probe with 15 mm (0.6 in.) long solenoid
▲ = right angle probe with 15 mm (0.6 in.) long solenoid

The data show that, for example, a 5 mm (0.2 in.) deep discontinuity in a good surface should be detectable through more than 10 mm (0.4 in.) of coating.

The maximum coating thickness through which a discontinuity should be detectable depends on the size of the probe solenoid. Figure 7 compares the performances of different probe designs.

For sizing of discontinuities under coatings, the sizing tables cover a range of liftoff values to compensate for the fact that the amplitude is reduced. The coating thickness needs to be known but only to the nearest millimeter (about 0.04 in.) because the effect is small.

The limitation above applies to nonconductive coatings. The alternating current field measurement technique can be used to test through thin conducting coatings (such as galvanizing, copper loaded grease, flame sprayed aluminum) but only if the coating thickness is small compared to the standard depth of penetration, about 1 mm (0.04 in.) at 5 kHz in the cases described above.

Deep Crack Limit

Any technique that uses induced currents to interrogate surface breaking discontinuities will, for sufficiently deep discontinuities, face the problem that any further increase in discontinuity depth has no effect on the current distribution on the face of the discontinuity. Therefore, no information can be gained about where the bottom of the crack is. This limiting discontinuity depth depends on the probe design — in particular, on the size of the inducing magnetic field. Figure 8 shows experimental results for the rate of change in B_x signal amplitude

FIGURE 8. Rate of increase in magnetic flux density B_x minimum with increasing slot depth.

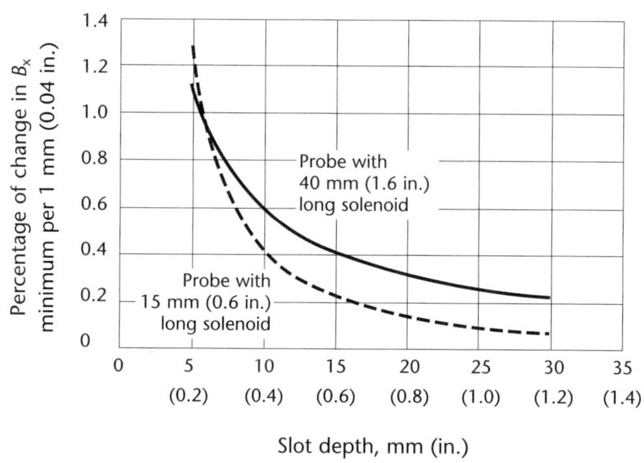

versus discontinuity depth. The points at which the curves fall below about 0.2 percent per millimeter (5 percent per inch) are the deepest points that can be determined with each particular probe type.

Although this limiting depth is larger than for standard eddy current probes, where the small input field usually gives a maximum distinguishable depth of about 5 mm (0.2 in.), it is important to know the limitation during testing. If a discontinuity is sized with a depth close to the limit, it should be recognized that this depth is an estimate and that the true depth may be larger.

Sensitivity to Small Discontinuities

A larger input field than in a conventional eddy current probe means that sensitivity to small discontinuities, particularly in nonferrous metals, is reduced. Sensitivity can be improved by using a higher operating frequency and smaller sensor coils but at the expense of noise. If uncorrected, the problems can give less accurate depth sizing. Using smaller sensor coils allows the coils to be deployed with centers closer to the metal surface, which improves sensitivity to shallow discontinuities. Also, smaller diameter coils give better detection of the ends of short discontinuities because when the coil is larger than about half the discontinuity length, the positive and negative B_z signals from the two ends tend to cancel each other out.

FIGURE 9. Comparison of 5 kHz and 50 kHz, 2.0 mm (0.08 in.) diameter coil probes on slots in ferrous steel.

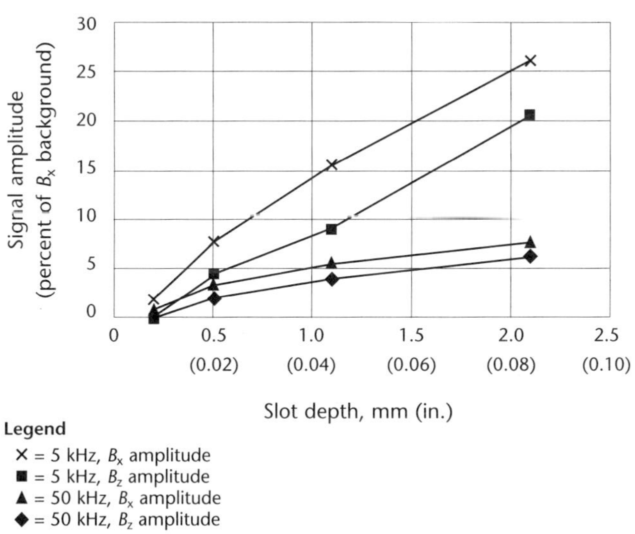

The smallest discontinuity detectable by alternating current field measurement is a function of many parameters. With sensitive probes on good surfaces, discontinuities as small as 2 mm (0.08 in.) long or 0.2 mm (0.008 in.) deep have been detected in ferritic steel. In nonferrous metals, the shallowest detectable discontinuity is around 0.5 mm (0.02 in.) deep.

Experimental data showing signal amplitudes for slots with conventional and high frequency probes are shown in Fig. 9. Signals from a small discontinuity in steel at both frequencies are shown in Fig. 10.

Plate Edges

Compared to a conventional eddy current probe, the larger size of the magnetic field inducer for an alternating current field measurement probe means that the induced currents spread farther out from the center of the probe into the test object. Nearby geometry changes can affect the current flow and so produce changes in the measured magnetic flux density. Features that can produce signals in this way include plate edges, holes and support plates.

FIGURE 10. Signals from 5 mm (0.2 in.) long, 0.2 mm (0.008 in.) deep slot using straight, pencil shaped probes of 2.0 mm (0.08 in.) coil diameter: (a) at 5 kHz; (b) at 50 kHz.

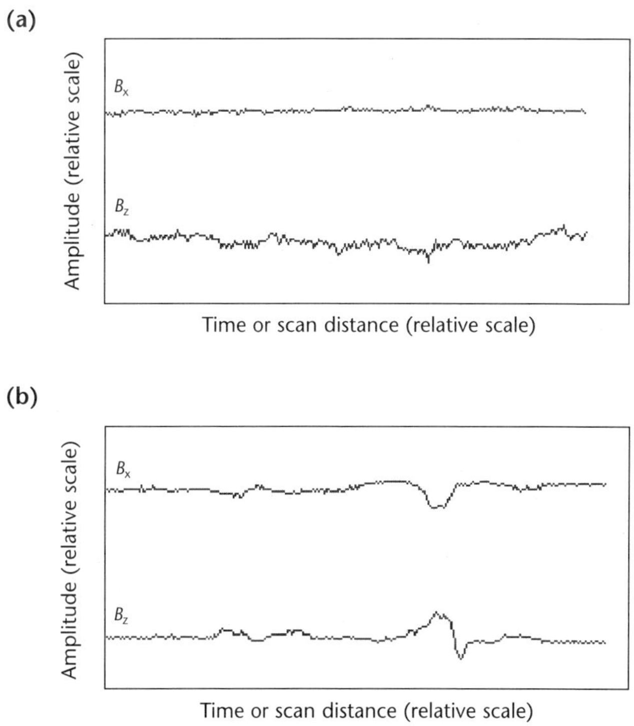

Figure 11 shows B_x signals from two probes scanning up to a plate edge in ferritic steel. It can be seen that B_x decreases as the probe approaches the edge before increasing rapidly to the in-air value as the sensor reaches the edge. Comparison of the two probes shows, however, that the effect on the probe with the smaller inducer is restricted more to the area near the edge. Another way to reduce the effect is to use a probe with two sets of sensors on a line parallel to the edge, connected differentially. In this way, the changing signal from the edge (seen equally by both sets of sensors) cancels out whereas a discontinuity signal (seen more strongly by the sensors passing over the discontinuity) still shows up. The drawback with using such a differential probe is that knowledge of the background value of B_x is lost, so it is not possible to size the discontinuity accurately.

To size a discontinuity that lies within the range of influence of a plate edge, the value of the background B_x magnetic flux density must be estimated at the point where the discontinuity is deepest. It must be estimated what value B_x would have had if the discontinuity had not been present. This value is obtained either by drawing a curve joining the two sections of plot on either side of the discontinuity (see Fig. 12) or by making a second probe scan parallel to the discontinuity but away from its influence.

Transverse Discontinuities

The simple picture of current perturbations producing the measured signals would suggest that discontinuities oriented transverse to the probe scan direction, thus being parallel to the uniform currents, would not be detected.

In fact, discontinuities in this orientation in ferrous steel generally produce measurable signals that arise from flux leakage effects rather than current perturbation. The signals are relatively short (roughly the length of the sensor coils). The B_x signal consists of an upward peak (caused by the increased flux density above the crack) whereas the B_z signal is a close peak-to-trough pair (caused by the flux going up, out of and then down into the metal on either side of the crack). This combination results in an upward loop in a butterfly shaped plot (Fig. 13c), a loop that is distinct from the normal longitudinal discontinuity signal but may be confused with the signal from a seam weld. The differences between the signals from a transverse discontinuity and a seam weld are that the transverse discontinuity gives shorter signals and that the signal from a seam weld is constant wherever the probe crosses it. The signals are strongest when crossing the deepest, or widest, part of the crack — no strong signals are produced at the crack ends.

Because the signal intensity is related as much to the crack opening as the depth, signal intensity cannot be used to calculate discontinuity depth. Also, no such signal is obtained in nonferrous metals. For these reasons, to guarantee detection of transverse discontinuities, test procedures should require the operator to make two sets of scans with the probe oriented in two orthogonal directions (or to use an array probe that continually switches between two orthogonal current inputs).

FIGURE 11. Change in normalized magnetic flux density B_x reading, approaching edge of steel plate.

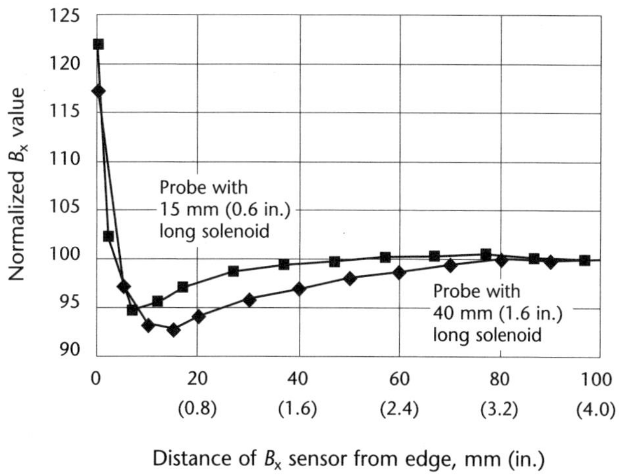

FIGURE 12. Estimation of background magnetic field density B_x near plate edge.

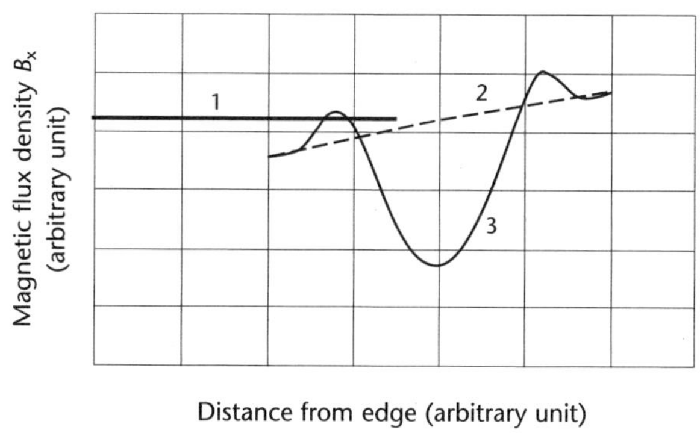

Legend
1. Background value for calibration.
2. B_x background with no discontinuity.
3. B_x signal with discontinuity.

For cracks in ferrous steel oriented somewhere between the purely longitudinal and the purely transverse, the signals lie between the two extremes (Figs. 4 and 13, respectively). For discontinuities within about 30 degrees of the longitudinal direction, the signals appear similar to a longitudinal discontinuity except that the amplitude of the B_x trough is reduced and that the B_z signal becomes asymmetric — the peak (or trough) at the leading end of the crack is larger than the corresponding trough (or peak) at the trailing end.

For cracks oriented within 30 degrees of the transverse direction, the signals look like those from a transverse discontinuity, except that the B_z signal is strongly asymmetric.

For cracks oriented at about 45 degrees, the B_x signal can practically disappear but B_z signals are obtained from both the center and the ends of the discontinuity (Fig. 14).

FIGURE 13. Magnetic flux density signals from transverse discontinuity compared to parallel discontinuity and seam weld: (a) chart recorder plot of B_x measurements; (b) chart recorder plot of B_z measurements; (c) butterfly shaped plot of magnetic flux density.

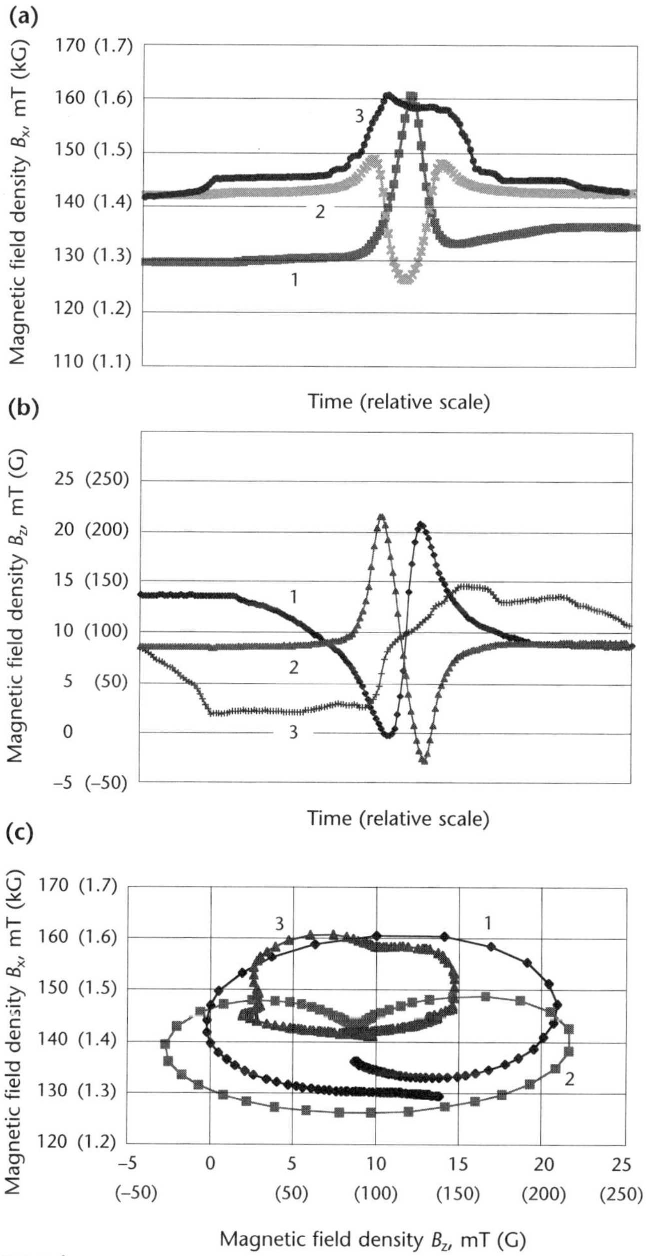

Legend

1. Transverse discontinuity.
2. Parallel discontinuity.
3. Seam weld.

Restrictions in Theoretical Model

The theoretical model used to produce the sizing tables is based on a number of assumptions. One assumption is that the input current is unidirectional and of uniform intensity. It is also assumed that

FIGURE 14. Magnetic flux density from cracks oriented at different angles to scan direction: (a) B_x measurement; (b) B_z measurement.

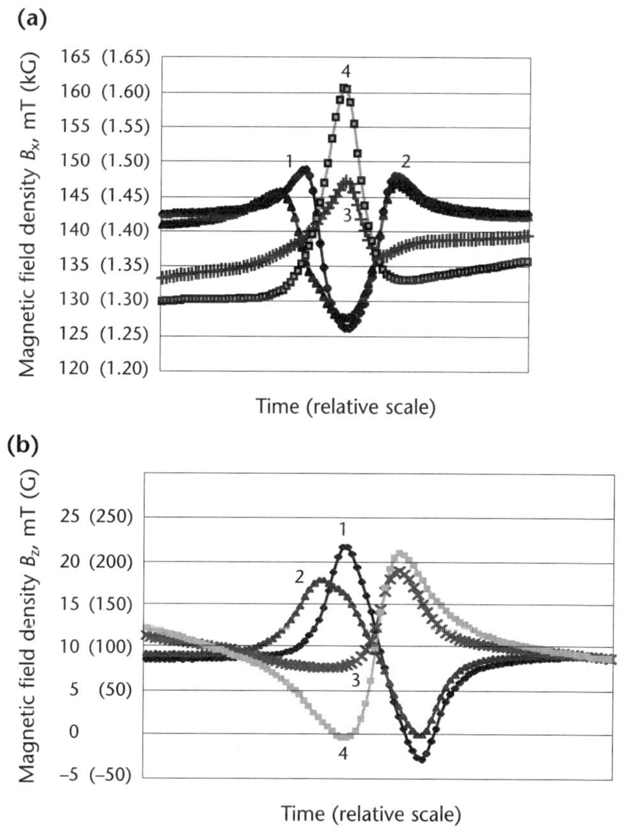

Legend

1. 0 degrees.
2. 30 degrees.
3. 45 degrees.
4. 90 degrees.

the standard depth of penetration is small compared to the dimensions of the discontinuity and that the discontinuity has a semielliptical shape with a length at least twice as large as the depth.[9]

There are also restrictions in the parameter space covered by the sizing tables for practical reasons (time needed to generate each datum, memory required for storage and other software functions). Consequently, there are limits to the minimum and maximum length of discontinuity that can be sized and to the maximum liftoff that can be compensated.

As stated above, the assumption of a uniform input field is required to simplify the modeling of the interaction between the current and a planar discontinuity. Practical alternating current field measurement probes are designed to have a uniform field but there is inevitably some nonuniformity caused by the finite size of the inducer, particularly for the smaller probes. The effect of this nonuniformity, together with any direct induction between the induction solenoid and the sensors, is compensated for during manufacturing setup. Also, the models have been extended to cover the effects of nonuniformity in a real probe[18] and to improve accuracy.

The restriction to a thin standard depth of penetration means that the problem becomes two-dimensional where the test object surface and the crack face can be considered as one continuous two-dimensional surface. This assumption simplifies the problem but means that the results from the model cannot be used to size discontinuities in nonferrous, low conductivity metals such as stainless steel, titanium and nickel alloys. Even in high conductivity metals such as aluminum and copper, the standard depth of penetration is often comparable to the discontinuity depth. In these materials, estimating discontinuity depth requires calibration (although each probe is calibrated once, at the manufacturing stage).

A further consequence of assuming a small standard depth of penetration is that the inclination of the crack plane to the surface has no effect on the results. Therefore, no information on crack inclination can be obtained in practice. The depth values obtained are the distances measured down the crack face, which for an inclined crack will be greater than the through-thickness penetration of the crack. If the standard depth of penetration is not small compared to the discontinuity, there is likely to be some asymmetry in the signal from a scan made across the discontinuity. Measurement of this asymmetry could give information on the inclination of the discontinuity to the surface (as has been done with voltage measurements in the alternating current potential drop technique[19]). The asymmetry in alternating current field measurement, however, is much smaller.

Finally, a small standard depth of penetration means that no signal perturbation is produced by a discontinuity that does not break the surface. In thick skin materials, it is possible to detect subsurface discontinuities but theory does not allow the submerged depth or size of subsurface discontinuities to be calculated from alternating current field measurements.

The restriction to semielliptical crack shapes is again a practical restriction. Sizing tables can be produced for other shapes (such as circular arc or rectangular) if required but semielliptical shapes are chosen because they best fit the real shape of fatigue cracks. The restriction that the crack must be shallower than semicircular is a limit of the transformation used for a semielliptical coordinate system. However, in practice, it is unusual for cracks to grow deeper than this. Also, for a semicircular crack, the currents already flow predominantly around the ends of the crack rather than underneath. Any further increase in depth for the same surface length then has very little effect on the current distribution, so it is not possible to accurately measure the depth of such discontinuities — any estimate obtained will be less than the true depth.

The shortest crack length measurable with the technique is determined by the physical size of the B_z sensor coil because, when the crack length is less than about twice the coil diameter, the distance between the peak and trough in the B_z signal is related to the coil diameter rather than the crack length. For this reason, sizing tables in the 1990s were restricted to lengths above about 5 mm (0.2 in.).

As crack length gets long compared to the size of the probe, the effect of length on the B_x signal amplitude (and hence on the calculated depth) is reduced. Above a certain limit, the current density at the middle of a long discontinuity will be independent of exactly where the ends of the crack are. Therefore, the sizing tables are truncated at an upper length limit, usually around 300 mm (12 in.).

The signal intensity reduces with height above the discontinuity (liftoff) so the liftoff tables also need to cover a range of liftoff to give accurate sizing. For reasons of space, the tables are truncated at an upper limit of about 5 mm (0.2 in.).

Nonuniform Field Effects

The original model assumed a uniform input field and probes are designed to provide a uniform input. However, a uniform field would mean that no signals would be obtained from a long crack of uniform depth, making such a crack undetectable by alternating current field measurement. In fact, such cracks are readily detected by a strong dip in the B_x reading as a probe crosses the crack. The size of this dip depends on the crack depth (Fig. 15).

To quantify this effect, some modeling work was carried out on the effects of nonuniformity in the magnetic flux densities actually generated by finite sized solenoids in real probes.[18] This model was able to show the change in B_x signal amplitude with crack depth and also accounts for the direct induction between the solenoid and the sensor coils, induction required for accurate sizing when the probe liftoff is high.

The nonuniform model requires more parameters (the size, shape and turn distribution of the solenoid) than does the uniform field model. Because these parameters are specified, any set of results is specific to a particular design of probe.

FIGURE 15. Amplitude of magnetic flux density B_x obtained from long slots with 15 mm (0.6 in.) long solenoid probe.

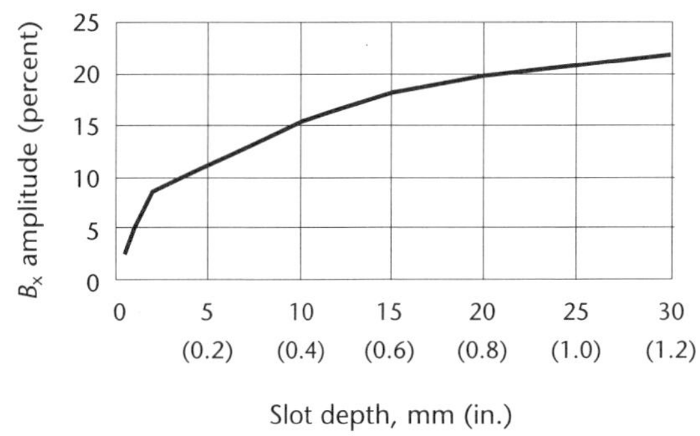

Part 3. Alternating Current Field Measurement Accuracy

Probability of Detection and Probability of Sizing

As with any nondestructive test technique, it is necessary to understand the capabilities and reliability of alternating current field measurement in order to properly use the information it provides. Reliability can be determined only through extensive trials carried out on realistic discontinuities in realistic test objects. The results of such trials are then usually expressed in terms of probability of detection, probability of sizing or receiver operating characteristic. Equipment using the alternating current field measurement technique has undergone a number of such trials, both separately and with other techniques.

Trials with alternating current field measurement equipment were carried out during technique development.[20-22] A library of welded tubular nodes (K, T, X and Y shaped joints) was produced and were fatigued to produce real fatigue cracks of varying length and depth. About 200 fatigue cracks located in various geometries were produced and were tested using underwater equipment by the alternating current field measurement technique together with other techniques, for a comparison of performance. Probability of detection curves were produced for all of the techniques. Underwater alternating current field measurement proved to have detection capabilities similar to those of underwater magnetic particle testing, both when calculated against length (Fig. 16a) and depth (Fig. 16b) but alternating current field measurement had fewer false calls (10 compared to 39 for magnetic particle testing, out of 120 real discontinuities).

It should be noted that the limiting probability of detection of 90 percent shown in Fig. 16a is a lower bound estimate resulting from the finite number of discontinuities in the trial. The discontinuities were arranged in order of characterized length and then assigned to four groups of 29 discontinuities. To be conservative, each group was assigned to the length of the longest crack in the group. Binomial statistics dictate that if all 29 discontinuities in a group are detected, there is a 95 percent confidence level that, of all discontinuities of the same length, the test technique would detect 90 percent. In reality, neither technique missed any discontinuities longer than 20 mm (0.8 in.) In other words, experimental probability of detection was 100 percent for discontinuities longer than 20 mm (0.8 in.).

Another independent evaluation of the reliability of the technique was carried out for an array probe system deployed on a remotely operated vehicle. The remotely operated vehicle test system was subjected to blind trials where a series of cracked

Figure 16. Probability of detection for underwater alternating current field measurement and magnetic particle testing from 1991 trials: (a) versus length; (b) versus depth.

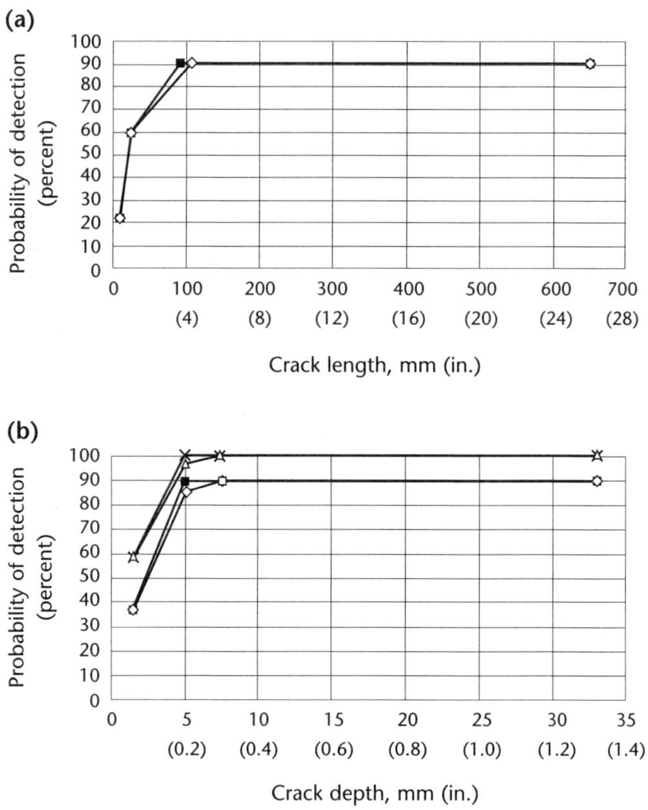

Legend

◇ = magnetic particle testing (95 percent confidence level)
■ = alternating current field measurement (95 percent confidence level)
Δ = magnetic particle testing, experimental probability of detection
✕ = alternating current field measurement, experimental probability of detection

and uncracked plates were tested.[23] All data were transferred via the remotely operated vehicle's umbilical to the surface where interpretation was carried out. Although there were insufficient discontinuities for a meaningful probability of detection measure, the trial included 47 discontinuities ranging in size from 15 to 200 mm (0.6 to 8 in.) long with depth from 2 to 10 mm (0.08 to 0.4 in.). A detection percentage of 98 was achieved and two false calls were recorded. The false calls were both called as discontinuities smaller than the target size of 15 mm × 2 mm (0.6 × 0.08 in.) whereas the one discontinuity missed was close to this limit. The performance of the alternating current field measurement system deployed by the remotely operated vehicle was comparable with that of manual alternating current field measurement but with a lower false call rate.

Further tests were carried out blind on real fatigue cracks in realistic geometries. The project included a wider range of test techniques (including some deployed by remotely operated vehicles), a wider range of test sites (including one in seawater near shore) and a wider range of test objects. The large number of test objects (almost 200) and discontinuities (more than 300) also allowed the project to study operator variability for various techniques.[24,25]

Some experimental results from these trials are shown in Fig. 17. Figure 17a shows the range of experimental probability of detection versus crack length obtained in diver deployed tank trials on tubular welded joints. The total number of fatigue cracks included in the tank trials was 89. Also shown are results from a more limited sea trial offshore. Figure 17b shows the same results plotted against crack depth.

Although it is important for a nondestructive test technique to have a high probability of detection, it is also important that it does not produce too many false calls. As well as measuring probability of detection, the trials also counted false calls and combined the results as a receiver operating characteristic.

As part of a program to obtain approvals for alternating current field measurement in the United Kingdom rail industry, a comparative blind trial was carried out under normal workshop conditions on 15 railroad car axles.[16] All of the axles had discontinuities present (fatigue cracks or corrosion pits) produced during service, which previously would have caused them to be scrapped by the overhauler using magnetic particle and ultrasonic testing. Alternating current field measurement before cleaning produced an experimental probability of detection of 84 percent on discontinuities more than the target size of 0.5 mm (0.020 in.) deep. This result compared to 44 percent for the same discontinuities with magnetic particle testing, even though the magnetic particle testing was carried out after cleaning. The alternating current field measurement system, deployed on a lathe, was able to detect discontinuities down to 0.2 mm (0.008 in.) in depth.

FIGURE 17. Experimental probability of detection for alternating current field measurement by divers on tubular joints underwater in tanks: (a) versus length; (b) versus depth.

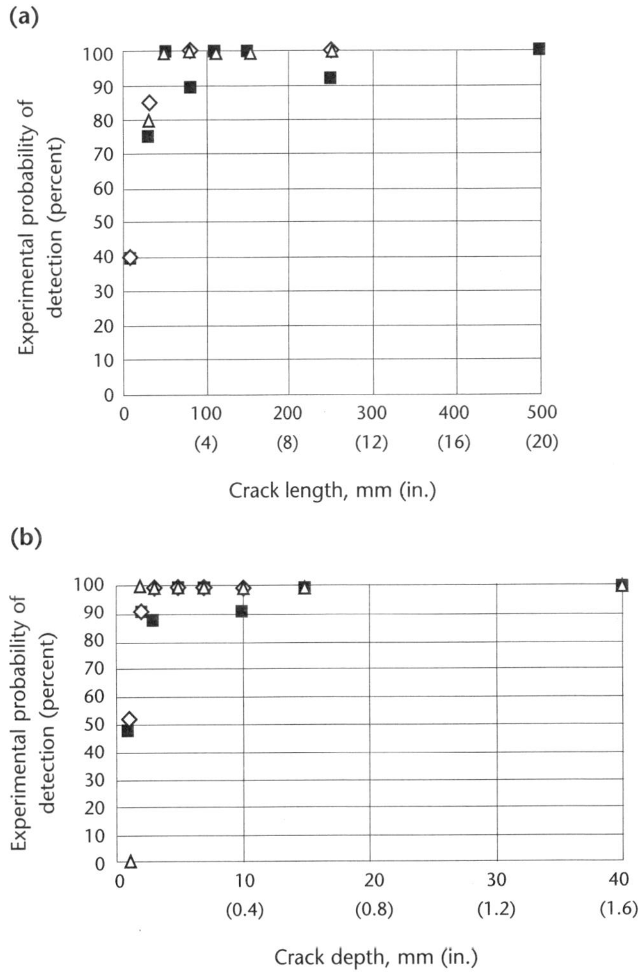

Legend
- △ = tank trial with most discontinuities detected
- ■ = tank trial with fewest discontinuities detected
- ◇ = sea trial

Influences on Sizing Accuracy

The accuracy of length sizing is expected to be good for alternating current field measurement because the physical locations of the B_z peak and trough are closely related to the discontinuity ends. However, there are instances where the crack length measured is shorter than that measured by magnetic particle or liquid penetrant testing. After sectioning of some discontinuities in the underwater trials mentioned above,[20-22] it was noticed that there were instances where the discontinuity had wing shaped ends too shallow to be picked up by alternating current field measurement. Instead, the B_z signal was responding to the points where these wings ended and the crack depth suddenly increased.

The accuracy of depth sizing, on the other hand, can be affected by a number of factors including crack inclination, crack shape and morphology, geometry effects and material property changes.

Crack Inclination

As mentioned above, alternating current field measurement testing measures (as does alternating current potential drop testing) the crack depth down the crack face. If the crack is inclined to the surface, this distance will be greater than the through-thickness penetration of the crack (the important parameter for calculating the remaining mechanical strength). The test technique will also overestimate discontinuity depth if the crack branches under the surface.

Morphology

On the other hand, some discontinuities can be discontinuous under the surface. In this case, alternating current field measurement will only measure the depth of the discontinuity connected to the surface and so will underestimate the depth of the deepest, unconnected, part of the discontinuity. All these factors need to be kept in mind when depths from alternating current field measurement are compared with depths from ultrasonic measurements that locate the crack tip relative to the surface.

A situation where simple interpretation of alternating current field measurement signals can incorrectly size discontinuities is undergoing testing for fatigue cracks in railroad rail heads. The stress conditions in the head of a rail mean that these discontinuities tend to grow sideways as they propagate, making the length under the surface greater than the surface breaking length. With this shape, there is a tendency for more of the current to flow around the ends of the discontinuity on the surface than would normally flow around a semielliptical fatigue crack of the same length. In this case, however, the discontinuities tend to grow in well defined patterns, so there is a close relationship between discontinuity shape and depth. Crack depth can then be measured accurately by calibration.

Geometry Effects

Geometric effects need to be taken into account when sizing discontinuities. The effect of geometry on current flow and how to compensate for it are described above (in the discussion of plate edges). In other situations, such as discontinuities at plate ends or in grooves, it is best to measure the background B_x signal at the same place on a similar geometry rather than immediately outside the discontinuity.

Another situation where a discontinuity is not semielliptical is when a crack grows to a plate edge. The crack may start from a corner, for example, or may run the full width of a plate. In these situations, where there are not two discontinuity ends to measure between, the normal sizing procedure cannot be followed.

A crack growing from a corner is a problem. If the crack is symmetric around the edges, the current perturbation will also be symmetric, so using twice the distance from the crack end to the plate corner as the crack length should give the correct answer.

For cracks that are highly nonsymmetric, most of the current flows around the short part of the crack, so the signal perturbation is independent of the length of the long part of the crack.

Sizing a full width crack requires the nonuniform current model mentioned above. Otherwise, use of plate width as crack length results in a reasonable estimate of discontinuity depth.

Sensor Coverage and Lateral Displacement

A large, uniform input field in alternating current field measurement means that the current perturbation from a discontinuity extends some distance away from the line of a discontinuity. However, there is a limit beyond which a probe will no longer be able to detect a given discontinuity. This limiting distance is larger for deeper (and, to a lesser extent, longer) discontinuities and determines the test width covered by a probe in one scan. This width coverage in turn determines the number of passes needed to inspect a given weld cap, for example, or the optimum spacing between sensors in an array probe.

Detectability of a discontinuity is itself dependent on surface roughness, background signal variations and other factors but it is reasonable to expect that a discontinuity will be detected if the B_x signal amplitude is 1 percent or more. On this basis, Fig. 18 shows the lateral displacement at which discontinuities at three depths can still be detected with a variety of probe types. The discontinuities all have a length around ten times the depth, typical for fatigue cracks at welds, but the results are relatively insensitive to discontinuity length. The plot shows that different pencil shaped probes perform similarly but that, because of its large solenoid coil measuring 40 mm (1.6 in.) in length, the weld probe offers better coverage for the deeper discontinuities. For example, a 5 mm (0.2 in.) deep discontinuity could be detected from a distance of 18 mm (0.7 in.) by using this weld probe.

The minimum discontinuity size reliably detected by alternating current field measurement in blind trials at welds is usually found to be around 1 mm (0.04 in.) deep. Figure 18 shows that this size discontinuity could be detected in the trial from 5 to 9 mm (0.2 to 0.4 in.) away. Because the detection range is symmetric around the center line of the sensor, this detectability implies that a probe adequately tests a band between 10 and 18 mm (0.4 and 0.7 in.) wide. There will be some variation with results from different probes and from discontinuities in different geometries. A coverage of 15 mm (0.6 in.) is typical. If the test is required to find only deeper discontinuities, this coverage will be wider.

Because the B_x amplitude decreases with lateral displacement, discontinuity depths will be underestimated if the lateral displacement is higher than the value assumed in the theoretical sizing tables. This value is zero for pencil shaped probes (expected to be scanned directly along the line of the discontinuity) and 2.5 mm (0.1 in.) for weld probes with 40 mm (1.6 in.) long solenoids (where it is assumed that the discontinuity is at the weld toe whereas the sensors are set back from the front of the probe).

Figure 19 shows the experimental effect of lateral displacement on depth sizing accuracy. It can be seen that a discontinuity will be sized around 70 percent of the true depth at a lateral

FIGURE 18. Lateral displacement at which amplitude of magnetic flux density B_x drops to 1 percent.

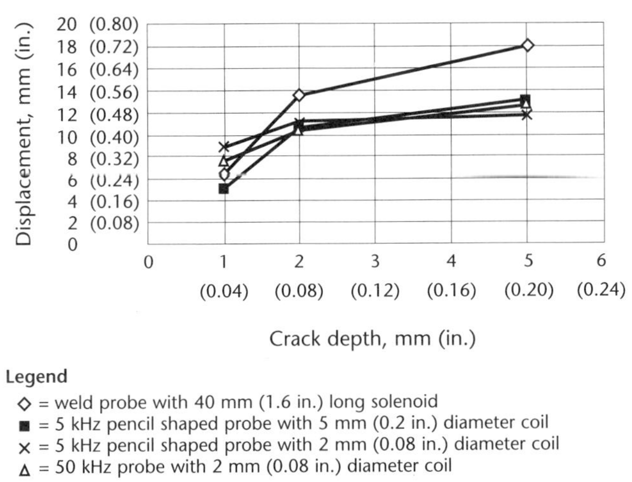

FIGURE 19. Decrease in predicted depth with lateral displacement: (a) weld probe with 40 mm (1.6 in.) long solenoid; (b) probe with 20 mm (0.08 in.) diameter coil.

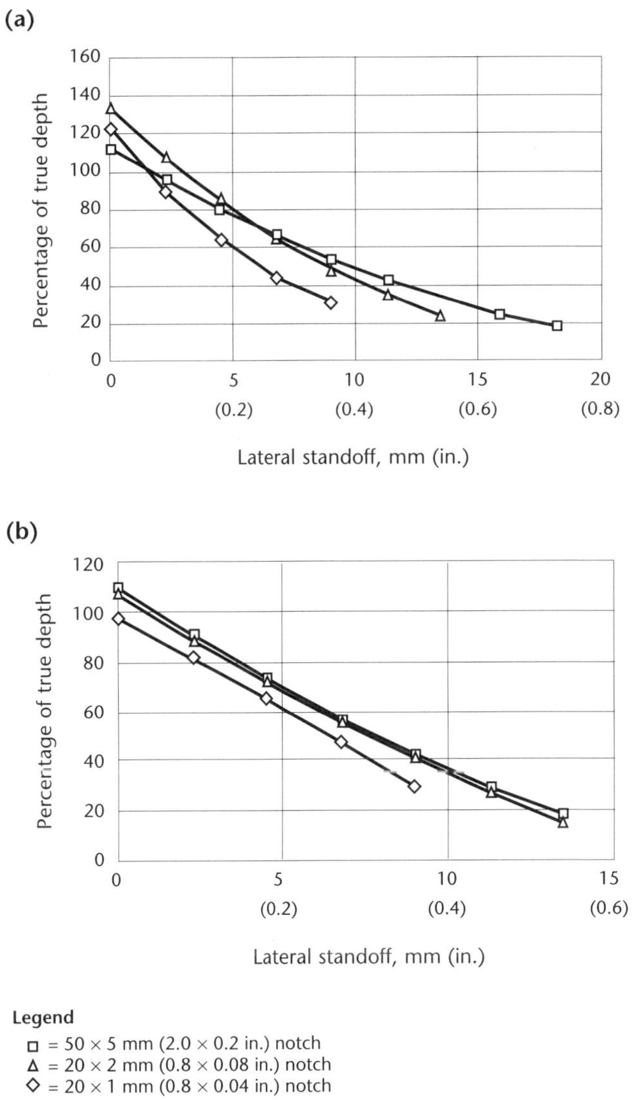

displacement of 5 mm (0.2 in.). The weld probe overestimates discontinuity depth at zero displacement because of the assumption in the sizing model that the front edge of the probe, not the sensors, runs along the weld toe.

False Calls

The false call rate for alternating current field measurement is generally low unless frequency is turned too high in an effort to detect very small discontinuities. Nevertheless, large signals can be confused with discontinuities. Such situations occur mainly when there is a material property change transverse to the scanning direction. This situation can be caused by a seam weld perpendicular to the weld being tested, especially where it is ground off and so not visible to the operator. A seam weld will usually give a peak in B_x (as in Fig. 13a) but can give a trough shaped indication if the base material provides an opposite property difference. A similar effect can also arise where a discontinuity has previously been ground out and refilled with weld metal. These signals can usually be differentiated from discontinuity indications by taking parallel scans some distance away: a discontinuity indication will drop rapidly in amplitude whereas a weld seam signal will not.

One material where particular care has to be used to avoid false calls is duplex steel. This is a mixture of ferritic and nonferritic steels in which the permeability can vary across the surface and where shallow localized grinding of the surface greatly changes the local permeability.[26]

PART 4. Alternating Current Field Measurement Indications

Cracks

Fatigue Cracks

Alternating current field measurement was designed for the detection and sizing of fatigue cracks. Three considerations make the technique well suited for finding such discontinuities.

1. Fatigue cracks are generally surface breaking discontinuities.
2. Fatigue cracks tend to grow at defined stress concentrations well suited for the linear scanning path of alternating current field measurement probes.
3. Fatigue cracks tend to grow in a semielliptical shape and at right angles to the surface, as assumed in the theoretical model used for sizing.

There are situations where fatigue cracks are not semielliptical, however. One such situation occurs on large tubular welded intersections where cracks often initiate at multiple sites. The curved shape of the weld in this case means that, as the separate cracks grow, they are not coplanar. This means that the ends of neighboring cracks often grow past each other, resulting in crack overlaps or bridges of metal between the cracks. While the cracks remain separated, they can be treated as two separate semielliptical discontinuities. Eventually, however, the bridge of metal between the cracks breaks and the cracks connect. At this stage, the crack has a W shape. The alternating current field measurement signals from such a discontinuity are distinctive but accuracy of depth sizing is reduced. As the crack grows deeper, it rapidly becomes semielliptical again.

Stress Corrosion Cracking

Stress corrosion cracking can take the form of a series of parallel cracks acting as a colony. In other cases, it can be present as crazed cracking. The orientation of the cracking and the proximity of individual cracks can lead to problems in interpretation of alternating current field measurement signals. The large scale input field means that the signal from one discontinuity is superimposed on signals from neighboring discontinuities. It is difficult to isolate discontinuities closer together than the distance over which each indication extends. When there are many discontinuities, it is also difficult to match the two discontinuity end signals together correctly.

Some work has been carried out on quantifying these effects.[27] In general, it has been found that detection of clusters of stress corrosion cracking is reliable and that depth values obtained by treating isolated clusters as single discontinuities agrees reasonably well with the typical discontinuity depth.[15]

Hydrogen Induced Cracking

The alternating current field measurement technique has also been used to detect sulfide stress concentration cracking, hydrogen induced cracking, hydrogen sulfide cracking and stress orientated hydrogen induced cracking in the base metal adjacent to the heat affected zone.[28]

Hydrogen cracks are different from fatigue cracks: they are not mechanically induced but result from a combination of internal or external chemical reactions, usually resulting in the production of hydrogen. If these pockets of hydrogen are beside inclusions or very hard areas, cracking will occur.

The cracks tend to have similar features in that they are parallel to the surface and can occur at the sites of inclusion clusters, especially elongated inclusions and in areas of hard metallurgical structures such as martensite or bainite found in heat affected zones. Sulfide stress concentration cracks normally occur in clusters; the other types of hydrogen cracks are lenticular, occurring in parallel bands, and may be shallow. They do not have the normal elliptical shape of fatigue cracks because they are metallurgical rather than mechanically associated discontinuities and thus can be affected by the metallographic structure. Although alternating current field measurement can detect these cracks, their complicated subsurface structure (branching and splitting) makes depth sizing difficult.

Fatigue Cracks in Rail Heads

Nonsemielliptical discontinuities occur also in railroad rails. Head checking (also called *gage corner cracking*) is cracking that initially grows into the top surface of a rail at a highly inclined angle (typically

25 degrees to the surface). As they grow below a certain depth, they turn to a steeper angle but also start to grow sideways, so their length is greater beneath the surface.[29] Such complicated shapes are difficult to model, so depth sizing relies on empirical curves or on calibration with reference standards.

Corrosion Pitting

The unidirectional currents used in alternating current field measurement are most strongly perturbed by planar discontinuities. However, surface corrosion pitting also perturbs current flow to some extent and can also be detected. The degree of current perturbation is much lower than for a crack of the same depth and length, so on an initial scan, a corrosion pit looks like a shallow crack. However, the distinguishing feature of a pit is that, unlike a crack, it will produce the same signal regardless of the orientation of the interrogating current. Systems designed to distinguish cracks and pits therefore use two orthogonal current inducing coils, usually with an array of sensors to speed up tests.[30]

Some modeling work has also been carried out on the perturbation of uniform currents by hemispherical pits.[31]

References

1. Brearley, T., E.A. Pedersen and J. Krol. "Electromagnetic Array Inspection Technology." *Proceedings of International Offshore Contracting and Subsea Engineering* [Aberdeen, United Kingdom, October 1992]. London, United Kingdom: International Marine Contractors Association (1992).
2. "Potential Drop Nondestructive Testing." *Nondestructive Testing Handbook,* second edition: Vol. 9, *Special Nondestructive Testing Methods.* Columbus, OH: American Society for Nondestructive Testing (1996): p 378-397.
3. Collins, R., D.H. Michael and K.B. Ranger. "The AC Field around a Plane Semi Elliptical Crack in a Metal Plate." *Proceedings of the 13th Symposium on Non-Destructive Evaluation.* San Antonio, TX: Nondestructive Testing Information Analysis Center with the South Texas Section of the American Society for Nondestructive Testing (1981): p 470-479.
4. Dover, W.D., F.D.W. Charlesworth, K.A. Taylor, R. Collins and D.H. Michael. "The Use of AC Field Measurements to Determine the Shape and Size of a Crack in Metal." *Eddy-Current Characterization of Materials and Structures.* Special Technical Publication 722. West Conshohocken, PA: ASTM International (1981): p 401-427.
5. Michael, D.H., R.T. Waechter and R. Collins. "The Measurement of Surface Cracks in Metals by Using A.C. Electric Fields." *Proceedings of the Royal Society of London:* Series A, *Mathematical and Physical Sciences.* Vol. 381. London, United Kingdom: Royal Society (1982): p 139-157.
6. Michael, D.H. and R. Collins. "The AC Field around a Plane Crack in a Metal Surface When the Skin Depth Is Large." *Journal of Nondestructive Evaluation.* Vol. 3, No. 1. New York, NY: Plenum Press (March 1982): p 19-24.
7. Mirshekar-Syahkal, D., R. Collins and D.H. Michael. "The Influence of Skin Depth on Crack Measurements by the A.C. Field Technique." *Journal of Nondestructive Evaluation.* Vol. 3, No. 2. New York, NY: Plenum Press (June 1982): p 65-76.
8. Collins, R., D.H. Michael, D. Mirshekar-Syahkal and H.G. Pinsent. "Surface Electromagnetic Fields around Surface Flaws in Metals." *Journal of Nondestructive Evaluation.* Vol. 5, No. 2. New York, NY: Plenum Press (June 1985): p 81-93.
9. Lewis, A.M., D.H. Michael, M.C. Lugg and R. Collins. "Thin-Skin Electromagnetic Fields around Surface-Breaking Cracks in Metals." *Journal of Applied Physics.* Vol. 64, No. 8. Melville, NY: American Institute of Physics (1988): p 3777-3784.
10. Marques, F.C.R., M.V.M. Martins and D.A. Topp. "Experiences in the Use of ACFM for Offshore Platform Inspection in Brazil." *15th World Conference on NDT Proceedings* [Rome, Italy, October 2000]. Brescia, Italy: Italian Society for Nondestructive Testing and Monitoring Diagnostics (2000).
11. Bajula, D.R. "ACFM in Lieu of WFMT Inspections for HIC Damage." *ASNT Spring Conference and 10th Annual Research Symposium 2001 Abstracts Book.* Columbus, OH: American Society for Nondestructive Testing (March 2001): p 72.
12. Raine, G.A. and N. Smith. "NDT of On and Offshore Oil and Gas Installations Using the Alternating Current Field Measurement (ACFM) Technique." *Materials Evaluation.* Vol. 54, No. 4. Columbus, OH: American Society for Nondestructive Testing (April 1996): p 461, 462, 464, 465.
13. LeTessier, R., R.W. Coade and B. Geneve. "Sizing of Cracks Using the Alternating Current Field Measurement Technique." *International Journal of Pressure Vessels and Piping.* Vol. 79. Amsterdam, Netherlands: Elsevier Science (2002): p 549-554.
14. Gaynor, T.M., D.L. Roberts, E. Holman and W.D. Dover. "Reduction in Fatigue Failures through Crack Detection by Alternating Current Field Measurements." Paper IADC/SPE 35033. Presented at IADC/SPE Drilling Conference [New Orleans, LA, March 1996]. Houston, TX: International Association of Drilling Contractors (1996).

15. Raine, G.A. and C. Laenan. "Additional Applications with the ACFM Technique." *Insight.* Vol. 40, No. 12. Northampton, United Kingdom: British Institute of Non-Destructive Testing (December 1998): p 860-863.
16. Howitt, M. "Bombardier Brings ACFM into the Rail Industry." *Insight.* Vol. 44, No. 6. Northampton, United Kingdom: British Institute of Non-Destructive Testing (June 2002): p 379-382.
17. Mirshekar-Syahkal, D., D.H. Michael and R. Collins. "Parasitic Voltages Induced by Artificial Flaws When Measured Using the A.C. Field Technique." *Journal of Nondestructive Evaluation.* Vol. 2. New York, NY: Plenum Press (1981): p 195-202.
18. Zhou, J., M.C. Lugg and R. Collins. "A Nonuniform Model for Alternating Current Field Measurement of Fatigue Cracks in Metals." *International Journal of Applied Electromagnetics and Mechanics.* Vol. 10, No. 3. Amsterdam, Netherlands: IOS Press (1999): p 221-235.
19. Lugg, M.C., H.M. Shang, R. Collins and D.H. Michael. "The Measurement of Surface Crack Inclination in Metals Using A.C. Electric Fields." *Journal of Physics D: Applied Physics.* Vol. 21. London, United Kingdom: Institute of Physics (1988): p 1814-1821.
20. Dover, W.D. and J. Rudlin. "Crack Sizing Trials." *Proceedings of International Offshore Contracting and Subsea Engineering* [Aberdeen, United Kingdom, October 1992]. London, United Kingdom: International Marine Contractors Association (1992): p 13-16.
21. Rudlin, J. and W.D. Dover. "Results of Probability of Detection Trials." *Proceedings of International Offshore Contracting and Subsea Engineering* [Aberdeen, United Kingdom, October 1992]. London, United Kingdom: International Marine Contractors Association (1992).
22. Raine, G.A., W.D. Dover and J. Rudlin. "Trials on Coated Nodes." *Proceedings of International Offshore Contracting and Subsea Engineering* [Aberdeen, United Kingdom, October 1992]. London, United Kingdom: International Marine Contractors Association (1992).
23. Lugg, M.C., D. Cooke and D. Topp. "ROV Deployed Inspection Using ACFM Arrays." Presented at International Offshore Conference and Exhibition [Aberdeen, United Kingdom, 1994]. Malden, United Kingdom: Spearhead Exhibitions.
24. Visser, W. *POD/POS Curves for Non-Destructive Examination.* H.S.E. Offshore Technology Report 2000/018. London, United Kingdom: Health and Safety Executive (2002).
25. Rudlin, J. and W.D. Dover. "Performance Trends for POD As Measured in the ICON Project." *Offshore Mechanics and Arctic Engineering.* Vol. 2. New York, NY: ASME International (1996): p 509-513.
26. Lugg, M.C. and A. Raine. "Inspection of Duplex and Superduplex Steels Using the Alternating Current Field Measurement Technique." *Insight.* Vol. 37, No. 6. Northampton, United Kingdom: British Institute of Non-Destructive Testing (June 1995): p 436-439.
27. Carroll, L.B. and C.C. Monaghan. "Detection and Classification of Crack Colonies Using ACFM Technology — Phase I." *NDE Performance Demonstration, Planning and Research.* PVP-Vol. 352; NDE-Vol. 16. New York, NY: ASME International (1996).
28. Lugg, M.C. "The Use of the ACFM Technique for Detection of Environmental Cracking." *Proceedings of the Joint EC-IAEA Specialists Meeting* [Petten, Netherlands, March 1999]. Brussels, Belgium: European Commission, Joint Research Centre (1999): p 206-214.
29. RT/PWG/001, *Rolling Contact Fatigue in Rails: A Guide to Current Understanding and Practice.* No. 1. London, United Kingdom: Railtrack PLC (2001).
30. Lugg, M.C. "Recent Advances in ACFM Array Systems." *Studies in Applied Electromagnetics and Mechanics.* Vol. 12. Amsterdam, Netherlands: IOS Press (1997): p 165-170.
31. McKurdy, D.M. and A.M. Lewis. "ACFM above a Hemispherical Pit in an Aluminum Block." *Review of Progress in Quantitative Nondestructive Evaluation* [La Jolla, CA, July 1992]. Vol. 12A. New York, NY: Plenum Press (1993): p 265-270.

11

CHAPTER

Reference Standards for Electromagnetic Testing

Donald M. Bailey, Roseville, California

Timothy E. Kinsella, Dassault Falcon Jet, Wilmington, Delaware

Walter R. Matulewicz, Huntsville, Alabama

PART 1. Introduction to Reference Standards for Electromagnetic Testing

This chapter is an updated version of the corresponding section in the second edition.[1]

Development and Use of Reference Standards

Success in electromagnetic testing depends on the proper use of available reference standards.

The development and use of eddy current reference standards requires a thorough understanding of the test to be performed. Considerations include (1) the materials tested, (2) their size and shape, (3) discontinuities of interest, (4) means of producing artificial discontinuities, (5) indications from artificial discontinuities, (6) noise that might be encountered, (7) instrument limitations and (8) criteria for relevant indications.

In view of the many types of artificial discontinuities that can be produced, an understanding of their relationship to the discontinuities of interest is critical. For example, if transverse discontinuities in a tube are of interest, it might seem obvious to use a transverse notch. However, that notch could be either straight, which is much easier to fabricate, or curved to match the radius of the tube (see Fig. 1). Each notch will produce different signals that must be understood to properly relate them to the condition to be detected. Another similar case is that of cracks in a fastener hole, which can be simulated by several notches as shown in Fig. 2, all of which will produce different signals.

One of the primary considerations in designing a reference standard is the ability to make more than one identical reference standard according to the same specifications. The reference standard and artificial discontinuities can be faithfully reproduced so that consistent tests can be performed at more than one location, a test can be repeated at another facility for evaluation and analysis and an identical reference standard can be produced if the original is damaged or lost. Therefore, the fabrication method must be carefully considered. For example, a notch made in a typical machine shop will not be as consistent as an electric discharge machined notch made by a vendor who specializes in them and provides certification. The lack of repeatability is a drawback to the use of natural discontinuities as references because no two natural discontinuities are identical.

Another important concern is the purpose or function of the reference standard. It could be used to establish or

FIGURE 1. Transverse notches in tube: (a) straight; (b) curved.

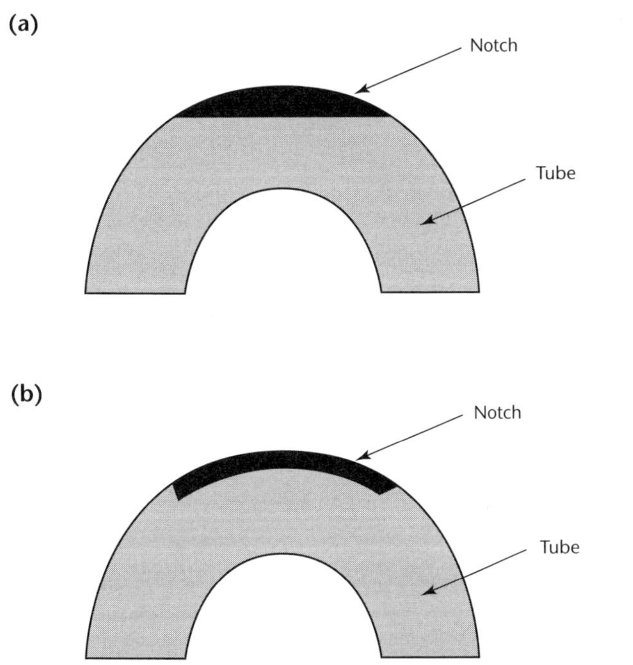

FIGURE 2. Notches to simulate cracks as might be found in fastener holes.

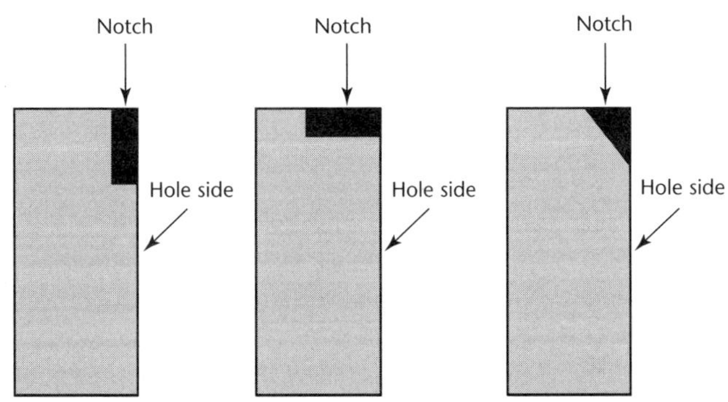

confirm the proper setup, to duplicate types of discontinuities that might be encountered or to help interpretation by discriminating between noise and discontinuities. Finally, if the test is sufficiently controlled, the reference standard can be used to produce the same response as the discontinuity of interest.

The variables found in electromagnetic testing are almost always nonlinear. There are some situations where the variables are not even monotonic. Figure 3, for example, shows that the thickness impedance curve reverses direction even though the changes in thickness continue in the original direction. Actually, there are test points on all impedance plane diagrams where signal reverses occur. Therefore, intermediate variable reference standards will often be required in addition to reference standards that cover the end points of the tested variables.

In conductivity tests, minimum reference standards used for calibration are often provided with the equipment. For greater accuracy, however, additional reference standards that cover the specific range and material being tested are recommended.

Alloy sorting, heat treatment verification, hardness determination and thickness measurement must each have reference standards that properly match all the changes in the variables that might exist in the test objects. Standardization and setup are very important for locating

FIGURE 3. Expanded, relative scale of thickness variations for tests of three metals at 120 kHz.

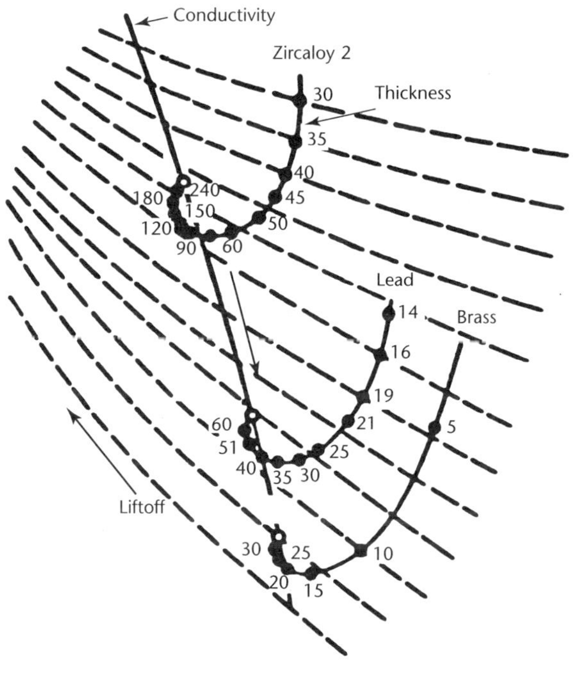

discontinuities. To obtain sufficient information, test choices might include different frequencies, different probes, different orientations or different procedures.

Requirements of Codes and Specifications

An important requirement for successful electromagnetic testing is the use of an accurate reference standard for equipment calibration. The reference standard is used to adjust the electromagnetic equipment's sensitivity to various specimen parameters (cracks, surface roughness, conductivity and permeability variations and other material conditions). Consistency of calibration is maintained with procedural documents such as those issued by ASTM International, the American Petroleum Institute, government military offices and other organizations.

As an example, the *ASME Boiler and Pressure Vessel Code* of the American Society of Mechanical Engineers (ASME) is widely used for testing of tubing or pipe in mills (as required by various code sections) and for in-place testing in nuclear power plant steam generators (Section XI). The reference standards described in Section V, Article 8, Appendices I and II, are used for testing of installed nonferromagnetic heat exchanger tubing and are required when specified by a governing party or referenced by another code section. Article 8 is also voluntarily used for installed tube testing in many industries because there are few governing standards for it. The reference standards described there are used to establish and verify system response and can also provide a means of establishing depth-versus-phase curves for evaluation of signals from discontinuities. Depth-versus-phase calibration is made using a series of flat bottom holes in increments of 20 percent of wall thickness. A plot of the measured phase angle versus the depth of each flat bottom hole is made and used as a reference to estimate the depth of signals from the inspected tube.[2] Other calibration or reference standards should be included to represent specific types of discontinuities to be detected.

As a means for controlling the accuracy of discontinuities in a standard, ASTM E 215 calls for a tolerance of ±0.025 mm (±0.001 in.) for flat bottom hole depths. Several holes are drilled around the circumference of the tubular reference at depths of one-third to two-thirds of the tube wall thickness. To be considered as an acceptable primary reference standard, the responses of any of the one-third thickness holes must be within ±20 percent of the mean of the three one-third thickness holes. Also, the two-thirds thickness response must be within ±10 percent of the mean indication for the two-thirds thickness holes.[3]

The procedural document often specifies the type of reference standard to be used for a given test, depending on the material under test and its geometry.

Figure 4 shows several types of reference standards. These reference standards and the procedural documents that govern their use are often the key to successful electromagnetic testing.

FIGURE 4. Types of reference standards; (a) notched tubes; (b) calibration block; (c) block with graduated holes.

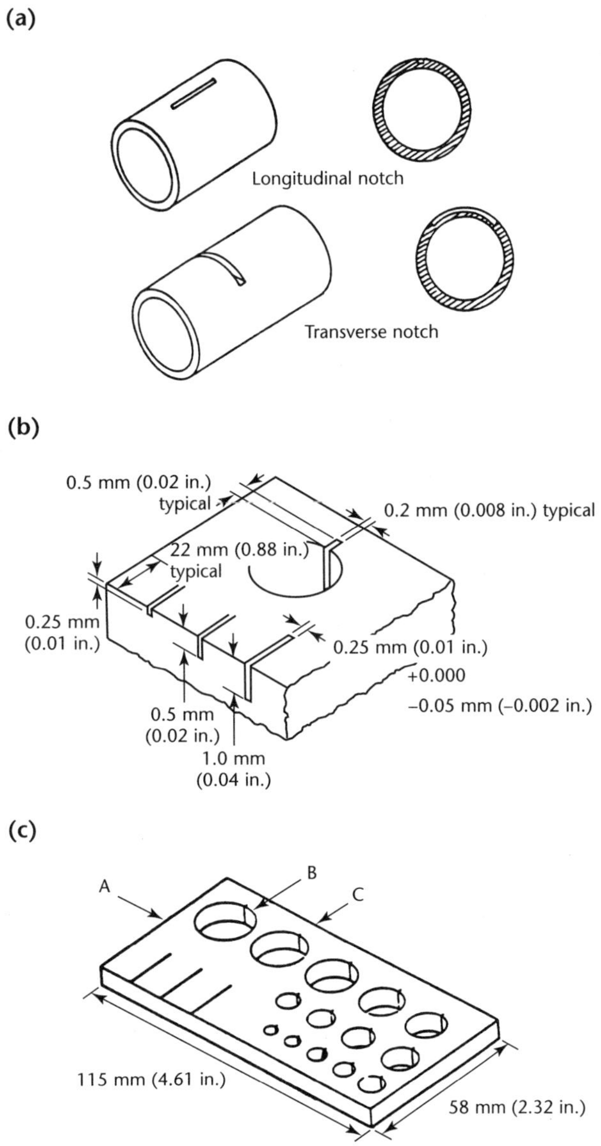

Legend

A. Made of same material as part to be inspected, 6.4 mm (0.25 in.) thick, 32 root mean square finish.
B. Holes reamed to ±0.05 mm (± 0.002 in.) tolerance and 32 root mean square finish.
C. Vibration etched hole sizes and material.

Part 2. Types of Reference Standards

Conductivity Reference Standards

Eddy current testing is widely used to determine electrical conductivity.[4] Both primary and secondary reference standards can be used to determine conductivity.

1. Primary reference standards have values assigned through direct comparison with a standard calibrated by the National Institute of Standards and Technology[5] or have been calibrated by an agency that has access to reference standards calibrated in fundamental units. For example, some primary reference standards have values assigned through direct comparison with standards calibrated according to test technique ASTM B 193.[6] Primary reference standards are usually kept in a laboratory environment and are used only to calibrate secondary reference standards.
2. Secondary reference standards are those reference standards supplied with the instrumentation or reference standards constructed by the user for a specific test. These reference standards are used to calibrate the test instrument.

Conductivity reference standards should be tested with a relatively small coil to determine the uniformity of electrical conductivity over the surface of the standard; both the front and back surface should be tested for conductivity differences. If possible, scanning the surfaces at several input signal frequencies is recommended.

Each time the reference standards are used, the probe coil is placed at the same position within 6.4 mm (0.25 in.) of the center of the standard.

Conductivity reference standards are precise electrical materials and should be treated as such. Any scratching of the standard surface could introduce error in measurement. Avoid dropping and other rough handling and keep the surface of the standard as clean as possible with a nonreactive liquid and a soft cloth or tissue. Reference standards are stored where the temperature is relatively constant; placing the reference standards where large temperature variations occur may cause thermal shock and should be avoided.

Coating Thickness Reference Standards[7]

Each instrument should be calibrated in accordance with the manufacturer's instructions before use. Calibration should be checked at frequent intervals during use.

Calibration reference standards of uniform thickness are available in two types: (1) foils or shims of known thicknesses laid on an appropriate substrate and (2) actual coatings affixed to prepared substrates as supplied or recommended by the instrument manufacturer or standardizing organization (Table 1).[5]

Calibration Shims

Calibration foils or shims are placed on the surface of an uncoated basis metal when calibrating electromagnetic testing instruments. Shims are well suited for calibrating on curved surfaces and are often more readily available than a coated standard. To prevent measurement errors due to poor contact between shim and substrate, there must be intimate contact between them. Calibration shims are subject to indentation and should be replaced when damaged.

Table 1. Standard reference materials from National Institute of Standards and Technology for calibration of instruments used in measurement of organic and nonmagnetic inorganic coatings over steel. Each 45 × 45 mm (1.8 × 1.8 in.) block consists of fine grained copper electrodeposited on low carbon steel substrate.[5]

Material Number	Coating Thickness µm	Coating Thickness (in. × 10^{-3})
1358a	80, 255, 1000	(3.1, 9.8, 39)
1359b	48, 140, 505, 800	(2.0, 5.5, 20, 32)
1361b	6, 12, 25, 48	(0.2, 0.5, 1.0, 2.0)
1362b	40, 80, 140, 205	(1.6, 3.1, 5.5, 7.9)
1363b	255, 385, 505, 635	(9.8, 16, 20, 26)
1364b	800, 1000, 1525, 1935	(32, 39, 59, 79)

Nonmagnetic shims may be used to calibrate magnetic thickness gages for measurement of nonmagnetic coatings. Plastic shims can be used to calibrate electromagnetic testing instruments for measurement of nonconductive coatings.

Shims should should be made of materials that will not change shape (strain or bend) when pressed on with the probe and that will not change in thickness with variations in moisture or temperature. Two or more shims on top of each other should be avoided unless flexibility of thin shims is required for a curved surface.

Coated Substrate Reference Standards

Each coating specimen used as a reference standard has a coating of a uniform thickness permanently bonded to a substrate material. Calibration reference standards of several known coating thicknesses may be used for a single application. For calibration, the thickness of one reference coating should be as close as possible to the upper limit, one as close as possible the lower limit and another as close as possible to the desired coating thickness.

For instruments that measure coatings on magnetic substrates, calibration reference standards should have the same magnetic properties as the coated test specimen. For electromagnetic testing instruments, such as eddy current liftoff gages, the calibration standard should have the same electrical and magnetic properties as the coated test specimens.

To determine calibration validity, a reading should be made on a bare specimen identical to the test object in magnetic and electrical properties. If the coating process is changed after previous calibration, the calibration may no longer be valid, especially for magnetic coating gages and eddy current thickness gages, and the initial reading must again be established.

In some cases, calibration of instruments with two-pole probes is checked with the poles rotated 0, 90, 180 and 270 degrees.

The substrate thickness for testing and calibration should be the same if their thickness can be sensed by the instrument (that is, if their thickness is less than about four or five standard depths of penetration). Often, it is possible to back up the substrates of standard and test specimens with sufficient thicknesses of the same material (to exceed the critical thickness) and to then make readings independent of substrate thickness.

If the curvature of the coating is so extreme as to preclude calibration on a flat surface, then the curvature of the coated standard (or of the substrate on which the calibration foil is placed) should have the same contour.

Magnetic Thickness Gages

Reference standards are used for the calibration of gages that measure the thickness of nonmagnetic coatings on magnetic materials. In such reference standards, for example, the steel substrates may have the magnetic properties of low carbon steel and the nickel coatings may have the magnetic properties of an annealed watts nickel electrodeposit free of cobalt and iron.

These reference standards are often used to measure the thickness of paint and other organic coatings on steel, as well as galvanized zinc and other nonmagnetic metallic coatings. The number of different thicknesses required for these calibrations depends on the type of gage and the coating thicknesses to be measured.

Magnetic thickness gages may also be used to estimate the magnetic properties of austenitic stainless steel weld metal. Because the magnetic properties of the weld metal are closely related to the ferrite content of the weld, magnetic thickness gages are used to estimate the weld's ferrite content.

Sorting Reference Standards[8]

When sorting with the absolute encircling coil technique, a known acceptable calibration standard and a known unacceptable standard are required. When using the comparative encircling coil technique, usually two known acceptable specimens of the test object and one known unacceptable specimen are required. For a three-way sort, it is best to have three calibration reference standards, including either two for the high and low limits of acceptability for one group or one each for two unacceptable groups. The third reference standard represents the acceptable lot of material.

Calibration and Standardization

Electromagnetic testing is used to sort objects by comparing them to standard specimens. Empirical data and physical tests on samples representing properties to be separated determine the validity of the sorting. The calibration and standardization procedure is based on the properties of the sample requiring separation. The sorting may require more than one test operation.

When using the absolute encircling coil technique, the known acceptable calibration standard is inserted in a fixed position in the coil and the test instrument is adjusted to achieve an on-scale meter reading, an oscilloscope reading or both. The acceptable standard is replaced in the exact position with a known unacceptable standard and the sensitivity of the instrument is adjusted to maximize the indicated difference reading without exceeding 90 percent of the available scale range.

When using the comparative encircling coil technique, a reference standard is selected (usually one that falls within the acceptable limits of the specimens being tested) and placed in the reference coil. This coil and the reference standard are placed in a location where they will not be accidentally disturbed during the sorting operation.

When used with a two-way mix, two calibrated reference standards are chosen: one represents the acceptable group and the other represents the unacceptable group. The acceptable calibration standard is placed at a fixed position in the test coil coinciding with the position of the reference standard in the reference coil. Then the operator balances the instrument. This acceptable calibration standard is replaced with one representing the unacceptable group and the test instrument's phase, sensitivity and coil current are adjusted to maximize the indicator reading without exceeding 90 percent of the available scale range. The acceptable standard is reinserted and the instrument controls are alternately adjusted to retain a null value for the acceptable standard and maximum indication for the unacceptable standard.

A typical case of using reference standards for the high and low limits of acceptability is in the measurement of maximum and minimum acceptable hardness. In this instance, the reference standard representing the acceptable lot is placed in the test coil and the instrument is adjusted for a null or zero reading. The controls are then adjusted to maximize the indications without exceeding ±90 percent of the available scale range from the null for each of the maximum and minimum reference standards. Alternate readjustment of the controls may be necessary to retain the null reading, as well as the maximum and minimum limits for acceptance.

For a three-way sort, when three dissimilar grades of material become mixed, the third acceptable reference standard is placed in the test coil and the instrument is nulled. The two reference standards representing the other two grades are successively inserted into the test coil and the instrument's controls are adjusted to maximize the indications without exceeding 90 percent of the available scale range from the null for each of the other two reference standards. Alternate readjustment of the controls may be necessary to retain the null reading as well as the indication for the other two reference standards.

The procedure with probe coils is similar to that used with encircling coils. Instead of placing reference standards in the coils, however, the probe is positioned in a consistent, suitable location on the reference standard.

PART 3. Functions of Reference Standards

One purpose of nondestructive testing is to ensure that all test objects containing critical discontinuities are rejected. Critical discontinuities are those that cause unsatisfactory performance. More importantly, all parts with anomalies that are unsafe and could cause bodily injury or death should be segregated from the acceptable parts.

Simulation of Acceptable Parts

The favored technique for ensuring that electromagnetic tests will reject anomalous parts is to set up and calibrate the test equipment by using reference standards. Reference standards simulate the parts to be inspected except they contain known discontinuities. The physical, electrical and magnetic characteristics of a reference standard must represent (1) what is expected of the population of parts and (2) what the test process is sensitive to. The reference standard should be a stable device with stable characteristics from which the performance of the electromagnetic test can be established and evaluated.

Often, a single reference standard can be used to both verify the performance of the electromagnetic test equipment (establish test variables such as frequency, gain, balance and gate threshold), as well as to check the thoroughness of the test coverage (accuracy of the threshold, stability of the electronic equipment, stability of holding fixtures for part and probe and other details). Therefore, the same reference standard can be used both to set up and to calibrate test equipment.

Calibration reference standards are used to verify the accuracy of an electromagnetic test before a group of tested and accepted parts are released.

If the total surface of each part is to be tested, then the reference standard must duplicate the total part. If only a portion of a part is to be tested, then the reference standard need only duplicate that portion of the part, providing a reliable means to properly locate the tested portion in a holding fixture. If the dimensions of the parts have tolerances, then the reference standard should be of average size. If there is a possibility that an out-of-tolerance surface may mask a discontinuity, then a preliminary test should ensure that those out-of-tolerance parts are removed before testing or the holding fixture may be made to reject the out-of-tolerance parts. If the surface finish of a part influences the test results, then the reference standard should have an average surface finish of the specified tolerance. If the conductivity of a metal varies, the reference standard should have an average value of electrical conductivity. If the value of magnetic permeability varies, then the reference standard should have an average value of magnetic permeability.

Notice that specific metal alloys may not need to be duplicated; only some of their physical, electrical and magnetic characteristics need to be. In some applications, there is no need for acceptance reference standards. The integrity of the electromagnetic test equipment is depended on to accept good parts. In these cases, the purpose of the reference standard is to establish and maintain sensitivity to discontinuities. Therefore, only rejection reference standards for both setup and calibration functions are needed.

It has been observed that contractors who use nondestructive testing equipment often have different criteria for establishing rejection thresholds. Some contractors set the threshold so that the reference standards are just barely rejected (the discontinuity signal from the reference standard just penetrates the threshold). Other contractors set the threshold so that the discontinuity signal from the reference standard is twice as high as the threshold. Still others set the discontinuity signal to go 10 percent, 25 percent and even 50 percent beyond the threshold. These differing criteria are another reason why acceptance reference standards are not used.

Simulation of Discontinuities

There are many varieties of natural discontinuities, even of a particular type such as cracks. No two are identical and each will therefore produce a different eddy current signature. If a reference standard is to be used to represent actual discontinuities, then extreme care must be taken to ensure that, however a discontinuity is simulated, it must represent the entire range of discontinuities to be encountered. An extreme example of this is the manufacture of steel bars containing cracks, inclusions, internal bursts, laps, scratches, seams and tears. Manufacture of reference standards representing these types in all their natural variations would be prohibitive.

This dilemma can be addressed during technique development by studying many natural discontinuities, by identifying discontinuity mechanisms and by developing procedures to detect the discontinuities of interest. A reference standard can then be developed that uses artificial discontinuities, not merely to represent rejectable objects, but rather to verify that the equipment is oriented properly and that the test setup adequately reproduces the parameters established in the laboratory to find discontinuities of interest.

If some form of cracking is the discontinuity of interest, then the next consideration is the dimensions of a rejectable crack. Fracture mechanics can provide these limits in the form of critical crack size.[9]

The most significant parameter in critical crack size is crack depth. Assuming the length of a crack will be at least ten times the depth (a very conservative assumption), the critical crack depth is the minimum value, under given load and environmental conditions, where catastrophic failure (brittle fracture) can occur. Once the critical crack depth is known, a safety factor is added to determine a smaller crack depth that is acceptable. Product safety is achieved by rejecting the deepest acceptable crack depth.

Natural Cracks

Natural cracks display a great deal of variation that can be visible to various eddy current instruments. Natural cracks are not flat or parallel, their ends may or may not be tapered and they may contain corrosion products or other foreign material. All these variables can affect the signal produced and make interpretation difficult.

Some natural cracks can be grown in reference standards. Artificial fatigue cracks have been available in primary reference standards from the National Institute of Standards and Technology.[10]

When detected by electromagnetic testing, a grown fatigue crack normally will not respond the same way as a quench crack. The reason is conductivity. A tension fatigue crack has shiny faces that rub together as the crack grows; when at rest, these faces often touch one another. Thus, fatigue cracks are usually more conductive than oxide coated or carbon coated quench cracks. A fatigue crack is an inefficient simulation for a quench crack.

Machined Cracks

It is also possible to machine very narrow slots to simulate cracks. Electromagnetic test equipment is sensitive to the depth, length and width of cracks. In eddy current tests, the size of a crack is revealed by the magnitude of the eddy current signal. If the detection equipment uses an analog meter to reveal the presence of discontinuities, then simulated cracks may be used (typical meters respond only to the magnitude of a signal and thus the size of a crack). These detection meters ignore phase angles related to crack depth.

Electric discharge machined slots can be cut with faces 0.15 mm (0.006 in.) apart. If possible, the phase angle of separation between natural cracks and simulated cracks should be oriented on a cathode ray tube so that the points of intersection of simulated and typical natural cracks are equidistant from a line that is normal to the threshold and that passes through the origin of both signals. With this arrangement, minimum detection error occurs. Figure 5 illustrates the proper angular arrangement of signals from the reference standard (with a

FIGURE 5. Angle of display for horizontal alarm threshold to minimize rejection error: (a) improper angle; (b) proper angle.

(a)

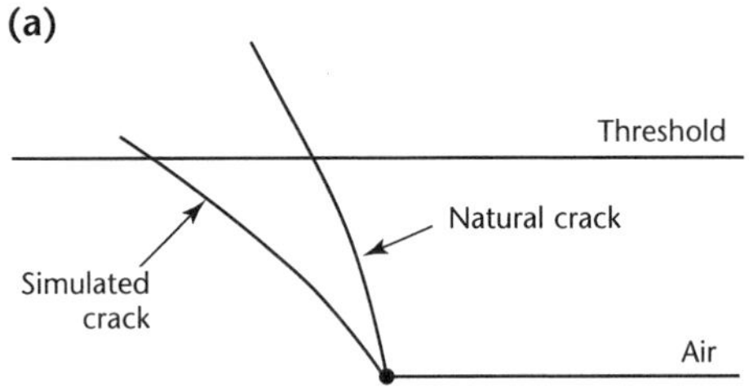

(b)

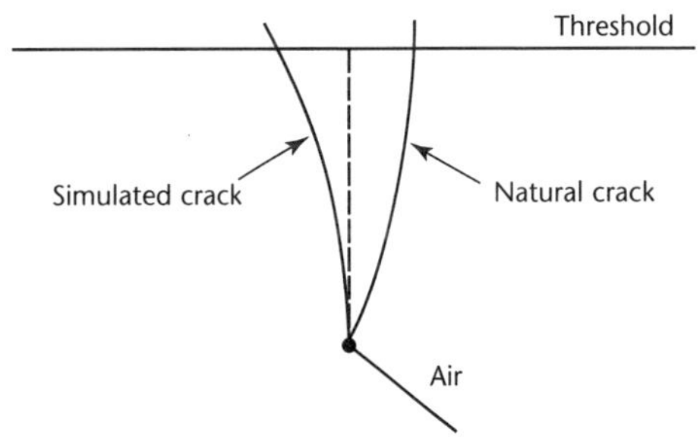

simulated crack) and an anomalous part (with a natural crack).

Simulated cracks should be placed in predetermined locations where they can cause the greatest weakness or where there is a high probability of crack occurrence. Orientations most difficult to detect should be included to ensure that natural cracks in those orientations are not overlooked.

Several techniques have been developed for fabricating electromagnetic reference standards. The advantages and limitations of these fabrication techniques may vary with the application but excellent dimensional tolerances and repeatabilities are possible.

Part 4. Techniques of Reference Standard Fabrication

Reference standards are made to represent the conditions evaluated in the test part. Variables to be considered and controlled during reference standard fabrication include geometry, conductivity, permeability, surface finish and coating. How best to simulate a discontinuity, inclusion or crack has been a topic for discussion and debate since the standardization of reference test objects in the 1950s. The ideal case might be to locate an actual part that contains a natural discontinuity of known size and then to use it as a reference standard. Reference standards have been obtained this way but it is difficult to find several identical parts containing identical discontinuities for use as natural and known reference standards.

Drilled Holes

One of the earliest techniques for making electromagnetic reference standards was the drilling of holes. The reference standard was fabricated from material of the same conductivity and permeability as that of the object to be tested. The reference standard's geometry matched that of the component and small drilled holes were located at sites that would simulate natural discontinuities. These drilled holes were typically much smaller than the probe coil and the depth of the hole was much greater than the penetration depth of the eddy currents. In this way, equipment response to the reference standard was controlled by the drilled hole diameter alone.

The following example of a drilled hole reference standard illustrates this reference standard type, along with its advantages and limitations. The test incorporated the following specifications for fatigue crack detection. Holes of 0.8 mm (0.03 in.) diameter were drilled into aircraft aluminum alloy having a conductivity of about 32 percent of the International Annealed Copper Standard. Test frequency was about 200 kHz.

The test was conducted with a metered probe having a single coil with a diameter of about 2.5 mm (0.1 in.). The response from the hole was a minimum of 4 percent of full scale deflection. Background noise (due primarily to test surface roughness) was typically 2 percent of full scale. Eddy current indications found equal to or exceeding the reference standard hole response were further evaluated as probable fatigue cracks. This test system resulted in the detection of confirmed cracks with measured surface lengths as small as 0.33 mm (0.013 in.).

Advantages of Drilled Hole Reference Standards

The primary advantage of the drilled hole reference standard is its simplicity. Such a reference standard can be produced and reproduced at minimal cost without expensive tools or machinery. Uniformity of response from one reference standard to the next is quite high, typically within a few percent. When used in the manner described above, this reference standard provides an indication of signal sensitivity, background noise level and, from these, signal-to-noise ratio. In summary, this reference standard is economical to produce with repeatable results for a sensitive discontinuity detection setup.

Disadvantages of Drilled Hole Reference Standards

The main drawback of a drilled hole is that it does not always behave like a crack. This difference is present in the test response produced by the reference standard. Although the impedance plane response for the reference standard is similar to that of a small crack, that similarity is lost when larger holes are used. Large drilled holes (compared to the probe size) result in modified and larger reference standard responses. These responses resemble the impedance plane response of a specimen edge (edge effect) and not the response of a crack. Effective use of the drilled hole reference standard is restricted to a narrow range of drill sizes that are closely related to the size of the probe coil being used.

Accuracy and Repeatability

The accuracy of the drilled hole reference standard is relative and not absolute. Crack length can be estimated or predicted from a reference standard hole size only by correlations established with experience. New correlations are required when significant changes occur in test condition or component geometry. Test system repeatability with the drilled hole reference standard is excellent when test conditions and component geometry are held constant. Care should be taken during drilling on thin materials to avoid distortions of the test object and the hole.

Notches

Notches are the first choice for producing simulated crack reference standards. The shape of the notch can generally be controlled during fabrication and the shape of the expected crack can be accurately simulated in the reference standard. Several techniques for producing simulated crack reference standards are discussed below.

Electric Discharge Machining

Electric discharge machining (EDM) is a metal removal technique suitable for a wide range of materials. The technique uses a controlled electrical discharge between the reference standard material and a preshaped electrode. The electrode shape determines the shape of the notch, a simulated crack. Electrodes are generally cut from thin foil, graphite, brass or copper tungsten and notch widths of only a few hundredths of a millimeter (a few thousandths of an inch) are commonly produced. Figure 6 shows an apparatus used to position and move the electrode relative to the reference standard.

One caution should be noted when using electric discharge machining on alloys containing significant amounts of the elements iron, nickel and cobalt. The permeability of such alloys (including the nonmagnetic stainless steels and some of the high temperature nickel base alloys) can be modified by the electric discharge machining process. When high power, high speed electric discharge machine cutting is used, a layer of recast material is produced at the base and on the walls of the notch and this recast material can have greatly altered magnetic permeability. For small notches intended to simulate small cracks, the eddy current response can be dominated or greatly affected by the magnetic recast material. Slower cutting speeds and fluid flow during electric discharge machining will reduce or eliminate this problem. Electrode feed rates less than 0.013 mm (0.0005 in.) per minute are common.

The primary advantage of electric discharge machining is accuracy. It is possible to have electrodes with widths as small as 0.05 mm (0.002 in.) and to produce slots with very short surface lengths, as small as 0.13 mm (0.005 in.). Errors between the desired dimensions of a notch and the resulting dimensions after the electric discharge machining process are often less than 0.013 mm (0.0005 in.). Because of these small dimensions, electric discharge machining can produce simulated discontinuities whose electromagnetic test indications closely resemble those of actual cracks. By designing the electrode according to the shape of the discontinuity, many different widths and length-to-depth ratios are possible. The electrode may also be placed inside a fastener hole, on a radius or on a flat surface. Figure 7 shows cross sections of finished electric discharge machined notches in carbon steel (Unified Numbering System G10260).

One disadvantage of electric discharge machining is expense. It may cost hundreds of dollars per slot to have accurate reference standards made.

In summary, the accuracy of electric discharge machining fabrication can be very good but care must be taken so that recast material does not accumulate in the notch. The operator also must carefully

FIGURE 6. Feed mechanism for electric discharge machining of reference standards: (a) outside surface cutting; (b) inside surface cutting.

(a)

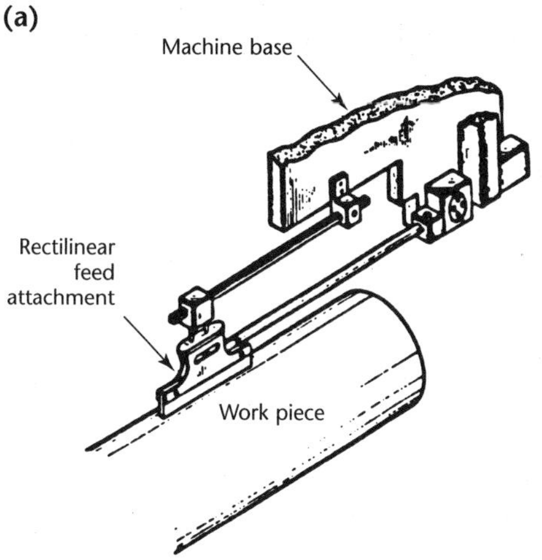

(b)

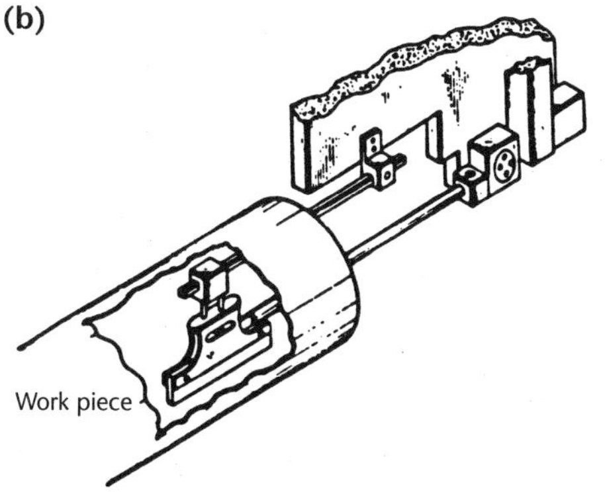

choose the electrode material, voltage, current and the right cutting speed for compatibility with the reference standard material.

Planing Fabrication

The planing technique for fabricating simulated crack reference standards is also a metal removal technique. A very small tool is used to precisely gouge the surface, producing a very narrow slot. Typically, a tool maker will hand grind a small tip on a rod of tool steel about 0.2 to 0.25 mm (0.008 to 0.01 in.) wide. The tool is placed in the chuck of a planer and aligned with the reference standard. After one pass, the tool is indexed deeper into the material and another pass is made. Care must be taken not to move the tool too quickly because heat or chatter will develop. Heat may alter the conductivity of the reference standard and chatter will cause surface roughness on the walls of the slot. Slow cutting speeds and adequate fluid flow during the machining process will reduce this problem.

A derivation of the planing technique is a manually filed notch mentioned in ASTM E 243.[11] The notch is made using a 6.4 mm (0.25 in.) diameter, number four cut, round file.[12] The reference standard material is seamless copper and copper alloy tubing. The required tolerance of the notch depth is ±0.013 mm (±0.0005 in.).

FIGURE 7. Notch replicas (magnified 50×) representing the stabilizing effect of rectilinear feed guides on notch profiles when machine is subject to lateral vibration: (a) notch made with radial feed guide; (b) notch made with rectilinear feed.

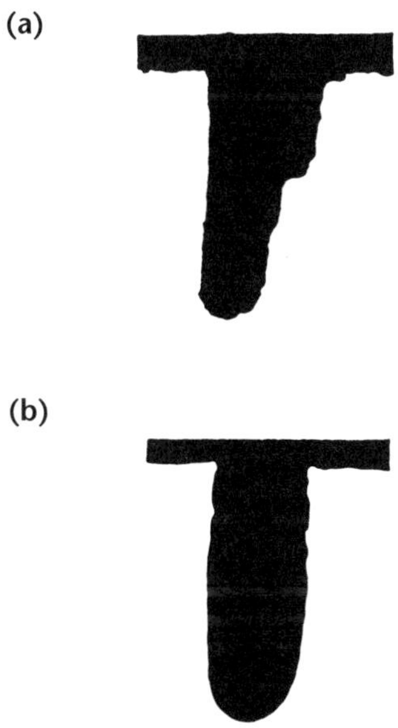

The primary advantage of the planing technique is economy. Slots can be machined in a short time with conventional shop equipment.

The major disadvantage of the process is its inability to produce narrow slots. It is very difficult to make a strong, sharp cutting tool less than 0.25 mm (0.01 in.) wide. In addition, because of the tool's size and the large forces acting on it as it travels across the reference standard's metal surface, the tool often breaks.

The accuracy and repeatability of this process is good when the same operator is using the same machine and the same tool. Because the tools are usually ground by hand, no two will be alike.

Jet Abrasives

Another metal removal technique for fabricating simulated crack reference standards is the jet abrasive technique. A high pressure stream of fluid containing abrasive compounds is forced through a small opening, hitting the metal surface with enough force to cause erosion. Slot width is controlled by the orifice size and slot depth is controlled by the number of passes over the material.

The major advantage of the jet abrasive process is its repeatability. After the equipment settings are known for a given type of reference material and slot geometry, they can be accurately reset at another time. There are no moving parts or ablative surfaces (as in the electric discharge machining or planing processes) that contribute to inaccuracies and nonrepeatability.

A disadvantage is the lack of control over slot geometry, due to the spreading nature of the fluid. Slot widths tend to be wide because of this spreading. Another disadvantage is that, after the process is complete, the surface of the slot has a texture related to the size of the abrasive compound. This pitting may have an effect on eddy currents at high frequencies.

Natural Cracks

The cost of obtaining or producing natural cracks for use as electromagnetic reference standards continues to be prohibitive. It must be noted, however, that the electromagnetic response from an electric discharge machined notch or other simulated discontinuity is not the same as that from a natural crack. Care should be taken when selecting reference standards and test systems to ensure that natural crack sensitivity is maintained.

The obvious advantage of the natural crack is that it most closely resembles an actual discontinuity. The gap between faces of the crack will be very small.

The main disadvantage in the natural crack technique is the expense of acquiring specimens. If the cracks are grown from fatigue specimens, the cost could be very high. If the cracks are from an inservice part, the availability of the specimens will be a limitation. Another problem is that of size. There is no control over the size of actual inservice specimens. After the reference standard is obtained, some other form of nondestructive testing will be needed to measure the discontinuity's size.

The accuracy of a natural crack reference standard will always be in question because its exact size is not directly known until it is broken open. Repeatability can be controlled somewhat by closely monitoring the growing of the crack and then machining away portions to make it smaller.

Transverse Notches

The guidelines for producing simulated crack reference standards with longitudinal notches apply equally well to transverse notches. However, changes in geometry must be compensated for by adapting the electrode in electric discharge machining or by developing sophisticated machining motion for planer techniques. It may be less difficult to fabricate notches that travel along the direction of the tube or radius rather than against it. For that reason, it could be more advantageous to fabricate transverse notches using the electric discharge machining or jet abrasive techniques. In general, the problems inherent in the fabrication of simulated crack reference standards are similar to those of longitudinal notches.

References

1. Bailey, D. and P. McEleney. Section 10, "Reference Standards for Electromagnetic Testing." *Nondestructive Testing Handbook,* second edition: Vol. 4, *Electromagnetic Testing*. Columbus, OH: American Society for Nondestructive Testing (1986): p 251-264.
2. *ASME Boiler and Pressure Vessel Code:* Section V, *Nondestructive Examination.* New York, NY: American Society of Mechanical Engineers (2001).
3. ASTM E 215, *Standard Practice for Standardizing Equipment for Electromagnetic Examination of Seamless Aluminum-Alloy Tube.* West Conshohocken, PA: ASTM International (1998).
4. ASTM E 1004, *Standard Practice for Determining Electrical Conductivity Using the Electromagnetic (Eddy-Current) Method.* West Conshohocken, PA: ASTM International (2002).
5. *Standard Reference Materials Online Catalog*. Gaithersburg, MD: National Institute of Standards and Technology (2003).
6. ASTM B 193, *Standard Test Method for Resistivity of Electrical Conductor Materials.* West Conshohocken, PA: ASTM International (2002).
7. ASTM E 376, *Standard Practice for Measuring Coating Thickness by Magnetic-Field or Eddy-Current (Electromagnetic) Test Methods.* West Conshohocken, PA: ASTM International (1996).
8. ASTM E 703, *Standard Practice for Electromagnetic (Eddy-Current) Sorting of Nonferrous Metals.* West Conshohocken, PA: ASTM International (1998).
9. Liebowitz, H. *Fracture, An Advanced Treatise*. Vol. 5. New York, NY: Academic Press (1968).
10. Document 303.2, *Artificial Flaw for Eddy Current NDE.* Reference Material 8458. Gaithersburg, MD: National Institute of Standards and Technology (1991).
11. ASTM E 243, *Standard Practice for Electromagnetic (Eddy-Current) Examination of Copper and Copper-Alloy Tubes.* West Conshohocken, PA: ASTM International (1997).
12. BS 498, *Specification for Rasps and Engineers' Files.* London, United Kingdom: British Standards Institution (1990).

12

CHAPTER

Microwave Testing

Pradeep Ramuhalli, Michigan State University, East Lansing, Michigan

Nasser Qaddoumi, American University of Sharjah, Sharjah, United Arab Emirates (Part 3)

Ronald J. Botsko, Huntington Beach, California (Part 2)

Part 1. Introduction to Microwave Testing

Background

Microwave testing is accomplished by transmitting electromagnetic waves at a relatively high frequency toward a structure and using some of the transmitted or reflected wave properties to obtain information about the composition and the integrity of the structure. Microwave radiators have been used for the nondestructive testing of various materials since the technique's introduction in the 1960s.[1-6]

The advent of new materials and structures calls for innovation to meet the increasing demands of technology. The increased use of composite materials for industrial, medical and military applications challenges many nondestructive test methods. Difficulties arise from the inherent anisotropy and physical property heterogeneities of these materials, as well as the relatively high absorption and scattering of the radiated signals. The ability of microwaves to penetrate deeply inside dielectric materials suits them for interrogating composite materials.[5-29] One exception is the interior of advanced carbon fiber composites whose fibers are electrically conductive.

Also, microwave signals reflect completely from metallic objects, so surface features and discontinuities of metallic surfaces (including rust, cracks and surface blemishes) may be detected and evaluated.[6,30-40]

Principles of Operation

The interaction of the microwave signal with the material system is best described in terms of the interaction of waves with the material system. When a wave impinges on a material system, the following takes place.

1. Part of the incident signal reflects back. This reflection is influenced by the wave impedance of the material, its geometry, its size and the parameters of its constituents.
2. Part of the signal travels into the material system. As a signal travels into a material, the signal maintains the same frequency but the wave velocity and wavelength change. The variation of the velocity and wavelength suggests that a certain phase change must take place. Additionally, the material absorbs some of the energy in the signal, which causes the amplitude to change.

In microwave testing, the magnitude or phase of a transmitted microwave signal is used to create a two-dimensional image of an object, to perform line scans or to obtain a single-point measurement.

Microwave testing measures properties that reveal information about the test material's parameters and physical dimensions. Each material is characterized by a unique, complex dielectric constant ε_r relative to free space or a vacuum. The real part (permittivity) of the relative complex dielectric constant is a measure (farad per meter) of the material's ability to store the incident electric energy whereas the imaginary part (loss factor) indicates the material's ability to absorb the incident electric energy. In microwave testing, a subsurface inclusion is announced by a change in the complex dielectric constant for the inclusion's location. Microwave testing shows this change, either with or without contact between probe and test object.

Microwaves are electromagnetic signals with frequencies between 300 MHz and 300 GHz, corresponding to vacuum wavelengths between 1 m (3.3 ft) and 1 mm (0.04 in.), respectively. Because of the short wavelengths, circuit theory generally cannot be used directly to describe the interaction between the microwave signals and the material system interrogated by these signals. Radiation from the circuit affects the analysis because, when the dimensions of circuit elements become comparable to the wavelength, the voltage and current relation used to represent the properties of a load (that is, impedance) changes to a relation between the electric and magnetic fields.

Microwave testing can be performed in either the near field or the far field. The near field approach uses simple probes such as open ended waveguides and coaxial lines whereas the far field approach requires an antenna for focusing the microwave energy. Furthermore, far field testing does not offer good spatial resolution because the footprint is relatively large. Focusing lenses are often used to remedy this problem.[41] Measurements are conducted using either phase or amplitude information. Far field techniques generally use amplitude information. Near field testing is more versatile because phase information is easily produced and often contains more than amplitude information. Additionally, in near field testing, the footprint of a sensor is close to its physical aperture (for example, the waveguide opening).

Limitations and Advantages

Limitations of microwave testing include the following.

1. Microwave testing is ill suited for detecting volumetric or subsurface discontinuities in metals and other electrically conductive materials.
2. The laboratory equipment needed to develop a technique for a particular application may be expensive.
3. A power source is needed.
4. Technicians who have been trained to conduct microwave tests are fewer than for other nondestructive test methods.

Generally, microwave testing has the following advantages.

1. Microwave testing is fast.
2. In some techniques, the sensor does not require contact with the test surface.
3. Discontinuities may be discovered through intervening dielectric layers.
4. Microwave testing is applicable to the surface of nonferromagnetic as well as ferromagnetic metals or alloys and coarse grained materials.
5. Microwave testing may be applied to curved and other complicated surfaces.
6. Discontinuity orientation can be determined.
7. No special operator skill in microwaves or signal interpretation is needed for discontinuity detection.
8. Little or no surface preparation is required.
9. Microwave testing is environmentally compliant and operator friendly.
10. The required microwave power is in a safe, low milliwatt range.
11. A microwave test system may be battery operated and portable.
12. The results are obtained in real time.

Closing

Microwave testing has been used for nondestructive testing since the 1960s, most often in a research setting. In the twenty-first century, industrial applications of microwave testing are becoming more widespread.

Part 2. Theory of Microwave Testing

Microwave propagation, like other types of electromagnetic radiation, depends on the interaction between time varying electric and magnetic fields. Such fields oscillate in waves that are called *traveling waves* because energy is transported from one position to another. The velocity of propagation of the microwave depends largely on the medium through which the wave propagates. For instance, in free space or vacuum, the propagation velocity of the wave is the speed of light, $c = 2.998 \times 10^8$ m·s^{-1} (6.706×10^8 mi·h^{-1}).

Basic Differential Equations for Electromagnetic Fields

The basic theory of microwaves can be understood with Maxwell's equations. Maxwell's equations are given in differential form for time varying fields:[42]

$$(1) \quad \nabla \times \boldsymbol{E} = -\frac{\partial \boldsymbol{B}}{\partial t}$$

$$(2) \quad \nabla \times \boldsymbol{H} = \boldsymbol{J} + \frac{\partial \boldsymbol{D}}{\partial t}$$

$$(3) \quad \nabla \bullet \boldsymbol{B} = 0$$

$$(4) \quad \nabla \bullet \boldsymbol{D} = \rho$$

where $\boldsymbol{B}$ is magnetic flux density (tesla), $\boldsymbol{D}$ is electric flux density (coulomb per square meter), $\boldsymbol{E}$ is electric field intensity (volt per meter), $\boldsymbol{H}$ is magnetic field intensity (ampere per meter), $\boldsymbol{J}$ is current density (ampere per square meter), t is time (second), ρ is the volume charge density (coulomb per cubic meter) and the bold letters indicate vector quantities. These equations contain the equation of continuity:

$$(5) \quad \nabla \bullet \boldsymbol{J} = -\frac{\partial \rho}{\partial t}$$

In addition to these five equations, the constitutive relations given below are necessary for solving Maxwell's equations:

$$(6) \quad \boldsymbol{D} = \varepsilon \boldsymbol{E}$$

$$(7) \quad \boldsymbol{B} = \mu \boldsymbol{H}$$

$$(8) \quad \boldsymbol{J} = \sigma \boldsymbol{E}$$

Here, σ is the conductivity (siemens per meter) of the medium, permittivity $\varepsilon = \varepsilon_0 \varepsilon_r$ and permeability $\mu = \mu_0 \mu_r$, where ε_r is the relative dielectric constant, μ_r is relative permeability, μ_0 is the permeability (henry per meter) of vacuum or free space and ε_0 is the permittivity of vacuum or free space (farad per meter) and is the absolute value of the dielectric constant:

$$(9) \quad \varepsilon_0 = 8.854 \times 10^{-12} \cong \frac{1}{36\,\pi} \times 10^{-9} \text{ F·m}^{-1}$$

$$(10) \quad \mu_0 = 4\pi \times 10^{-7} \text{ H·m}^{-1}$$

For nonferromagnetic materials, $\mu_r \cong 1$, meaning that the permeability of the material can bc approximated as the permeability of a vacuum.

The wave equations governing the behavior of the electric and magnetic fields are obtained by solving Maxwell's equations:[42,43]

$$(11) \quad \nabla^2 \boldsymbol{E} = \mu\varepsilon \frac{\partial^2 \boldsymbol{E}}{\partial t^2}$$

$$(12) \quad \nabla^2 \boldsymbol{H} = \mu\varepsilon \frac{\partial^2 \boldsymbol{H}}{\partial t^2}$$

Plane waves are a special category of traveling waves with a planar wave front. Similarly, spherical waves have a spherical wave front. Only the case of plane waves is considered here. For plane waves traveling along the X direction, the wave equation is given by:

$$(13) \quad \frac{\partial^2 \boldsymbol{E}}{\partial x^2} = \mu\varepsilon \frac{\partial^2 \boldsymbol{E}}{\partial t^2}$$

Note that, in general, the vector $\boldsymbol{E}$ can have three components — E_x, E_y and E_z — along the X, Y and Z directions. It is easy to show that, for plane waves traveling in the X direction, $E_x = 0$. Thus, plane waves have electric field and magnetic field components that are not only transverse to each other (as in all electromagnetic waves) but also transverse to the direction of propagation. Such electromagnetic waves are called *transverse electric magnetic* waves, a term more specific than *transverse electromagnetic waves*. In addition, the phase of a plane wave is constant across the wave front. If the magnitude of the wave is held constant, the special case of a uniform plane wave is obtained.[42,44]

Another important aspect of plane electromagnetic waves is that in a vacuum the electric and magnetic fields are in temporal and spatial phase with each other. In other words, the intensities of both the electric field and the magnetic field change or oscillate at the same rate and reach a maximum value at the same time (and at the same point along the beam axis) for a traveling wave in a lossless medium. Because no energy is lost during transmission in a lossless medium, the maximum intensities of the ***E*** and ***H*** fields remain constant as the electromagnetic wave travels along its beam axis.

However, if the medium absorbs electromagnetic energy at the frequency of propagation, the time phase of the electric field is advanced with respect to the phase of the magnetic field if the losses are electrical as in the case of dielectric media. Alternatively, the phase of the electric field is retarded with respect to that of the magnetic field if the energy loss results from magnetic effects, as in the case of conducting metallic test materials or ferromagnetic materials. Sometimes the losses can be of both electrical and magnetic origins.

The electric and magnetic fields are also shifted out of phase with respect to the incident wave during reflection of electromagnetic waves, even if the transmission medium is lossless. The reflection phase shift can thus indicate certain properties of the test material from which waves are reflected. These material interactions are discussed in more detail below.

In most practical cases, plane waves are sinusoidally varying as a function of time. In these cases, the wave equations are reduced:[42,43]

$$\nabla^2 \boldsymbol{E} = \gamma^2 \boldsymbol{E} \tag{14}$$

and:

$$\nabla^2 \boldsymbol{H} = \gamma^2 \boldsymbol{H} \tag{15}$$

The intrinsic propagation constant γ can be expressed:

$$\gamma = \alpha + j\beta = \sqrt{j\omega\mu\left(\sigma + j\omega\varepsilon\right)} \tag{16}$$

where $j = \sqrt{(-1)}$, α and β are attenuation and phase shift constants, respectively, and ω is angular frequency (radian per second). The solutions to these equations for traveling waves in a lossless, homogeneous medium are of the form:

$$E_y\left(x,t\right) = C_1 \cos\left(\omega t - \beta x\right) + C_2 \cos\left(\omega t + \beta x\right) \tag{17}$$

Similar expressions can be obtained for the E_z, H_y and H_z components.

The wavelength λ (meter) of such waves is defined as the distance over which the sinusoidal waveform passes through a full cycle of 2π rad. The wave varies sinusoidally as a function of βx:

$$\beta\lambda = 2\pi \tag{18}$$

or:

$$\beta = \frac{2\pi}{\lambda} \tag{19}$$

The wavelength λ is also related to the frequency f (hertz) by means of the relationship:

$$\lambda = \frac{v}{f} \tag{20}$$

where v is the propagation velocity (meter per second).

Electromagnetic Plane Wave Properties

In the following discussion of several properties of electromagnetic waves in general and plane waves in particular, it is assumed that the plane wave is traveling in the X direction. The first property relates ***E*** and ***H*** in a uniform plane wave. When electromagnetic plane waves travel in a vacuum (or free space), then the ratio of electric field intensity to magnetic field intensity is always constant:[42-44]

$$\frac{E_y}{H_z} = -\frac{E_z}{H_y} = \eta_0 = \sqrt{\frac{\mu_0}{\varepsilon_0}} \cong 377\ \Omega \tag{21}$$

This quantity, which has the dimensions of impedance, is called the *intrinsic impedance* of a vacuum. In a more general case, where the wave is traveling through a medium with permittivity and permeability equal to ε and μ respectively, the intrinsic impedance η of the medium is given by:

$$(22)\quad \eta = \frac{E_x}{H_y} = \sqrt{\frac{\mu}{\varepsilon}} = \sqrt{\frac{\mu_r \mu_0}{\varepsilon_r \varepsilon_0}} \cong 377\sqrt{\frac{\mu_r}{\varepsilon_r}}$$

Because $\mu_r \cong 1$ for most nonferromagnetic materials, Eq. 22 reduces to:

$$(23)\quad \eta \cong \frac{377}{\sqrt{\varepsilon_r}}\ \Omega$$

Further, ***E*** and ***H*** in a traveling plane wave are at right angles to each other and the electric field vector crossed by the magnetic field vector gives the direction in which the wave travels.

Dual-Frequency Modulation of Beat Frequency[42]

Modulation is an important means by which information can be transmitted. Various modulation techniques are used in nondestructive testing. For instance, frequency modulation, where the frequency of a microwave is changed according to the amplitude of a second wave, is frequently used in microwave ranging. When electromagnetic plane waves contain two oscillation frequencies f_1 and f_2, each frequency is characterized by its own angular frequency, ω_1 or ω_2, and its own phase shift, β_1 or β_2:

$$(24)\quad \beta_1 = \frac{2\pi}{\lambda_1}$$

$$(25)\quad \beta_2 = \frac{2\pi}{\lambda_2}$$

For the specific case in which these two frequencies differ by equal but opposite increments from a central frequency f_0, the corresponding angular frequencies can be written as:

$$(26)\quad \omega_1 = \omega_0 + \Delta\omega$$

with a corresponding phase shift:

$$(27)\quad \beta_1 = \beta_0 + \Delta\beta$$

and:

$$(28)\quad \omega_2 = \omega_0 - \Delta\omega$$

with a corresponding phase shift:

$$(29)\quad \beta_2 = \beta_0 - \Delta\beta$$

For the special case in which the amplitudes are equal for the waves of these two frequencies, the total electric field ***E*** for the combined waves can be given by:

$$(30)\quad E_y = 2E_0 \sin(\omega_0 t - \omega_0 x) \times \sin(\Delta\omega t - \Delta\beta x)$$

Because two sinusoidal terms appear in Eq. 30, the dual frequency wave will exhibit beats, or a cyclic variation in amplitude of oscillations, as sketched in Fig. 1.[1,45] The beat frequency is equal to the difference of the two component frequencies ($f_1 - f_2$) and is the frequency of modulation of the envelope enclosing the sine waves. The actual frequency of the individual sinusoidal waves within the modulation envelope is the average of the two different frequencies of excitation:

$$(31)\quad f_0 = \frac{f_1 + f_2}{2}$$

The effect is comparable to two sideband frequencies caused by the modulation of a central carrier frequency f_0 by a modulation frequency Δf, the carrier frequency then being suppressed from the signal.

Phase Velocity

Figure 2 illustrates the propagation of electric waves in accordance with Eq. 17. A constant phase point P moves along the X axis of propagation with the sinusoidal waves of electric field ***E***. The velocity of a constant phase point is the phase velocity *v* of the traveling wave. The phase point

FIGURE 1. Modulation and beat frequency.[1,5,45]

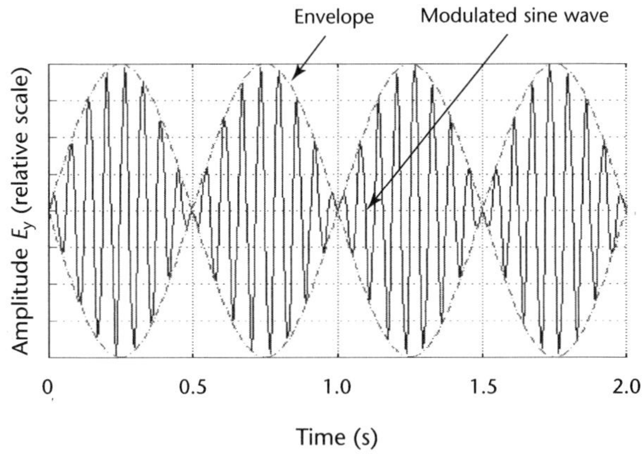

remains constant when the term $(\omega t - \beta x)$ of Eq. 17 has a constant value. The phase velocity is given (in meter per second) by:

$$(32)\quad v = \frac{dx}{dt} = \frac{\omega}{\beta} = \lambda f$$

The phase velocity of electromagnetic waves in a vacuum is equal to the velocity of light c. For media other than vacuum, the phase velocity p relative to the velocity of light is given by:

$$(33)\quad p = \frac{\sqrt{\mu_0 \varepsilon_0}}{\sqrt{\mu \varepsilon}} = \frac{1}{\sqrt{\mu_r \varepsilon_r}}$$

Group Velocity

In Eq. 30, the argument of the second sine term can be made equal to a constant:

$$(34)\quad \Delta\omega t - \beta_0 x = \text{Constant}$$

By differentiating to obtain $dx \cdot (dt)^{-1}$, an expression is obtained for the velocity u (meter per second) of the wave envelope:

$$(35)\quad u = \frac{dx}{dt} = \frac{\Delta\omega}{\Delta\beta} = \Delta f \, \Delta\lambda$$

This velocity, with which the envelope containing the individual waves moves forward along the X axis of microwave beam propagation, is called the *group velocity.* The group velocity is the phase velocity of the wave envelope. Figure 2

FIGURE 2. Phase and group velocities.

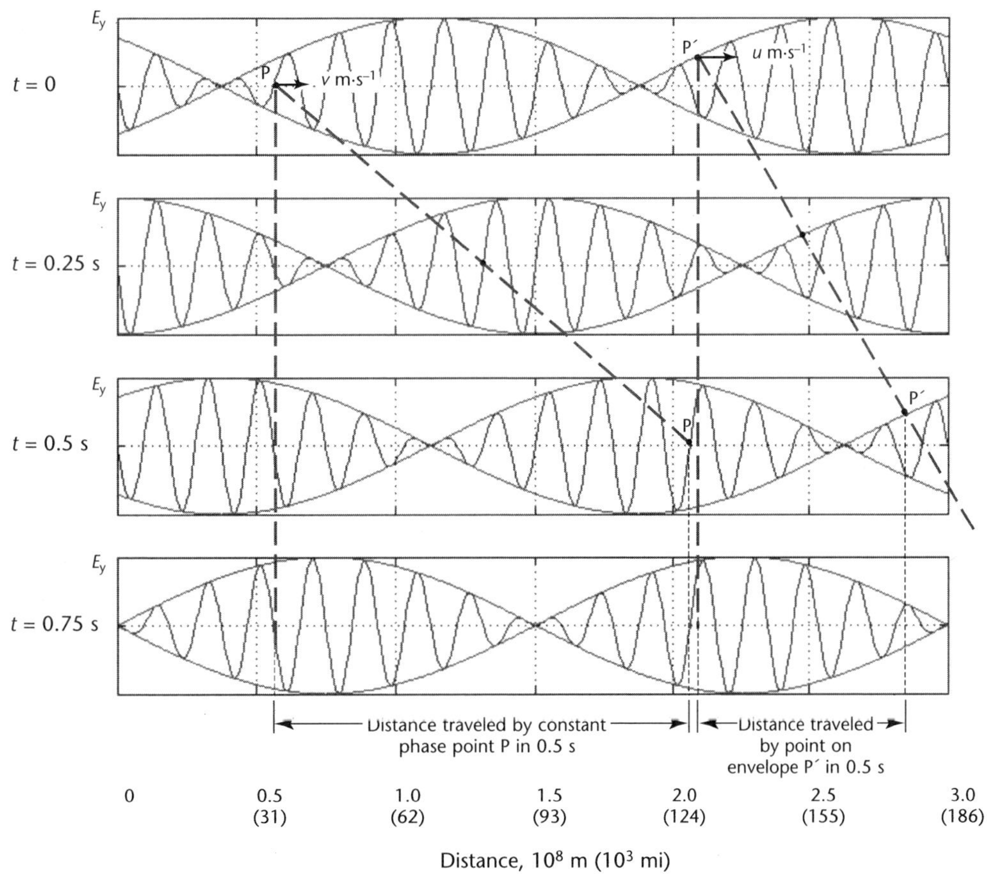

Legend

E_y = electric field intensity in Y direction
t = time (s)
u = velocity of point P
v = velocity of point P´

illustrates the movement of a point P′ on the envelope that moves with the group velocity u.

By comparison, the individual waves within the envelope move forward along the X axis of wave propagation at their own phase velocity v:

$$(36)\quad v = \frac{\omega_0}{\beta_0} = f_0 \lambda_0$$

In Fig. 2, the phase velocity v is indicated by the movement of point P for the case in which the group velocity u is half of the phase velocity v.

Conductors and Dielectrics

Most materials can be classified into either of two major classes: (1) conductors and (2) dielectrics, or insulators. Generally speaking, this distinction is not clear cut, some materials being conductors in one part of the frequency spectrum and dielectrics in another. The dissipation factor is:[42]

$$(37)\quad D = \frac{\sigma}{\omega \varepsilon}$$

The dissipation factor is a commonly used parameter to characterize materials, with good dielectrics having a dissipation factor close to zero. Conductors typically have very high dissipation factors.

A lossless medium is one where the conductivity σ is zero. Under such conditions, the attenuation factor α in Eq. 16 disappears and the wave can propagate without suffering any attenuation or loss. Most materials are not truly lossless although many may be approximated as lossless. A nonzero conductivity results in some attenuation of the electromagnetic wave. At the other end of the scale, materials with very high values of conductivity are called *conductors*. Typical conductors have conductivities of $\sigma \geq 1 \times 10^6$ S·m^{-1}. At the limit of $\sigma = \infty$, the attenuation factor is infinity and the material is considered a perfect electric conductor. A byproduct of the high conductivity (and consequently high attenuation) is the fact that electromagnetic waves cannot travel for great distances inside conductors. In fact, the depth of penetration of an electromagnetic wave in conductors is so small that, for all practical purposes, the wave is considered to be restricted to the surface of the conductor.

Indeed, almost all of the energy is reflected from the surface of a conductor and a perfect conductor reflects all of the energy. This reflection is the opposite of what happens when microwaves impinge on a dielectric. In fact, the energy is attenuated to $1 \cdot e^{-1}$ of the surface energy when the wave reaches a depth δ inside the conductor. This depth is called the *standard depth of penetration* (sometimes called *skin depth*):[42-44]

$$(38)\quad \delta = \frac{1}{\sqrt{\pi f \mu \sigma}} = \frac{1}{\alpha} = \frac{1}{\beta}$$

The standard depth of penetration depends on the frequency f of the incident wave. At a frequency of 30 GHz, the standard depth of penetration for copper ($\sigma = 5.8 \times 10^7$ S·m^{-1}) is 0.38 µm (about 1×10^{-5} in.). The skin effect at microwave frequencies has a severe effect on measurements even in moderately conducting media, so microwaves cannot be used to test inside materials that have a conducting boundary. However, microwaves can be used for detection of surface breaking cracks in metals.

Polarization

The term *polarization* refers to the direction of the electric field and how it changes with respect to time. The most general, two-dimensional case for a transverse electric magnetic wave is elliptical polarization, where the electric field rotates as it progresses along the propagation axis X. The field figuratively traces out an elliptically shaped envelope on a plane perpendicular to the propagation axis. As the wave propagates, the electric field traces out a helix or spiral along the propagation axis.

Figure 3 illustrates elliptical polarization, together with its limiting conditions of linear and circular polarization. As shown in Fig. 3a, a linearly polarized electromagnetic wave has its electric field oriented at all times in the same direction (along the Y axis in Fig. 4).[46]

Cross field waves are waves whose electric field is tilted in the direction of beam propagation. This tilt leads to the most generalized, three-dimensional form

FIGURE 3. Polarization of microwaves: (a) linear polarization; (b) elliptical polarization; (c) circular polarization.[46]

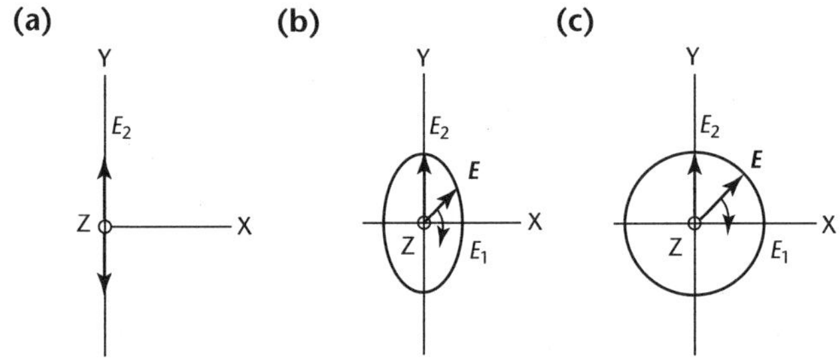

of polarization, which is elliptical cross field polarization. This mode is commonly referred to as *space polarization.* In contrast, an unpolarized electromagnetic wave is one for which the rotating behavior of the electric field is random and has no predictable pattern with respect to time. For example, the radiation from a common light bulb is unpolarized. Polarization can play an important role in microwave nondestructive tests of anisotropic test materials whose properties vary as a function of direction.

Poynting Vector and Energy Flow

It is useful to define the amount of power delivered by a microwave. The poynting vector $\boldsymbol{P}$ gives a measure of the rate of energy flow per unit area at any point:[47]

$$(39)\quad \boldsymbol{P} = \boldsymbol{E} \times \boldsymbol{H}$$

The direction of flow of energy is perpendicular to both $\boldsymbol{E}$ and $\boldsymbol{H}$ and is in the direction given by $\boldsymbol{E} \times \boldsymbol{H}$. This statement is referred to as *Poynting's theorem.*

The total energy density per unit volume (or the stored energy per unit volume) of an electromagnetic wave is equal to the sum of the energy densities of the electrical field and the magnetic field. The *electric field energy density* (stored electric energy per unit volume in the electric field) w_e (joule per cubic meter) is determined by the electrical field intensity $\boldsymbol{E}$ and the dielectric properties of the medium:

$$(40)\quad w_e = \frac{1}{2}\varepsilon|\boldsymbol{E}|^2$$

Similarly, the stored energy density w_m of the magnetic field is given by:

$$(41)\quad w_m = \frac{1}{2}\varepsilon|\boldsymbol{H}|^2$$

The total power applied to a system is related to the energy density of the electric and magnetic fields:

$$(42)\quad -\int_{V_{vol}} \boldsymbol{E}\bullet\boldsymbol{J}dV = \int_{V_{vol}} \sigma|\boldsymbol{E}|^2 dV + \frac{\partial}{\partial t}\int_{V_{vol}}\left(\frac{\mu|\boldsymbol{H}|^2}{2} + \frac{\varepsilon|\boldsymbol{E}|^2}{2}\right)dV + \oint_S \boldsymbol{E}\times\boldsymbol{H}\bullet d\boldsymbol{S}$$

Equation 42 can be analyzed as having four parts, where applied power P is the sum of three rates: $P = R_1 + R_2 + R_3$. Equations 43 to 46 define these parts of Eq. 42.

Equation 43 expresses the total instantaneous power P dissipated in volume V_{vol}:

$$(43)\quad P = -\int_{V_{vol}} \boldsymbol{E}\bullet\boldsymbol{J}dV$$

Equation 44 expresses the rate R_1 at which energy is converted into heat in volume V_{vol}:

$$(44)\quad R_1 = \int_{V_{vol}} \sigma|\boldsymbol{E}|^2 dV$$

Equation 45 expresses the rate R_2 at which the total stored electric and magnetic energy is changing with time:

$$(45)\quad R_2 = \frac{\partial}{\partial t}\int_{V_{vol}}\left(\frac{\mu|\boldsymbol{H}|^2}{2} + \frac{\varepsilon|\boldsymbol{E}|^2}{2}\right)dV$$

Equation 46 expresses the rate R_3 of the flow of energy outward through the surface enclosing the volume V_{vol}:

$$(46)\quad R_3 = \oint_S \boldsymbol{E}\times\boldsymbol{H}\bullet d\boldsymbol{S}$$

FIGURE 4. Traveling plane electromagnetic wave in loss free isotropic medium: (a) instantaneous values of E_y and H_z along X axis at time $t = 0$; (b) same values at later time $t = 0.25$ cycle (in this interval, wave has traveled 0.25 wavelength to right).[46]

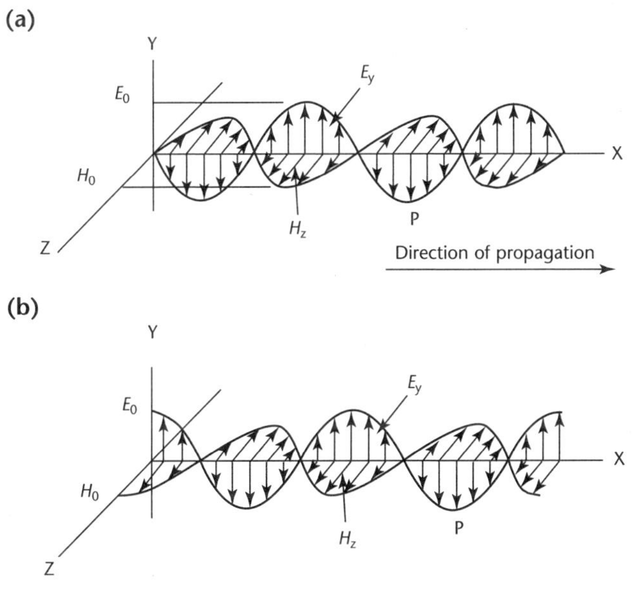

Legend
E = electric field intensity
H = magnetic field intensity
P = reference point

Microwaves at Interfaces between Different Media

The behavior of microwaves at the interface between two media is of considerable interest in nondestructive testing. The properties of microwaves at interfaces determines the inspectability of many materials. In all cases, the electromagnetic field must obey the boundary conditions given below.[42,43]

1. The tangential component of the electric field $\boldsymbol{E}$ is continuous across the boundary.
2. The normal component of the magnetic flux density $\boldsymbol{B}$ is continuous across the boundary.
3. The normal component of the electric flux density $\boldsymbol{D}$ is discontinuous across the boundary by an amount equal to the free surface charge density ρ_S at the boundary.
4. The tangential component of the magnetic field intensity $\boldsymbol{H}$ is discontinuous across the boundary by an amount equal to the free surface current density $\boldsymbol{J}$ at the boundary.

Reflection from Conductors

As already explained, almost all of the incident energy on a conducting surface is reflected. In the present discussion, it is assumed that the conductor is a perfect electric conductor. Cases of normal incidence can be distinguished from cases of oblique incidence. In the case of normal incidence, the incident plane wave is directed along the X axis at a conductor parallel to the $x = 0$ plane (Fig. 5). The electric field intensity $\boldsymbol{E}_i$ of the incident wave will be:

$$\textbf{(47)} \quad \boldsymbol{E}_i = E_i e^{-j\beta x}$$

FIGURE 5. Reflection from conductor.

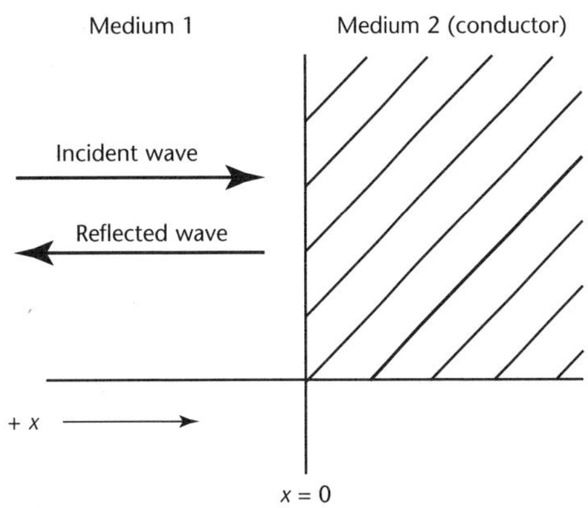

where E_i is amplitude, x is distance (meter) on the X axis and β is the propagation constant defined in Eq. 19. The electric field $\boldsymbol{E}_r$ in the reflected wave is given by:

$$\textbf{(48)} \quad \boldsymbol{E}_r = E_r e^{j\beta x}$$

The boundary conditions indicate that the tangential component of the electric field is continuous across the boundary. Because the field inside a conductor is zero, the tangential component of the electric field just outside the conductor must also be zero. Furthermore, there is no loss of power in a perfect conductor. Thus, the field intensities in the reflected wave will be the same as those in the incident wave, with the direction of power flow being the only difference. All of the energy must be reflected:

$$\textbf{(49)} \quad E_r = -E_i$$

If the incident wave is normally directed at the interface, the reflected wave, which is of the same frequency as the incident wave, will interfere with the incident wave. This causes the creation of a standing wave:

$$\textbf{(50)} \quad E_{\text{total}}(x,t) = 2E_i \sin(\beta x) \sin(\omega t)$$

Such a wave does not travel in space. However, the magnitude of the electric field $\boldsymbol{E}$ varies sinusoidally as a function of distance from the reflecting plane, with the total field being zero at the plane and a maximum value of $2E_i$ at distances that are multiples of a half wavelength (0.5 λ) from the surface. Likewise, the magnetic field intensity also has a standing wave pattern. However, the total magnetic field lags behind the total electric field by 90 degrees.

Figure 6 shows the standing wave pattern for the case in which Eq. 50 applies.[46] The envelope represents a wave stationary in space, extending along the axis of the incident and reflected beams. The value for the total field E of the standing wave at any particular point along the beam axis is a sinusoidal function of the distance x from the metallic reflector of the electromagnetic waves normal to the plane of incidence.

If the metallic reflector were moved from its initial position, the standing wave pattern would move with it. Thus, if the microwave source were stationary, movements of a metallic reflector could be readily detected. Microwave resonators can be constructed to exploit this principle. In a similar manner, a detector placed in a microwave waveguide can be used to detect distances to a metallic reflector surface from the waveguide exit.

Whenever there is a standing wave, the standing wave ratio (SWR) can be defined:[42,47]

$$(51)\quad \text{SWR} = \frac{\text{Maximum value of envelope of standing wave}}{\text{Minimum value of envelope of standing wave}}$$

For the case of reflection from a perfect conductor, the ratio is infinite:

$$(52)\quad \text{SWR} = \infty$$

The other extreme is when there is no reflection. In this case, the standing wave ratio is zero.

Interface of Two Dielectrics

In the most general case, any microwave energy propagating in one dielectric medium and incident on a second dielectric medium is partially reflected at the boundary and partially transmitted through to the second medium. This statement assumes that there are no losses in the different media (perfect dielectrics). The proportion of energy reflected depends on the physical properties of the two media. Specifically, consider the case of an electromagnetic plane wave traveling in the X direction and incident on an interface parallel to the plane defined by $x = 0$. Further, assume that the incident wave is in a medium with permittivity ε_1 and permeability μ_1. The second medium is of a different permittivity ε_2 and a different permeability μ_2. Let E_i represent the incident electric field, E_r the reflected field and E_t the transmitted field. Now define the intrinsic impedance η of medium 1:

$$(53)\quad \eta_1 = \sqrt{\frac{\mu_1}{\varepsilon_1}} = \frac{E_i}{H_i} = -\frac{E_r}{H_r}$$

where H_i is the incident magnetic field intensity (ampere per meter), H_r is the reflected magnetic field intensity (ampere per meter) and the negative sign in the last expression is caused by the fact that the reflected wave is traveling in the direction opposite to the direction of the incident wave. Similarly, the intrinsic impedance η of medium 2 can be defined:

$$(54)\quad \eta_2 = \sqrt{\frac{\mu_2}{\varepsilon_2}} = \frac{E_t}{H_t}$$

The tangential components of $\boldsymbol{E}$ and $\boldsymbol{H}$ are continuous across the boundary:

$$(55)\quad E_i + E_r = E_t$$

$$(56)\quad H_i + H_r = H_t$$

Based on Eqs. 53 to 56, Eqs. 57 and 58 give the reflection coefficient Γ and the transmission coefficient T:

$$(57)\quad \Gamma = \frac{E_r}{E_i} = \frac{\eta_2 - \eta_1}{\eta_2 + \eta_1} = -\frac{H_r}{H_i}$$

$$(58)\quad T = \frac{E_t}{E_i} = \frac{2\eta_2}{\eta_2 + \eta_1} = \frac{\eta_2}{\eta_1}\frac{H_t}{H_i}$$

For the case of oblique incidence, the situation is more complex. Here, the direction of the transmitted wave is altered when it passes from medium 1 to medium 2. This phenomenon is called *refraction.* The angle of incidence θ_i that the incident field makes with the normal is always equal to the angle of reflection θ_r (angle between the reflected field and the normal):

$$(59)\quad \theta_i = \theta_r$$

This is Snell's law, illustrated in Fig. 7a. The relationship between the refracted wave and the incident wave depends on the polarization of the incident wave (Figs. 7b and 7c). With horizontal or perpendicular polarization, the electric field direction is perpendicular to the plane of incidence and parallel to the

FIGURE 6. Standing wave patterns for three magnitudes of reflection coefficient, $R' = 0$, 0.5 and 1.0. Note that envelopes of standing waves are not typically sinusoidal.[46]

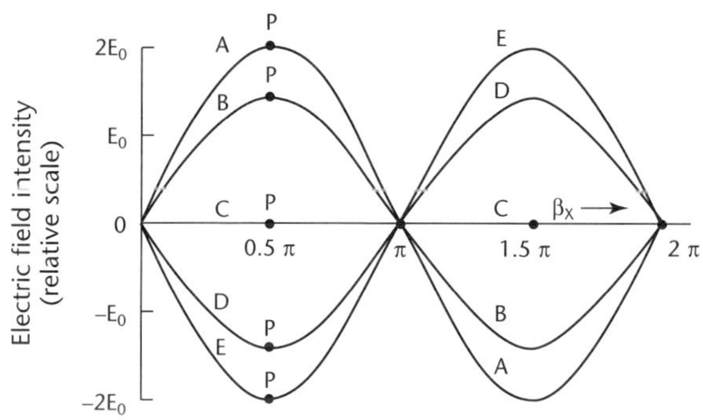

Legend
A. $t = 0$
B. $t = 0.125\ T$, where T is an arbitrary unit of time
C. $t = 0.25\ T$
D. $t = 3 \times 0.125\ T$
E. $t = 0.5\ T$

FIGURE 7. Reflection and refraction: (a) at interface of one dielectric medium to another; (b) at interface of air to dielectric medium; (c) at interface of dielectric medium to air.

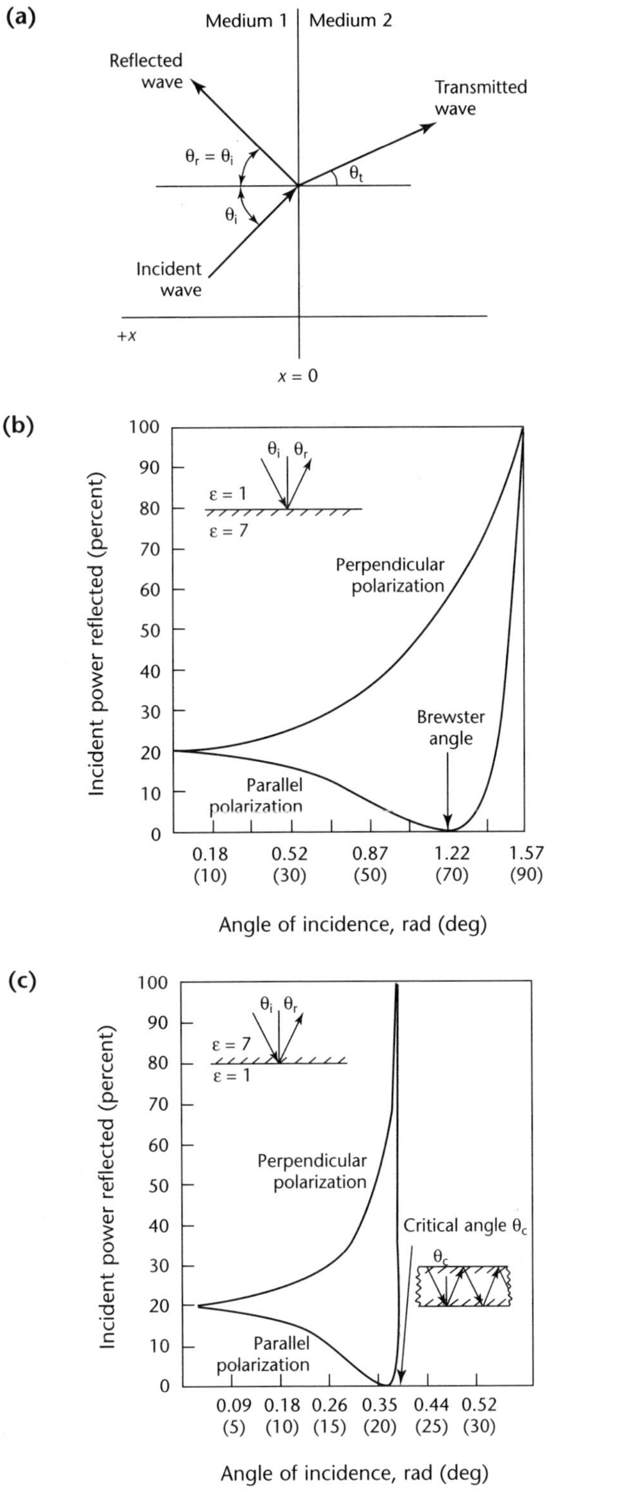

Legend

ε = permittivity (ε = 1 for air)
θ_c = critical angle
θ_i = angle of incidence
θ_r = angle of reflection
θ_t = angle of refraction
x = increment of wave propagation on X axis

reflecting surface. In this case, it can be shown that:[42]

$$(60)\quad \frac{E_r}{E_i} = \frac{\cos\theta_i - \sqrt{\dfrac{\varepsilon_2}{\varepsilon_1} - \sin^2\theta_i}}{\cos\theta_i + \sqrt{\dfrac{\varepsilon_2}{\varepsilon_1} - \sin^2\theta_i}}$$

The other case is that of parallel or vertical polarization, where the electric field vector is parallel to the plane of incidence. Here, the reflection coefficient is given by:[42]

$$(61)\quad \frac{E_r}{E_i} = \frac{\left(\dfrac{\varepsilon_2}{\varepsilon_1}\right)\cos\theta_i - \sqrt{\dfrac{\varepsilon_2}{\varepsilon_1} - \sin^2\theta_i}}{\left(\dfrac{\varepsilon_2}{\varepsilon_1}\right)\cos\theta_i + \sqrt{\dfrac{\varepsilon_2}{\varepsilon_1} - \sin^2\theta_i}}$$

Of particular interest is the case in Eq. 61, where no reflection occurs. This phenomenon occurs when θ_i satisfies the following condition:

$$(62)\quad \tan\theta_i = \sqrt{\frac{\varepsilon_2}{\varepsilon_1}}$$

This angle is called the *brewster angle.* At this angle, shown in Fig. 7b, there is no reflected wave when the incident wave is parallel polarized. In contrast, there is no corresponding brewster angle for perpendicular polarization. Finally, total internal reflection occurs if all of the incident energy (irrespective of the polarization) is reflected:

$$(63)\quad \sin\theta_i > \sqrt{\frac{\varepsilon_2}{\varepsilon_1}}$$

As mentioned above, electromagnetic waves are refracted when they travel from one medium to another. The index of refraction, defined as the ratio of the velocity of light in a vacuum to the phase velocity in the dielectric medium, is given by:

$$(64)\quad n = \sqrt{\varepsilon_r}$$

where n is the refractive index and ε_r is the relative dielectric constant of the dielectric medium. A common example of refraction is the bending of light when it propagates through water.

Just as in the case of reflection from a perfect conductor, reflection of a portion of the energy from a dielectric material can create a standing wave. As above, the standing wave ratio can be computed for partial reflection from a dielectric. The standing wave ratio in this case can also be related to the reflection coefficient:

$$(65)\quad |\Gamma| = \frac{\text{SWR} - 1}{\text{SWR} + 1}$$

or:

$$(66)\quad \text{SWR} = \frac{1 + |\Gamma|}{1 - |\Gamma|}$$

where $|\Gamma|$ represents the magnitude of the quantity. For a perfect conductor, Γ is unity and the standing wave ratio is infinity.

S Parameter Theory[44]

At lower frequencies, any two-port network can be represented in terms of the voltages and currents at the two ports. The relationships between the voltages and currents are represented in terms of the H, Y or Z parameters. However, such an approach does not work at microwave frequencies because of the difficulty of measuring the total voltage and current at these frequencies. As such, a different set of parameters called *scattering parameters,* or *S parameters,* is used to characterize microwave devices. Consider a negative port (n port) microwave device, each port of which is assumed to be lossless (Fig. 8). It is assumed that a_i is the incident wave at port i and b_i is the reflected wave at port i:

$$(67)\quad b_i = \sum_{j=1}^{n} S_{ij} a_i$$

or equivalently:

FIGURE 8. Four-port microwave junction.

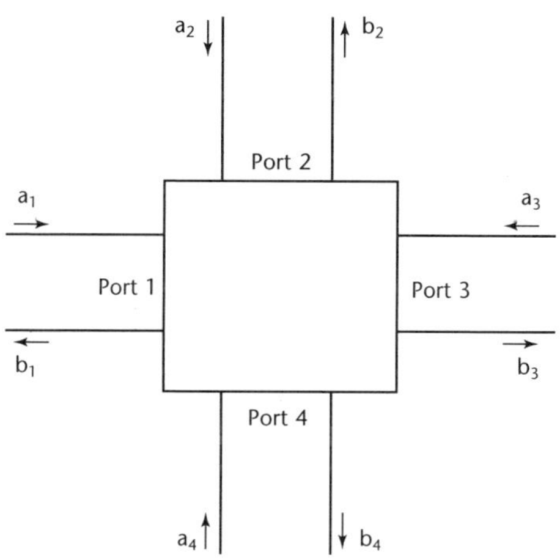

$$(68)\quad \begin{aligned} b_1 &= S_{11}a_1 + S_{12}a_2 + \ldots S_{1n}a_n \\ b_2 &= S_{21}a_1 + S_{22}a_2 + \ldots S_{2n}a_n \\ &\ldots \\ b_n &= S_{n1}a_1 + S_{n2}a_2 + \ldots S_{nn}a_n \end{aligned}$$

In matrix notation:

$$(69)\quad \boldsymbol{b} = \boldsymbol{Sa}$$

where:

$$(70)\quad \boldsymbol{b} = \begin{bmatrix} b_1 \\ b_2 \\ \vdots \\ b_n \end{bmatrix}$$

$$(71)\quad \boldsymbol{a} = \begin{bmatrix} a_1 \\ a_2 \\ \vdots \\ a_n \end{bmatrix}$$

and:

$$(72)\quad \boldsymbol{S} = \begin{bmatrix} S_{11} & S_{12} & \ldots & S_{1n} \\ S_{21} & S_{22} & \ldots & S_{2n} \\ \vdots & & & \\ S_{n1} & S_{n2} & \ldots & S_{nn} \end{bmatrix}$$

The S matrix is called the *scattering matrix* and its elements are the scattering parameters, or S parameters. These parameters are related to the reflection and transmission coefficients for three conditions:

1. $S_{ij} = \Gamma_{ij}$, the reflection coefficient of the ith port if $i = j$, with all other ports terminated in matched loads;
2. $S_{ij} = T_{ij}$, the forward transmission coefficient (transmission from the ith port to the jth port) of the jth port if $i > j$, with all other ports terminated in matched loads;
3. $S_{ij} = T_{ij}$, signifying in this case the reverse transmission coefficient (transmission from the jth port to the ith port) of the jth port if $i < j$, with all other ports terminated in matched loads.

The S parameters provide a simple means to completely characterize a device or medium under test. The S parameters can be determined quickly by using modern microwave equipment such as network analyzers. The reflection and transmission coefficients can then be computed using the S parameters, completely characterizing the material or structure. Examples of using the reflection coefficient for nondestructive testing are provided below.

Scattering by Small Reflectors

Scattering is a generalized term for the reflection phenomenon.[48-50] In its most general form, the direction, nature and degree of scattering depend on a highly complex interrelationship of many factors, including wavelength, polarization, near or far field conditions, shape and orientation of object and the index of refraction between the object and its surrounding medium (accounting for the dielectric and conductive properties of the two media). Excellent treatment of the physics and mathematical equations involving the more generalized descriptions of the scattering phenomenon are given in the literature.[1-3,48-50]

As a simplified example, the scattering behavior (normalized scatter cross section) of a metal sphere in space (under far field sensing conditions) is portrayed in Fig. 9.[51] As shown, with very large wavelengths (corresponding to low frequency microwaves), the degree of scattering is low and varies with the fourth power of the wavelength. This wavelength region is known as the *rayleigh region.* As the wavelength decreases and approaches the order of magnitude of the sphere's circumference, an oscillating type of scatter takes place. This wavelength range is known as the *resonance region.* The maximum normalized scattering cross section occurs when the microwave wavelength and the sphere circumference are equal.

If the wavelength is further decreased so that it is very small compared to the circumference of the sphere, the scattering cross section becomes equal to the physical cross section. The region in which this behavior exists is identified as the *optical region.*

FIGURE 9. Fundamental microwave backscattering relationship in far field for spherical conductors.

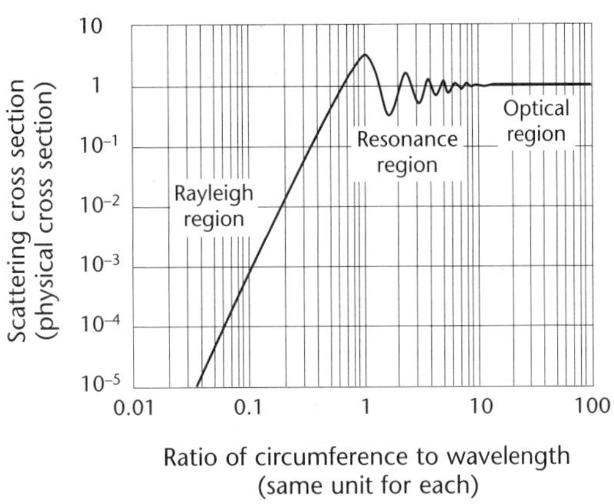

Some simple equations can be derived that approximate scattering from metal or dielectric spheres surrounded in a dielectric medium when rayleigh scattering conditions apply (at wavelengths much larger than the sphere radius a):

$$(73)\quad \frac{2\pi a}{\lambda} < 0.4$$

For a conductive sphere located in an infinite surrounding medium, the far field scattering cross section ψ (square meter) for plane wave illumination in the rayleigh region is given for backscatter:[52]

$$(74)\quad \frac{\psi}{\lambda^2} = \frac{9}{4\pi}\left(\frac{2\pi a}{\lambda}\right)^6$$

and for forward scatter:

$$(75)\quad \frac{\psi}{\lambda^2} = \frac{1}{4\pi}\left(\frac{2\pi a}{\lambda}\right)^6$$

Note that backscattering is almost an order of magnitude greater than forward scattering. The quantity $\psi \cdot \lambda^{-2}$ represents a fraction of the total incident power that would be scattered if the total incident power were confined to a circular area of λ^2.

Under the same rayleigh conditions, the backscatter from a low loss dielectric sphere can be approximated:[52]

$$(76)\quad \frac{\psi}{\lambda^2} = \frac{1}{\pi}\left(\frac{\varepsilon_r - 1}{\varepsilon_r + 2}\right)^2\left(\frac{2\pi a}{\lambda}\right)$$

where ε_r is the dielectric constant of the sphere relative to the surrounding medium.

Analogy of Atmospheric Scattering of Light

The scattering relationships of Fig. 9 are true throughout the electromagnetic spectrum. For example, the daytime sky can be blue and the sunset red because of light scattering. Air molecules, small in circumference compared to light wavelengths, fall deep into the rayleigh region. The shorter blue wavelengths are scattered more than the red wavelengths. Hence, light transmitted directly through the Earth's atmosphere from the sun appears redder than that scattered by the atmosphere, so that in the northern hemisphere light from the sky overhead is bluer.

These effects can be contrasted with the optical scattering of a strong beam of light by large chalk particles. In this case, the scattered light is white because the wavelengths for the entire visible light spectrum are many times smaller than the size of the chalk particles. In this optical region (see Fig. 9), the scattering cross section is the same for all the visible light wavelengths.

Scattering from Reflector Shapes and Arrays

The same general scattering relationships shown in Fig. 9 also hold true for other shapes of reflectors, including conical, cylindrical and ellipsoidal reflectors. Their scattering graphs are understandably different from that of Fig. 9 but the same three scattering regions occur.

Scattering by periodic structures such as gratings and arrays has created considerable interest and has been the basis for some specialized microwave test applications. When a linearly polarized microwave beam impinges normally on a flat screen or grating formed of wires, the microwave scattering and transmission are strongly affected by the angle between the direction of microwave polarization and the direction of the parallel wires in the reflector.

When the microwave polarization axis and the direction of the wires are orthogonal (perpendicular to each other), most of the microwave beam energy is transmitted through the wire grating. When the microwave beam is polarized in a direction parallel to the wires in the grating, most of the microwave beam energy is scattered or reflected. Proportional amounts of the microwave radiation are either scattered or transmitted for angles (of microwave polarization with respect to the condition of the wires) between these two extremes.

FIGURE 10. Acoustic model of radiation pattern, showing near field and far field. Source is at left, with energy propagating from left to right. Photograph shows acoustic waves propagating in illuminated liquid (schlieren technique).[53]

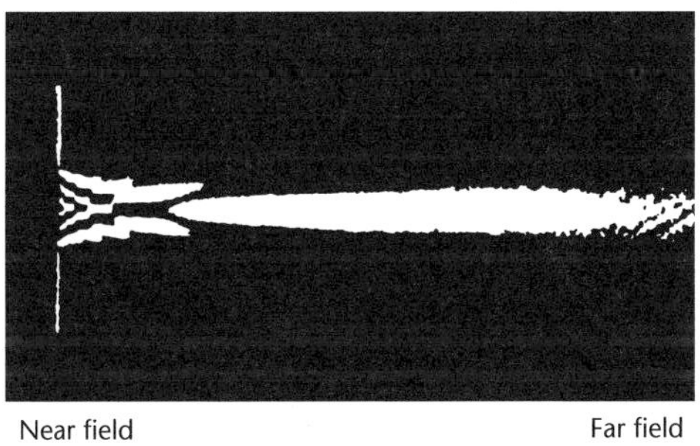

Radiation Patterns

According to Huyghen's principle, any form of radiation (electromagnetic or acoustic) can be analyzed by the addition of contributions from some given distribution of simple sources, properly selected in phase and amplitude to represent the physical situation. This principle forms the basis for analyzing the microwave intensity distribution as a function of any point in space in front of the radiator.

The intensity distribution and shape of a radiation pattern is determined by the ratio of the dimensions of the radiator and the wavelength in the propagating medium. The diffraction pattern is formed by the phase and amplitude relationships of the huyghen wavelets propagating from the face of the radiator. Figure 10 illustrates a radiation pattern from a circular source with a diameter of 6.7 wavelengths: $D \cdot \lambda^{-1} = 6.7$.[53] The light areas represent high intensity whereas dark areas represent low intensity. The complex diffraction pattern closer to the radiator is called the *near field* (or *fresnel field*) whereas the more uniform pattern further from the radiator is called the *far field* (or *fraunhofer field*).

Figure 11 shows the distribution of relative intensity I along the Y axis of the radiation pattern from a circular continuous wave radiator with a radius-to-wavelength ratio of $R \cdot \lambda^{-1} = 9.5$.[54] For continuous wave radiation in a lossless medium, the maxima along the axis all have the same intensity whereas the minima are all zero. The far field or near field boundary is located at the last (farthest) axial maximum from the face of the radiator y_{o+}. The location of the various maxima and minima along the beam axis in the near field is a function of $D \cdot \lambda^{-1}$.

FIGURE 11. Distribution of axial intensity I for circular radiator $R \cdot \lambda^{-1} = 9.5$.[45]

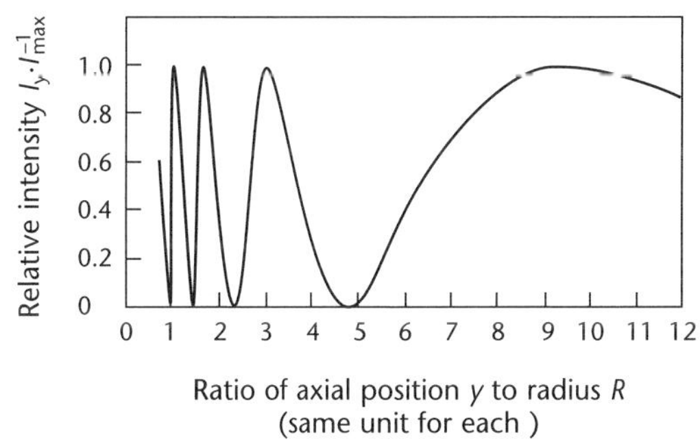

Near Field (Fresnel Field)

The near field intensity pattern transverse to the radiation axis (radiator axis) also consists of a series of maxima and minima. The distribution and number of these maxima or minima occurring across a given transverse axis depend on the transverse axis location along the longitudinal Y axis (its distance from the face of the radiator). The number of maxima occurring immediately in front of a radiator is given by the ratio $D{\cdot}\lambda^{-1}$. The near field contains *cross polarization,* where a component of $\boldsymbol{E}$ lies in the propagation direction.

Far Field

The far field (fraunhofer field) begins at $y_{o+} \geq R^2{\cdot}\lambda^{-1}$ and extends to infinity, with the axial intensity uniformly decreasing with y^{-2} (inverse square law function). Beam spread occurs in the far field because of diffraction, producing a main lobe and side lobes in the radiation pattern (Fig. 12).[54] As illustrated, the shape or directivity of the main lobe and the number of side lobes are functions of $D{\cdot}\lambda^{-1}$. The directivity or beam sharpness of the main lobe for a circular radiator is given by the fraunhofer formula:

$$\text{(77)} \quad \sin\theta = 0.61\frac{\lambda}{D}$$

where θ is the half angle of beam spread, or beam divergence. The fraunhofer formula for beam spread from a square wave radiator is given by:

$$\text{(78)} \quad \sin\theta = \frac{\lambda}{\ell}$$

where ℓ is the side of the square radiator.

The wavefronts in the main lobe are spherical, following intensity reduction in accordance with the inverse square law. At great distances in the far field, the wavefront characteristics closely approximate those of plane waves.

FIGURE 12. Radiation patterns in far field for circular radiators with various ratios of radiator diameter D to wavelength λ: (a) diagram; (b) $D{\cdot}\lambda^{-1} = 0.5$; (c) $D{\cdot}\lambda^{-1} = 1.2$; (d) $D{\cdot}\lambda^{-1} = 2$; (e) $D{\cdot}\lambda^{-1} = 4$.

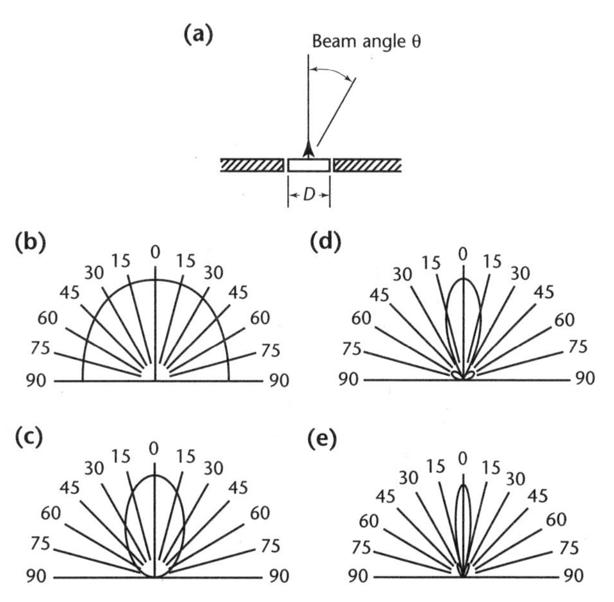

Guided Waves and Waveguides[44]

A waveguide is a dielectric filled metallic tube, usually either rectangular or circular in cross section, that can guide an electromagnetic wave. Such waveguides are usually used only at microwave frequencies. In many cases, the dielectric material within the waveguide is air. The electric and magnetic fields are confined to the space within the waveguide, with negligible power loss at the walls of the waveguide. Dielectric losses are also usually small because in most cases air is used as the dielectric material. In microwave testing, waveguides are essential in instrumentation because they couple the microwave energy to the test object. The exact type of waveguide also depends on its operating frequency.

In general, it is possible to propagate several modes of electromagnetic waves in a waveguide, such as transverse electric, transverse magnetic and transverse electric and magnetic. However, each waveguide has a cutoff frequency for each mode, above which energy can be transmitted down the waveguide without attenuation. Below the cutoff frequency, the energy is attenuated to a negligible value in a very short distance. The mode with the lowest cutoff frequency is called the *dominant mode.* The allowable modes, as well as the cutoff frequency, are related to the cross sectional dimensions of the waveguide. Usually, waveguides are selected such that only the dominant mode can propagate.

For rectangular waveguides, a propagation constant can be defined in the waveguide:

$$\text{(79)} \quad \gamma_g^2 = \gamma^2 + k_c^2$$

where γ is the intrinsic propagation constant of the dielectric and k_c is the cutoff wave number. For a lossless dielectric material:

$$\text{(80)} \quad \gamma_g = \pm j\sqrt{\omega^2 - k_c^2}$$

No wave propagation occurs when:

$$\gamma_g = 0 \quad (81)$$

or:

$$\omega_c^2 \mu\varepsilon = k_c^2 \quad (82)$$

This condition occurs when the frequency of the impressed electromagnetic wave is equal to the cutoff frequency. Thus, the cutoff frequency can be derived as:

$$f_c = \frac{k_c}{2\pi\sqrt{\mu\varepsilon}} \quad (83)$$

For cases where the frequency is above the cutoff, wave propagation in the guide occurs; for cases where the frequency is below the cutoff, the wave is rapidly attenuated in the guide. The special case of $f = f_c$ is also called *evanescence.* Similar expressions can be derived for waveguides with other cross sections.

PART 3. Applications of Microwave Testing

The ability of microwaves to penetrate deeply inside dielectric materials and composites makes microwave testing very attractive for interrogating such materials.[2,5,6,9,55] Table 1 lists some application areas.[56-83]

Generally, microwave measurement setups are divided into three categories: reflection, transmission and scattering setups. As in monostatic radar, reflection measurements (Fig. 13a) are conducted using one antenna for transmitting and receiving signals. Different antennas may be used for transmission and reception. The transmission approach (Fig. 13b) is similar to bistatic radar and is conducted by using an antenna on each side of the test object. Transmission approaches require access to both sides of the test object. Scattering measurements (Fig. 13c) are obtained using several transmitters or several receivers positioned at certain locations.

Microwave Testing Using Far Field Approach

As mentioned above, microwave nondestructive test techniques are applied either in the far field or in the near field. The mathematical derivations of the far field approach are relatively simple and easy to implement in a code format.

Detection of Disbonds in Concrete Structures

The presence of a disbond or a crack can be modeled as an additional layer of air within a concrete structure. When a plane wave is incident on a dielectric structure, part of it will be reflected at the boundary between any two layers and the rest will travel through. A wave that moves within a dielectric material attenuates and undergoes more rapid phase changes than a wave traveling in a vacuum. Waves reflect and refract whenever they travel

TABLE 1. Applications of microwave testing.

Medium	Test Purpose
Concrete	testing of concrete for constituent determination, reinforcing bar location, chloride detection and safety evaluation[18,56-62]
	detection of grout in masonry[63,64]
Layered media	detection and evaluation of corrosion under paint and thick stratified composite laminate coatings[22,32]
	detection of minute thickness variations in each layer of stratified dielectric medium[26,65]
	accurate thickness measurements of coatings, single dielectric slabs (paint) and layered dielectric composites[11,15,66]
	detection of disbonds, delaminations and voids in stratified media and depth estimation of these discontinuities[66,67]
Nonmetals	inspection of thick plastics and composites for interior discontinuities, fiber bundle orientation and moisture content[4,68]
	detection and estimation of porosity in ceramics, thermal barrier coatings, plastics and glass[69]
	detection and measurement of moisture content in wood, grains, textiles and other materials[63]
	impact damage detection and evaluation for reinforced composite structures, including graphite composites[70]
	fiber orientation determination in graphite and glass reinforced composites[71]
	characterization of constituents (for example, volume contents and dielectric constants) in dielectric mixtures[28,63,72,73]
	evaluation of cure states in chemically reactive materials (carbon loaded rubber, resin binder and cement)[74,75]
Metals	sizing of surface cracks (stress and fatigue) in metals and in cracks filled or covered by paint, rust or dirt[21,30,31,33-40]
	profiling and roughness evaluation of metal surfaces[63]
Subsurface media	imaging of localized and extended area interior and surface discontinuities[23,67,76]
	heating lossy dielectric materials for medical or thermographic imaging[25,77,78]
	imaging of buried objects[79,80]
	determination of ratio of water to cement in fresh and hardened concrete with microwave reflection techniques[81]
	test for segregation in concrete structures in service[75,82]
	inspection of luggage for contraband[83]
	liquid state monitoring

from one medium or layer to another. The reflection coefficients between the layers and the attenuation through them are calculated until the wave reaches an infinite half space or a conducting layer.[11] By comparing the reflected wave from a layer to the incident wave on the layer, the reflection coefficient is obtained at that layer. The reflection coefficient at the first layer (effective reflection coefficient) can be calculated and measured as well. The presence of a disbond or a void is detected through measuring variation in the phase of the effective reflection coefficient Γ. To demonstrate this approach, a concrete structure with relative dielectric constant $\varepsilon_r = 6 - j\,0.6$ was assumed,[81] where the relative dielectric constant is a constant unique for each material.

Figure 14 shows the calculated phase of the reflection coefficient as a function of disbond thickness for a disbond at the depth of either 10 mm (0.4 in.) or 50 mm (2.0 in.) and at two frequencies of 3 GHz and 6 GHz. The results indicate that, if the disbond thickness is zero (that is, if there is no disbond), all frequencies see an infinite half space of concrete. This is indicated by a common phase to all graphs of 180 degrees. As the thickness of the disbond increases, the phase begins to change from that of an infinite half space. The figure also shows that at a depth of 50 mm (2 in.) and at 6 GHz, the variation in the phase becomes very small because the depth of penetration of this frequency is limited compared to that at 3 GHz. The fact that a constant phase difference is present at 6 GHz for any disbond thickness exceeding 80 mm (3.15 in.) is an indication that this phase difference can be used to either locate or estimate thicknesses of disbonds.

Simulations have indicated that, as the depth of the disbond increases, frequencies around 6 GHz seem useless. Consequently, the phase of the reflection coefficient is calculated for different disbond depths and thicknesses at different frequencies. Then, the phase difference between the bonded and disbonded areas was calculated and plotted as a function of frequency. Figure 15 shows the difference in the phase of reflection coefficients between the bonded case and a 5 mm (0.2 in.) thick disbond at different depths. It is clear that if a disbond is deep, maximum phase difference is obtained at lower frequencies. For example, a disbond at a depth of 60 mm (2.4 in.) gives maximum difference at a frequency of about 2 GHz whereas a disbond at a depth of 20 mm (0.8 in.) gives maximum difference at a frequency of 6 GHz.

It seems that deciding on an optimum frequency is hard, especially if the disbond depth and thickness are unknown as in most applications. However, test results show that frequencies around 3 GHz are optimal in obtaining high differences for all depths. Figure 15 also shows that, at some combinations of disbond depth, thickness and frequency of operation, the reflection coefficient's phase does not change from the case without disbond. This result indicates that a test at a single frequency does not guarantee the detection of a disbond. Even if the disbond is detected, two unknowns need to be determined, namely the disbond thickness and depth. Consequently, two different measurements are needed to find disbond thickness and depth.

The easiest way to obtain more tests is to change the frequency of operation. Conducting two or more measurements will ensure the detection and evaluation of the thickness and depth because, if there were no disbond in the structure, the two tests will result in same phase of the reflection coefficient (that is, infinite half space of concrete).

FIGURE 13. Measurement setups: (a) reflection; (b) transmission; (c) scattering.

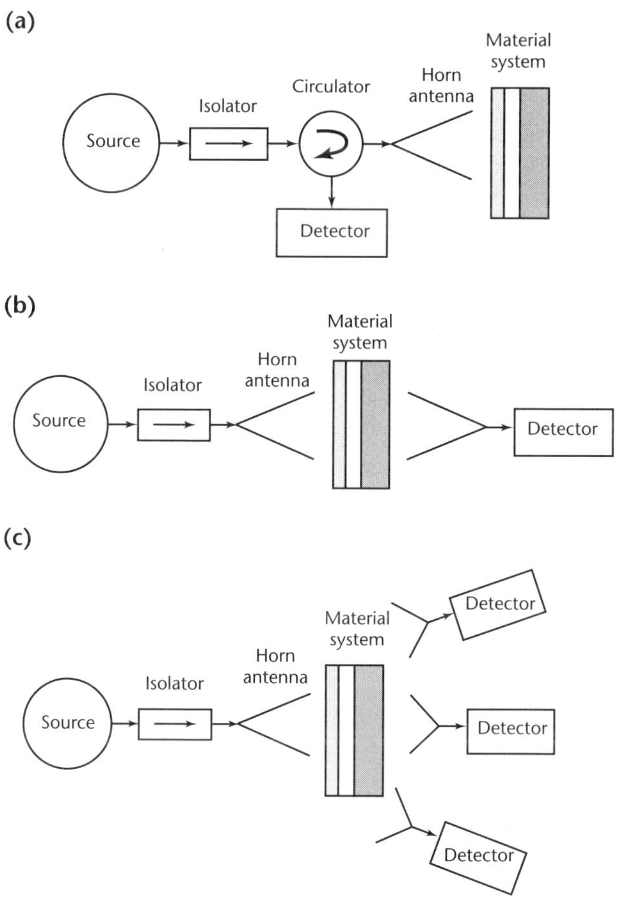

Usually when tests are done using microwaves, the first step is to calibrate the microwave system. This step may be eliminated if the phase difference of two tests at two different frequencies is taken into account. Because phase difference does not change, waves travel the same distance all the time. This means that if

FIGURE 14. Calculated phase versus disbond thickness for 10 mm (0.4 in.) and 50 mm (2.0 in.) deep disbonds at 3 and 6 GHz.

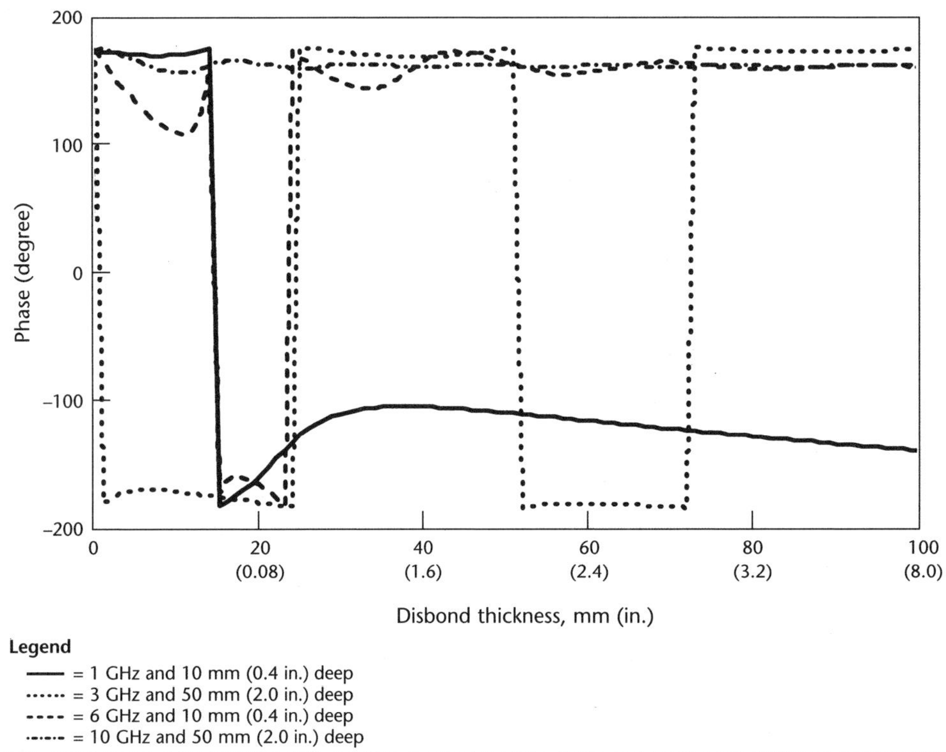

FIGURE 15. Phase difference between no disbond case and 5 mm (0.2 in.) thick disbond.

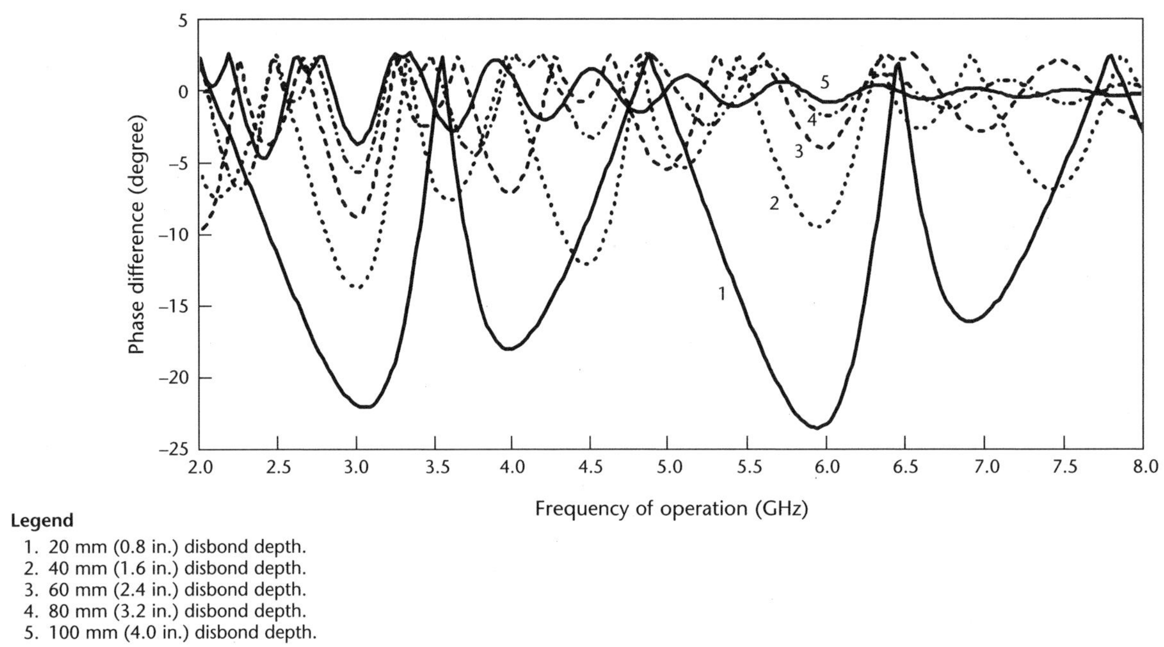

the test is performed at three different frequencies and the phase difference of the three tests is calculated, the disbond will be detected and its depth and size will be evaluated as well. Figure 16 shows that the phase difference changes with the changes in disbond thickness. This correlation was true also for a varying disbond depth and a fixed thickness.

In the case of no disbond, the phase difference will remain at zero. For a disbond of any thickness at any depth, using two values of phase difference between phases at three frequencies reveals the thickness and depth of any disbond. In most applications, the phase of the reflection coefficient can be measured at different frequencies. Consequently, a lookup table may be used with a simple program to search the data generated by using the theoretical code for the phase difference combinations closest to the measured ones.

In an experiment, a phase detection system using a horn antenna was built and used. A 1 m × 1 m × 50 mm (39 × 39 × 2.0 in.) concrete block was made and a frame that can be used to adjust the height (that is, to control the disbond thickness) of the concrete block was built. Phases caused by 10 mm (0.4 in.) disbonds, 20 mm (0.8 in.) disbonds and cases of no disbonds were measured at several frequencies. Results at different phases show that disbonds can be detected (Fig. 17). The differences, shown in Fig. 18, were also calculated and used to estimate the depth and thickness of the disbonds. For example, if tests are done at 3.8 and 3.2 GHz and the phase difference measured is 10 degrees, the disbond thickness will be 10 mm (0.4 in.) at a depth of 50 mm (2.0 in.). Phase differences at three frequencies will give the thickness and depth of a disbond.

Microwave Testing Using Near Field Approach

Near field microwave testing of structures with discontinuities has received considerable attention. Near field microwave measurements are influenced by several factors such as probe type (such as rectangular waveguide, circular

FIGURE 16. Phase difference of two frequencies versus disbond thickness at depth of 50 mm (2.0 in.).

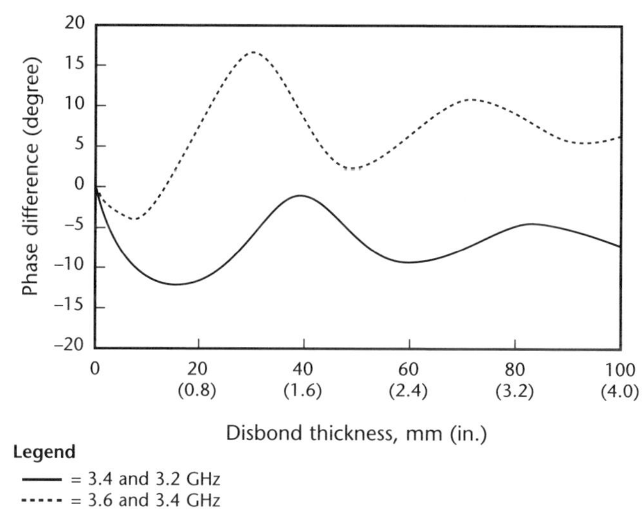

FIGURE 17. Measured changes in phase of reflection coefficient versus measured changes in disbond thickness.

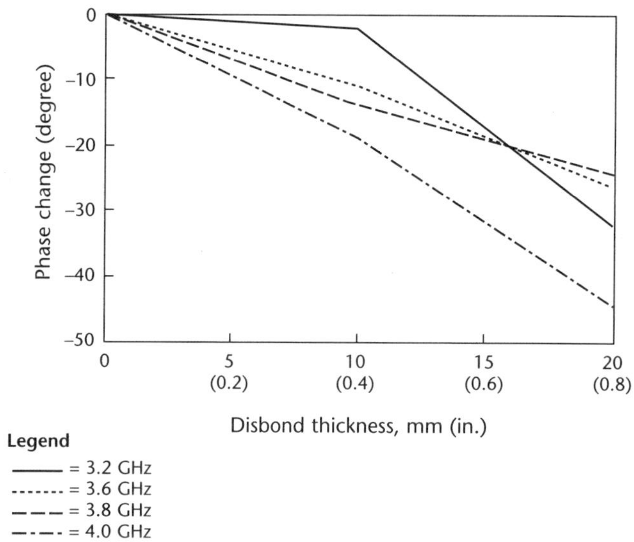

FIGURE 18. Measured phase difference of two frequencies versus measured disbond thickness at depth of 50 mm (2.0 in.).

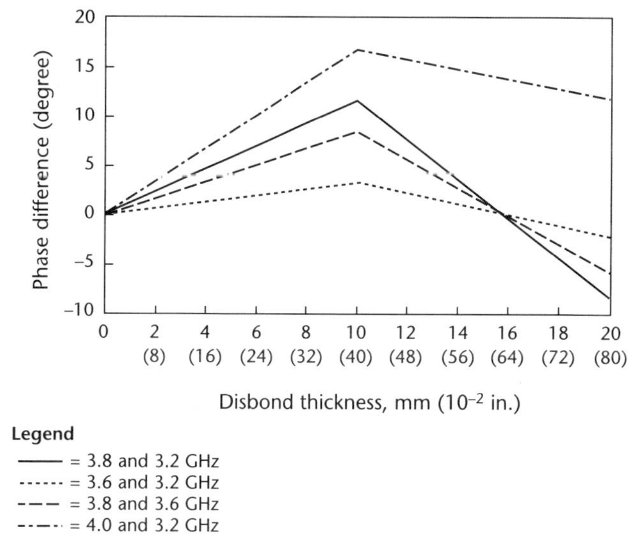

waveguide or coaxial line) and field properties (such as main lobe, side lobes and half power beam width), geometrical and physical properties of both the discontinuity and the material under test.

Near Field Imaging of Structures

Inspection and detection of discontinuities in opaque media is a classical problem. Over many years, this problem has motivated scientists and engineers to develop probes that can characterize such entities. Most imaging systems are intended to address this problem in a wide range of applications, including medical imaging, microscopy and material nondestructive testing. There are many test techniques for imaging purposes and they share the basic principle of image formation: an image of a test object is formed by the interaction between the imaging sensor and certain physical or chemical properties of the test object. In essence, this applies for near field microwave imaging of discontinuities in various test objects.

A microwave image is obtained by arranging detected microwave signal data, gathered by performing a raster scan over a test object, to produce a visual impression of structural geometry and discontinuities. The microwave data may include information such as the phase or magnitude of either the reflection coefficient or the transmission coefficient. Also, attenuation information can be used to produce a microwave image of a test object as well as any combination of the above. The general geometry of a composite panel illuminated by a waveguide sensor is shown in Fig. 19. The sensor operates at a certain frequency and at a certain standoff distance, that is, at a certain distance between the waveguide aperture and the test surface.

The use of an open ended rectangular waveguide as a sensor for measuring material properties at microwave frequencies has received considerable attention.[7-9,13-16,21-23,71] An open ended rectangular waveguide offers advantages such as a relatively small footprint (sensing area) compared with an antenna operating from the far field. Table 2 lists some of the most used frequency bands with the corresponding frequency range and waveguide aperture size.

Near field waveguide systems are simple yet powerful imaging tools. They are based on the basic idea that microwaves launched into a medium are very sensitive to discontinuities in the material space. A discontinuity could be practically realized as a crack in metal, a void in concrete structure, a foreign inclusion in composite structure or interfaces between different dielectric layers. This discontinuity, defined as change in the spatial distribution of the dielectric properties, causes some change in the properties of the reflected and transmitted waves. The power and the phase properties of the forward and backward traveling waves contain valuable information about the type, location and orientation of discontinuities. Any microwave system is intrinsically a nondestructive test system if it is outside the test material space and measures microwaves to evaluate the material.

Generally, a near field waveguide based microwave imager is an imaging system that uses a waveguide as an imaging probe to measure the reflection coefficient in a certain imaging plane as a function of the spatial coordinates and then provides that information as a two-dimensional image. Usually, the imaging plane is taken to be parallel to the test object's surface. Imaging for nondestructive testing has two fundamental objectives.

FIGURE 19. Composite structure is scanned by near field waveguide based imager.

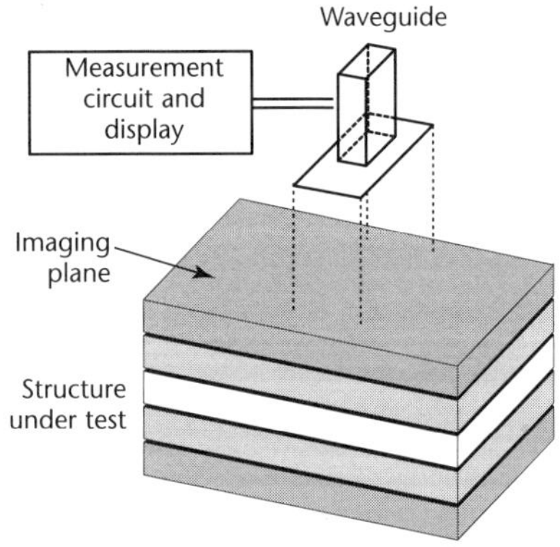

TABLE 2. Commonly used frequency bands and corresponding waveguide apertures.

Band	Frequency Minimum (GHz)	Frequency Maximum (GHz)	Aperture Diameter Minimum mm	Aperture Diameter Minimum (in.)	Aperture Diameter Maximum mm	Aperture Diameter Maximum (in.)
S	2.6	3.95	34.04	(1.34)	72.14	(2.84)
G	3.95	5.85	22.15	(0.87)	47.55	(1.87)
J	5.85	8.2	15.80	(0.62)	34.84	(1.37)
X	8.2	12.4	10.16	(0.40)	22.86	(0.90)
Ku	12.4	18.0	7.90	(0.31)	15.80	(0.62)
K	18.0	26.5	4.32	(0.17)	10.67	(0.42)
Ka	26.5	40.0	3.56	(0.14)	7.11	(0.28)
V	50.0	75.0	1.88	(0.15)	3.75	(0.15)

1. The detection problem — imaging equipment should detect a subsurface inclusion such that, from the captured image, an inspector can document the existence of the inclusion.
2. The inverse problem — the imaging equipment should assess the inclusion through quantitative interpretation of the captured image to extract the inclusion's physical and electrical properties.

Finally, two important metrics are used to assess the performance of the imager itself.

1. Sensitivity is a measure of the minimum detectable variation in the material properties.
2. The resolution of the imager is its ability to distinguish two close material variations.

A general near field microwave measurement setup is shown in Fig. 20. A single-frequency transmitter (sweep oscillator or gunn diode) generates a microwave signal that is transmitted through an open ended rectangular waveguide probe terminated into a large metallic flange. As a standard practice, a square flange with sides greater than the vacuum wavelength λ_0 is used to terminate the aperture of the waveguide to approximate an infinite ground plane. As the signal reaches the aperture, part of it gets reflected into the waveguide according to the effective dielectric properties of the medium in front of the aperture. A receiver (a detector diode or a mixer) is used to measure an electric potential related to the properties of the reflected signal.

As the scan progresses, the measured voltage values are recorded in a matrix. If the scan starts at a spot devoid of any discontinuity, the detector measures a certain, almost constant, voltage. When an indication appears (that is, when a discontinuity appears in the sensing range of the probe) by the detector aperture, the effective dielectric property of the structure in front of the aperture varies. This variation results in a change in the reflected signal and consequently the measured voltage. Images produced in this fashion are referred to as *contrast images* (that is, the presence of a heterogeneity is indicated by a different color, or intensity level, in the image).

After fulfilling the requirements of the test process, mentioned above, the type of images to be captured should be decided. Then, an optimization procedure should be carried out to facilitate the inverse problem solution — that is, image interpretation — in the later stages.

FIGURE 20. General near field microwave imaging experimental setup.

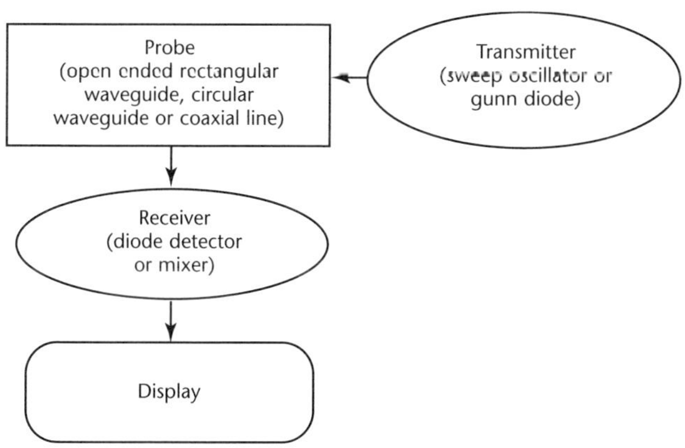

Types of Captured Images

The reflection coefficient is a complex parameter that depends directly on the properties of the material discontinuities. Using special microwave circuitry, the phase and the magnitude of reflection coefficient can be measured. According to the measured parameter, there are three types of images and hence three types of realizations of the near field waveguide based imagers: phase images, magnitude images and complex images.

Phase Images. Phase images are images in which the intensity at every spatial point is proportional to the phase of the reflection coefficient at that point. The waveguide is scanned in the imaging plane and the phase of the reflection coefficient, or a signal proportional to it, will be measured at each scanned point. A microwave circuit that measures scatter may be used for this purpose, capturing images within a dynamic range of 180 degrees. These images may reveal surface and subsurface discontinuities in material structures. For this type of test, the sensitivity of a phase image is quantified by the ratio of the change of the phase of the reflection coefficient to the amount of change in the electrical property of the material.

Magnitude Images. In a magnitude image, the intensity at each scanned point is proportional to the magnitude of the reflection coefficient. The implementation of the magnitude imagers is simpler than that of the phase imagers. Because the magnitude of the reflection coefficient has a range from zero to one, the dynamic range of the magnitude images span a portion of this range.

Complex Images. Complex images are the simplest to capture. The intensity at each point in the captured image is proportional to the complex reflection coefficient as a whole. These images do not convey phase or magnitude information exclusively.

Image Optimization

One of the most attractive features of near field microwave imaging systems is that they lend themselves to systematic optimization to enhance their capability in various applications. In the imaging context, this is achievable when the sensitivity (physical contrast) and the lateral resolution of the captured images are high enough for the image to be interpreted quantitatively.

Optimization Parameters. There are two main optimization parameters to be used to enhance the captured image for a given detection scenario: the operation frequency and the standoff. These parameters can be applied to any microwave test.

1. The optimum frequency of operation (the waveguide band) should be determined to meet the detection requirements. Special optimization computer codes written to find these parameters are based on mathematical models developed to describe the interaction between microwaves and material systems. For waveguide probes particularly, the frequency of operation and the waveguide dimensions not only determine the sensitivity but also restrict the lateral resolution to a ratio of the waveguide footprint. Sometimes, there may be a tradeoff between the sensitivity and the lateral resolution. In this case, the other optimization parameters and tactics help to resolve the tradeoff.
2. The standoff distance, or liftoff distance, is the mechanical distance between the waveguide aperture and the surface of the test material. Many investigations have proved the impact of this optimization parameter on the detection capability of near field microwave test systems. Changing the standoff distance by small fractions of the operating wavelength can optimize the detection sensitivity.

Optimization Tactics. Optimization tactics are used for special cases or critical applications when the optimization parameters alone fail to meet the detection requirements.

1. Backing the structure under test with a conducting sheet enhances the detection sensitivity. This measure works especially for low loss, low permittivity structures.
2. Adding a layer of a dielectric material to replace the liftoff focuses the field, shifts the phase and maintains constant liftoff.
3. Filling the waveguide with a coupling material can increase the amount of power penetrating into the structure. Waveguide filling constitutes a powerful optimization parameter to test material of very high loss tangent.

Laminated composite materials have become prominent engineering materials in a wide range of applications. For critical applications, these materials are engineered to have specific attributes in which accidental variations can cause catastrophic failures. It is required that the nondestructive test technique provide quantitative measures about the type, location and orientation of subsurface inclusions. Such information about these inclusions could be extracted from two-dimensional images. Near field microwave imaging systems with open ended rectangular waveguides as imaging probes have shown promising results in detecting subsurface discontinuities in such opaque media.[8,22,23,71]

Theoretical Image Formation

To form and optimize an image theoretically, it is crucial to have the mathematical model that describes the interaction between microwaves and the laminated composite structures. The model is used to calculate the phase and magnitude of the reflection coefficient at each point in the imaging plane. The developed model has been applied to obtain images of subsurface inclusions in a five-layer laminated composite structure (Table 3).

The structure was backed by a conducting sheet. An air inclusion, or air pocket (relative dielectric constant $\varepsilon_r = 1$), measuring at least $2 \times 3 \times 0.3$ mm ($0.08 \times 0.12 \times 0.012$ in.) in the X, Y and Z directions, respectively, may be introduced in the first or second adhesive layers. To detect and assess the effect of each inclusion independently, they need to be imaged with acceptable

TABLE 3. Model for microwave testing of layered composite structure, where $j = \sqrt{(-1)}$.

Layer	Relative Dielectric Constant ε_r	Thickness mm	(in.)
Skin	$4.5 - j\,0.04$	2.5	(0.10)
Adhesive	$3.1 - j\,0.01$	0.3	(0.012)
Inner core	$1.1 - j\,0.003$	40.0	(1.60)
Adhesive	$3.1 - j\,0.01$	0.3	(0.012)
Skin	$4.5 - j\,0.04$	2.5	(0.10)

nonoverlapping dynamic ranges. The frequency of operation, standoff (liftoff) distance and the waveguide band that yield the required images must be determined before application.

By using the developed model, the requirements were met by operating at a frequency of 32 GHz in the K_a band at a standoff distance of 1 mm (0.04 in.). Figure 21a shows the raster phase image when the inclusion is present in the first adhesive layer. The spatial extent of the inclusion is apparent from the image with a dynamic range of 40 degrees. In Fig. 21b, the inclusion extent in the second adhesive layer is also evident with a narrower yet still acceptable dynamic range of 25 degrees. The dynamic ranges are not overlapping and the two locations of the inclusion can be distinguished at the same frequency. The two images also show that, as the discontinuity location becomes deeper, its image becomes larger. This change occurs mainly because the beam becomes wider as the distance from the radiator increases.

Near field waveguide based microwave imagers are suited for detection and assessment of foreign inclusions in laminated composite materials. Imaging of various inclusions can be achieved through model based optimization procedures. These procedures result in theoretical images to be used with practical images to extract subsurface discontinuity properties.

FIGURE 21. Phase image for flat inclusion in first adhesive layer: (a) dynamic range of 40 degrees; (b) dynamic range of 25 degrees.

(a)

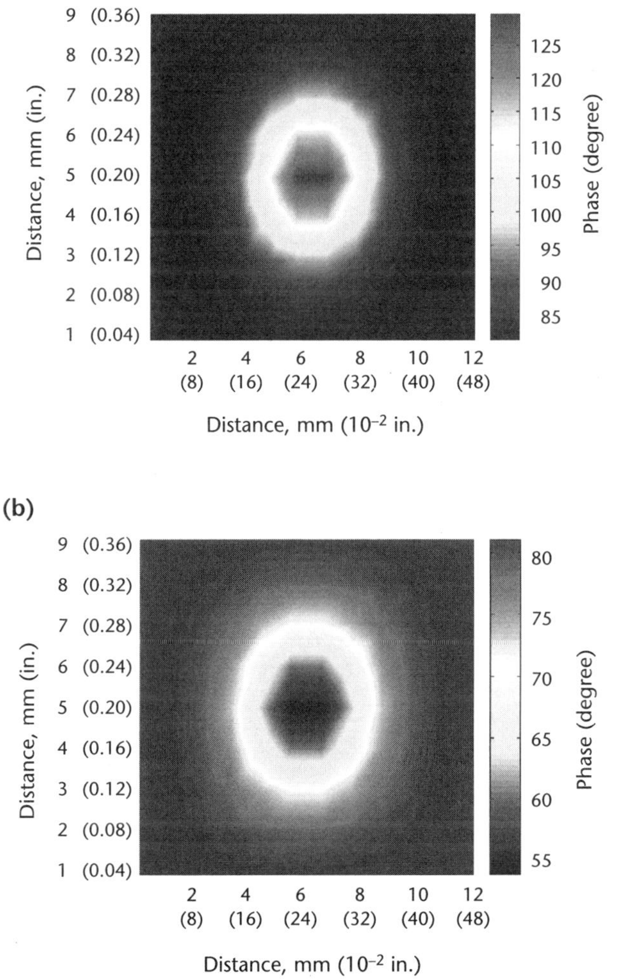

Experimental Results

Numerous examples of experimentally obtained images have been published in the literature. The following examples demonstrate the applicability of microwave testing.

Moisture Permeation Detection. Figure 22 shows an image obtained using a real time handheld microwave imaging system operating at 10.5 GHz and 2 mm (0.08 in.) standoff distance for 0.5 cm^3 (0.03 $in.^3$) of water injected at the bottom of a 20 mm (0.8 in.) thick composite panel. The two-dimensional intensity image indicates that such discontinuities may be clearly detected and their extent assessed.

Metal Bar in Structure. Figure 23 shows a two-dimensional image of a steel bar inserted at an angle in a relatively dense composite structure at a frequency of 10.5 GHz and a standoff distance of 2 mm (0.08 in.). As the depth changes, the intensity level caused by the bar changes as well.

FIGURE 22. Real time complex image of water permeation in thick composite structure.

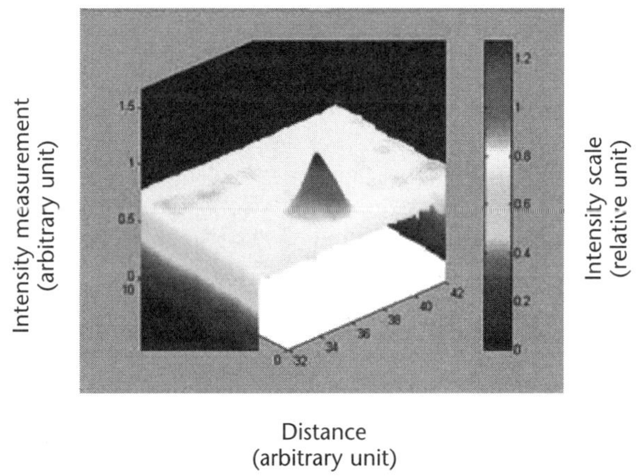

Corrosion Detection

Detection of corrosion under paint and composite laminate coatings is crucial in many industrial, civil and military applications. In many countries, detection of rust and other corrosion is an important practical concern. Because the presence of rust or corrosion may be considered as an additional new thin dielectric layer under the paint or coating, microwave testing is well suited for evaluating this type of damage.[22,32]

One study evaluated a microwave test system that exploited the phase and magnitude information of the reflection coefficient. For these measurements, a direct current voltage proportional to the magnitude and phase of the reflection coefficient at the waveguide aperture was used to generate raster scan images of rusted steel specimens. The experimental results illustrate the practicality of near field microwave imaging for corrosion detection. Figure 24a shows a photograph of a steel specimen having an area of rust in the middle. This specimen was produced from a relatively flat piece of steel covered by a thin layer of naturally occurring rust. The rusted area was masked by a piece of tape and the remaining exposed surface was sand blasted to remove the rust.

Subsequently, this specimen was painted as uniformly as possible with common spray paint as shown in Fig. 24b. Figure 24c shows a 130 × 70 mm (5.2 × 2.8 in.) image of the rusted steel specimen with a rust thickness of about 0.5 mm (0.02 in.) under a paint thickness of 0.1 mm (0.004 in.) at a frequency of 25.5 GHz and at a standoff distance of 1 mm (0.04 in.) of air. The rusted area is clearly visible. To demonstrate the use of a dielectric cover for optimization, a layer of rubber was used to replace the standoff distance (Fig. 24d). The electromagnetic waves become more intense and the spatial extent of the rusted area in the image closely matched the rusted area in the object.

Figure 25a shows a rusted steel flange used in the petrochemical industry. The flange was painted uniformly. Figure 25b shows a 200 × 350 mm (8 × 14 in.) image at a frequency of 25.5 GHz and at a standoff distance of 8 mm (0.32 in.). This standoff distance was picked to avoid touching the bolt shown on the picture. The rust was clearly detected as well as other surface features of the flange.

Monitoring of Liquid States and Properties

Determination of the dielectric properties of liquids has high practical potential in many industrial, medical, as well as environmental sectors. The properties of liquids (for example, engine oil, transformer oil and antifreeze liquids) change when they are used in different environments. Change of properties is usually associated with a change of the percentages of the constituents of a liquid as well as the addition of new particles — for example, metallic particles in oil, percentage of carbon and others. Dielectric properties of mixtures are usually influenced by the dielectric properties of the individual constituents and their percentages. Consequently, monitoring the dielectric properties of a liquid with a sensitive probe can yield information about the state of the liquid.

FIGURE 23. Complex image of 130 × 120 mm (5.2 × 4.8 in.) metal bar in thick composite structure.

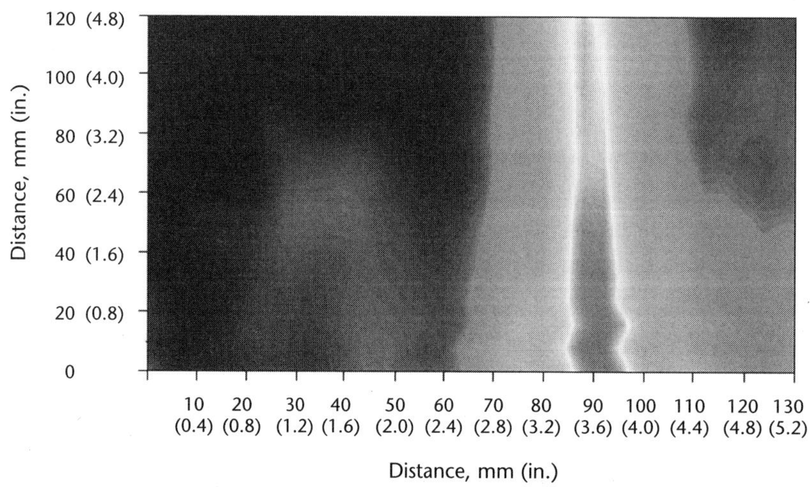

An experiment was conducted using a microwave testing system with a monopole antenna (Fig. 26). To use such a probe, the dimensions indicated on the drawing must be found. Additionally, to alleviate any confusion in the dielectric properties of the liquid under investigation, multiple-frequency measurements and calculations must be used. Frequencies of 0.7, 0.8, 1.0 and 1.2 GHz with monopole dimensions of inner radius a = 0.925 mm (3.64×10^{-2} in.), outer radius b = 3.13 mm (0.123 in.) and height h = 120 mm (4.72 in.) were found to be optimal.

Mathematical operations called *fuzzy logic systems*, developed to describe the operation of uncertain systems, were applied in this case. A fuzzy logic system alone and a fuzzy logic system with a maximum likelihood algorithm were used to determine the dielectric properties of liquids ranging in relative complex dielectric constant from $\varepsilon_r = 20 - j10$ to $\varepsilon_r = 70 - j60$. The systems were exposed to 50 random experiments in blind tests — that is, 50 random dielectric constants. Each experiment resulted in four reflection coefficients at the four optimum frequencies chosen above. From these reflection coefficients, the system is supposed to estimate the liquid dielectric properties — real and imaginary parts of

FIGURE 24. Rusted steel specimen: (a) before painting; (b) after painting; (c) experimentally obtained image; (d) experimentally obtained image with same specifications as in Fig. 24c but standoff distance medium is rubber.

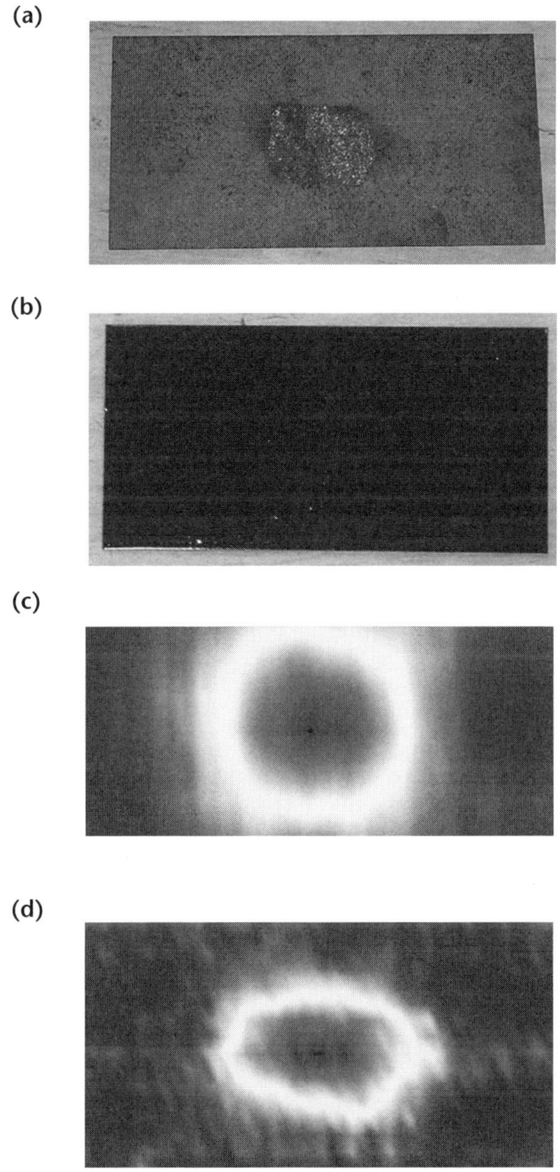

FIGURE 25. Rusted steel flange; (a) photograph; (b) experimentally obtained image at 25.5 GHz.

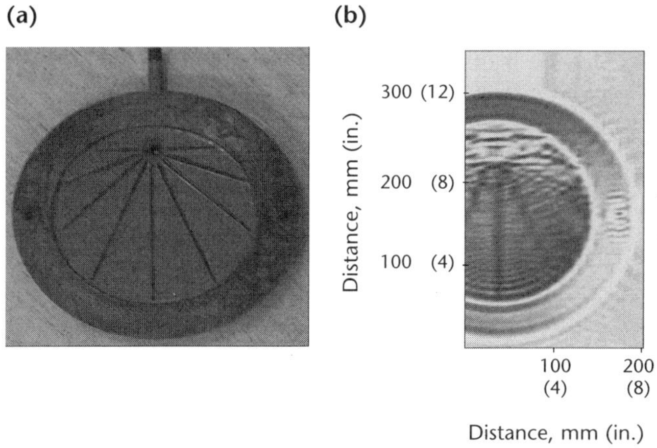

FIGURE 26. Monopole antenna.

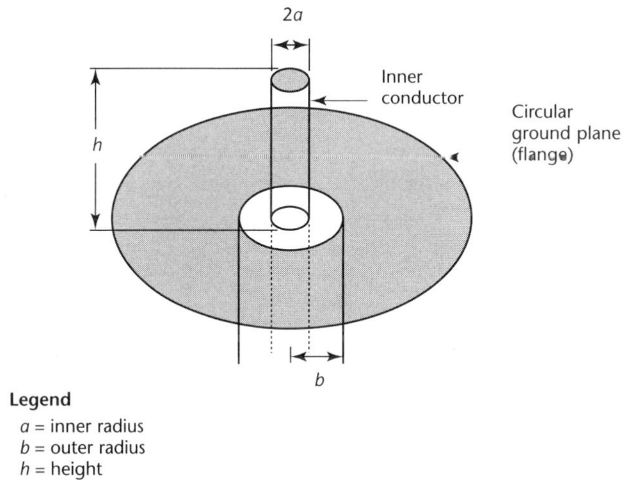

the dielectric constant. Figure 27 shows the output of the system for the real and the imaginary parts of the dielectric constant, respectively. As shown in the figures, the system using the maximum likelihood algorithm outperformed the fuzzy system algorithm and was capable of determining the dielectric properties with high accuracy. The results demonstrate the potential of using such a system for online liquid state monitoring.

Surface Crack Detection

Metal surface crack detection is another important application area of many nondestructive test techniques. Cracks may be detected underneath a layer of paint or rust or both, as well as cracks filled with a dielectric. Crack evaluation using microwave test techniques has been successfully conducted by using either open ended rectangular waveguide probes[21,33-40] or coaxial line sensors.[30,31] Each sensor has its own advantages and disadvantages as they relate to crack detection. For example, coaxial line sensors have wider operating frequency bandwidth and smaller footprints and can go into areas, such as bore holes, that rectangular waveguides cannot. On the other hand, coaxial sensors are much weaker radiators: they have to be close to the structure under test. They are also sensitive to minute standoff distance variation and their properties change with big changes in temperature.

Figure 28 shows a line scan of a steel specimen with a 0.08 mm (0.003 in.) wide crack under a 0.05 mm (0.002 in.) thick paint layer. The scan was obtained using a coaxial line sensor operating at a frequency of 10.5 GHz and at a standoff distance of zero (that is, contact). The crack is detected and the line scan indicates that information about the width of the crack may be obtained by correlating the dip at the middle of the line scan to the dimensions of the coaxial line and the width of the crack.

FIGURE 27. Dielectric constant: (a) estimation of real part; (b) estimation of imaginary part.

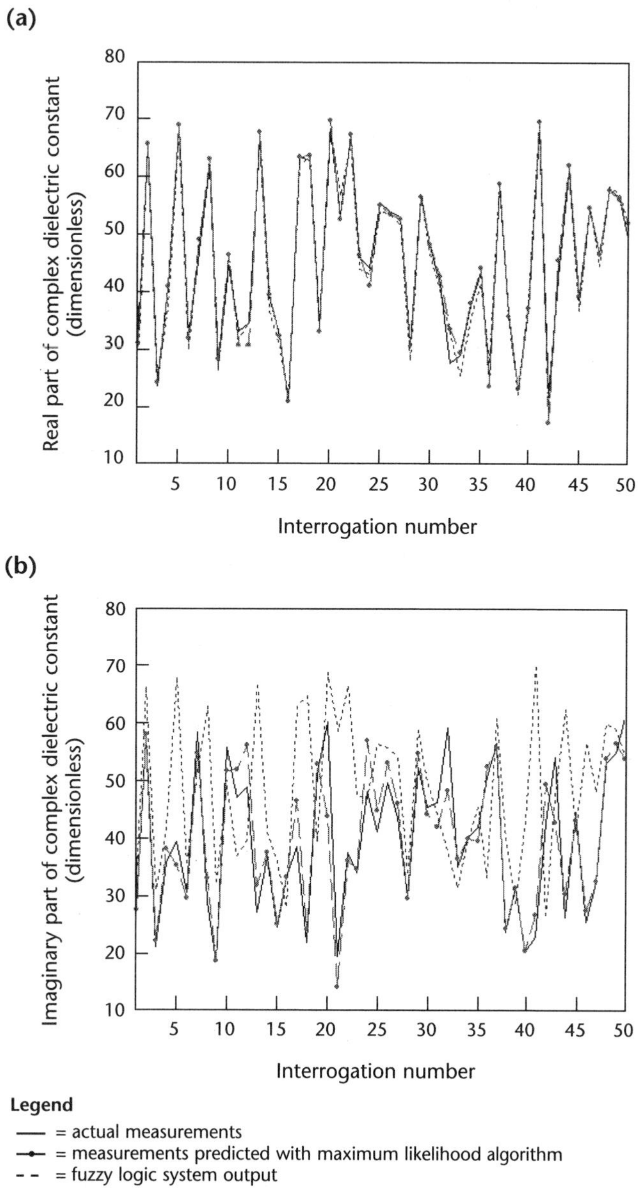

FIGURE 28. Line scan of 0.08 mm (0.003 in.) wide crack using coaxial line sensor.

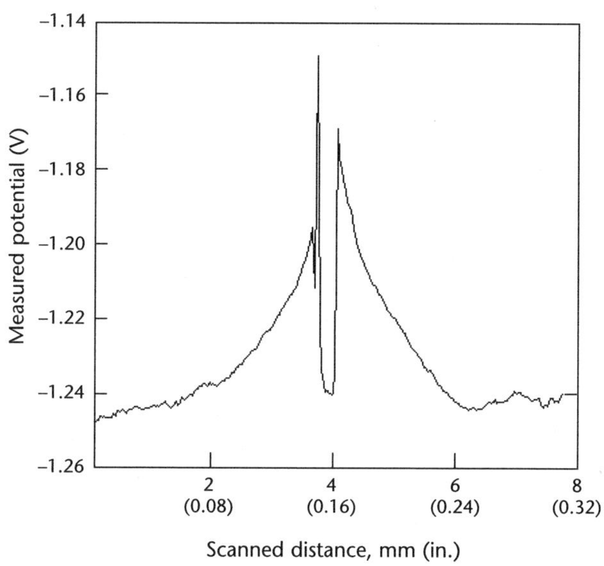

Complex Impedance Plane Testing

In addition to the microwave test techniques described above, a relatively simple spot-to-spot (or sample-to-sample) impedance plane technique can be used for various applications.[3,51,84]

The microwave impedance technique is similar to the eddy current impedance plane technique, where a flying dot signal is presented on a two-dimensional screen, such as a liquid crystal display, plasma display or cathode ray tube monitor. The flying dot represents the tip of the impedance phasor. The display can be rectilinear, showing in-phase signal changes along one axis and quadrature (1.57 rad or 90 deg out of phase) signal amplitude changes along the other axis. Or the display can be polar (Fig. 29), showing amplitude changes in the radial direction and phase changes in the circumferential direction.

The advantage of the simple complex impedance plane display is that various physical or chemical variables often move the flying dot in recognizable loci, permitting an irrelevant variable to be sorted from the variable of interest. Signal response can be interpreted as with the eddy current impedance plane technique. Either single-frequency or multifrequency techniques can be used. The practicality of the technique depends on the material and the application.

FIGURE 29. Microwave impedance plane presentation (flying dot) with phase changes in circumferential direction and amplitude changes in radial direction. Movement of signal dot from P_1 to P_2 defines direction and magnitude of change for particular material variable (affecting both phase and amplitude).[84]

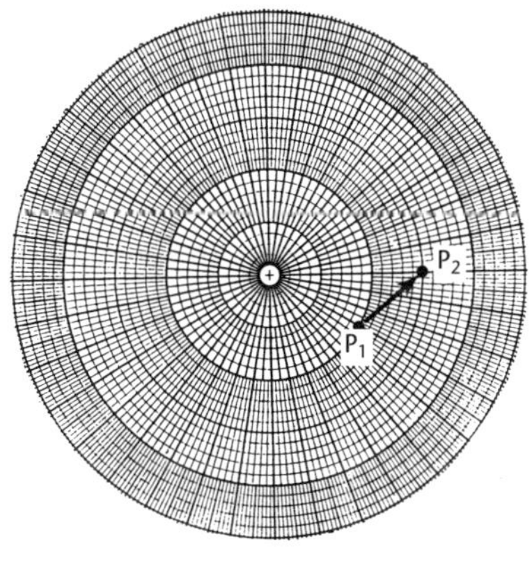

Conclusion

The examples provided in this discussion and references indicate the applicability of microwave inspection and imaging techniques for nondestructive testing. Microwave nondestructive test systems can be designed and built to be operator friendly, handheld, battery operated and portable. Microwave imaging systems are used to obtain real time images; quantifying the severity and extent of a discontinuity may require additional data and signal processing.

Quantifying may be viewed as an inverse problem: once it is solved, all the information related to the characteristics of a discontinuity may be obtained. The inverse problem is defined as reading or extracting the hidden entity's properties from the captured data. The solution to this problem is essentially a fulfillment of the fundamental objectives of the imager, such as assessment.

It is important to optimize the image theoretically before application. The optimization process solves the inverse problem.

References

1. [Botsko], R.J. and R.C. McMaster. Section 16, "Microwave Theory." *Nondestructive Testing Handbook,* second edition: Vol. 4, *Electromagnetic Testing.* Columbus, OH: American Society for Nondestructive Testing (1986): p 461-488.
2. [Botsko], R.J. and R.C. McMaster. Section 17, "Microwave Properties of Dielectric Materials." *Nondestructive Testing Handbook,* second edition: Vol. 4, *Electromagnetic Testing.* Columbus, OH: American Society for Nondestructive Testing (1986): p 489-504.
3. [Botsko], R.J., R.W. Cribbs, R.J. King and R.C. McMaster. Section 18, "Microwave Methods and Applications in Nondestructive Testing." *Nondestructive Testing Handbook,* second edition: Vol. 4, *Electromagnetic Testing.* Columbus, OH: American Society for Nondestructive Testing (1986): p 505-560.
4. NT 93-02, *Benchmarking of State-of-the-Art in NDE for Applicability of the Composite Armored Vehicle (CAV) Advance Technology Demonstrator (ATD) Program.* Austin, TX: Nondestructive Testing Information Analysis Center (1993).
5. Bahr, A.J. *Microwave Nondestructive Testing Methods.* Langhorne, PA: Gordon and Breach Science Publishers (1982).
6. Lavelle, T.M. "Microwaves in Nondestructive Testing." *Materials Evaluation.* Vol. 21, No. 11. Columbus, OH: American Society for Nondestructive Testing (November 1967): p 254-257.
7. Zoughi, R. and S. Bakhtiari. "Microwave Nondestructive Thickness Monitoring of Dielectric Slabs Backed by Conducting Plates." *International Advances in Nondestructive Testing.* Vol. 17. Langhorne, PA: Gordon and Breach Science Publishers (1994): p 301-317.
8. Ganchev, S.I., R. Runser, N. Qaddoumi, E. Ranu and G.W. Carriveau. "Microwave Nondestructive Evaluation of Thick Sandwich Composites." *Materials Evaluation.* Vol. 53, No. 4. Columbus, OH: American Society for Nondestructive Testing (April 1995): p 463-467.
9. Qaddoumi, N., T. Bigelow, R. Zoughi, L. Brown and M. Novack. "Reduction of Sensitivity to Surface Roughness and Slight Standoff Distance Variations in Microwave Testing of Thick Composite Structures." *Materials Evaluation.* Vol. 60, No. 2. Columbus, OH: American Society for Nondestructive Testing (February 2002): p 165-170.
10. Bahr, A.J. "Nondestructive Evaluation of Ceramics." *IEEE Transactions on Microwave Theory and Techniques.* Vol. MTT-26. New York, NY: Institute of Electrical and Electronics Engineers (1978): p 676-683.
11. Bakhtiari, S. and R. Zoughi. "Microwave Thickness Measurement of Lossy Layered Dielectric Slabs Using Incoherent Reflectivity." *Research in Nondestructive Evaluation.* Vol. 2, No. 3. Columbus, OH: American Society for Nondestructive Testing (1990): p 157-168.
12. Bakhtiari, S., S.I. Ganchev and R. Zoughi. "A Novel Technique for Microwave Thickness Measurement of Dielectric Slabs Using an Open-Ended Rectangular Waveguide." *Review of Progress in Quantitative Nondestructive Evaluation.* Vol. 11A. New York, NY: Plenum Press (1992): p 529-535.
13. Bakhtiari, S. *Open-Ended Sensors for Microwave Nondestructive Evaluation of Layered Composite Media.* Ph.D. dissertation. Fort Collins, CO: Colorado State University (1992).
14. Bakhtiari, S., S.I. Ganchev and R. Zoughi. "Open-Ended Rectangular Waveguide for Nondestructive Thickness Measurement and Variation Detection of Lossy Dielectric Slabs Backed by a Conducting Plate." *IEEE Transactions on Instrumentation and Measurement.* Vol. 42, No. 1. New York, NY: Institute of Electrical and Electronics Engineers (February 1993): p 19-24.
15. Bakhtiari, S., S.I. Ganchev and R. Zoughi. "Microwave Swept-Frequency Optimization for Accurate Thickness or Dielectric Property Monitoring of Conductor-Backed Composites." *Materials Evaluation.* Vol. 51, No. 6. Columbus, OH: American Society for Nondestructive Testing (June 1993): p 740-743, 748.

16. Bakhtiari, S., S.I. Ganchev and R. Zoughi. "A General Formulation for Admittance of an Open-Ended Rectangular Waveguide Radiating into Stratified Dielectrics." *Research in Nondestructive Evaluation.* Vol. 7, No. 2/3. Columbus, OH: American Society for Nondestructive Testing (1995): p 75-87.
17. Baribaud, M. and M.K. Nguen. "Maximum Entropy Image Reconstruction from Microwave Scattered Field Distribution." *18th European Microwave Conference* [Stockholm, Sweden, September 1988]. Louvain-la-Neuve, France: European Microwave Association (1988): p 891-896.
18. Mubarak, K., K.J. Bois and R. Zoughi. "A Simple, Robust and On-Site Microwave Technique for Determining Water-to-Cement (w/c) Ratio of Fresh Portland Cement-Based Materials." *IEEE Transactions on Instrumentation and Measurement.* Vol. 50, No. 5. New York, NY: Institute of Electrical and Electronics Engineers (October 2001): p 1255-1263.
19. Bolomey, J.C., L. Jofre and G. Peronnet. "On the Possible Use of Microwave Active Imaging for Remote Thermal Sensing." *IEEE Transactions on Microwave Theory and Techniques.* Vol. MTT-31, No. 9. New York, NY: Institute of Electrical and Electronics Engineers (September 1983): p 777-781.
20. Bolomey, J. and C. Pichot. "Microwave Tomography: From Theory to Practical Imaging Systems." *International Journal of Imaging Systems and Technology.* Vol. 2. New York, NY: John Wiley and Sons (1990): p 144-156.
21. Qaddoumi, N., E. Ranu, J.D. McColskey, R. Mirshahi and R. Zoughi. "Microwave Detection of Stress-Induced Fatigue Cracks in Steel and Potential for Crack Opening Determination." *Research in Nondestructive Evaluation.* Vol. 12, No. 2. Columbus, OH: American Society for Nondestructive Testing (2000): p 87-103.
22. Qaddoumi, N., L. Handjojo, T. Bigelow, J. Easter, A. Bray and R. Zoughi. "Microwave Corrosion Detection Using Open-Ended Rectangular Waveguide Sensors." *Materials Evaluation.* Vol. 58, No. 2. Columbus, OH: American Society for Nondestructive Testing (February 2000): p 178-184.
23. Qaddoumi, N. and R. Zoughi. "Preliminary Study of the Influences of Effective Dielectric Constant and Nonuniform Probe Aperture Field Distribution on Near Field Microwave Images." *Materials Evaluation.* Vol. 55, No. 10. Columbus, OH: American Society for Nondestructive Testing (October 1997): p 1169-1173.
24. D'Ambrosio, G., R. Messa, M. Migliore, A. Cilibetro and C. Sabatino. "Microwave Defect Detection on Low-Loss Composites." *Materials Evaluation.* Vol. 51, No. 2. Columbus, OH: American Society for Nondestructive Testing (February 1993): p 285-289.
25. Edrich, J. "Centimeter- and Millimeter-Wave Thermography — A Survey of Tumor Detection." *Journal of Microwave Power.* Vol. 14, No. 2. Edmonton, Canada: International Microwave Power Institute (June 1979): p 95-104.
26. Edwards, J. and R. Zoughi. "Microwave Sensitivity Maximization of Disbond Characterization in Conductor Backed Dielectric Composites." *Journal of Nondestructive Evaluation.* Vol. 12, No. 3. New York, NY: Plenum Press (September 1993): p 193-198.
27. Ermert, H., G. Fuller and D. Hiller. "Microwave Computerized Tomography." *11th European Microwave Conference* [Amsterdam, Netherlands, September 1981]. Louvain-la-Neuve, France: European Microwave Association (1981): p 421-426.
28. Bois, K.J., L. Handjojo, A. Benally, K. Mubarak and R. Zoughi. "Dielectric Plug-Loaded Two-Port Transmission Line Measurement Technique for Dielectric Property Characterization of Granular and Liquid Materials." *IEEE Transactions on Instrumentation and Measurement.* Vol. 48, No. 6. New York, NY: Institute of Electrical and Electronics Engineers (December 1999): p 1141-1148.
29. Bois, K.J., A. Benally and R. Zoughi. "An Exact Multimode Solution for the Reflection Properties of an Open-Ended Rectangular Waveguide Radiating into a Dielectric Half-Space: Forward and Inverse Problems." *IEEE Transactions on Instrumentation and Measurement.* Vol. 48, No. 6. New York, NY: Institute of Electrical and Electronics Engineers (December 1999): p 1131-1140.
30. Wang, N. and R. Zoughi. "Moment Method Solution for Modeling the Interaction of Open-Ended Coaxial Probes and Surface Cracks in Metals." *Materials Evaluation.* Vol. 60, No. 10. Columbus, OH: American Society for Nondestructive Testing (October 2002): p 1253-1258.

31. Wang, Y. and R. Zoughi. "Interaction of Surface Cracks in Metals with Open-Ended Coaxial Probes at Microwave Frequencies." *Materials Evaluation*. Vol. 58, No. 10. Columbus, OH: American Society for Nondestructive Testing (October 2000): p 1228-1234.
32. Qaddoumi, N. and R. Zoughi. "Microwave Detection of Corrosion under Paint and Composite Laminate Coatings." *Nondestructive Evaluation of Aging Aircraft, Airports and Aerospace Hardware* [Scottsdale, AZ, November 1996]. SPIE Proceedings, Vol. 2945. Bellingham, WA: International Society for Optical Engineering (1996): p 346-352.
33. Zoughi, R., S.I. Ganchev and C. Huber. "Microwave Measurement-Parameter Optimization for Detection of Surface Breaking Hairline Cracks in Metals." *Nondestructive Testing and Evaluation*. Vol. 14. Langhorne, PA: Gordon and Breach Science Publishers (1998): p 323-337.
34. Huber, C., H. Abiri, S.I. Ganchev and R. Zoughi. "Modeling of Surface Hairline Crack Detection in Metals under Coatings Using Open-Ended Rectangular Waveguides." *IEEE Transactions on Microwave Theory and Techniques*. Vol. 45, No. 11. New York, NY: Institute of Electrical and Electronics Engineers (November 1997): p 2049-2057.
35. Huber, C., S.I. Ganchev, R. Mirshahi, J. Easter and R. Zoughi. "Remote Detection of Surface Cracks/Slots Using Open-Ended Rectangular Waveguide Sensors: An Experimental Investigation." *Nondestructive Testing and Evaluation*. Vol. 13. Langhorne, PA: Gordon and Breach Science Publishers (1997): p 227-237.
36. Huber, C., H. Abiri, S.I. Ganchev and R. Zoughi. "Analysis of the Crack Characteristic Signal Using a Generalized Scattering Matrix Representation." *IEEE Transactions on Microwave Theory and Techniques*. Vol. 45, No. 4. New York, NY: Institute of Electrical and Electronics Engineers (April 1997): p 477-484.
37. Ganchev, S.I., R. Zoughi, C. Huber, R. Runser and E. Ranu. "Microwave Method for Locating Surface Crack Tips in Metals." *Materials Evaluation*. Vol. 54, No. 5. Columbus, OH: American Society for Nondestructive Testing (May 1996): p 598-603.
38. Yeh, C. and R. Zoughi. "Sizing Technique for Surface Cracks in Metals." *Materials Evaluation*. Vol. 53, No. 4. Columbus, OH: American Society for Nondestructive Testing (April 1995): p 496-501.
39. Yeh, C. and R. Zoughi. "Microwave Detection of Finite Surface Cracks in Metals Using Rectangular Waveguide Sensors." *Research in Nondestructive Evaluation*. Vol. 6, No. 1. Columbus, OH: American Society for Nondestructive Testing (1994): p 35-55.
40. Yeh, C. and R. Zoughi. "A Novel Microwave Method for Detection of Long Surface Cracks in Metals." *IEEE Transactions on Instrumentation and Measurement*. Vol. 43, No. 5. New York, NY: Institute of Electrical and Electronics Engineers (October 1994): p 719-725.
41. Gopalsami, N., S. Bakhtiari, S.L. Dieckman, A.C. Raptis and M.J. Lepper. "Millimeter-Wave Imaging for Nondestructive Evaluation of Materials." *Materials Evaluation*. Vol. 52, No. 3. Columbus, OH: American Society for Nondestructive Testing (March 1994): p 412-415.
42. Jordan, E.C. and K.G. Balmain. *Electromagnetic Waves and Radiating Systems*. Upper Saddle River, NJ: Prentice Hall (1991).
43. Balanis, C.A. *Advanced Engineering Electromagnetics*. New York, NY: John Wiley and Sons (1989).
44. Liao, S.Y. *Microwave Devices and Circuits*. Upper Saddle River, NJ: Prentice Hall (1980).
45. Harrington, R.F. *Time-Harmonic Electromagnetic Fields*. New York, NY: McGraw-Hill (1961).
46. Kraus, J.D. *Electromagnetics*, third edition. New York, NY: McGraw-Hill (1984).
47. Chatterjee, R. *Advanced Microwave Engineering*. Chichester, United Kingdom: Ellis Horwood Limited (1988).
48. Stratton, J.A. *Electromagnetic Theory*. New York, NY: McGraw-Hill (1941).
49. Harvey, A. *Microwave Engineering*. New York, NY: Academic Press (1963).
50. Kerker, M. *The Scattering of Light and Other Electromagnetic Radiation*. New York, NY: Academic Press (1969).
51. Hochschild, R. "Microwave Nondestructive Testing in One (Not-So-Easy) Lesson." *Materials Evaluation*. Vol. 26, No. 1. Columbus, OH: American Society for Nondestructive Testing (January 1968): p 35A-42A.
52. Bahr, A.J. Report No. AMMRC CTR-77-29, *Microwave Techniques for Nondestructive Evaluation of Ceramics*. Contract No. DAAG46-76-C-0048 with SRI International, Menlo Park, CA. Watertown, MA: Army Materials and Mechanics Research Center (1977).

53. Krautkrämer, J. and H. Krautkrämer. *Ultrasonic Testing of Materials,* fourth edition. Berlin, Germany: Springer-Verlag (1990): p 59.
54. Hueter, T. and R. Bolt. *Sonics,* third edition. New York, NY: John Wiley and Sons (1962).
55. Zoughi, R. and S.I. Ganchev. *Microwave Nondestructive Evaluation: State-of-the-Art Review.* Austin, TX: Texas Research Institute, for the Nondestructive Testing Information Analysis Center (1994).
56. Peer, S., J.T. Case, R. Zoughi and K.E. Kurtis. "Microwave Evaluation of Accelerated Chloride Ingress in Mortar Subjected to Cyclical Loading." *ASNT Spring Conference and 11th Annual Research Symposium* [Portland, Oregon, March 2002]. Columbus, OH: American Society for Nondestructive Testing (2002): p 80.
57. Zoughi, R., G. Cone and P. Nowak. "Microwave Nondestructive Detection of Rebars in Concrete Slabs." *Materials Evaluation.* Vol. 49, No. 11. Columbus, OH: American Society for Nondestructive Testing (November 1991): p 1385-1388.
58. Bois, K.J., A. Benally and R. Zoughi. "Near-Field Microwave Non-Invasive Determination of NaCl in Mortar." *IEE Proceedings — Science, Measurement and Technology.* Vol. 148, No. 4. London, United Kingdom: Institution of Electrical Engineers (July 2001): p 178-182.
59. Bois, K.J., A. Benally, P.S. Nowak and R. Zoughi. "Microwave Nondestructive Determination of Sand-to-Cement Ratio in Mortar." *Research in Nondestructive Evaluation.* Vol. 9, No. 4. Columbus, OH: American Society for Nondestructive Testing (1997): p 227-238.
60. Muñoz, K. and R. Zoughi. "Influence of Cyclical Soaking in Chloride Bath and Drying of Mortar on Its Microwave Dielectric Properties: The Forward Model." *Review of Progress in Quantitative Nondestructive Evaluation* [Bellingham, WA, July 2002]. Vol. 22. AIP Proceedings, Vol. 657. Melville, NY: American Institute of Physics (2003): p 470-477.
61. Case, J.T., E. Gallaher, S. Peer, R. Zoughi and K.E. Kurtis. "Microwave Reflection and Dielectric Properties of Mortar Exposed Periodically to 2% Chloride Salt Solution and Compression Force." *Review of Progress in Quantitative Nondestructive Evaluation* [Bellingham, WA, July 2002]. Vol. 22. AIP Proceedings, Vol. 657. Melville, NY: American Institute of Physics (2003): p 1132-1138.
62. Case, T., E. Gallaher, S. Peer, R. Zoughi and K.E. Kurtis. "Microwave Reflection and Dielectric Properties of Mortar Exposed Periodically to Chloride Solution with 1% Salinity and Compression Force." *Proceedings of the Eleventh International Symposium on Nondestructive Characterization of Materials* [Berlin, Germany, June 2002]. Berlin, Germany: Deutsche Gesellschaft für Zerstörungsfreie Prüfung (2002).
63. Zoughi, R., S.I. Ganchev, J. Bhattacharyya, S. Gray and D. Radford. "Microwave Diagnosis of Carbon Black Rubber Compounds and Porosity Estimation in Composites." *Advanced Metallization for Devices and Circuits — Science, Technology, and Manufacturability* [San Francisco, CA, April 1994]. Pittsburgh, PA: Materials Research Society (1994).
64. Bois, K.J., H. Campbell, A. Benally, P.S. Nowak and R. Zoughi. "Microwave Noninvasive Detection of Grout in Masonry." *Masonry Society Journal.* Vol. 16, No. 1. Boulder, CO: Masonry Society (June 1998): p 49-54.
65. Zoughi, R. and M. Lujan. "Nondestructive Microwave Thickness Measurement of Dielectric Slabs." *Materials Evaluation.* Vol. 48, No. 9. Columbus, OH: American Society for Nondestructive Testing (September 1990): p 1100-1105.
66. Gray, S. and R. Zoughi. "Dielectric Sheet Thickness Variation and Disbond Detection in Multilayered Composites Using an Extremely Sensitive Microwave Approach." *Materials Evaluation.* Vol. 55, No. 1. Columbus, OH: American Society for Nondestructive Testing (January 1997): p 42-48.
67. Qaddoumi, N., R. Zoughi and G.W. Carriveau. "Microwave Detection and Depth Determination of Disbonds in Low-Permittivity and Low-Loss Thick Sandwich Composites." *Research in Nondestructive Evaluation.* Vol. 8, No. 1. Columbus, OH: American Society for Nondestructive Testing (1996): p 51-63.
68. Venugopalan, P., K. Jose, K. Nair, P. Chaturvedi and V. Ravindran. "Microwave Technique for Locating Inhomogeneities in Cured Rocket Propellant Samples." *NDT International.* Vol. 19, No. 6. Oxford, United Kingdom: Elsevier Science Limited (December 1986): p 395-397.

69. Gray, S., S.I. Ganchev, N. Qaddoumi, D. Radford and R. Zoughi. "Porosity Level Estimation in Polymer Composites Using Microwaves." *Materials Evaluation.* Vol. 53, No. 3. Columbus, OH: American Society for Nondestructive Testing (March 1995): p 404-408.
70. Radford, D.W., S.I. Ganchev, N. Qaddoumi, G. Beauregard and R. Zoughi. "Millimeter-Wave Nondestructive Evaluation of Glass Fiber/Epoxy Composites Subjected to Impact Fatigue." *Advanced Microwave and Millimeter-Wave Detectors.* SPIE Proceedings, Vol. 2275. Bellingham, WA: International Society for Optical Engineering (1994): p 21-26.
71. Qaddoumi, N., G.W. Carriveau, S.I. Ganchev and R. Zoughi. "Microwave Imaging of Thick Composites with Defects." *Materials Evaluation.* Vol. 53, No. 8. Columbus, OH: American Society for Nondestructive Testing (August 1995): p 926-929.
72. Ganchev, S.I., N. Qaddoumi, D. Brandenburg, S. Bakhtiari, R. Zoughi and J. Bhattacharyya. "Microwave Diagnosis of Rubber Compounds." *IEEE Transactions on Microwave Theory and Techniques.* Vol. 42, No. 1. New York, NY: Institute of Electrical and Electronics Engineers (January 1994): p 18-24.
73. Ganchev, S.I., N. Qaddoumi, S. Bakhtiari and R. Zoughi. "Calibration and Measurement of Dielectric Properties of Finite Thickness Composite Sheets with Open-Ended Coaxial Sensors." *IEEE Transactions on Instrumentation and Measurement.* Vol. IM-44, No. 6. New York, NY: Institute of Electrical and Electronics Engineers (December 1995): p 1023-1029.
74. Paoloni, F.J. "Implementation of Microwave Diffraction Tomography for Measurement of Dielectric Constant Distribution." *IEE Proceedings:* Series H, *Microwaves, Antennas, and Propagation.* Vol. 134, No. 1. London, United Kingdom: Institution of Electrical Engineers (February 1987): p 25-29.
75. Bois, K.J., A.D. Benally, P.S. Nowak and R. Zoughi. "Cure-State Monitoring and Water-to-Cement Ratio Determination of Fresh Portland Cement Based Materials Using Near Field Microwave Techniques." *IEEE Transactions on Instrumentation and Measurement.* Vol. 47, No. 3. New York, NY: Institute of Electrical and Electronics Engineers (June 1998): p 628-637.
76. Zoughi, R. and S. Bakhtiari. "Microwave Nondestructive Detection and Evaluation of Voids in Layered Dielectric Slabs." *Research in Nondestructive Evaluation.* Vol. 2, No. 4. Columbus, OH: American Society for Nondestructive Testing (1990): p 195-205.
77. Perronet, G., C. Pichot, J.C. Bolomey, L. Jofre, A. Izadnegahdar, C. Szeles, Y. Michel, J.L. Guerquin-Kern and M. Gautherie. "A Microwave Diffraction Tomography System for Biological Applications." *13th European Microwave Conference* [Nürnberg, Germany, September 1983]. Louvain-la-Neuve, France: European Microwave Association (1983): p 529-533.
78. Pichot, C., L. Jofre, G. Peronnet and J.C. Bolomey. "Active Microwave Imaging of Inhomogeneous Bodies." *IEEE Transactions on Antennas and Propagation.* Vol. AP-33, No. 4. New York, NY: Institute of Electrical and Electronics Engineers (April 1985): p 416-425.
79. Michiguchi, Y., K. Hiramoto, M. Nishi, T. Ootaka and M. Okada. "Advanced Subsurface Radar System for Imaging Buried Pipes." *IEEE Transactions on Geoscience and Remote Sensing.* Vol. GE-26, No. 6. New York, NY: Institute of Electrical and Electronics Engineers (November 1988): p 733-740.
80. Richards, P.J. and A.P. Anderson. "Microwave Imaging of Sub-Surface Utilities in an Urban Environment." *8th European Microwave Conference* [Paris, France, September 1978]. Sevenoaks, United Kingdom: Microwave Exhibitions and Publishers Limited, for the European Microwave Association (1978): p 33-37.
81. Bois, K.J., A. Benally and R. Zoughi. "Microwave Near-Field Reflection Property Analysis of Concrete for Material Content Determination." *IEEE Transactions on Instrumentation and Measurement.* Vol. 49, No. 1. New York, NY: Institute of Electrical and Electronics Engineers (February 2000): p 49-55.
82. Zoughi, R. *Microwave Non-Destructive Testing and Evaluation.* Dordrecht, Netherlands: Kluwer Academic Publishers (2000).
83. Watters, D., D. Falconer, K. Harker, R. Ueberschaer and A. Bahr. "Microwave Inspection of Luggage for Contraband Materials Using Imaging and Inverse-Scattering Algorithms." *Research in Nondestructive Evaluation.* Vol. 7, No. 2/3. Columbus, OH: American Society for Nondestructive Testing (July 1995): p 153-168.

84. [Botsko], R.J. "Nondestructive Testing of Plastics with Microwaves." *Materials Evaluation*. Vol. 17, No. 6. Columbus, OH: American Society for Nondestructive Testing (June 1969): p 25A-32A.

Bibliography

Altschuler, H.M. "Dielectric Constant." *Handbook of Microwave Measurements*. Vol. 2, Chapter 9. Brooklyn, NY: Polytechnic Press (1963).

Anderson, A.P. and S.J. Mawani. "A Microwave Eye Can Be Almost Human." *6th European Microwave Conference* [Rome, Italy, September 1976]. Sevenoaks, United Kingdom: Microwave Exhibitions and Publishers Limited, for the European Microwave Association (1976): p 105-111.

Arcone, S.A. and R.W. Larson. "Single-Horn Reflectometry for in Situ Dielectric Measurements at Microwave Frequencies." *IEEE Transactions on Geoscience and Remote Sensing*. Vol. 26, No. 1. New York, NY: Institute of Electrical and Electronics Engineers (January 1988): p 89-92.

Bahr, A.J., R. Zoughi and N. Qaddoumi. "Microwave." *Nondestructive Evaluation: Theory, Techniques and Applications*. New York, NY: Marcel Dekker (2002).

Baker-Jarvis, J., C. Jones, B. Riddle, M. Janezic, R. Geyer, J. Grosvenor, Jr. and C. Weil. "Dielectric and Magnetic Measurements: A Survey of Nondestructive, Quasi-Nondestructive, and Process-Control Techniques." *Research in Nondestructive Evaluation*. Vol. 7, No. 2/3. Columbus, OH: American Society for Nondestructive Testing (1995): p 117-136.

Bakhtiari, S., N. Gopalsami and A.C. Raptis. "Characterization of Delamination and Disbonding in Stratified Dielectric Composites by Millimeter Wave Imaging." *Materials Evaluation*. Vol. 53, No. 4. Columbus, OH: American Society for Nondestructive Testing (April 1995): p 468-471.

Belhadj-Tahar, N.-E., A. Fourrier-Lamar and H. de Chanterac. "Broad Band Simultaneous Measurement of Complex Permittivity and Permeability Using a Coaxial Discontinuity." *IEEE Transactions on Microwave Theory and Techniques*. Vol. MTT- 38. New York, NY: Institute of Electrical and Electronics Engineers (January 1990): p 1-7.

Broquetas, A., M. Ferrand, J.M. Rius, L. Jofre, E. de los Reyes, A. Cardama, A. Elias and J. Ibanez. "Temperature and Permittivity Measurements Using a Cylindrical Microwave Imaging System." *17th European Microwave Conference* [Rome, Italy, September 1987]. Louvain-la-Neuve, France: European Microwave Association (September 1987): p 892-895.

Campbell, H., F. Bastidas, E. Winkler, R. Zoughi, D. Raymond and D. Williams. "Optimization of Microwave Inspection Techniques for Fatigue Crack Detection in Bridges." *ASNT Fall Conference and Quality Testing Show 2000 Paper Summaries Book* [Indianapolis, IN, November 2000]. Columbus, OH: American Society for Nondestructive Testing (2000): p 183-184.

Caorsi, S., G.L. Gragnani and M. Pastarino. "A Numerical Approach to Microwave Imaging." *18th European Microwave Conference* [Stockholm, Sweden, September 1988]. Louvain-la-Neuve, France: European Microwave Association (1988): p 892-895.

Chaloupka, H. "Imaging of Objects in a Half-Space with Unknown Permittivity." *Electronics Letters*. Vol. 20, No. 13. London, United Kingdom: Institution of Electrical Engineers (June 1987): p 570-572.

Chu, T.H. and K.Y. Lee. "Wideband Microwave Diffraction Tomography under Born Approximation." *IEEE AP-S International Symposium Digest*. New York, NY: Institute of Electrical and Electronics Engineers, Antennas and Propagation Society (June 1987): p 1042-1045.

Ciocan, R. and N. Ida. "A Transmission Line Matrix Model for Microwave NDT." *ASNT Spring Conference and 11th Annual Research Symposium* [Portland, Oregon, March 2002]. Columbus, OH: American Society for Nondestructive Testing (2002): p 86.

Decreton, M.C. and F.E. Gardiol. "Simple Non-Destructive Method for Measurement of Complex Permittivity." *IEEE Transactions on Instrumentation and Measurement*. Vol. IM-23. New York, NY: Institute of Electrical and Electronics Engineers (December 1974): p 434-438.

Diener, L. "Microwave Near-Field Imaging with Open-Ended Waveguide — Comparison with Other Techniques of Nondestructive Testing." *Research in Nondestructive Evaluation*. Vol. 7, No. 2/3. Columbus, OH: American Society for Nondestructive Testing (1995): p 137-152.

Ghodgaonkar, D.K., V.V. Varadan and V.K. Varadan. "A Free-Space Method for Measurement of Dielectric Constant and Loss Tangent at Microwave Frequencies." *IEEE Transactions on Instrumentation and Measurement.* Vol. 37, No. 3. New York, NY: Institute of Electrical and Electronics Engineers (June 1989): p 789-793.

Goth, M. and D. Raymond. "Surface Crack Detection on Bridge #E-16-HI in Golden, Colorado Using an Open-Ended Rectangular Waveguide Device." *ASNT Spring Conference and 11th Annual Research Symposium* [Portland, Oregon, March 2002]. Columbus, OH: American Society for Nondestructive Testing (2002): p 77.

Han, H. and E. Mansueto. "Thin Film Inspection with Millimeter-Wave Reflectometer." *Research in Nondestructive Evaluation.* Vol. 7, No. 2/3. Columbus, OH: American Society for Nondestructive Testing (1995): p 89-100.

Hughes, D., C. Behrens, R. Zoughi, G. Green and P. Campbell. "Microwave and Millimeter Wave Inspection of Impact Damage in GFRP Composite and Pitted Corrosion in Steel." *ASNT Spring Conference and 11th Annual Research Symposium* [Portland, Oregon, March 2002]. Columbus, OH: American Society for Nondestructive Testing (2002): p 85.

Kaiser, J. "Reflection and Absorption of Microwaves (2-100 GHz) by Carbon-Fiber-Reinforced Composite Surfaces." *Research in Nondestructive Evaluation.* Vol. 5, No. 4. Columbus, OH: American Society for Nondestructive Testing (1994): p 275-283.

Lundien, J.R. Technical Report M-72-4, *Determining Presence Thickness and Electrical Properties of Stratified Media Using Swept Frequency Radar.* Vicksburg, MS: United States Army Corps of Engineers, Waterways Experimental Station (November 1972).

Orme, R.D. and A.P. Anderson. "High Resolution Microwave Holographic Technique." *IEE Proceedings: Series H, Microwaves, Antennas, and Propagation.* Vol. 134, No. 1. London, United Kingdom: Institution of Electrical Engineers (February 1987): p 25-29.

Otto, G.P. and W.C. Chew. "Microwave Inverse Scattering — Local Shape Function Imaging for Improved Resolution of Strong Scatters." *IEEE Transactions on Microwave Theory and Techniques.* Vol. MTT-42, No. 1. New York, NY: Institute of Electrical and Electronics Engineers (January 1994): p 137-141.

Paoloni, F.J. and M.J. Duffy. "Microwave Imaging with Digital Reconstructions." *Journal of Electrical and Electronics Engineering, Australia.* Vol. 4, No. 1. Barton, Australia: Institution of Engineers, Australia (March 1984): p 54-59.

Qaddoumi, N., A. Shroyer and R. Zoughi. "Microwave Detection of Rust under Paint and Composite Laminates." *Research in Nondestructive Evaluation.* Vol. 9, No. 4. Columbus, OH: American Society for Nondestructive Testing (1997): p 201-212.

Qaddoumi, N., S.I. Ganchev and R. Zoughi. "Microwave Diagnosis of Low-Density Fiberglass Composites with Resin Binder." *Research in Nondestructive Evaluation.* Vol. 8, No. 3. Columbus, OH: American Society for Nondestructive Testing (1996): p 177-188.

Raymond, D.M. and M.L. Goth. "Design and Development of a Portable Microwave System for Detection of Surface Cracks in Metals." *ASNT Spring Conference and 11th Annual Research Symposium* [Portland, Oregon, March 2002]. Columbus, OH: American Society for Nondestructive Testing (2002): p 76.

Rhim, H.C., O. Büyüköztürk and D.J. Blejer. "Remote Radar Imaging of Concrete Slabs with and without a Rebar." *Materials Evaluation.* Vol. 53, No. 2. Columbus, OH: American Society for Nondestructive Testing (February 1995): p 295-299.

Schlegel, J.L., J.W. Wagner and R.E. Green, Jr. "Microwave Dielectrometry Measurements of Glass Reinforced Polyester Resins." *Materials Evaluation.* Vol. 57, No. 10. Columbus, OH: American Society for Nondestructive Testing (October 1999): p 1091-1094.

Shalaby, W. and R. Zoughi. "Microwave Compressive Strength Estimation of Cement Paste Using Monopole Probes." *Research in Nondestructive Evaluation.* Vol. 7, No. 2/3. Columbus, OH: American Society for Nondestructive Testing (1995): p 101-105.

Tabib-Azar, M. "Applications of an Ultra High Resolution Evanescent Microwave Imaging Probe in the Nondestructive Testing of Materials." *Materials Evaluation.* Vol. 59, No. 1. Columbus, OH: American Society for Nondestructive Testing (January 2001): p 70-78.

13

CHAPTER

Electromagnetic Techniques for Material Identification

Timothy E. Kinsella, Dassault Falcon Jet, Wilmington, Delaware

Donald J. Hagemaier, Huntington Beach, California (Part 2)

Johann H. Hinken, Hochschule Magdeburg-Stendal, Magdeburg, Germany (Part 6)

George Mordwinkin, Sensor Corporation, Scottdale, Pennsylvania (Parts 5 and 6)

Andrew P. Washabaugh, Jentek Sensors, Waltham, Massachusetts (Part 7)

Markus Zahn, Massachusetts Institute of Technology, Cambridge, Massachusetts (Part 7)

Part 1. Introduction to Electromagnetic Identification of Materials

Absolute identification of chemical composition of microstructure uses techniques such as chemical spot testing, metallography, spectroanalysis and X-ray fluorescence. These techniques are covered in other volumes of the *Nondestructive Testing Handbook*.[1,2]

Electromagnetic tests can nondestructively provide useful information for characterization or sorting of materials. Electromagnetic techniques include impedance plane analysis, conductivity testing, electrical resistivity measurements, hysteresis curve analysis, thermoelectric measurements and dielectric measurements. The discussion in this chapter is based on information in the previous edition[3] and has additional information on dielectrometry.

Eddy current testing is well suited to the task of sorting materials. When a conductive material is placed in the field of an eddy current probe, a specific impedance value is established in the test circuit. When a different material is placed in the field, the resistance and inductive reactance of the circuit will change. In many cases these material variations can be analyzed by examining the impedance plane display of an eddy current instrument.

Metals are by definition good conductors of electricity. The addition of alloying elements or changes in microstructure caused by processing can cause changes in a material's electrical conductivity that can be measured and analyzed. The effects of microstructure can be measured either with special purpose conductivity testers or with conventional eddy current equipment, preferably equipment having an impedance plane display.

Conventional eddy current measurement of electrical conductivity is an indirect measurement negatively influenced by surface conditions (roughness, curvature and others). Instruments, such as a four-point probe, are available that measure resistivity directly. These instruments are significant because they require no reference standard, because measurements can be made on bulk material in the as manufactured state and because measurements are made directly without intermediate calculation.

Magnetic materials are characterized by their hysteresis curve, which relates magnetic field intensity and the resulting flux density. The variables shown in the hysteresis loop (saturation, retentivity, coercive force and permeability) are influenced to some extent by basic characteristics and properties of magnetic materials, including their chemical composition, metallurgical structure and heat treatment.

Another useful approach to material characterization or sorting uses thermoelectric effects. The most commonly used is the seebeck effect: if two conductors are joined and the junctions are maintained at different temperatures then an electromotive force will be developed around the closed loop. This force is the thermocouple voltage. Seebeck coefficients vary for different materials. Changing the probe tip materials and temperatures can optimize detectable differences.

Although dielectric materials are not metals, they can also be analyzed by using electromagnetic techniques that measure resistivity or permittivity. A dielectric material is a poor conductor of electric charge in which an applied electric field causes a displacement of charge but little or no flow of charge.

PART 2. Eddy Current Impedance Plane Analysis

Impedance Plane

When an alternating current voltage at angular frequency ω (radian per second) is applied to an eddy current circuit, the current flows through the inductive reactance X_L and the resistance R where $X_L = L_\omega$ and L is inductance (henry). The inductive reactance voltage is identified as E_1 in Fig. 1 and the resistance voltage is identified as E_2. The specific value of the voltage is the product of the current I (ampere) and either the inductive reactance or the resistance:

$$E_1 = I X_L \tag{1}$$

and

$$E_2 = I R \tag{2}$$

The voltage E_1 across the inductive reactance is 90 degrees out of phase with the voltage E_2 across the resistance. These two voltages can be represented as shown in Fig. 1.

Because the current through both the inductive reactance and the resistance at any given time is at the same value, the voltage values on the voltage plane diagram may be divided by the current value to give the values of inductive reactance and resistance in the circuit. The resulting diagram is called a *phase vector* or *phasor diagram* and is used to show the amplitude and phase relationship of alternating current signals having the same frequency (see Fig. 1).

FIGURE 1. Voltage plane and impedance plane diagrams: (a) voltage plane; (b) resistance inductance circuit; (c) impedance plane.

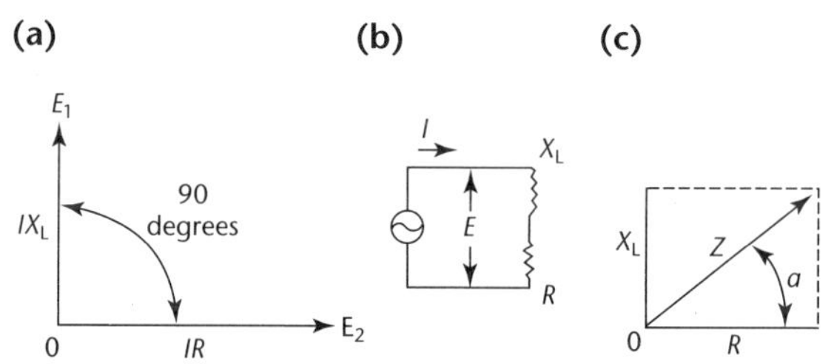

Legend

- a = phase angle
- E = voltage
- I = current
- R = ohmic resistance
- X_L = inductive reactive of coil with self inductance L
- Z = impedance phasors

When the values of R and X_L are varied, the voltage drop across the circuit varies and, depending on these values, the voltage drop is represented by different impedance phasors Z on the impedance plane. The total voltage drop E_r is the phasor or vector sum of E_1 and E_2.

$$Z = jX_L + R \tag{3}$$

where $j = \sqrt{(-1)}$, or:

$$Z = \sqrt{X_L^2 + R^2} \tag{4}$$

The values of these voltage components depend on the value of circuit resistance, reactance and frequency. The voltage drop across the resistance is proportional to the current times the resistance: IR.

The voltage drop across the reactance is proportional to both the inductance and the frequency.

Vector addition of the values of inductive reactance and resistance, plotted 90 degrees apart, will indicate the impedance value $\boldsymbol{Z}$ and the lag of phase angle. A similar addition can be applied to the current values.

In general, the test coil is characterized by two electrical impedance quantities: (1) the inductive reactance X_L (where the frequency of the alternating current field is in hertz and coil self-inductance L is in henry) and (2) the ohmic resistance R.

It is common practice to plot reactance X_L as the ordinate and resistance R as the abscissa in the impedance plane. In this way, the test coil impedance $\boldsymbol{Z}$ is represented by a point P formed by two perpendicular components X_L and R on the impedance plane. In the absence of a test object, the empty test coil has a characteristic impedance with coordinates X_{L0} and R_0 shown on the impedance plane by the coil in air, point P_0 of Fig. 2. If the probe is placed on the test object, the original field of the coil in air is modified by the superimposed field of the eddy currents. This field modification has exactly the same effect as would be obtained if the characteristics of the test coil itself had been changed. The influence of the test object can be described by a variation in the test coil characteristics. The apparent impedance P_0 of the coil in air is displaced to P_1 (corresponding to new values of X_L and R) under the influence of the test object (Fig. 2).

The magnitude and direction of the displacement of the apparent impedance from P_0 to P_1 under the influence of the test object are functions of the properties of the test object and the characteristics of the instrumentation. Significant properties of the test object include: (1) electrical conductivity σ, (2) dimensions of the test object, (3) magnetic permeability μ, (4) standard depth of penetration δ and (5) presence of discontinuities such as cracks. Significant instrument characteristics include (1) frequency f of the alternating current field in the test coil, (2) size and shape of the test coil and (3) distance of the test coil from the test object (liftoff).

Liftoff and Edge Effects on Impedance Plane

The spacing between the probe and the test object surface is referred to as the *liftoff*. Figure 3 shows the impedance plane response that occurs when the spacing is increased between the probe and the test object surface. The upper portion of the impedance plane is the magnetic domain, where responses occur from ferromagnetic materials. The lower portion is the domain where responses are obtained from nonmagnetic materials.

FIGURE 2. Representation of test coil characteristics on impedance plane.

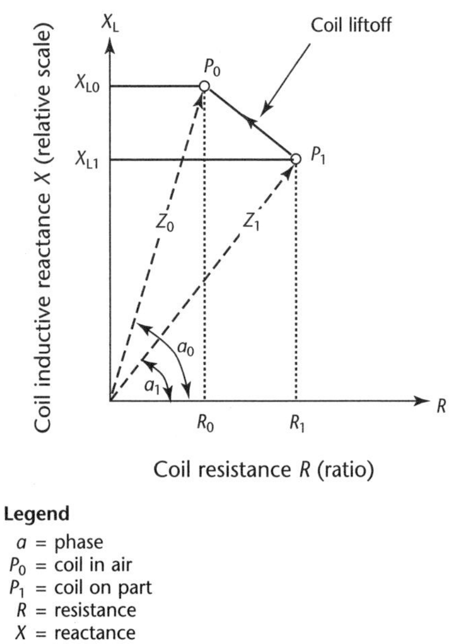

Note the nonlinear (logarithmic) changes among the liftoff loci for equal increments of spacing.

When the probe is moved near the edge of the test object, an edge effect occurs because a part of the magnetic field is outside the test object. For nonmagnetic materials, the resultant effect produces a response similar to the liftoff response (compare Fig. 3a). For ferromagnetic materials, the edge effect response curves to the left of the liftoff locus line. This evidently occurs because the magnetic field becomes distorted at the edge of the test object; the magnetic field wants to remain in the material. The angle of curvature A (Fig. 3b) increases or decreases as a function of operating frequency and coil diameter.

FIGURE 3. Liftoff and edge effect loci on impedance plane: (a) liftoff loci; (b) edge effect loci.

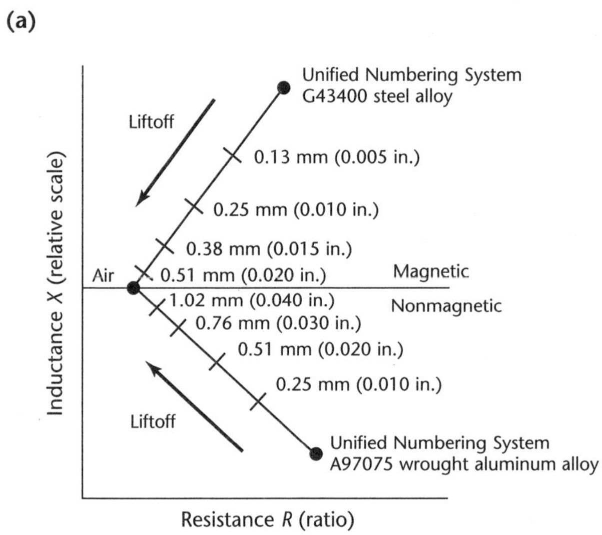

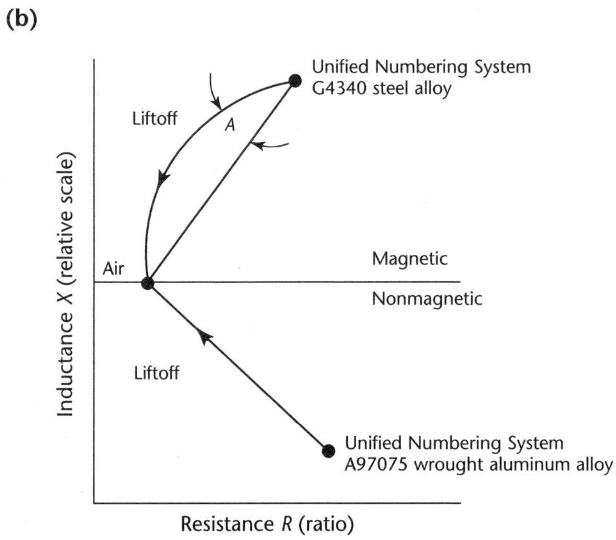

Conductivity and Permeability Loci on Impedance Plane[4]

The impedance plane response to the different conductivities of various nonmagnetic alloys is shown in Fig. 4. The material points trace out a characteristic comma shaped curve with conductivity increasing in a clockwise direction. The coil liftoff loci are shown for the different metals. Note that the separation angle between the conductivity locus and the liftoff locus is much smaller for titanium than it is for copper. Hence the unwanted liftoff variable will affect test results less when testing copper or aluminum alloys at 100 kHz than it will when testing titanium or graphite.

The material points are spaced around the conductivity locus in a nonlinear fashion. For example, the spacing between titanium and stainless steel at the top of the curve is much greater than it is between bronze and aluminum at the bottom of the curve.

Figure 5 shows the effect of test frequency on the conductivity and liftoff curves for nonmagnetic alloys. Frequency changes shift the points along the conductivity locus in a nonlinear fashion. This phenomenon, also true for other impedance curves, can be used advantageously because it allows the material points to be located for optimum response or suppression. Specifically a frequency should be chosen that causes the material points for the variables to be measured to move in a substantially different direction from those points to be suppressed.

FIGURE 4. Conductivity and liftoff loci on impedance plane.

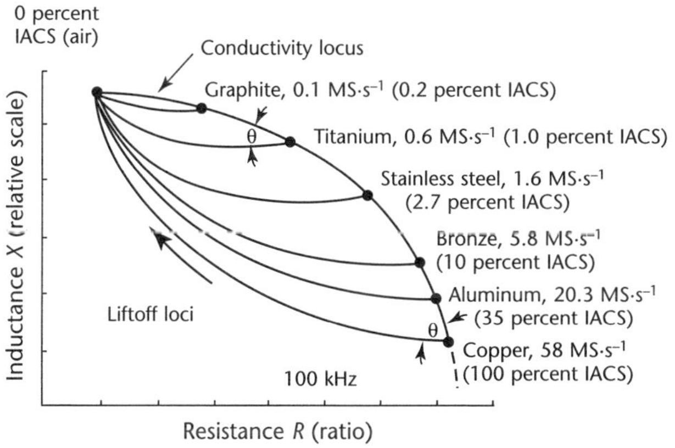

Legend

IACS = International Annealed Copper Standard

θ = curve of separation between liftoff curve and conductivity curve

FIGURE 5. Movement of material points by frequency changes: (a) low frequency, 20 kHz; (b) medium frequency, 100 kHz; and (c) high frequency, 1 MHz.

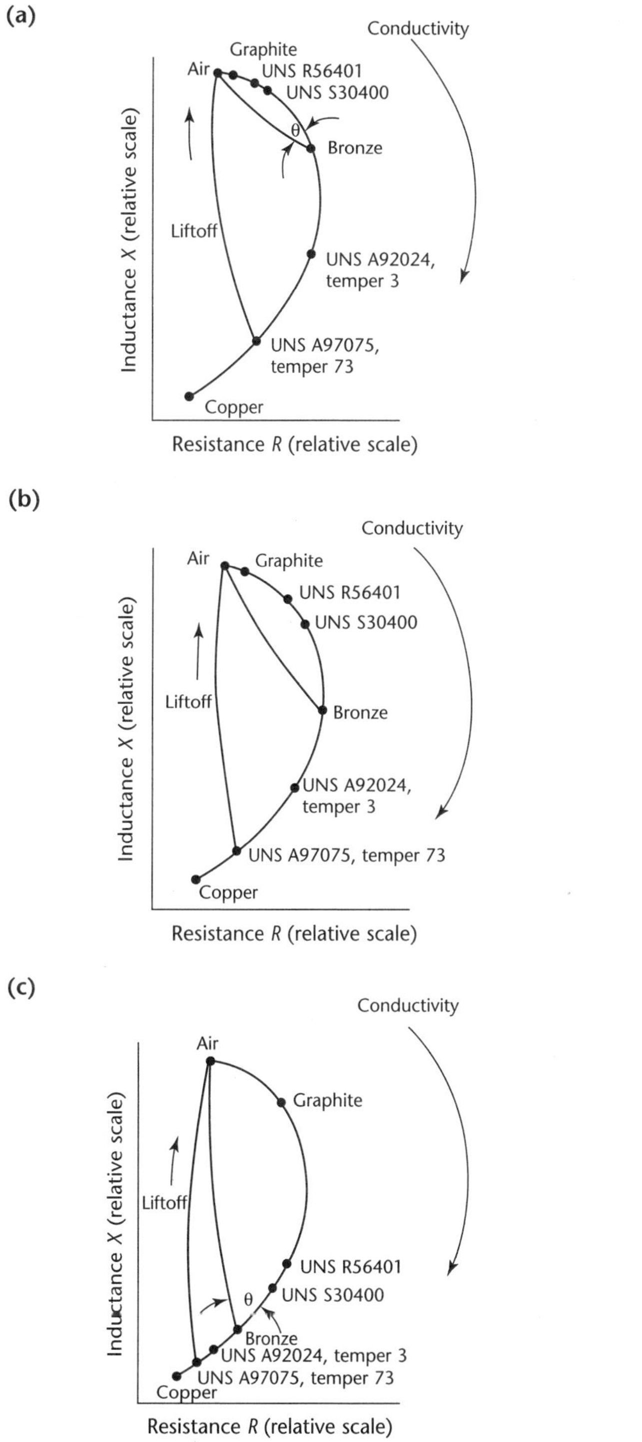

Legend

UNS A92024 = Unified Numbering System A92024 heat treatable wrought aluminum alloy

UNS A97075 = Unified Numbering System A97075 heat treatable wrought aluminum alloy

UNS R56401 = Unified Numbering System R56401 titanium alloy

UNS S30400 = Unified Numbering System S30400 austenitic chromium nickel stainless steel

At low frequencies (Fig. 5a), the separation angle θ between the liftoff curve and the conductivity curve for bronze is quite small. Thus, it becomes more difficult to obtain liftoff suppression. If a higher frequency is chosen (Fig. 5c), the separation angle for bronze is large, allowing liftoff suppression and good sensitivity to conductivity variations. For sorting titanium alloys, a frequency between 500 kHz and 1 MHz would be chosen but, to sort aluminum alloys, a frequency between 20 kHz and 100 kHz would be chosen.

Generally for conductivity measurements (alloy sorting, heat treat determination and others) and for surface crack detection, a frequency should be chosen that places the material point just below the knee in the conductivity curve. At this point, a large separation angle exists between the liftoff and conductivity curves.

For magnetic materials, the liftoff and magnetic permeability loci curves are virtually superimposed (Fig. 6a) but their respective values increase in opposite directions. Figure 3 shows that the reactance component of the test coil impedance is decreased by the presence of nonmagnetic materials. This reactance reduction occurs because induced currents flow in the conductive and nonmagnetic object and set up a secondary field that partially cancels the primary field of the coil. The opposite is true when a magnetic material such as iron or ferrite is placed within the field of the coil. This happens because the presence of the magnetic field intensity of the primary coil field causes atomic magnetic elements of the magnetic material to become aligned with the field, increasing the flux density. The magnetic permeability μ is the ratio of flux density *B* to magnetic field intensity *H*:

$$(5) \quad \mu = \frac{B}{H}$$

where B = magnetic flux density (tesla) and H = magnetizing force or magnetic field intensity ($A \cdot m^{-1}$). In free space, magnetic permeability $\mu_0 = 4\pi \times 10^{-7}\ H \cdot m^{-1}$.

The nickel zinc ferrite cores (Fig. 6b) were chosen as examples because they have a low conductivity and two different values for permeability. The effect of the increased flux density gives a greater induced voltage in the test coil that in turn raises the impedance. The increase in impedance is in the reactance direction except for the effect of a small amount of energy loss resulting from hysteresis. The nickel zinc ferrite cores may have an initial permeability of 850, high on the permeability line of Fig. 6a. Usually practical engineering materials also have an associated electrical conductivity that affects the impedance as shown for 422 steel and 4340 steel in Fig. 6b. The relative relationship of the permeability

FIGURE 6. Permeability, liftoff and conductivity loci on impedance plane: (a) permeability and liftoff locus; (b) permeability loci for different materials; (c) loci for permeability μ and conductivity σ.

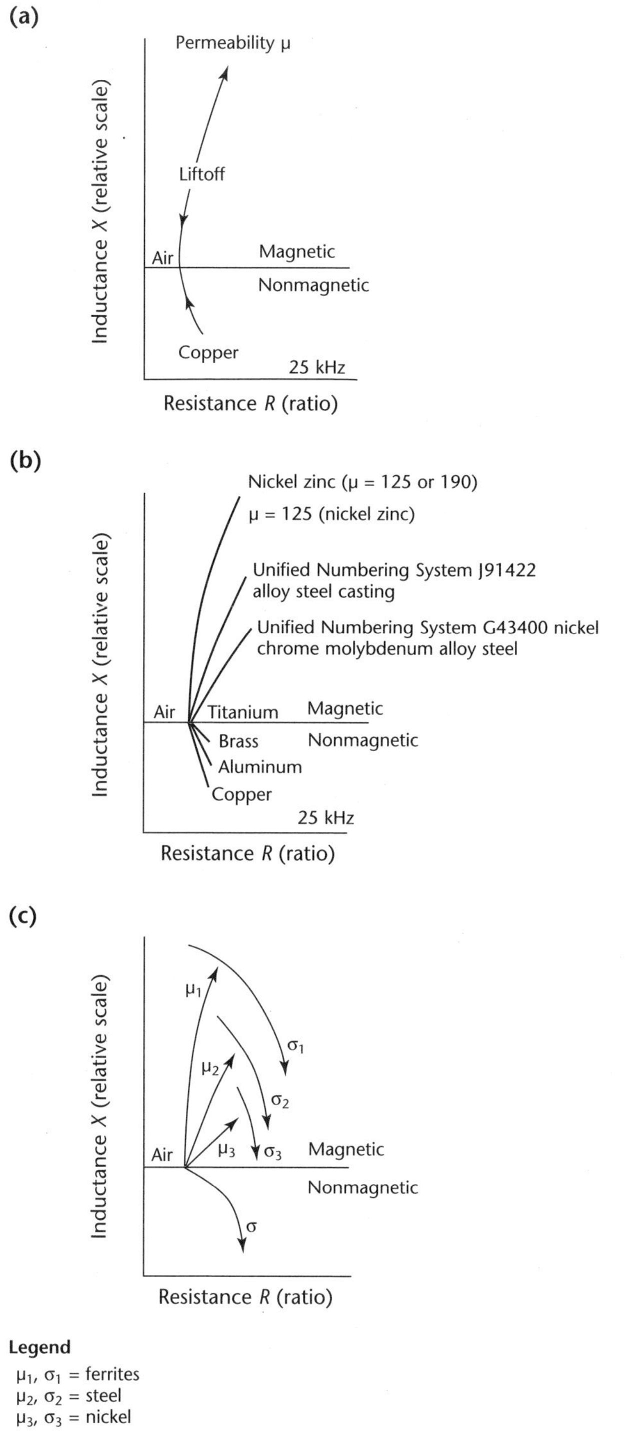

loci lines and the conductivity curves for three materials are shown in Fig. 6c.

The vector or phasor values of inductive reactance and resistance for different material conditions yield unique loci or phasor plots on the impedance plane at particular operating frequencies. The phase angle of the impedance vectors will change at different frequencies because the inductive reactance value is a function of inductance and frequency. Hence vector points may move relative to one another along the conditional loci curves when the operating frequency is changed. This shift in phase is shown in Fig. 5 for the conductivity values of nonmagnetic materials. Similar phase angle changes for the permeability of 4340 steel are shown in Fig. 7 as the frequency changes from 75 to 300 kHz. These changes in phase shift at different frequencies do not interfere with impedance plane analysis, provided that the operator is aware of this factor. In some cases, test results may be improved by changing the frequency to cause phase shifts.

With phase analysis eddy current instruments, an operator can produce impedance plane loci plots or curves automatically on a flying dot oscilloscope or integral cathode ray tube. Such impedance plane plots can be presented for the following material conditions (as shown in Fig. 8): (1) liftoff and edge effects, (2) cracks, (3) material separation and spacing, (4) permeability, (5) specimen thinning, (6) conductivity and (7) plating thickness.

Evaluation of these plots shows that ferromagnetic material conditions produce higher values of inductive reactance than values obtained from nonmagnetic material conditions. Hence the magnetic domain is at the upper quadrant of the impedance plane whereas nonmagnetic materials are in the lower quadrant. The separation of the two domains occurs at the inductive reactance values obtained with the coil removed from the conductor (sample); this is proportional to the value of the coil's self-inductance L.

Linear material values do not produce linear responses on the impedance plane loci. With the eddy current probe balanced on the metal specimen, the loci values for linear material conditions are displayed as follows.

1. Magnetic and nonmagnetic liftoff conditions are displayed logarithmically (in X).
2. Magnetic and nonmagnetic edge effects are displayed logarithmically.
3. Magnetic and nonmagnetic conductivities vary with test frequency.
4. Magnetic permeability varies with test frequency.
5. Metal thinning varies exponentially.

FIGURE 7. Phase angle changes on impedance plane caused by frequency changes.

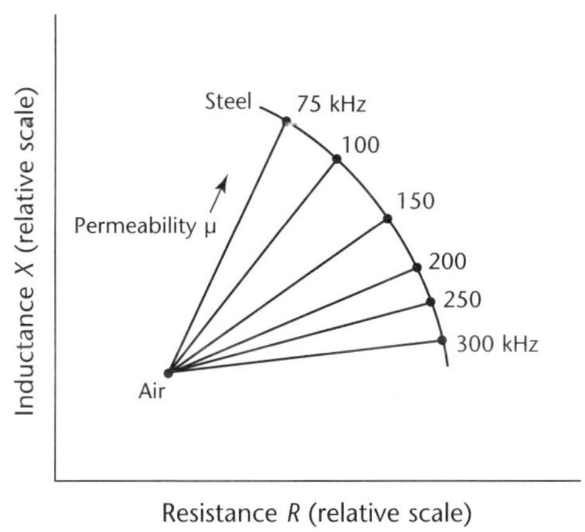

FIGURE 8. Impedance changes in relation to one another on impedance plane.

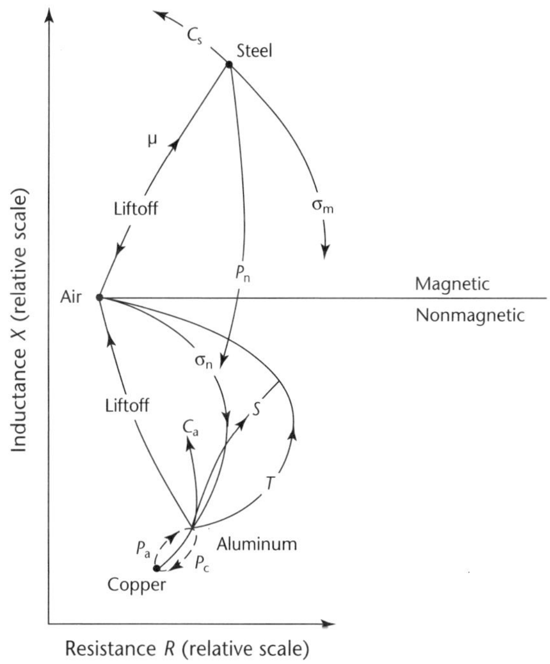

Legend

C_a = crack in aluminum
C_s = crack in steel
P_a = plating (aluminum on copper)
P_c = plating (copper on aluminum)
P_n = plating (nonmagnetic)
S = spacing between aluminum layers
T = thinning in aluminum
μ = permeability
σ_m = conductivity for magnetic materials
σ_n = conductivity for nonmagnetic materials

6. Nonmagnetic plating thickness is displayed logarithmically.
7. Material spacing or separation varies exponentially.

Electromagnetic induction effects are not easy to understand. Neither the magnetic fields nor the eddy currents can be seen.

In a problem solving situation, impedance plane analysis is a useful tool because it improves the ability to detect various conditions and provides a better understanding and interpretation of the eddy current test results.

PART 3. Conductivity Testing[1]

Effect of Alloys

Metals are by definition good conductors of electricity. Conductivity depends on the arrangement of atoms in each metal lattice and the distribution and energy of the electrons surrounding each atom. Any variation in the structure of metals that affects the electronic structure and energy of the atoms decreases the conductivity of the metal. For simplicity, a decrease in conductivity may be associated with obstacles in the path of electron flow through a metal. The obstacles to electron flow may be caused by lattice distortions resulting from dislocations, missing atoms (lattice vacancies), foreign atoms or grain boundaries. The presence of particles of different composition also restricts the flow of electrons; a greater number of smaller particles offers more resistance than fewer larger particles.

Most engineering metals are alloys. An alloy is formed by adding one or more metals or nonmetals to a base metal. Alloying elements are usually added during melting of a base metal and the quantities added are normally specified over a percentage range.

The alloying elements can take one or more forms in the solidified state depending on the amount added and the rate of cooling from the melting temperature. Some elements may occupy lattice positions normally occupied by atoms of the base metal. The alloy thus formed is called a *substitutional solid solution.* Very small atoms such as those of carbon, nitrogen and hydrogen take up positions between the base metal atoms to form an *interstitial solid solution.* This action can actually change the lattice structure as when carbon is added to iron to make steel.

Alloying elements can also form new lattice structures continuous throughout the metal or distributed as small particles of various sizes throughout the metal. The distribution of the alloying elements depends on (1) the amount of alloying elements added in relation to the amount that can be tolerated in the lattice of the base metal and (2) their change in solubility with temperature.

Alloy Effects on Mechanical Properties

All of the alloying element distributions increase the resistance of a metal to deformation. Increased strength results from the interference of particles formed by the alloying atoms with the movement of dislocations or by the generation of new dislocations. The distribution can often be modified by heat treatment.

Alloy Effects on Conductivity

The conductivity of a metal is decreased as increasing amounts of alloying elements are added. Even small amounts of foreign atoms can greatly reduce conductivity (Figs. 9 and 10). Some alloying elements have a much greater effect on conductivity than others. Generally atoms that most severely differ from the base metal in size and electron distribution cause the greatest decrease in conductivity. The lattice distortion caused by the alloying atoms (and particles of different chemical composition) inhibits the flow of electrons through the lattice. Because of variations in chemical composition resulting from the tolerances in alloy additions, a conductivity range rather than a specific conductivity value is obtained for each alloy (Fig. 11).

FIGURE 9. Decrease in conductivity of copper caused by various impurities.

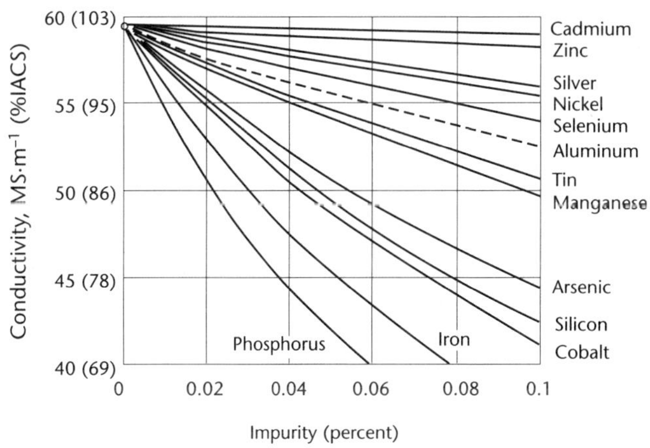

Legend
%IACS = percentage of International Annealed Copper Standard

Heat Treatment

The properties of metals can be altered by changing the number and distribution of dislocations, alloying atoms and particles of different composition. These changes can be accomplished through various types of heat treatment. The three principal types of heat treatment are: (1) annealing, (2) solution heat treatment and (3) precipitation heat treatment or artificial aging.

In annealing, the metal is heated to a sufficiently high temperature to remove the effects of cold working by redistribution of dislocations and in some instances by the formation of a new stress free grain (recrystallization). During the annealing of alloys, the temperature is selected sufficiently high to permit the alloying atoms to migrate readily. However, this selected temperature is sufficiently below that of maximum solubility to favor the formation of separate particles and compounds by the alloying atoms. Slow cooling from the annealing temperature encourages even more alloying atoms to move from their random position in the base metal lattice to aid in the growth of larger secondary compounds.

Annealing removes many of the obstacles to plastic flow such as interacting dislocations and the numerous individual alloying atoms and fine particles that normally resist plastic deformation. These processes generally result in metals of lower strength and greater ductility after annealing.

The annealing process reduces obstacles to electron flow. Therefore annealing improves the conductivity of a metal. Increased annealing times favor more complete diffusion and greater coalescence and growth of particles with associated increases in conductivity.

The maximum number of alloying atoms will occupy lattice sites of the base metal when a temperature slightly below melting point is reached. In interstitial solid solutions, the maximum number of atoms will occupy interstitial positions. As temperatures are lowered, the atoms of many alloying elements will tend to diffuse together and form separate compounds or regions with a different lattice. If the metal is cooled rapidly enough, the atoms do not have time to diffuse and are held in their original lattice positions (retained in solution). The process is called *solution heat treatment*. Any delay in rapid cooling (delayed quench) or a slow rate of cooling will permit an increased amount of diffusion and reduce the number of alloying atoms held in solution.

The alloying atoms retained in base metal lattice positions by solution heat treatment present obstacles to dislocation movement. The resistance to plastic deformation increases the strength of the metal. In many instances, more than one alloying element contributes to the higher strength of alloys. Slow rates of cooling from solution heat treatment temperatures or a low solution temperature can reduce the strength of the heat treated alloy.

The distortion and stresses established by the substitution of alloying atoms for those of the base metal reduce the conductivity of the metal. The greater the number of solute atoms of a specific material, the greater the reduction in conductivity. The presence of lattice vacancies caused by solution also disrupts the electronic structure of an alloy and contributes to lower conductivity. The conductivity is not lowered as much if solution heat treatment temperatures are low or if cooling from solution heat treatment temperatures is excessively slow; poor solution heat treatment practices such as these permit too many atoms to come out of solution or form secondary particles.

If an alloy has been solution heat treated to retain atoms in the same lattice occupied at high temperature, properties can be further modified by a precipitation or aging treatment. During a precipitation treatment, an alloy is heated to a temperature that will allow alloying atom diffusion and coalescence to form microscopic particles of different

FIGURE 10. Influence of metallic additives on conductivity of aluminum.

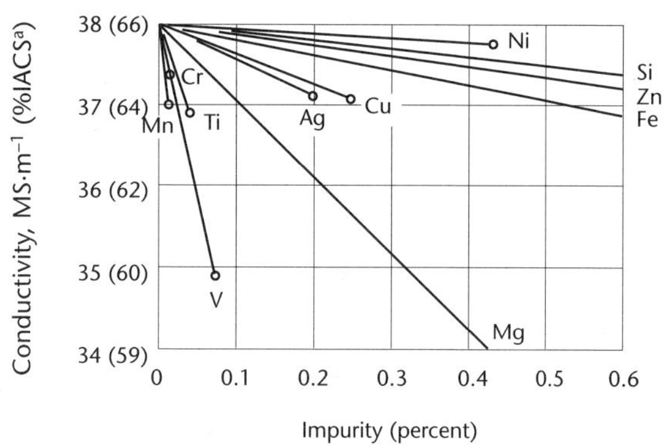

Legend

%IACS = percentage of International Annealed Copper Standard
Ag = silver
Cr = chromium
Cu = copper
Fe = iron
Mg = manganese
Mn = magnesium
Ni = nickel
Si = silicon
Ti = titanium
V = vanadium
Zn = zinc

composition and lattice structure in the metal. The number, size and distribution of the particles are controlled by the time and temperature of the aging process. Temperatures are much lower than those required for solution heat treatment or annealing. Lower temperatures and shorter times result in smaller particle sizes. Higher temperatures favor the formation of fewer but larger particles.

Precipitation or aging treatments are generally designed to increase the strength of alloys, particularly the yield strength. The strengthening is accomplished by the formation of small particles of different composition and lattice structure from the original lattice. The small particles provide obstacles to the movement of dislocations in which planes of atoms slip over each other, causing plastic deformation. Greatest strengthening usually occurs at a specific range of particle sizes for a particular alloy system. In many cases, aging is performed under conditions designed to provide a specific combination of strength and ductility or corrosion resistance. As aging increases beyond the optimum time or temperature, the particle size increases and gradual softening occurs. When material has been aged for excessive times or at too high a temperature, then it is said to be *overaged.*

The removal of foreign atoms from the parent lattice during precipitation hardening removes much of the distortion of the electron distribution in the lattice. This action favors the movement of electrons through the metal and results in higher conductivity. As increased amounts of foreign atoms are removed from solution and particle growth occurs during overaging, conductivity continues to increase.

Conductivity of Aluminum Alloys

Conductivity measurement is often applied to aluminum alloys.[5] This application results from the extensive use of aluminum alloys in the aerospace industry and the wide variation in the electrical conductivity and mechanical properties between different alloys and heat treatment. For most common usage, specific conductivity ranges have been established for each alloy and temper.

FIGURE 11. Electrical conductivity values for various metals and alloys.

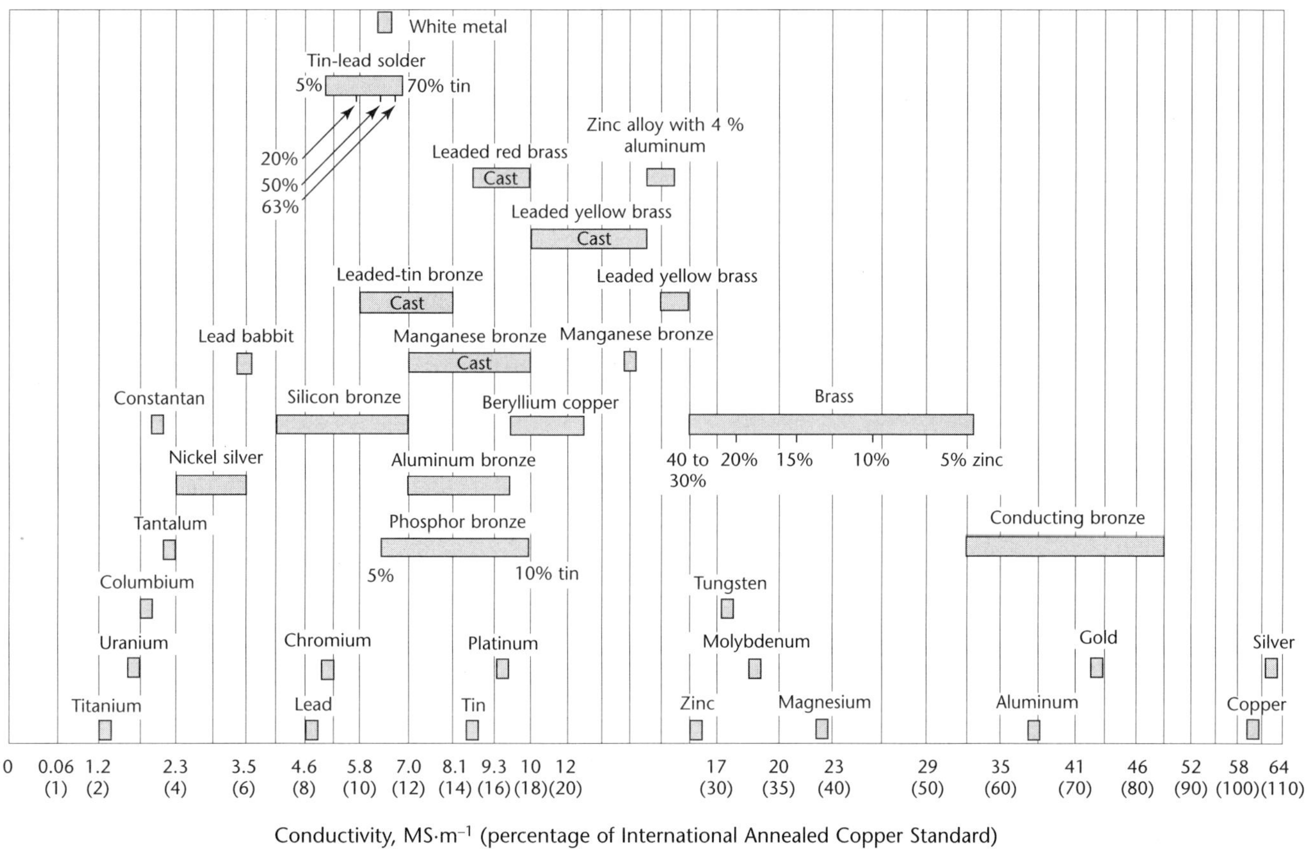

Table 1 lists the conductivity ranges for many of the aluminum alloys commonly used in aircraft structural applications. These data represent a composite of values from various airframe manufacturers and government agencies. The ranges include all values obtained for standard heat treatments except for extreme values obtained from one or two sources clearly outside the ranges of all other lists. Any time a conductivity value is obtained for an aluminum alloy and temper outside of the applicable range, its mechanical properties should be considered suspect.

An aluminum alloy has the highest conductivity and lowest strength when it is in the fully annealed temper. After quenching from the solution heat treatment temperature, the strength is increased and the conductivity is decreased. Many aluminum alloys are unstable after solution heat treatment even if held at room temperature. During this time, a certain amount of atom migration takes place to initiate the formation of submicroscopic particles. This process, sometimes called *natural aging,* increases the strength of the alloy but either has no effect on conductivity or causes a slight decrease in conductivity. Some aluminum alloys remain unstable for such long periods after quenching that they are never used in the solution heat treated condition — for example, Unified Numbering System A97075 wrought aluminum alloy.

If a solution heat treated alloy is precipitation hardened by heating at relatively low temperature, between 93 and 232 °C (200 and 450 °F), alloying atoms form small particles. At a critical size and distribution of particles, the strength of the aluminum alloy reaches a maximum. Conductivity increases during the precipitation hardening or artificial aging process. If aging is carried beyond the point where optimum strength is obtained, the strength will decrease but conductivity will continue to increase. Figure 12 shows the relationship between conductivity and strength for a typical structural aluminum alloy.

Variations from specified heat treatment practice can result in aluminum alloys with strengths below required levels. Heat treatment discrepancies

TABLE 1. Ranges of electrical conductivity for aluminum alloys.

Alloy and Temper[a]		Electric Conductivity			
		Minimum		Maximum	
Unified Numbering System (UNS)	Aluminum Association (AA)	$MS \cdot s^{-1}$	(%IACS[b])	$MS \cdot s^{-1}$	(%IACS[b])
UNS A91100	AA 1100	33.1	(57.0)	36.0	(62.0)
UNS A92014, untempered	AA 2014-0	28.1	(48.5)	29.9	(51.5)
UNS A92014, temper 3XX	AA 2014-T3XX	18.3	(31.5)	20.3	(35.0)
UNS A92014, temper 4XX	AA 2014-T4XX	18.3	(31.5)	20.3	(35.0)
UNS A92014, temper 6XX	AA 2014-T6XX	21.5	(37.0)	24.1	(41.5)
UNS A92019, untempered	AA 2219-0	24.9	(43.0)	26.7	(46.0)
UNS A92019, temper 3XX	AA 2219-T3XX	15.7	(27.0)	18.0	(31.0)
UNS A92019, temper 62X	AA 2219-T62X	18.0	(31.0)	20.6	(35.5)
UNS A92019, temper 8XX	AA 2219-T8XX	18.0	(31.0)	20.6	(35.5)
UNS A92024, untempered	AA 2024-0	26.4	(45.5)	29.0	(50.0)
UNS A92024, temper 3XX	AA 2024-T3XX	16.2	(28.0)	19.1	(33.0)
UNS A92024, temper 4XX	AA 2024-T4XX	16.5	(28.5)	18.9	(32.5)
UNS A92024, temper 6XX	AA 2024-T6XX	20.3	(35.0)	23.8	(41.0)
UNS A92024, temper 8XX	AA 2024-T8XX	20.9	(36.0)	24.7	(42.5)
UNS A93003	AA 3003	25.8	(44.5)	29.0	(50.0)
UNS A96061, untempered	AA 6061-0	27.3	(47.0)	29.6	(51.0)
UNS A96061, temper 4XX	AA 6061-T4XX	20.6	(35.5)	24.1	(41.5)
UNS A96061, temper 6XX	AA 6061-T6XX	23.2	(40.0)	26.1	(45.0)
UNS A97075, untempered	AA 7075-0	25.5	(44.0)	27.8	(48.0)
UNS A97075, temper 6XX	AA 7075-T6XX	17.4	(30.0)	20.3	(35.0)
UNS A97075, temper 73X	AA 7075-T73X	22.0	(38.0)	24.7	(42.5)
UNS A97075, temper 76X	AA 7075-T76X	20.9	(36.0)	22.6	(39.0)
UNS A97178, untempered	AA 7178-0	24.9	(43.0)	27.3	(47.0)
UNS A97178, temper 6XX	AA 7178-T6XX	16.8	(29.0)	19.7	(34.0)
UNS A97178, temper 76	AA 7178-T76	20.3	(35.0)	22.6	(39.0)

a. *X*s represent numerals unspecified in this list.

b. Percentage of International Annealed Copper Standard.

include deviations or misapplication of the following processes: (1) solution heat treatment temperature, (2) solution heat treatment time, (3) quenching practice, (4) aging temperature, (5) aging time, (6) annealing temperature and time and (7) uncontrolled temperature application.

Solution heat treatment temperatures for each aluminum alloy vary over a range. Wider ranges can lead to greater variations in conductivity and mechanical properties. For instance, Unified Numbering System 97075 wrought aluminum alloy can be solution heat treated at temperatures from 460 to 500 °C (860 to 930 °F) depending on product form and size. Solution heat treatment at the higher temperatures results in greater numbers of atoms in solution and in vacancies. These conditions provide slight increases in hardness and decreases in conductivity after quenching. Solution heat treatment at the lower end of the temperature range causes slightly higher conductivity. The variations in permissible heat treatment temperature partially contribute to the ranges in conductivity obtained.

During solution heat treatment, a minimum time at the solution heat treatment temperature is usually specified. The minimum time is required to ensure that the entire cross section of the part reaches solution heat treatment temperature and the soluble elements are uniformly diffused into the base metal lattice. Insufficient time at temperature can result in too few alloying atoms in solution and a consequent decrease in strength and hardness and an increase in conductivity. Too long a time at temperature promotes grain growth in the material and can result in excessive diffusion in clad aluminum. Unless reaction with the heat treatment atmosphere occurs or extremely large grains are formed, little effect is noted on mechanical properties or conductivity because of excessively long times at temperature.

FIGURE 12. Relationship between yield strength and conductivity for 2024 aluminum alloy. In saturation, the magnetic permeability $\mu_0 = 4\pi \times 10^{-7}$ H·m^{-1}.

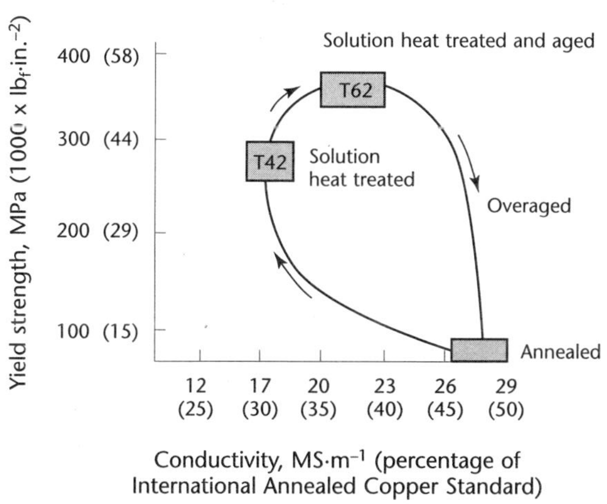

The solubility of alloying elements decreases rapidly with decreasing temperature. If an aluminum alloy is cooled slowly from the solution heat treatment temperature, the alloying atoms coalesce to form large particles. These larger particles form preferentially at the grain boundaries. Grain boundary precipitation increases susceptibility to intergranular corrosion. The fewer number of alloying atoms retained in solution by slow quenching reduces the hardness and strength and increases conductivity. Slow quenching also prevents adequate response to subsequent aging: specified hardness and strength cannot be attained.

Aging temperatures for aluminum alloys are chosen to obtain a desired combination of strength, ductility and corrosion resistance. Standard aging temperatures provide the optimum combination of these properties. In some cases, the aging temperatures are varied to obtain properties other than strength. For instance, temper 6 and temper 73 of Unified Numbering System A97075 wrought aluminum alloy are obtained by different artificial aging treatments of solution heat treated and quenched that alloy. Temper 6 provides higher strength but temper 73 has greater resistance to intergranular corrosion and stress corrosion cracking. Each of these tempers has a different conductivity range. The lower strength of temper 73 is accompanied by higher conductivity. Conductivity measurement is widely used as a process control for determining the adequacy of aging for temper 73. Excessive aging temperatures for aluminum alloys result in a greater particle growth and cause conductivity values higher than specified. Low aging temperature may not result in optimum strength and resulting conductivities are lower than usual for a particular alloy and temper.

As precipitation heat treatment times at a specific temperature increase for an aluminum alloy, the alloy's strength and hardness increase to a maximum and then decline. Ductility is usually at a minimum when strength reaches a maximum. As the aging time increases, electrical conductivity continually increases until maximum precipitation and growth of precipitate particles occur. With prescribed aging times, the conductivity should lie within the conductivity range for the appropriate alloys and tempers as listed in Table 1.

Aged tempers are identified as 5XX, 6XX, 7XX and 8XX following the alloy number, the *X*s representing unspecified numerals. Aging times less than those specified can result in conductivities below the specified range.

Annealing of structural aluminum alloys can be performed at more than one temperature, depending on previous heat treatment and processing history. When effects of previous solution heat treatment are to be removed, heating to about 413 °C (775 °F) followed by slow cooling to about 260 °C (500 °F) or lower is necessary to obtain adequate precipitation and coalescence of alloying elements. To remove the strain hardening effects from forming operations on previously annealed material, heating is performed at 340 °C (650 °F) for a time sufficient to ensure temperature uniformity. Strength is lowest after annealing and ductility is highest. Conductivity reaches the highest value for each aluminum alloy following a full anneal. Excessive annealing times have little effect on properties or conductivity.

Occasionally during processing or in service, aluminum alloys may be subjected to the application of uncontrolled temperatures for indefinite periods of time. An example of a processing application would be the heating of a heat treated and aged alloy for a forming operation where heating is applied without specific controls. During service, heat may be applied in the form of jet exhaust by aerodynamic heating or as a result of an aircraft fire. Depending on temperature and time of exposure, the strength and corrosion resistance can be significantly reduced. These effects are usually accompanied by changes in electrical conductivity. The direction and amount of change in electrical conductivity depends on the temperature applied, the time of application and the rate of cooling.

If heat is applied in a nonuniform manner, as might be the case in a fire, a wide range in strength and conductivity may be obtained. Areas heated near solution heat treatment temperatures and rapidly cooled by water or adjacent cold metal may have low conductivity. Adjoining areas may be heated to temperatures above the aging temperature and close to the annealing temperature. This heating results in low strength and relatively high conductivity.

Magnetic Permeability

Measurement of electrical conductivity is applicable to materials with a relative magnetic permeability of one or nearly one when using most general purpose instruments. If the permeability exceeds one, it will affect the conductivity measurement and erroneous readings will be obtained. Some stainless steels may be essentially nonmagnetic in the annealed condition but slight amounts of cold working or exposure to extremely low temperature can cause transformation to a magnetic structure.

Saturation techniques provide meaningful eddy current data about some magnetic materials. A direct current field is applied to the material until it reaches saturation on the hysteresis curve. At this point, the material has an effective permeability of one and behaves like a nonferromagnetic material.

Geometry Effects on Measurement

Any change in part configuration that affects distribution or penetration of eddy currents will result in erroneous electrical conductivity readings. The following sources of error are included in these categories: (1) proximity to test object edges or adjoining structure, (2) metal thickness less than the effective depth of penetration and (3) excessive curvature of the test object surface.

If conductivity readings are taken too close to the edge of a test object, a deviation from true conductivity will be obtained. This deviation increases as the probe approaches the edge of the test object. Generally edge effects can be entirely eliminated if the coil stays at least 6 mm (0.25 in.) away from the edge of the test object. Proximity of other materials, particularly steel, can also affect the measured conductivity values.

As the radius of curvature of a test object decreases, the measured conductivity value decreases below the true value. A curve demonstrating this decrease is shown in Fig. 13. Because the deviation from true value will change with various conductivity measuring instruments and probes, with base metal conductivity and with liftoff adjustment, curves should be established for each application as necessary. In general, convex radii of curvature greater than 100 mm (4 in.) and concave radii greater than 500 mm (20 in.) in aluminum alloys do not require any adjustment or special calibration curves. Deviation from true value increases with increasing conductivity of the test object.

Conductivity measuring eddy current instruments often have a preset liftoff adjustment. For some models, liftoff compensation is set at 0.08 to 0.1 mm (0.003 to 0.004 in.) by an internal adjustment. Another model has a preset liftoff adjustment of about 0.05 mm (0.002 in.). The liftoff adjustment is usually set during calibration of the instrument. Maintenance manuals describe the procedures that can be performed by trained nondestructive test personnel.

With probe wear and changes in the instrument's electrical components over a period of time, liftoff adjustment can change. When conductivity measurements are performed on rough surfaces or through thin nonconductive coatings, therefore, liftoff adjustment should be checked before the measurements. After calibrating an instrument against the conductivity standards, the liftoff adjustment should be checked against a specimen with conductivity representative of the test object. Subsequent to determining the conductivity of a bare specimen, nonconductive shims should be inserted between the specimen and the probe. The amount of liftoff adjustment is the maximum shim thickness at which the conductivity begins to differ from that obtained on the bare surface.

Liftoff greater than the amount of preset liftoff adjustment (if any) results in errors in conductivity readings. Causes of excessive liftoff can be heavy nonconductive coatings of paint or plastic, local waviness, minor changes in contour and excessively rough surfaces.

When using general purpose equipment, the effect of liftoff depends on the amount of liftoff adjustment used and the sensitivity used. The effects of excessive liftoff on conductivity measurements are shown in Fig. 14.

FIGURE 13. Decrease of measured conductivity with decreasing radius for Unified Numbering System A92024 wrought aluminum alloy, temper 42.

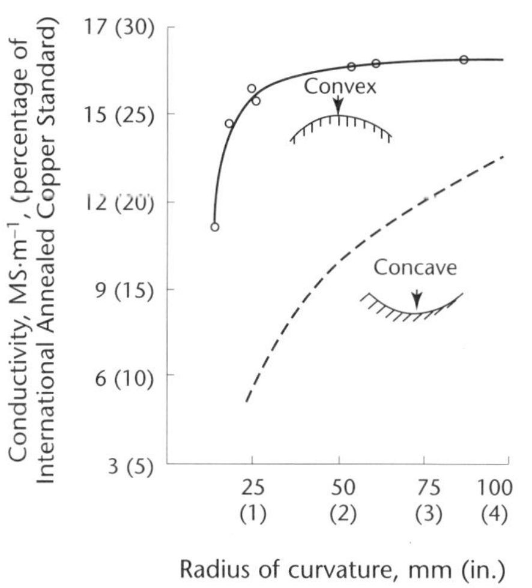

Temperature Effects on Measurement

Higher temperature increases the thermal activity of the atoms in a metal lattice. The thermal activity causes the atoms to vibrate around their normal positions. The thermal vibration of the atoms increases the resistance to electron flow and thereby lowers the conductivity of the metal. Lower temperature reduces thermal oscillation of the atoms resulting in increased electrical conductivity.

The conductivity of reference standards is usually determined at a specific temperature; 20 °C (68 °F) is commonly used. Typical conductivity values are allowable conductivity ranges also established at about this temperature. If all instrument calibration and conductivity measurement could be performed at this temperature, then errors in conductivity measurement related to temperature variation would not occur and temperature compensation would not be required. In field applications, however, test temperatures can range from –20 to 50 °C (0 to 120 °F). Unless precautions are taken in selection of reference standards in instrument calibration and in testing, there can be errors in the measured conductivity values. Two ways in which erroneous readings occur are (1) difference in temperature between reference standards and test object and (2) difference between the temperature at which the conductivity of the reference standard was originally established and the temperature at which instrument calibration and conductivity measurements are performed.

To prevent errors from differences in temperature between reference standards and test object, the instrument and

FIGURE 14. Effects of liftoff on conductivity measurements.

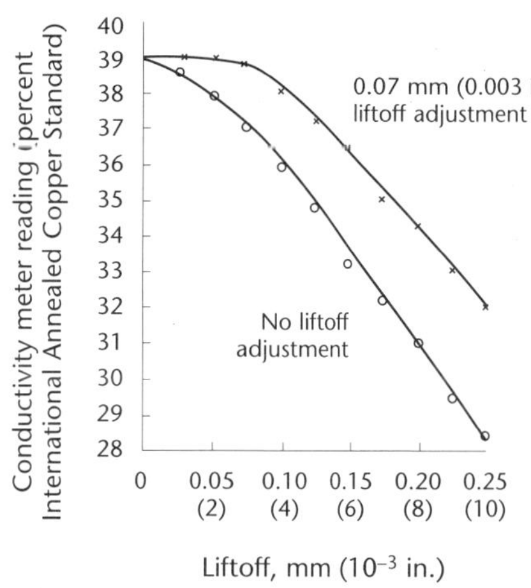

reference standards should be allowed to stabilize at the test object temperature before calibration and conductivity measurements are performed. In no instance should measurements be taken if the temperatures of the test object and the reference standard differ by more than 5 °C (10 °F). Even though reference standards and test objects are at the same temperature, error in determining conductivity occurs when the measuring temperature differs from the temperature at which the conductivity of a reference standard was originally established. The magnitude of the error becomes larger as this difference in temperature increases.

Two other factors also contribute to errors caused by temperature differences: (1) increased difference in conductivity between the upper and lower reference standards and (2) differences in the temperature coefficient of electrical resistivity between reference standards and the test object. These two sources of error can be reduced by decreasing the range between the conductivity standards and using standards of the same or about the same temperature coefficient of electrical resistivity as the test object. Because all aluminum alloys have about the same rate of change of electrical resistivity with temperature change, aluminum conductivity reference standards are preferred for aluminum alloys.

Conductivity measurements should not be performed under conditions where the relative humidity exceeds 85 percent.

Conductivity Reference Standards

For calibration of eddy current conductivity requirements, the number of reference standards needed may vary depending on the test purpose and the accuracy required. The conductivity range of the reference standards used for calibration (calibration blocks) must be within the range of the instrument and must cover the range of conductivity values to be measured. Preferably the calibration reference standards should have the same change in resistivity with temperature as the test objects. It is desirable for the difference in conductivity between the calibration standard representing the low end of the calibration range and the calibration standard representing the high end of the range to be at least 10 percent International Annealed Copper Standard (IACS) but not to exceed 25 percent. When a calibration curve is established for a general purpose instrument, a sufficient number of calibration standards are necessary to develop a smooth continuous curve over the range of interest.

For convenience of transportation and storage, conductivity standards are usually kept relatively small. Reference standards must have sufficient size to prevent edge effects or thickness from having a bearing on conductivity readings. These requirements can be satisfied by requiring length and width to be 25 mm (1 in.) greater than the probe diameter and the thickness greater than 3.5 times the standard depth of penetration at the test instrument frequency. Calibration standards should be flat, have a smooth surface and be free of coatings. Reference standards used for calibrating instruments immediately before measuring conductivity should be accurate within ±0.5 percent International Annealed Copper Standard of the nominal value. A second set of calibration standards accurate within ±0.6 $MS{\cdot}m^{-1}$ (±0.35 percent of the International Annealed Copper Standard) should be periodically made available for checking the performance of instruments and field calibration standards.

Calibration standards should be traceable to the National Institute of Standards and Technology, Gaithersburg, Maryland. Such calibration standards are available from some manufacturers of eddy current conductivity instruments.

Many reference standards, particularly those of aluminum alloys, are subject to metallurgical changes if exposed to temperatures of 65 °C (150 °F) or greater. Surfaces of reference standards can also corrode if exposed to moisture or other hostile environments. Damage caused by rough handling can also lead to error in conductivity readings. For these reasons, standards should be transported and stored in dry, clean, protected areas not subject to excessive temperatures.

PART 4. Hysteresis Loop Characteristics[1]

Magnetic Hysteresis[6]

The magnetic behavior of a ferromagnetic material is characterized by its hysteresis loop (magnetization curve). In ferromagnetic materials, there is no linear relationship between magnetic field intensity H and flux density B. A very small change in H may produce a large change in B. The relationship is best shown in the magnetization or B,H curve (Fig. 15). As H increases, the flux density increases rapidly up to the knee of the curve. Beyond the knee, a further increase in H causes no useful increase in B and the material is said to be *saturated.*

It is evident from Fig. 15 that the magnetic permeability μ (the ratio of magnetic flux density B to magnetic field intensity H) up to the saturation magnetization I_s is not constant. The magnetic permeability's variation as a function of B is shown in Fig. 16. The largest value on this curve is the maximum permeability μ_{max}. Another important quantity is the initial permeability μ_0, measured at very low fields at the toe of the hysteresis loop.

It is a familiar phenomenon that some ferromagnetic materials when magnetized by an external field do not return to a completely unmagnetized state when removed from that field. In fact, these materials must be subjected to a reversed field of a certain intensity to demagnetize them. Other ways to demagnetize are to heat the material to a characteristic temperature called the *curie point* (above which ferromagnetic ordering of atomic moments is thermally destroyed) or to work the material mechanically to reduce the magnetization. If an external field varied in a controlled manner is applied to a completely demagnetized (virgin) specimen and if the magnetic induction in the specimen is measured, the magnetization curve of the material may be determined. Figure 17 shows a representative hysteresis loop for a ferromagnetic material.

As shown in Fig. 17a, starting at the origin O with the specimen in the unmagnetized condition and increasing the magnetizing force H in small increments, the flux density B in the material increases quite rapidly at first and then more slowly until it reaches a point beyond which any increase in the magnetic field intensity does not increase the flux density. This is shown by the dashed curve OA. In this condition the specimen is said to be *magnetically saturated.*

When the magnetic field intensity is gradually reduced to zero, the curve AB results (Fig. 17b). The amount of magnetism that the steel retains at point B is called *residual magnetism* or *remanence* and is represented by B_r.

When the magnetizing current is reversed and gradually increased in magnitude, the flux continues to diminish. The flux does not become zero until point C is reached, at which time the magnetic field intensity is represented by OC (see Fig. 17c), which graphically

FIGURE 15. Magnetization or BH curve showing relation between flux density B and magnetic field intensity H in ferromagnetic materials. The intensity of magnetization at saturation is I_s.

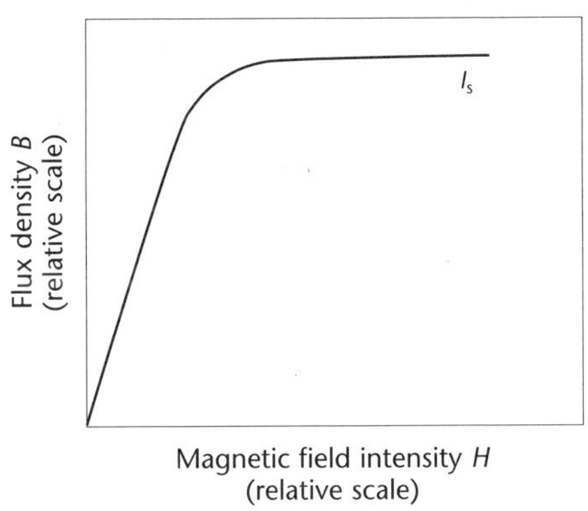

FIGURE 16. Variation of magnetic permeability with flux density.

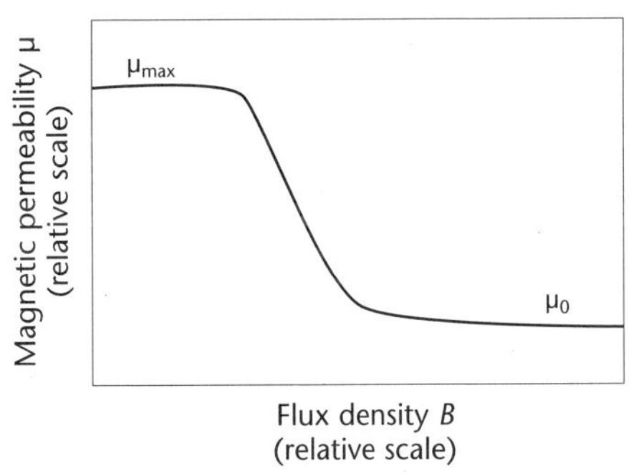

designates the coercive force H_c in the material.

As the reversed field is increased beyond C, point D is reached (Fig. 17d). At this point, the specimen is again saturated but in the opposite polarity. The magnetic field intensity is now decreased to zero and the DE line is formed and retains reversed polarity residual magnetism B_r in the specimen. Again increasing the magnetic field intensity in the original direction completes the curve EFA. Now the cycle is complete and the *hysteresis curve* (ABCDEFA) is called the *hysteresis loop*. In alternating current applications, the ferromagnetic material goes through this cycle for every reversal in current, 60 times per second on a 60 Hz power line. The area enclosed by the loop is proportional to the energy dissipated per cycle.

The lag throughout the cycle between the magnetic field intensity and the flux is called *hysteresis*. It is significant that the various factors of the hysteresis loop (saturation, remanence, coercive force and permeability) are influenced in different ways by material properties such as alloy composition, structure and internal stress.

Magnetic saturation is mainly a function of chemical composition and crystal structure. The coercive force is influenced by (1) internal stress and magnetostriction; (2) number, magnitude and distribution of foreign bodies embedded in or between ferromagnetic crystals; and (3) the energy and size of the ferromagnetic crystals. The magnetostriction, crystal energy, saturation magnetization and curie point are independent of the structural condition. However, the coercive force, permeability and rayleigh constant are influenced by internal stress and structural condition. The magnetic behavior depends to a large degree on selection of the initial materials and on melting, foundry, rolling and annealing processes.

FIGURE 17. Representative magnetization (hysteresis) curve for ferromagnetic material: (a) magnetic saturation; (b) reduction in magnetic force; (c) coercive force; (d) hysteresis loop.

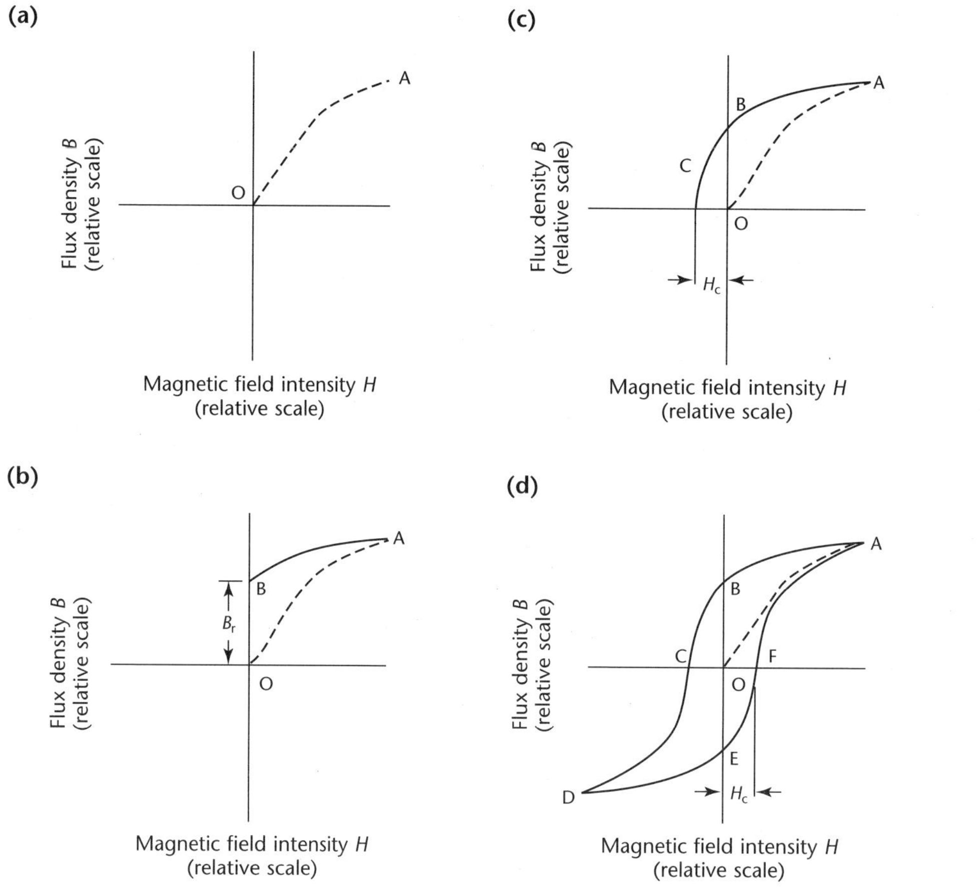

Legend

A, B = reference points
B_r = residual magnetism
H_c = coercive force

Hysteresis Loop Tests

The hysteresis loop can be represented for steel samples produced easily from standard semifinished parts. Only the infinitely long sample or the ring sample without magnetic poles permit representation of the hysteresis loop without distortion. The tube shown in Fig. 18 has been used as a standard sample for magnetic investigations. For magnetizing, this sample is placed on a conducting rod (Fig. 19) that in turn is placed in a high current fixture.

An alternating current passing through the conductor produces circular magnetization without free poles in the sample. A winding is used to measure the test piece induction and the central conductor is equipped with an air ring coil for measuring the applied field intensity. These measurements are then applied across the vertical and horizontal sweeps of an oscilloscope.

FIGURE 18. Standard test object for investigating ferromagnetic properties of material.

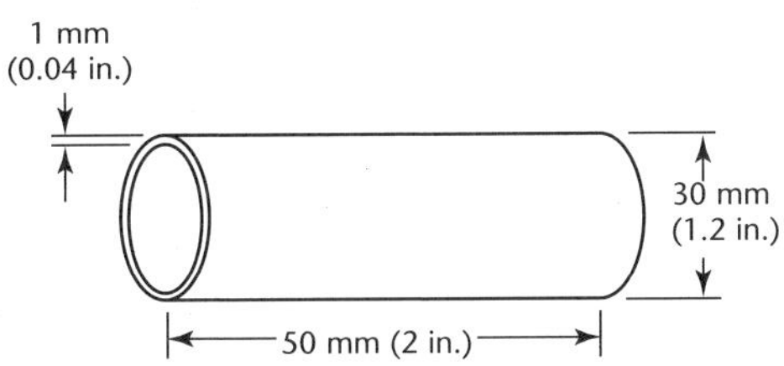

FIGURE 19. Device for magnetizing standard ring samples with axial current and measurement of magnetic field intensity with air ring coil.

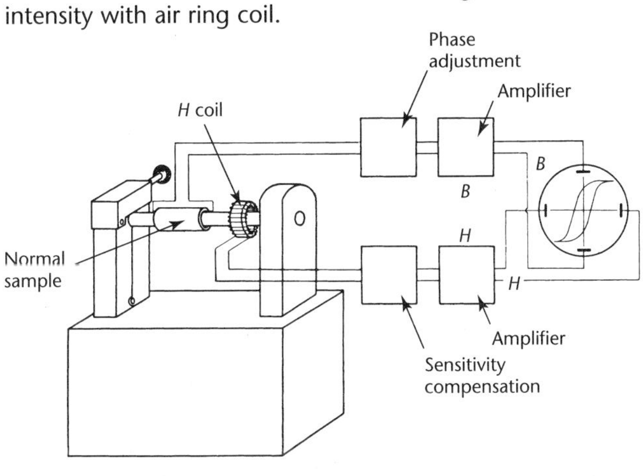

Legend

B = magnetic flux
H = magnetic field intensity

Loop Pattern Analysis[6]

By subjecting various standard test pieces to a wide variety of treatments, a large number of cathode ray tube screen pictures can be developed illustrating the effects of cold drawing, annealing, carburizing, quenching and tempering. Such pictures illustrate the capability of loop pattern analysis in nondestructive testing. Three sets of loop patterns are shown in Fig. 20. The family of loop patterns in Fig. 20a is for a sample of cold

FIGURE 20. Three sets of hysteresis loop patterns for samples of Unified Numbering System G10150 carbon steel: (a) patterns for sample as cold drawn; (b) patterns for same sample after stress relief annealing; (c) patterns for samples carburized and then either quench hardened or slowly cooled.

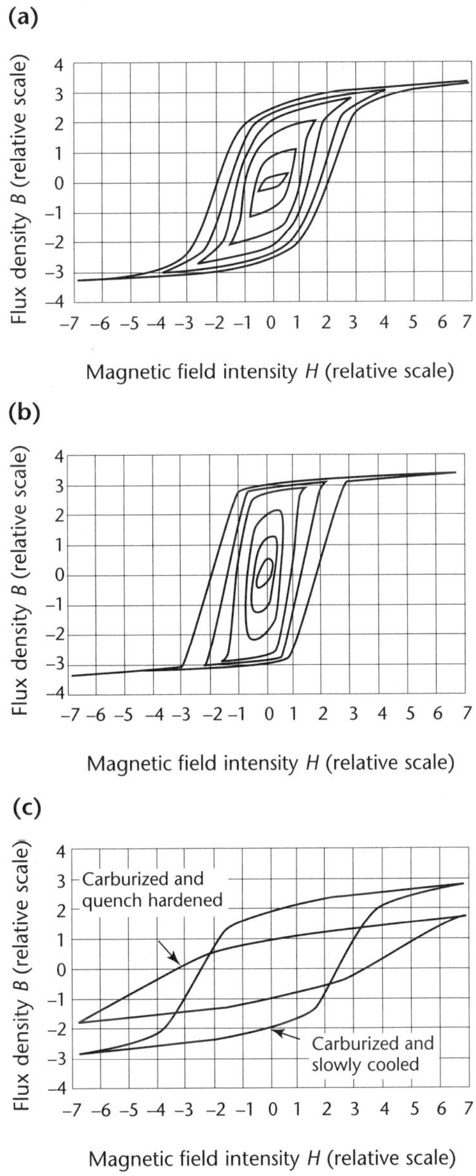

drawn Unified Numbering System G10150 carbon steel tested at six different magnetic field intensities (*H* values). When the same sample was stress relief annealed, the loop patterns were altered, appearing as shown in Fig. 20b. Figure 20c compares loop patterns of standard samples of carbon steel that were gas carburized for 1 h at 900 °C (1650 °F) and then cooled from the carburizing temperature at markedly different rates. The larger loop, with higher magnetization values, represents the slowly cooled sample; the smaller loop, with lower magnetization values and higher coercive force, represents the sample that was quench hardened.

Comparator Bridge Tests[6]

For sorting, the hysteresis loop is not the best representation of magnetic material variables. An improved presentation of the electromagnetic characteristics of a material occurs in test equipment that can indicate (1) the magnetization variations, (2) the permeability variations or (3) the variations of the curvature of the hysteresis loop of the test object with reference to a standard, depending on the selection of the presentation.

Using such an instrument, the procedures followed in sorting two steels of different chemical composition A and B are outlined in Fig. 21. The principal objective in this instance is to obtain the greatest possible difference in amplitude and shape between the patterns displayed on the screen for the two different steels. This differentiation can be used as a basis for sorting. In Fig. 21, the shaded symbols represent the controls that are adjusted during each step of setup.

As shown in Fig. 21a, a sample of steel A placed in the test coil causes a pattern to appear on the screen. The phase shifter and coarse sensitivity are adjusted until the entire pattern is displayed and centered. A fine sensitivity adjustment is used to bring the curve to a specific indication height called the *absolute value.* This adjustment lets all positions on the sensitivity selector be expressed as percentages of absolute value. If a second sample A_1 of steel A (Fig. 21b) is placed in the unoccupied difference coil and the sensitivity control is adjusted to a higher level, then a difference between the magnetic properties of samples A and A_1 will appear on the screen as indicated by the solid line in Fig. 21b. If A is then replaced by sample B_1, a curve for this sample might be as indicated by the broken line in Fig. 21b. Next, sample A_1 is reinserted in the difference coil, replacing B_1, and the compensator is adjusted so that curve A_1 becomes a horizontal,

FIGURE 21. Operating procedures and screen displays obtained when sorting two steels of composition A and composition B: (a) after phase and sensitivity adjustments; (b) second sample B_1 introduced; (c) sample A reintroduced; (d) curve B_1 centered; (e) curve A_1 symmetrically displaced; (f) amplification. Shaded symbols represent adjusted controls. Screen images show variation of permeability with magnetization (relative scale).

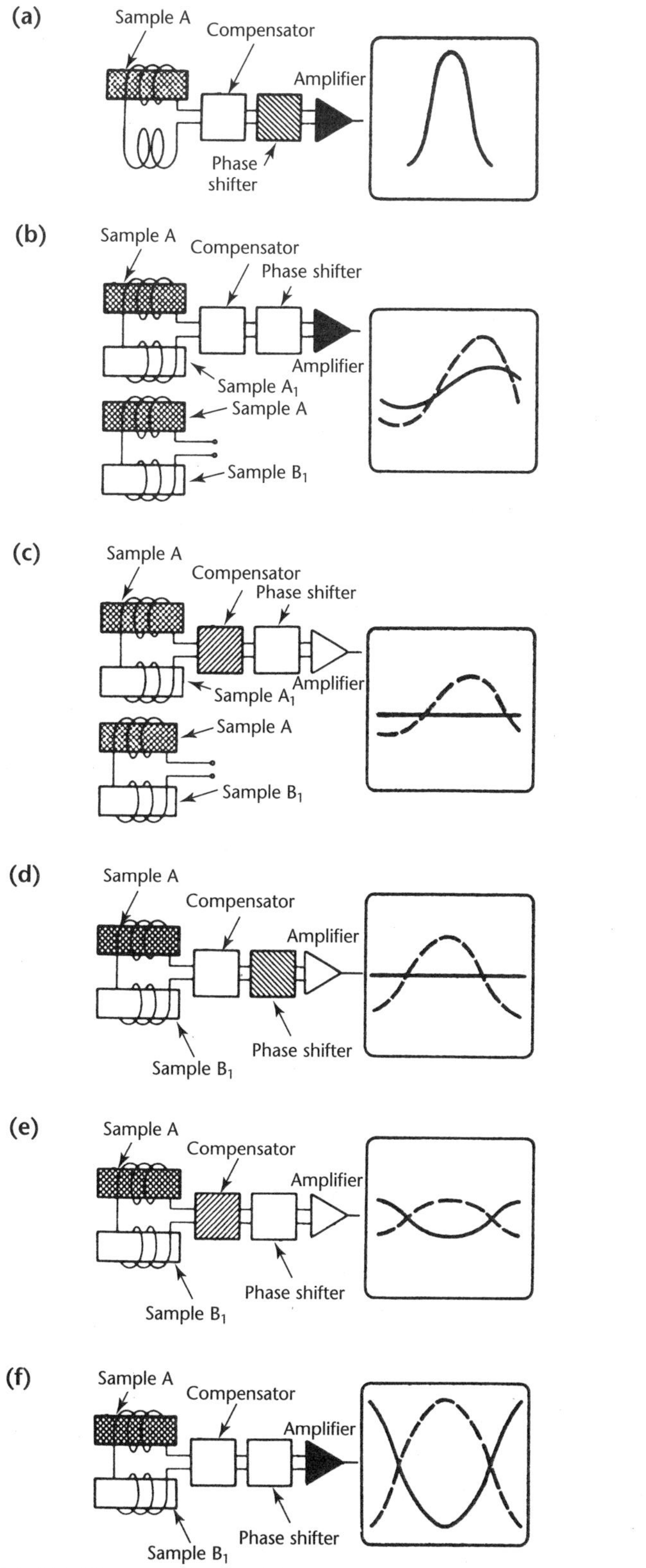

straight line (Fig. 21c). In the next step, curve B_1 is displayed by the phase shifter so that its peak or maximum lies in the center of the screen (Fig. 21d). In Fig. 21e, curve A_1 is symmetrically displaced downward with reference to curve B_1 by using the compensator. Finally the amplifier is turned up so that the area of the screen is fully used (Fig. 21f).

Spread Bands

When a large number of samples representing two steels of different chemical composition are sorted, minor variables in samples of the same steel result in a spread band or slight variations to either side of the principal curve. Slight differences in internal stresses or surface decarburization are among the factors that account for these variations. Typical spread bands made with identical instrument settings are shown in Fig. 22 for carbon steel and a sulfurized free machining steel. The bands shown in Fig. 22 reflect the spread obtained when 1000 samples of each steel were tested.

FIGURE 22. Spread bands obtained when 1000 samples each of two steels of different composition (Unified Numbering System G10150 carbon steel and free machining steel) were comparator bridge tested at identical instrument settings.

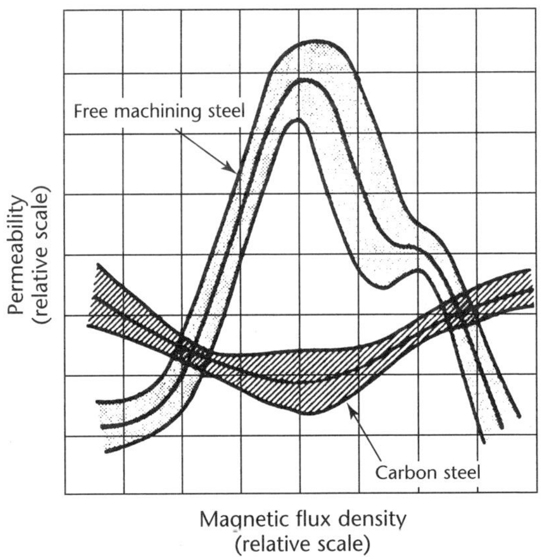

PART 5. Electrical Resistivity Measurements[1]

The resistivity of metals has been studied for many years and resistivity values for most materials are readily available. The reason that this parameter has not been widely used for metal identification or sorting, however, is the difficulty in determining the resistivity without special laboratory techniques.

Resistivity testing is also called *potential drop testing* and can be used for discontinuity detection.

Conductivity measurements are often made with eddy current devices. These techniques determine the conductivity or resistivity indirectly by measuring its effect on a coil in a high frequency alternating current test circuit. Consequently surface roughness, surface curvature and trace impurities can significantly affect the results. Another drawback is the requirement for a standard sample to compare with the unknown.

Instruments are now available that permit measuring resistivity directly. The advantages of direct measurement are that measurements can be made on bulk material in its manufactured state, that no reference standards are required and that resistivity values do not require intermediate calculation.

FIGURE 23. Probe arrangement and schematic diagram of resistivity measuring instrument.

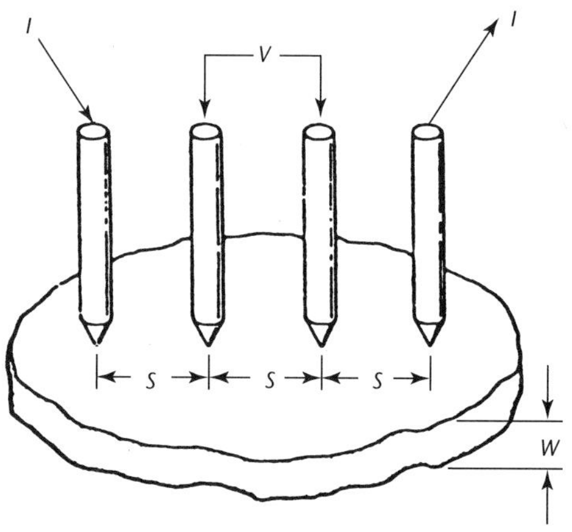

Legend
- I = current (A) through sample
- S = distance (meter) between probes
- V = potential difference (volt)
- W = sample thickness (meter)

Principles of Resistivity Tests

A typical instrument consists of two parts: a four-point probe and an electronics package to supply current, determine the voltage drop and convert it to a resistivity value. The four-point probe has been widely used for studying semiconductor materials; the relationship between probe geometry, voltage drop and sample resistivity has been established for many common cases. Although this relationship cannot be solved in closed form for a sample of arbitrary geometry, two important cases lead to very simple solutions. Using the notation of Fig. 23 I is the current through the sample, S is the distance (in meter) between the probes, V is the voltage detected across the inner probe and W is the thickness (in meter) of the sample. For samples of length and width several times the overall probe spacing, the resistivity ρ for sheets with thickness $W < 0.5\ S$ is:

$$(6) \qquad \rho = \frac{VW\pi}{I(\ln 2)} = 4.53\,\frac{VW}{I}$$

and

$$(7) \qquad \rho = \frac{2\pi SV}{I} = 6.28\,S\,\frac{V}{I}$$

for sheets with $W > 3S$. Resistivity ρ is usually expressed in microohm centimeter or microohm meter. Note that for thin sheets resistivity ρ is determined independently of the probe spacing. For thick sheets the determination is independent of thickness. An important implication of Eq. 7 is that the resistivity can be determined for massive samples like ingots or bars, provided only that

FIGURE 24. Typical resistivity measuring system.

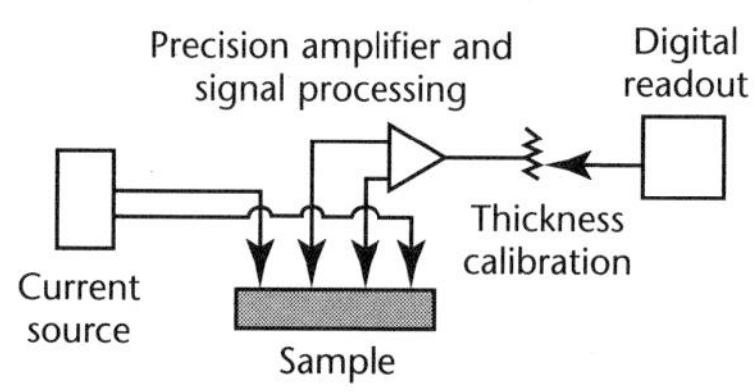

there is one flat surface and that the overall sample size is large compared to the probe spacing.

Figure 24 shows a block diagram of a typical instrument. The current source puts a known current through the sample, the signal processing electronics remove interfering voltages and the sample thickness (or probe spacing) is accounted for by an analog multiplier. The circuit output voltage is read by a digital volt meter. The calibration is adjusted so that the output reading is directly in microohm meter or microohm centimeter. A minor circuit change could make the output read in a conductivity unit.

Applications of Resistivity Measurements

This technique can be very useful when absolute resistivity readings are required. There may be cases however when test object geometry, material thickness or probe spacing will not be within the constraints of the theory. In these instances for which no theoretical correction factors are available, the application may be solved empirically. Sorting for mixed material is a good example of an application that usually requires only relative readings. Theoretical correction factors for some applications are available from several sources.

Resistivity changes arising from the heat treatment (and subsequent precipitation and recrystallization) of various alloys are frequently large enough to use as a determination of whether the heat treatment has led to the desired microstructure. Resistivity measurements for this purpose are particularly useful on steels because such measurements on steel do not suffer any effects from permeability variations as do eddy current tests. Another important advantage of resistivity measurements over other techniques is that a significant volume of the sample is measured (to a depth of about the probe spacing) rather than just the surface. This distinction can be important in view of the effects of surface oxidation, segregation and cold work in making the surface different from the bulk.

So far it has been assumed that the material being studied is homogeneous. Heterogeneities such as voids, cracks, slag or flux inclusions are generally nonconductive and will sharply raise the observed resistivity if they are in the vicinity of the voltage probes. It is estimated that voids greater in diameter than 10 percent of the probe spacing S could be detected if they are within a distance S of the surface. Cracks perpendicular to the surface and the line made by the probe points can be most easily detected whereas those perpendicular to the surface but along the line of the probes would have a very much smaller effect. This difference allows some localization and measurement of the size of discontinuities.

PART 6. Thermoelectric Sorting[1]

Though not strictly electromagnetic, another useful approach to material sorting involves thermoelectric effects. There are three thermoelectric effects: the seebeck effect, the peltier effect and the thompson effect.

The seebeck effect is observed when two conductors are joined: if the junctions are maintained at different temperatures, an electromotive force (EMF) will be developed around the closed loop. This force is the thermocouple voltage or seebeck voltage θ. It is easily measured by inserting a voltmeter in the loop.

The peltier effect is the evolution (or absorption) of heat when an electric current passes from one material to a different material at the same temperature. The rate at which heat is evolved at the junction when a current J is passing from material 1 to material 2 is $\pi_{12}J$ where π_{12} is the peltier coefficient.

The thompson effect is the evolution (or absorption) of heat when an electric current passes from a material at one temperature to the same material at a different temperature (in a material with a temperature gradient imposed). The rate at which heat is evolved per unit temperature difference is $\mu{\cdot}J$ where μ is the thompson coefficient and J the current flowing from the higher to the lower temperature. The thompson heat like the peltier heat is reversible and is in addition to any (irreversible) joule heat.

The mechanisms can be visualized that give rise to thermoelectricity and they can be loosely described as the tendency of heat to drag along electricity. The tendency of electricity to drag along heat could also be described but is of no particular interest here because most commercially available instruments use the first approach.

Imagine a piece of material as a long box of electrons with a uniform concentration of charge carriers. Make one end hot and the other end cold. The electrons at the hot end will diffuse more rapidly than those at the cold end, so they will move around a little more and more of the hot electrons will move to the cold end. This flow of electric current would keep up indefinitely if the piece of material under consideration were not electrically insulated. In this case the initial current produces a pile up of electrons at one end, which gives rise to a back electromotive force that prevents further flow of charge.

In the equilibrium situation, hot electrons keep flowing to the cold end while cold electrons flow to the hot end because of the voltage gradient. Thus, there is no net flow of particles or charge. There is however a flow of energy, a heat current. Under the influence of a temperature gradient, there exists then a heat flow together with a tendency for electricity to flow. This tendency is balanced by the back electromotive force.

Application of the Thermoelectric Effect

There exist two relationships between the seebeck, peltier and thompson quantities.

$$\theta_{12} + \pi_{12}(T_c) - \pi_{12}(T_h) + \int_{T_c}^{T_h} (\mu_1 - \mu_2)\,dT = 0 \tag{8}$$

$$\frac{\pi_{12}(T_c)}{T_c} - \frac{\pi_{12}(T_h)}{T_h} + \int_{T_c}^{T_h} \frac{\mu_1 - \mu_2}{T}\,dT = 0 \tag{9}$$

These equations are obtained by considering a thermocouple with an infinitesimal current flowing through it. The thermocouple is made of material 1 and material 2 with one junction at a higher temperature T_h and the other at a cooler temperature T_c. The main point to be made is that there are three physical effects and two relationships between them. If all three measurements are carried out, the two relationships are tested and there remains one parameter to describe the material. In other words, the measurement of any one effect determines the other two. This is made clearer by the introduction of the seebeck coefficient, the differential coefficient of the seebeck voltage:

$$(10)\quad \frac{d\theta_{12}}{dT} = \alpha_1(T) - \alpha_2(T)$$

$$(11)\quad \pi_{12}(T) = \left[\alpha_1(T) - \alpha_2(T)\right]T$$

$$(12)\quad \int_0^T \frac{\mu(T)}{T}\,dT = \alpha_1(T)$$

where $\alpha_1(T)$ is the (absolute) seebeck coefficient of material 1 at the temperature T. The relative seebeck coefficient of the two materials is $\alpha_1(T) - \alpha_2(T)$. Equations 8, 9 and 10 taken together are equivalent to Eqs. 8 and 9 with a defining equation for the seebeck coefficient.

The thompson coefficient is determined for a single material rather than a pair of materials and so is the seebeck coefficient. This rather difficult measurement, however, needs to be made only on one material. The seebeck coefficient of any other material may be obtained by measuring the temperature dependence of the seebeck voltage of a thermocouple constructed of the unknown and a reference standard.

The magnitude of the seebeck coefficient for different kinds of systems may be estimated using the following relationship:

$$(13)\quad q\alpha \cong \frac{C_v}{N}$$

where C_v = the specific heat of the system of charge carriers, N = the number of conducting particles and q = the charge of each particle. The usual unit of $C_v \cdot N^{-1}$ is in terms of Boltzmann's constant k whereas α is in microvolt per degree.

FIGURE 25. Variation in seebeck coefficients for four hypothetical materials.[7]

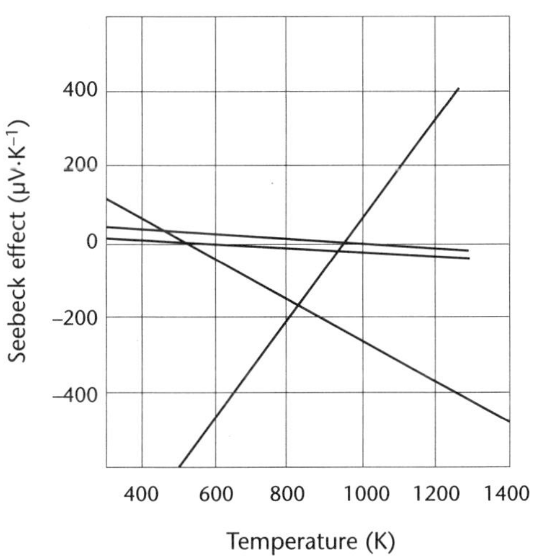

Most thermoelectric instruments currently do not actually measure the seebeck coefficient but rather the seebeck voltage. Copper is commonly the hot contact although some instruments can work with other metals. As is evident from Eq. 10, the voltage produced depends on the difference in the seebeck coefficients of the two materials.

Figure 25 illustrates how seebeck coefficients vary for different materials.[7] Consequently a greater (and more easily measured) voltage may be produced by judicious selection of the hot probe's material and temperature. Because of the nature of the circuitry, the temperature is seldom variable although probe tips may be interchanged on some instruments. Although not as effective as in identifying alloy variations, thermoelectric techniques may be used to identify heat treatment and possibly other metallurgical variations.

To measure the electromotive force in terms of the thermoelectric voltage, normally two electrodes are brought into direct contact with the sample at two points maintained at different temperatures and the electric potential between the electrodes is measured. This measurement unavoidably includes the thermoelectric voltage generated in the sample, the voltage of the electrodes and the possible contribution from the contact potential caused by oxidation or other contaminants on the sample surface. The requirement of direct contact with the surface and the dependence on surface conditions limit the application of thermoelectric testing mostly to raw materials or specially prepared samples.[22] The small value of the thermoelectric voltage makes measurement of small variations extremely difficult.

To overcome the surface contamination problems, a technique can be applied to use noncontacting readout of the thermoelectric signals by most sensitive magnetometers.[23] This technique has potential for detection of segregations,[23,24] regions with fatigue damage before cracking,[25-27] texture[28] and stress[29] and for further microstructural analysis.

PART 7. Dielectrometric Techniques for Material Characterization

Dielectrometry involves the measurement and characterization of weakly conducting or insulating dielectric materials. This is accomplished with electrode structures that apply an electric field to the test material so that measurements of the effective impedance or admittance between the electrodes reflects the complex dielectric permittivity of the test material. These measurements can also be expressed in terms of the dielectric permittivity, electrical conductivity and loss tangent. Typically these measurements are then correlated to other properties of interest, such as the cure state, moisture content, concentration of impurities and additives, coating thickness, density and aging status. The electric fields in these sensors are analogous to magnetic fields in eddy current sensors and magnetometers. Different electrode structures and multiple excitation frequencies allow spatial and temporal variations of the dielectric and material properties to be determined; the variations lend valuable insights into physical phenomena in materials and equipment. These dielectric property sensors provide instrumentation for system monitoring and diagnostics and can be used for optimization of design and performance characteristics.[8-11]

Dielectrometers have been used to measure the properties of many liquid and solid dielectrics, composites and surface coatings.[8] Such applications are: cure state monitoring of polymer composites, adhesives and epoxies; porosity measurements of ceramics, such as thermal barrier coatings and microporous alkaline battery separators; surface conductivity determination of thin semiconducting coatings; monitoring of toxic vapor absorption in chemical protection garments; detection of discontinuities in glass fiber epoxy composites; moisture diffusion monitoring in oil pressboard systems used in power transformers; measurement of interfacial electrochemical parameters of debye length, zeta potential, molecular diffusion coefficients and ion mobilities; and detection of buried objects such as plastic and metallic land mines.[9] Other applications involve capacitive sensors for system diagnostics such as proximity sensors and liquid level sensors.[10]

A typical dielectrometry system is shown in Fig. 26. The sensor is connected to probe electronics for signal amplification and buffering. The electronics in turn are connected to instrumentation for measuring the terminal impedance or admittance of the sensor electrodes. Most systems stand alone and provide front panel access for instrument control and data display but some also use a computer, which then provides greater flexibility for data display and processing. Buffers in the probe electronics allow the sensor itself to be at a distance from the instrumentation. This separation is particularly useful for manufacturing settings where it is convenient to have all of the control systems and diagnostic information in a central location.

The goal of dielectrometry is to relate physical and electrical characteristics of an unknown material to electrical admittance or impedance measurements made at electrical terminals. This is done most easily for materials with homogeneous electrical properties using parallel plate electrodes. When a voltage is placed across the electrodes (Fig. 27), the electric field is spatially uniform in the central region between electrodes and points in a single direction perpendicular to the electrode surface. An equivalent circuit for the response is shown in Fig. 28. For a homogeneous material of permittivity ε and ohmic conductivity σ (inverse of resistivity ρ), the capacitance C

FIGURE 26. Dielectrometry system including impedance measurement instrumentation, probe electronics and interchangeable sensors for single sided property measurements.

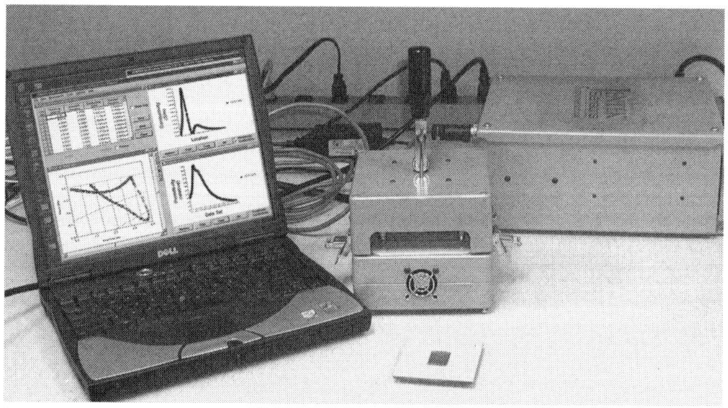

and conductance G (inverse of resistance R) can be expressed as:

$$(14)\quad C = \frac{\varepsilon A}{d}$$

and

$$(15)\quad G = \frac{1}{R} = \frac{\sigma A}{d}$$

where A is the electrode area and d is the gap thickness between the electrodes. The permittivity of free space is $\varepsilon_0 = 8.854\ \text{pF}\cdot\text{m}^{-1}$. Note that for a homogeneous ohmic dielectric:

$$(16)\quad RC = \frac{\varepsilon}{\sigma}$$

This result is true, independent of electrode geometry.[12] When excited by a sinusoidal voltage in time with frequency f and angular frequency $\omega = 2\pi f$, the corresponding equivalent admittance Y (inverse of impedance Z) can be expressed as:

$$(17)\quad Y = \frac{1}{Z} = G + j\omega C$$

FIGURE 27. Parallel plate electrode sensor: (a) basic sensor; (b) sensor with guard electrodes.

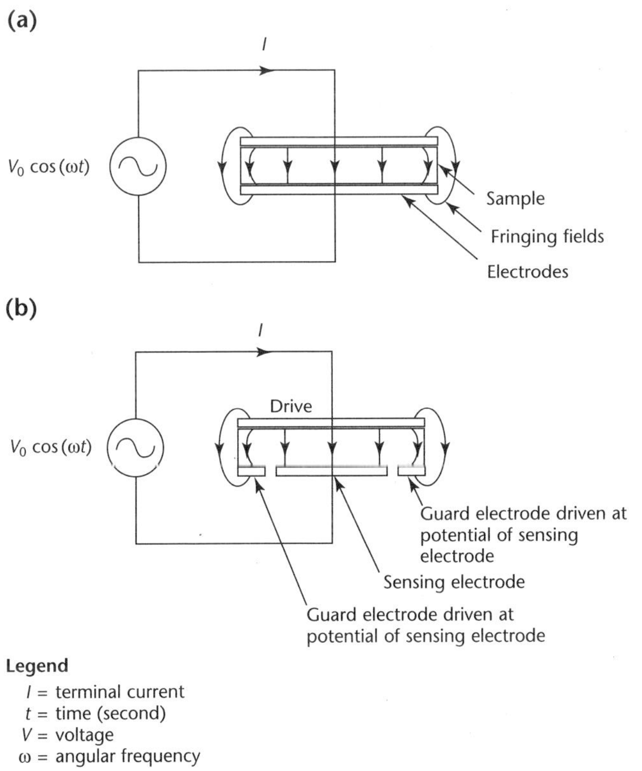

where $j = \sqrt{(-1)}$. The dielectric material properties can also be expressed in terms of the complex permittivity ε^*:

$$(18)\quad \varepsilon^* = \varepsilon' - j\varepsilon'' = \varepsilon - \frac{j\sigma}{\omega}$$

Typical excitation frequencies range from less than 1 mHz for insulating materials up to several megahertz for semiconducting materials. The sensor geometry is accurately captured by the ratio of the area to the gap ($A \cdot d^{-1}$) when the electrode widths are much larger than the gap so that the fringing fields at the electrode sides can be neglected. Placing guard electrodes around the sense electrode with their voltage the same as the sense electrode voltage helps minimize the effects of the fringing fields as illustrated in Fig. 27b. A guard electrode can also be placed behind the sense electrode to further reduce extraneous coupling from the fringing fields. The final use of these dielectrometry measurements is to infer related physical properties such as moisture content, density, porosity and impurities. Empirical measurements then can generally map values of the physical variable to values of the material permittivity and conductivity.

For simple systems (as in Fig. 28), measurements at a single excitation frequency can be used to determine both the permittivity and conductivity of the material. Most materials are dispersive, however, so the effective properties depend on the excitation frequency. Dispersiveness can be attributed to heterogeneous material properties, such as particles embedded in a matrix or multiple layers of different material properties, or can be attributed to multiple physical processes, such as multiple conduction mechanisms. Different techniques or models are then used to determine the properties of interest. As examples for multiple layered materials, equivalent circuits and expressions for the terminal capacitance or conductance (Eqs. 14 and 15) that account for the properties of each layer

FIGURE 28. Equivalent circuit for parallel plate electrode sensor and homogeneous dielectric material.

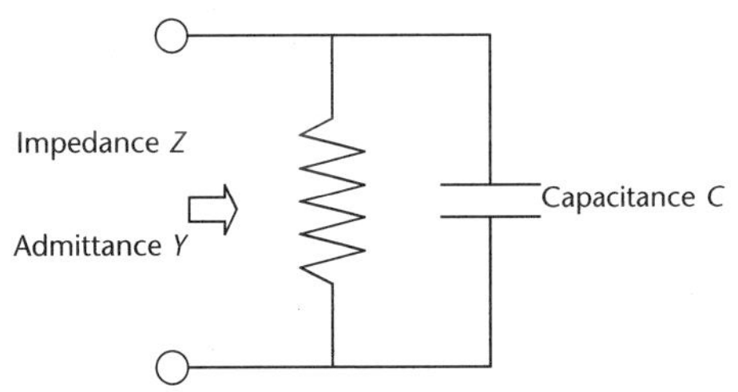

can be used.[12] For two-phase composite materials, the effective dielectric properties of the composite can be related to the dielectric properties and geometry of the constituent materials.[13]

The effects of material heterogeneity can also be displayed graphically. If excited by a sinusoidal voltage in time, as the frequency is varied from zero to infinity, a plot of the imaginary part of the impedance or admittance versus the real part of the impedance or admittance traces out a semicircle for the equivalent circuit of Fig. 28. Such plots are called *cole-cole plots* and when semicircular provide a quick verification that the material is homogeneous and frequency nondispersive. If the dielectric properties of permittivity and conductivity are spatially dependent or dependent on frequency, then the cole-cole plots will deviate from the semicircular shape.

To illustrate this effect, consider Fig. 29, which shows the effective properties for a relatively insulating liquid dielectric.[14] In this case, a parallel plate sensor was immersed in a liquid dielectric so that no air gaps would be present between the test material and the electrodes. At high frequencies, the effective permittivity is constant and the effective conductivity is unmeasured because the response is dominated by the capacitive effects. At intermediate frequencies, both the permittivity and conductivity can be determined and are constant with frequency. At very low frequencies, the effective properties show a dependence on frequency and reflect the existence of an electrical double layer at the electrode interfaces. The cole-cole plot of Fig. 30 shows this frequency dependence: the semicircular shape of the response is lost at low frequency and at the elevated temperatures because of the electrical double layer capacitance.

In many circumstances, only one side of the test material is accessible or the spatial variations of the material properties may be of interest. In these situations, it is generally impractical to use a parallel plate sensor and one-sided measurement schemes are required. These single-sided sensor schemes can be visualized as parallel plate capacitors that have been opened so that the sensor responds to the material properties through changes in the fringing electric fields between the electrodes. Guard electrodes can also limit unintentional coupling of the fields to the sense electrode.

One single-sided electrode format that has an advantage of providing a relatively large sensing response within a given sensor footprint is an interdigitated electrode structure as shown in Fig. 31a.[15-17] The spatial periodicity of the electrodes is determined by the wavelength λ. One set of electrode fingers is driven by a sinusoidally time varying signal V_D with known amplitude and frequency whereas the second set of

FIGURE 29. Representative plots of effective material properties for transformer oil: (a) at 15° C (59° F); (b) at 70° C (158° F).

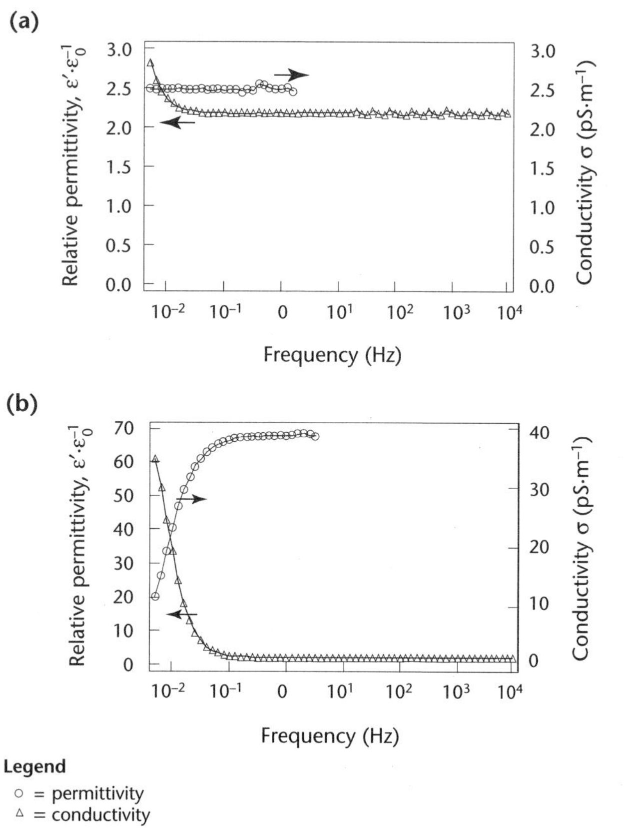

FIGURE 30. Representative cole-cole plots of real and imaginary parts of complex impedance Z corresponding to data of Fig. 30.

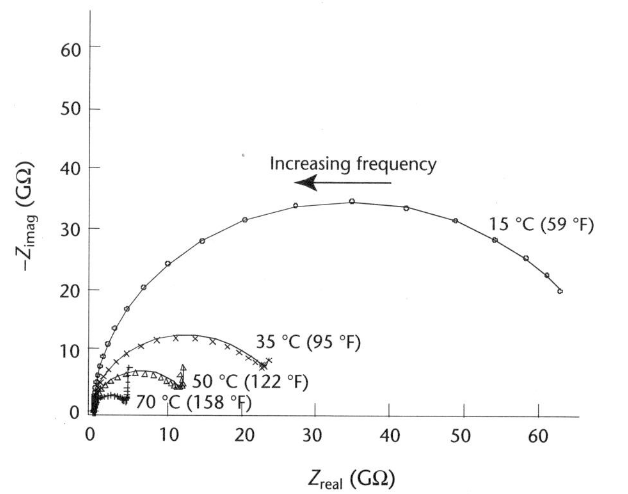

interdigitated fingers is capacitively loaded and floats to a sensed voltage V_S or is virtually grounded with a terminal current I_S. The sensed signal amplitude and phase with respect to the driven voltage depend on the complex permittivity of the adjacent dielectric. With these interdigitated structures, the inversion between the measured sensor terminal variables and the dielectric properties is more complicated than with the parallel plate structures.[17-19] The spatially periodic variation of electric potential along the surface (in the Y direction) produces an electric field that penetrates into the medium (in the Z direction). The potential obeys Laplace's equation and can be represented using a fourier series.

This model shows that the electric field for each spatial mode at a given frequency decays exponentially (in the Z direction) with a decay proportional to the spatial wavelength of the periodic electrodes (Fig. 31b). Consequently, sensors with longer wavelengths have larger penetration depths and will respond to changes of material properties far from the interface of sensor to dielectric material. Smaller wavelength sensors will primarily respond to changes near the interface. Thus, multiple wavelength sensors provide spatial profile information about the test material and permit, for example, simultaneous measurement of the dielectric properties and thickness of any air gaps that may be present between the test material and the sensor. These multiple wavelength sensors can use multiple sets of interdigitated spatially periodic electrodes attached to a common substrate or colocated designs that interweave the electrodes for a multiple wavelength measurement in a single sensor footprint.[20] Segmenting of the electrodes into arrays also permits wide area imaging of dielectric properties.

Multiple frequency techniques are also used to study interfacial electrochemical processes such as surface corrosion and energy storage in battery components and for electrochemical impedance spectroscopy.[21] In addition to the bulk response of the materials, these techniques allow interfacial processes, such as reaction rates and transport of charged species, to be isolated. These measurements provide insight into physical phenomena in the materials and are used to optimize performance characteristics.

FIGURE 31. Single-sided electrode format: (a) interdigital dielectrometry sensor in one-sided contact with material under test; (b) electrode spatial wavelength limits electric field intensity depth of penetration into test material.

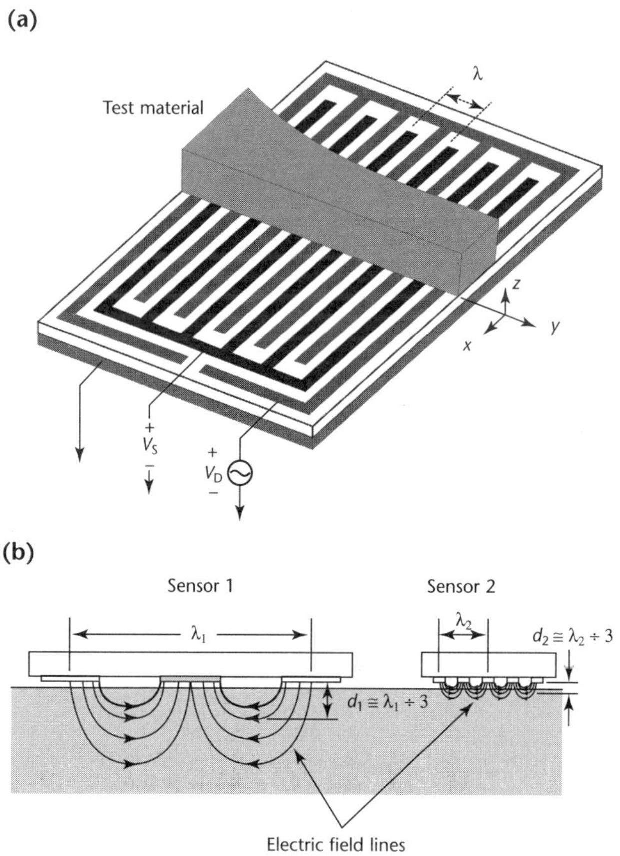

Legend

d = electric field penetration depth
V_D = sinusoidally varying time signal
V_S = sensed voltage
x,y,z = directional coordinates
λ = wavelength of spatially periodic electrodes

References

1. "Metal and Alloy Identification." *Nondestructive Testing Handbook*, second edition: Vol. 9, *Special Nondestructive Testing Methods*. Columbus, OH: American Society for Nondestructive Testing (1995): p 363-376.
2. ."X-Ray Diffraction and X-Ray Fluorescence." *Nondestructive Testing Handbook*, third edition: Vol. 4, *Radiographic Testing*. Columbus, OH: American Society for Nondestructive Testing (2002): p 427-435.
3. "Electromagnetic Sorting Techniques." *Nondestructive Testing Handbook*, second edition: Vol. 4, *Electromagnetic Testing*. Columbus, OH: American Society for Nondestructive Testing (1986): p 216-249.
4. Hagemaier, D.J. "Eddy Current Impedance Plane Analysis." *Materials Evaluation*. Vol. 41, No. 2. Columbus, OH: American Society for Nondestructive Testing (February 1983): p 211-218.
5. Rummel, W.D. "Theory of the Use of Eddy Current Conductivity Devices to Monitor Aluminum Alloys." *Materials Evaluation*. Vol. 24, No. 9. Columbus, OH: American Society for Nondestructive Testing (September 1966): p 507-511.
6. Section 42, "Eddy Current Test Indications." *Nondestructive Testing Handbook*, first edition: Vol. 2. Columbus, OH: American Society for Nondestructive Testing (1959).
7. Egli, P.H., ed. *Thermoelectricity; Including the Proceedings of the Conference on Thermoelectricity, September, 1958*. New York, NY: John Wiley and Sons (1960).
8. Von Hippel, A. *Dielectric Materials and Applications*. Boston, MA: Artech House (1954).
9. Zahn, M. "Optical, Electrical, and Electromechanical Measurement Methodologies of Field, Charge, and Polarization in Dielectrics." *IEEE Transactions on Dielectrics and Electrical Insulation*. Vol. 5, No. 5. New York, NY: Institute of Electrical and Electronics Engineers (October 1998): p 627-650.
10. Carr, J.J. *Sensors and Circuits*. Upper Saddle River, NJ: PTR Prentice Hall (1993).
11. Auld, B.A., A.V. Clark, S.R. Schaps and P.R. Heyliger. "Capacitive Probe Array Measurements and Limitations." *Review of Progress in Quantitative Nondestructive Evaluation* [La Jolla, CA, July 1992]. Vol. 12A. New York, NY: Plenum (1993): p 1063-1077.
12. Zahn, M. *Electromagnetic Field Theory: A Problem Solving Approach*. Malabar, FL: Krieger Publishing (1987).
13. Neelakanta, P.S. *Handbook of Electromagnetic Materials*. New York, NY: CRC Press (1995).
14. Washabaugh, A.P., A. Mamishev, Y. Du and M. Zahn. "Dielectric Measurements of Semi-Insulating Liquids and Solids." *12th International Conference on Conduction and Breakdown in Dielectric Liquids* [Rome, Italy, July 1996]. New York, NY: Institute of Electrical and Electronics Engineers (1996): p 381-384.
15. Sheppard, N.F., Jr., D.R. Day, H.L. Lee and S.D. Senturia. "Microdielectrometry." *Sensors and Actuators*. Vol. 2. Lausanne, Switzerland: Elsevier Sequoia (1982): p 263-274.
16. Senturia, S.D., N.F. Sheppard, Jr., H.L. Lee and D.R. Day. "In-Situ Measurement of the Properties of Curing Systems with Microdielectrometry." *Journal of Adhesion*. Vol. 15. Langhorne, PA: Gordon and Breach (1982): p 69-90.
17. Zaretsky, M.C., L. Mouayad and J.R. Melcher. "Continuum Properties from Interdigital Electrode Dielectrometry." *IEEE Transactions on Electrical Insulation*. Vol. 23, No. 6. New York, NY: Institute of Electrical and Electronics Engineers (December 1988): p 897-917.
18. Melcher, J.R. and M.C. Zaretsky. United States Patent 4 814 690, *Apparatus and Methods for Measuring Permittivity in Materials* (March 1989).
19. Lesieutre, B.C., A.V. Mamishev, Y. Du, E. Keskiner, M. Zahn and G.C. Verghese. "Forward and Inverse Parameter Estimation Algorithms of Interdigital Dielectrometry Sensors." *IEEE Transactions on Dielectrics and Electrical Insulation*. Vol. 8, No. 4. New York, NY: Institute of Electrical and Electronics Engineers (August 2001): p 577-588.

20. Goldfine, N.J., M. Zahn, A.V. Mamishev, D.E. Schlicker and A.P. Washabaugh. United States Patent 6 380 747, *Methods for Processing, Optimization, Calibration, and Display Of Measured Dielectrometry Signals Using Property Estimation Grids.* (April 2002).
21. Macdonald, J.R., ed. *Impedance Spectroscopy: Emphasizing Solid Materials and Systems.* New York, NY: Wiley-Interscience (1987).
22. Morgner, W. "Fundamentals of Nondestructive Materials Characterization." *NDT&E International.* Vol. 27, No. 5. Kidlington, Oxford, United Kingdom: Elsevier Science Limited (October 1994): p 263-268.
23. Hinken, J.H. and Y. Tavrin. "Thermoelectric SQUID Method for the Detection of Segregations." *Review of Progress in Quantitative Nondestructive Evaluation* [Montreal, Canada, July 1999]. Vol. 19B. Melville, NY: American Institute of Physics (2000): p 2085-2092.
24. Morgner, W. "Introduction to Thermoelectric Nondestructive Testing." *Materials Evaluation.* Vol. 49, No. 9. Columbus, OH: American Society for Nondestructive Testing (September 1991): p 1081-1087. Errata, Vol. 49, No. 12 (December 1991): p 1505.
25. Tavrin, Y., G.S. Krivoy, J.H. Hinken and J.-P. Kallmeyer. "A Thermoelectric Method to Detect Weak Local Structural Changes and Inclusions in Metal Parts." *Review of Progress in Quantitative Nondestructive Evaluation* [Ames, IA, July 2000]. Vol. 20B. Melville, NY: American Institute of Physics (2001): p 1710-1716.
26. Nagy, P.B. and J. Hsu. "Thermoelectric Detection of Early Fatigue Damage in Metals." *Review of Progress in Quantitative Nondestructive Evaluation* [San Diego, CA, July-August 1997]. Vol. 17B. New York, NY: Plenum Press (1998): p 1573-1580.
27. Maslov, K. and V.K. Kinra. "Noncontact Thermoelectric NDT for Alloy Microstructural Analysis." *Materials Evaluation.* Vol. 59, No. 9. Columbus, OH: American Society for Nondestructive Testing (September 2001): p 1081-1084.
28. Carreon, H., A.H. Nayfeh and P.B. Nagy. "Anisotropic Effects in Noncontacting Thermoelectric Material Characterization." *Review of Progress in Quantitative Nondestructive Evaluation* [Brunswick, ME, July-August 2001]. Vol. 21B. Melville, NY: American Institute of Physics (2002): p 1455-1462.
29. Carreon, H., P.B. Nagy and M.P. Blodgett. "Thermoelectric Nondestructive Evaluation of Residual Stress in Shot-Peened Metals." *Research in Nondestructive Evaluation.* Vol. 14, No. 2. Columbus, OH: American Society for Nondestructive Testing (June 2002): p 59-80.

Bibliography

ASTM E 566, *Standard Practice for Electromagnetic (Eddy-Current) Sorting of Ferrous Metals.* West Conshohocken, PA: ASTM International (1999).

ASTM E 703, *Standard Practice for Electromagnetic (Eddy-Current) Sorting of Nonferrous Metals.* West Conshohocken, PA: ASTM International (1998).

ASTM E 977-84, *Standard Practice for Thermoelectric Sorting of Electrically Conductive Materials.* West Conshohocken, PA: ASTM International (1999).

ASTM E 1476, *Standard Guide for Metals Identification, Grade Verification, and Sorting.* West Conshohocken, PA: ASTM International (1997).

ASTM E 1916, *Standard Guide for Identification and/or Segregation of Mixed Lots of Metals.* West Conshohocken, PA: International (1997).

"Electromagnetic Sorting and Testing." *Metals Handbook*, eighth edition: Vol. 11, *Nondestructive Inspection and Quality Control.* Materials Park, OH: ASM International (1976): p 93-105.

IEC 60028, *International Standard of Resistance for Copper.* Geneva, Switzerland: International Electrotechnical Commission (2001).

Mordwinkin, G. United States Patent 4 059 795. *Digital Eddy Current Apparatus for Sensing and Analyzing Characteristics of an Electrically Conductive Material* (November 1977).

Mordwinkin, G. United States Patent 4 230 987, *Digital Eddy Current Apparatus for Generating Metallurgical Signatures and Monitoring Metallurgical Contents of an Electrically Conductive Material* (October 1980).

14

CHAPTER

Primary Metals Applications of Electromagnetic Testing

Ram P. Samy, The Timken Company, Canton, Ohio

Tatsuo Hiroshima, Marktec Corporation, Taiei, Chiba, Japan (Part 4)

Joseph M. Mandula, Seven Hills, Ohio (Part 3)

Masashi Mizuno, Tohoku Steel, Murata, Japan (Part 5)

Takahide Sakamoto, Sumitomo Metal Technology, Amagasaki, Japan (Part 4)

Wolfgang Stumm, Erbach, Germany (Part 6)

PART 1. Electromagnetic Testing in Primary Metals Industries

The basic principles of electromagnetic testing were known in the nineteenth century and practical applications of electromagnetic techniques have been commonplace in the metals industry since 1930. Eddy current testing has evolved from relatively simple devices for metal sorting to complex, automated test systems as part of manufacturing processes.[1] Electromagnetic techniques offer simplicity, low cost, noncontact and couplant free operation, high speed and high temperature capabilities. They are widely used in all types of industries to evaluate the quality of materials and components, including both ferritic and nonferritic metals. The total number of tests performed annually by these techniques may exceed that of all other nondestructive test techniques.

Because of the skin effect, which limits the depth of penetration, eddy current testing is limited to surface and near surface evaluation of materials and products. Both eddy current and magnetic techniques are preferred in the steel industry throughout the world for inline surface testing of bars, billets and tubes at production speeds. The eddy current technique is used by the metals industry for the inline testing of hot wires at high production speeds, often in excess of 120 $m \cdot s^{-1}$ (265 $mi \cdot h^{-1}$).

Electromagnetic test techniques find their application in all stages of forming, shaping and heat treating of metals and alloys, where the effectiveness of processing steps can be quickly evaluated. Materials damaged during processing can be detected and removed from production without incurring further processing costs. Thermal treatments such as annealing, normalizing, hardening, case hardening and other heat treating processes can be monitored directly in many instances.

Eddy Current versus Magnetic Testing

Metals are said to be *ferromagnetic* if, like iron (*ferro-* comes from the Latin word for iron), they can be magnetized. Ferromagnetic metals include iron, cobalt and nickel. Nonferromagnetic metals include copper and aluminum. Magnetic techniques such as magnetic flux leakage and magnetic particle testing can be used only on metals that can be magnetized — on steel, not aluminum.

On the other hand, eddy current testing can be used on all electrically conductive materials — magnetic and nonmagnetic. For this reason, eddy current testing is extensively used in factories that manufacture nonmagnetic but electrically conductive materials and components, such as aluminum and copper. The anisotropic and highly nonlinear behavior of ferromagnetic materials during magnetization tends to generate eddy current signals that are difficult to interpret. For this reason, ferromagnetic objects tested with the eddy current technique are often magnetically saturated, making them behave like nonmagnetic objects.

Developments in Eddy Current Testing

Eddy current techniques are widely used to test materials and components at high temperature. A differential technique with encircling coils is used to perform continuous process testing of hot rolled wires and rods in many steel mills around the world.

Testing of hot steel billets (round and square) is also performed using eddy current probes. Both nonrotating and rotating probes scan the hot surface.

Technique Developments[2]

Pulsed and Multifrequency Techniques. Many complex eddy current test problems can be solved through the application of pulsed and multifrequency eddy current techniques, as in the nondestructive measurement of case carburized and case hardened thickness in bearing components.

Inversion of Eddy Current Data. Separating a desired variable from eddy current signals influenced by many undesirable variables is difficult despite advances in signal processing. Improvements in the reliability of online eddy current tests have been implemented mainly through eddy current signal processing techniques.

Phase Measurement. Many industrial applications use the amplitude of an eddy current signal to judge the product quality and do not use the equally valuable phase information from the same signal. Discontinuity depth in steel mills has typically been sized through amplitude only. The reliability of eddy current testing is improved if both amplitude and phase are used in developing accept/reject criteria.

Probe Designs

Multiple-sensor eddy current test heads have been developed to detect surface discontinuities in continuously cast slabs at high temperature. Phase discrimination can be used to suppress liftoff variations.

The need for high speed data acquisition and probes contoured to fit complicated shapes has led to arrays of electronically scanned eddy current sensors rather than mechanical scanning with solitary sensors.

Probes are designed to optimize operating parameters such as excitation frequency, test speed and liftoff. Successful design of eddy current tests requires accurate knowledge of electrical and magnetic characteristics of the test object and of the materials chosen for the construction of the probe. Even the best mathematical model is of little value without these material characteristics, which are often (depending on the application) nonlinear functions of magnetic flux density, space and temperature. Measurement of these characteristics in materials is important for optimization of eddy current tests.[2]

Numerical Modeling

Most practical eddy current problems are three-dimensional; some of them involve anisotropic materials and materials with nonlinear electrical and magnetic characteristics. Numerical models have shown success in handling nonlinear situations and awkward discontinuity boundaries, especially in two-dimensional or axisymmetric test geometries.[2]

Thanks to advances in digital computer technology since 1975, numerical modeling techniques have overcome most drawbacks associated with analytical techniques. Numerical techniques are not limited by material nonlinearities and complex discontinuity shapes but by computer memory.

The numerical model has much in common with the experimental approach. Numerical analysis techniques, unlike analytical models, do not produce any equation as the solution but rather produce flux density, current density, phase and impedance plane trajectory plots. Numerical modeling techniques can predict the complex interactions between fields and discontinuities, interactions important for discontinuity characterization. Better eddy current tests and sensors have been designed with the help of advanced computer models.

Eddy Current Applications

Presented below are case histories representing typical applications of electromagnetic test techniques in the primary metals industry for the surface testing of bars, square billets and hot wires, rods and tubes.

Most of these techniques are eddy current tests for hot metal. For efficient process control, eddy current testing is performed on steel while it is still hot because it is more efficient to detect discontinuities early in the process before the material undergoes further working.

Eddy currents are induced by probe coils driven at medium to high frequencies. The coverage of the test surface depends on the probe motion relative to the test object. This relative motion can be achieved either by moving the test object or by moving the probe during testing. The applications in the rest of this chapter each illustrate an arrangement for this relative motion.

Part 2. Rotating Probe Testing of Hot Rolled Bars[3]

Test Requirements

A rotating probe eddy current technique has been used for surface testing of hot steel bars as they are rolled.

There are several requirements for an effective bar testing system: (1) high discontinuity sensitivity, (2) the ability to classify bar quality, (3) fully automatic operation, (4) rugged construction for use in mills, (5) the ability to test bars as received without special preparation. That is, bars would not have to be pickled or shot blasted to remove surface scale before testing. Likewise, bars that meet commercial straightness criteria would not have to be straightened before testing.

The rotating probe eddy current system costs less than alternative systems. Unlike ultrasonic test systems, an eddy current system does not require a liquid couplant and can achieve a higher sensitivity to surface discontinuities.

Magnetic flux leakage testing requires magnetization of the test object. Disadvantages of the magnetic particle version of such systems are that accurate discontinuity depth discrimination is not possible, that full automation is usually impractical and that magnetic particles are expensive and cannot be fully reclaimed after testing. A significant improvement over magnetic particle testing for this application, magnetic flux leakage testing has not only the disadvantage of requiring magnetization but also the drawback that the probe must be very close to the bar surface, preferably riding on it.

In the rotating probe eddy current technique, the equipment rotates an eddy current probe around an advancing bar, the probe being held a preset minimum distance from the bar surface. The equipment maintains the selected level of test sensitivity regardless of changes in discontinuity signal amplitude caused by varying surface spacing from probe to bar and marks only those discontinuities that exceed a preselected length and depth. Bars with discontinuities are automatically separated from discontinuity free bars. Automatic paint marking is also available if visual discontinuity identification is required for subsequent reconditioning.

Mechanical Equipment

The major items of the mechanical equipment developed for the eddy current system are (1) bar handling equipment for supplying bars for testing and (2) a test station for rotating the test probe around an advancing bar.[4]

Bar Handling Equipment

Though fairly conventional in design, the bar handling equipment is designed (1) to automatically supply one steel bar at a time for testing, (2) to operate at a line speed up to 0.76 $m{\cdot}s^{-1}$ (1.7 $mi{\cdot}h^{-1}$) and (3) to separate bars into two classifications, those to be accepted in one cradle and those to be scrapped or reconditioned in a second.

Test Station

Testing of bars occurs while the bars are inside the test station. To carry out testing, the test station must perform several functions.

1. The station must rotate the eddy current probe around the bar at a preset rotational speed. Top rotational speed is 26.7 cycles per second (1600 rotations per minute). Speed depends on bar diameter.
2. The station must extend the probe to the test position when a bar is in the unit and retract the probe when the bar leaves.
3. For all bar diameters, the station must maintain a minimum preset spacing from probe to bar.
4. The station must transfer signals from the rotating probe to the electronic instrumentation.
5. The station must supply probe position signals for subsequent discontinuity identification and marking.

There are three major components of the test station.

1. A floating head can move up to 13 mm (0.5 in.) vertically or horizontally to accommodate testing of nonstraight bars. The test probe used to scan the bar surface is inside the head.

2. A rotary transformer transfers signals between the rotating probe and the electronic instrumentation.
3. A variable speed motor rotates the transformer and bore of the head assembly.

The test station operates with a constant center line while testing bars of different diameters; however, the conveyor line operates with a fixed base pass line. For this reason, the test station is on a platform whose height can be adjusted when the test is changed to a different bar diameter.

During testing, the probe follows a helical test path on the bar. The spacing between successive helical scans is determined by both the rotational speed of the test probe and the forward speed of the bar. The helical pattern permits 100 percent surface coverage except for end loss. Testing is carried out to within about 0.2 m (8 in.) of either end of each bar.

Electronic Instrumentation

Electronic instrumentation for the bar tester consists of the following circuits: (1) eddy current circuits for detecting the presence of discontinuities, (2) automatic gain control circuits for maintaining a constant test sensitivity, (3) signal processing circuits that function separately and together to discriminate between noise and actual discontinuity signals and (4) marking circuits.

Discontinuity Detection

Figure 1 is a block diagram of the electronic instrumentation. The discontinuity detection circuits are coupled to the rotating probe through one channel of the three-channel rotary transformer.

Electrical signals developed by the high frequency discontinuity oscillator and applied to coils in the test probe generate eddy currents on the bar surface. When a discontinuity is present on the bar surface, the orderly flow of eddy currents is disrupted; the deeper the discontinuity, the greater its effect on the eddy current flow.

Detection, amplification and filtering circuits then develop electrical signals that indicate the presence of discontinuities and provide a signal amplitude proportional to discontinuity depth for processing by subsequent circuits to assess discontinuity severity.

Automatic Gain Control

One drawback of eddy current testing is that the amplitude signals for a given discontinuity will decrease with increasing spacing between the test probe and bar surface. Although several techniques have been developed to minimize this effect, they have limitations such as the narrow range in spacing from probe to bar over which they are effective and poor signal correction accuracies. To overcome these shortcomings, the instrumentation includes circuits for measuring variations in distance from probe to bar and developing a liftoff signal for use in other circuits for correcting eddy current signal amplitude. These circuits are included in the block diagram of Fig. 1. As is the case for the discontinuity detection circuits, the low signal amplitude circuits for measuring distance from probe to bar are

FIGURE 1. Circuits of eddy current bar testers.

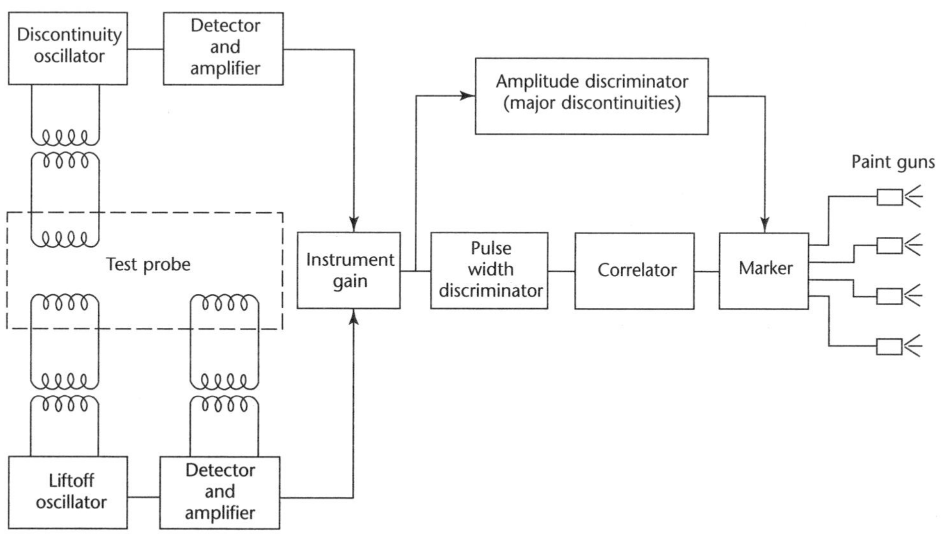

coupled through the rotary transformer to coils in the probe assembly. The liftoff signal provided by this equipment is a function of the distance between the bar surface and the discontinuity sensing coils. This signal automatically varies the instrument gain, thereby providing a constant test sensitivity under conditions of varying distance from probe to bar. Although the automatic gain control thus maintains a constant test sensitivity, the desired level of sensitivity is obtained by adjusting the manual gain control.

Signal Processing Circuits

In the case of eddy current testing, noise signals can be caused by conditions such as surface roughness, scale and electrical interference, as well as short and shallow discontinuities that are not causes for rejection. These problems can be overcome through two techniques: pulse width discrimination and signal correlation. Gross discontinuity signals are processed by conventional amplitude discrimination. Finally, the system provides marking for subsequent visual location of discontinuities if required.

Pulse Width Discrimination

The pulse width discriminator identifies as a discontinuity indication any eddy current signal that (1) exceeds a preset amplitude and (2) decreases in amplitude from its peak value to half its peak value in less than a predetermined time. The corollary is that the discriminator disregards other signals. For simplicity, the period of time in which the amplitude decreases is referred to as *signal width.*

Figure 2 can be used to illustrate the way in which pulse width discrimination works. There are three typical eddy current signals applied to the discriminator circuit. Signal A, typical of conditions such as shallow surface scratches or surface roughness, does not exceed the preset amplitude value and is therefore disregarded by the circuitry. Although signal B, characteristic of conditions such as loose surface scale, exceeds the amplitude value, the signal width D exceeds the preselected time. This signal is likewise disregarded by the circuitry. Signal C is characteristic of a surface discontinuity because it exceeds the amplitude value and because its amplitude decreases (in width E) from the peak to half the peak amplitude in less than the predetermined time. Pulse C is identified by the pulse width discriminator as an indication of a discontinuity. The discriminator generates a digital output signal in such cases.

FIGURE 2. Identification of discontinuity signals by pulse width discriminator.

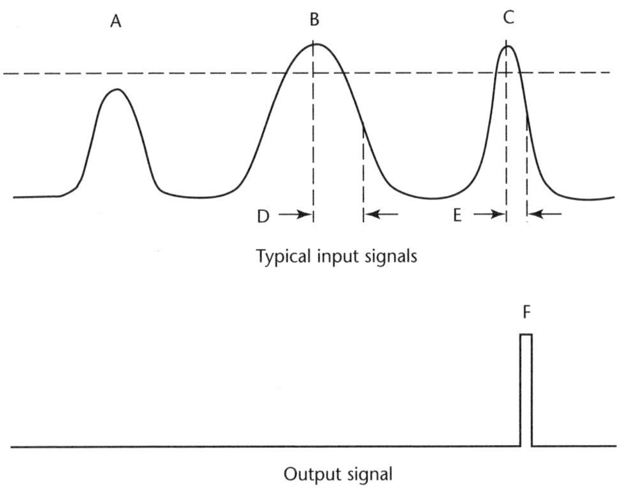

Legend

A = pulse less than preset amplitude threshold
B, C = pulses in excess of amplitude threshold
D = amplitude decrease longer than preset time
E = amplitude decrease briefer than preset time
F = signal triggered by abrupt amplitude change

Signal Correlation

Signal correlation not only discriminates between discontinuities and harmless surface imperfections such as short, shallow scratches but also serves to further suppress the effects of random noise signals. Correlation is used in the bar tester as a type of discontinuity pattern recognition based on the principle that harmful surface discontinuities such as laps and seams are longitudinal or continuous. Laps and seams will, therefore, be detected by the rotating probe at about the same circumferential position on the bar for successive probe scans.

As seen in Fig. 3, the correlation of discontinuity signals is performed as follows. As the eddy current probe rotates around the bar, an encoder in the test station is synchronously driven with the

FIGURE 3. Signal correlation for identification of continuous discontinuity.

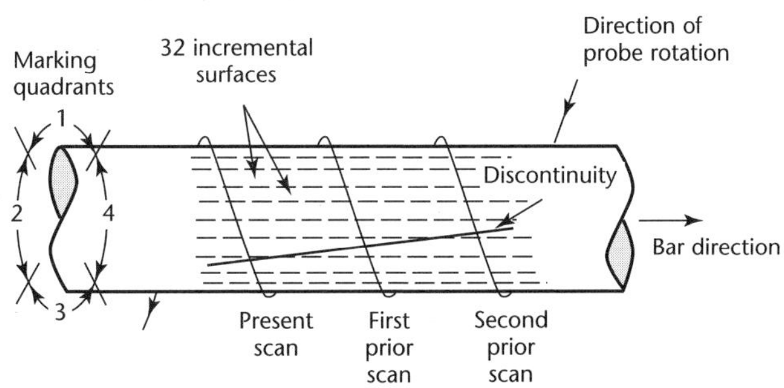

probe, dividing the bar circumference into 32 incremental surfaces. When a discontinuity signal is provided by the pulse width discriminator, information identifying the incremental surface and scan number is stored in the correlation electronic circuits. Using this information, the electronic circuits correlate discontinuity signals for the same position in a preselected number of prior scans. Because the equipment can, depending on instrument setting, compare the present scan and incremental position with up to two adjacent incremental positions for prior scans, it has the additional capability of allowing for bar turning and discontinuity skew. Finally, the tester can be adjusted to correlate data for up to five probe scans.

Major Discontinuity

Discontinuities that are deep but short, such as scabs and slivers, occur within only one probe scan and are processed in a conventional manner by amplitude discrimination. Figure 1 shows the block diagram location of these circuits. Because discontinuities of this type usually greatly exceed the amplitude of noise signals as well as correlated signals, they can be readily detected by amplitude level discrimination techniques. Signals of this type bypass the pulse width discriminator and correlation circuits.

Marking

The system includes provision for accurate discontinuity marking so that the discontinuities identified by the electronic circuits can be visually located for purposes such as subsequent bar reconditioning.

The essentials of the marking equipment are as follows. Four paint guns are arranged around the circumference of the bar, each gun being centered in a quadrant of the bar (Fig. 3). The marking circuits (Fig. 1) process signals from the major discontinuity and correlator circuits and provide a paint mark in the quadrant where the discontinuity is located. An electrical signal from an encoder driven by the conveyor line delays marking until the anomalous area of the bar reaches the marking guns. In Fig. 4, the gun in the foreground is seen spraying a white stripe in a quadrant where the electronic circuits identify a discontinuity.

Calibration

Before the test is begun, the operator adjusts the equipment for bar diameter, metallurgical grade of the bars and the surface test requirements. To simplify adjustments for bar diameter, the controls and readouts for test station height, probe rotational speed and conveyor line speed are calibrated in terms of bar diameter. Adjusting for metallurgical grade is simple, usually requiring only a slight readjustment of two potentiometers to provide readings for *on bar* and *off bar,* respectively.

Surface test requirements can be stated in terms of minimum depth of discontinuity to be marked and for continuous or seam type discontinuities, the minimum length to be marked. Thus, the setting for the desired test is obtained by adjusting the sensitivity potentiometer and the switch for discontinuity length.

At the beginning of each turn, an overall check of equipment operation can be made by inserting a calibration fixture into the test station. This fixture has an accurately machined 50 mm (2 in.) diameter round section shifted off the center line of the test station. On the surface are accurately machined artificial discontinuities. The off center round section and the artificial discontinuities make it easier for the operator to quickly check a number of mechanical and electronic features of the equipment.

FIGURE 4. Marking guns.

PART 3. Eddy Current Device for Total Surface Testing of Square Billets[5]

Integration of Test in Rolling Process

Eddy current testing can automatically inspect 100 percent of the surface of steel billets (having a square cross section) without the need of an operator's judgment for interpreting test results. The system described here can detect seams, evaluate their severity and mark the location of those that exceed an acceptable depth. The key component is a scanning head assembly that keeps an eddy current probe in contact with and tangent to the billet surface at all locations around the periphery, including the corners. The machine is designed to test round cornered, square billets as they are rolled. This integration with the manufacturing process is an important step in the development of integrated automatic testing and conditioning systems.

The inherent shortcomings of visual testing motivated the development of electronic techniques for measuring seam dimensions. One seam evaluation device uses a probe coil that causes eddy currents to flow in the test surface of the metal. Any discontinuity in the surface affects the electrical loading of the probe coil; in the case of seams, coil loading is inversely proportional to seam depth.

For the detection of such discontinuities, absolute coil loading has little meaning. However, the changes in the loading of a coil as it crosses a discontinuity are significant. Therefore, relative movement between the probe coil and the product is required. In a manual test, this movement is achieved by moving the probe while holding the test object stationary. In an automatic test, the test object, the probe or both are moved.

Application to Rounds

The diameter of the test object dictates the means of obtaining the relative movement necessary between the search probe and the test object. Machines designed to test small diameter bars rotate and propel the product while the probe is held stationary. For large diameter bars, pipe and billets, the probes are rotated around the product as it moves forward through the machine.

Using the second means of obtaining relative motion, an installation for testing of rounds consists of two machines designed for 75 to 250 mm (3 to 10 in.) diameter, straightened solid product. These machines have been installed in a mill finishing line that also includes facilities for grit cleaning, straightening and grinding.

Square Products

To test products having a square cross section, a prototype billet test machine has been built. In this machine, a search probe is reciprocated across the face of the billet by an air cylinder while the billet moves forward.

The next step is the designing and building of production equipment. It includes two machines, each testing two of the four faces of a billet on one pass. The search probes are reciprocated by a hydraulic cylinder with servo controlled reversing. Although these probes test the face of the billet adequately, they do not test the corners, also susceptible to seams.

Design Requirements

One way to test the entire circumference of the surface, including corners, is for the probe to revolve continuously around the billet. This motion, however, presents challenges in system design.

1. The probe must be maintained tangential to and at a close but constant distance from the surface. The probe may bounce rapidly when rounding the corners, causing an electronic signal similar to that from a discontinuity. Excessive bounce also changes the electrical coupling and varies the sensitivity to discontinuities.
2. The probe holding assembly has to be mechanically rugged to withstand the shock encountered while passing each corner and an occasional very rough surface.

3. The speed of the probe relative to the test surface must be maintained within limits. Experimental studies have established that the optimum range is 0.4 to 0.7 $m \cdot s^{-1}$ (0.9 to 1.6 $mi \cdot h^{-1}$). A slower probe speed would result in an inconsistent measurement of discontinuity depths. Speeds above this range cause objectionable electronic noise.

Probe Assembly

These problems are addressed by the search probe scanning assembly (Figs. 5 and 6), similar to a roller chain with the search probe in place of a center roll. The assembly is flexibly mounted so that the rolls smoothly guide the search probe around the billet surface.

The search probes, at the center of the assembly, are encapsulated in nylon housings mounted in a titanium holder. Attached to the integral parts of the holder are tungsten carbide wear shoes that bear on the billet surface. Two coils are used in each search probe assembly to increase the area of surface tested.

The entire probe assembly is held against the surface by a spring that exerts constant force. The test surface forces the outside rolls to open the pivot mounted arms, causing the chainlike arrangement to become taut and conform to the test surface. The four small springs hold the various elements against the surface at different positions around the square section.

Most of the assembly is constructed from a titanium alloy selected for high strength and lightness. Weight must be minimized because centrifugal forces resulting from rotation and inertia tend to lift the assembly from the billet surface.

FIGURE 5. Components of billet search probe scanning assembly.

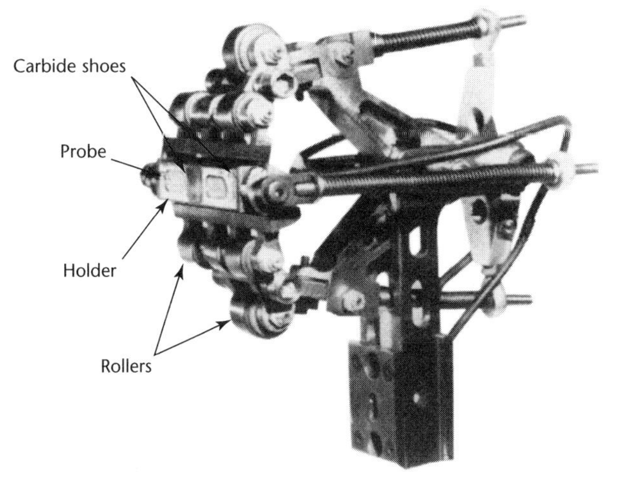

The high strength of this material permits the assembly to withstand forces resulting from surface roughness and the directional changes necessary during testing. The rolls are tungsten carbide for maximum wear resistance and use bearings but require no lubrication.

The dimensions of the probe assembly components affect the instantaneous velocity of the probe relative to the billet surface. The most significant dimensions are the spacing between the rolls, the diameter of the rolls and the distance of the probe pivot from its contact surface. Figure 6 shows the relative speed of the probe at different positions on a quarter section of a billet with a 0.1 m (4 in.) square cross section. The element shape for this assembly gives the probe a velocity that is maximum on the flat part of the surface, where bounce is least likely, and minimum at the corner, where

FIGURE 6. Probe at various positions on quarter section of 100 mm (4 in.) billet: (a) cross section; (b) speed.

(a)

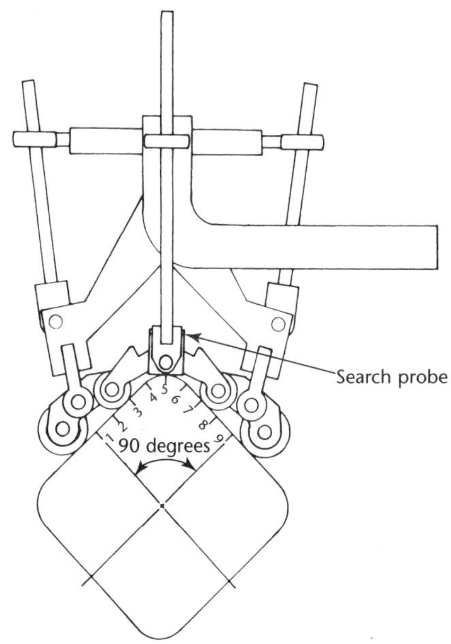

(b)

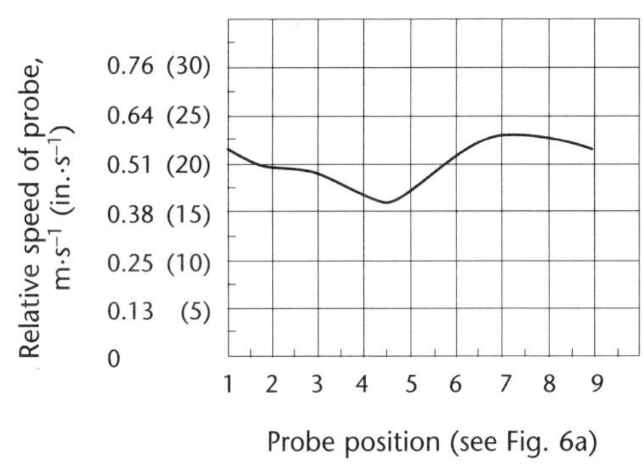

the coil has the greatest tendency to leave the surface. The speed fluctuates smoothly within the allowable testing range as the probe assembly is rotated around a square billet. The probe assembly dimensions are optimized for 102 and 127 mm (4 and 5 in.) billets but this same assembly can test a greater range.

Discontinuity Marking System

A marking system is selected that consists of spray markers, an air compressor, a paint reservoir and valves to direct paint to the billets. The spray markers and two retractable search assemblies are attached to pivot arms on a rotating face plate. The face plate in turn is attached to a rotating, open ended drum through which the billet passes. The spray markers are spaced to mark the exact location of the seam anywhere around the billet periphery. The other components of the marking system are attached to the exit end of the drum and also rotate.

Electrical energy is transmitted by means of slip rings to the search probes, to the solenoid valves for the marking system and to the compressor.

Machine Details

The drum that encircles the billet moves freely in all directions perpendicular to its axis, thus enabling the drum to follow twists and bends in billets. Camber up to 114 mm (4.5 in.) in overall length can be accommodated.

The machine elements that permit free movement of the drum are shown in Fig. 7. Guide rolls ensure that the center

FIGURE 7. Machine elements that permit free movement of drum and enable testing of cambered product: (a) diagram; (b) photograph.

(a)

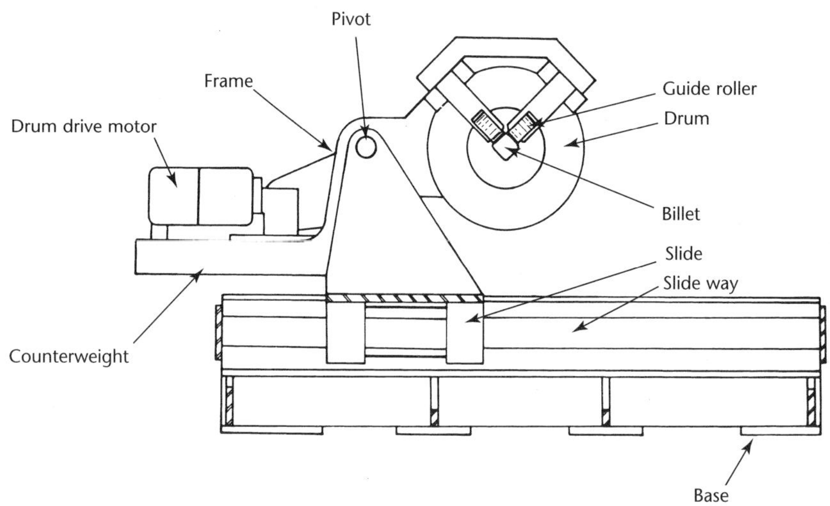

(b)

line of the drum corresponds to the billet center line. The drum assembly is counterweighted by the drum motor and steel weights. This entire assembly is attached to an A shaped member through a pivot that permits vertical motion of the drum. The A shaped member is mounted on the base by a low friction slide assembly that permits the drum to move horizontally. Any movement of the billet is transferred by the guide rolls to the drum and the combination of the pivot and slide allows the drum to move readily in the planes perpendicular to its axis. An extension of the slide for horizontal movement of the drum is used to remove the entire machine from the production line for calibration and easy access for service.

Production Installation

The installation includes a grit blaster. The turnover mechanism moves the billet from the grit blast conveyor to the testing conveyor. The forward speed of the conveyor is adjusted for various sizes of billets. To ensure the detection of all discontinuities longer than 25 mm (1 in.), the forward speed of the conveyor for 127 mm (5.0 in.) billets, for example, is 0.17 $m \cdot s^{-1}$ (0.4 $mi \cdot h^{-1}$). With no delays, this would provide a test rate of 64 000 kg (140 000 lb) per hour.

Calibration

To calibrate the electronic circuitry, the entire drum assembly is moved out of the line of billet travel. A small billet section is swung into the test position (Fig. 7b).

FIGURE 8. Chart recording of electronic signals from sample 100 mm (4 in.) billet.

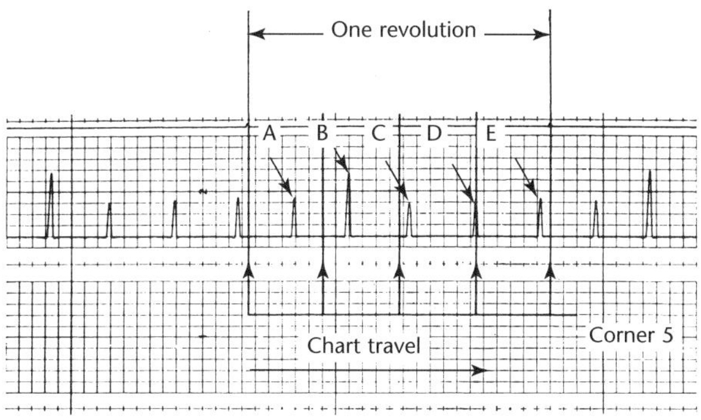

Legend

A. Artificial seam on flat surface.
B. Natural seam 1.5 mm (0.06 in.) on flat surface.
C. Artificial seam 0.75 mm (0.03 in.) deep lags corner.
D. Artificial seam 0.75 mm (0.03 in.) deep directly on corner.
E. Artificial seam 0.75 mm (0.03 in.) deep leads corner.

The drum is then rotated with the probes contacting the test sample and the signals are recorded. The electronic circuit is calibrated by simple dial adjustments.

Figure 8 shows a chart recording of the signals obtained from the calibration billet specimen. The billet has four discontinuities milled about 0.75 mm (0.03 in.) deep by a 0.15 mm (0.006 in.) wide cutter. The seams are on the flat surface before, on and after the corner. A natural discontinuity, 1.5 mm (0.06 in.) deep on the flat, is also present in this billet.

The calibration procedure provides for adjustment of the electronic circuitry and supplies the operator with information on equipment performance. It can also be used to verify that the probes are riding the billet properly, that no elements in the probe assembly are binding and that both markers are operating properly.

Part 4. Rotating Machine to Test Hot Steel Rods and Wires[6]

Integration of Test before Coiling

Wires and bars are usually coiled immediately after they are hot rolled. When discontinuity detection is performed on cold products, they must be uncoiled for testing. If discontinuity detection can be accomplished during the rolling process, while wires or bars are still hot, costs can be reduced by saving electric power and by omitting the uncoiling process.

Eddy current testing using an encircling coil has been applied to hot rolling of bars.[7] Steel mills use encircling coil eddy current systems to test hot wires. Generally, encircling coils in a differential configuration can detect short discontinuities such as scabs and roll marks. It is, however, difficult to detect harmful longitudinal discontinuities such as seams and cracks.

When the eddy current tester uses a rotating probe for wires and bars in hot rolling, there are technical problems.

1. The rotating machine must be centered mechanically at the pass line.
2. Electronic correction is required to detect discontinuities without disturbance of liftoff variation.
3. The probe must be cooled to protect against heat damage.

Testing Machine

To apply the rotating probe technique to hot wires and bars, it is important to develop the techniques to suppress the liftoff variation and correct discontinuity signal amplitude by measurement of liftoff variation. The liftoff variation is caused by wobble of wires or bars and by difference of centering between the rotating test machine and the pass line.

Components have been developed for the technique. An electronic controller can eliminate the signal caused by the residual liftoff variation and correct the corresponding discontinuity signal. The influence of liftoff variation is also minimized by several means.

1. Two pairs of pinch rolls can keep the tested wire near the center of the rotary machine. Also, they can reduce the vibration of the wire.
2. The mechanical stage supporting the eddy current tester can be driven by stepping motors to position the machine precisely at the pass line.
3. An air jet emitted from the probe case to the surface of the wire lifts the probe for noncontact scanning. The air continues along the surface of the wire to minimize the liftoff variation. The air jet also serves to cool the probe.

FIGURE 9. Rotating probe machine for eddy current testing of hot steel wires and rods.

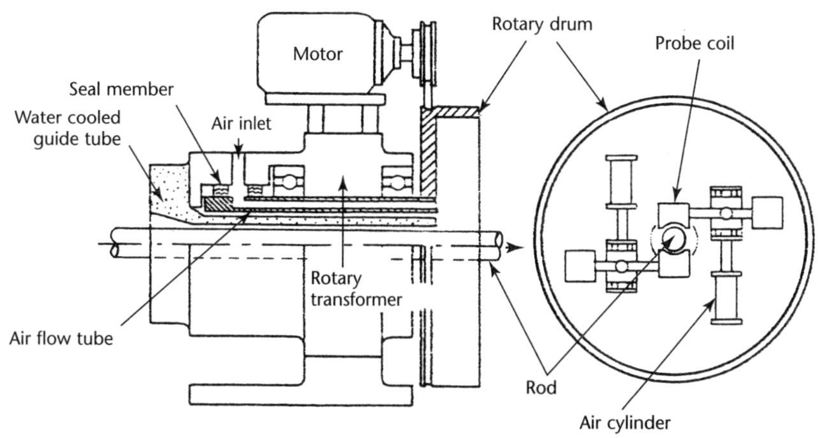

4. Multifrequency signal processing eliminates the liftoff variation signal caused by high frequency components of the wobble.
5. The liftoff variation is measured by a displacement probe and the discontinuity signal change caused by the liftoff variation can be compensated electronically.

By these means, the rotating probe eddy current system performs quantitative evaluation of discontinuities on the surface of wire.

Rotating Probe Machine

Figure 9 schematically shows the eddy current testing machine with rotating probes. It consists of a rotary drum that has two sensor holders and actuators for sensor holders, a cylindrical housing and a motor that drives the rotary drum.

The cylindrical housing has a rotary transformer, an air flow tube of duplex steel and a water cooled guide tube. The rotary drum is mounted on the downstream side of the housing. The motor can drive the rotary drum at the maximum rotation speed of 16.7 cycles per second (1000 rotations per minute).

Within the rotary drum, the sensor holders are mounted with probes and air cylinders for retracting the probe. The probe is moved away from the center of the rotary drum by the air cylinder to keep the probe from rubbing against the top or bottom of the material. The mechanism for retracting the probe successfully operates at up to 16.7 cycles per second (1000 rotations per minute) even though centrifugal force acts against the mechanism.

The air flow tube is made of duplex steel and comprises an inner tube and an outer sleeve. The air conduit extends into the rotary drum at the downstream side end. To supply compressed air to the rotary drum, sealing members are fitted between rotor and stator as shown in Fig. 9. The compressed air has three functions: (1) noncontact scanning using air jets, (2) cooling the probe coil and (3) driving the air cylinder.

The air cylinder is controlled by solenoid valves mounted on the rotary drum. The water cooled guide tube is fitted into the air tube. The guide tube protects the rotary transformer and bearings against heat damage. The signals from the noncontacting probe coil are sent or received by the rotary transformer.

Air Flotation Scanning

In the air flotation scanning station, the probe case has an air inlet and several air jet outlets. An air jet flows to the surface of wire through these outlets, lifting the probe. The air flotation force keeps the liftoff constant because air force becomes stronger when liftoff decreases and becomes weaker when liftoff increases (Fig. 10).

Air flotation scanning closely follows its intended path until it reaches a scanning speed of 1.05 $m{\cdot}s^{-1}$ (2.35 $mi{\cdot}h^{-1}$), a rotation speed of 16.7 cycles per second (1000 rotations per minute) at the wire's diameter of 20 mm (0.8 in.), when the value of eccentricity is 0.4 mm (0.016 in.). For this application, the eccentricity is considered to be the distance from the center of the wire to the center of the rotating machine.

The probe opposite to the tested material of temperature over 1000 °C (1800 °F) can be cooled by the air jet. The temperature rise of the probe coil is only 50 °C (120 °F) and does not affect test results in practice.

Centering Machine

The eddy current testing machine is set on a truck with a specially designed mechanical centering machine as shown in Fig. 11. The truck can be pulled out from the pass line to calibrate and maintain the test machine.

The centering machine consists of two pairs of pinch rolls and a mechanical stage. Two horizontal and two vertical pinch roll assemblies hold the wire. The centering machine is used to set the rotating drum in the center of pass line.

The mechanical stage is moved by Y and Z axis actuators driven by stepping motors as shown in Fig. 12. A displacement sensor arranged near the probe picks up the liftoff variation during probe rotation. The value of eccentricity is computed from both signals measured by the displacement sensor and a probe position sensor. The probe position sensor mounted on the cylindrical housing generates one pulse for each revolution of the drum; the rotating position of the probe is obtained by noting the pulse interval.

FIGURE 10. Air flotation scanning technique.

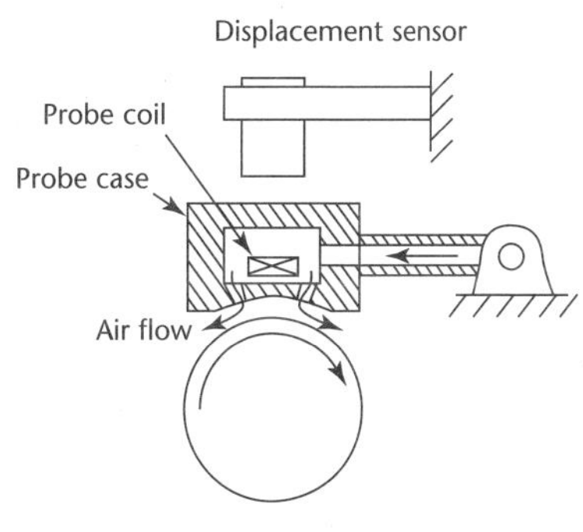

A set of digital panel meters shows the vertical and horizontal components of eccentric value as shown in Fig. 12. The cathode ray tube displays an actual trace of the probe. An operator drives the stepping motors to minimize the eccentricity. The setting is accurate to ±0.2 mm (±0.008 in.).

The test for hot wires or bars is performed as follows.

1. Upstream and downstream, hot metal detectors at the sides detect the hot wire.

FIGURE 11. Layout of eddy current machine for hot wires and rods.

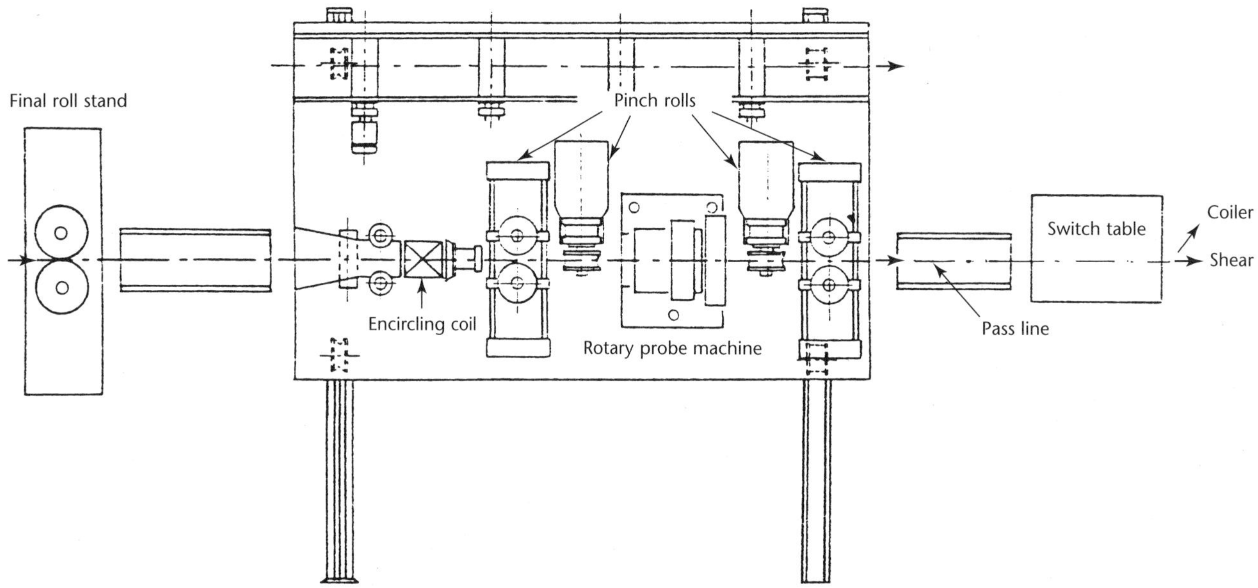

FIGURE 12. Block diagram for automatic alignment of rotating probe machine.

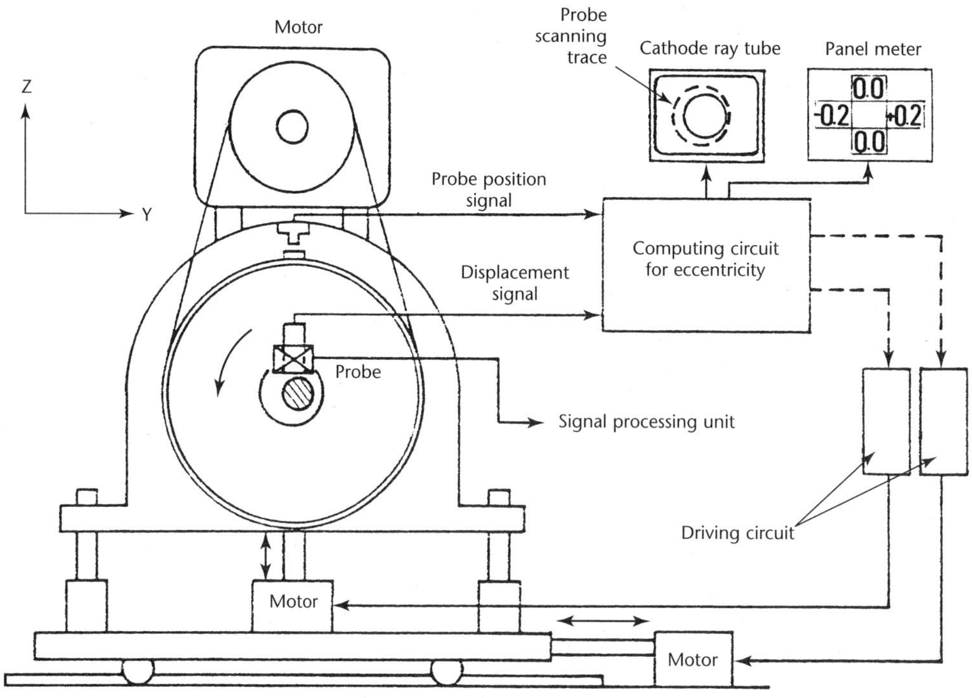

2. Pinch rolls are actuated by the hot metal detector signals and hold the hot wire.
3. The air cylinder on the rotary drum is actuated. The rotating probe approaches the wire and the test begins.
4. After the hot wire passes at the hot metal detector point, the air cylinder is actuated to make the probe retract.
5. Pinch rolls release the hot wire.

To shorten the untested length of wire's ends, the rotary drum is always rotated during the test.

Signal Processing

Multifrequency Signal Processing

Phase analysis and frequency analysis are used to suppress the liftoff variation signal. Because the phase angle difference between the discontinuity signal and the liftoff variation signal in hot steel testing is usually slight, such as 0.18 to 0.35 mrad (10 to 20 deg), it is difficult to discriminate the liftoff variation by phase analysis.

On the other hand, the frequency analysis technique serves to suppress an undesirable signal by means of a frequency difference between the undesirable signal and the discontinuity signal. Because the liftoff variation caused by the vibration of wire has frequency components similar to those of the discontinuity signal, however, this technique cannot distinguish the signals.

The wobble of wire is caused by vibrations from a roll stand or a coiling machine. The air floating sensors cannot follow the surface of wire whose vibration frequency is over 20 cycles per second.

To make it possible to suppress the signal caused by this vibration and to detect a discontinuity with higher ratio of signal to noise, the multifrequency eddy current technique is applied. Multifrequency eddy current testing of hot wires with encircling and probe coils has been reported.[6,8]

Single-Frequency versus Dual-Frequency Testing

The single-frequency and dual-frequency techniques have been compared for detectability by using an austenitic stainless steel bar with artificial discontinuities. Electromagnetic properties of the stainless bar are similar to those of hot steel over the magnetic curie temperature. The liftoff variation of the tested bar is given by the vibrator.

The relation between the vibration amplitude of the tested bar and detectability of artificial discontinuities whose depth is 0.3 mm (0.012 in.) is shown in Fig. 13a. The data are obtained under the condition without eccentricity. For detecting artificial discontinuities, the single-frequency technique is better than the dual-frequency technique when there is no eccentricity and no vibration. However, detectability with the

FIGURE 13. Effect of dual-frequency eddy current testing of 24.3 mm (0.96 in.) diameter stainless steel: (a) without eccentricity and with 0.3 mm (0.012 in.) deep notch; (b) with 0.4 mm (0.016 in.) eccentricity and with 0.5 mm (0.02 in.) deep notch.

(a)

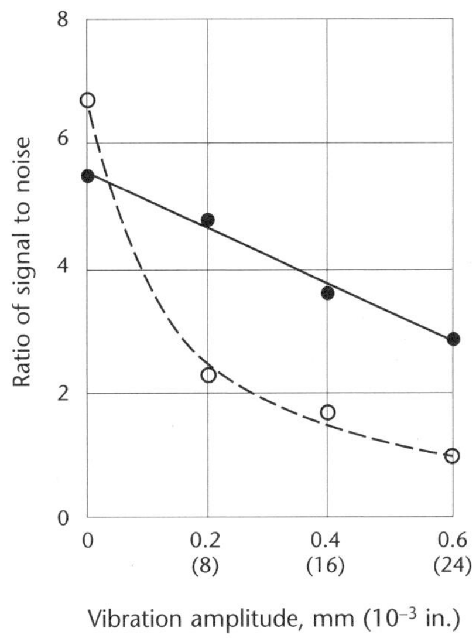

(b)

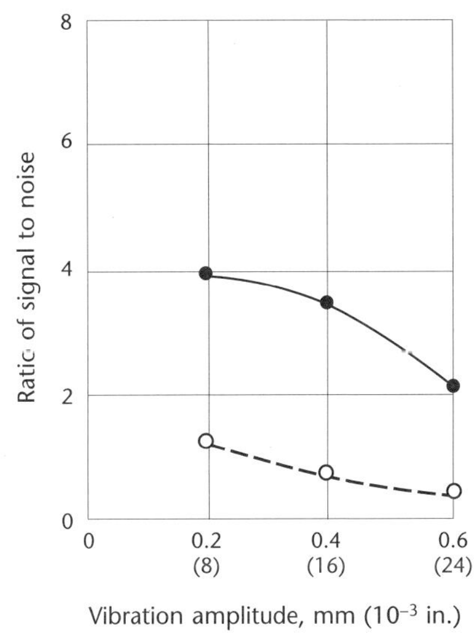

Legend

—— = dual-frequency probe, 32 and 128 kHz
— — = single-frequency probe, 128 kHz

single-frequency technique decreases markedly with the increase of vibration amplitude. On the other hand, discontinuity detectability decreases slowly with the dual-frequency technique. The dual-frequency technique is superior to the single-frequency technique when there is vibration.

Where the eccentricity is greater than the vibration of the tested bar, the difference in detectability between the two techniques becomes even greater, as shown in Fig. 13b. Both vibration and eccentricity are present in the testing of hot wires, making the dual-frequency technique very efficient for discontinuity detection.

Signal Processing for Liftoff Compensation

The liftoff variation changes the amplitude of the discontinuity signal itself. When liftoff varies, it is impractical to measure a discontinuity size by the amplitude of discontinuity signals. In such cases, the liftoff compensation circuit can be adopted.

The compensation circuit consists of a function converter and a calculator. The liftoff signal is converted to a compensated value by the function converter. The outputs of the function converter and discontinuity signal are multiplied by the calculator and provide a signal in proportion to discontinuity depth. For example, when the liftoff varies from 0.7 mm to 1.3 mm (0.03 to 0.05 in.), the signal amplitude nearly doubles. This circuit ensures that signal amplitude varies by less than ±10 percent.

PART 5. Seam Testing in Hot Steel Rods[9]

Surface Testing in Rods

Surface discontinuity testing is essential in the quality assurance of iron and steel products. In many mills, quality control of hot rolled rods is provided through eddy current and magnetic flux leak testing carried out after the rolling, shearing and cooling processes. If the test is made during hot rolling, information about surface quality could be rapidly fed back to the rolling process, thus minimizing the quantity of surface discontinuities in future products. To achieve this, the encircling coil technique[10] has been put into practice. However, the orientation of encircling coils is unsuited for detection of long discontinuities, such as seams.

Rotary Probe System Features

The rotary probe eddy current technique has been widely used for cold rods and is described here for the detection of long surface discontinuities during the hot rolling of iron and steel rods. The probe has been modified for the temperature of the rods. Rotary probe techniques, when installed immediately after the finishing stand in the rolling process, are suited for detecting seams with depths over 0.3 mm (0.012 in.).

In addition to functions needed for cold rods, the rotary probe discontinuity detector must perform the following functions for hot rods: (1) provide heat resistance for discontinuity detection in high temperature materials, (2) compensate for the vibration of the rod under hot rolling and (3) provide for probe retraction to protect the probe from the deformed ends of the hot rod.

The modifications made to the probe to achieve these additional functions are described below.

Heat Resistance

The rotating mechanism, exposed directly to radiant heat from hot rods, is equipped with a water cooled sleeve and the probe itself is put into a heat resistant case. The water cooled sleeve is the outer part of a cast iron pipe with a double structure to allow the circulation of water inside. This pipe is installed on the inside of the rotating mechanism to absorb radiant heat to protect the probe rotating mechanism (but not the probe) from the radiant heat (Fig. 14).

The probe case protects the probe from the radiant heat outside. The interior of the probe case is air cooled. To keep discontinuity detection performance from deteriorating because of temperature, the material at the bottom of the probe case is carefully selected for minimum eddy current loss, for heat resistance and for mechanical strength. Materials such as ceramics and austenitic stainless steel satisfy these requirements. A 0.2 mm (0.008 in.) thick Unified Numbering System S30400 austenitic chromium nickel stainless steel plate is selected because of its easy machinability. The actual eddy current loss caused by this plate is found to be about 20 percent, too small to limit discontinuity detection.

Compensation for Rod Vibration

High speed rolling produces large vibrations of the rods. Measures are required to compensate for vibratory effects. The vibrations could be either

FIGURE 14. Probe rotating mechanism.

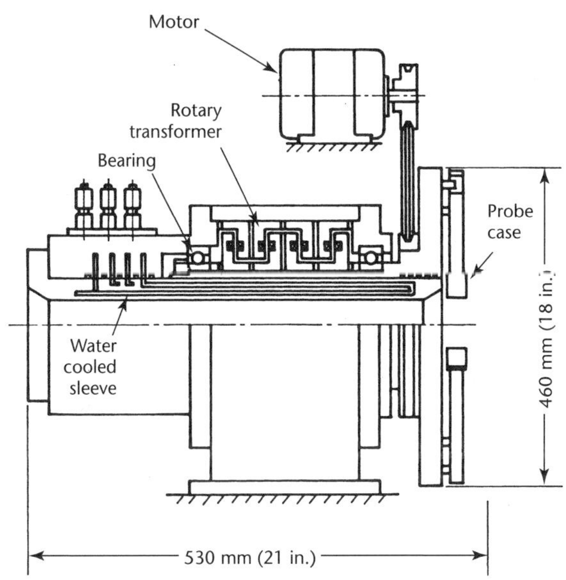

high or low frequency, calling for different countermeasures.

High frequency vibration causes fluctuation in the liftoff, thus deteriorating the ratio of signal to noise. On the other hand, the frequency of the discontinuity signal is proportional to the probe rotation speed. If the discontinuity signal is of much higher frequency than the liftoff fluctuation noise, it may be distinguished by means of a high pass filter. It is therefore necessary to increase the probe rotation speed as a measure against vibration. The vibration frequency of rods under hot rolling is found to be about 10 cycles per second (600 rotations per minute). Hence, probe rotation is set to three times the vibration frequency, 30 cycles per second (1800 rotations per minute).

The low frequency vibration refers to the fluctuation of the passing position of the rods (pass line fluctuation), a phenomenon that rarely occurs at low speed. When this does occur, the probe rotational orbit becomes eccentric with the rods, causing liftoff fluctuation as the probe rotates even if the rod itself does not rotate. Being proportional to the probe rotational speed, this noise cannot be distinguished from the discontinuity signal even if the rotational speed of the probe is increased. To eliminate this noise, a servo mechanism is added to make the probe rotating mechanism follow the pass line fluctuation.

Liftoff compensation is desirable and may be provided by an optical or other sensor designed for the application.

Probe Retraction

Rods under hot rolling generally have both ends distorted and are not completely circular. Because probes generally have liftoffs of 1 to 2 mm (0.04 to 0.08 in.), set on the premise that the rod is circular throughout, a probe may get damaged on contact with the largely distorted ends of the rod. Hence, it is necessary to retract the probe at the ends of the rod.

Figure 15 shows the technique used for retracting the probe at 30 cycles per second (1800 rotations per minute) when the end of a rod appears and then reapproaching the rod once that end passes. The inner and outer disks in Fig. 15 have the same rotational speed. However, when the outer disk is retarded by means of the brake, the difference in speeds between the inner and outer disks creates a force large enough to retract the probe. Release of the brake will allow the probe to go back to its normal position. This technique is an alternative to retraction techniques that use a hydraulic cylinder or an electromagnetic solenoid.

Experimental Work

Experiments using the rotary eddy current probe have been made by installing the probe rotating mechanism after the finishing stand in the hot rolling process.

Artificial discontinuities are produced on billets before rolling by cutting slits along the length of the billets with a thin blade grinder. These slits turn into seams with depths of 0.1 to 1.0 mm (0.004 to 0.040 in.) after rolling. These artificial discontinuities are of almost the same shape as the natural discontinuities found in practice in the billets after rolling (Fig. 16). The billets with the artificial discontinuities are inserted into the reheating furnace before the rolling process in a manner similar to that used in normal practice.

FIGURE 15. Probe retracting technique.

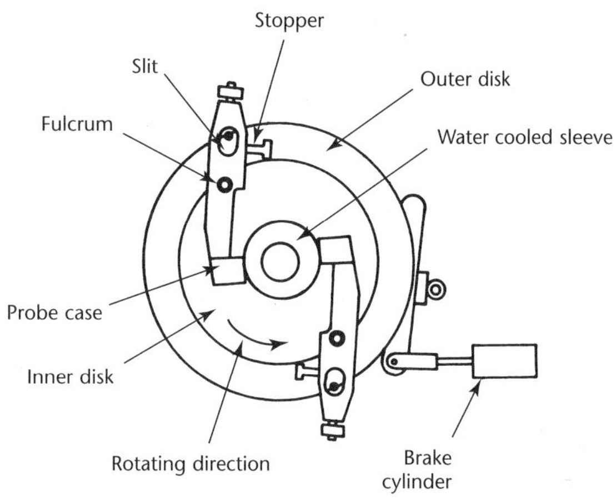

FIGURE 16. Cross section of artificial discontinuity produced in billet.

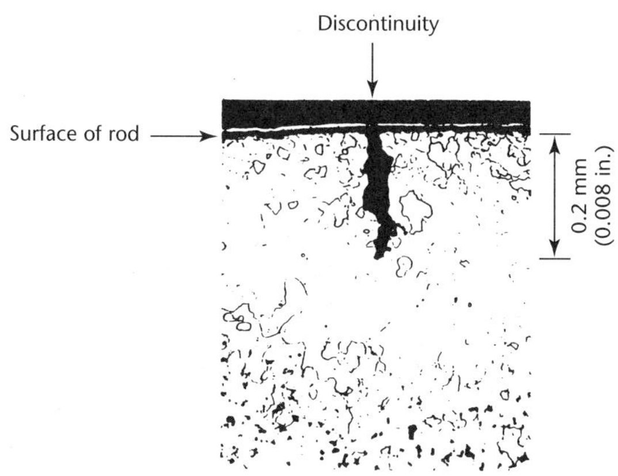

The conditions for the hot eddy current test are as follows: multiple-coil induction differential probe (Fig. 17), 32 Hz discontinuity detection frequency, 30 cycles per second (1800 rotations per minute) motion frequency, 1 to 2 mm (0.04 to 0.08 in.) probe liftoff between bottom of probe and rod, 800 to 1000 °C (1470 to 1830 °F) rod temperature, 5 $m \cdot s^{-1}$ (11.2 $mi \cdot h^{-1}$) rolling speed, 40 mm (1.6 in.) rod diameter and carbon steel rods.

Signals have indicated the dimension and position of real and artificial discontinuities on billets 150 × 150 mm (6 × 6 in.) and 12 m (39 ft) long rolled into 40 mm (1.6 in.) diameter rods over a length of about 230 m (750 ft).

The rods used in the experiment have been sheared and cooled. Fluorescent magnetic particle testing has then been used to detect the discontinuities and the rods have been ground to measure the discontinuity depth (Fig. 18). These discontinuity signals are found to have a pitch of 167 mm (5 m at 30 rotations per 1 s). Measurements at distance intervals of 500 mm (20 in.) have shown that indication amplitude correlates well with discontinuity depth.

FIGURE 17. Schematic diagram of multiple-coil induction differential probe: 50 turns in exciting coil, 70 turns in receiving coil.

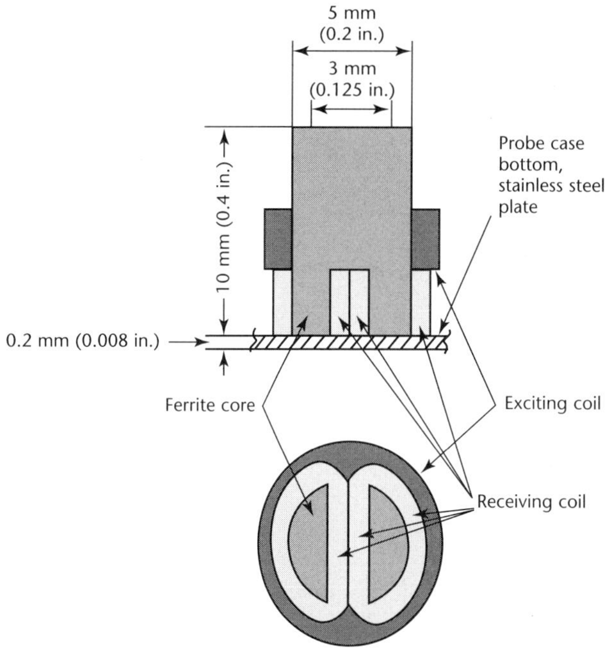

FIGURE 18. Discontinuity depths and signals in sections of steel rod made from billets that had 3 mm (0.12 in.) wide artificial discontinuities machined into them before rolling: (a) 4 mm (0.16 in.) prerolling depth; (b) 3 mm (0.12 in.) prerolling depth; (c) 5 mm (0.20 in.) prerolling depth.

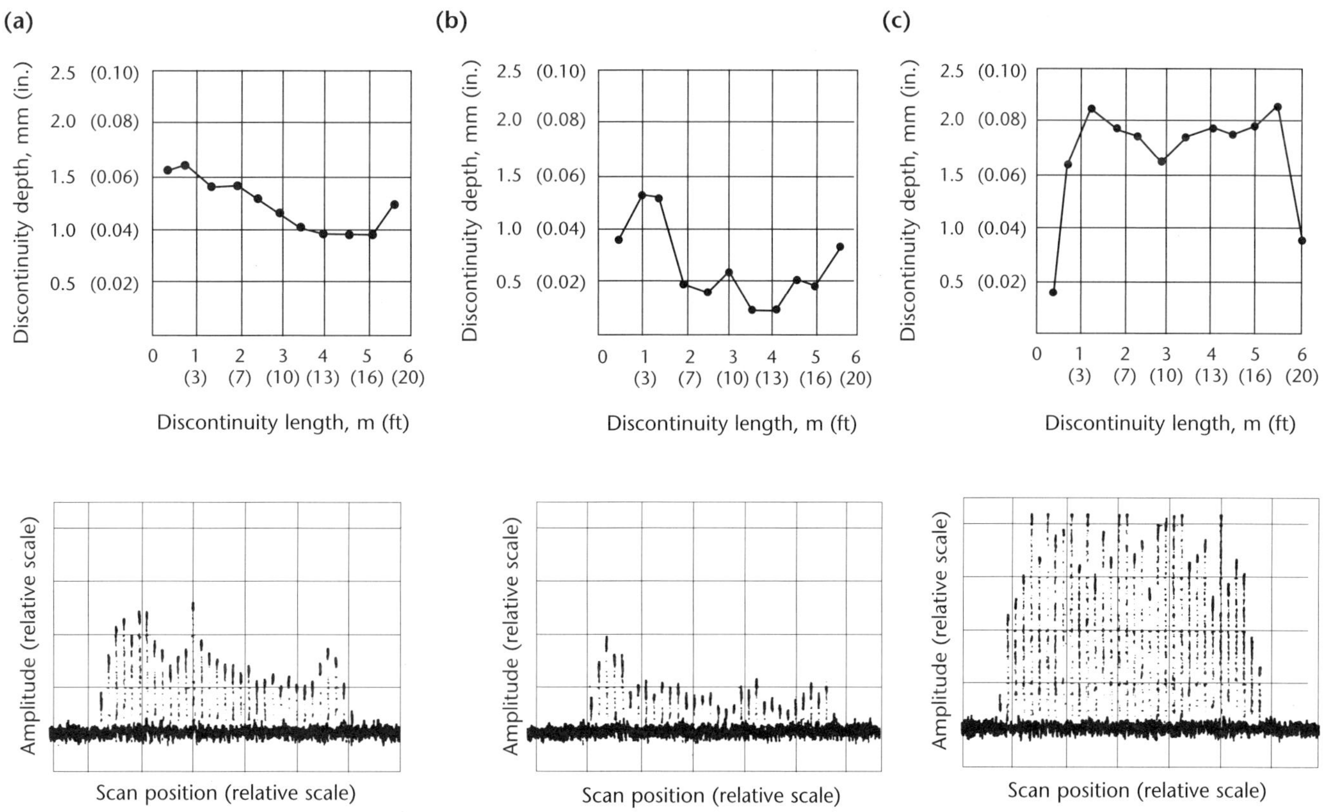

FIGURE 19. Results of hot discontinuity detection experiment: (a) relation between discontinuity depth and ratio of signal to noise; (b) relation between discontinuity depth and detection rate. Detection rate is number of detected discontinuities expressed as percentage of total discontinuities.

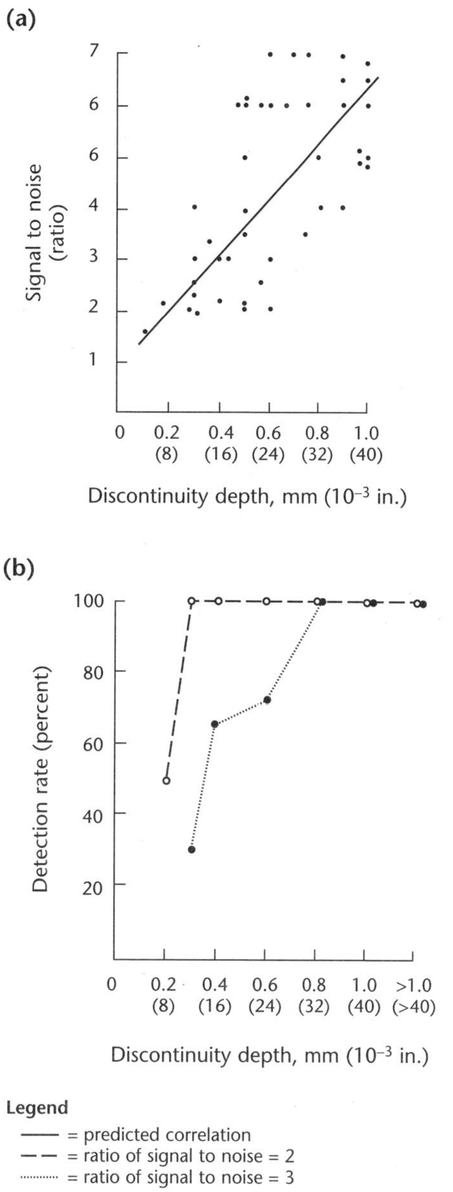

Reliability

The variation of the signal-to-noise ratio and the detection rate with the discontinuity depth are shown in Fig. 19. If a discontinuity is considered detectable at a signal-to-noise ratio of 2, the probability of detection for discontinuities over 0.3 mm (0.012 in.) is 100 percent. On the other hand, if a discontinuity is considered detectable only at a signal-to-noise ratio of 3, the probability of detection of discontinuities 0.3 mm (0.012 in.) and deeper drops to an average of 83 percent.

Part 6. Online Testing of Hot Metal Products

Process Control of Hot Metal Rods[10]

Quality control of hot metal is difficult in view of the material's high temperature and the production environment, which is unsuitable for sensitive electronic measurements. The problem is compounded in single-strand rod mills by the speed of the production line, which reaches 120 m·s^{-1} (268 mi·h^{-1}).

Eddy current testing is well suited for the quality evaluation of hot rods because of its fundamental characteristics: (1) measurement without contact, which allows quality control at high temperature and high speed, and (2) the rapid response time of the sensor, which permits quality testing in real time with a suitable computer.

Statistical Process Control

A steel coil is normally 5 or 6 km (3 or 4 mi) long and can be as long as 9 km (6 mi), depending on the size of the billet used and the diameter of the rod produced. One particular eddy current technique concentrates on determining a statistical distribution of the discontinuities on the coil length rather than on locating each discontinuity precisely.

There are four main stages of quality control.

1. Discontinuities are detected, located and sized.
2. Discontinuities are counted in each 10 mm (0.4 in.) length. This gives a count, for example, of 600 000 data over a 6 km (4 mi) coil.
3. These data are gathered in segments or windows of a predetermined length.
4. A final quality report is based on an algorithm specific to the mill. This includes an overall quality index and depends on the quality criteria desired. For example, the operator may decide the discontinuity density that must not be exceeded.

Figure 20 shows the four main stages in the eddy current test.

1. The sensor gathers raw analog data of the eddy current discontinuity detection.
2. The analog signal processing system displays analog discontinuity signal data.
3. The digital preprocessing system sorts the discontinuity signals according to amplitude and location.
4. The results are printed out using software for a final quality report of the coil.

Sensor. The sensor is an encircling coil located just after the finishing stand. Photoelectric cells on either side signal the beginning and end of rods passing through at high speeds. Guides at either end of the sensor feed the rod through the exact center of the coil.

Arranged as a self-comparison differential coil, the sensor has separate injection and reception functions. The

FIGURE 20. Diagram of single-strand steel rod control system, showing four stages: (1) signal acquisition, (2) data display, (3) data sorting and (4) reporting.

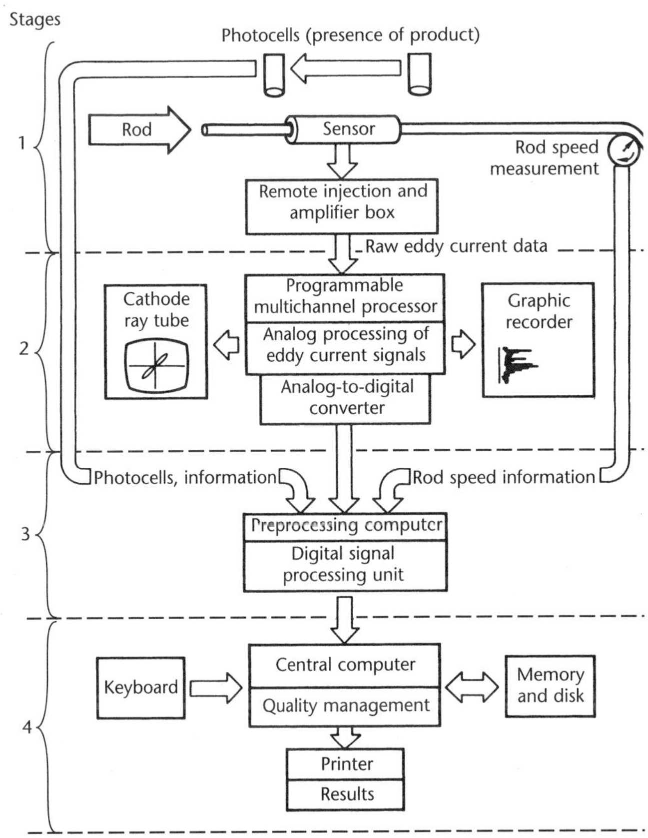

sensor injection coil is powered by a high frequency current (about 100 kHz) from the programmable multichannel processor (an analog unit, described below), through an injection box near the sensor. The reception coils deliver an eddy current signal immediately as a discontinuity passes through their magnetic fields.

Analog Signal Processing. The raw signals are sent to the programmable multichannel system through the amplification box near the injection box and the sensor. The programmable multichannel system is an analog unit that can be controlled remotely. It processes the discontinuity signals, with processes such as filtering, dephasing and expanding.

The signals are then (1) displayed in phase and amplitude on a normal cathode ray tube (for checking or adjustment of the system) and (2) presented on a multichannel graphic recorder (one channel per strand). This information gives the rod mill operator an immediate, initial idea of rod quality.

Digital Preprocessing. A preprocessing computer receives the product presence and speed signals in addition to data from the programmable multichannel processor. Then, with special software, the preprocessing computer performs data reduction. The eddy current signals are digitized, are integrated for 10 mm (0.4 in.) unit lengths of rod and are sorted according to three amplitude levels, one of which is background noise.

Results. Finally, the preprocessed information is collected by the central processing unit that manages the whole system. This computer produces an analysis report of each window over the entire coil as well as an overall quality index. The report is produced according to a quality program in the computer. The user can select or customize programs according to the quality level desired for the application. Of course, the central processing unit may be hooked up to the central quality management computer of the rod mill or the plant.

Signal Processing

The discontinuity signals are processed and shaped before they go to the preprocessing computer and central processing unit to produce an overall quality report (Fig. 21). They must pass from the eddy current detector through the remote amplifier, the programmable multichannel processor unit and the analog-to-digital converter before reaching the preprocessing computer.

Multichannel Processing of Raw Signals. The amplitude of the imbalance signals from the coil's electrical circuit is largely a function of the size of the discontinuities. The amplified raw signal is processed in the programmable multichannel processor in three ways: (1) a phase projection (along the Y axis) corrects the rod vibration influence by treating it as liftoff, (2) analog filtering improves the ratio of signal to noise and (3) automatic balancing eliminates the drifts that can occur, for example, with temperature variations or changes in the structure of the metal.

Sampling, Discontinuity Counting and Digitizing. A pulse generator at the rod drive wheel delivers a pulse for every 10 mm (0.4 in.) of the passing rod. This pulse will group the discontinuities detected in each unit length of rod. The programmable multichannel processor counts and memorizes the passing peaks. A 6 km (4 mi) coil, for example, will have 600 000 data after sampling and counting. This information, in the form of analog voltages, is then converted into digital values.

Signal Interpretation

The preprocessing computer receives the sampling signals — the digital eddy current signals as well as the signals that

FIGURE 21. Signal processing of eddy current signals.

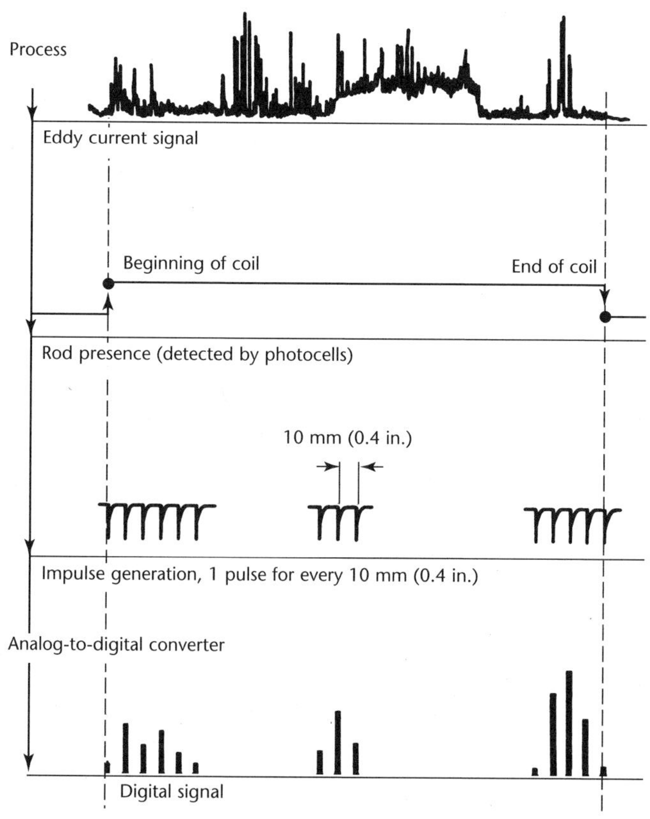

the steel rod has started and finished passing. The quality assessment of the coil begins once these data are assembled.

Amplitude Classification. The digitized signal is sorted (Fig. 22) according to three preset amplitude levels: two preselectable levels S_1 and S_2 for signal detection and a background threshold level S_{bkg} for detecting the ratio of signal to noise. This sorting stage is fundamental because it can help reveal the nature of the discontinuity detected. The application described below shows how long and short discontinuities are recognized.

Summarizing the Information. Having 600 000 measurements for a coil of 6 km (4 mi) is difficult to manage. In practice, the information is assembled into image segments called *windows*. The number and length of these windows depends on the user's quality control needs.

Data processing software is flexible. At one plant, for example, a coil measuring 5 mm (0.2 in.) diameter and 6 km (4 mi) long is broken down into 200 windows of 30 m (100 ft) each. Another plant, however, uses only 12 windows (including one head and one tail window). The format of the quality report may be slightly different but the basic principles of quality control are the same.

At this stage, the system has gathered enough information to produce a quality distribution report for the coil.

Discontinuity Length Measurement. The encircling sensor does carry with it certain limitations.

With short discontinuities, there is no problem. A short discontinuity generates a high signal that is perfectly identifiable. It will be detected above the S_1 threshold (Fig. 23).

FIGURE 22. Sorting of eddy current indications.

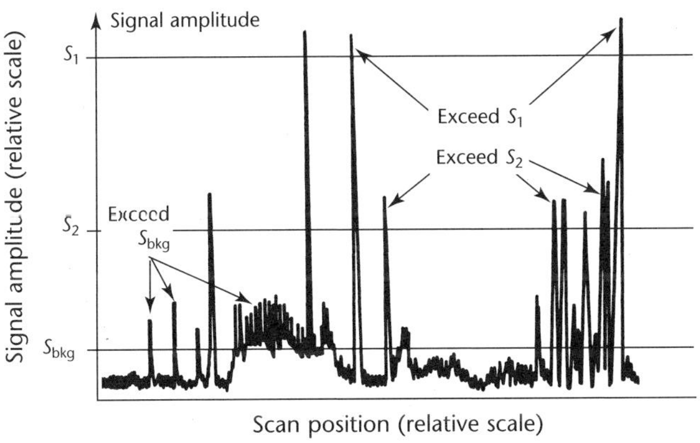

Legend

S_1 = high signal threshold for short discontinuities
S_2 = medium signal threshold for long discontinuities
S_{bkg} = threshold for detection of elevated background noise

In theory, a long discontinuity that is perfectly straight (like a razor blade cut) is undetectable because of the very nature of the encircling sensor. At best, the sensor will signal the beginning and the end of the discontinuity as it would indicate two short discontinuities (Fig. 24).

Fortunately, most of the long discontinuities on a rod have uneven edges that generate an elevation in the level of background noise as well as significant signals above S_2. Pulse density and noise density increase. Thus, they can be detected with the S_2 and S_{bkg} levels adjusted correctly.

FIGURE 23. Indications from short discontinuities: (a) test object; (b) high amplitude signals exceeding preset level at threshold S_1 with normal background.

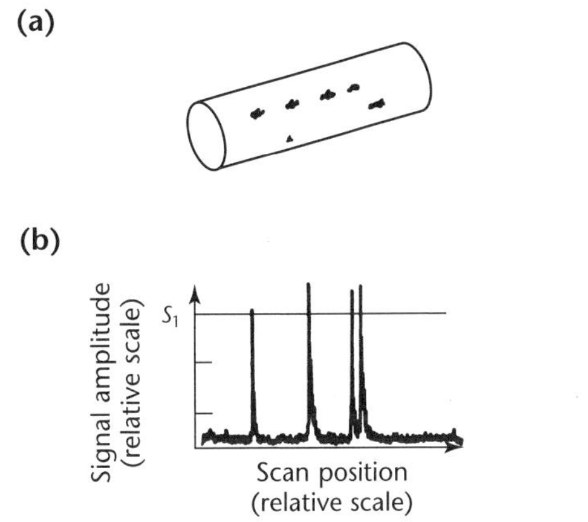

FIGURE 24. Indications from long discontinuities: (a) test object; (b) significant signals exceeding preset level at threshold S_2 with noise rising above background signal S_{bkg}.

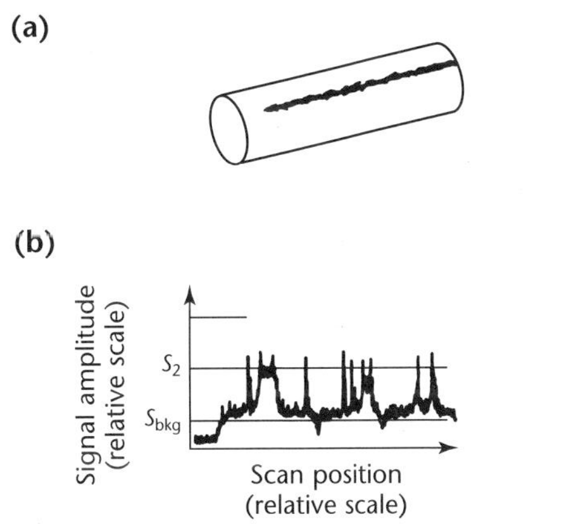

High Speed Testing of Hot Wire and Hot Tubes[11]

Nondestructive testing of hot wire differs from other test processes in one essential aspect: production speed. Speeds up to 50 m·s^{-1} (112 mi·h^{-1}) are routine and speeds up to 100 m·s^{-1} (224 mi·h^{-1}) are not unusual. Material flow and test data flow from 100 percent surface testing are extremely high. Online nondestructive testing in the production line is a cybernetic process. Different feedback mechanisms with different time responses occur in such a process, depending on the type of discontinuities and the information about detected discontinuities.

Hot Aluminum Products

Hot Rolled Aluminum Wire. Aluminum wire having a temperature of about 300 °C (570 °F) and a diameter of about 15 mm (0.6 in.) is continuously tested with the eddy current unit at 10 m·s^{-1} (22 mi·h^{-1}). The discontinuities are marked immediately with a paint that resists heat up to 500 °C (930 °F). The marked discontinuities are controlled visually and cut out before cold drawing. The eddy current test record and the micrographs of some typical discontinuities are shown in Fig. 25.

Extremely small discontinuities are of less interest because they are removed by the cold drawing process. The representation of dangerous, hard ferrous inclusions is out of proportion to their occurrence. Because of the high magnetic permeability of iron, even small iron inclusions are detected.

Hot Formed Aluminum Tubes. Another successful application of eddy current testing is in aluminum alloy tube production. Aluminum alloy tubes are produced at about 1 m·s^{-1} (2.2 mi·h^{-1}) at about 450 °C (840 °F) with an extrusion press. Directly after the press, the material flow through the eddy current test is nearly continuous.

Figure 26 shows eddy current indications of an extruded tube of 16 mm (0.63 in.) diameter with drilled holes according to the specification and indicated with a good ratio of signal to noise. Because only small differences exist between hot and cold aluminum, these typical test results, which have been attained by a test of a cold tube, are also valuable for hot material.

Welded Steel Tubes

In one installation, welded steel tubes are tested by an eddy current system just behind the calibration mill in a welded steel tubing line. Transverse welds and typical discontinuities are indicated with a

FIGURE 25. Hot testing of 10 mm (0.4 in.) diameter aluminum wire at speeds of about 10 m·s^{-1} (22 mi·h^{-1}) and at temperature of about 300 °C (570 °F): (a) overlap; (b) mechanical damage; (c) ferrous inclusion.

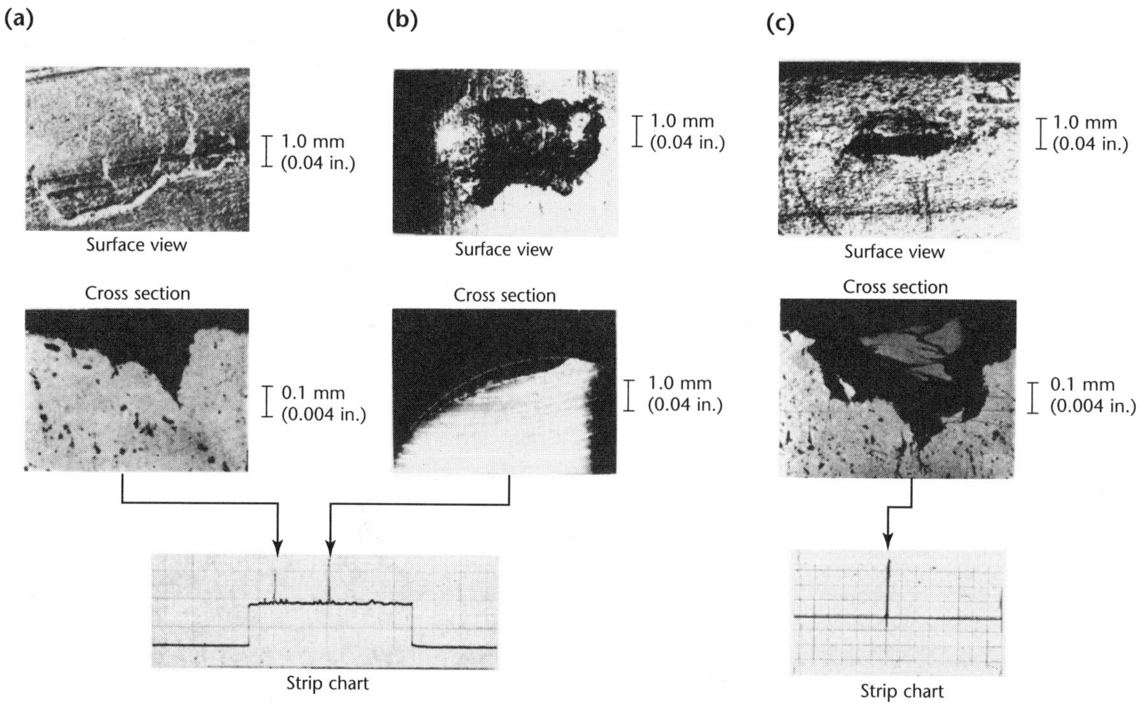

high ratio of signal to noise by the test instruments and an automatic saw is activated by the output signals. This eddy current test obviates hydrostatic testing for this product.

Hot Steel Wire

Steel wire production is not only the widest field of application for hot testing equipment but also the field with the highest demands because of the physical conditions for the detectability of small discontinuities. The following investigations have been carried out in seven European rolling mills.

As a first step, single discontinuities and their eddy current indications have been compared. To this end, holes with diameters between 2 and 5 mm (0.08 and 0.20 in.) and depths from 5 to 9 mm (0.20 to 0.35 in.) have been drilled in the cold billets to produce artificial discontinuities in the rolled wire. The difficulty with this technique is the fact that the discontinuities in the wire will have large depths, 1 to 3 mm (0.04 to 0.12 in.). It is difficult to attain smaller discontinuities in the wire by this technique because smaller holes in the billet will be burnt out during the heating process. There are many ways to produce discontinuities in the wire. One is to overheat the billet and roll it in scale. This technique will cause discontinuities of various sizes.

In the test record and micrographs of a wire rolled from a billet with drilled holes, all the large artificial discontinuities are well indicated.

FIGURE 26. Eddy current test of 16 mm (0.63 in.) diameter extruded aluminum tubes with 1.5 mm (0.06 in.) thick walls: (a) strip chart; (b) schematic.

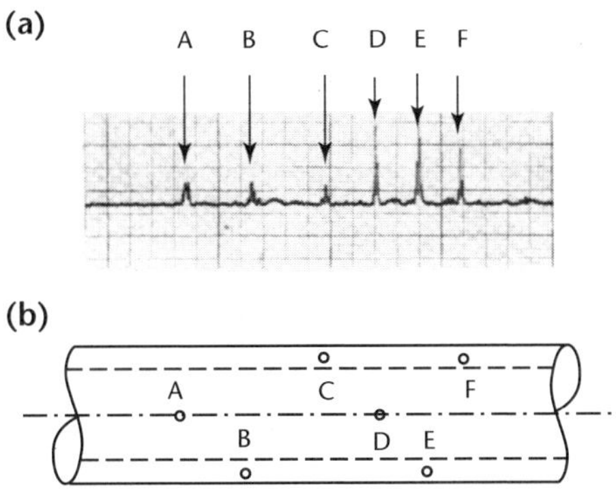

Legend

A, B, C = drilled holes of 0.94 mm (0.037 in.) diameter and 0.41 mm (0.016 in.) deep
D, E, F = drilled holes of 0.94 mm (0.037 in.) diameter and 0.84 mm (0.033 in.) deep

Conclusion

Online applications to metals demonstrate the speed and effectiveness of electromagnetic testing for the control of process and product quality. The method's widespread use in primary metals production saves that industry many millions of dollars annually and provides the world with stronger and safer products.

In addition to the detection of cracks, seams, inclusions and other discontinuities that occur randomly, it is possible to detect repetitive discontinuities caused by broken or cracked rollers. Overheated billets and other problems that occur as a function of time are also detectable.

References

1. *Nondestructive Testing Handbook*, second edition: Vol. 4, *Electromagnetic Testing*. Columbus, OH: American Society for Nondestructive Testing (1986).
2. [Samy], R.[P.] "The Future of Electromagnetic Testing." *Materials Evaluation*. Vol. 49, No. 9. Columbus, OH: American Society for Nondestructive Testing (September 1991): p 1158–1159, 1161.
3. Hoffman, J.P. "A New Rotating-Probe Eddy Current Method for Inspecting Bar Surface." *Materials Evaluation*. Vol. 33, No. 10. Columbus, OH: American Society for Nondestructive Testing (October 1975): p 237-242.
4. Hoffman, J.P. and R.C. Booth. United States Patent 3 673 493, *One-Probe Method and Apparatus for Detecting, Correlating, and Classifying Defects in Test Members* (June 1972).
5. Mandula, J.M. "Billetscan — A New Eddy Current Device for Total Surface Inspection of Square Billets." *Materials Evaluation*. Vol. 30, No. 3. Columbus, OH: American Society for Nondestructive Testing (March 1972): p 49-54.
6. Sakamoto, T. and T. Hiroshima. "Rotating Eddy Current Machine for Hot Steel Rods and Wires." Paper 5. *QualTest-2 Conference Proceedings* [Dallas, TX, October 1983]. Columbus, OH: American Society for Nondestructive Testing (1983): p 1-11.
7. Arnelo, A. and A. Von Heijne. "Eddy Current Inspection of Hot Rod." *Materials Evaluation*. Vol. 26, No. 11. Columbus, OH: American Society for Nondestructive Testing (November 1968): p 230-235.
8. Labbe, G. and C. Maeder. "Multifrequency Testing of Hot Steel Wires." Paper 1B-5. *Eighth World Conference on Nondestructive Testing* [Cannes, France, September 1976]. Paris, France: Confédération Française pour les Essais Non Destructifs [French Society for Nondestructive Testing], for the International Committee on Nondestructive Testing (1976).
9. Mizuno, M. "Rotary-Probe Eddy Current Testing of Hot Steel Rods." *Materials Evaluation*. Vol. 49, No. 6. Columbus, OH: American Society for Nondestructive Testing (June 1991): p 691-694.
10. Cousin, M. "An On-Line Eddy Current System for the Quality Evaluation of Hot Rods." *Materials Evaluation*. Vol. 43, No. 13. Columbus, OH: American Society for Nondestructive Testing (December 1985): p 1649-1654.
11. Stumm, W. "New Developments in the Eddy Current Testing of Hot Wire and Hot Tubes." *Materials Evaluation*. Vol. 29, No. 7. Columbus, OH: American Society for Nondestructive Testing (July 1971): p 141-147.

Bibliography

Burley, C.E. and J.E. Duarte. "Quality Control of Heat-Treated Aluminum Plate by Electrical Conductivity Testing." *Materials Evaluation*. Vol. 42, No. 12. Columbus, OH: American Society for Nondestructive Testing (November 1984): p 1487-1491.

Cecco, V.S., J.R. Carter and S.P. Sullivan. "Eddy Current Technique for Detecting and Sizing Surface Cracks in Carbon Steel." *Materials Evaluation*. Vol. 51, No. 5. Columbus, OH: American Society for Nondestructive Testing (May 1993): p 572-577.

Dodd, C.V. and W.A. Simpson, Jr. "Thickness Measurements Using Eddy Current Techniques." *Materials Evaluation*. Vol. 31, No. 5. Columbus, OH: American Society for Nondestructive Testing (May 1973): p 73-79, 84.

Förster, F. "The Nondestructive Inspection of Tubings for Discontinuities and Wall Thickness Using Electromagnetic Test Methods: Part 1." *Materials Evaluation*. Vol. 28, No. 4. Columbus, OH: American Society for Nondestructive Testing (April 1970): p 21A-25A, 28A-31A.

Förster, F. "The Nondestructive Inspection of Tubings for Discontinuities and Wall Thickness Using Electromagnetic Test Methods: Part 2." *Materials Evaluation*. Vol. 28, No. 5. Columbus, OH: American Society for Nondestructive Testing (May 1970): p 19A-23A, 26A-28A.

Henry, E.B. "Lester Honor Lecture 1988: The Role of Nondestructive Testing in the Production of Pipe and Tubing." *Materials Evaluation.* Vol. 47, No. 6. Columbus, OH: American Society for Nondestructive Testing (June 1989): p 714-715, 718, 720, 722-724.

Kubota, J., S. Sasaki, I. Sato, S. Ito, T. Kadowaki, H. Yamaguchi, K. Fujisawa and R. Murayama. "An Improved Electromagnetic Ultrasonic Testing Technique for Flaw Detection for Hot Steel." *Materials Evaluation.* Vol. 46, No. 4. Columbus, OH: American Society for Nondestructive Testing (March 1988): p 523-527.

Kwun, H. and C.M. Teller. "Detection of Fractured Wires in Steel Cables Using Magnetorestrictive Sensors." *Materials Evaluation.* Vol. 52, No. 4. Columbus, OH: American Society for Nondestructive Testing (April 1994): p 503-507.

Mandula, J.M. and E.S. Monks. "NDT Systems for Steel Billets, Bars and Tubes." *Materials Evaluation.* Vol. 34, No. 10. Columbus, OH: American Society for Nondestructive Testing (October 1976): p 230-236.

Prince, J.M., L.D. Reid and D.L. Lessor. "Two-Frequency Eddy Current Instrument for Measuring the Thickness of Zircaloy Cladding on Uranium under Conditions of Varying Lift-Off." *Materials Evaluation.* Vol. 43, No. 12. Columbus, OH: American Society for Nondestructive Testing (November 1985): p 1562-1565.

Rummel, W.D. "Characterization and Evaluation of 2014 Aluminum Alloy by Eddy Current Conductivity Techniques." *Materials Evaluation.* Vol. 24, No. 6. Columbus, OH: American Society for Nondestructive Testing (June 1966): p 322-326.

Rummel, W.D. "Theory of the Use of Eddy Current Conductivity Devices to Monitor Aluminum Alloys." *Materials Evaluation.* Vol. 24, No. 9. Columbus, OH: American Society for Nondestructive Testing (September 1966): p 507-511.

Sather, A. "Pulsed Eddy Current Testing Apparatus for Use on Smooth and Ribbed Tubing." *Materials Evaluation.* Vol. 35, No. 12. Columbus, OH: American Society for Nondestructive Testing (December 1977): p 55-59.

Sengupta, A.K. and W.A. Theiner. "Nondestructive Evaluation of Stresses in Welds by Micromagnetic Method." *Materials Evaluation.* Vol. 53, No. 5. Columbus, OH: American Society for Nondestructive Testing (May 1995): p 554-558, 561.

Shaffer, R.D. "Eddy Current Testing, Today and Tomorrow." *Materials Evaluation.* Vol. 52, No. 1. Columbus, OH: American Society for Nondestructive Testing (January 1994): p 28-32.

Smith, J.H., C.V. Dodd and L.D. Chitwood. "Multifrequency Eddy Current Examination of Seam Weld in Steel Sheath." *Materials Evaluation.* Vol. 43, No. 12. Columbus, OH: American Society for Nondestructive Testing (November 1985): p 1566-1572.

Sun, Y.S., S.S. Udpa, W. Lord and D. Cooley. "A Remote Field Eddy Current NDT Probe for the Inspection of Metallic Plates." *Materials Evaluation.* Vol. 54, No. 4. Columbus, OH: American Society for Nondestructive Testing (April 1996): p 510-512.

Standards

ASTM B 483/B 483M, *Standard Specification for Aluminum and Aluminum-Alloy Drawn Tubes for General Purpose Applications.* West Conshohocken, PA: ASTM International (2003).

ASTM B 491/B 491M, *Standard Specification for Aluminum and Aluminum-Alloy Extruded Round Tubes for General-Purpose Applications.* West Conshohocken, PA: ASTM International (2000).

ASTM E 215, *Standard Practice for Standardizing Equipment for Electromagnetic Examination of Seamless Aluminum-Alloy Tube.* West Conshohocken, PA: ASTM International (1998).

ASTM E 243, *Standard Practice for Electromagnetic (Eddy-Current) Examination of Copper and Copper-Alloy Tubes.* West Conshohocken, PA: ASTM International (1997).

ASTM E 309-95, *Standard Practice for Eddy-Current Examination of Steel Tubular Products Using Magnetic Saturation.* West Conshohocken, PA: ASTM International (2001).

ASTM E 426, *Standard Practice for Electromagnetic (Eddy-Current) Examination of Seamless and Welded Tubular Products, Austenitic Stainless Steel and Similar Alloys.* West Conshohocken, PA: ASTM International (1998).

ASTM E 570, *Standard Practice for Flux Leakage Examination of Ferromagnetic Steel Tubular Products.* West Conshohocken, PA: ASTM International (1997).

ASTM E 571, *Standard Practice for Electromagnetic (Eddy-Current) Examination of Nickel and Nickel Alloy Tubular Products.* West Conshohocken, PA: ASTM International (1998).

ASTM E 1033, *Standard Practice for Electromagnetic (Eddy-Current) Examination of Type F–Continuously Welded (CW) Ferromagnetic Pipe and Tubing above the Curie Temperature.* West Conshohocken, PA: ASTM International (1998).

ASTM E 1312, *Standard Practice for Electromagnetic (Eddy-Current) Examination of Ferromagnetic Cylindrical Bar Product above the Curie Temperature.* West Conshohocken, PA: ASTM International (1999).

ASTM E 1606, *Standard Practice for Electromagnetic (Eddy-Current) Examination of Copper Redraw Rod for Electrical Purposes.* West Conshohocken, PA: ASTM International (1999).

MT-003-0, *Module #12 — Detection, Classification, and Elimination of Rod and Bar Surface Defects.* Washington, DC: American Iron and Steel Institute (1996).

SAE AMS 4071-L, *Aluminum Alloy, Drawn, Round, Seamless Hydraulic Tubing 2.5Mg 0.25Cr (5052-0) Annealed.* Warrendale, PA: SAE International (2002).

SAE AMS 5556, *Steel Corrosion and Heat-Resistant, Seamless or Welded Tubing 18Cr 11Ni 0.70Cb (SAE 30347) Solution Heat Treated.* Warrendale, PA: SAE International (2003).

SAE AMS 5568-F, *Steel, Corrosion and Heat Resistant, Welded Tubing 17Cr 7.1Ni 1.1Al Solution Heat Treated, Precipitation-Hardenable.* Warrendale, PA: SAE International (2001).

SAE AMS 5589-D, *Nickel Alloy, Corrosion and Heat Resistant, Seamless Tubing 52.5Ni 19Cr 3.0Mo 5.1Cb 0.90Ti 0.50Al 18Fe Consumable Electrode or Vacuum Induction Melted, 1775Mdf (968Mdc), Solution Heat Treated.* Warrendale, PA: SAE International (2000).

SAE AMS S 7420, *Steel Bars, Alloy, Chromium, High Carbon E52100 (Aircraft Quality).* Warrendale, PA: SAE International (2003).

SAE AMS WWT 700, *Tube, Aluminum and Aluminum Alloy, Drawn, Seamless, General Specification for.* Warrendale, PA: SAE International (2001).

SAE J 349, *Detection of Surface Imperfections in Ferrous Rods, Bars, Tubes, and Wires.* Warrendale, PA: SAE International (1991).

SAE J 425, *Electromagnetic Testing by Eddy Current Methods.* Warrendale, PA: SAE International (1991).

SAE J 2281, *Selecting and Specifying Hot-Rolled Steel Bar Products.* Warrendale, PA: SAE International (1997).

15

CHAPTER

Chemical and Petroleum Applications of Electromagnetic Testing

David R. Bajula, Longview Inspection, La Porte, Texas (Part 1)

David M. Amos, MFE Enterprises, Humble, Texas (Part 2)

Lawrence O. Goldberg, Sea Test Services, Merritt Island, Florida (Part 4)

Roderic K. Stanley, NDE Information Consultants, Houston, Texas (Part 3)

PART 1. Electromagnetic Testing of Process Tubing and Heat Exchangers

Tubing

Tube testing is an important part of maintenance for the refining and petrochemical industry. Heat exchangers and condensers are designed to keep products in the tubes separate from products in the vessel (see Fig. 1). A leaking tube not only could cause a significant impact on production but also could cause a catastrophic failure and loss of life.

Tube testing techniques include magnetic flux leakage testing, remote field testing, conventional eddy current testing, ultrasonic testing, laser profilometry and remote visual testing. The present discussion concentrates on electromagnetic techniques; ultrasonic and laser methods complement the electromagnetic techniques and often are used in parallel.

Tube testing is typically broken down into two categories: ferrous and nonferrous. Ferrous metals are metals such as carbon steel, 400 series stainless steel and metals with similar magnetic properties; nonferrous metals are nonmagnetic and include copper, brass and most stainless steels. Table 1 lists techniques used for tubes made of various materials.

The choice of technique is mainly influenced by the type of service damage to be detected but often the technique is dictated by tube cleanliness. For example, rotary ultrasonic testing and laser profilometry require very clean interior surfaces whereas electromagnetic tests do not. Often, electromagnetic techniques are used as screening tools before cleaning for ultrasonic or laser techniques.

Several damage mechanisms and discontinuities can occur. Some are volumetric and not connected with either surface. However, the primary discontinuities are either outside diameter or inside diameter surface breaking discontinuities. Table 2 lists various discontinuities that can be detected with the various techniques for both nonferrous and ferrous tubing materials.

FIGURE 1. Cutaway image of typical heat exchanger, showing tube bundle.

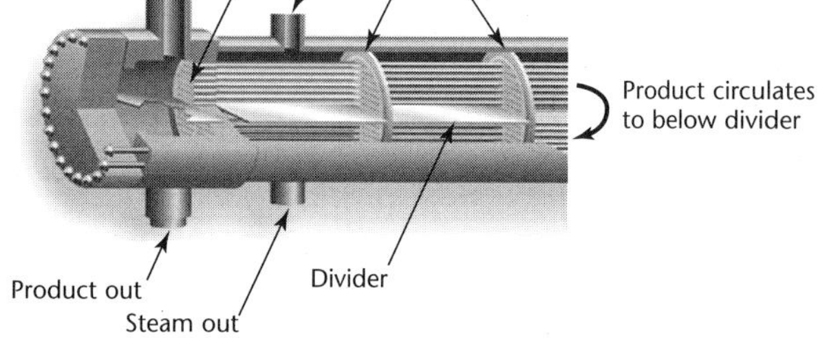

Eddy Current Testing

The eddy current technique works by inducing electrical currents (eddy currents) in electrically conductive materials as detailed elsewhere. Bobbin probes containing coils are used for tube testing (Fig. 2a). In theory, any discontinuities in the material such as cracks, pitting or wall loss will disrupt the flow of the eddy currents and thus be detected by the instrumentation.[1,2] Saturation or special probes can be used for thin walled ferromagnetic tubing. Most tube exchanger bundles contain supports susceptible to damage in service. Multiple-channel systems are capable of suppressing or mixing out the signal responses from supports to closely interrogate the material under and near the supports.

Conventional eddy current testing is used mainly on nonferrous (nonmagnetic) materials because of the effects from permeability with ferrous materials. In many cases, the owners and users of the exchangers prefer eddy current testing to internal rotary ultrasonic testing because the cleanliness of the tubes is less critical.

TABLE 1. Applicability of nondestructive tests to ferrous and nonferrous metals.

Technique	Applicability to Metals: Ferrous	Nonferrous
Eddy current testing	no	yes
Magnetic flux leakage testing	yes	no
Remote field testing	yes	yes
Laser techniques	yes	yes
Ultrasonic testing	yes	yes

Additionally, eddy current testing can be several times faster than internal rotary ultrasonic testing.

Remote Field Testing

Remote field testing was developed for ferrous or carbon steel materials and requires a special remote field eddy current probe in which the exciter coil is separated from the pickup coil by a distance of two to three times the tube diameter (Fig. 2b). The receiving or pickup coil then detects the generated flux lines that cross the tube wall twice. Because of the highly magnetic properties of ferrous materials, meaningful eddy current testing requires higher power fields. Other eddy current techniques for ferrous tubing require complete magnetic saturation of the tube material but remote field testing does not. The remote field testing amplifier provides the higher power output levels needed for ferrous tube testing and remote field probe coils are designed to handle the increased power levels.[3]

Because remote field testing is transmitted through the tube wall, it is equally sensitive to discontinuities on the inside surface and outside surface of the tube. However, much like eddy current testing, the factor having the greatest effect on the signal is change in the cross sectional area. Without the proper instrumentation, a 10 percent wall reduction for 360 degrees of tube surface could have a response similar to that for a 90 to 100 percent pinhole. The owners and users of the exchangers prefer remote field testing to internal rotary ultrasonic testing because the cleanliness of the tubes is less critical. Additionally, remote field testing can be three times faster than internal rotary ultrasonic testing. Remote field testing is somewhat slower than conventional eddy current testing and the speed of travel must be as constant as possible to obtain accurate responses.

Magnetic Flux Leakage Testing

Magnetic flux leakage testing uses a strong magnet inside the probe to magnetize the test object (Figs. 2c and 3). As the probe encounters a wall reduction or a sharp discontinuity, the flux distribution varies around that area and is detected with either a hall effect sensor or an inductive pickup coil.[4,5] Magnetic flux leakage response is sensitive to discontinuities such as isolated pitting. Magnetic flux leakage testing has been used successfully on air cooled, finned, heat exchanger tubes of carbon steel. Magnetic flux leakage testing is less sensitive to signal effects from the aluminum fins coiled around the carbon steel tubes than remote field testing is.

Complementary Methods

Internal Rotary Ultrasonic Testing. Internal rotary ultrasonic testing is well suited for petrochemical and refinery tube tests. The technique uses an ultrasonic beam to scan the tube internal surface in a helical pattern to ensure that the full circumference of the tube is tested. The system monitors the front wall and the back wall echoes to measure the tube wall thickness precisely. Essentially, a radial

TABLE 2. Discontinuity detection by nondestructive tests for ferrous and nonferrous metals in used components.

Damage Mechanism	Eddy Current Testing	Magnetic Flux Leakage Testing	Remote Field Testing	Laser Profilometry	Ultrasonic Testing
Nonferrous Materials					
Pitting, inside surface	yes	no	tube and pipe	yes	yes
Pitting, outside surface	yes	no	tube and pipe	no	yes
Stress corrosion cracking, inside surface	yes	no	no	limited	no
Stress corrosion cracking, outside surface	yes	no	no	no	no
Volumetric discontinuities, embedded and other	yes	no	tube and pipe	no	limited
Wall loss, inside surface	yes	no	tube and pipe	limited	yes
Wall loss, outside surface	yes	no	tube and pipe	limited	yes
Ferrous Materials					
Pitting, inside surface	no	yes	tube and pipe	yes	yes
Pitting, outside surface	no	yes	tube and pipe	no	yes
Stress corrosion cracking, inside surface	no	no	no	limited	no
Stress corrosion cracking, outside surface	no	no	no	no	no
Volumetric discontinuities, embedded and other	no	no	no	no	limited
Wall loss, inside surface	no	limited	tube and pipe	limited	yes
Wall loss, outside surface	no	limited	tube and pipe	limited	yes

B-scan of the tube profiles total wall thickness and pitting on the inside or outside of the tube. A drawback of ultrasonic testing is that the tubes are required to be extremely clean and typically are sandblasted with silicon grit or soda ash. Also, ultrasonic testing can require three times as much time as an electromagnetic technique would.

Laser Profilometry of Tubing. Laser profilometry is based on the principle of optical triangulation. A laser source similar to a standard laser pointer is directed at the surface whose height is to be measured. An imaging lens collects the light reflected from the surface and focuses it onto a position sensitive detector. As the surface height changes, the position of the focused laser spot on the detector moves. The output from the detector is processed electronically to convert the detector positions to accurate height measurements that can be stored on a computer for display and analysis.

Essentially, laser techniques provide information regarding the nearside surface by profiling the tube wall. Pitting and wall losses can be detected as a diameter

FIGURE 2. Bobbin coil probes for electromagnetic testing: (a) probe for eddy current testing; (b) probe for remote field testing; (c) probe for magnetic flux leakage testing.

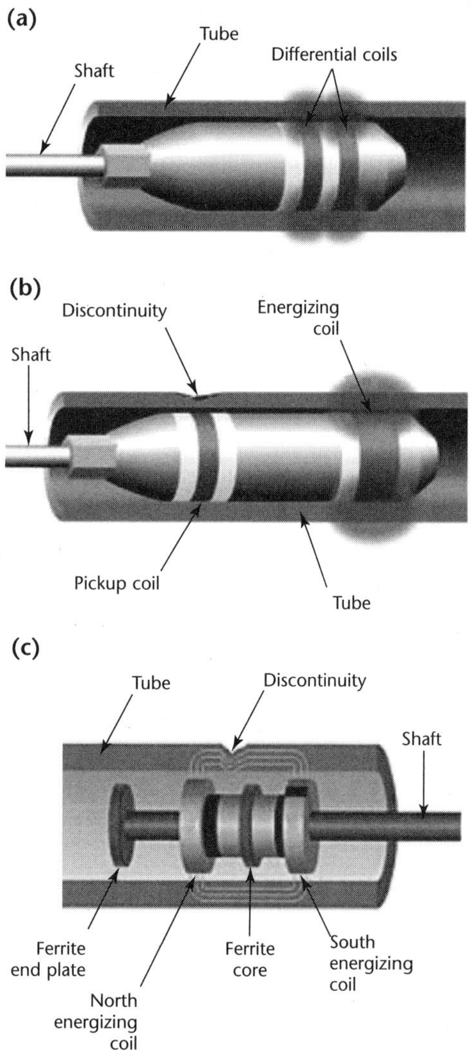

FIGURE 3. Magnetic flux leakage probe inserted in carbon steel tube bundle of crude petroleum processing unit.

FIGURE 4. Gasoline processing plant: (a) external view, showing distillation columns; (b) interior view of chamber in distillation column.

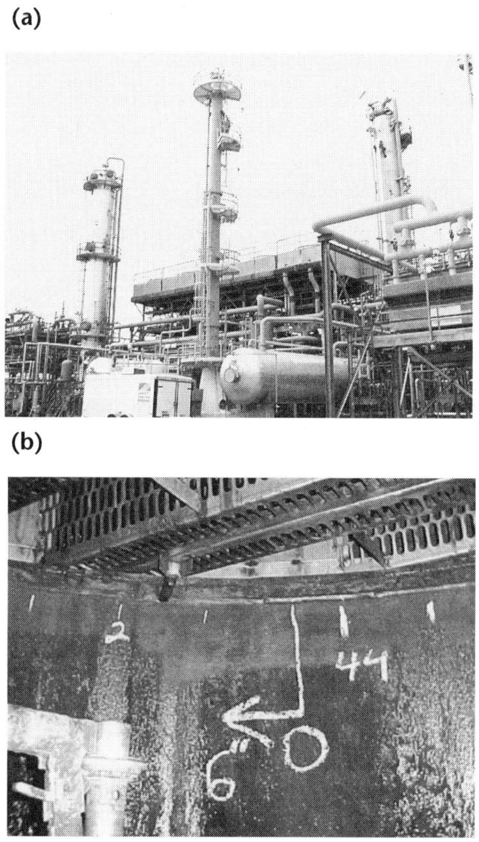

change with a high degree of accuracy. Additionally, laser techniques are used for the detection of cracking, which appears as a disruption or distortion in the optical field.

Pressure Vessels

Pressure vessels are continually subject to testing and are considered one of the most critical pieces of equipment in a petrochemical plant or refinery (Fig. 4). Traditional preservice tests include radiographic and ultrasonic testing during fabrication. Traditional inservice tests include visual, ultrasonic, magnetic particle and more recently electromagnetic techniques such as eddy current testing and alternating current field measurement. Electromagnetic testing can be used for the detection, sizing and evaluation of damage mechanisms such as cracking.

Industry practices for inservice tests of pressure vessels have specified visual testing, ultrasonic testing, wet fluorescent magnetic particle testing and electromagnetic testing. Electromagnetic techniques such as eddy current testing and alternating current field measurement offer distinct advantages. To perform wet fluorescent magnetic particle testing, the vessel surfaces must be prepared by sandblasting. Eddy current testing and alternating current field measurement techniques do not require sandblasting and, unlike magnetic particle testing, can also provide depth sizing information.

Alternating Current Field Measurement

Alternating current field measurement (Fig. 5) was developed from the alternating current potential drop technique. Potential drop testing has been used for crack sizing and crack growth monitoring for underwater applications such as offshore platforms. The alternating current field measurement technique is simple, relying on the measurement of surface magnetic fields instead of surface electric fields, thus requiring no electrical contact. This reliance on magnetic fields allows the technique to be used through coatings up to 6 mm (0.25 in.).[6]

Eddy current techniques and probes can be dramatically influenced by probe liftoff but alternating current field measurement, with its unidirectional fields, is not. Another benefit of alternating current field measurement is that the technique requires no calibration. The technique relies on field values compared with a theoretical model and database of known crack responses.

FIGURE 5. Equipment for alternating current field measurement.

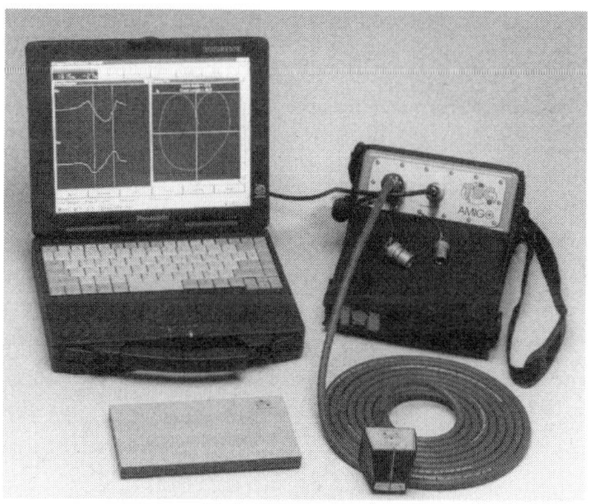

Part 2. Electromagnetic Testing of Transmission and Storage Systems

Pipelines

Pipelines connect field production (gas and oil extraction) with refineries and petrochemical plants where gas and crude petroleum are processed into usable products (Fig. 6). Because pipelines cross state lines in the United States, they are governed by the Department of Transportation. The construction, maintenance and testing of these pipelines are critical to the safety of the environment and the general public. Buried pipelines not only have the potential for catastrophic failure but could contaminate lakes, rivers and underground water sources if leakage occurs.

Traditional preservice tests include radiographic and ultrasonic testing during fabrication to ensure the quality of the welding. Once a pipeline is in service, the pipeline companies depend largely on inservice testing to assess corrosion.

Test strategies before 1970 included leak detection systems. Since the late 1960s, flux leakage testing tools have been inserted into the pipelines and propelled by product flow. This expedient offers a test technique without significant interruption in pipeline production.

In magnetic flux leakage testing, changes in the material mass such as corrosion or pitting cause a localized flux leakage to occur at the discontinuity. These perturbations in the magnetic field are detected by the sensors within the magnetic circuit, are recorded and later are analyzed and reviewed. Much like the baffles or supports in a tube exchanger bundle, the pipeline circumferential welds provide abrupt signals and easy landmarks when the data are evaluated for discontinuity locations.[7-9]

FIGURE 6. Carbon steel, 0.75 m (30 in.) outside diameter, gas transmission pipeline.

Smart pigs are test vehicles that product flow pushes through a pipeline (Fig. 7). The technique got its name from a squealing sound from the pig moving through the pipe. At the end of the line or run, the pig is retrieved and the onboard data are then processed and analyzed. The pigs are similar to the magnetic flux leakage probes used in tube testing but the pigs are constructed to propel themselves down pipelines and collect the required test data.

FIGURE 7. Equipment for magnetic flux leakage testing of pipes and tubes: (a) pig tool; (b) data acquisition from pig sensors.

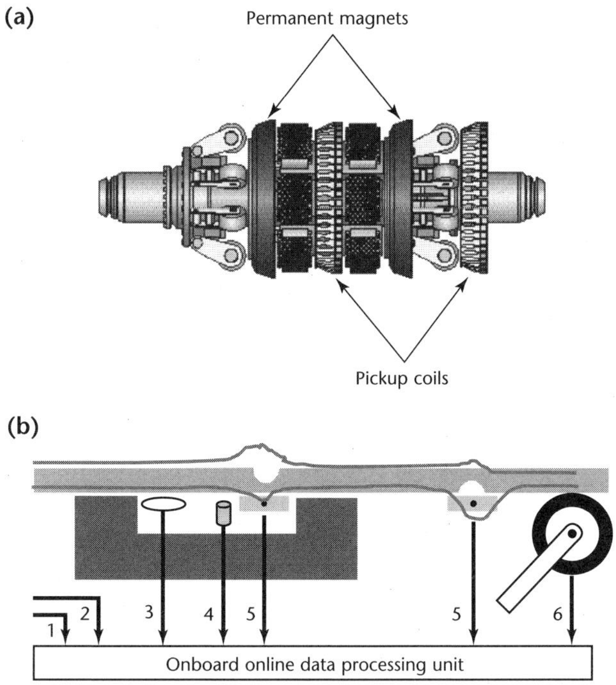

Legend

1. Pressure.
2. Ambient temperature.
3. Magnetic field (magnetization).
4. Surrounding magnetic flux.
5. Magnetic flux leakage (stray flux).
6. Odometer (distance and speed).

Magnetic Flux Leakage Testing of Aboveground Storage Tank Floors

Tank floors of aboveground storage tanks (Fig. 8) are subject to corrosion where they touch the ground. In the 1970s, ultrasonic testing was being performed on tank floors — spot ultrasonic testing using transducers on large wheels and automated ultrasonic techniques such as C-scanning. One destructive technique was to randomly cut out 0.3 × 0.3 m (12 × 12 in.) square coupons, to visually test them and then either to weld them back in place or to replace them with new patch plates.

Magnetic flux leakage test techniques have been widely used in the oil field industry since the 1970s for the testing of pipe, tubing and casing, both new and used. During the 1980s, magnetic flux leakage testing for tank floor applications was introduced to the petrochemical and refining industry. Since 1990, this technique has been applied to aboveground storage tank floors to provide a reliable indication of overall floor condition within an economical time frame.[10,11]

Magnetic flux leakage floor scanners provide reliable tests at a fraction of the time and cost associated with ultrasonic thickness gaging. A tank floor test at regular intervals is required by some specifications.[12] As with other techniques, evaluation by ultrasonic testing is required when magnetic flux leakage testing is specified. Generally, the evaluation is accomplished by ultrasonic thickness gaging and sometimes by B-scan or C-scan ultrasonic testing.

For tank floor testing, a magnetic bridge is used to introduce as near a saturation of flux as is possible in the test material between the poles of the bridge.

FIGURE 8. Aboveground storage tank for petroleum products.

Any significant reduction in the thickness of the plate will force some of the magnetic flux into the air around the reduction area. Sensors that can detect this flux *leakage* are placed between the poles of the bridge (Fig. 9a). To create leakage fields from corrosion or pitting, it is necessary to achieve near saturation of the magnetic field in the material. Near saturation is accomplished with powerful rare earth magnets, which offer more stability than electromagnets. The sensor can detect the magnetic flux leakage field caused by corrosion and pitting but cannot reliably determine if the flux

FIGURE 9. Magnetic flux leakage test: (a) schematic of bridge; (b) tank floor scanner incorporating magnetic flux leakage test bridge.

(a)

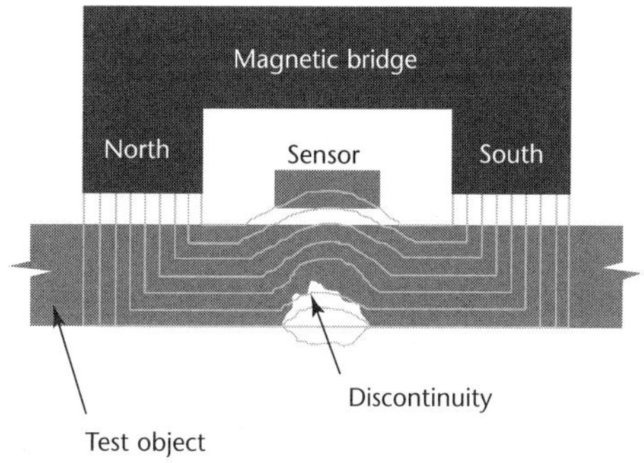

(b)

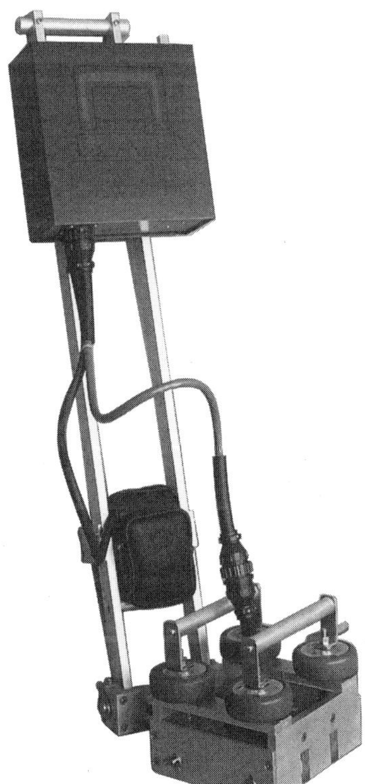

leakage is caused by top or bottom indications. For uncoated materials, the top discontinuities can be verified during a simple visual test. Other methods such as ultrasonic testing are performed for coated floors.

Floor scanning has problems not evident in the testing of tubular goods, where certain parameters can be closely controlled. Probably the greatest problem is that tank floors are never flat whereas tubes are always round. The unevenness of tank floors makes it hard to get reasonably consistent quantitative information. The application of rigid accept/reject criteria based on signal amplitude thresholds is also very unreliable for quantitative information. A realistic approach is required in the application of this test technique and in the design of the test equipment to ensure that fewer significant discontinuities are missed.

Test Conditions

To optimize the test, it is necessary to consider the environment and address the physical restrictions imposed by the actual conditions found when testing most tank floors.

Climate. The range of temperature and humidity conditions varies enormously during the year and around the world. The effect on both operator and equipment must be taken into consideration.

Cleanliness. Most aboveground storage tanks are dirty and sometimes dusty places to work. The conditions vary widely and depend on how much the tank operator cleans the floors in preparation for magnetic flux leakage scanning. As an absolute minimum, a good water blast is necessary and all loose debris and scale must be removed from the test surface. The surface does not have to be dry but puddles of standing water need to be removed. The cleaner the floor, the better the test.

Surface Condition. Significant top surface corrosion and buckling of the floor plates represent serious limitations to both the achievable coverage in the areas concerned and also the achievable sensitivity. Although very little can be done to improve this situation before testing, it must be considered in the design of the equipment. The effect of corrosion and buckling on the sensitivity of the test must be appreciated by both the tank operator and the inspector. Any physical disturbance of the scanning system as it traverses the floor will result in the generation of noise. The rougher the surface, the greater the noise and therefore the more difficult it is to detect small indications.

Equipment

It is important that magnetic flux leakage equipment produced for this particular application be designed to handle the environmental and practical problems always present. Figure 9b shows a mobile floor scanning unit.

Powerful rare earth magnets are well suited for introducing the required flux levels into the material under test. Electromagnets by comparison are excessively bulky and heavy. They do have an advantage in that the magnetic flux levels can be easily adjusted and turned off if necessary for cleaning. Permanent magnet heights can be adjusted to alter flux levels but the bridge requires regular cleaning to remove ferritic debris. The buildup of debris can impair system sensitivity significantly.

It is virtually impossible for this technique to achieve 100 percent coverage because physical access is limited. The equipment should be designed so that it can scan as close as possible to the lap joint and shell. The wheel base of the scanner is an important consideration on floors that are not perfectly flat. Smaller scanning heads can be used in confined spaces to increase coverage.

Sensors

Two types of sensors are used for magnetic flux leakage of aboveground tanks: coils and hall effect sensors. Both can detect the flux leakage fields caused by corrosion on tank floors. There is a fundamental difference, however, in the way that they respond to leakage fields and generate a response.

Coils are passive devices and follow Faraday's law in the presence of a magnetic field. As a coil passes through a magnetic field, a voltage is generated in the coil. The level of this voltage depends on the number of turns in the coil and the rate of change of the flux leakage. Scanning speed has a direct effect on the rate of change of the magnetic flux leakage passing through the coils (scanning speed needs to be constant.)

Hall effect sensors are solid state devices that form part of an electrical circuit. When passed through a magnetic field, the voltage in the circuit varies with the flux density. It is necessary to carry out some cross referencing and canceling with this type of sensor so that true signals can be separated from other causes of large variations in voltage levels generated by the test.

Hall effect sensors are more sensitive than coils and so result in false calls when surface conditions are imperfect. For tube testing, on the other hand, coils are adequately sensitive and are more stable and reliable than hall sensors.

Interpretation of Indications

Surface Differentiation. Magnetic flux leakage testing cannot differentiate between indications from the top and bottom of the test object. Some attempt has been made to use the eddy current signals from top discontinuities for the purposes of surface differentiation. Such discrimination is unreliable on real tank floors because the test surface is uneven and dirty. In most cases, visual testing is adequate. Contrary to what is expected, the flux leakage response from a top indication is significantly lower in amplitude than that from an equivalent bottom indication. To some degree, the influence of the top indications can be tuned out to assess the bottom indications.

Quantitative Assessment. Magnetic flux leakage testing is not quantitative but is a reliable, qualitative detector of corrosion on tank floors. Because of environmental and physical restrictions during tests, no reliable quantification of indications is possible. Amplitude alone does not indicate remaining wall thickness because it depends on volume loss. Discontinuities exhibiting various combinations of volume loss and through-wall dimension can give the same amplitude signal. This difficulty plus the continually changing spatial relationship of magnets, sensor and test surface makes an accurate assessment of remaining wall thickness virtually impossible. Quantitative results can be obtained by using ultrasonic testing as a followup test.

Misuse of Signal Threshold. Expediency has sometimes motivated accept/reject criteria using a signal threshold but signal amplitude alone is not a reliable indicator of remaining wall thickness. Significant indications can be completely missed where there is a single threshold or where the equipment does not provide a real time display to the operator during the test. To carry out a reliable test, the operator must have as much information as possible available in a real time display that is easy to interpret.

Computerized Signal Mapping. Mapping of flux leakage signals to tank floor layout is available on some systems. These maps can be used to plan further tests, for corrosion surveys and for hard copy reporting. The usefulness of this equipment must be weighed against the risk of electrical equipment inside storage tanks.

Training and Qualification. Training available to inspectors using magnetic flux leakage testing on tank floors is limited. Training must be specific to the equipment. The ultrasonic test must be carried out by personnel who are adequately trained and qualified. It must be remembered that this is not just *thickness measurement* but rather corrosion evaluation and the technician must have a full understanding of the damage mechanisms and the test technique.

Conclusions

Magnetic flux leakage testing is a reliable and economical means of qualitatively assessing the condition of tank floors.

The environment and physical restrictions must be addressed in the design of the equipment. Despite the greater sensitivity of hall effect sensors, coils are more reliable for this application.

Amplitude of flux leakage signals is an unreliable indicator of remaining wall thicknesses. Quantitative information can be obtained by applying ultrasonic testing to the areas indicated by magnetic flux leakage.

PART 3. Electromagnetic Testing of Drill and Coil Pipe

Drill Pipe Testing

Drill pipe is manufactured from various grades of seamless carbon steel tube, with tool joints of different steel grades friction welded on at either end. Details of chemistry, sizes and grades for both the pipe body and tool joints can be found in various specifications.[13-17] Drill pipe has also been manufactured from aluminum, a material not addressed here.

Testing in Manufacturing Plants

To detect material discontinuities, the carbon steel tube body is 100 percent tested by magnetic flux leakage testing, by shear wave ultrasonic testing or by both and may be repaired by discontinuity removal to leave a remaining wall of at least 87.5 percent of the specified wall thickness. The test sensitivity for new tube body is determined from API SPEC 5D.[13] Acceptable pipe then is upset at each end and the box and pin tool joints are friction welded in place. Tool joints are threaded, one end as a pin connection and the other as a box connection (Fig. 10).[3] The excess metal on the inner and outer surfaces from this process is removed by machining. Then, the friction welds between the pipe body and the tool joints may be tested with conventional shear wave ultrasound for material discontinuities.

Inservice Testing

Table 3 shows criteria applied to used drill pipe according to API RP 7G.[15] Several points need particular attention.

1. In the case of outside surface cuts and gouges, the remaining wall thickness shall be (a) not less than 80 percent for premium pipe, (b) not less than 70 percent for class 2 pipe with longitudinal discontinuities and (c) not less than 80 percent for class 2 pipe with transverse discontinuities. Discontinuities may be ground out provided (a) the remaining wall is not reduced to less than 80 percent for premium pipe, (b) the remaining wall is not reduced to less than 70 percent for class 2 pipe and (c) the removal by grinding is approximately *faired,* or smoothed, into the outer contour of the pipe.
2. Where cracks are found, the pipe is considered unfit for further drilling.
3. The average adjacent wall is determined by measuring the wall thickness on each side of the cut or gouge adjacent to the deepest penetration.
4. *String shot* refers to discontinuities caused by expansion of the pipe wall after a controlled explosion to dislodge drill pipe stuck in the hole.

Figure 10 shows the various parts of a drill pipe. Table 4 lists problems that occur with different parts of used drill pipes.

Tube Body Testing

The original form of drill pipe testing was visual, with an optical gage for the critical areas at the ends. Then, for many years, drill pipe tube bodies were tested by magnetizing the tube longitudinally and scanning it with rings of inductive coil or

FIGURE 10. Diagram for drill pipe testing. (See Table 4.)

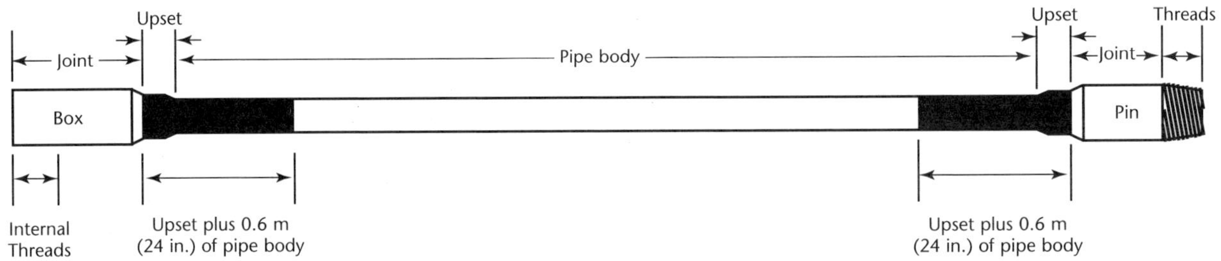

TABLE 3. Acceptance criteria for three classes of inservice drill pipe according to API RP 7G.[15]

Pipe Condition	Premium Class	Class 2	Class 3
Exterior Conditions			
Corrosion	≥ 80 percent[a]	≥ 70 percent[a]	< 70 percent[a]
Cracks	none	none	none
Cuts and gouges			
Longitudinal	≥ 80 percent[a]	≥ 70 percent[a]	< 70 percent[a]
Transverse	≥ 80 percent[a]	≥ 80 percent[a]	< 80 percent[a]
Dents and mashes	diameter reduction ≤ 3 percent[b]	diameter reduction ≤ 4 percent[b]	diameter reduction > 4 percent[b]
Slip area mechanical damage			
Crushing and necking	diameter reduction ≤ 3 percent[b]	diameter reduction ≤ 4 percent[b]	diameter reduction > 4 percent[b]
Cuts and gouges	depth ≤ 10 percent[c]	depth ≤ 20 percent[c]	depth > 20 percent[c]
Stress induced diameter variations			
Stretched	diameter reduction ≤ 3 percent[b]	diameter reduction ≤ 4 percent[b]	diameter reduction > 4 percent[b]
String shot	diameter reduction ≤ 3 percent[b]	diameter reduction ≤ 4 percent[b]	diameter reduction > 4 percent[b]
Wear	≥ 80 percent[a]	≥ 70 percent[a]	imperfections that exceed those of Class 2
Interior Conditions			
Corrosive pitting	≥ 80 percent[a,d]	≥ 70 percent[a,d]	< 70 percent[a,d]
Erosion and wear	≥ 80 percent[a]	≥ 70 percent[a]	< 70 percent[a]
Cracks	none	none	none

a. Percent of specified wall.
b. Percent of specified outside diameter.
c. Percent of average adjacent wall.
d. Measured from base of deepest pit.

TABLE 4. Testing of areas of used drill pipe.[15] (See Fig. 10 for parts of pipe.)

Location	Test Description	Discontinuity or Detail of Interest
Basic Test		
Tube body, transverse	full length, magnetic flux leakage test	internal and external fatigue cracks, gouges and pitting
Tube body, longitudinal	full length, magnetic flux leakage test	internal and external seams, slip area cracks, wire line cuts
Body wall measurement	ultrasonic and gamma ray tests	wear and conformance to specifications
Tool joints and upsets	transverse external magnetic particle testing (threads excluded)	external fatigue cracks and pits
	longitudinal external magnetic particle testing (threads excluded)	external seams, heat check cracks
Permanent marking	metal stencil applied to pin tool joint shoulder	date, test company, joint number, object classification
Optional Features		
Upsets and 0.6 m (24 in.) of pipe body	critical area internally tested by magnetic particle testing	internal transverse fatigue cracks
Tool joint outside diameter	caliper measurement to specification	undersized (worn) tool joints
Shoulder (face)	visual testing for discontinuities affecting pressure holding capacity	galls, nicks, washes, fins
Joint	random check of 10 percent of tool joints for manufacturers' markings	determination of reworking of tool joints
Shoulder width	mechanical gaging of tool joint shoulders	nonuniform wear caused by crooked drill pipe
Box swell	measurement of inside diameter of box counterbore	dimensional change caused by excessive torque
Thread profile	mechanical gaging of threads	galled or stretched threads, flank wear, sharp crowns
Thread	box and pin threads are cleaned and tested with ultraviolet radiation or magnetic particle testing	transverse crack in thread roots

hall sensors (Fig. 11). This simple magnetic flux leakage test is sensitive to discontinuities with a transverse or volumetric component, such as internal pitting, external pitting, drilling rig slip marks and fatigue cracks. Inflatable magnetic rubber balloons were also used to detect fatigue cracks on the inside surface.

A conductor would also be passed through the pipe, a shot of current fired and the tube magnetized circularly before magnetic flux leakage or magnetic particle testing over the tube body. Figure 12 shows a typical signal from such a test. Additionally, the tool joints could be tested with the residual circular field for longitudinal imperfections such as cracks from string shot and longitudinal heat check cracking on the tool joints. Circular magnetization has been accomplished by using capacitive discharge units.[18-20]

Spinning gamma ray pipe wall thickness gages are used to measure the wall thickness of the tube in a spiral pattern. This test does not cover the tool joints and scans only a limited part (2 to 30 percent) of the tube wall but is extremely effective in detecting wall thinning from wear and pipe eccentricity.

In magnetic flux leakage testing for transverse discontinuities, a coil magnetizes the pipe wall longitudinally to saturation and detector coils or hall effect sensors ride inside the magnetizing coil as close to the tube outside surface as they can be placed. Usually, they are mounted in brass shoes that have the same radius as the pipe and have a layer of tungsten carbide as a wear plate. The axis of the coil is generally perpendicular to the pipe axis, as in Fig. 11. This form of sensor is somewhat tuned to short range magnetic flux leakage and so is effective in reducing longer range noise. It should be noted that magnetic flux leakage induces eddy currents in brass shoes. The signals are filtered and the largest is fed to a chart recorder. This simple system has been effective for many years.

In operation, the inspector first backs the test head as far as it will go onto the pipe tool joint connection area, then reverses the direction and scans the entire pipe to the other upset. Often, signals from the internal and external upsets can interfere with imperfection signals and interpretation is difficult in these areas. Ultrasound represents a better option for end regions.

FIGURE 11. Tube is scanned by ring of transverse sensors: (a) longitudinal section diagram; (b) photograph of equipment.

(a)

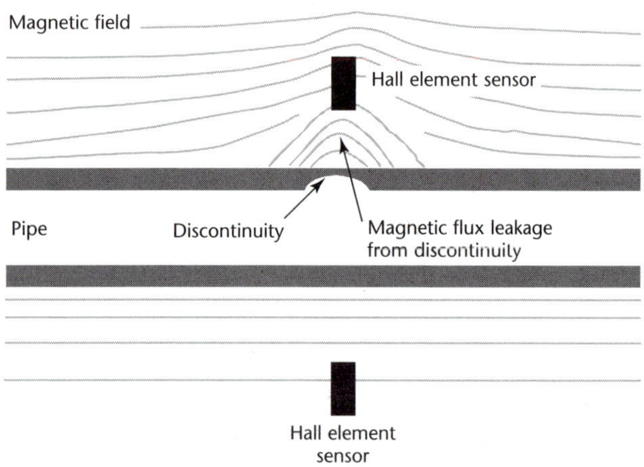

(b)

FIGURE 12. Typical signal for flat, parallel coil.

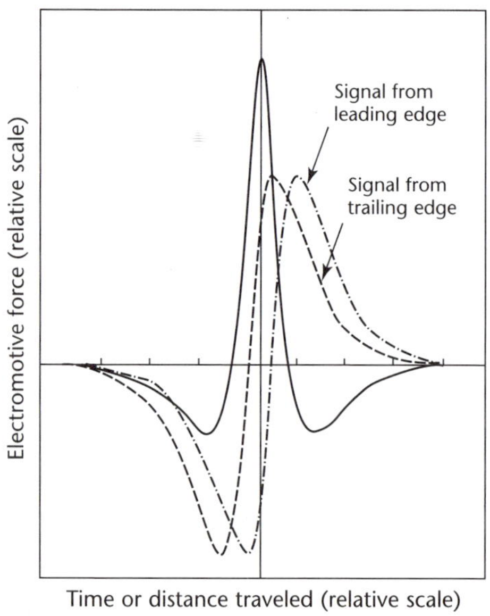

For the longitudinal discontinuity test, two forms of magnetization occur. In one, an internal conductor rod is passed through the pipe, which is then magnetized by one or many shots, often from a capacitor discharge system. It is then scanned by rotating sensors and tested by the resulting residual circular induction. This test is relatively problematic because signals from permeability variations tend to mask serious discontinuities unless signal processing is performed. Of more value is a rotating pole yoke so that the test can be performed in an active induction, where the magnetic flux leakage from imperfections can be many times in magnitude what they were in residual induction.

Magnetic Measurement of Wall Thickness

For the simple drill pipe field test that uses only longitudinal magnetization, one problem has been the lack of a measurement of the wall thickness, especially on worn drill pipe. Erosion on the inside and outside surfaces is a common problem. The tube body wall is measured with an ultrasonic thickness gage at sample points, which may not include the thinnest part of the wall.

No wall thickness measurement is taken by magnetic flux leakage testing, so one improvement effected in the 1990s was the inclusion of an encircling coil inside the magnetizing coil and connected to an integration circuit.

Magnetic wall thickness has also been measured by adding a pickup coil and an integration circuit to measure the total magnetic flux Φ in the magnetizing coil (Fig. 13).[21,22] With the air term (the flux in the air between the coil and the pipe and inside the pipe) subtracted by calibration, the total flux Φ_{steel} in the steel is measured:

FIGURE 13. Encircling coil total flux system for magnetic flux leakage testing of tubes.

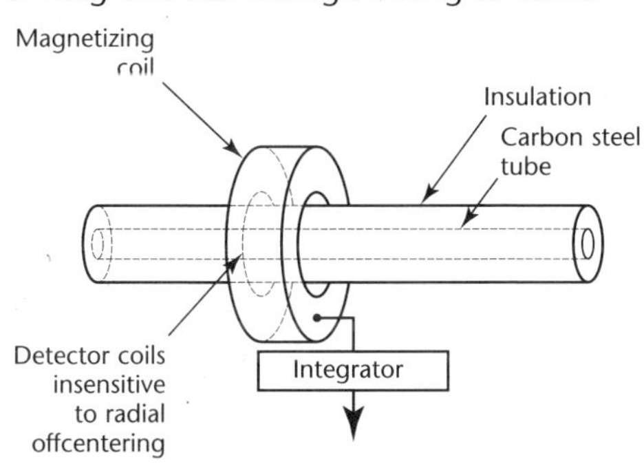

$$\text{(1)} \quad \Phi_{steel} = B_{steel} \times A_{steel}$$

If the longitudinal saturation flux density B_{steel} in the steel is a constant, the cross sectional area A_{steel} of the steel is calculated:

$$\text{(2)} \quad A_{steel} = \pi t_{av} \left(D_{meas} - t_{av} \right)$$

The average wall thickness t_{av} can also be calculated from Eq. 2.

The measured outside diameter D_{meas} of the drill pipe might not include the *minimum* wall thickness in the case of eccentric seamless tubular goods and eccentrically worn tubes. Outside diameter measurement, however, does help to determine wall thickness of drill pipe by a magnetic noncontact technique and is very effective in locating the thinner regions of the pipe. This technique is then used to measure the average wall thickness of oilfield tubing as it is pulled from a well.

An improvement to the outside diameter measurement technique is to add hall sensors to the magnetic flux leakage shoes and measure the tangential magnetic field immediately above the pipe wall. This technique effectively places the encircling coil close to the pipe surface and collects localized signals (Fig. 14).

Tool Joint Testing

Because drill pipe threads suffer tension and consequently often stretch, measurement with a mechanical lead gage is performed to determine the amount of stretch. If the connection is made loose and the pipe is used in deviated wells, there is a possibility of fatigue cracking in

FIGURE 14. Tangential magnetic field immediately above pipe wall is measured by hall sensors in magnetic flux leakage shoes. This technique places encircling coil close to pipe surface to collect localized signals.

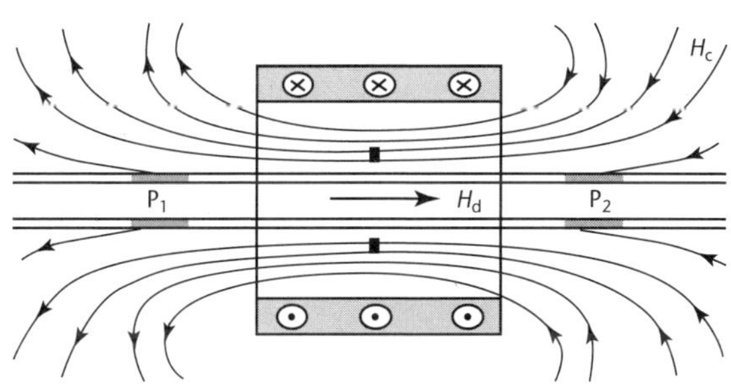

Legend
H_c = magnetic field
H_d = demagnetizing field
P_1 and P_2 = sensor locations

the last engaged thread region. The end of the tool joint can be tested by two techniques.

The tool joint is wrapped in a coil, magnetized and then tested by the wet fluorescent magnetic particle method. Formulas for the magnetization have been given by Moyer.[20] Careful cleaning of the threads is essential before applying the particle suspension. The critical area may be searched for transverse cracks with multiple-transducer ultrasonic systems.[23]

Pipe Threads[24]

An area that requires special attention during the testing of used drill pipe is the threaded region of the pin and box connections (see Fig. 10). Common problems in these regions include fatigue cracking from overtorquing at the pin thread roots and stretching of the thread metal. Automated systems that use both active and residual magnetic flux leakage techniques can be used for detecting such discontinuities.

The stretching and cracking of threads is a common problem. For example, when tubing, casing and drill pipe are overtorqued at the coupling, the threads are in their plastic region. Metallurgical changes in the metal can create regions where stress corrosion cracking takes place in highly stressed areas at a faster rate than in areas of less stress. Couplings between tubes may become highly stressed. Drill pipe threads are a good example of where such stress can cause plastic deformation and thread root cracking.

Analysis of Magnetic Flux Leakage Signals from Coils

In one study,[4] flat coil signals from internal pitting in drill pipe and other oil country tubular goods were analyzed. It was found that the amplitude of the magnetic flux leakage signal, defined as only the upper part of Fig. 12, (1) generally increases as the pit deepens or the remaining wall above it lessens, (2) often decreases as the pit becomes longer and (3) increases as the pit widens. These competing variations lead to scatter diagrams such as the one in Fig. 15, which indicate that trying to assess discontinuity depth from such amplitude based curves is relatively pointless. The plot in Fig. 15 was actually performed for American Petroleum Institute grade J-55 tubing, a carbon steel with 380 MPa (5.5×10^4 lbf·in.$^{-2}$) yield strength.[25] The general trend is for deeper pits to give bigger amplitude signals because deeper pits also tend to be wider. Such spreads have also been reported by others in assessing pipeline pigging signals.[26]

Better assessments are generally made by treating the signal amplitude as the maximum distance between the peak and valley on a signal such as the one in Fig. 12. Such signals generally have one valley deeper than the other because the discontinuity is asymmetrical.

Standardization of Magnetic Flux Leakage Units

Drill pipe test units are often standardized by running the sensors over a 1.6 mm (0.06 in.) diameter through drilled hole in a test standard at the same speed as the pipe to be tested (such coil based signals are speed dependent) and setting the resulting amplitude at some convenient height on a moving chart. For testing of new casing and tubing, electric discharge machined notches are generally used and the test units are expected to show some signal amplitude consistency when the reference indicator is located at twelve, three, six and nine o'clock positions on the pipe. The signals confirm that the pipe is running centrally through the testing unit. Data such as those shown in Fig. 15 indicate the danger of using the amplitude from magnetic flux leakage reference indicators to decide whether to perform further evaluation of the indication.

Coiled Tubing[27]

Coiled tubing and line pipe are made from electric welded carbon steel manufactured in various grades for use in oil and gas well servicing, coiled tubing drilling and installed well tubing. Tubing sizes are typically 25 mm (1 in.) to 89 mm

FIGURE 15. Scatter plot of magnetic flux leakage test signals for 73 mm (2.9 in.) outside diameter pipe.

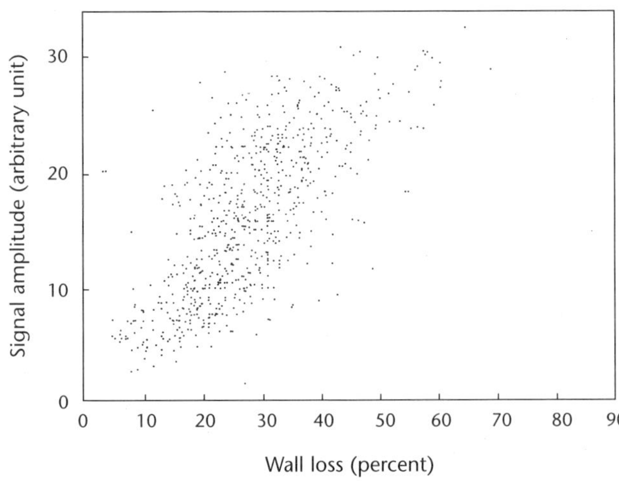

(3.5 in.) outside diameter with wall thicknesses varying from 2 mm (0.08 in.) to 6.3 mm (0.25 in.). Outside diameters of coiled line pipes range from 13 to 165 mm (0.5 in. to 6.5 in.). Strips are spooled onto drums. Strings are made by welding strips together, end to end, and then passing the strip through a high frequency induction electric resistance welding mill. The result is often a coiled string of length 6 to 10 km (4 to 6 mi).

The testing of new coiled product is usually conducted according to American Petroleum Institute specifications.[16,28,29] Conventional nondestructive test methods are used: radiographic, ultrasonic and liquid penetrant testing, as well as electromagnetic techniques. The testing of used coiled tubing is different because the anticipated discontinuities differ from those for new product. However, the equipment for inservice testing follows from the desirability of noncontact testing of wall thickness, ovality, pitting, erosion and other damage. Five types of electromagnetic testing are important for the tubing body.

1. An electromagnetic sensing system for diameter measurement detects ballooning, necking and ovality. Standard eddy current standoff measurement sensors are mounted in a ring to detect changes in outside diameter (Fig. 16). Such ovality measurements are used in collapse pressure calculations. This measurement requires an unpitted outside surface or much compensation and averaging circuitry.

FIGURE 16. Eight eddy current liftoff sensors measure ovality of coiled tubing from fixed distance.

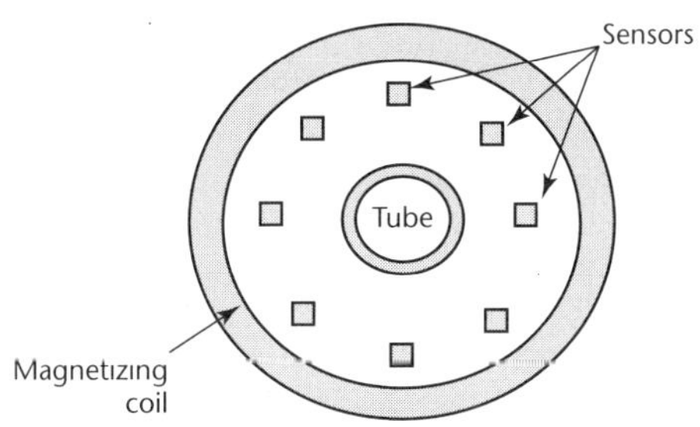

2. A magnetic reluctance wall thickness measurement system enables the thickness of the ambient wall (not localized pits) to be measured along the string, with thin areas caused by erosion, general corrosion and rubbing against the side of the well. These results can then be used for cross sectional area computations and maximum tensile forces for each section of a string. In this method, the field intensity measured with rings of hall effect sensors, placed next to the tube wall, is related to the wall thickness immediately below it as shown in Fig. 14. In principle, the number of poles at locations P_1 and P_2 affects the demagnetizing field H_d, which in turn affects the tangential field in the sensor ring.
3. The same rings of sensors are also used for detection of pitting, gouges and transverse discontinuities by measuring their magnetic flux leakage. Signals from localized pitting can be electronically removed. Sensitivity to small surface imperfections depends on the liftoff from the tubing surface to the electromagnetic center of the sensor.
4. A standard eddy current (3 kHz) system can be used to detect longitudinal discontinuities and areas of heavy cycling in the tubing surface.
5. Because tubing stretches, it is important to know where the highly fatigued areas are, irrespective of where a length indicator says they are. Such areas may be removed if the fatigue life is higher than that of the rest of the tubing. The same eddy currents that respond to damage within the metal through changes in the electrical conductivity are used.

Electromagnetic test results may be confirmed with visual, liquid penetrant, magnetic particle, radiographic or ultrasonic testing.

PART 4. Eddy Current Testing of Offshore Welds[30]

Eddy current testing can be used for manual inservice nondestructive testing of welds in marine environments. Eddy current testing of underwater welds has become common for oil and gas companies in Europe and the United States.

Method Selection

Nondestructive Test Methods

Nondestructive testing methods each have advantages and limitations for detecting various types of weld indications.

1. Magnetic particle testing is used for detecting short length and shallow surface breaking indications. Its sensitivity, however, is reduced in detection of indications through coatings of 0.2 to 0.4 mm (0.008 to 0.016 in.). Magnetic particle testing is difficult to use on wet surfaces.
2. Ultrasonic testing is used for detecting volumetric indications. It is generally not as sensitive as magnetic particle testing for detection of fine, surface breaking indications.
3. Radiographic testing is used for volumetric detection of indications. It cannot, however, detect laminar indications. Radiographic testing requires special safety precautions.
4. Eddy current testing is used for detecting surface breaking indications through coating thicknesses as great as 2 mm (0.08 in.) and can be used on wet surfaces. However, because only the area under the probe is being tested at one moment, several scans must be used for complete coverage.

Eddy Current Testing and Magnetic Particle Testing

Consideration must be given to the component being tested and to the type and size of indication requiring detection — during fabrication, in service or during repair tests. For example, for inservice tests on offshore structures, the predominant indications are surface breaking, mostly in the toe of the weld. Because magnetic particle testing is ideally suited for detection of this type of indication, it has been the method most widely used. Its main drawback, however, is its inability to see through certain coating thicknesses; nor can magnetic particle testing be used on wet surfaces — for example, on surfaces wet from rain.

Eddy current testing has the ability to overcome both of these disadvantages.[31] Magnetic particle testing loses its sensitivity when applied through most coatings,[32] so the coating must be removed and reapplied if magnetic particle testing is to be used. In contrast, eddy current testing can be reliably performed through 2 mm (0.08 in.) of nonconductive coating.

Both wet and dry magnetic particle testing techniques are difficult or impossible to implement in wet or windy environments. Portable eddy current instruments can be placed into lightweight, waterproof enclosures. Eddy current probes are inherently waterproof and can be used on wet surfaces.

Magnetic particle testing is a two-handed operation. This constraint does not matter for most applications but is difficult for projects where the inspector must hold the yoke overhead. In contrast, lightweight eddy current probes can be held for scanning with one hand. Using a lightweight instrument of about 3 kg (6 lb), the eddy current technique is suitable for rope access and for overhead applications (Fig. 17).

Eddy current testing can be used with minimal visibility as in, for example, the underwater testing of jack supports. To verify any eddy current indications, however, visibility must return for magnetic particle testing to be performed (Fig. 18).

Magnetic particle testing produces a residue of particles in the environment. Although particles (wet and dry) may be nontoxic, they may require workers to wear protective equipment to reduce airborne particle inhalation. This may be an important consideration for nuclear applications.

Limitations of Eddy Current Testing

Eddy current testing has distinct limitations compared with other test methods.

1. Compared to other surface breaking indication detection methods (primarily magnetic particle testing), eddy current testing requires a higher inspector skill level for accurate interpretation of signals.
2. Eddy current testing requires the probe to be close to the indication for detection. Specific scanning patterns must be used for the heat affected zone, for the toe of the weld and for weld surface tests. Careful attention must be given to geometry, access and full testing of the part.
3. If equal surface preparation, normal access and a need to test the entire weld (not just one weld toe) are assumed, eddy current testing is slow.
4. Unlike magnetic particle testing, eddy current testing does not produce a visible indication on the test object. Eddy current indications require verification with magnetic particle testing. Typically, the eddy current indication is cleaned to bare metal by using hand tools or a needle gun (an electric, handheld descaling tool) before testing.[33]
5. On extremely corroded, rough surfaces, eddy current test performance is degraded by low ratio of signal to noise.
6. Eddy current testing is not suitable for evaluation of indications by grinding because detection is unreliable for indications that are extremely shallow, less than 0.5 mm (0.02 in.).

FIGURE 17. Rope access for eddy current testing through coatings.

Before being allowed to perform tests, eddy current inspectors should be independently qualified by performing practical demonstrations on test specimens having indications in the range of sizes and geometries of those to be found in the field.

Other Considerations

As part of a joint industry project in the 1990s, a procedure using lightweight commercial eddy current equipment and weld testing probes was developed.[31] Using qualified recommended practices and personnel, results of eddy current testing were found to be in agreement with results of magnetic particle testing.

FIGURE 18. Eddy current testing of welds in marine environment: (a) eddy current scanning through coatings and on damp surfaces; (b) magnetic particle testing verifies eddy current results (arrows point to magnetic particle indication).

(a)

(b)

Speed of Testing

With efficient practices and inspectors, magnetic particle testing is faster than eddy current testing on bare metal. However, operational factors, surface condition and cost are important. Magnetic particle testing works on the assumption that the area between the yoke legs, about 150 mm (6 in.) wide and 75 mm (3 in.) long, is fully tested in one yoke placement. To test a weld completely, the yoke must be placed in two directions. The scanning rate for magnetic particle testing is about 5 $mm \cdot s^{-1}$ (1 $ft \cdot min^{-1}$) for transverse indication scans and about 2.5 $mm \cdot s^{-1}$ (0.5 $ft \cdot min^{-1}$) for indications parallel with the weld, such as toe cracks, centerline cracks and cracks parallel in the base metal.

Eddy current testing interrogates only the area directly under the probe. Five eddy current scans are typically used for weld testing: two for the base metal (parallel and transverse), one specifically targeted for the weld toes and two for the weld face (parallel and transverse). Additionally, rough weld faces typically will decrease scan speed because of increased signal complexity. Eddy current scanning rates vary with the size and profile of the weld face.

Access

Eddy current testing is easier to use from rope access. If most of the testing is in the overhead position, the one-handed eddy current technique is ergonomically easier than magnetic particle testing.

Sensitivity

Of the two methods (magnetic particle testing and eddy current testing), magnetic particle testing has a slightly greater sensitivity for indication detection. One recommended practice[34] gives the sensitivity of magnetic particle testing as 6 mm (0.25 in.) long and 1 mm (0.04 in.) deep. According to a European standard,[35] current testing of welds has a sensitivity of 5 mm (0.2 in.) long and 1 mm (0.06 in.) deep. The slight difference in sensitivity between magnetic particle testing and eddy current testing is sometimes not critical. Eddy current testing should be considered for tests of intact coatings in the following circumstances: (1) where magnetic particle testing would require coating removal and reapplication, (2) for wet or damp surfaces (bare metal or painted) when the surface would have to be dried to perform magnetic particle testing, (3) for operations using rope access and (4) for underwater operations where visibility limits the use of magnetic particle testing.

Conclusion

A combination of eddy current and magnetic particle testing has been successfully used on a number of applications, including the top structural testing of painted offshore oil rigs, large aboveground storage tanks and the testing of painted ship details. The inspector should select the best technique to achieve safety, the required sensitivity and the desired cost effectiveness. Written practices for eddy current weld testing and qualification of inspectors should be specified so that the eddy current test procedures are documented.

References

1. Datt, P.L. and T.R. Schmidt. "Some Applications of Eddy Current Testing in the Petroleum Industry." *Nondestructive Testing.* Vol. 20, No. 5. Columbus, OH: American Society for Nondestructive Testing (September-October 1962): p 335-338.
2. Wehrmeister, A.E. "Eddy Current System for Maintenance Inspection of Chemical Reformer." *Materials Evaluation.* Vol. 31, No. 2. Columbus, OH: American Society for Nondestructive Testing (February 1973): p 21-24.
3. Schmidt, T.R. and M.J. Maulucci. Section 8, "Eddy Current Tests of Ferromagnetic Tubes." *Nondestructive Testing Handbook,* second edition: Vol. 4, *Electromagnetic Testing.* Columbus, OH: American Society for Nondestructive Testing (1986): p 205-215.
4. Stanley, R.K. "Basic Principles of Magnetic Field Flux Leakage Inspection of Oilfield Tubular Materials." *National Fall Conference: American Society for Nondestructive Testing* [Houston, TX]. Columbus, OH: American Society for Nondestructive Testing (1980): p 157-166.
5. Stanley, R.K. "Simple Explanation of the Theory of the Total Magnetic Flux Method for the Measurement of Ferromagnetic Cross Sections." *Materials Evaluation.* Vol. 53, No. 1. Columbus, OH: American Society for Nondestructive Testing (January 1995): p 72-75.
6. Raine, G.A. and N. Smith. "NDT of On and Offshore Oil and Gas Installations Using the Alternating Current Field Measurement (ACFM) Technique." *Materials Evaluation.* Vol. 54, No. 4. Columbus, OH: American Society for Nondestructive Testing (April 1996): p 461, 462, 464, 465.
7. Shannon, R.W.E. and L. Jackson. "Flux Leakage Testing Applied to Operational Pipelines." *Materials Evaluation.* Vol. 46, No. 12. Columbus, OH: American Society for Nondestructive Testing (November 1988): p 1516-1518, 1520-1522, 1524.
8. Mackintosh, D.D., D.L. Atherton, P.C. Porter and A. Teitsma. "Test Rigs for Magnetic Flux Leakage Inspection Tools for Pipelines." *Materials Evaluation.* Vol. 50, No. 1. Columbus, OH: American Society for Nondestructive Testing (January 1992): p 13-17.
9. Beuker, T.M.E., B.W. Brown, A.E. Crouch and M. Afzal. *Advanced Magnetic Flux Leakage Signal Analysis for Detection and Sizing of Pipeline Corrosion: Field Evaluation Program.* Final Report, GRI-00/0109. Chicago, IL: Gas Technology Institute (1999).
10. Amos, D.M. "Magnetic Flux Leakage as Applied to Aboveground Storage Tank Flat Bottom Tank Floor Inspections." *Materials Evaluation.* Vol. 54, No. 1. Columbus, OH: American Society for Nondestructive Testing (January 1996): p 26-28.
11. You, Z. and D. Bauer. "Combining Eddy Current and Magnetic Flux Leakage for Tank Floor Inspection." *Materials Evaluation.* Vol. 52, No. 7. Columbus, OH: American Society for Nondestructive Testing (July 1994): p 816-818.
12. API STD 653, *Tank Inspection, Repair, Alteration, and Reconstruction,* third edition. Washington, DC: American Petroleum Institute (2001).
13. API SPEC 5D, *Specification for Drill Pipe,* fifth edition. Washington, DC: American Petroleum Institute (2001).
14. API SPEC 7, *Specification for Rotary Drill Stem Elements,* fortieth edition. Washington, DC: American Petroleum Institute (2002).
15. API RP 7G, *Recommended Practice for Drill Stem Design and Operating Limits,* sixteenth edition. Washington, DC: American Petroleum Institute (2000).
16. API RP 5A5, *Recommended Practice for Field Inspection of New Casing, Tubing, and Plain End Drill Pipe,* sixth edition. Washington, DC: American Petroleum Institute (1999).
17. API RP 574, *Inspection Practices for Piping System Components,* second edition. Washington, DC: American Petroleum Institute (1998).

18. Moake, G.L. and R.K. Stanley. "Inspecting Oil Country Tubular Goods Using Capacitive Discharge Systems." *Materials Evaluation*. Vol. 41, No. 7. Columbus, OH: American Society for Nondestructive Testing (June 1983): p 779-782.
19. Bray, D.E. and R.K. Stanley. *Nondestructive Evaluation: A Tool in Design, Manufacturing, and Service,* second edition. Boca Raton, FL: CRC Press (1996).
20. Moyer, M.C. "Magnetic Requirements for Oilfield Tubulars." *Materials Evaluation*. Vol. 44, No. 6. Columbus, OH: American Society for Nondestructive Testing (May 1986): p 616-617, 620-624.
21. Bradfield, J.E., J.E. Kahil, M.S. Jaynes, G.L. Moake, M. Milewits, W. Curtis, C.C. Lam and R.K. Stanley. United States Patent 4 710 712, *Method and Apparatus for Measuring Defects in Ferromagnetic Elements* (December 1987).
22. Stanley, R.K. "Magnetic Measurements in NDT." *British Journal of Non-Destructive Testing*. Vol. 36, No. 5. Northampton, United Kingdom: British Institute of Non-Destructive Testing (May 1994): p 360.
23. Stanley, R.K. and L. Wells. "Recent Advances in Used Drill Pipe Inspection by Ultrasonic Method." *Materials Evaluation*. Vol. 52, No. 11. Columbus, OH: American Society for Nondestructive Testing (November 1994): p 1282-1285.
24. Stanley, R.K., T. Hiroshima and M. Mester. Section 24, "Diverted Flux Applications." *Nondestructive Testing Handbook,* second edition: Vol. 4, *Electromagnetic Testing*. Columbus, OH: American Society for Nondestructive Testing (1986): p 633-651.
25. API SPEC 5CT/ISO 11960, *Specification for Casing and Tubing,* seventh edition. Washington, DC: American Petroleum Institute (2002).
26. Cueto, A.G., B.A. Toledo and A.P. Soberanes. "Development of Ultrasonic Techniques to Confirm the Kind of Discontinuities Detected by Automated Inspection Systems (Pig) in Oil and Gas Pipelines in Mexico." Presented at *Second Pan-American Conference for Nondestructive Testing (PACNDT) and ASNT's International Chemical and Petroleum Industry Inspection Technology (ICPIIT) VII Topical Conference* [Houston, TX]. Columbus, OH: American Society for Nondestructive Testing (June 2001).
27. Stanley, R.K. "Testing of Coiled Oilfield Tubing — An Update." *Materials Evaluation*. Vol. 58, No. 8. Columbus, OH: American Society for Nondestructive Testing (August 2000): p 970-975.
28. API RP 5C7, *Recommended Practice for Coiled Tubing Operations in Oil and Gas Well Services*, first edition. Washington, DC: American Petroleum Institute (1996).
29. API SPEC 5LCP, *Specification for Coiled Line Pipe*. Washington, DC: American Petroleum Institute (1999).
30. Goldberg, L.O. "Eddy Current Testing: An Emerging NDT Method for Ferritic Weld Inspection." *Materials Evaluation*. Vol. 56, No. 2. Columbus, OH: American Society for Nondestructive Testing (February 1998): p 149-152.
31. *Ferritic Weld Inspection Using Eddy Current*. Report. Merritt Island, FL: Sea Test Services (1996).
32. NP-5951, *Reliability of Magnetic Particle Inspection Performed through Coatings*. Palo Alto, CA: Electric Power Research Institute (July 1988).
33. *Project #90-61 RP for Underwater Magnetic Particle Inservice Weld Inspection of Offshore Fixed Platforms and Guidelines for Qualification of Magnetic Particle Inspector Divers*. Research Report. Washington, DC: American Petroleum Institute (1991).
34. API RP 2X, *Recommended Practice for Ultrasonic and Magnetic Examination of Offshore Structural Fabrication and Guidelines for Qualification of Technicians*. Washington, DC: American Petroleum Institute (1996).
35. BSEN 1711, *Non-Destructive Examination of Welds — Eddy Current Examination of Welds by Complex Plane Analysis*. Brussels, Belgium: European Committee for Standardization (2000).

16

CHAPTER

Electric Power Applications of Electromagnetic Testing

Jack C. Spanner, Jr., Electric Power Research Institute, Charlotte, North Carolina (Parts 1 and 3)

Thomas U. Bipes, Progress Energy, Raleigh, North Carolina (Part 3)

Anmol S. Birring, NDE Associates, Webster, Texas (Part 2)

Gary P. Rogles, AmerenUE, Saint Louis, Missouri (Part 4)

Lawrence O. Goldberg, Sea Test Services, Merritt Island, Florida (Part 5)

Part 1. Introduction to Electromagnetic Testing of Electric Power Components

Components that are electromagnetically tested in nuclear and fossil fueled power plants include main steam heat exchangers, condensers, feedwater heaters, lubrication oil coolers, boiler tubes, instrumentation tubing and main generator hydrogen coolers. Surfaces are electromagnetically tested to detect cracking in welds, coated surfaces and other places difficult to test with liquid penetrants.

The most common use of electromagnetic testing in the power industry is tube testing in heat exchangers. Tubes in a nuclear steam generator are important for safety because they carry the primary cooling water.

Steam Generators

The inspection of the steam generator tubing is an important task during an outage, so a variety of strategies are used to reduce the electromagnetic test time, including the simultaneous use of two robotic probe positioners in each steam generator and over 30 data analysts working in shifts around the clock. The requirements for examination plans and processes to maintain the steam generators are provided in guidelines prepared by the industry.[1]

Degradation of steam generator tubing results from corrosion and mechanical mechanisms: intergranular corrosion, stress corrosion cracking, thinning, pitting, denting, mechanical wear, impingement and fatigue. Most tubes are fabricated from iron nickel chromium and nickel chromium alloys (Unified Numbering System N06600, N06690 and N08800).

For electromagnetic testing, multifrequency bobbin coil techniques are used to test the full length of the tubes. It is recognized that the bobbin coil is not qualified for all tube locations, so other techniques such as rotating coil technology, array probes and ultrasonic search units are also used. The U bend, tube sheet and expanded tube regions are difficult to test; both differential and absolute coil techniques are used.

The test program for steam generators is based on performance. The techniques must be demonstrated to have a probability of detection of 80 percent at a 90 percent confidence level for each damage mechanism. To meet this qualification requirement, all the rejectable grading units must be detected in data sets that contain a minimum of 11 to 17 rejectable grading units. One miss is allowed for data sets containing 18 to 24 rejectable grading units and two misses are allowed for sets of 25 to 31

Table 1. Heat exchanger materials and damage mechanisms.

Heat Exchanger	Tube Material	Potential Damage Mechanisms
Feedwater heat exchanger	carbon steel nickel copper alloy stainless steel	axial cracking circumferential cracking condensate grooving loose part damage pitting roll transition cracking steam erosion vibrational wear
Generator stator cooling system	alloy of 90 percent copper and 10 percent nickel	inside surface pitting
Generator hydrogen cooler	aluminum brass; alloy of 90 percent copper and 10 percent nickel	inside surface pitting and outside surface damage
Lubricating oil cooler	admiralty brass	inside surface pitting
Main condenser	stainless steel copper nickel alloys; admiralty brass; titanium	inside surface pitting and vibrational wear

rejectable grading units. Sizing accuracy is based on correlation of the eddy current parameters and on a standard error of regression at an interval where there is a 90 percent probability that discontinuity sizing is accurate to within 50 percent of discontinuity depth. The demonstration results and techniques are peer reviewed by a panel of experts.

The industry guidelines specify a number of personnel qualification requirements.[1] Before becoming qualified data analysts, personnel must be certified as a Level II or Level III (as defined in the guideline) and attend additional data analysis training. They must pass written and practical examinations and attend annual training. In addition, utilities have established a site specific performance demonstration program for all prospective analysts. To prepare for the next steam generator inspection, personnel should attend training that will provide practice analyzing indications of interest for that site. A site specific practical examination must be passed for each planned outage and written examinations may be required. The sampling plan for the inspection may be prescriptive or performance based: inspections and additions to the test object sets are partly determined by previous inspections. The inspection results and operating history of other power plants are also considered.

Balance-of-Plant Heat Exchangers

In a nuclear power plant, heat exchangers that do not carry reactor cooling water from the containment vessel are commonly referred to as *balance-of-plant heat exchangers.* Most of them have counterparts in fossil fuel power plants and are tested with similar techniques. Some of the tubing materials are ferromagnetic, so techniques other than conventional eddy current are required. These other techniques include magnetically biased eddy current, remote field eddy current and magnetic flux leakage testing. Table 1 lists routinely inspected heat exchangers, potential damage mechanisms and the tubing material.

Nonferromagnetic tubing is usually tested with multifrequency eddy current techniques in the differential and absolute modes using bobbin coil probes. In most cases, phase angle analysis is used and relies on a flat bottom hole reference standard. During calibration, the operator may decide to set the 100 percent hole response at about 40 degrees. As the outside surface discontinuities get deeper, from 20 percent to 100 percent through the wall, the eddy current lissajous figures rotate counterclockwise. The changing phase angles can be correlated to discontinuity depth and used to establish a calibration curve of phase angle to discontinuity depth (Fig. 1). However, this setup is not ideal for measuring depths of inside surface discontinuities, such as pits.

Amplitude signals are directly proportional to the volume of wall loss.[2] As a consequence, amplitude analysis is usually recommended for volumetric wall losses such as fretting wear at tube support intersections, inside surface pitting and erosion. Sometimes, it is advantageous to use calibration reference standards with reference discontinuities that look similar to the discontinuities of interest. Tapered elongated notches can be machined to replicate wear damage at a support plate, for instance.

FIGURE 1. Curves obtained with calibration reference tube: (a) calibration curve; (b) depth curve. Percentages express ratio of discontinuity depth to total wall thickness.

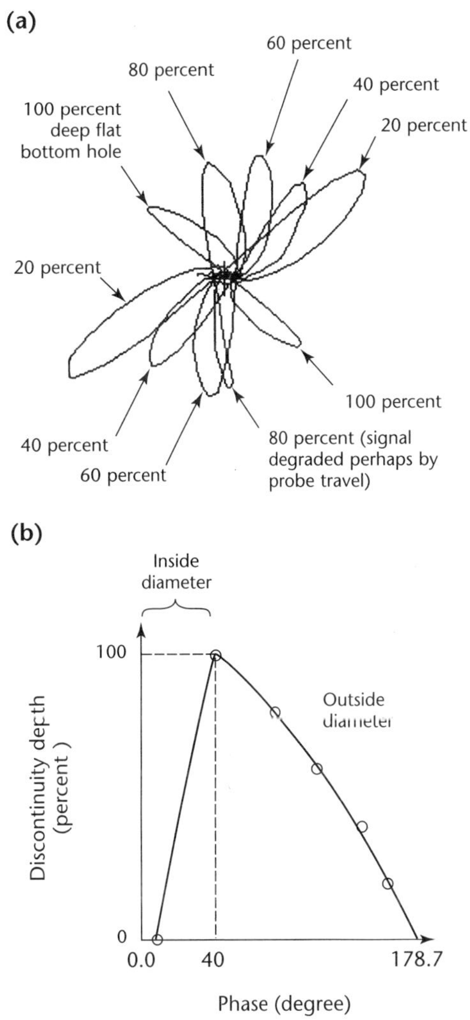

Surface Tests

Surface tests are also conducted using electromagnetic techniques to detect cracking in a variety of components that are difficult to test with other surface methods such as liquid penetrant, magnetic particle and visual testing. Eddy current techniques tend to be used more on nonferromagnetic materials or with a remote technique, such as automated ultrasonic testing. Components typically tested include turbine rotors, turbine blades, socket welds, expansion joint bellows, coated surfaces, control rod drive mechanisms and stainless steel pipe susceptible to corrosion cracking of the outside surface. These applications typically require flat (pancake) coils or flexible array coils. The coils are usually operated in the absolute mode, in either a cross wound or a transmit/receive configuration. Depth sizing is inaccurate in some surface tests.

PART 2. Electromagnetic Techniques for Heat Exchanger Tubing[3]

A variety of corrosion resistant heat exchanger tube materials is encountered in power and petrochemical plants. These include stainless steel, admiralty brass, nickel alloys, titanium, nickel copper alloys, nickel, duplex steel, ferritic stainless steel and carbon steel. The tubes are inspected periodically to locate and size discontinuities such as pits, cracks, erosion and wear. Selection of a nondestructive test technique depends on the tube material and on the type of discontinuity expected in the tubing.

Nondestructive test techniques for tubes include conventional eddy current, full saturation eddy current, remote field eddy current, magnetic flux leakage, internal rotary ultrasonic testing and laser profilometry.

No single nondestructive test technique can be applied to all of the tube materials. Each nondestructive test technique has advantages and limitations. For example, conventional eddy current testing is very sensitive to pits and cracks but its application is limited to nonferromagnetic materials. Internal rotary ultrasonic testing is accurate in measuring wall thickness but will miss small discontinuities such as pin holes and cracks. Optical techniques are limited to inside surface discontinuities. Proper selection of nondestructive test techniques is therefore a key to testing heat exchangers.

Technique Selection

Several nondestructive test techniques for testing heat exchanger tubing are discussed below. The selection of technique depends on the material and type of discontinuity. Table 2 lists tube materials and the recommended techniques.

Eddy Current Testing

The conventional eddy current technique is based on measuring the impedance of the test coil.[4] The impedance of the coil

TABLE 2. Selection of nondestructive test techniques for tube testing.

Discontinuity Type	Technique	Comments
Nonferromagnetic Metals		
Pits, cracks, wall loss	conventional eddy current	reliable for discontinuity detection; accurate for pit sizing; maximum speed 1.8 $m \cdot s^{-1}$ (6 $ft \cdot s^{-1}$)
Cracks in finned tubes	multiple-coil pancake probe	detects cracks greater than 50 percent deep
Cracks in tubesheet expansion zone	rotating pancake probe	slow but reliable
	multichannel transmit/receive	fast; special probe required
Partial and Thin Ferromagnetic Metals[a]		
Pits, cracks, wall loss	full saturation eddy current	same as conventional eddy current; inside surface pit sizing inaccurate; maximum speed 1.8 $m \cdot s^{-1}$ (6 $ft \cdot s^{-1}$)
Thick Ferromagnetic Metals (Carbon Steel)		
Wall loss	remote field eddy current	only for wall loss, limited pit sizing accuracy; maximum speed 0.25 $m \cdot s^{-1}$ (0.9 $ft \cdot s^{-1}$)
	partial saturation eddy current	only for wall loss, limited accuracy for pits and pin holes
Pits and wall loss	magnetic flux leakage	cannot size discontinuities; more sensitive to inside surface discontinuities; rust on inside surface affects sensitivity; maximum speed 0.9 $m \cdot s^{-1}$ (3 $ft \cdot s^{-1}$)
All Materials		
Wall loss and large pits	internal rotary ultrasonic testing	accurate for thickness measurement; slow; requires clean tubes; cannot detect cracks; may miss pin holes and small pits; maximum speed 0.075 $m \cdot s^{-1}$ (0.25 $ft \cdot s^{-1}$); minimum thickness about 0.9 mm (0.035 in.)
Inside surface	laser profilometry	limited to inside surface testing; requires clean tubes; maximum speed 0.075 $m \cdot s^{-1}$ (0.25 $ft \cdot s^{-1}$)

a. Unified Numbering System S31803, S44627 and S44736 duplex stainless steels.

changes as the electromagnetic field interacts with the material being tested. Initially, the coil is placed in the tube and the instrument is calibrated on a reference standard having known, machined discontinuities. The probe is pulled through the tube and variations in coil impedance are recorded. These impedance changes are related to types and sizes of discontinuities. Conventional eddy current testing is fast and can be conducted at speeds up to 1.8 $m \cdot s^{-1}$ (6 $ft \cdot s^{-1}$).

Conventional eddy current testing is limited to nonferromagnetic materials: stainless steel, admiralty brass, copper nickel alloys, titanium, nickel alloy and others. The conventional eddy current test can be performed in either of two modes: differential or absolute. Differential mode detects small discontinuities such as pitting and cracking whereas absolute mode detects wall loss. A mixed channel detects discontinuities under supports — the mixed channel is a sum of two frequency channels added to cancel out the support plate signal (Fig. 2).

Discontinuities are measured as percentages of wall thickness and are detected with a frequency that produces a phase spread of 180 degrees between the 0 percent inside diameter and 0 percent outside diameter. Calibration is first performed on a tube machined according to the applicable code.[5] The phase of the signal from the 100 percent calibration hole is set to 40 degrees. Signals that fall below 40 degrees represent inside diameter discontinuities whereas signals greater than 40 degrees represent outside diameter discontinuities (Fig. 1).

FIGURE 2. Eddy current test display, showing detection of discontinuity under support plate by using digitally mixed channel.

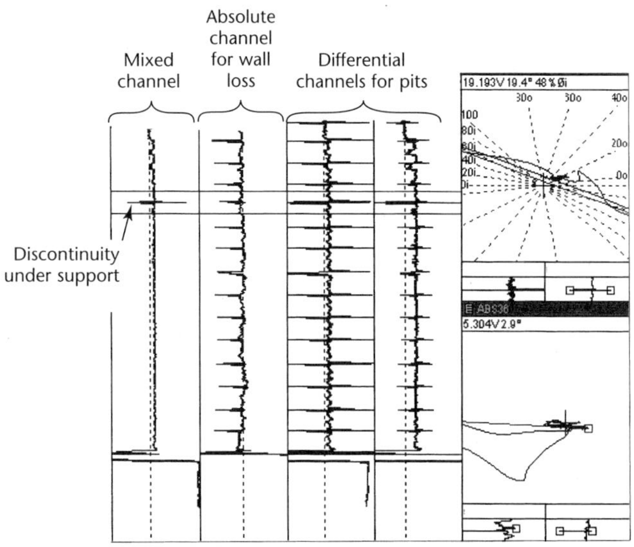

The interpretation of discontinuities is not simple: other anomalies can produce signals that also look like discontinuity indications. One such case is ferrite deposits in stainless steel tubes. These ferrite deposits produce signals representative of inside surface discontinuities. The phase of these signals is in the 20 to 30 degree range when the frequency for 180 degree phase is used. This range represents a 50 to 75 percent deep inside surface discontinuity. The discrimination of metallic deposits from inside surface pits requires additional frequencies. Data should be taken at four frequencies and the signals checked for consistency. If the signal is from a discontinuity, then all signals will correspond to either an inside or outside surface discontinuity. Mismatches among the four frequencies will occur when the signal is from a metallic deposit. Careful selection of the four frequencies is vital for this test.

Sizing of the discontinuities is performed using a depth curve generated on an eddy current calibration tube. Typically, this tube has five pit type discontinuities and two grooves, including one 100 percent hole and outside surface flat bottom holes with depths of 80, 60, 40 and 20 percent of the tube wall thickness (Fig. 1). A depth curve is generated by taking measurements on the five outside surface pits. Because there are no inside surface pits in the reference standard, a straight line is usually drawn between the 100 percent hole at 40 degrees and 0 percent at 0 degrees.

Inaccuracies in sizing result from two causes, the first being that the depth curve does not account for the effect of discontinuity diameter on the phase. An increase in diameter reduces the phase angle of the discontinuity. It is therefore not uncommon to obtain a phase angle of 50 to 60 degrees from pinholes in thin titanium tubes and 25 to 30 degrees from large diameter holes in admiralty brass tubes. Errors in sizing can result if only the phase is used for sizing. These errors can be reduced if both amplitude and phase are considered for sizing. The second reason for sizing inaccuracy is connecting the 100 percent hole to 0 percent (at 0 degrees) by a straight line. This error is significantly influenced by the frequency. The 40 degree spread reduces with reduction in frequency. The error in sizing inside surface pits can be minimized by proper frequency selection.

Rotating and Multicoil Eddy Current Probes

Special eddy current probes are used for applications where the bobbin coils can miss discontinuities. These may include detection of circumferential cracks in finned tubes and cracking next to the tube sheet. Bobbin coils cannot detect circumferential cracks in finned tubes because the eddy currents are also in the circumferential direction and parallel to the crack. In such a case, multichannel pancake coil probes should be used. A typical probe would have four to eight pancake coils placed around the circumference.

Another application where bobbin coils cannot be used is cracking near the tube sheet. This area is also referred to as the *expansion zone*. Simple multichannel coils can be used but the liftoff caused by the expansion zone produces a large signal that can mask the crack signal. In such a case, there are two options: to use a motorized rotating pancake coil probe or to use a multichannel send/receive probe. The former contains a single pancake coil that is rotated as the probe is pulled slowly through the testing area. This probe is very sensitive to cracking in the transition zone. One multichannel transmit/receive probe design is called the *cecco probe*,[6] in which the transmit/receive coils are placed around the circumference of the probe. This transmit/receive design is less sensitive to liftoff than a coil operated in impedance mode but can detect smaller discontinuities than the differential multichannel probe can. Testing with the multichannel transmit/receive coil is significantly faster than with the motorized rotating pancake coil probe.

Full Saturation Eddy Current Technique

The principle of full saturation eddy current is the same as that of conventional eddy current. The technique is applicable to partially ferromagnetic materials such as nickel copper alloy, Unified Numbering System S31803 duplex (austenitic ferritic) chromium nickel molybdenum stainless steels and thin ferromagnetic materials such as Unified Numbering System S44627 or S44736 ferritic chromium molybdenum stainless steel. The full saturation probe contains a conventional eddy current coil and a magnet. The magnetic field of the magnet saturates the material. Once saturated, the relative permeability of the material drops to one and the principles of conventional eddy current testing are then applicable.

The main problem with this technique is ensuring that the material has been fully saturated. Saturation can be confirmed by running a test on a calibration tube. A fully saturated tube produces a normal phase spread on outside surface discontinuities in the calibration tube. It is therefore very important to machine the calibration tube from the same material as that of the test object. Consistency in material chemistry is especially important for nickel copper alloys whose permeability can vary from batch to batch. In addition, the strength of the magnets in a full saturation probe can vary from vendor to vendor. Weaker magnets will not saturate the material and will produce a noisy signal. It is imperative to check the quality of the probe before using it. The application of a full saturation eddy current technique depends on the permeability of the material, tube thickness and tube diameter. Larger diameter tubes will allow placement of larger magnets, whereby slightly thicker tubes could be saturated.

Sizing of outside surface discontinuities is done in a similar manner as in conventional eddy current testing. Inside surface discontinuities cannot be sized with signal phase because the depth of the discontinuity does not influence the phase. The sizing of inside surface pits is, therefore, done purely on the basis of signal amplitude.

Partial Saturation Eddy Current Technique

The partial saturation technique is applied on ferromagnetic tubes too thick to be fully saturated. A conventional eddy current instrument monitors variations in impedance caused by changes in permeability. Permeability measurements vary with thickness. A loss of thickness increases the intensity of the magnetic field and hence reduces the permeability. The reduction in permeability affects the coil impedance, measured by the eddy current instrument.

Because this technique depends on gross changes in permeability, it is limited to detecting variations in wall loss. Because the magnetic field becomes weaker as the tube becomes thicker, this technique is insensitive to small discontinuities such as pits that have minimal influence on the total magnetic field. The technique is also limited to a maximum thickness of about 2.5 mm (0.1 in.) in a carbon steel tube measuring 25 mm (1 in.) in outside diameter.

Remote Field Testing

The remote field test technique is based on the transmission of an electromagnetic field through the tube material.[7] The exciter coil generates eddy currents at a low frequency in the circumferential direction. The electromagnetic field transmits through the thickness and travels on the outside surface. A receiver coil placed in the remote field zone of the exciter picks up this field. In this zone, the wall current source dominates the primary field directly from the exciter. The separation between the two coils is two to four times the tube's inside diameter.

The remote field technique is effective in the testing of carbon steel tubes and pipes, especially for large area discontinuities such as steam erosion and baffle wear. For pit sizing, remote field results can be affected by scale in the tube, causing small jerks when probes get momentarily stuck. Accurate sizing of a pit depends on the volume of metal loss at the pit.

The test speed with remote field testing is lower than with conventional eddy current testing. Conventional eddy current can easily be performed at speeds of 1.8 m·s^{-1} (6 ft·s^{-1}) whereas remote field testing takes place at about 0.25 m·s^{-1} (10 in.·s^{-1}).

Discontinuity sizing with remote field eddy current is done using the voltage plane curves (Fig. 3). These curves measure wall loss. The curves relate discontinuity depth, discontinuity length and discontinuity circumference to the phase of the remote field signal. Inaccuracies result because encountered discontinuities differ in geometry from calibration discontinuities. Internal rotary ultrasonic testing can be used to verify the remote field eddy current measurements.

FIGURE 3. Voltage plane curves for remote field sizing of wall loss in carbon steel tubes. Curved lines show circumferential extent of discontinuity; straight lines show discontinuity depth.

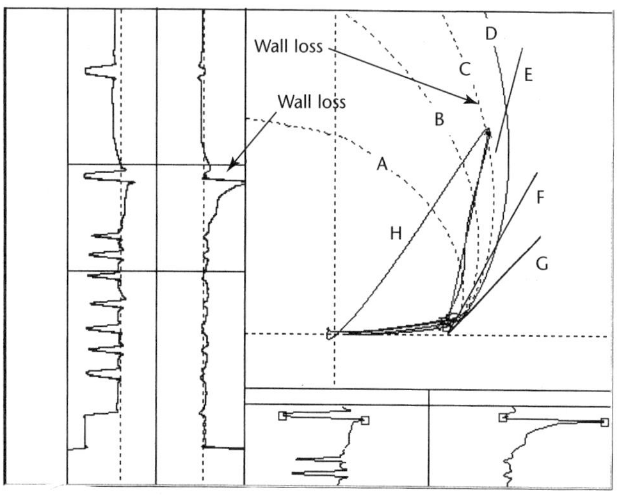

Legend

A. Curve for 90 degree extent.
B. Curve for 180 degree extent.
C. Curve for 270 degree extent.
D. Curve for 360 degree extent.
E. Line for 40 percent depth.
F. Line for 20 percent depth.
G. Line for 0 percent depth.
H. Voltage plane curve (loop).

Magnetic Flux Leakage Testing

Magnetic flux leakage testing is based on the influence of discontinuities on a magnetic field. The technique is limited to ferromagnetic materials. The magnetic flux leakage probe consists of a magnet with two types of magnetic pickups: coil and hall element. The coil picks up the flux rate of change whereas the hall effect detector picks up absolute flux. The coil detects small discontinuities that cause perturbations in the flux. Following Faraday's law, the flux rate of change induces an output voltage read by the magnetic flux leakage instrument. Because the output voltage is directly proportional to the flux rate of change, a constant pull speed is required. Sudden changes in speed induce electromagnetic voltage that can be misinterpreted as discontinuities. Magnetic flux leakage coils are more sensitive to sharp discontinuities than to gradual wall loss. In fact, the coils can totally miss long areas of wall loss if the changes in wall thickness are gradual. A hall effect detector is therefore used to detect gradual wall loss.

The output of the magnetic flux leakage coils is related to the change of flux caused by the discontinuity but not to the discontinuity size. The technique therefore cannot size discontinuities. A small diameter, 25 percent pit will produce a larger signal output than a large diameter 75 percent deep discontinuity with a gradual change in depth. In addition, rust at the inside surface will also produce signal noise that can overshadow discontinuity signals.

Hall effect detectors measure the absolute flux and can be used for sizing wall loss, not pits. But the output of the hall effect detector depends on the orientation of the sensor in the probe relative to the discontinuity and whether the location of the discontinuity is on the inside or outside surface. Inside surface discontinuities will produce larger signals than outside surface discontinuities because the field strength on the inside surface is higher than on the outside surface.

Nonelectromagnetic Techniques

Laser Profilometry. Laser profilometry uses a rotating laser beam that scans the inside surface as the probe is pulled out of the tube. The reflected beam is picked up by a lateral detector that measures changes in proximity caused by variations on the inside diameter. The information received by the detector is processed to create an image of the inside surface. Limited to inside surface testing with a speed of up to 75 mm·s^{-1} (15 ft·min^{-1}), the technique also requires the tube to be cleaned to minimize optical scattering. Because of the slow speed and cleaning requirements, the technique is used not alone but to complement other techniques such as eddy current testing.

Internal Rotary Ultrasonic Testing. The internal rotary technique of ultrasonic testing measures thickness with ultrasound. The probe is pulled through the tube and ultrasonic data are evaluated in a C-scan display of the tube's cross section. The rotary ultrasonic test technique is used mostly for carbon steel tubes and is used sometimes for nonferromagnetic tubes to validate or quantify results of other techniques. The technique is accurate for thickness measurement as well as detecting inside and outside surface pits. Rotary ultrasonic testing will, however, miss pinholes and cracks. The technique is slow, with test speeds limited to about 75 mm·s^{-1} (15 ft·min^{-1}). Because of the inability of maintaining water coupling in the entire tube length, the technique does not result in 100 percent coverage.

Technique Selection

Which technique is best depends on the tube material

Nonferromagnetic Tubes

Conventional eddy current is the most suitable technique for the testing of nonferromagnetic tubes. These include stainless steel, admiralty brass, copper nickel alloys and titanium. The testing can be done with regular bobbin probes for the tube length and under the supports. Special probes are required for detection of circumferential cracks in finned tubes and cracks next to the tube sheet.

Partially Ferromagnetic and Thin Ferromagnetic Tubes

Thin and partially ferromagnetic materials include nickel copper alloy, duplex stainless steel (Unified Numbering System S31803) and thin ferritic stainless steel (Unified Numbering System S44627 and S447366). Full saturation eddy current can be applied if full saturation can be demonstrated in the calibration tube. There is, however, a maximum thickness limit for application of this technique. The thickness limit depends on the tube diameter. In general, for 19 mm (0.75 in.) diameter and 25 mm (1 in.) diameter tubes, the thickness limit is about 1.7 mm (0.07 in.) for partial ferromagnetic materials and 0.8 mm (0.03 in.) for ferritic stainless steel.

Ferromagnetic Tubes

Carbon steel tubes, with and without aluminum fins, are classified as ferromagnetic. Three electromagnetic techniques that can be applied for these tubes are remote field, partial saturation eddy current and magnetic flux leakage testing. The first two techniques are limited to detecting wall loss and will miss small pits. Magnetic flux leakage testing is sensitive to both pitting and wall loss but cannot measure discontinuity depth. The sensitivity of the flux leakage technique is impaired by noise signals produced from oxides (rust) in the tube's inside surface. Internal rotary ultrasonic testing is the most reliable technique for the testing of ferromagnetic tubes.

Reliability of Techniques

A study was conducted to assess the reliability of nondestructive test techniques for heat exchanger tubes.[8] The study was done on samples with both service induced and artificial discontinuities. Discontinuities were representative of corrosion and mechanical wear initiating on the inside and outside surfaces of tubes.

Tube Material

Table 3 shows the detection performance for three techniques. The table shows the high reliability of eddy current on nonferromagnetic tubes and the high reliability of rotary ultrasonic testing on carbon steel tubes. However, ultrasonic method reliability drops to 28 percent on

TABLE 3. Discontinuity detection performance by tube material and test technique.

Technique	Carbon Steel	Stainless Steel[a]	Titanium	Copper Nickel[b]	Admiralty Brass
Eddy current testing	—	91 percent	98 percent	91 percent	92 percent
Remote field testing	77 percent	—	—	—	—
Ultrasonic testing	83 percent	28 percent	68 percent	—	—

a. Unified Numbering System 30400 austenitic chrome nickel stainless steel.
b. 90 percent copper, 10 percent nickel alloy.

TABLE 4. Discontinuity detection performance by two inspectors using eddy current testing.

Test	Stainless Steel[a]	Titanium	Copper Nickel[b]	Admiralty Brass
Inspector 1	91 percent	98 percent	91 percent	92 percent
Inspector 2	58 percent	52 percent	83 percent	89 percent

a. Unified Numbering System 30400 austenitic chrome nickel stainless steel.
b. 90 percent copper, 10 percent nickel alloy.

stainless steel because ultrasonic scanning is insensitive to small pits and cracks.

Inspector Training

Variation in reliability is caused by differences among test inspectors. Table 4 shows results of conventional eddy current tests performed by two different inspectors. In the case of stainless steel, discontinuity detection dropped from 91 percent for one inspector to 58 percent for the other. These results show the importance of training for eddy current inspectors. The training should be done on samples with service induced discontinuities and each inspector's performance with samples should be evaluated.

Conclusions

No single nondestructive test technique can be applied for testing all heat exchanger tubing materials: several techniques are needed. Inspectors should be aware of the limitations of each technique. Selection of the proper technique makes reliable tests and accurate sizing possible.

In addition, technicians performing the test should be properly trained on representative samples. The training should also include education on distinguishing false calls from discontinuities and knowledge of factors that can affect discontinuity sizing.

PART 3. Eddy Current Detection of Forging Laps in Pressurized Water Reactor Tubing

Testing before Outages

During every refueling outage, utilities are required to test a sampling of the high temperature, nickel chromium alloy tubing in their steam generators. The tubing is susceptible to eight damage mechanisms: intergranular corrosion, stress corrosion cracking, thinning, pitting, denting, mechanical wear, impingement and fatigue.

Before their installation into the steam generator, tubes are given a baseline test. After installation, they are tested again during preservice tests before initial operation of the system to ensure that the tubing is free of discontinuities. Innocuous manufacturing discontinuities are accurately located and characterized before the steam generator begins service. This information is necessary for inservice tests to ensure that eddy current indications are properly evaluated.

Many significant improvements have been made in the testing of steam generator tubes. As a result, manufacturing discontinuities that in the twentieth century would not have been reported before plant operation are detected and reported during preservice tests. A study conducted to evaluate the detectability of these discontinuities and their impact on structural integrity is described below.[9]

FIGURE 4. Photomicrograph of typical lap in high temperature nickel chromium alloy tubing.

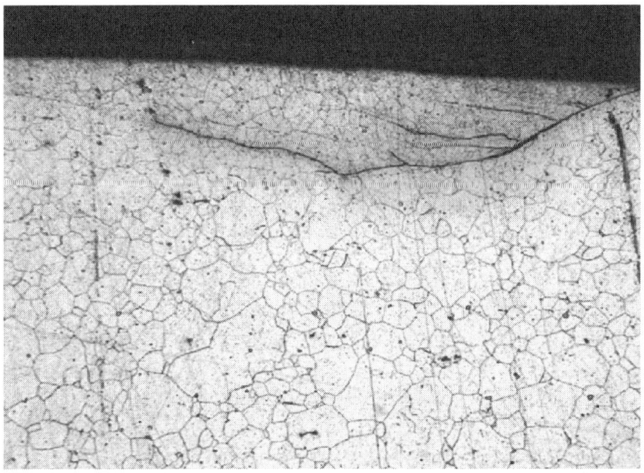

200 µm (0.008 in.)

Forging Laps

One type of manufacturing discontinuity that has been detected in steam generator tubing is called a *lap* or *forging lap* (Fig. 4). A lap is produced when an imperfection on the tube surface is folded during the forming process. It is generally oriented parallel to the surface and not perpendicular to the primary stresses.

In the study, the detection of laps during manufacturing was compared with their detection during service. Tests at the tube mill are generally performed using ultrasonic equipment and electromagnetic encircling coils. A review of the manufacturer's test reports showed that the laps were detected but were below the reporting level. Multifrequency bobbin coil techniques using the latest digital equipment detected the laps during preservice or inservice tests.

Signal Interpretation

A study was conducted including nondestructive testing and metallurgical analysis of the indications. Two tubes with laps were pressurized until they leaked. The burst pressure was typical for new tubing, indicating that laps do not affect the structural integrity of the tubing.

A lap signal was detected with a bobbin coil using multifrequency techniques. The signal in Fig. 5 is the response from a lap at 300 kHz, about half the prime frequency of 630 kHz. The indication was confirmed with a rotating coil probe also operating at 300 kHz as shown in Fig. 6.

Figure 5 is a typical screen presentation for multifrequency bobbin coil analysis and is divided into three major areas. The left side shows strip chart data from the entire length of the tube. The chart signal on the left is the high frequency (630 kHz) differential vertical signal that includes data from the full length of the U bend tube. The probe entry and exit from the tube are visible at the bottom and top of the signal, respectively.

The chart signal second from the left is the low frequency (150 kHz) absolute vertical signal. The horizontal excursions locate the nine support plates that the probe detects as it is pulled from the hot leg tube end, through the hot leg tube sheet, past the flow baffle hot leg, past the first support plates (01H to 09H), over the U bend region where the antivibration bars (AV1 to AV8) are located, past the cold leg side of the support plates (09C to 01C) and out the cold leg tube end. During calibration, the 150 kHz absolute phase angle has been adjusted so that metal losses make excursions to the right and metal gains to the left. The offset in the data in the U bend region is typically a result of the thermal stress relief the U bends receive or may be caused by liftoff and the probes' not being centered.

In the upper right part of Fig. 5, the lissajous signal is 631.2 mm (24.85 in.) above the seventh support plate (07H). This signal is selected for display by scrolling the chart's horizontal line through the tube data; the line is between 07H and 08H in the strip chart on the left part of Fig. 5. The signal is from the 300 kHz differential output. It is noted on the screen that the signal is interpreted as a nonquantifiable indication (NQI) detected with the half frequency of 300 kHz.

The lower right part of Fig. 5 is an expanded version of the data selected for display with the horizontal scroll line. It shows strip chart data — the vertical output on the left and the horizontal on the right. Combining the horizontal and vertical signals results in the lissajous figure, which also provides phase angle information.

Bobbin coil techniques perform well to detect various manufacturing imperfections. Careful analysis of the data is necessary to avoid making false positive calls, that is, to avoid rejecting good material. Experience has shown that, particularly for volumetric discontinuities, depth sizing by using the amplitude information tends to be more accurate than phase information alone. The laps were evaluated to be insignificant in terms of operational and structural concerns. The lap signals should be tracked in future outages and monitored for changes. This practice is normal for benign indications detected during baseline tests. Such signals should be included in a site specific eddy current training session by the data analysts immediately before each outage.

FIGURE 5. Lap indication with bobbin coil at 300 kHz.

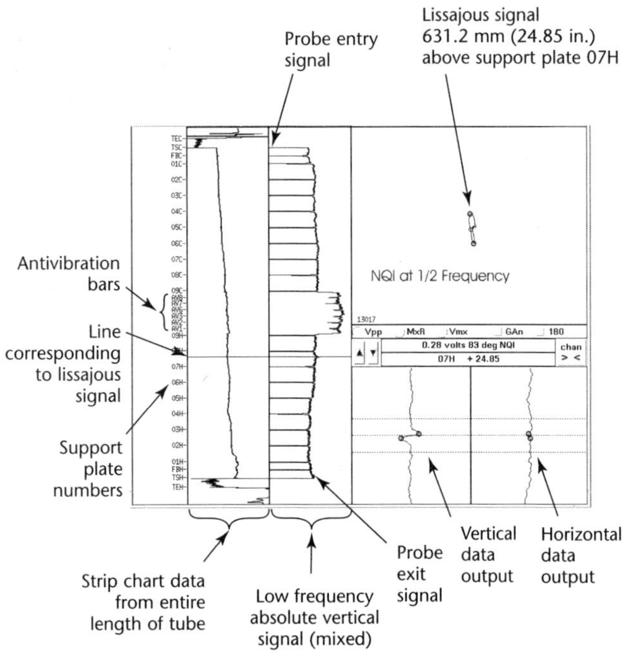

FIGURE 6. Lap indication with rotating coil at 300 kHz.

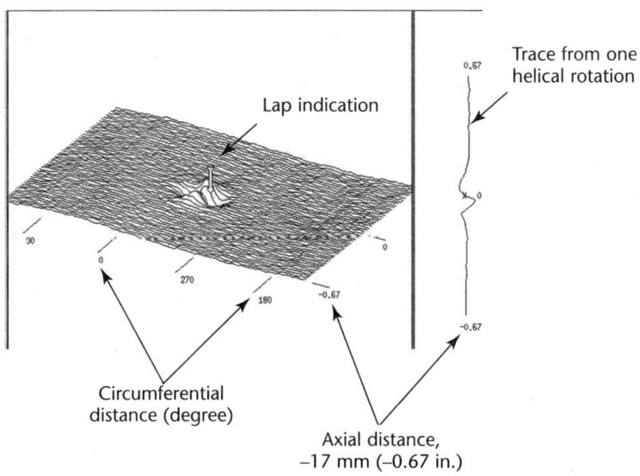

Part 4. Dezincification in Inhibited Admiralty Brass Tubing

Tubing Corrosion

Tubing corrosion is a subject of concern for fossil plants.[10-15] Condensers are a source of silica contamination in the boiler and on turbine blades, resulting in loss of efficiency and a need to flush the steam drum. Condenser tube leak incidents increased in the 1990s, resulting in lost megawatt hours because of silica load limits and downtime for leak checking.

Dezincification

A major reason for condenser tubing failure is dezincification, a chemical process in which zinc is dealloyed from the copper alloy.

Many condensers in fossil power plants have tubing made of admiralty brass (71 percent copper, 28 percent zinc and 1 percent tin). For resistance to dezincification, tube material is generally *inhibited,* that is, formulated to be resistant to dezincification by the addition of 0.02 to 0.10 percent arsenic, antimony or phosphorus (producing brass corresponding to Unified Numbering System C44300, C44400 or C44500, respectively).

However, dezincification occurs even in inhibited brass. Figure 7 shows tubes corroded through dezincification. The cross section in Fig. 8 shows consequent loss of material on the inside surface of a tube. In many instances, the cause of failure is unknown and may be a combination of failure mechanisms.

Figure 7. Split section of corroded tube, showing tube roughness on inside surface.

Eddy Current Testing

Eddy current testing is regularly performed on heat exchangers in fossil plants. Remote field testing has been used for carbon steel in low pressure heaters on supercritical units and conventional eddy current testing has been useful in assessing the overall condition of the tube bundles. Eddy current testing provides indications of cracking, wear, erosion, denting, small inside surface pitting and large, volumetric inside surface pitting.

Eddy current testing has been used to help minimize condenser tube leaks and to help determine failure mechanisms. Although the testing provides an overall assessment of the condenser tubes, it has not completely eliminated condenser tube leaks. Many tubes leak because debris gets lodged in them and from other causes.

Figure 8. Color energy dispersive display map showing lack of zinc (light area) in pit.

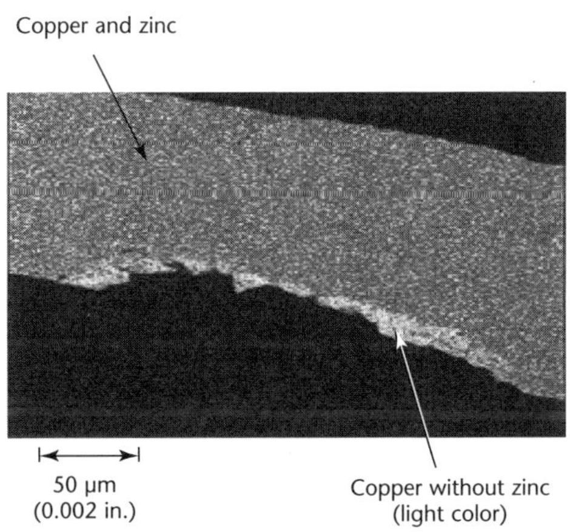

Case Histories

Case histories of four plants are described here.

1. Plant 1 had the first condenser identified with dezincification. The damage was isolated to only one of the four passes. This pass was a first pass bundle in the low pressure shell where the high energy piping, drains and dumps are connected through the side wall.
2. All of the six condensers in plant 2 had some form of dezincification combined with stress corrosion cracking. The cracking led to massive tube failures and eventual tubing replacement.
3. In plant 3, the damage occurred in one of the upper passes but had not been a problem until, during summer operation, the plant was unable to maintain a tight condenser for extended periods of time because of multiple tube leaks.
4. In plant 4, the dezincification resulted in large inside surface pits in the top half of the tube.

These case histories are discussed in detail below.

In each plant, all of the dezincification damage occurred in original tubing, made of admiralty brass in accordance with ASTM B 111.[16] Material analysis confirmed that the material for the condensers was inhibited but damage caused by dezincification was identified in all four.

As part of the normal maintenance outage work, eddy current testing was performed on tubes in the main condensers. Where suspect indications were detected, tube samples were removed to determine the cause of the indications and to verify eddy current results. Failure analysis and material evaluation were performed on the tubes and the results were used to help design new eddy current calibration reference standards made from tubing removed from service.

Plant 1

Plant 1 was a 600 MW, horizontal, dual-pressure unit with twin shells. The first pass was on the low pressure side and the second pass was on the high pressure side. This condenser had a condensing surface of 11 729 m^2 (126 250 ft^2) per shell. The original tubing was inhibited admiralty brass measuring 25 mm (1 in.) in outside diameter, 1.2 mm (0.049 in.) in thickness and 11 m (36 ft) in length.

The damage was identified in the north outside pass along the north wall near the high energy drains. Eddy current testing showed severe pitting and wall loss consistently between the fifth and sixth tube supports from the test end. The main steam drains between these two tube supports were suspected of causing the damage.

A review of the operational history showed problems caused 20 years earlier by high temperature, main steam drains impinging on the tubes in the north outside pass. Analysis at that time by an outside contractor had determined the tubes had been overheating and causing dezincification. Although these drains were later routed inside the condenser to spray over the top of the tube bundle, the tubes continued to deteriorate.

Twenty years later, several tubes had small pinholes between the fifth and sixth tube supports. The inside surface was rough. Wall thickness measurements indicated the average tube wall was 0.75 mm (0.030 in.) with a reduced wall of 0.50 mm (0.020 in.) between the fifth and sixth tube supports. Several pits were examined with microscopes; intergranular corrosion and dezincification were found.

The dezincification had occurred in tubing known to have direct impingement by main steam drains. Although the drains had been rerouted, the corrosion and erosion of the tube wall continued for nine years until the tubes were replaced. The microstructure of the tubing in the corroded area had a larger grain size than tubing away from the damage area, indicating a change in material properties. Grain growth could have decreased the corrosion resistance, thereby letting the erosion and corrosion continue.

Plant 2

The condenser tested in plant 2 was one of six in a shared, common header system serving eight boilers. The turbine sizes ranged from 40 to 60 MW. The condenser was a horizontal, twin shell with two passes of tubes, in an upper bank and a lower one. The condenser had a condensing surface of 7178 m^2 (77 250 ft^2) and used river water for cooling. The original tubing was inhibited admiralty brass and measured 25 mm (1 in.) in outside diameter, 1.24 mm (0.049 in.) in thickness and 8.0 m (26.25 ft) in length.

This condenser had never been eddy current tested. Testing was initiated to assess the current condition of the tube bundle. An example test is shown in Fig. 9. Although testing was difficult because many of the tubes were plugged with silt and sand, initial indications identified possible through-wall cracks. Because of the severe debris in the tubes, a cleaning contractor was hired to clean the tubes. A flood check was then performed on the condenser and nearly 2100 tubes were found leaking. Tubes were pulled to determine the cause of failure but only parts of the tubes could be extracted because of their deterioration. Analysis of the tubes revealed a silt deposit on both the outside and inside surfaces. The condenser had not been operated or maintained much in the previous 15 years. A contributing factor was the inability to isolate the condenser because of leaking gate valves.

All of the tube sections pulled from this condenser had both longitudinal and circumferential cracks 360 degrees around the tubes but the cracking did appear to be worse on the top half of the tube. Testing showed that the cracking originated at the tube outside surface and propagated to the inside surface (Fig. 10). Energy dispersive spectroscopy revealed layers of dezincification (Fig. 11) on the tube's outside and inside surfaces. In addition, dezincification with pitting was occurring on the tube's outside surface, where cracks originated. The tubes ultimately failed because of stress corrosion cracking. The root cause of failure was corrosion under deposits after years of lack of maintenance. The outside surface of the tubes had been allowed to sit with a silt deposit in a moist atmosphere for years, thereby creating the environment needed for dezincification and stress corrosion cracking.

FIGURE 9. Eddy current display from test of condenser tube.

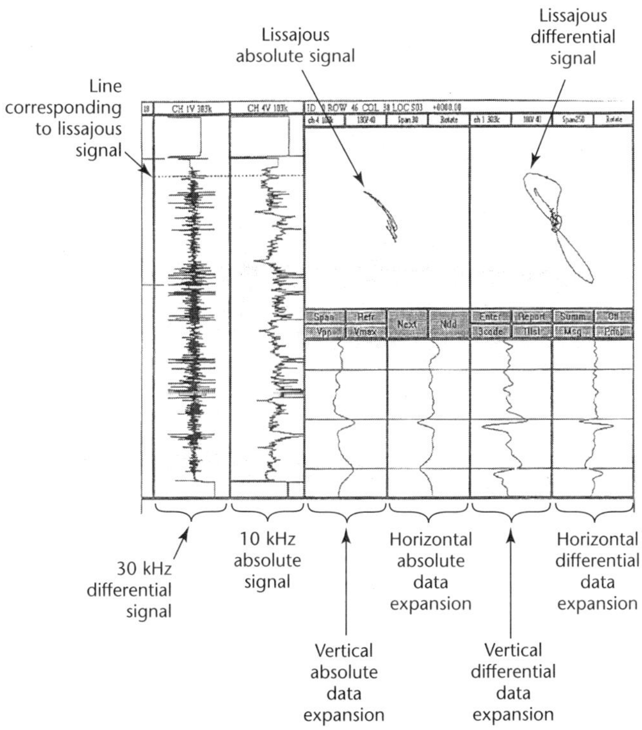

Plant 3

Plant 3 was rated at 275 MW and had a horizontal condenser with two shells and two passes of tubes, upper and lower. The condenser had a total condensing surface of 12 077 m^2 (130 000 ft^2). The original tubing measured 25 mm (1 in.) in outside diameter, 1.2 mm (0.049 in.) in thickness and 9.1 m (29.75 ft) in length.

Eddy current testing of this condenser was conducted after data showed that several tubes had inside surface pitting. Dezincification was occurring beneath the dark deposit but was inactive as long as the deposit remained intact. The wall under the deposit was 1.14 mm (0.045 in.) thick and the eroded and corroded tube wall was 0.66 mm (0.026 in.) thick.

The eddy current results showed that all of the 116 tubes tested had inside wall loss indications and that more than half of the tubes tested had wall loss greater than 75 percent in places. Most of the tested tubes had indications covering half

FIGURE 10. Cracking on inside surface of condenser tube.

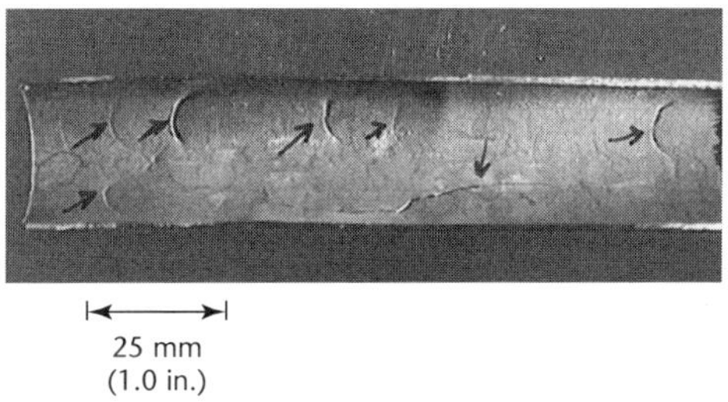

FIGURE 11. Color energy dispersive display map of condenser tube, showing lack of zinc in crack (light area).

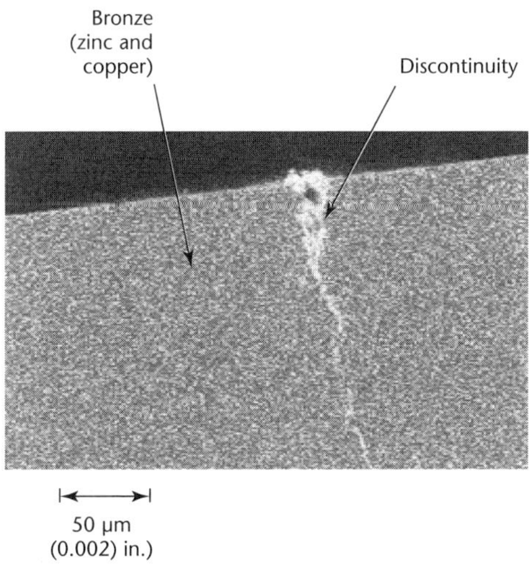

of the tube length in the outlet half of the tube. An analysis of pulled tubes showed pin holes where the tube wall had been thinned on its inside surface; the pits were approximately aligned.

A dark inside surface deposit had covered most of the inside wall and the thinning was occurring where this dark deposit had been removed. A comparison of the inside surface condition to tubes removed 12 years earlier showed an increase in the removal of this inside surface deposit. Energy dispersive spectroscopic analysis with the scanning electron microscope showed the inside surface deposit to be mostly silicon and iron.

The corrosion and erosion occurred in tubing near the top of the condenser. The cause for the removal of the dark inside surface deposit was unknown but the upper bank of tubes would have been subjected to high temperatures because of steam exhaust from the turbine. Dezincification was occurring under the deposit but, as long as the deposit remained intact, the tube wall was protected and not thinned by pitting. It was only when the deposit was removed that the tube wall would corrode and erode, leading to small holes.

The dezincification appeared to be caused by corrosion under deposits. The lack of oxygen under this deposit appeared to create an anodic site at which zinc was dissolved from brass. The thinning of the tube wall was continuing because of the weakening of the base metal through dezincification and metal's erosion by circulating water. This process continued until small holes formed tube leaks.

Plant 4

This plant was a 600 MW fossil plant whose horizontal, twin shell condenser had an upper bank of tubes and a lower bank. This condenser had a total condensing surface of 24 944 m^2 (268 500 ft^2) per shell. The tubing was inhibited admiralty brass and measured 38 mm (1.25 in.) in outside diameter, 1.2 mm (0.049 in.) in thickness and 10.4 m (34 ft) in length.

Eddy current testing did not identify any real concerns with the tubing. After several small areas showed an increase in the number of tube leaks, testing was limited to three areas. The eddy current test results showed one area had large wall loss indications at two tube supports similar to condensate grooving, that another had damage associated with the high energy drain connections and that a third had large inside surface indications throughout the tube length. This last area was in the second pass bundle along a side wall of the condenser. Tubes were pulled to evaluate the eddy current signals and determine the cause of wall loss.

The first tube showed condensate grooving adjacent to the tube supports and a visual test of the inside surface of the second tube showed large inside surface pits in the top half of the tube. There was also a well defined split between the top and bottom half of the tube that showed a dramatic difference in wall thickness. Along with this wall thickness difference was a dark inside surface deposit covering the thicker, top half of the tube. Measurements showed the remaining wall thickness was about 1.1 mm (0.042 in.) in the top half and about 0.7 mm (0.028 in.) in the bottom half. Sections were cut so that pits up to 38 mm (1.5 in.) long could be measured. The diameter of the pits varied from 3 mm (0.125 in.) to 13 mm (0.5 in.). Measurements also showed the wall thickness had been reduced to as low as 0.3 mm (0.012 in.) in some pits.

Energy dispersive spectroscopy with the scanning electron microscope showed the dark inside surface deposit to be very similar to the deposit in plant 3, mostly silicon and iron plus minor elements. Further analysis showed that the pitting was occurring because of dezincification. As with the tubing in plant 3, dezincification was occurring under the inside surface deposit but the wall remained intact where the deposit adhered tightly.

Evaluation of Results

The tubes failed because of cracking and inside surface pitting. One plant's condensers had longitudinal and circumferential cracks that required immediate retubing. Cracking occurred at only one plant whereas the pitting occurred at all the plants. The pitting was in the form of isolated pits and aligned pitting. Many tubes had inside surface pit indications at different locations down the tube length; some, the entire length. One tube had inside surface pitting that looked more like aligned cavities along half of the tube. These cavities were present where a dark inside surface deposit had been removed from the tube wall. Another condenser had inside surface cavities up to 38 mm (1.5 in.) long and 19 mm (0.75 in.) in diameter that were once again in the half of the tube with the dark inside surface deposit.

The stress corrosion cracking initiated at outside surface pits were under a silt deposit. The admiralty brass had both layer and plug types of dezincification and cracks had formed in both the axial and circumferential directions. Lack of maintenance by the plant led to eventual failure and replacement of the tubing.

Pitting appeared to grow by the combination of erosion and corrosion. The constant water flow did not allow the brass tubing to sufficiently form the protective oxide layer. The circulating water velocities were unknown but may have exceeded the recommended rate of 2.1 to 2.4 $m \cdot s^{-1}$ (7 to 8 $ft \cdot s^{-1}$). Corrosion occurred through dezincification, which resulted in a porous metal surface that was very weak. The site at which the pitting developed was typically on the top half of the tube. All of the tubes with dezincification pits developed them on the inside surface, where a dark deposit or scale was once on the surface. The deposit seemed to flake off in hot areas near the condensers. One location was where the high energy drains and dumps were connected. The other two areas were in the upper passes of the condenser that sits under the turbine. The last area was near a side wall that may also have had less water flow.

Dezincification could have occurred if the water velocity was too great, if temperatures were higher than expected or if water velocities were too slow, which will let deposits form. The formation of deposits created, under the deposit, an anodic site that was oxygen deprived. This site would then corrode, producing pits.

Conclusions

Eddy current testing has succeeded in detecting long term damage caused by dezincification in brass tubing. The final failure modes documented in these case histories was stress corrosion cracking in one condenser and pitting in the other condensers.

The mechanisms causing dezincification in fossil plant condensers may never be known completely. The material meets the required specifications but operating conditions may have altered the corrosion protection of the admiralty brass. The condition of the tubing needs to be monitored. In the twenty-first century, all admiralty brass tubing has been replaced with stainless steel in the installations described above.

Part 5. Eddy Current Testing of Ferritic Welds in Nuclear Transfer Casks

Background

Manual eddy current testing is used as a nondestructive test technique to supplement the load testing and visual testing of welds in nuclear fuel transfer casks in accordance with ANSI N14.6.[17]

In the 1980s, eddy current testing was introduced as a supplement to magnetic particle testing for welds in North Sea oil and gas fields.[18] Efforts were also made in the United States to perform magnetic particle testing through coatings.[19-21] The expense of removing marine growth or paint to perform underwater magnetic particle testing and topside tests requiring paint removal resulted in developmental projects for applications in both magnetic particle and eddy current testing.

Magnetic particle testing has been performed under water and through a thin film of black oxide 0.10 to 0.15 mm (0.004 to 0.006 in.) thick. Magnetic particle testing through black oxide is highly reliable and can detect indications as short as 1.5 mm (0.06 in.). In 1987, the code was changed for magnetic particle testing to allow performance demonstration for validating testing through coatings.[22] However, for many applications, magnetic particle testing through coatings for fabrication, repair and inservice tests is not feasible because of detection requirements. Additionally, most painted welds typically have greater than 0.15 mm (0.006 in.) of coating, the maximum paint thickness for magnetic particle testing typically cited in reports and standards.

In the early 1990s, a series of research projects and round robins were performed to evaluate eddy current testing through coatings where magnetic particle testing could not be used, primarily topside applications.[23] The round robin included computer aided electromagnetic testing (alternating current field measurement) for three-dimensional mapping of corrosion and discontinuities. Experienced operators using a commercial eddy current instrument and a weld probe detected discontinuities of about 0.4 mm (0.015 in.) through paint, similar to results from magnetic particle testing on bare metal. The eddy current instrument was small and lightweight, so it was used with industrial rope access instead of scaffolding.

Description of Method

In eddy current testing, a small probe containing cross wound coils is passed over the weld test object. A current is passed in the coil, which induces a magnetic field in the test object. The magnet field induces eddy currents within the test object. Surface discontinuities are indicated by perturbations on a display screen. Electric discharge machined notches can be used to calibrate as well as to reveal the system's performance sensitivity. The display signal produced on the eddy current instrument is set to a certain screen height (as with ultrasonic testing). To account for paint, previously measured using an eddy current absolute probe, plastic shims of the same thickness as the area under test are placed over the electric discharge machined notch and gain is adjusted accordingly. The eddy current instrument is set up so that the eddy current signal produced from the electric discharge machined notch produces a vertical trace that can be distinguished from normal background noise, liftoff and signals produced from the toe of the weld and various weld geometries.

In eddy current testing, only the area under the weld probe is interrogated; several scans must be used to cover the full weld, making it inherently slower than magnetic particle testing unless paint must be removed and reapplied. Separate scans that test the base metal, the weld toe and the weld face are performed. The eddy current inspector moves the probe over the test surface, looking for signals produced by weld geometries and for crack signals.

A major advantage of eddy current testing is that it can be used on a wet surface and on rainy days. The equipment is lightweight and battery operated, making it ideal for applications requiring rope access.

Technique Development

A project was conducted in the petroleum industry to establish a confidence level for eddy current testing of ferritic welds.[23] The project used test objects with discontinuities of the shape and size required. A discontinuity size of 6 mm (0.24 in.) was chosen for the minimum size for detectability; other test objects had indications as small as 1.5 mm (0.06 in.). Samples (Fig. 12) with service induced cracks and others with artificial cracks were selected on the basis of discontinuity orientations (longitudinal and transverse), joint shapes (butt, fillet and cruciform joints) and discontinuity locations (on weld face, weld toe and base metal and near access ports).

In the study, manual eddy current test results showed an 87 percent agreement with magnetic particle test results. Subsequent eddy current test results with the test bed showed a 100 percent agreement with magnetic particle testing. Causes of improved reliability included fine tuning the initial recommended practice and using highly qualified eddy current inspectors.

Computer assisted results showed 62 percent agreement with the magnetic particle test results and significant false alarms. One factor impairing the results was that the eddy current probes were difficult to apply at weld toes and small access ports.

Eddy Current Testing of Transfer Casks

Background

Transfer cask shells (Fig. 13) fabricated from a specified alloy steel[24] are used for transferring transportable storage canisters containing spent nuclear fuel bundles. Transfer casks are required to be tested on a scheduled basis in accordance with ANSI N14.6.[17]

Annual testing of special lifting devices such as a transfer cask can be a significant disruption to an operation that runs all day, every day. This disruption becomes even more severe if the components need to have paint stripped and reapplied for the purposes of inspecting the load bearing welds. Paint stripping is further exacerbated by the fact that a transfer cask is considered to be contaminated unless a significant effort is expended to prove otherwise. Previous annual testing resulted in out-of-service time for these components of about one month.

Magnetic particle testing has been done on transfer cask load bearing welds, requiring removal of all paint in those areas. Although magnetic particle testing is ideal for detection of surface cracks on ferritic material, magnetic particle testing loses its sensitivity when applied through coatings thicker than 200 to 300 µm (0.008 to 0.012 in.). Eddy current testing obviates removal and reapplication of coatings.

Ferritic Weld Trials

A nuclear engineering firm was asked by a nuclear operator to explore eddy current testing to replace magnetic particle testing. The application was unique in that a procedure was specified to detect a 1.6 mm (0.063 in.) discontinuity. Previous written practices for inservice testing specified a minimum detectability of 6 mm (0.25 in.) long and 0.8 mm (0.03 in.) deep by nonvisual testing.

Field and manufactured samples were used to qualify the technique but there was no reference standard for calibration or for validation of discontinuity size. A calibration block with an electric discharge machined notch measuring 1.6 mm (0.063 in.) in length, 0.51 mm (0.020 in.) in depth and 0.015 mm (0.006 in.) in width was used because it was already referenced in an applicable standard.[25] The written procedure was revised to reflect this sensitivity.

FIGURE 12. Transfer cask samples for training and qualification of personnel.

FIGURE 13. Transfer cask: (a) photograph; (b) eddy current test of access port weld; (c) eddy current test of trunnion weld.

(a)

(b)

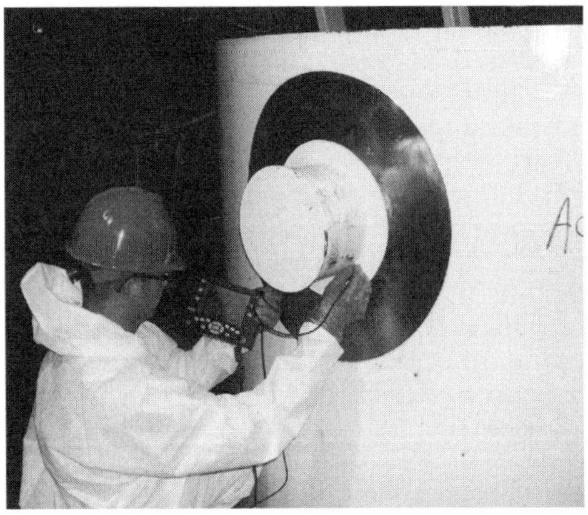

(c)

Additionally, a calibration block having electric discharge machined notches with depths of 0.20 mm (0.008 in.), 0.51 mm (0.020 in.) and 1.02 mm (0.040 in.) was used to validate other test parameters such as permeability. The written procedure was then validated in a set of blind trials witnessed by the nuclear power company's quality assurance, inspection and engineering personnel.

In addition to qualifying the procedure, eddy current personnel were then qualified according to the written procedure by having to achieve, in blind trials, a minimum 90 percent detection on test objects with 25 or more indications of varying sizes and types in various geometries. The equipment consisted of a commercially available eddy current instrument, an eddy current weld probe, an absolute probe, electric discharge machined calibration blocks and plastic shims. All materials and test equipment were calibrated to reference standards traceable to the National Institute of Standards and Technology.

The eddy current procedure consisted of a series of calibration checks including (1) setting discontinuity sensitivity, (2) checking the symmetry of the cross wound weld probe, (3) checking that the permeability of the steel under test matched that of the calibration block and (4) measuring the steel through a coating to compensate for gain adjustments on the discontinuity calibration block.

In the field tests, it remained to confirm whether the permeability of the calibration block agreed with that of the transfer cask. The calibration blocks were made of low alloy steel and carbon steel (Unified Numbering System G43400 and G10180, respectively). No problems were encountered and permeability checks were within acceptable tolerances.

If an eddy current test indication was detected, it was proof tested by magnetic particle testing. Magnetic particle testing showed a visual display of the indication and provided additional confidence in the eddy current test result.

Conclusion

Eddy current is a proven and accepted technique for detecting short length indications through paint on ferritic welds, such as those in nuclear transfer casks.

References

1. *Pressurized Water Reactor Steam Generator Examination Guidelines,* Revision 6: *Requirements.* Palo Alto, CA: Electric Power Research Institute (2002).
2. TR-110 392, *Eddy Current Testing of Service Water Heat Exchangers for Engineers Guideline.* Palo Alto, CA: Electric Power Research Institute (1999).
3. Birring, A.S. "Selection of NDT Techniques for Heat Exchanger Tubing." *Materials Evaluation.* Vol. 59, No. 3. Columbus, OH: American Society for Nondestructive Testing (March 2001): p 382-391.
4. "Eddy Current Inspection." *ASM Handbook,* ninth edition: Vol. 17, *Nondestructive Evaluation and Quality Control.* Materials Park, OH: ASM International (1989): p 164-194.
5. *ASME Boiler and Pressure Vessel Code:* Section V, *Nondestructive Examination.* Article 8, "Eddy Current Examination of Tubular Products." New York, NY: American Society of Mechanical Engineers (2001).
6. Mark, J.T. and D. Anges. "Steam Generator Inspections with the Cecco Probe." Presented at *16th EPRI Steam Generator NDE Workshop* [Palm Beach, FL, July 1997]. Palo Alto, CA: Electric Power Research Institute (1997).
7. Fisher, J.L. "Remote Field Eddy Current Inspection." *ASM Handbook,* ninth edition: Vol. 17, *Nondestructive Evaluation and Quality Control.* Materials Park, OH: ASM International (1989): p 195-201.
8. Krzywosz, K. Report GC 111 672, *Flaw Detection and Characterization in Heat Exchanger Tubing.* Palo Alto, CA: Electric Power Research Institute (1998).
9. Bipes, T. and G.P. Pierini. "Tube Manufacturing Indications in the Shearon Harris Replacement Steam Generator." Presented at *21st Steam Generator NDE Workshop* [Berkeley, CA, July 2002]. Palo Alto, CA: Electric Power Research Institute (2002).
10. Rogles, G.P. "Investigation of Dezincification in Inhibited Admiralty Brass Tubing." *6th Balance of Plant Heat Exchanger NDE Symposium* [Scottsdale, AZ, June 2000]. Palo Alto, CA: Electric Power Research Institute (2000).
11. Reynolds, S.D. and F.W. Pement. "Corrosion Failures of Tubing in Power Plant Auxiliary Heat Exchangers." Presented at *Corrosion/74* [Chicago, IL, March 1974]. Katy, TX: National Association of Corrosion Engineers (1974).
12. Boffardi, B.P. "Control the Deterioration of Copper-Based Surface Condensers." *Power.* Vol. 129, No. 7. New York, NY: McGraw-Hill (July 1985): p 57-60.
13. Khartak, H.S., J.B. Gnanamoorthy and P. Rodriguez. "Failure of Admiralty Brass Condenser Tubes." *ASM Handbook of Case Histories in Failure Analysis,* Vol. 2. Materials Park, OH: ASM International (1993): p 91-94.
14. Beavers, J.A., A.K. Agrawal and J.H. Payer. "Dealloying and Stress-Corrosion Cracking (Waterside)." RD-2282-SR, *Prevention of Condenser Failures — The State of the Art.* Palo Alto, CA: Electric Power Research Institute (1982).
15. Beavers, J.A., A.K. Agrawal and W.E. Berry. NP-1468, *Corrosion-Related Failures in Power Plant Condensers.* Palo Alto, CA: Electric Power Research Institute (1980).
16. ASTM B 111M, *Standard Specification for Copper and Copper-Alloy Seamless Condenser Tubes and Ferrule Stock (Metric).* West Conshohocken, PA: ASTM International (1998).
17. ANSI N14.6, *Radioactive Materials — Special Lifting Devices for Shipping Containers Weighing 10,000 Pounds (4500 kg) or More.* New York, NY: American National Standards Institute (1993).
18. Goldberg, L.O. "Eddy Current Testing: An Emerging NDT Method for Ferritic Weld Inspection." *Materials Evaluation.* Vol. 56, No. 2. Columbus, OH: American Society for Nondestructive Testing (February 1998): p 149-152.
19. API Research Report, *Project #90-61 RP for Underwater Magnetic Particle Inservice Weld Inspection of Offshore Fixed Platforms and Guidelines for Qualification of Magnetic Particle Inspector Divers.* Washington, DC: American Petroleum Institute (1991).
20. Report NP-5951, *Reliability of Magnetic Particle Inspection Performed through Coatings.* Palo Alto, CA: Electric Power Research Institute (July 1988).

21. Goldberg, L.O. *Reliability of Magnetic Particle Inspection through Coating, Phase II.* Houston, TX: Exxon Production Research (1985).
22. *ASME Boiler and Pressure Vessel Code:* Section V, *Nondestructive Examination.* 1986 Edition, 1987 Addenda. New York, NY: American Society of Mechanical Engineers (1987).
23. *Ferritic Weld Inspection Using Eddy Current.* Report. Merritt Island, FL: Sea Test Services (1996).
24. ASTM A 588/A 588M, *Standard Specification for High-Strength Low-Alloy Structural Steel with 50 ksi (345 MPa) Minimum Yield Point to 4-in. (100-mm) Thick.* West Conshohocken, PA: ASTM International (2001).
25. MIL-STD 271F, *Requirements for Nondestructive Testing Methods.* Washington, DC: United States Department of Defense (1986).

Bibliography

Dodd, C.V., J.R. Pate and J.D. Allen. "Advancement in Eddy Current Test Technology for Steam Generator Tube Inspection." *Materials Evaluation.* Vol. 53, No. 1. Columbus, OH: American Society for Nondestructive Testing (January 1995): p 49-50, 52-54.

Frankfurt, V.I. and D.S. Kupperman. "Review of Electromagnetic NDT Methods for Monitoring the Degradation of Nuclear Reactor Components." *Materials Evaluation.* Vol. 59, No. 9. Columbus, OH: American Society for Nondestructive Testing (September 2001): p 1053-1057.

Ida, N., H. Hoshikawa and W. Lord. "Finite Element Prediction of Differential EC Probe Signals from Fe_3O_4 Deposits in PWR Steam Generators." *NDT International.* Vol. 18, No. 6. Kidlington, United Kingdom: Elsevier Science Limited (December 1985): p 331-338.

Mahmoud, S.E., P.I. Abrams and L.O. Goldberg. "Reliability of Techniques for Nondestructive Testing over Coatings for MODU Inspections."

Obrutsky, L.S., V.S. Cecco, S.P. Sullivan and D. Humphrey. "Transmit-Receive Eddy Current Probes for Circumferential Cracks in Heat Exchanger Tubes." *Materials Evaluation.* Vol. 54, No. 1. Columbus, OH: American Society for Nondestructive Testing (January 1996): p 93-98.

Sabbagh, H.A., E.H. Sabbagh, R.K. Murphy and J. Nyenhuis. "Assessing Thermal Barrier Coatings by Eddy Current Inversion." *Materials Evaluation.* Vol. 59, No. 11. Columbus, OH: American Society for Nondestructive Testing (November 2001): p 1307-1312.

Sullivan, S.P., S.P. Smith and F.L. Sharp. "Simultaneous Absolute and Differential Operation of Eddy Current Bobbin Probes for Heat Exchanger Tube Inspection." *Materials Evaluation.* Vol. 58, No. 5. Columbus, OH: American Society for Nondestructive Testing (May 2000): p 634-638.

Yi, J.K. and D.L. Atherton. "Remote Field Gap Evaluation for Nuclear Fuel Channel Tubes." *Materials Evaluation.* Vol. 56, No. 6. Columbus, OH: American Society for Nondestructive Testing (June 1998): p 771-773.

17

CHAPTER

Infrastructure Applications of Electromagnetic Testing

Glenn A. Washer, Federal Highway Administration, Turner Fairbank Highway Research Center, McLean, Virginia (Part 1)

Margarit G. Lozev, Edison Welding Institute, Columbus, Ohio (Part 1)

Michael L. Scott, Wiss, Janney, Elstner Associates, McLean, Virginia (Part 2)

Herbert R. Weischedel, NDT Technologies, Incorporated, South Windsor, Connecticut (Part 3)

Reza Zoughi, University of Missouri, Rolla, Missouri (Part 4)

Part 1. Eddy Current Detection of Cracks in Steel Bridges

Background

There are almost 590 000 bridges, tunnels and culverts in the United States. Steel superstructures are in more than 185 000 bridges, more than any other superstructure material. Fatigue cracking in steel bridges can result from cyclical loading caused by traffic.

Many fatigue cracks under field conditions develop at welds, which commonly have both complex geometry and complex metallurgy. Fabrication discontinuities in welds, such as inclusions or lack of fusion, can provide stress risers that act as initiation sites for fatigue crack growth. The geometry of a particular weld, intersecting welds and high residual stress levels can also provide initiation sites. Hydrogen embrittlement and hydrogen assisted cracking can result in cracks in welds that may initially be small but can propagate under the cyclical loading conditions of a bridge.

The propagation of fatigue cracks in bridges can lead to structural collapse if the cracks are not detected. Because many cracks begin in welds, crack detection near them must be effective so that cracks can be detected before they grow significantly. This requirement has been an impediment to the application of eddy current testing on highway bridges: the welding geometry and varying magnetic properties caused by the complex metallurgy result in a low signal-to-noise ratio for some traditional eddy current technologies.

Historically, the eddy current technique has been used to test nonferrous structural items in the aerospace, chemical and processing industries.[1] Since about 1990, the technology has been extended to include ferromagnetic materials and even ferromagnetic welds. Sensor designs address the variations in magnetic properties that have led to difficulties in the past. The discussion below reports on the effectiveness of eddy current testing for detecting cracks in weldments typical of highway bridge construction.

Test Equipment

Probe

The probe used in this application was a differential probe with bidirectional sensitivity. This probe can detect surface breaking cracks in both base and weld metal of steel bridge members. The probe consists of two circular coils with axes parallel to the surface and perpendicular to one another. A sketch of the probe configuration is shown in Fig. 1. Probes of similar design are available from several manufacturers.

The probe operates in a differential configuration in which the impedance of one coil is compared with the impedance of the other, a common technique in eddy current testing. For this probe, each coil is very near the specimen surface, so both coils are simultaneously affected by the test specimen. The eddy currents generated in the surface of the specimen are an image of the coils in the probe, so each coil generates current with a significant current density aligned with its winding. Under this geometry, orthogonal currents in the surface of the specimen are generated.

In the differential mode, assuming negligible magnetic anisotropy of the material, each coil can have impedance consistent with that of the other as the probe is scanned along the specimen. The presence of a crack aligned with one coil and orthogonal with the other results in a

Figure 1. Bidirectional probe arrangement: (a) oblique view; (b) side view.

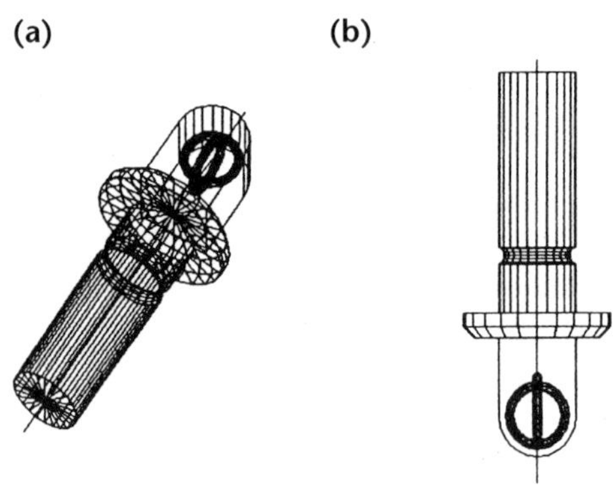

change in relative impedance of the coils and consequently creates a detectable signal. In fact, any electromagnetic condition not common to each coil will be detected. Because the coils are located very near each other, gradual changes caused by conductivity or permeability variation are limited. Liftoff effects are also minimized but not suppressed, as both coils are simultaneously affected when distance from the test surface to the probe changes.

The maximum probe response is observed when the crack is perpendicular to the direction of either coil winding. Under this scenario, the crack interrupts the currents generated by the coil oriented orthogonal to the crack and has minimum influence on currents oriented parallel to the crack. Minimal responses occur when the crack is at 45 degrees from the direction of the coil winding because each coil experiences similar impedance changes and hence no relative change.

Instrument

The research was performed using a commercially available instrument that displays and stores eddy current test data collected by a probe. The manner in which the test signal is processed and displayed depends on the instrument settings. The values of drive levels (voltage peak) as well as frequencies can be selected independently. The instrument voltage ranges from 0 to 9 V peak and the frequency ranges from 5 Hz to 10 MHz. Instruments with similar characteristics are available from several manufacturers.

Impedance plane mode and voltage plane mode were used during the testing. In both modes, the drive signal is applied across an alternating current bridge. One arm of the bridge consists of a resistor connected in series with one of the two coils of the probe. The second arm consists of a second resistor connected in series with the second coil in the probe. The voltage difference across the center of the bridge between the two coils is amplified and passed through a demodulator, which detects the in-phase and out-of-phase signal. The signal is then digitized to 16 bits and made available to the computer bus.

In the impedance plane mode, the eddy current signal is displayed as a calibrated percent of change in impedance. In the voltage plane mode, the signal is also generated from an impedance bridge but the data are displayed as detected in-phase and out-of-phase voltages.

Reference Standard Calibration Blocks and Crack Specimens

Reference standard calibration blocks were produced from ASTM A 588 and ASTM A 36 structural steels, materials commonly used in bridge construction.[3,4] Each block had three electric discharge machined (EDM) notches with depths of 0.2, 0.5 and 1.0 mm (0.008, 0.02 and 0.04 in.). The specimens were measured with no attempt to anneal or demagnetize them and the surface of the material was in its as rolled condition with mill scale removed. To study the effects of typical bridge coatings, both an inorganic zinc based coating and a lead based coating were applied to the surface of an ASTM A 588 calibration block.

Three ASTM A 588 steel specimens with surface breaking cracks were tested. These specimens had weld details typical of steel bridge construction with implanted cracks. Two specimens had T joints that represent web-to-stiffener or web-to-flange connections. One specimen contained a butt joint, typically used for flange or web splices. The butt joint specimen and one of the T specimens had complete penetration groove welds. The other T joint was connected with a fillet weld. These specimens were subjected to corrosive environmental conditions such that the surface was pitted and covered with iron oxide. Testing was conducted without changing the surface conditions of the specimens.

A single specimen of ASTM A 36 steel with a groove weld containing a transverse crack was tested. The implanted crack had a depth of 0.90 mm (0.035 in.).

Experimental Testing, Results and Discussion

Frequency Response

Tests were conducted to determine the frequency response of the eddy current probe and supporting instrumentation. Calibration blocks of ASTM A 588 and ASTM A 36 steel were used for the measurements. No significant differences in the results were found between the two materials, which are similar in chemistry and manufacturing process. The data points in Figs. 2 and 3 represent the maximum signal magnitude generated by scanning the probe along a path perpendicular to the long axis of an electric discharge machined notch or crack. A separate scan was required for each frequency datum shown in the figures.

The voltage plane mode scan of the reference standard ASTM A 588 steel specimen is shown in Fig. 2a. This figure indicates the variation in probe coil voltage versus frequency for 0.2, 0.5 and 1.0 mm (0.008, 0.02 and 0.04 in.) deep electric discharge machined notches. Maximum signal magnitudes for a given discontinuity depth are observed for frequencies ranging from 200 to 250 kHz. The general shape of these curves is a characteristic of the bridge output sensitivity. The voltage amplitude varied as a function of electric discharge machined notch depth as shown in Fig. 2a.

The impedance plane mode probe response of two electric discharge machined notches, 0.5 and 1.0 mm (0.02 and 0.04 in.) deep, and a 0.9 mm (0.035 in.) deep implanted crack specimen are shown in Fig. 2b. The figure indicates that the probe response from a 0.5 mm (0.02 in.) deep electric discharge machined notch is higher than the response from the 0.9 mm (0.035 in.) deep implanted fatigue crack, regardless of the test frequency. This verifies that the crack indication signal depends not only on discontinuity depth but also on discontinuity length and crack tightness. Electrical contact between crack faces because of crack geometry or oxides can cause a short circuit of the flow of the eddy currents around the crack opening and can change the response drastically.[1] Additionally, some flow of current can circumvent the crack because of the semielliptical geometry of the crack. Consequently, smaller signals are expected from cracks than from open electric discharge machined notches.

The effect of typical coatings are illustrated in Fig. 3. The figures show the variation in the inductive reactance component versus frequency for a 1.0 mm (0.04 in.) electric discharge machined notch covered by either zinc (Fig. 3a) or red lead based coating (Fig. 3b). In these figures, the inductive reactance amplitude is observed to attenuate progressively because of the liftoff effect and

Figure 2. Fatigue crack signal frequency responses from electric discharge machined notches: (a) with change in probe coil voltage for ASTM A 588 structural steel; (b) with change in probe coil inductive reactance for ASTM A 36 structural steel.

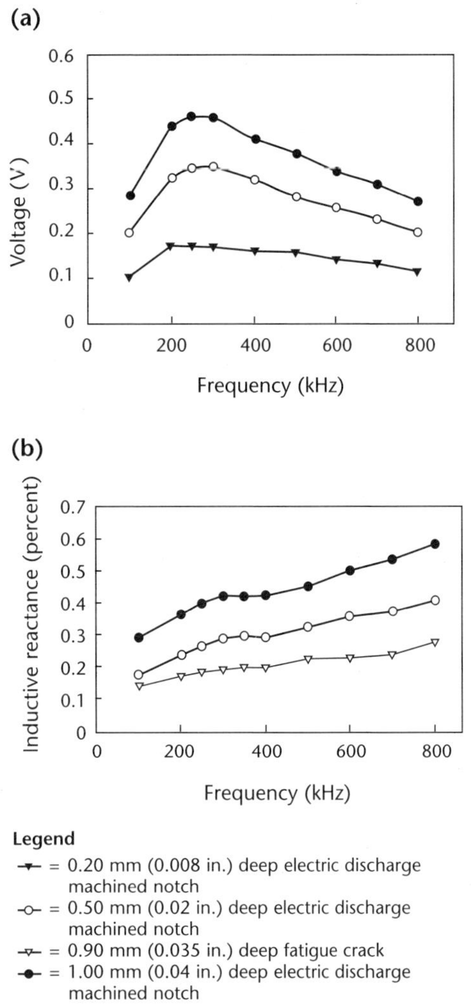

Figure 3. Effects of paint on eddy current signal on electric discharge machined block of ASTM A 588 structural steel: (a) zinc based paint; (b) lead based paint.

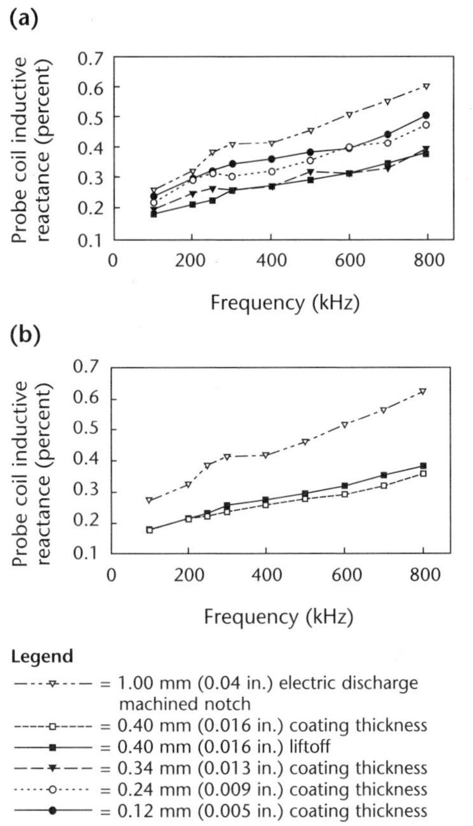

distribution of the eddy currents in the coating layer. This attenuation is proportional to the coating thickness. The low electrical conductivity of the coating layer ensures the penetration of the eddy currents into the base metal. In fact, no significant variations are observed between signals affected by 0.4 mm (0.016 in.) red lead base coating and 0.4 mm (0.016 in.) liftoff (Fig. 3b). Similar results were found on the zinc based system as shown in Fig. 3a. These data indicate that these coating systems can be penetrated by the magnetic field of the coils and that the effect of the coatings can be represented effectively by probe liftoff.

Eddy Current Response to Cracks and Machined Notches

This section illustrates the response of the eddy current probe to electric discharge machined notches and implanted cracks. The frequency and drive level used during scanning was 240 kHz and 3 V, respectively. The probe response is illustrated primarily as a reactance component, with only small changes in the resistive component of the impedance. The abscissa of these figures plots successive data points, that is, as dimensionless time. This abscissa can be interpreted as distance, given a constant scanning velocity. Data were collected by manually scanning the probe along a linear path that traversed the feature of interest, either a crack or a notch. Scanning was conducted at constant rate of about 25 mm·s^{-1} (5 ft·min^{-1}).

Figure 4 illustrates the eddy current response to scanning over electric discharge machined notches in the ASTM A 588 steel specimen. Scanning was performed from left to right along the surface of the specimen shown schematically in Fig. 4a. As expected, the magnitude of response increases with increasing notch depth. However, the ability of the eddy current techniques to accurately define the actual depth of a discontinuity is limited, as discussed in the next section.

Figures 5 to 7 illustrate the eddy current response to cracks in different geometric conditions. In each case, scanning direction was coincident with the weld axis. Therefore, a transverse crack in a weld, as shown in Fig. 5, is oriented orthogonal to the scanning direction. Figure 5b shows the response of the eddy current probe to a transverse crack. Of note in this figure is the low noise in areas adjacent to the discontinuity and the increasing magnitude of response as the probe approaches the crack and decreasing magnitude as the probe is scanned away from the crack. The maximum response corresponds to the probe positioned directly over the discontinuity.

FIGURE 4. Reactive and resistive component of signal response for electric discharge machined notches in ASTM A 588 structural steel: (a) cross section; (b) data.

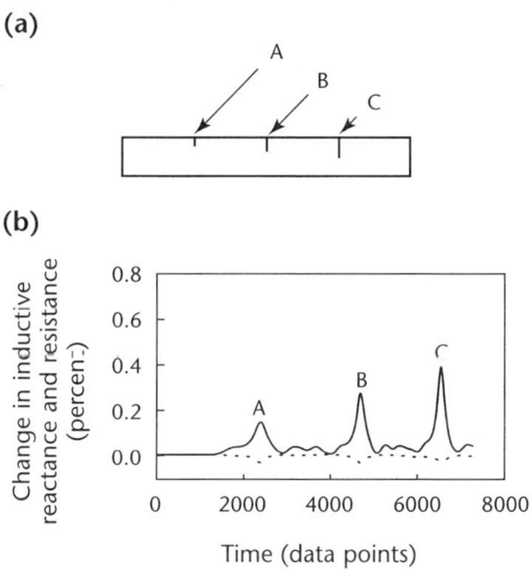

Legend

A = 0.20 mm (0.008 in.) deep
B = 0.50 mm (0.02 in.) deep
C = 1.00 mm (0.04 in.) deep
—— = inductive reactance
- - - = resistance

FIGURE 5. Reactive and resistive component of signal response for transverse crack: (a) cross section of butt or groove weld; (b) data.

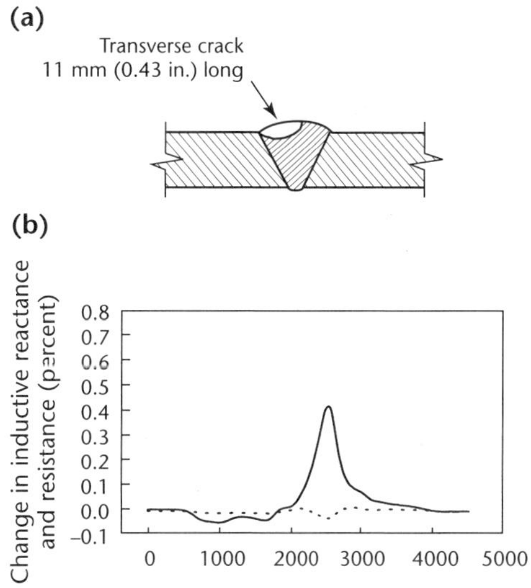

Legend

—— = inductive reactance
- - - = resistance

Figure 6 indicates the eddy current response to a longitudinal crack in the weld crown. Figure 7 indicates the responses to a toe crack (Fig. 7b) and root crack (Fig. 7c). Of note in these figures is the characteristic eddy current response of a longitudinal crack, increasing in magnitude as the probe approaches the crack, maintaining a magnitude displacement as the probe is scanned directly over the crack and diminishing in magnitude as the probe is scanned beyond the crack. Compared qualitatively to the response to a transverse crack, eddy current response shows crack orientation and can be easily interpreted. Additionally, the bidirectional sensitivity of the probe enables it to detect both longitudinal and transverse cracking as it is scanned along a weld. As mentioned previously, the probe is least sensitive to a discontinuity oriented at 45 deg to the probe coil axes.

Case Study: Steel Box Girder

The eddy current technique previously described was field tested on a welded steel box girder. The fracture critical member supported the south span of a double-swing bridge. The pivot girder supporting the north span of the bridge had time delayed cracking following production and shipping to the bridge site. Magnetic particle testing had revealed more than 150 cracks in this member during tests at the fabrication plant, following shipment to the bridge site and following placement. The magnetic particle testing procedure required the removal of a zinc based coating to facilitate contact between the prods and the member — a procedure undesirable to the bridge owner. The eddy current technique was used to eliminate the need for paint removal during testing of the south girder. At the time of the test, both pivot girders were in place on the structure.

The member tested was constructed of ASTM A 709 grade 70W steel,[5] with multipass fillet welds at the intersection of each corner of the box and along the stiffeners. Plate thickness ranged from 32

FIGURE 6. Impedance components of signal response for longitudinal crack: (a) T or fillet weld; (b) data from probe as probe is passed along length of crack.

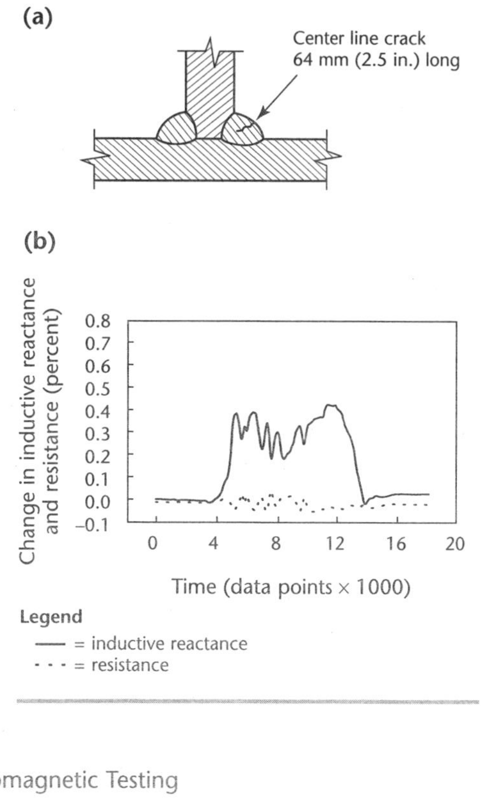

Legend
— = inductive reactance
· · · = resistance

FIGURE 7. Impedance components of signal response for longitudinal crack: (a) cross section of T or groove weld; (b) data for crack in toe; (c) data for crack in root.

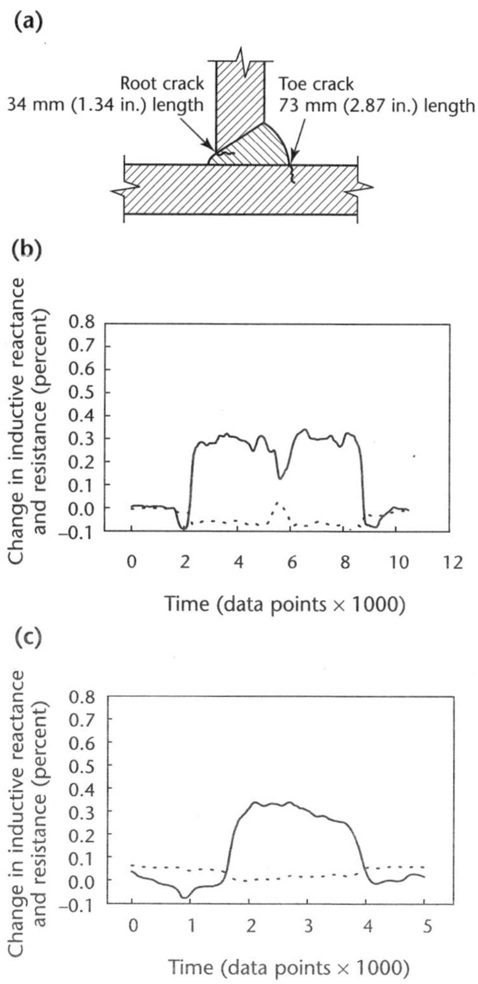

Legend
— = inductive reactance
· · · = resistance

to 76 mm (1.25 to 3.0 in.). The zinc based coating on the structure had a thickness of about 0.2 mm (0.01 in.), based on measurements taken on the north girder. The coating system included both an intermediate and finish layer, for a total coating thickness of about 0.4 to 0.5 mm (about 0.02 in.). Surface scanning of about 250 m (820 ft) of fillet weld included all flange-to-web and web-to-stiffener welds. The testing of the weld bead required between three and five passes of the probe for complete coverage of the weld surface and the heat affected zone.

Known fatigue cracks located on the north girder by magnetic particle testing were used as test specimens to determine frequency and drive levels for the probe. A calibration specimen was made from a piece of ASTM A 709 steel containing electric discharge machined notches with depths of 0.5, 1.0 and 2.0 mm (0.02, 0.04 and 0.08 in.). Liftoff due to coating was assumed to reduce signal amplitudes about 40 percent, which compensated for variations in the calibration procedure. Indications were determined by a simple, reactance threshold equivalent to about 20 percent full screen height. All signals exceeding this threshold were marked as indications.

Tests of the south girder revealed the following: three cracks, seven slag inclusions, seven undercuts and one rollover. Indications were excavated with a mechanical grinder to confirm the source of the indication. Of the 18 indications noted during the test, three correlated with cracks. It was found that the balance of the indications could be differentiated from cracks by visual testing in the case of geometric indications and by signal response in the case of slag inclusions.

Because of the time dependent nature of the cracking in the north girder, periodic tests of the girder at three-month intervals were scheduled. Based on the speed and simplicity of eddy current testing and the success on the south girder in detecting cracks without removal of coating, the eddy current technique was included in these tests. The instrument was a hand held battery operated eddy current instrument with a 4 MHz absolute probe having a probe footprint much smaller than the bidirectional probe of Fig. 1. This small footprint had the advantage of reducing the sensitivity to geometric indications. The hand held battery operated instrument displayed indications as normalized impedance and was self-balancing.

Conclusion

The discussion above reports on the effectiveness of eddy current testing for detecting cracks in weldments typical of highway bridge construction. The laboratory tests showed that the bidirectional probe configuration can be used to detect cracks in weld metal. The probe coil design suppresses the effects of spatially varying magnetic properties typically associated with the weld area in ferromagnetic steel. When the weld bead is scanned, irregular surface geometry is the main cause of noise.

The technique is effective for detecting cracks under paint. The laboratory tests showed that the eddy currents could penetrate both conductive and nonconductive coatings typically used on bridges. The effects of these coatings were proportional to the coating thickness and similar to liftoff effects. The attenuation caused by these effects can result in increased sensitivity to geometric indications that may mask smaller discontinuities; however, the cracks could generally be distinguished from welding discontinuities such as rollover or undercut.

The bidirectional probe responded well to field variables such as conductive coating thickness, irregular weld surfaces and weld joint geometry. The eddy current technique used has several advantages over magnetic particle testing. First, removal of coating to ensure good electrical contact is not required. Second, the bidirectional probe is sensitive to both longitudinal and transverse cracks simultaneously in weld metal.

Probes of this type are available from several manufacturers. It is recommended that the probe be calibrated for optimal frequency before performing any test. Interpretation of the eddy current signals can be accomplished with a moderate amount of training and experience and inspectors should have the appropriate qualifications.

Part 2. Applications of Ground Penetrating Radar to Bridge Decks

Background

Radar (*ra*dio *d*etection *a*nd *r*anging) is a technique that has been applied traditionally to problems related to object detection in air.[6] Since the 1960s, the technology has developed to detect and locate objects below the surface of the ground as well.[7] Ground penetrating radar systems typically transmit microwave pulses into the ground and measure reflected energy through a receiving antenna. Boundaries between materials with different dielectric properties (such as concrete and air) produce reflections when a radar pulse encounters them.[7] In the 1970s, the resolution of ground penetrating radar systems was suitable only for geophysical applications. Since then, ground penetrating radar systems with significantly higher resolution have been developed that can detect features of interest in civil infrastructure materials and components.

Civil infrastructure applications to roads,[8] bridges, dams and a variety of other infrastructure facilities have produced useful ground penetrating radar results. Ground penetrating radar applications to bridges are challenging and interesting because they typically require systems with high resolution to detect features of interest. Among bridge applications, testing of bridge decks has potential because of the large, accessible surface area and a high level of maintenance interest focused on them. Bridge decks deteriorate significantly faster than most other bridge components[9] and the dominant deterioration mechanism, corrosion induced delamination, is well understood. Ground penetrating radar is currently a useful tool for investigating many bridge deck conditions that include concrete cover depth, poorly consolidated concrete, voids and several other features of interest.[10] As technology advances, test results indicate that ground penetrating radar may also allow delamination to be directly addressed, as well.

Equipment

Ground penetrating radar systems typically include a radar pulse generator, transmitter, receiver, antenna and equipment for data acquisition and storage. Depending on the application, systems described as either ground coupled or air coupled may be used. Air coupled systems use antennas positioned at or above a height related to the center frequency the antenna transmits, a height clearance above the ground equivalent to a minimum of the wavelength at the radar center frequency. Most ground coupled antennas are positioned such that the antenna housing is directly in contact with the ground.

Typically, air coupled antennas are mounted in a vehicle to collect data from a moving platform in traffic. Conversely, ground coupled antennas are typically hand held (Fig. 8) where the antenna is manually positioned during data collection. For bridge deck applications, both of these ground penetrating radar systems may be used, depending on the survey needs.

The ground penetrating radar system functions by transmitting a radar pulse from the transmitting antenna into the ground and uses the receiving antenna to collect the reflected radar response from the ground. This process occurs very quickly, allowing the radar pulse to be transmitted and the waveform to be acquired with negligible distortion from movement, even at traffic speeds.

Data collection is usually triggered by a mechanism connected to a survey wheel. As the survey wheel turns, it initiates data

FIGURE 8. Hand held ground penetrating radar instrument in use on bridge deck.

acquisition at consistent intervals. The series of individual waveforms collected along the path of the radar antenna can be viewed collectively as a synthetic aperture. Raw data collected using a ground coupled ground penetrating radar are presented over a two-dimensional synthetic aperture (Fig. 9).

In this format, the radar response is presented in color maps that indicate high positive amplitude responses as light gray to white and low negative amplitude responses as dark gray to black. Basic features observed in these data include an example of a parabolic shaped response to reinforcing steel (identified with a circle) and a local area where reflection amplitudes are low, encircled by a horizontally oriented oval. Details of Fig. 9 are interpreted below.

The center frequency of a system is another important consideration when ground penetrating radar equipment is selected for a given application. A ground penetrating radar that operates at high frequencies can resolve smaller features than ground penetrating radar operating at lower frequencies. This resolution issue must be carefully considered when ground penetrating radar equipment is selected for an application like bridge deck evaluation.

Detection and Measurement of Deck Conditions

Several parameters that bridge engineers use can be measured with ground penetrating radar. In addition, there are distress modes that can be detected using this technology. The available measurements include concrete cover depth, bridge deck thickness and reinforcing steel location. Frequently detected discontinuity or distress modes include voids, poorly consolidated

FIGURE 9. Bridge deck: (a) data locations indicated by horizontal lines on deck surface; (b) numbered station locations with color or grayscale coded bar indicating confirmed delamination locations along data collection line [solid = delaminated, white = without delaminations]; (c) raw synthetic aperture radar data collected from bridge deck using commercial ground coupled radar system.

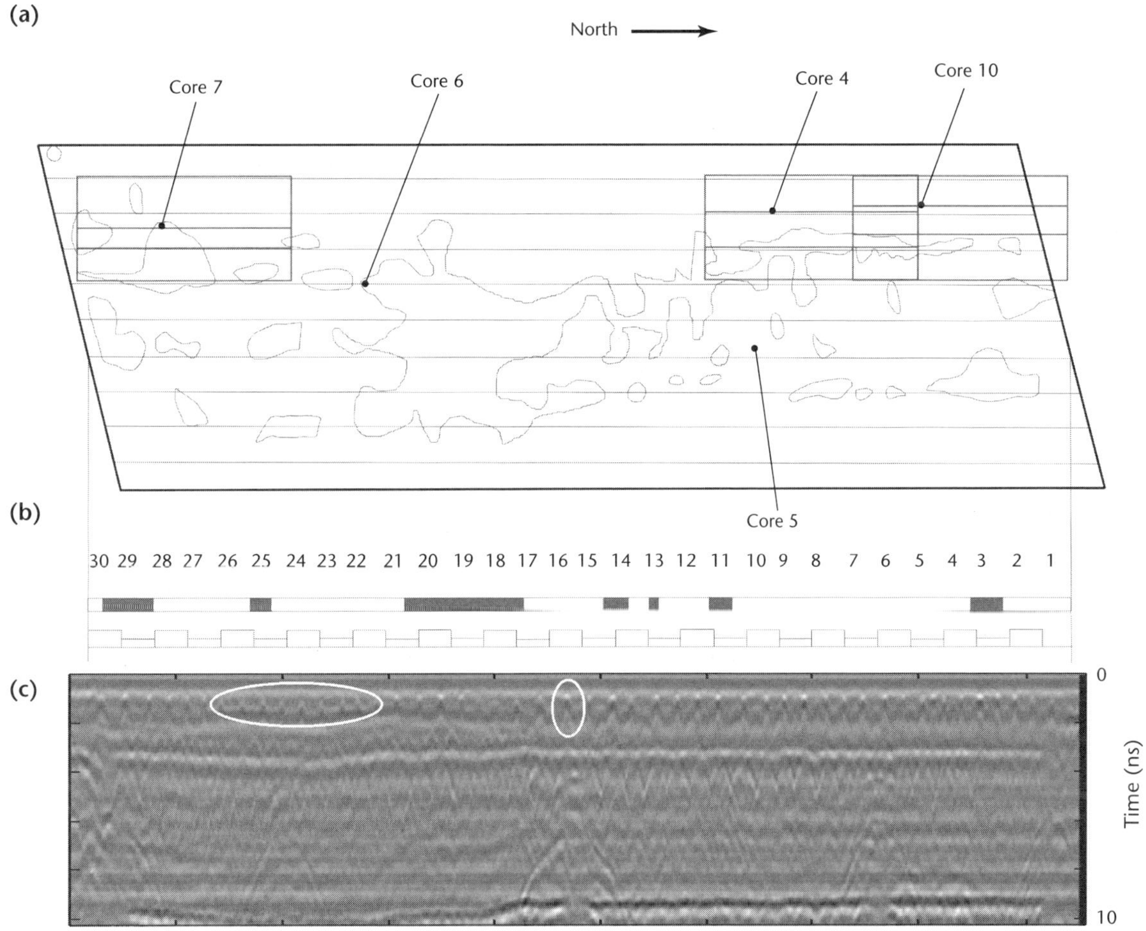

concrete and material property differences.

Measured parameters, such as concrete cover depth, must be determined using a calibration core or equivalent calibration data. The core or calibration data allow the propagation velocity and some dielectric properties of bridge deck materials to be determined for electromagnetic waves. After reference values have been determined at known locations, material properties of homogeneous areas around them can be inferred. Once these material property reference values are known, cover depth and bridge deck thickness can be calculated directly. In addition, migration algorithms and other data reconstruction techniques can be applied to raw data to produce radar images, given that dielectric material properties can be inferred from core data or equivalent test information. These radar images are often useful for locating of reinforcing steel.

Discontinuity or distress modes such as voids, poorly consolidated concrete and material property differences are all features that ground penetrating radar is generally useful in detecting. Air voids that can result from inadequacies in physical mixing are detected using ground penetrating radar because of the dielectric property contrast between concrete and the air in the void. In a similar way, poorly consolidated concrete produces a response where a portion of the concrete volume within the bridge deck scatters radar energy more than adjacent, properly placed concrete. Finally, material property differences in the concrete, caused by the presence of chlorides, water, deteriorated concrete or contaminants can affect both velocity and dispersion properties of concrete.

Two areas of ground penetrating radar research for bridge deck applications are imaging delaminations and penetrating asphalt overlays, although many systems already produce useful results through asphalt overlays. Delamination imaging is an important goal for bridge engineers but the resolution of commercial ground penetrating radar systems has made direct imaging of these discontinuities difficult.[9] Research indicates that high frequency ground penetrating radar systems can image a significant proportion of delamination in laboratory and field bridge decks.[10] As this technology develops, the capability of many ground penetrating radar systems to penetrate asphalt overlays may also improve, making it more useful for locating bridge deck problems without removing the asphalt layer. The additional thickness of the asphalt layer has contributed to dispersion and attenuation of the ground penetrating radar response — something that has needed to be compensated for.[11]

Ground Truth Data

The term *ground truth testing* generally refers to examination of features on the ground to confirm satellite telemetry. In the present discussion, *ground truth testing* refers to coring and acoustic tests for comparison with ground penetrating radar results. Ground truth data from several different techniques can be used to enhance the reliability of ground penetrating radar data. Implementation of these techniques through comprehensive testing of a bridge deck or local testing in areas of interest on a bridge deck can provide useful confirmation for ground penetrating radar data interpretation.[12]

Among ground truth techniques, ASTM D 4580 describes chain dragging, a qualitative acoustic method commonly applied to bare concrete bridge decks.[13] The technician drags a chain across the surface of the bridge deck and listens for hollow sounds. Other techniques that can be used for confirmation include hammer sounding, infrared imaging and coring.

The wide range of conditions in field bridge decks in terms of materials, property variations and design configurations has a role in determining the amount and type of additional testing that should accompany ground penetrating radar testing. Bridge decks with wide reinforcing steel spacing that exhibit material property homogeneity and generous clearances at boundaries, among other design features, are most conducive to ground penetrating radar testing and require minimal verification testing through ground truth techniques. However, bridge decks with material property or designs that impede ground penetrating radar testing require additional verification testing.

Example Applications

Two applications of bridge deck ground penetrating radar are presented to illustrate routine measurements and results obtained from emerging ground penetrating radar technology. The first example presents data collected from the Van Buren Road Bridge deck, Dumfries, Virginia. The second example presents data collected from a bridge over Lake Anna, near Fredericksburg, Virginia. Tests on the Van Buren Road Bridge provide a broad range of corroborating test data that accompany the ground penetrating radar data. The Lake Anna tests provide data for a more basic comparison between ground penetrating radar technology and typical acoustic[13] and coring tests.

Van Buren Road Bridge Deck

Figure 9 presents results obtained from a ground penetrating radar scan along the 18 m (59 ft) center span of the Van Buren Road Bridge deck. The path of the scan is indicated by a sequence of numbers on a plan view drawing of this deck (Fig. 9). Subsurface delamination cracking was identified in the bare concrete deck by using an acoustic chain drag method[13] before ground penetrating radar testing and is indicated by rectilinear outlines in the plan view diagram (Fig. 9a). In addition, a corresponding color coded bar identifies acoustically detected delamination locations along the path of the collected data (Fig. 9).

The ground penetrating radar scan along the designated path in Fig. 9 was conducted using a multichannel data acquisition system in a hand held configuration (Fig. 8). The results of the Fig. 9 scan illustrate one of the many ground penetrating radar responses to reinforcing steel at an individual circled location. The circled response is a characteristic parabolic shaped feature, typical of a point reflector in two-dimensional ground penetrating radar data. In this case, the reinforcing steel appears to be a point reflector because the steel orientation is orthogonal to the path of the ground penetrating radar. As the radar approaches this steel feature along a linear path, the travel time for the radar response to the steel changes from a relatively long duration down to a minimum travel time, when the antenna is directly above the steel.

As the antenna continues along its linear path beyond the steel feature, the travel time of the radar pulse begins to increase again. If the concrete's radar velocity and loss properties (numerically defined by dielectric properties) are known, then calculations can be performed to focus these parabola shaped features to point images. These images therefore represent the originally detected feature. There are several approaches to accomplishing this imaging task,[14,15] depending on assumptions about the problem at hand. One of these approaches, known as *wave field backpropagation,* has been implemented in Fig. 10 and is discussed below.

In addition to the commonly observed response to reinforcing steel, the ground penetrating radar scan presented in Fig. 9 illustrates an anomaly that has two possible interpretations. The anomaly is encircled by a horizontally oriented oval, which identifies several low amplitude reinforcing steel responses. This collection of responses identifies an area where the steel is either physically deeper below the deck surface than adjacent steel or the dielectric properties of the concrete are exhibiting variations. This type of anomaly may indicate deteriorating concrete.

In this particular bridge deck, the ground penetrating radar results and the acoustic sounding results presented in the plan view drawing (Fig. 9a) identify different problem areas. Comparing the locations of delaminations identified by acoustic testing of the entire bridge deck with anomalies in the ground penetrating radar data shows that the two test methods identify different phenomena. In this particular test, thin concrete delamination cracks were the feature of interest for an engineering research project. An analysis of the data that included concrete coring indicates that the acoustic testing was more effective than ground penetrating radar for identifying delamination in this bridge deck.

Figure 10 illustrates a range of different bridge deck tests conducted on the same span of the Van Buren Road Bridge deck presented in Fig. 9. Figures 10a to 10c each present results from the same span, overlaid on a separate diagram for clarity. This testing is described in detail elsewhere.[12]

In brief, these additional test data indicate many of the advantages and limitations of the ground penetrating radar test. Figure 10a shows the test results obtained by 21 different inspection teams using the acoustic chain drag test. In this figure, light gray areas indicate delamination locations identified by fewer inspection teams while dark areas indicate locations where more inspection teams identified delaminations. A reference chain drag survey result, obtained using unusual attention to detail, is indicated in Fig. 10a to Fig. 10c by line enclosed areas. Significant differences between inspection team results are observed. In Fig. 10b, results from the acoustic impact echo test are presented. Here, delaminated areas are indicated with closed circles while solid concrete is indicated with open circles. Agreement between the reference chain drag survey and the impact echo testing is evident.

Figure 10c presents test results from two ground penetrating radar systems. A Federal Highway Administration ground penetrating radar system was used to collect data in the plan view presentation. Here, images from the reinforcing steel level of the bridge deck indicate features such as the reinforcing steel itself and local variations in the response magnitude. The ground penetrating radar images were derived from raw data

FIGURE 10. Data collected using ground penetrating radar: (a) plan view; (b) data location grid; (c) plan view showing scan paths 1 and 2; (d) scan path images.

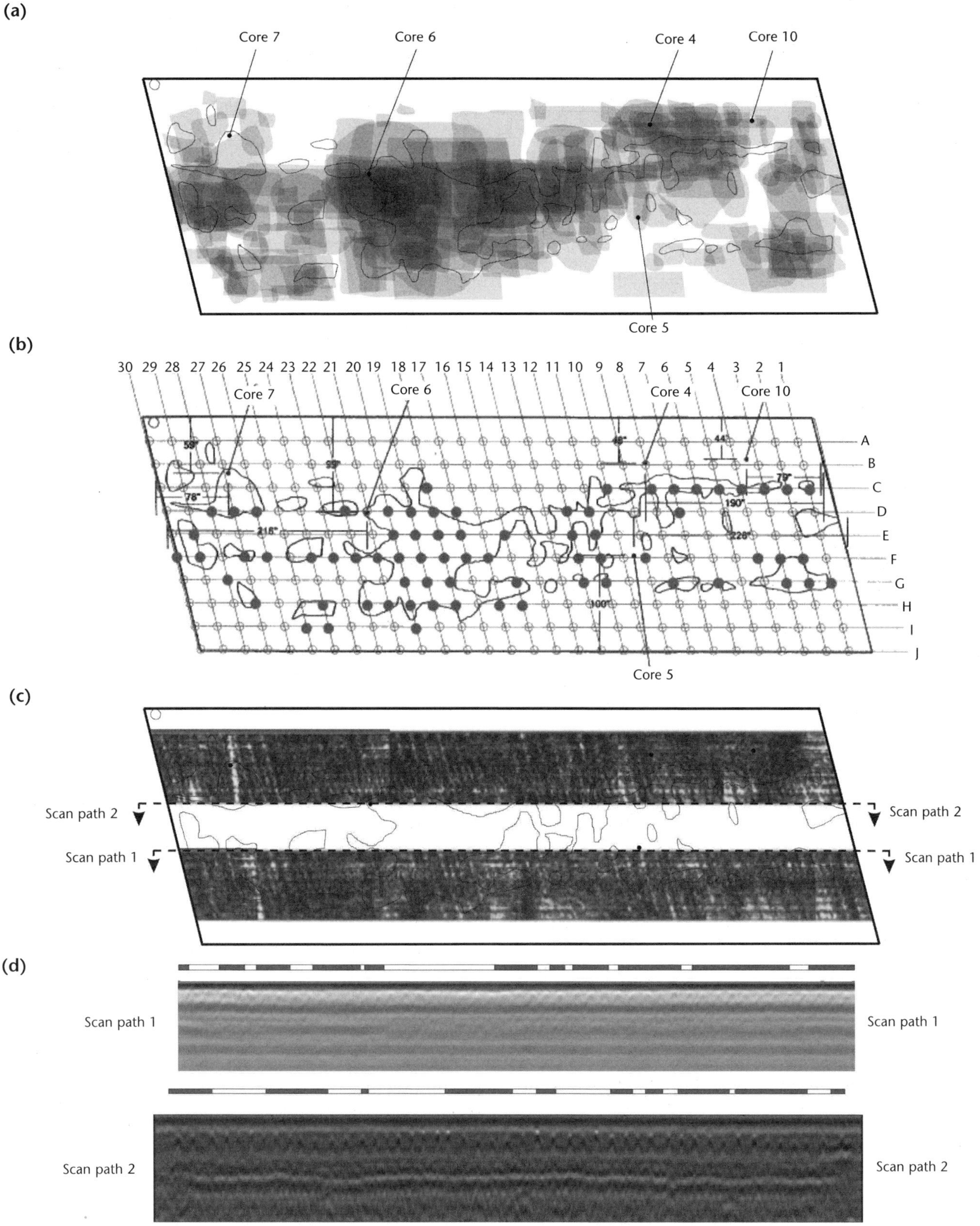

processed with wave field backpropagation[14] and were collected by traversing the bridge deck twice with an array of radar antennas at 8 $km \cdot h^{-1}$ (5 $mi \cdot h^{-1}$). This imaging indicates few areas of agreement between acoustic chain drag test results and ground penetrating radar system test results for delamination detection. Raw radar data from a ground penetrating radar antenna along scan path 1 in Fig. 10c are shown with raw data collected along scan path 2. These raw data illustrate typical radar responses from both systems; corresponding color bars indicate delamination detection results for both tests. Figure 10c radar results are consistent with the Fig. 9 results.

Lake Anna Bridge Deck

Figure 11 presents results from a Federal Highway Administration ground penetrating radar system.[10] These results were obtained from the Lake Anna Bridge deck. A diagram in Fig. 11a illustrates a portion of the bridge deck in plan view on a 0.6 × 0.6 m (24 × 24 in.) grid. An area is highlighted to indicate the delaminated areas detected using an acoustic chain drag technique and cored locations are indicated with black circular symbols. Figures 11b to 11d illustrate images derived from system data by using wave field backpropagation. These results are all presented to scale in a plan view (Fig. 11a). Figure 11b illustrates data from the deck surface, Fig. 11b shows data from the top reinforcing steel layer and Fig. 11c shows data from the bottom reinforcing steel layer. These images indicate strong radar reflections in light colors up to white and lower magnitude responses in dark colors down to black.

An area of interest in Fig. 11b is the bright response, to the right and below the middle of the image. This location corresponds to surface patch material that was used to repair the bridge deck. Figure 11c illustrates a response to a delaminated area indicated by an undulating boundary traversing the width of the image. On one side of the boundary, reinforcing steel is imaged clearly and sound concrete areas produce low magnitude radar reflections. On the other side of the boundary, reinforcing steel imaging is degraded and delaminated areas produce a significant radar reflection between reinforcing steel locations. The boundary between these areas has a geometry similar to that of the boundary identified in the chain drag testing but is shifted. Core testing indicates that shift can be accounted for by inaccuracies in data registration along the curved bridge section from which the data were acquired. Figure 11d illustrates the response at the bottom mat of the reinforcing steel, where the reflection from the delamination in the top mat of reinforcing steel has occluded features that would have otherwise been imaged in the bottom reinforcing steel mat.

Based on the ground penetrating radar data presented in Figs. 9 to 11, additional information can be derived, such as concrete cover depth and bridge deck thickness. These engineering measurements are subject to measurement errors caused by heterogeneities in concrete properties but their information is still useful, particularly when the measurements are made over an entire bridge deck. Delaminations, voids and other discontinuities and distress in bridge decks can be detected by ground penetrating radar when the resolution limits of a system are high enough. The results indicate that test methods that

FIGURE 11. Ground penetrating radar results from Lake Anna Bridge deck: (a) plan view of part of bridge deck; (b) backpropagation image from surface of bridge deck; (c) backpropagation image from top steel layer; (d) backpropagation image from bottom steel layer.

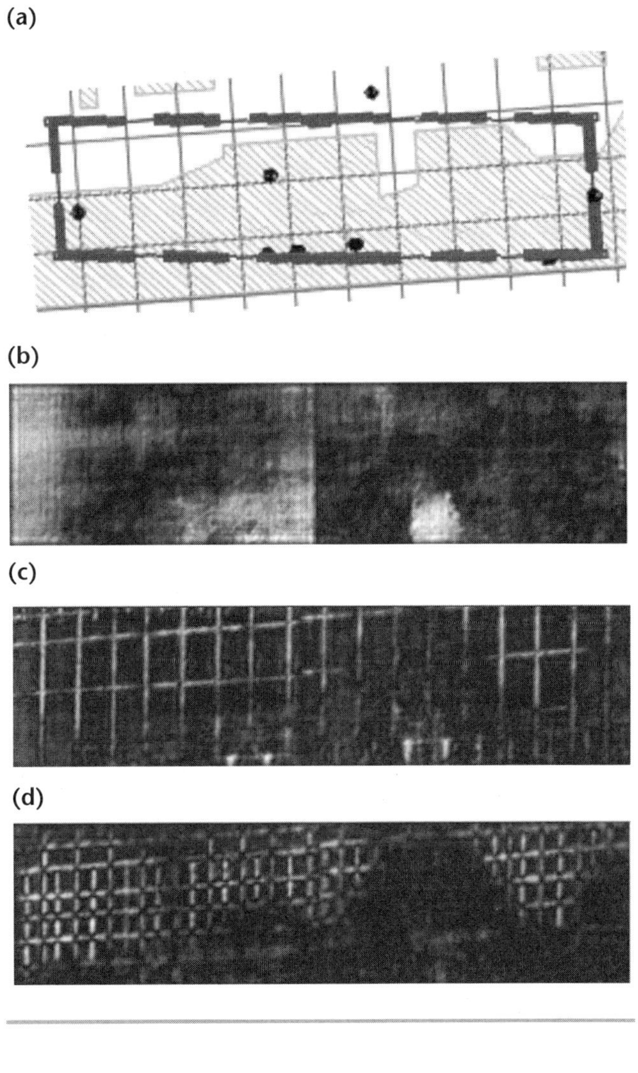

complement ground penetrating radar can be used to confirm the types of features a specific ground penetrating radar system can detect in a particular bridge deck.

Recommended Practices

Ground penetrating radar references that provide useful guidance to bridge engineers and ground penetrating radar practitioners include the following: AASHTO TP36, *Standard Test Method for Evaluating Asphalt Covered Concrete Bridge Decks Using Pulsed Radar;*[16] NCHRP Synthesis 255, *Ground Penetrating Radar for Evaluating Subsurface Conditions for Transportation Facilities;*[17] ASTM D 4748-87, *Test Method for Determining the Thickness of Bound Pavement Layers Using Short-Pulse Radar;*[18] and ASTM D 6432-99, *Standard Guide for Using the Surface Ground Penetrating Radar Method for Subsurface Investigation.*[19]

Conclusions

Modern ground penetrating radar technology provides a tool to evaluate distress and to measure subsurface engineering parameters of bridge decks. Example results have been presented for ground penetrating radar applications to specific bridge decks. These test results illustrate an approach to extracting engineering information from ground penetrating radar data. Bridge engineers and managers can subsequently use these data to make informed decisions. Although this technology has matured significantly, engineering judgment is required to select appropriate equipment, calibration techniques and corroborating test methods. Bridge decks are a demanding application of ground penetrating radar. The rigor of the application requires refined ground penetrating radar technology.

PART 3. Magnetic Flux Leakage Testing of Wire Rope

Background

Because the reliability of wire ropes is crucial for the safety of many mining, oil industry, crane and ski lift operations, concern with their integrity is a constant preoccupation of users and safety authorities. In spite of these concerns, a frequent reluctance to apply appropriate wire rope test techniques and retirement criteria compromise safety in many cases although dependable test techniques and instrumentation are available. Effective procedures, combined with a good understanding of degradation mechanisms and discard criteria, can notably increase wire rope safety.

For example, advanced electromagnetic wire rope test equipment of the magnetic flux leakage type has been developed since the 1960s. These instruments provide an important and in many cases indispensable element of wire rope testing.

Standard for Electromagnetic Tests of Wire Rope

ASTM E 1571, *Standard Practice for Electromagnetic Examination of Ferromagnetic Steel Wire Rope,*[20] describes several electromagnetic techniques to detect discontinuities and changes in metallic cross sectional area in ferromagnetic wire rope products. ASTM E 1571 is for rope diameters up to 64 mm (2.5 in.); larger diameters may be included, subject to agreement by the users of this practice. The standard practice also covers reference standards for wire rope testing.

Of the techniques described in ASTM E 1571, one type of magnetic flux leakage testing has been applied widely for wire rope testing and is the subject of the present discussion.

Approaches

Two different and distinct magnetic techniques have evolved for the detection and measurement of rope discontinuities.

1. Testing for loss of metallic cross sectional area quantitatively measures external or internal loss of metal because of corrosion (due to environmental conditions or poor lubrication) and wear (caused by rubbing along floors, by nicking, by high pressures and by poor lubrication).
2. Localized discontinuity testing qualitatively detects a wide variety of external and internal discontinuities such as broken wires and corrosion pitting. Broken wires are usually caused by fatigue, interstrand nicking and martensitic embrittlement.

Modern dual-function electromagnetic rope testers (Fig. 12) allow simultaneous tests for loss of metallic cross sectional area and localized discontinuities.

Underlying Principles

For dual-function instruments, strong permanent magnets induce a magnetic flux at the saturation level in the rope in the axial (longitudinal) direction. Various types of sensors close to the rope — such as coils, hall sensors or flux gate sensors — sense and measure the magnetic flux.

Any discontinuity — such as a broken wire or corrosion pitting — distorts the magnetic flux in the rope and causes it to leak from the rope. For localized discontinuity tests, the radial component

FIGURE 12. Electromagnetic testing of mooring rope.

of the leakage flux is measured by sensors called *radial sensors*. Note that these sensors are also called *differential sensors* because they sense only changes of the magnetic flux in the rope and not the flux itself. Therefore, discontinuity detection depends on a rapid change of the magnetic flux, typically caused by broken wires and corrosion pitting. Differential sensors cannot detect more gradual changes of the rope flux, changes typically caused by wear and corrosion.

The axial component of the leakage flux can be measured by axial sensors. This axial leakage flux signal is frequently offered as a substitute for the loss of metallic cross sectional area signal. However, while the axial signal can be useful for qualitative discontinuity characterization, it is very complex and cannot be used directly to determine the loss of metallic cross sectional area of a rope. Therefore, instruments that use this approach are not of the dual-function type as defined above. Techniques for the quantitative measurement of the loss of metallic cross sectional area of a rope are discussed below.

When a rope is magnetically saturated, the axial magnetic flux in the rope is proportional to its cross sectional area. Therefore, any loss of metallic cross sectional area can be determined by measuring this magnetic flux. Two types of sensors can be used to measure magnetic flux: hall sensors (or flux gate sensors) and coils in combination with electronic integrator circuits.

To measure flux density, hall sensors (and flux gate sensors) must be physically inserted directly into the magnetic flux path. Thus, the flux to be measured must intersect the sensors. This is not possible when measuring the magnetic flux inside the rope. Therefore, instruments that use hall sensors or flux gate sensors must always resort to an indirect technique for determining the axial rope flux. They measure some flux density outside the rope and determine or estimate the longitudinal rope flux from the external flux measurement.

Alternatively coils with integrators can be used. Because coils must encircle the magnetic flux to be measured, they can directly measure the magnetic flux inside the rope.

In general, two techniques are used for the determination of loss of metallic cross sectional area: the main flux technique and the return flux technique (Fig. 13).

Main Flux Technique

The main flux technique uses an annular coil together with an electronic integrator circuit to determine the local magnetic flux inside the rope (Fig. 13a). Note that

FIGURE 13. Cross sectional area measurement techniques: (a) main flux technique; (b) modified main flux technique; (c) return flux technique with hall sensors; (d) return flux technique with flux gate sensors.

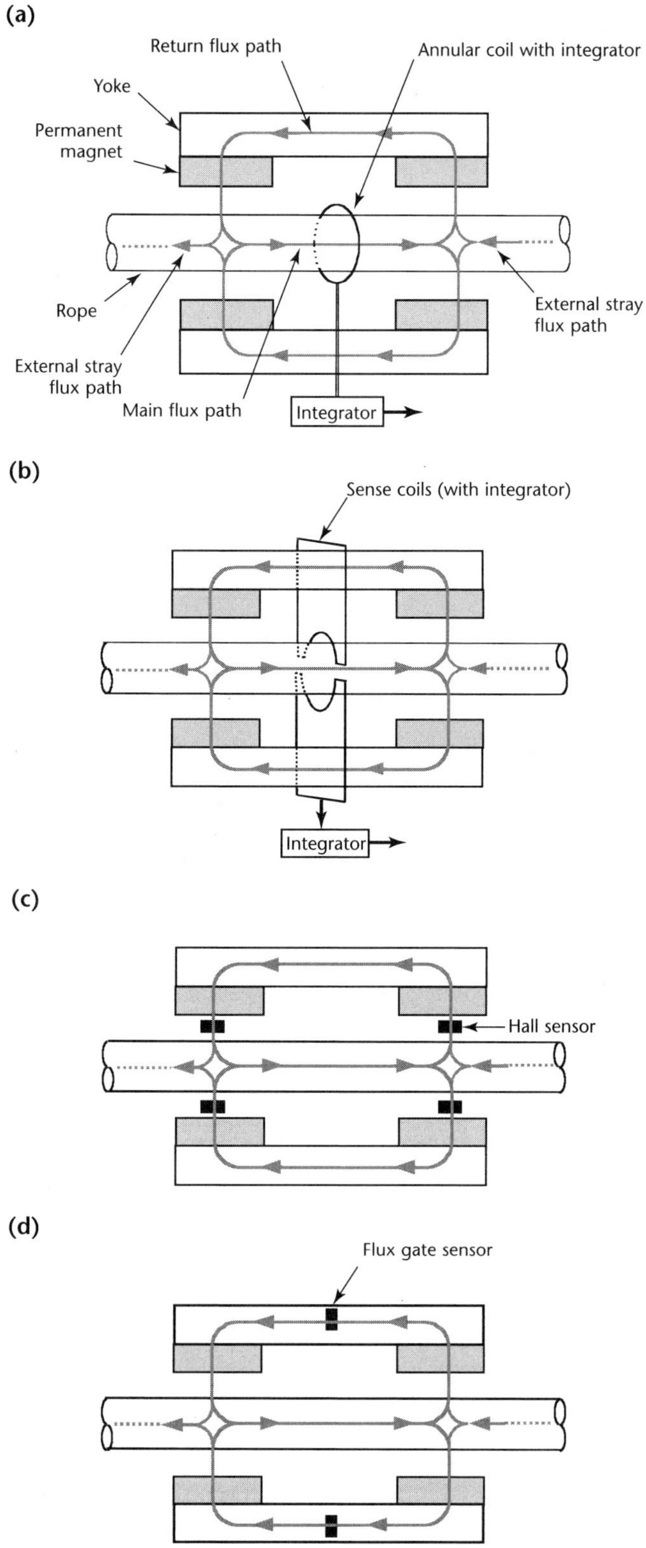

the coil must encircle the rope. Originally patented in the United Kingdom in the 1960s, this approach has been discussed in the literature.[21-23] Because it measures the magnetic flux inside the rope locally, the annular coil approach offers uncommon resolving power, signal fidelity and therefore test accuracy. The performance of this arrangement is unsurpassed and sets a standard by which all other techniques are measured.

Unfortunately, it is topologically impossible to implement a hinged annular coil with a large number of turns that can be opened and conveniently attached to the rope. Consequently, the practical implementation of this technique for inservice wire rope tests is seriously hampered by an inherent and insurmountable problem: an annular coil encircling the rope must be wound onto the rope in the field for each test. This cumbersome procedure allows only very few turns (say, 100) and hence only very small induced coil voltages. The coil voltages are of the same order of magnitude as the always present inherent offset voltages at the input of operational amplifiers used for the design of electronic integrator circuits. These inherent offset voltages make the long term, low drift integration of the coil voltages impossible. Hence, the annular coil approach is not feasible for inservice tests where loss of metallic cross sectional area measurements over longer time periods — say, over more than a few minutes — are required.

To illustrate the annular coil approach, Fig. 14 shows the loss of metallic cross sectional area and localized discontinuity traces of a laboratory test rope. The loss of metallic cross sectional area and localized discontinuity signals were acquired with an annular coil. Short pieces of wire, attached to the rope, simulate anomalies. The attached wires have different lengths as indicated. They typically represent a one percent increase of metallic cross sectional area. (The loss of metallic cross sectional area caused by the internal wire is unknown.) The two ends of the rope are welded together to form an infinite loop. The weld is also indicated in the chart.

Figure 14 shows the excellent results that could be obtained with annular coils. The increases of metallic cross sectional area caused by the attached wires are clearly indicated with their full magnitude for wires longer than about 50 mm (2 in.). The cross section changes caused by shorter wires are also indicated, albeit not to their full extent.

Modified Main Flux Technique

The modified main flux technique (Fig. 13b) tries to retain the superior performance of the main flux technique while allowing the use of hinged sensor heads shaped like a clamshell.[22,23] This design makes it easy to attach the sensor head to the rope, even under adverse field conditions.

The test signals of this modified main flux technique are a combination of two

FIGURE 14. Test results from main flux technique.

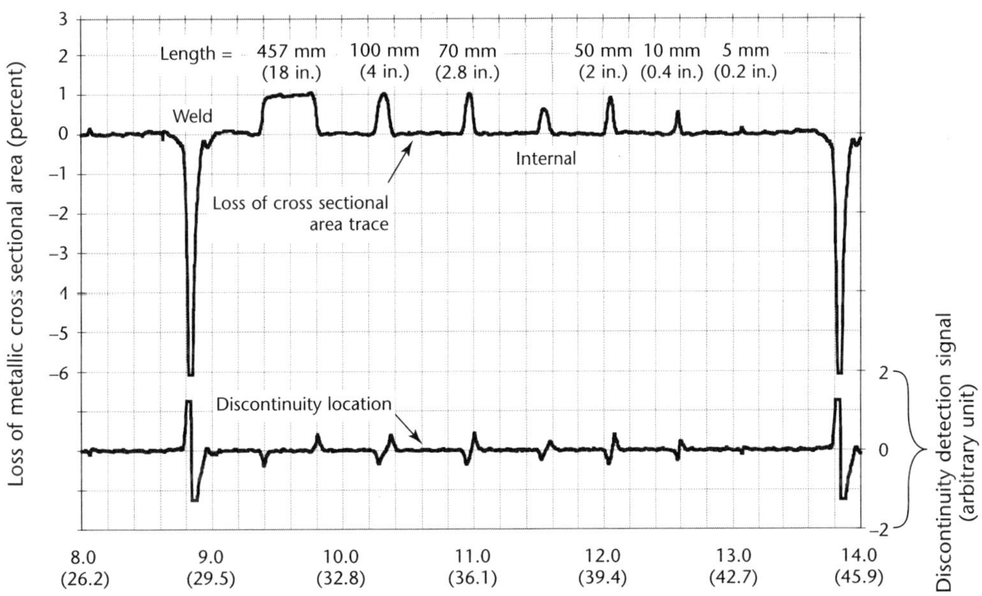

signal components: the main flux signal (Fig. 14) and a signal component caused by the outside stray flux, that part of the magnetic flux that flows along some external stray flux path outside the sensor head (Fig. 13).

Unfortunately, this outside signal component is significant. It compromises the quality of the test results and must be considered parasitic. Although this parasitic effect can be minimized, a true signal of the main flux type cannot be restored. Nevertheless, the modified main flux technique retains many features of the annular coil approach.

Return Flux Technique

The return flux technique uses hall sensors[24] (or more complex, flux gate sensors[25]) to measure the magnetic flux in the magnetic return path of the instrument.[26] Illustrated by Figs. 13c and 13d, the return flux is equal to the average value of the axial rope flux inside the sensor head plus the outside stray flux. Therefore, the return flux provides an estimate of the average cross sectional area of that section of the rope inside the sensor head. Flux sensors can either be inserted into the air gap between the permanent magnet poles and the rope or into the yoke of the magnetizer assembly. In contrast to the main flux approach, the return flux technique allows the sensor heads to be hinged like a clamshell. This shape makes it easy to attach the sensor head to the rope even in adverse field conditions.

Signal Generation and Evaluation[27]

For the interpretation of test results and for a correlation of test data with the actual rope condition, the rope inspector must understand the capabilities and limitations of the electromagnetic rope testing equipment. A proper appreciation of the signal generation process, together with an understanding of rope degradation mechanisms and discard criteria, is essential for making rational rope retirement decisions.

The functional block diagram of Fig. 15 illustrates the signal generation process. This figure shows the rope's cross sectional area — including variations caused by broken wires, corrosion, abrasion and other effects — as the input to an electromagnetic wire rope test system. From this input, the sensor head produces one or several electrical signals. These signals are electronically processed to produce the localized discontinuity and cross sectional signals, which are then recorded by a chart recorder or stored by a data acquisition system.

For the following discussion, step changes of metallic cross sectional area — caused by missing or added wires, for example — have particular significance. Because of its simple geometry, a step change can be called a *fundamental discontinuity*. Accordingly, the corresponding loss of metallic cross sectional area and localized discontinuity signals, caused by a fundamental discontinuity, can be called a *fundamental loss of metallic cross sectional area* or *localized discontinuity signals*, respectively. The fundamental signals can also be called an instrument's *step response*.

Any discontinuity can be represented as the sum of appropriately scaled and spaced fundamental discontinuities. Moreover, the process of signal generation is linear or almost linear. Hence linear superposition applies: if a discontinuity can be represented as the sum of several fundamental discontinuities, then the corresponding discontinuity signals are the sum of the corresponding fundamental loss of metallic cross sectional area and localized discontinuity signals. The concepts of fundamental discontinuities and signals, step response and linear superposition are discussed in the literature.[28] Determining and evaluating its step response is an excellent technique for characterizing the performance of an electromagnetic wire rope tester.

Filtering

The concept of filtering can be understood very loosely. For example, the sensor head of an electromagnetic rope tester, together with the signal electronics, may be viewed as a linear or nonlinear data filter. Figure 16 depicts the rope cross section as the input signal and the idealized corresponding loss of metallic cross sectional area and localized discontinuity output signals. Note that, for many rope testers, the localized discontinuity signal approximates the first

Figure 15. Functional block diagram of signal generation process.

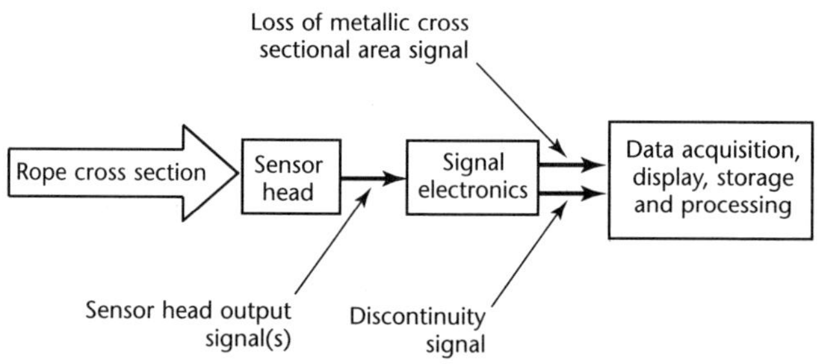

derivative of the rope cross section signal. For other instruments, the localized discontinuity signal is the second derivative of the rope cross section, shown as the alternative localized discontinuity signal in Fig. 16. Note that the signals generated with the main flux technique in Fig. 14 closely resemble the idealized results of Fig. 16.

Recognizing differentiation as the quintessential high pass filter operation, the localized discontinuity signal can be considered as the rope cross section input signal that has been high pass filtered. High pass filtering accentuates fast changes of signals and typically broken wires and corrosion pitting cause rapid variations of the rope cross section. Therefore, the high pass filtering feature makes the localized discontinuity signal useful for the detection of broken wires and corrosion pitting.

Any test equipment should present data in a form that facilitates interpretation by the human operator. Figure 16 shows that a perfect loss of cross sectional area signal could serve as an accurate and conceptually simple map of a rope's loss of metallic cross section — a map easy to interpret by the inspector.

Unfortunately, actual loss of metallic cross sectional area signals are far from this ideal. Although most rope testers can produce localized discontinuity signals with wave shapes very similar to those of Fig. 16, producing a loss of metallic cross sectional area signal that comes close to the idealized signal in Fig. 16 poses considerable problems.

To illustrate, some electromagnetic rope testers produce cross sectional step responses with considerable overshoot in both directions. The overshoot makes loss of metallic cross sectional area measurements complex, ambiguous and operator dependent. Chart interpretation for these instruments becomes especially problematic under actual field conditions. Because this type of performance is not amenable to analysis, it will not be discussed further.

FIGURE 16. Input and output signals of idealized rope test instrument.

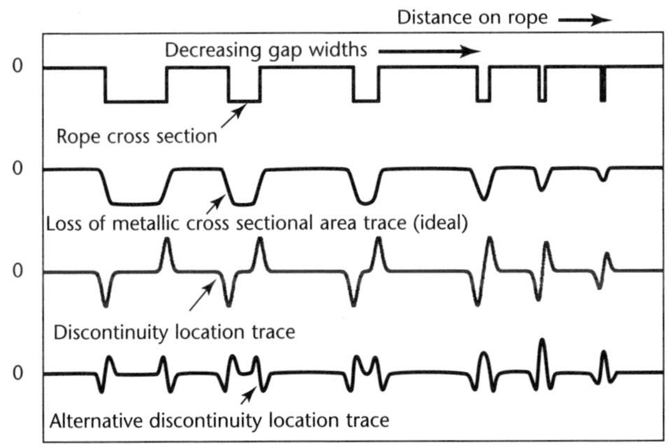

Resolution and Averaging Length

Resolution refers to the fineness of detail that can be distinguished in an image. For electromagnetic wire rope tests, resolution is always the foremost performance measure. In nondestructive testing, the terms *resolution* and *test accuracy* are often used synonymously.

In the discipline of electromagnetic wire rope testing, quantitative resolution[22] or averaging length (sometimes also called *scanning length*) is defined as the minimum length of a uniform anomaly for which the sensor provides an accurate measurement of a rope's loss of metallic cross sectional area.

To visualize the concept of averaging length, assume that, instead of measuring metallic cross sectional area directly, the rope tester continuously measures the metallic volume of consecutive rope sections with lengths that are equal to its averaging length. Figure 17 illustrates this concept.

Figure 17a shows a hypothetical rope with a uniform 10 percent loss of metallic cross sectional area extending over a length of 300 mm (12 in.). An instrument with a 50 mm (2 in.) averaging length will correctly measure this loss of metallic cross sectional area. As illustrated by Fig. 17b, a rope tester with an averaging length of 300 mm (12 in.) will also give a true indication of this anomaly.

Now consider a hypothetical rope with a 10 percent uniform loss of metallic cross sectional area extending over a length of 50 mm (2 in.). Figure 17c shows that an instrument with an averaging length of 50 mm (2 in.) can measure the loss of metallic cross sectional area caused by this anomaly. However, as can be seen from Fig. 17d, an instrument with an averaging length of 300 mm (12 in.) would indicate the same anomaly as a 1.7 percent loss of metallic cross sectional area extending over a length of 300 mm (12 in.) — a very inaccurate indication of the true rope condition. These examples show the importance of a short averaging length.

Note that signal averaging is a quintessential type of low pass filtering and that signals lose a significant amount of information (details) by low pass filtering. Figure 18 illustrates this situation. It shows how the quality of loss of metallic cross sectional area signals deteriorates as the loss of metallic cross sectional area averaging length increases.

An analogy can illustrate the problems associated with long loss of metallic cross sectional area averaging lengths: a chain is only as strong as its weakest link. Obviously, the strength of a chain is not determined by the average strength of some of its links. Similarly, the strength of a rope, which has lost metallic cross section by corrosion or wear, is determined by the minimum local metallic cross sectional area along the rope's length and not by some average value of the rope's cross sectional area.

Experience has shown that serious rope deterioration can occur over very short distances along the length of a rope. Hence, to determine and evaluate a rope's actual metal loss with acceptable accuracy, a short averaging length — of no more than a few centimeters (an inch or two) — is important.

Because all wire rope testers have a quantitative resolution or averaging length that is greater than zero, an accurate measurement of loss of metallic cross sectional area always requires minimum lengths of anomalies. As the above discussion shows and as illustrated by Fig. 18, the concept of quantitative resolution or averaging length is important for specifying and comparing the performance of rope testers.

FIGURE 17. Length of 10 percent loss of metallic cross sectional area versus averaging length of test system: (a) long loss, short averaging; (b) long loss, long averaging; (c) short loss, short averaging; (d) short loss, long averaging.

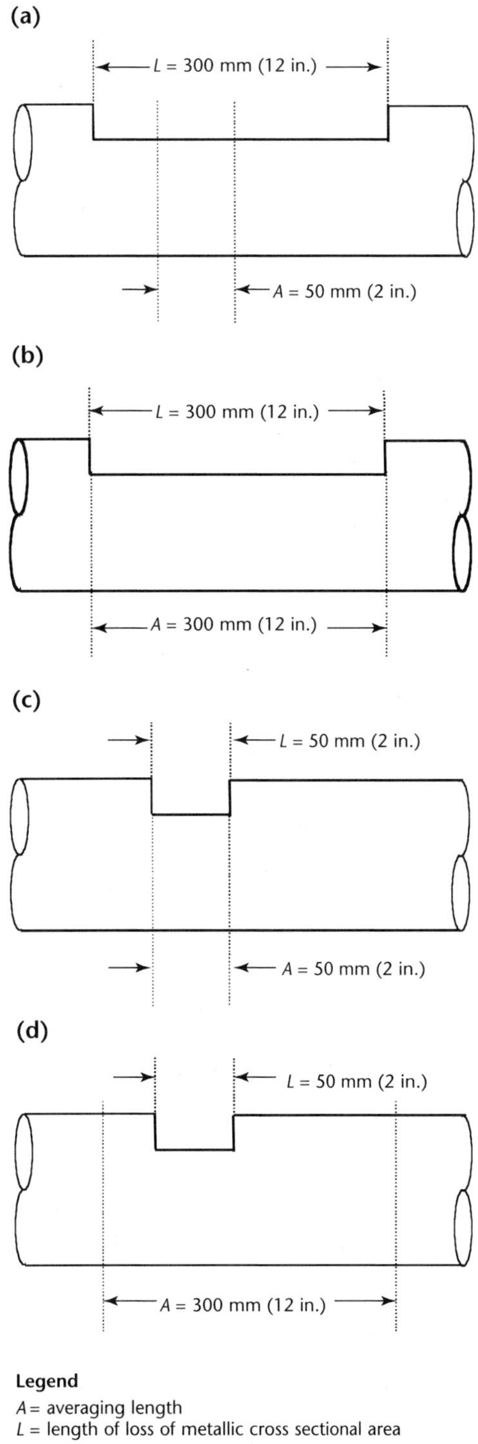

Wire Rope Testing and Retirement

Two different philosophies have been used to retire wire rope.

1. A *statutory life policy* mandates rope retirement at certain prescribed intervals. (The statutory life policy specifies a maximum time a rope can be in service).
2. *Retirement for cause* is based on retirement conditions evaluated periodically by nondestructive testing. (The approach requires that the rope must be retired when the deterioration exceeds a certain limit.)

Because a statutory life policy is inherently wasteful, regulators have

FIGURE 18. Loss of metallic cross sectional area signals as function of averaging length.

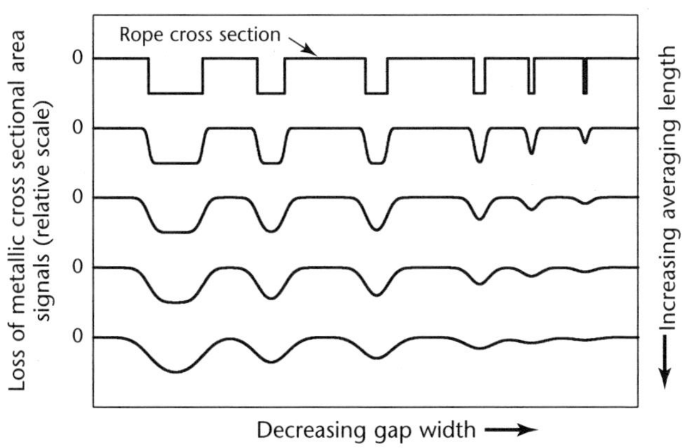

tended to adopt the retirement for cause approach wherever appropriate.

Wire rope deteriorates gradually throughout its service life. To keep abreast of deterioration, wire rope must be periodically tested. Because moderate deterioration is normally present, the mere detection of rope deterioration does not usually justify rope retirement.

There are two major nondestructive test methods for evaluation of rope degradation: visual testing and electromagnetic testing.

Visual Testing

The rag-and-visual technique is a simple yet useful way to detect a wide variety of external rope deteriorations. Using this approach, the inspector lightly grasps the rope — which moves at test speed — with a rag or cotton waste. External broken wires often become frayed and, as the rope moves, the broken wires snag the rag or cotton waste. The rope is then stopped at that point and the inspector assesses the rope condition by a visual test.

If broken wires do not fray, then a different test procedure must be used. The rope is moved 0.5 to 1 m (2 or 3 ft) at a time and visually tested at each stop. This procedure is tedious and, because the rope is often covered with grease, many external and internal discontinuities elude detection.

Another visual testing tool is measurement of the rope diameter. Rope diameter measurements compare the original diameter — when new and subjected to a known load — with the current reading under like circumstances. A change in rope diameter indicates external and internal rope damage. Inevitably, many sorts of damage do not cause a change of rope diameter.

Several visible signs can indicate distributed and cross sectional area losses due to corrosion, abrasion and wear. For example, corrosion products, flattening of outer wires and loss or, sometimes, increase of rope diameter frequently reveal external and internal corrosion. However, the extent of corrosion is often difficult to gage and its significance is even more difficult to assess.

Visual testing is inherently not well suited for the detection of internal rope deterioration. Therefore, it has limited value as a sole means of wire rope testing. However, visual testing is simple and does not require special instrumentation. When combined with the knowledge of an experienced rope examiner, visual testing can provide a valuable tool for evaluating many forms of rope degradation.

Electromagnetic Tests

Electromagnetic wire rope testing gives detailed insight into the condition of a rope. Its reliability has made electromagnetic techniques widely accepted for the testing of wire ropes in mining, for ski lifts and many other applications.

As discussed, two distinct electromagnetic test techniques have evolved to detect and classify discontinuities as either loss of metallic cross sectional area or localized discontinuity.

1. The localized discontinuity test, like the rag-and-visual method, is suited only for the detection of localized discontinuities, especially broken wires. Therefore, small hand held localized discontinuity instruments have been called *electronic rags*.
2. The loss of metallic cross sectional area test detects and measures changes of metallic cross section caused by wear and corrosion. More reliable than visual diameter checks, loss of metallic cross sectional area testing can replace diameter measurements made with a caliper. Therefore, loss of metallic cross sectional area instruments could be called *electronic calipers*.

Electromagnetic and visual wire rope tests complement each other. Both are essential for safe rope operation and both methods should therefore be used for maximum safety. The thrust of evolving regulations is clearly toward combined periodic electromagnetic and visual testing.

A thorough test must consider all aspects of a rope's condition, including (1) the findings of a visual test, (2) the results of an electromagnetic rope test, (3) the rope's operating conditions and related damage mechanisms and (4) the history of the rope under test and that of its predecessors.

Dependable test procedures, using combined visual and electromagnetic testing, can detect rope deterioration at its earliest stages. Wire rope users can use these techniques as an effective tool for preventive maintenance. There are several ways that these nondestructive tests can be used to prevent deterioration of wire ropes.

1. The early detection of corrosion allows immediate corrective action through improved lubrication.
2. Accelerating wear and interstrand nicking can indicate a need to reline sheaves to stop further degradation.

3. Careful testing can monitor the development of local damage at the crossover points of the rope on a winch drum. This way, the operator can determine the optimum time for repositioning the rope on the drum.

A program of periodic testing is especially effective. To establish baseline data for subsequent testing, such a program should commence with an initial test of the installed rope after a break-in period. Subsequent testing should then be performed at scheduled intervals. In particular, periodic electromagnetic testing allows the documentation of a rope's deterioration over its entire service life.

Figure 19 shows the schematic of a mine hoist including sections of mine hoist and balance ropes that are particularly susceptible to rope deterioration.[29,30]

Types of Rope Damage

Broken Wires

In running ropes, broken wires develop primarily in sections that move over sheaves, pulleys and winch drums. Typically, they are caused by bending over — sheave fatigue cycling. (See rope sections A, B and C in Fig. 19.)

Usually, breaks develop in segments of the rope surface that come into direct contact with the sheave. Here, various contact phenomena compound the fluctuating bending stresses. Breaks in these areas are external and usually visible. However, internal breaks can also develop, depending on the loading and especially on the rope construction. Once broken wires appear, a good many more are likely to develop soon.

Corrosion (Rust)

Corrosion is a serious hazard to a wire rope. Corrosion pitting causes stress concentrations. Furthermore, corrosion pitting inhibits the free movement of wires and strands, producing additional stresses in wires. The increased wire stresses combined with the above mentioned stress concentrations can drastically accelerate the development of fatigue breaks. Wires can also corrode uniformly over their entire surface, which may reduce their cross sectional area and cause loose wires.

Rust can cause shallow pitting on the working surfaces of a rope where the steady rubbing action of the sheave prevents deep cavities. This mechanism accelerates wear. Furthermore, deep corrosion pitting on the surfaces of internal wires can severely shorten service life.

The severity of corrosion often varies along the length of a rope. Frequently, corrosion is localized yet dangerous.

A corrosive environment often exists in certain zones along the length of a mine shaft. Corrosion often occurs in rope sections that stay in these corrosive zones over extended periods.

Corrosive areas are usually located where large and abrupt changes of temperature or humidity occur. Examples are below points where water or moist air can enter the shaft, humid shaft pits, fog areas in ventilation shafts and locations with increased air velocities (Fig. 19).

The extent of corrosion is often difficult to gage and — as shown by

FIGURE 19. Sections of hoist ropes susceptible to broken wires (A and B) and corrosion (C, D and E).

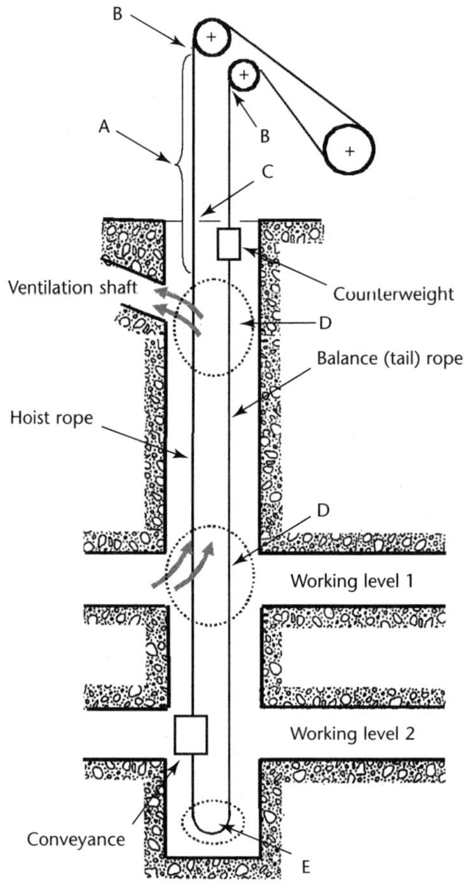

Legend

A. Acceleration section.
B. Rope sheave tangent points constitute nodes for all rope oscillations. (Oscillatory energy is absorbed and dissipated here.) Defective sheaves can cause corkscrew deformations.
C. Ropes can become wet at shaft collar.
D. Moist and warm air hits ropes in these positions.
E. Water and debris can accumulate inside loop of balance ropes.

experience — usually underestimated. Rust and dirt frequently clog up the rope surface and hide loose wires.

Wear

Wear results in loss of cross sectional area of the wires. The problems related to external and internal wear require special attention. External wear usually occurs on the working surface of a rope. Severe external wear can indicate that internal wires are similarly worn. Sometimes, severe wear can cause outer wires, or clusters of outside wires, to break abruptly. Rubbing between wires of a strand can cause internal wear.

Deformation and Mechanical Damage

Corkscrew type deformations can be caused by sheave grooves that are too tight, through manufacturing errors or as a result of severe wear. Corkscrew deformations can cause rope damage by increased exposure to wear. Furthermore, they increase the pressure between adjacent strands, which will eventually cause broken wires.

Kinks are permanent distortions caused by loops drawn too tightly. Loops, often precursors of kinks, are formed when a section of a rope under high torsion is allowed to become slack. Usually, ropes with kinks must be removed from service.

A common deterioration mode is peening, also called *plastic wear,* produced by localized impact or very high bearing pressure. Peening can occur by the slap of the rope at crossover points as the rope slips from layer to layer while winding on multilayer drums at high speeds. Peening sometimes gives the appearance of heavy wear although there is little loss of cross section. Peening can cause a fin on the edge of a worn wire that provides a ready site for the initiation of fatigue cracks.

Mechanical damage can have many causes such as a solid object hitting the rope, improper handling during rope installation, overloading or shock loading. Usually, mechanical damage is clearly visible and easy to detect. However, some mechanisms of mechanical damage, such as wire plucking, can be more difficult to locate. Mechanically equivalent to guitar string plucking, wire plucking is caused by lateral scraping of the rope at crossing points on a winch drum. It can lead to localized damage in the form of one or several broken wires at set intervals along the rope.

Martensitic Embrittlement

Martensite is a brittle phase of steel formed when the steel is heated above a critical temperature and then rapidly quenched. It occurs in wire rope as a result of frictional surface heating and the mass cooling effect of the cold metal beneath. Martensitic embrittlement can develop at rope crossover points while winding on multilayer drums. Here the rope can be heated by contact with the adjacent turn and then rapidly quenched by the surrounding metal. Martensite cracks easily and such cracks can propagate from the surface through the entire wire.

Combined Deterioration Modes

In practice, combined deterioration modes predominate. These include corrosion fatigue and corrosion assisted wear. Typically, one type of rope degradation can initiate and contribute to interactive deterioration mechanisms.

Corrosion assisted wear is probably the most common deterioration mode in stranded mine hoist ropes. Although broken wires do occur, most mine hoist ropes are retired because of unacceptable loss of metallic cross sectional area rather than broken wires.

Often, metal loss is caused by internal corrosion and abrasion that are invisible from the surface. As the various deterioration mechanisms progress, they will eventually cause broken wires or clusters of broken wires.

Electromagnetic tests of mine hoist rope show typical wear and fatigue chart patterns caused by bending over a sheave combined with acceleration and deceleration of the conveyance (Fig. 19). Typically, rope sections directly above the conveyance or counterweight show no deterioration because they never move over sheaves. In contrast, broken wires usually develop in rope sections that move over sheaves during acceleration or deceleration of the conveyance.

This typical pattern of rope damage can be explained as follows. When the conveyance starts at the bottom of the shaft, the rope conveyance arrangement develops rather slow longitudinal, lateral and torsional oscillations. When the conveyance reaches the top of the shaft, the initially low frequency oscillations become high frequency vibrations. The stopping and acceleration action of the conveyance at the shaft collar causes additional vibrations.

The tangent point of the rope and the sheave constitutes a node for all rope oscillations. These oscillations induce considerable tension, bending and torsional stresses at the tangent point. The oscillatory energy in the rope is dissipated and absorbed at this node. This action induces fatigue and abrasion at the tangent point.

Furthermore, as shown in Fig. 19, condensation water can form at the shaft collar. This, together with the fatigue damage, can cause fatigue corrosion and corrosion pitting in this section of the rope.

The localized discontinuity trace confirms the indications of the loss of metallic cross sectional area trace. In particular, the localized discontinuity trace indicates considerable corrosion pitting.

A common retirement criterion is that a rope must be retired when its loss of metallic cross sectional area exceeds 10 percent. For this reason, the condition of many mine hoist ropes is still considered acceptable although their charts indicate considerable degradation.

The above discussion shows that the best and worst sections of a rope are usually next to each other (Fig. 19). This proximity allows an easy comparison of the most degraded rope section with the rope essentially in its new condition. Fortunately, this proximity occurs in many cases and greatly facilitates rope evaluation.

Case Histories

Winch Ropes[31]

A locked coil rope was mounted on a mobile winch for about ten years after three months of earlier use on a friction winder installation (Fig. 20). The rope showed clear evidence of external corrosion, variable along the test length. Using retirement criteria appropriate for visual testing, this rope would have been rejected for further use. Because of its service history, the rope was not believed to contain any internal local discontinuities.

After dismantling, the rope showed severe corrosion on the outer layer and also significant corrosion on the second layer. The rope showed less corrosion in the third layer and appeared undamaged from the fourth layer inward because lubricants were still present. No local discontinuities were found.

Figure 20 shows an electromagnetic chart of this rope. The maximum measured loss of metallic cross sectional area is 5.1 percent compared to the best section on the rope covered by the chart. Note that the most convenient calibration technique for electromagnetic testing is to attach a calibration wire with known cross sectional area to the rope. In Fig. 20, a wire bundle that represents about a one percent increase in rope cross section is fastened to the rope and used for calibration.

Figure 20 shows variable corrosion, corrosion pitting and possibly broken wires. As discussed previously, the slap of the rope at the crossover points can cause peening, martensitic embrittlement and wire plucking with the associated rope damage as the rope crosses over from layer to layer on a drum. The deterioration pattern indicated to the left of the chart is typical for ropes that wind

FIGURE 20. Chart of locked coil rope.

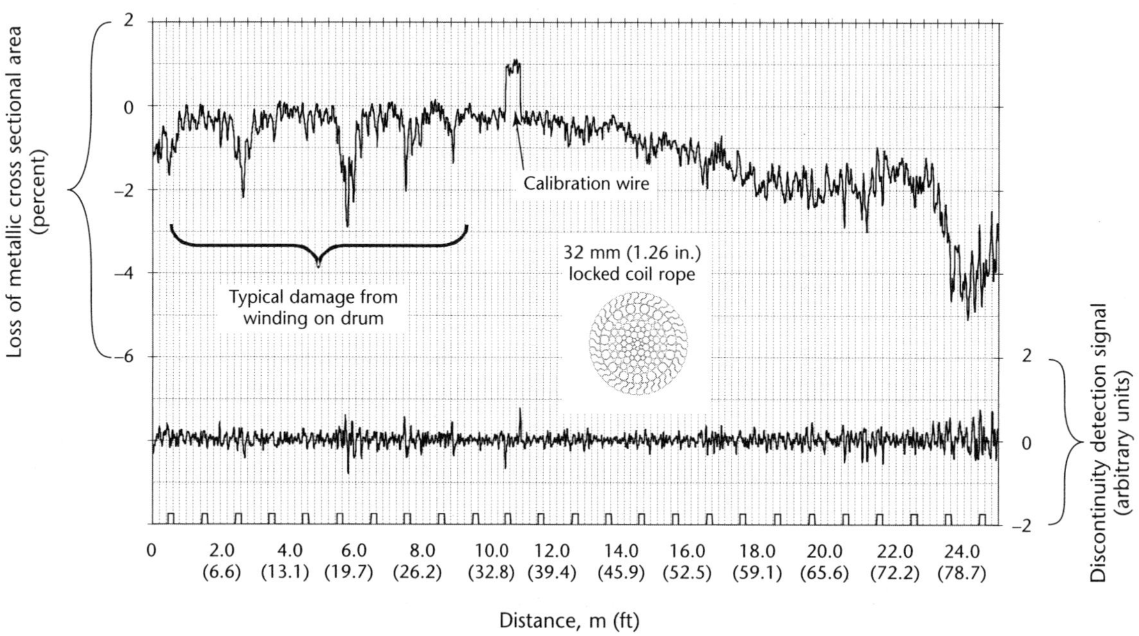

on a drum with the worst deterioration occurring at the crossover points as the rope slips from layer to layer while winding on a multilayer drum.

Multistrand Ropes[32,33]

Many ropes are of the torque balanced multistrand type comprising two or more layers of strands. Figure 21a shows a cutaway section of such a rope. Torque balance is achieved in multistrand ropes by laying outer and inner strands in opposite directions.

This type of rope construction limits axial rotation of the freely suspended rope under load. However, although multistrand ropes offer flexibility and a wear resistant surface profile, they have a tendency to deteriorate internally.

When multistrand ropes bend over sheaves or on a drum, they are subject to the combined effect of radial loading, relative motion between wires and bending stresses. This causes fretting wear or fatigue and interwire nicking across the interface between layers.

FIGURE 21. Multistrand rope: (a) construction; (b) cross section.

(a)

(b)

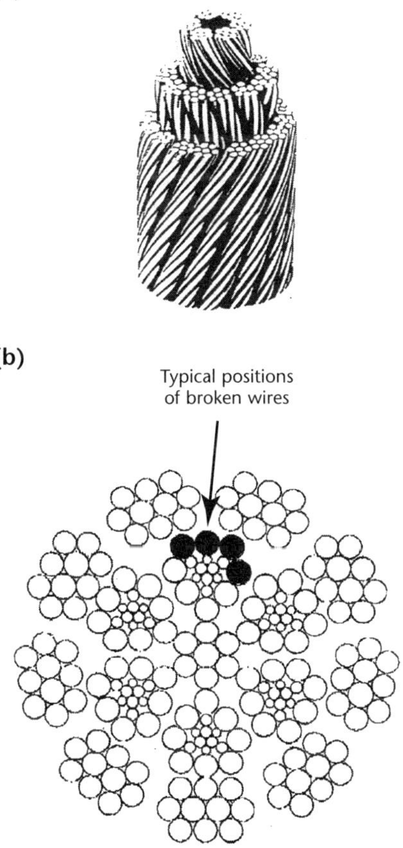

Therefore, multistrand ropes tend to develop internal broken wires. This breakup occurs primarily on the interface between the outer and second layer of strands, usually with no externally visible signs (see Fig. 21b). The wires in the second layer of strands typically show interstrand nicking and breaks caused by a combination of fluctuating axial wire stresses, motions between wires and fluctuating radial loads. The broken wires usually show squared off and Z shaped ends typical for fatigue breaks.

Many multistrand ropes are subject to corrosive environmental conditions. For example, offshore ropes are either immersed in the sea or continually wetted by salt water spray. In addition, heavy use in a marine environment can displace and degrade the rope lubricant. The combined effects of fatigue, corrosion and lubricant degradation can cause rapid internal deterioration where there is no effective form of protection.

Balance Ropes[34]

Figure 22 shows an electromagnetic test chart recording of a mine balance rope (see Fig. 19). Note that frequently torque balanced ropes of the multistrand type are used as balance ropes because regular stranded ropes develop internal torques during operation. Because balance ropes are under very little tension, these torques can cause loops, which are often precursors of kinks. Ropes with kinks must be removed from service. The use of multistrand ropes avoids this problem.

Figure 22b shows the typical corrosion chart pattern caused by the accumulation of water and debris inside the loop (see Fig. 19) of a multistrand balance rope. Water and debris tend to accumulate especially inside this loop while the conveyance is parked at the shaft collar for long periods. A cross sectional diagram of this rope is shown in Fig. 22b. Note that, as discussed above, the chart shows the best rope section — essentially with no rope deterioration — right next to the worst section. This feature greatly facilitates rope evaluation.

The chart shows a loss of metallic cross sectional area of 19 percent, which clearly calls for rope retirement. A subsequent destructive break test showed a loss of breaking strength of 48.3 percent for the worst rope section. On the other hand, the cross sectional area of the second layer of strands is about 47 percent of the total rope cross section. This suggests that the second layer of strands has completely lost all load bearing capability.

Experimental Multistrand Ropes[34]

This case study illustrates electromagnetic wire rope testing for the detection and quantitative characterization of internal broken wires and clusters of broken wires. The present experiment deals with the testing of a torque balanced multistrand rope with no corrosion and many broken wires. This rope had been used as a mine hoist rope on a trial basis and was known to contain numerous internal broken wires along its entire length.

The task at hand was to determine the number of broken wires in 100 mm (4 in.) segments along the length the rope. The difficulty of this quantitative discontinuity characterization problem was compounded by the fact that an undamaged rope section — usually the segment directly above the conveyance — was not available for comparison. Furthermore, at the time, the correlation of the typical deterioration modes of this and similar ropes with their electromagnetic test results was generally not well understood.

A cross sectional diagram of this multistrand rope is shown in Fig. 21b. As indicated in the figure, it is known that broken wires in multistrand ropes usually develop at the interface between the first layer and the second layer of strands. In addition, from this and similar ropes' service histories, it can be assumed that the rope under test has developed significant interstrand nicking together with numerous fatigue breaks of wires in the second layer of strands.

The detailed detection and quantitative characterization of internal broken wires in ropes with many breaks and clusters of breaks pose problems, as in this case. Difficulties are caused by the fact that, for electromagnetic wire rope testing, the indication of a broken wire is influenced by a number of parameters like broken wire cross sectional area and broken wire gap width. Difficulties are caused by the position of the broken wire within the cross section of the rope. For clusters of broken wires, an additional problem is caused by the fact that the relative position of broken wires with respect to

FIGURE 22. Electromagnetic test of mine hoist balance rope: (a) cross section; (b) chart.

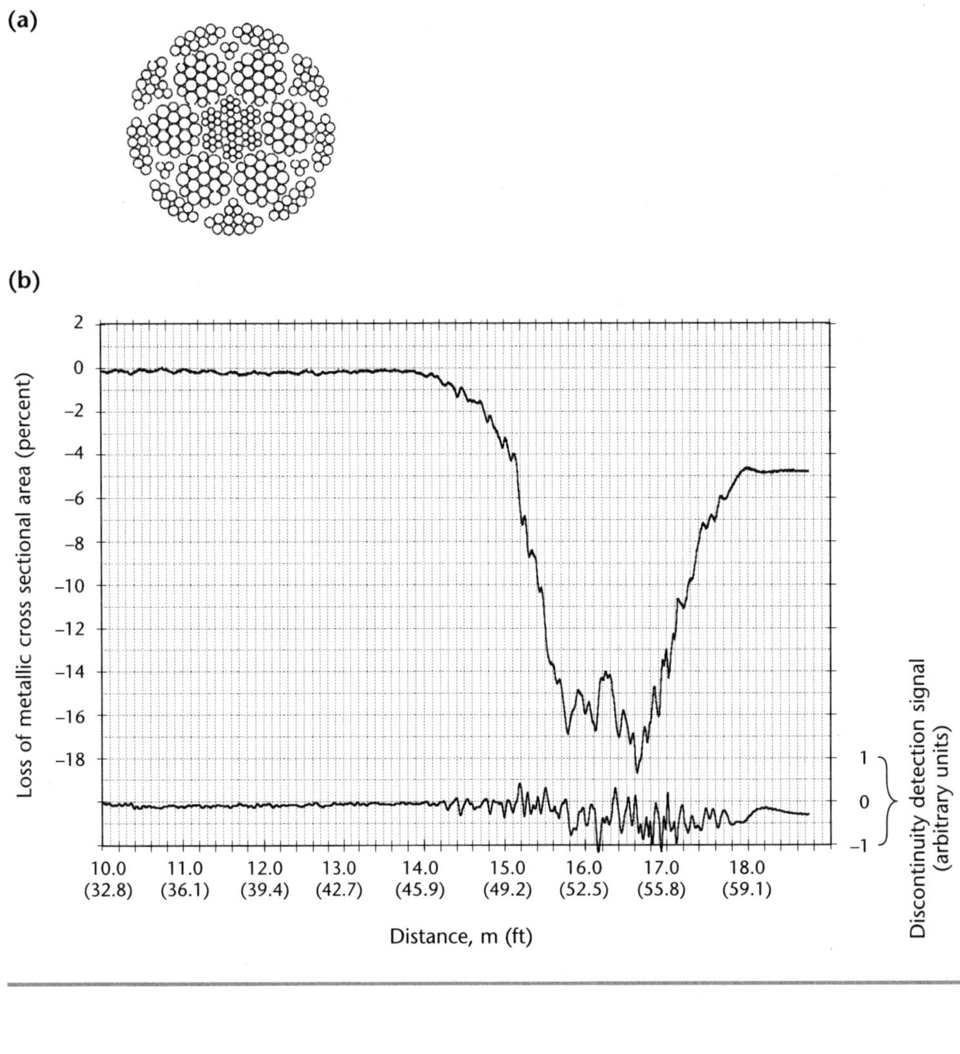

each other along the length of the rope is not known. For example, the gaps of broken wires could be aligned or staggered. Broken wires with zero or tight gap widths do not produce detectable magnetic leakage flux. For these reasons, only an estimate of the number of broken wires is possible.

Conventionally, the localized discontinuity trace is used for the detection of broken wires. However, the localized discontinuity signal is not quantitative and cannot be used for estimating the number of broken wires. On the other hand, the loss of metallic cross sectional area trace of the test chart in Fig. 23 shows rapid, relatively small variations of cross section. These variations are significant and can be used to estimate the number of broken wires per unit of rope length. Note, however, that the averaging length or quantitative resolution of the instrumentation must be sufficient to allow this quantitative discontinuity characterization.

The estimated number of broken wires per 100 mm (4 in.) of rope length N derived from the loss of metallic cross sectional area trace, is shown at the top of Fig. 23. Here, N_{max} denotes the maximum number of broken wires per 100 mm (4 in.) of rope length. Based on the operating history of this and similar ropes, a value of $N_{max} = 20$ can be estimated.

The rope was subsequently disassembled to determine the actual number of broken wires per unit of rope length. Broken wire estimates together with the actual number of broken wires along the length of the rope are shown in Fig. 24. Considering the fact that the loss

FIGURE 23. Test traces and relative number of broken wires ($N \cdot N_{max}^{-1}$).

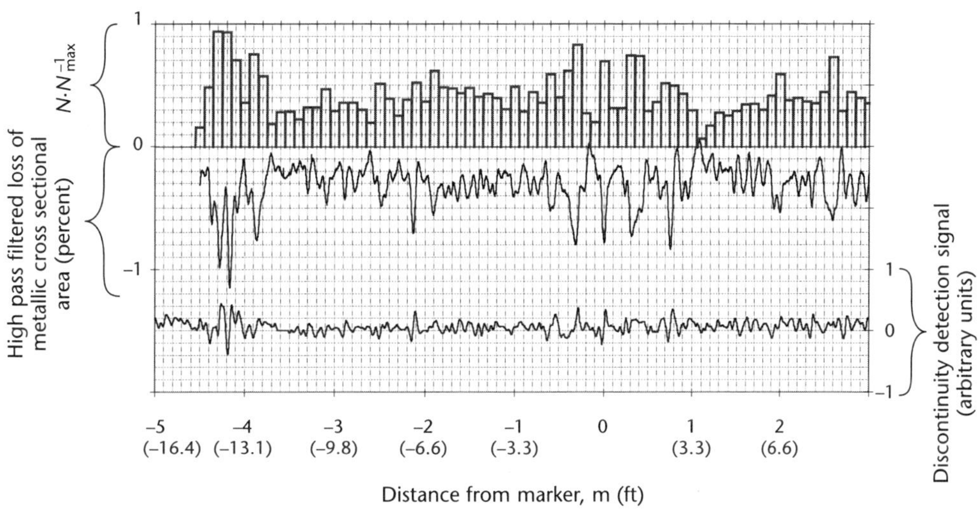

FIGURE 24. Broken wires at 100 mm (4 in.) intervals (smoothed data, three sample average).

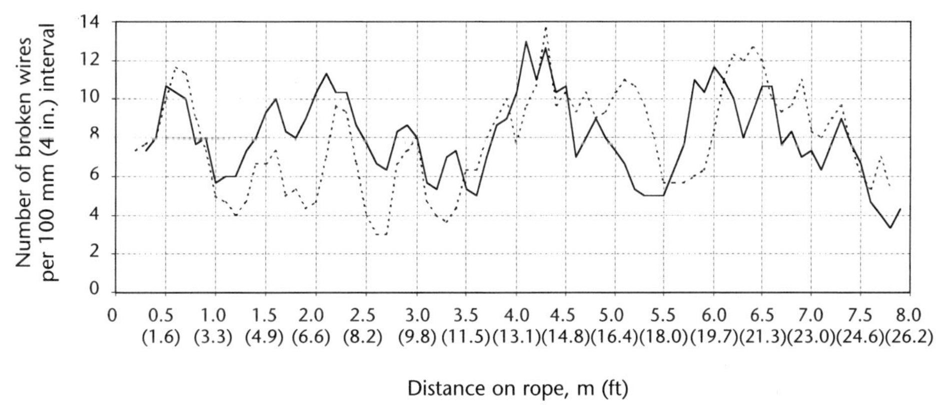

Legend

.... = actual

——— = estimate

of metallic cross sectional area trace not only indicates broken wires but also interstrand nicking, internal wear and other disturbances of the rope structure, there is a good correlation between the actual and estimated number of internal broken wires up to a rope distance of about 4.5 m (15 ft). Beyond this distance, there is an offset, which may be due to a distance measurement error during disassembly of the rope.

A subsequent destructive break test of the rope showed a 30.2 percent loss of breaking strength. Significantly, the second layer of strands of the rope (see Fig. 21) represents about 30 percent of the total rope cross sectional area. As with the balance rope test, this leads to the hypothesis that, for this rope, the second layer of strands has lost all load bearing capability.

The lack of sufficient information on the rope's operating history — and that of its predecessors — made this rope evaluation particularly difficult. Under normal circumstances, these details are known and must be considered when assessing the rope condition. Altogether, this evaluation shows that a quantitative discontinuity characterization for ropes with internal broken wires and clusters of broken wires is possible. The example illustrates the capabilities and limitations of electromagnetic wire rope test techniques for this particular discontinuity characterization problem.

Mooring Rope Test

This example deals with the testing of a mooring rope (Fig.12) having an independent wire rope core and having a diameter of 89 mm (3.5 in.).

Figure 25 shows a cross sectional diagram of such a rope. For independent wire rope core ropes, the outer wires of the outer strands have a larger diameter than the outer core strand wires. To minimize interstrand nicking between the outer strands and the independent wire rope core, these ropes are designed such that the wires of the outer strands and the independent wire rope core are almost parallel. (This construction is usually achieved by choosing a lang lay for the independent wire rope core and an ordinary lay for the outer strands.[29] In a lang lay, the strands of a rope are wound in the same direction as their constituent wires; in an ordinary lay, the strands are wound in the opposite direction from their constituent wires.) Typically, the wires of the outer strands are well supported by their neighbors whereas the outer wires of the independent wire rope core are relatively unsupported.

FIGURE 25. Typical six-strand rope with independent wire rope core.

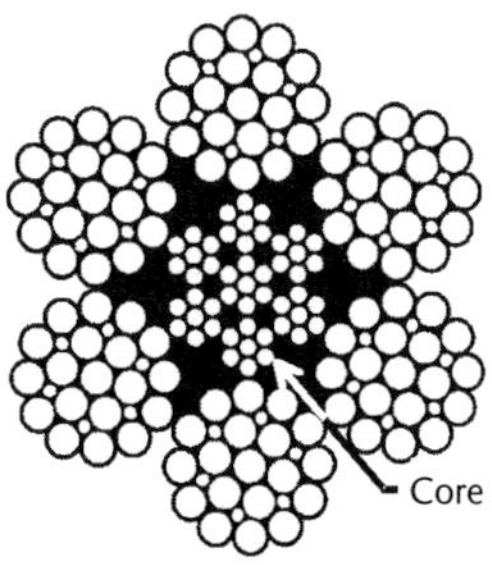

The result of these geometrical features is that, in response to tensile loads, the outer strands bear against the core and, in the localized contact areas, force the outer independent wire rope core wires into the valleys between the outer strand wires. Then, under continuing fluctuating tensile loads, this mechanism results in secondary bending stresses in the core wires, leading to large numbers of fatigue breaks. These breaks can be very close together and can form groups of breaks.

Test results (not shown) for this mooring rope have indicated a severe breakup of the independent wire rope core; the breakup is strong evidence of heavy fluctuating tensile fatigue loading. The loss of metallic cross sectional area and localized discontinuity traces show the typical patterns of broken and missing wires (compare Fig. 16). The missing wire patterns even indicate that short pieces of broken independent wire rope core wires might have fallen out of the rope. In addition, together with the findings of a visual test, electromagnetic testing indicates severe corrosion, including corrosion pitting. Corrosion products are clearly visible in Fig. 12.

Part 4. Near Field Microwave Testing of Cement Based Materials and Structures

Introduction

Near field microwave testing has shown great potential for evaluating electrically nonconducting (that is, insulating, dielectric) materials.[35] These techniques have been used to detect and evaluate the presence of voids and porosity, both local and distributed, in dielectric media and mixtures. They have also been used to measure accurately the thickness of various dielectric composites and ceramics and to evaluate disbonds and delaminations in thick sandwich composite structures.[36-39] Microwave techniques can also evaluate properties of materials composed of a mixture of several constituents involving both physical mixing (where there is no chemical interaction between constituents) and chemical reactions such as curing.[40-41]

Researchers have evaluated the potential of near field microwave test techniques, using open ended rectangular waveguide and monopole probes, for the purpose of testing and determining various properties of cement based materials and structures. Originally, a simple microwave nondestructive testing technique was used to detect the location of a reinforcing steel bar in a concrete slab.[42] In this preliminary investigation, the presence of a separation in the steel bar was also detected, in addition to demonstrating the sensitivity of microwave signals to aggregate size distribution.[42,43] Further investigations showed a strong correlation between the magnitude of reflection coefficient and the hardened paste's water-to-cement ratio, on one hand, to that paste's compressive strength on the other hand.[44] The potential of using monopole antennas for the same purpose has also been demonstrated.[45]

The objectives of several investigations were to measure, evaluate, analyze and correlate the near field microwave reflection properties of cement based materials to their sand-to-cement ratio, their coarse aggregate–to-cement ratio, their coarse aggregate size distribution and their compressive strength. Another objective was to detect and evaluate chloride ingress in these materials before and after mixing. The following is the summary of the findings.

Constituents of Cement

Cement paste — composed of cement powder, water and air — is the primary (matrix) phase of cement based materials such as concrete. The study of the interaction of microwave signals with cement paste is essential in understanding such aggregates. To this end, extensive study of the characteristics of the reflection coefficient of microwave signals from hardened cement paste blocks, when using an open ended rectangular waveguide probe, showed that the magnitude of reflection coefficient is directly related to the evaporation of free water from cement paste samples during the early days and to the curing process (that is, the water-to-cement ratio) in later days.[44,46] The latter correlation was also established with the compressive strength of these samples. A similar investigation was conducted using monopole antenna probes.[45] Similar results were obtained showing strong correlation between the magnitude of reflection coefficient, measured with this probe, and the water-to-cement ratio of hardened cement paste samples.

Mortar — composed of cement powder, water, fine aggregate (sand) and porosity (air) — was subsequently investigated, using open ended rectangular waveguides. Mortar is less homogenous than cement paste and its degree of homogeneity is a function of its compressive strength and the operating microwave frequency (that is, more homogenous at lower frequencies and less homogenous at higher frequencies). Subsequently, a simple relationship was established between the standard deviation of the magnitude of reflection coefficient at 10 GHz and the compressive strength of hardened mortar block samples. It was also shown that the average value of this parameter, when measured at lower microwave frequencies such as 3 GHz, is proportional to the water-to-cement ratio of mortar.[47] Finally, it was shown that simple dielectric mixing models can be used to estimate porosity of these mortar block samples.[48]

Concrete is the most widely used of cement based materials and is composed of cement powder, water, sand (fine aggregate), coarse aggregate (that is, rocks) and air. Therefore, concrete may be considered an heterogenous material at microwave frequencies. Consequently, when studying the reflection properties of concrete, it is necessary to consider the results from a statistical point of view. There are several important parameters of interest when using concrete: cure state (water-to-cement ratio), the directly related parameter of compressive strength, coarse aggregate size distribution and others. Extensive measurements using open ended rectangular waveguide probes and the magnitude of reflection coefficient show the potential for cure state monitoring of concrete with different constituent makeups.

It has also been shown that the water-to-cement ratio of hardened concrete can be determined for varying compressive strength and coarse aggregate size distributions.[49,50] The statistical properties (average and standard deviation) of the measured magnitude of reflection coefficient at relatively high frequencies (for example, 10 GHz) is shown to provide information about water-to-cement ratio and coarse aggregate size distribution. In addition, these statistics are correlated to the compressive strength of concrete through a maximum likelihood algorithm.[50,51] The extent of aggregate segregation in concrete placement is also evaluated using the statistics of the measured magnitude of reflection coefficient at relatively high microwave frequencies — for example, 10 GHz.[52]

Applications

Intrusion of chloride ions in concrete is responsible for corrosion of reinforcing steel bars. The presence of chloride ions and oxygen can initiate the process of corrosion. Subsequently, corrosion byproducts occupy a greater volume than concrete and result in the deterioration of the reinforced concrete structure. To this end, a preliminary study was conducted in which salt (that is, sodium chloride) was added to the mixing water of several mortar blocks. The reflection properties of these hardened mortar blocks were shown to be substantially different than those without sodium chloride added to their mixing water. Moreover, the results indicated a faster setting time for these samples compared to those without sodium chloride. Finally, a correlation between the magnitude of reflection coefficient and the compressive strength of these samples was obtained as a function of sodium chloride content and water-to-cement ratio.[53]

In another comprehensive study, many mortar samples were cyclically soaked in water and salt water (with varying salinity), while some of these samples were also loaded to promote microcracking and hence increased moisture permeation during the soaking cycles. The results showed that the properties of the reflection coefficient (magnitude and phase) at different frequencies can be used to distinguish between water soaked and salt water soaked samples. Additionally, a semiempirical electromagnetic model has been developed to simulate the reflection coefficient results and hence render the water content distribution in these samples as a function of days in a drying cycle and for several soaking cycles.[54-58]

Evaluation of water-to-cement ratio of early age cement based materials is an important practical issue. Monopole antenna probes are well suited for evaluating this parameter. Such a probe has been designed and optimized to give direct information about the early age water-to-cement ratio of many cement paste and concrete samples. Subsequently, a microwave measurement system (laboratory designed and assembled) incorporating this probe was used and successfully tested for this purpose.[59]

Detection of grout in masonry bricks and blocks is also an important quality control issue in the construction industry. Using a simple reflectometer and an open ended rectangular waveguide probe at 3 GHz, empty cavities can be distinguished from grout filled cavities in masonry blocks.[60]

Fiber reinforced polymer composite laminates are used for strengthening concrete structural members. These laminates are epoxy adhered to a structure. When the laminate becomes disbonded from the structure, no effective strengthening is provided. Thus, it is important to develop a nondestructive, quick, robust and portable technique to detect such disbonds and evaluate their spatial extent and their degree of severity. Using open ended rectangular waveguide probes, it has been demonstrated that near field microwave testing can provide the desired results.[61]

References

1. *Nondestructive Testing Handbook,* second edition: Vol. 4, *Electromagnetic Testing.* Columbus, OH: American Society for Nondestructive Testing (1986).
2. Lamtenzan, D., G. Washer and M.G. Lozev. "Detection of Cracks in Steel Bridges Using Eddy Currents." *Topics on Nondestructive Evaluation*: Vol. 2, *Nondestructive Testing and Evaluation of Infrastructure.* Columbus, OH: American Society for Nondestructive Testing (1998): p 13-23.
3. ASTM A 588/A 588M, *Standard Specification for High-Strength Low-Alloy Structural Steel with 50 ksi [345 MPa] Minimum Yield Point to 4-in. [100-mm] Thick.* West Conshohocken, PA: ASTM International (2001).
4. ASTM A 36/A 36M-01, *Standard Specification for Carbon Structural Steel.* West Conshohocken, PA: ASTM International (2001).
5. ASTM A 709/A 709M-01be3, *Standard Specification for Carbon and High-Strength Low-Alloy Structural Steel Shapes, Plates, and Bars and Quenched-and-Tempered Alloy Structural Steel Plates for Bridges.* West Conshohocken, PA: ASTM International (2001).
6. Skolnik, M.I. *Introduction to Radar Systems.* Boston, MA: McGraw-Hill (1980): p 581.
7. Daniels, D.J. *Surface Penetrating Radar.* London, United Kingdom: Institution of Electrical Engineers (1996): p 300.
8. Maser, K.R. "Pavement Characterization Using Ground Penetrating Radar: State of the Art and Current Practice." *Nondestructive Testing of Pavements and Backcalculation of Moduli: Third Volume* [Seattle, WA, June-July 1999]. ASTM Special Technical Publication 1375. West Conshohocken, PA: ASTM International (2000).
9. Bettigole, N. and R. Robison. *Bridge Decks.* Reston, VA: American Society of Civil Engineers, ASCE Press (1997): p 118.
10. Scott, M., A. Rezaizadeh and M. Moore. *Phenomenology Study of HERMES Ground Penetrating Radar Technology for Detection and Identification of Common Bridge Deck Features.* Report FHWA-RD-01-090. McClean, VA: Federal Highway Administration (2001).
11. Scott, M., M. Moore and G. Washer. *Testing and Evaluation of the PERES II Bridge Inspector.* Test and Evaluation Report DTFH61-C-98-00050. McClean, VA: Federal Highway Administration (2002): p 58.
12. Scott, M., A. Rezaizadeh, A. Delahaza, C.G. Santos, M. Moore, B. Graybeal and G. Washer. "A Comparison of Nondestructive Evaluation Methods for Bridge Deck Assessment." *Structural Materials Technology V: An NDT Conference* [Cincinnati, OH, September 2002]. Columbus, OH: American Society for Nondestructive Testing (2002): p 111-118.
13. ASTM D 4580, *Standard Practice for Measuring Delaminations in Concrete Bridge Decks by Sounding.* West Conshohocken, PA: ASTM International (2002).
14. Mast, J. *Microwave Pulse-Echo Radar Imaging for the Nondestructive Evaluation of Civil Structures.* Thesis. Urbana-Champaign, IL: University of Illinois (1993): p 105.
15. Armaghani, J. *Ground Penetrating Radar Signal Processing Enhancements for Geophysical Anomaly Identification.* Final Report, Contract HPR, Study 860, Work Phase 0510860. Gainesville, FL: Florida Department of Transportation (1999).
16. AASHTO TP36, *Standard Test Method for Evaluating Asphalt-Covered Concrete Bridge Decks Using Pulsed Radar.* Washington, DC: American Association of State Highway and Transportation Officials (2000).
17. NCHRP Synthesis 255, *Ground Penetrating Radar for Evaluating Subsurface Conditions for Transportation Facilities.* Washington, DC: National Academy of Sciences, Transportation Research Board (1998).
18. ASTM D 4748-87, *Standard Test Method for Determining the Thickness of Bound Pavement Layers Using Short-Pulse Radar.* West Conshohocken, PA: ASTM International (1999).

19. ASTM D 6432-99, *Standard Guide for Using the Surface Ground Penetrating Radar Method for Subsurface Investigation.* West Conshohocken, PA: ASTM International (1999).
20. ASTM E 1571, *Standard Practice for Electromagnetic Examination of Ferromagnetic Steel Wire Rope.* West Conshohocken, PA: ASTM International (2001).
21. Whitehead, E.A.N. *Method of Obtaining an Electrical Signal Proportional to the Cross-Sectional Area of a Magnetic Tube or Rod.* United Kingdom Patent 913 780 (December 1962).
22. Rieger, W. "Ein Beitrag zur Magnetinductiven Querschnittsmessung von Drahtseilen" ["A Contribution to the Magnetic Induction Cross Sectional Area Measurement of Wire Ropes"]. Doctoral dissertation. Stuttgart, Germany: Universität Stuttgart (1983).
23. Weischedel, H.R. "The Inspection of Wire Ropes in Service: A Critical Review." *Materials Evaluation.* Vol. 43, No. 13. Columbus, OH: American Society for Nondestructive Testing (December 1983): p 1592-1605.
24. Kitzinger, F. and G.A. Wint. United States Patent 4 096 437, *Magnetic Testing Device for Detecting Loss of Metallic Area and Internal and External Defects in Elongated Objects* (June 1978).
25. Marchent, B.G. United Kingdom Patent Application GB 2 012 966, *Apparatus for Nondestructive Testing of Elongate Objects* (December 1978).
26. Tomaiuolo, F.G. and J.G. Lang. United States Patent 4 495 465, *Method and Apparatus for Nondestructive Testing of Magnetically Permeable Bodies Using a First Flux to Saturate the Body and a Second Flux Opposing the First Flux to Produce a Measurable Flux* (January 1985).
27. Weischedel, H.R. "Electromagnetic Wire Rope Inspection: Signal Generation, Filtering, and Computer-Aided Rope Evaluation." *Non-Destructive Testing of Rope* [Krakow, Poland, September 1999]. OIPEEC Technical Meeting Proceedings, document 0685. Reading, United Kingdom: Reading Rope Research, for the International Organization for the Study of the Endurance of Wire Rope [OIPEEC] (1999).
28. Weischedel, H.R. "Quantitative In-Service Inspection of Wire Ropes, Applications and Theory." *International Advances in Nondestructive Testing.* Vol. 15. Langhorne, PA: Gordon and Breach Science Publishers (1990): p 83-118.
29. *Wire Rope Users Manual,* third edition. Woodstock, MD: Wire Rope Technical Board (1993).
30. Weischedel, H.R. "The Inspection of Mine Hoist Ropes." *Wire Rope News and Sling Technology.* Vol. 12, No. 5. Colonia, NJ: VS Enterprises (June 1991): p 14-25.
31. Smith, D.T. and P. McCann. *Evaluation of Instruments for the Non-Destructive Testing of Wire Ropes.* Report FE/02/07. Buxton, Derbyshire, United Kingdom: Health and Safety Laboratory (2002).
32. Weischedel, H.R. and C.R. Chaplin. "Inspection of Wire Ropes for Offshore Applications." *Materials Evaluation.* Vol. 49, No. 3. Columbus, OH: American Society for Nondestructive Testing (March 1991): p 362-367.
33. Weischedel, H.R. and C.R. Chaplin. "The Inspection of Offshore Wire Ropes: The State of the Art." Paper OTC 6969. *Offshore Technology Conference* [Houston, TX, May 1992]. Richardson, TX: Offshore Technology Conference (1992).
34. Dohm, M. *An Evaluation of International and Local Magnetic Rope Testing Instrument Defect Detection Capabilities and Resolution, Particularly in Respect to Low Rotation, Multilayer Rope Constructions.* Project Number GAP 503 and GAP 353. Johannesburg, South Africa: Safety in Mines Research Advisory Committee (1999).
35. Zoughi, R. *Microwave Non-Destructive Testing and Evaluation.* Dordrecht, Netherlands: Kluwer Academic Publishers (2000).
36. Zoughi, R. and S. Bakhtiari. "Microwave Nondestructive Detection and Evaluation of Disbonding and Delamination in Layered-Dielectric-Slabs." *IEEE Transactions on Instrumentation and Measurement.* Vol. 39, No. 6. New York NY: Institute of Electrical and Electronics Engineers (December 1990): p 1059-1063.
37. Zoughi, R. and S. Bakhtiari. "Microwave Nondestructive Detection and Evaluation of Voids in Layered Dielectric Slabs." *Research in Nondestructive Evaluation.* Vol. 2, No. 4. Columbus, OH: American Society for Nondestructive Testing (1990): p 195-205.
38. Gray, S., S. Ganchev, N. Qaddoumi, G. Beauregard, D. Radford and R. Zoughi. "Porosity Level Estimation in Polymer Composites Using Microwaves." *Materials Evaluation.* Vol. 53, No. 3. Columbus, OH: American Society for Nondestructive Testing (March 1995): p 404-408.

39. Bakhtiari, S., S. Ganchev and R. Zoughi. "Open-Ended Rectangular Waveguide for Nondestructive Thickness Measurement and Variation Detection of Lossy Dielectric Slabs Backed by a Conducting Plate." *IEEE Transactions on Instrumentation and Measurement.* Vol. 42, No. 1. New York, NY: Institute of Electrical and Electronics Engineers (February 1993): p 19-24.
40. Ganchev, S., N. Qaddoumi, D. Brandenburg, S. Bakhtiari, R. Zoughi and J. Bhattacharyya. "Microwave Diagnosis of Rubber Compounds." *IEEE Transactions on Microwave Theory and Techniques.* Vol. 42, No. 1. New York, NY: Institute of Electrical and Electronics Engineers (January 1994): p 18-24.
41. Qaddoumi, N., S. Ganchev and R. Zoughi. "Microwave Diagnosis of Low Density Glass Fibers with Resin Binder." *Research in Nondestructive Evaluation.* Vol. 8, No. 3. Columbus, OH: American Society for Nondestructive Testing (1996): p 177-188.
42. Zoughi, R., G.L. Cone and P.S. Nowak. "Microwave Nondestructive Detection of Rebars in Concrete Slabs." *Materials Evaluation.* Vol. 49, No. 11. Columbus, OH: American Society for Nondestructive Testing (November 1991): p 1385-1388.
43. Zoughi, R., T. Nast and P.S. Nowak. "Preliminary Results of Microwave Reflectometry As a Nondestructive Tool for Studying Concrete Properties." *Proceedings Nondestructive Evaluation of Civil Structures and Materials* [Boulder, CO, May 1992]. Boulder, CO Atkinson-Noland and Associates (May 1992): p 339-343.
44. Zoughi, R., S.D. Gray and P.S. Nowak. "Microwave Nondestructive Estimation of Cement Paste Compressive Strength." *ACI Materials Journal.* Vol. 92, No. 1. Farmington Hills, MI: American Concrete Institute (January-February 1995): p 64-70.
45. Bois, K.J. and R. Zoughi. "Analysis of Microwave Nondestructive Methods for Evaluating Compressive Strength of Cement Paste: An Optimization Study." *Proceedings 3rd Conference Nondestructive Evaluation of Civil Structures and Materials* [Boulder, CO, September 1996]. Boulder, CO: Atkinson-Noland and Associates (September 1996): p 109-120.
46. Shalaby, W. and R. Zoughi. "Analysis of Monopole Sensors for Cement Paste Compressive Strength Estimation." *Research in Nondestructive Evaluation.* Vol. 7, No. 2/3. Columbus, OH: American Society for Nondestructive Testing (1995): p 101-105.
47. Bois, K., A. Benally, P.S. Nowak and R. Zoughi. "Microwave Nondestructive Determination of Sand to Cement Ratio in Mortar." *Research in Nondestructive Evaluation.* Vol. 9, No. 4. Columbus, OH: American Society for Nondestructive Testing (1997): p 227-238.
48. Bois, K., R. Mirshahi and R. Zoughi. "Dielectric Mixing Models for Cement Based Materials." *Proceedings of the Review of Progress in Quantitative Nondestructive Evaluation* [Brunswick, ME, July-August 1996]. Vol. 16A. New York, NY: Plenum Press (1997): p 657-663.
49. Bois, K.J., A.D. Benally, P.S. Nowak and R. Zoughi. "Cure-State Monitoring and Water-to-Cement Ratio Determination of Fresh Portland Cement Based Materials Using Near Field Microwave Techniques." *IEEE Transactions on Instrumentation and Measurement.* Vol. 47, No. 3. New York, NY: Institute of Electrical and Electronics Engineers (June 1998): p 628-637.
50. Bois, K., A. Benally and R. Zoughi. "Microwave Near-Field Reflection Property Analysis of Concrete for Material Content Determination." *IEEE Transactions on Instrumentation and Measurement.* Vol. 49, No. 1. New York, NY: Institute of Electrical and Electronics Engineers (February 2000): p 49-55.
51. Bois, K. and R. Zoughi. "A Decision Process Implementation for Microwave Near-Field Characterization of Concrete Constituent Makeup." *Subsurface Sensing Technologies and Applications.* Vol. 2, No. 4. New York, NY: Kluwer Academic/Plenum Publishers (October 2001): p 363-376.
52. Bois, K., A.D. Benally, P.S. Nowak and R. Zoughi. "Application of Near-Field Microwave Sensing Techniques for Aggregate Segregation Detection in Concrete Members." *Review of Progress in Quantitative Nondestructive Evaluation* [Montreal, Canada, July 1999]. Vol. 19B. Melville, NY: American Institute of Physics (2000): p 1717-1722.
53. Bois, K., A. Benally and R. Zoughi. "Near-Field Microwave Non-Invasive Determination of NaCl in Mortar." *IEEE Proceedings — Science, Measurement and Technology.* Vol. 148, No. 4. New York, NY: Institute of Electrical and Electronics Engineers (July 2001): p 178-182.

54. Case, J.T., R. Zoughi, K. Donnell, D. Hughes and K.E. Kurtis. "Microwave Analysis of Mortar Prepared with Type I/II, III and V Cement and Subjected to Cyclical Chloride Exposure." *Review of Progress in Quantitative Nondestructive Evaluation* [Brunswick, ME, July-August 2001]. Vol. 21A. Melville, NY: American Institute of Physics (2002): p 498-505.
55. Peer, S., T. Case, K. Donnell, D. Hughes, R. Zoughi and K. Kurtis. "Investigation of Microwave Reflection Properties of Mortar Exposed to Wet-Dry Cycles of Tap Water and Chloride Bath." *Review of Progress in Quantitative Nondestructive Evaluation* [Brunswick, ME, July-August 2001]. Vol. 21B. Melville, NY: American Institute of Physics (2002): p 1269-1276.
56. Hu, C., T. Case, M. Castle, R. Zoughi and K. Kurtis. "Microwave Evaluation of Accelerated Chloride Ingress in Mortar." *Review of Progress in Quantitative Nondestructive Evaluation* [Ames, IA, July 2000]. Vol. 20A. Melville, NY: American Institute of Physics (2001): p 467-473.
57. Peer, S., J.T. Case, E. Gallaher, K.E. Kurtis and R. Zoughi. "Microwave Reflection and Dielectric Properties of Mortar Subjected to Compression Force and Cyclically Exposed to Water and Sodium Chloride Solution." *IEEE Transactions on Instrumentation and Measurement.* New York, NY: Institute of Electrical and Electronics Engineers (February 2003).
58. Peer, S. *Nondestructive Evaluation of Moisture and Chloride Ingress in Cement-Based Materials Using Near-Field Microwave Techniques.* M.S. thesis. Rolla, MO: University of Missouri–Rolla (2002).
59. Mubarak, K., K.J. Bois and R. Zoughi. "A Simple, Robust and On-Site Microwave Technique for Determining Water-to-Cement (*w/c*) Ratio of Fresh Portland Cement-Based Materials." *IEEE Transactions on Instrumentation and Measurement.* Vol. 50, No. 5. New York, NY: Institute of Electrical and Electronics Engineers (October 2001): p 1255-1263.
60. Bois, K., H. Campbell, A. Benally, P.S. Nowak and R. Zoughi. "Microwave Noninvasive Detection of Grout in Masonry." *Masonry Society Journal.* Vol. 16, No. 1. Boulder, CO: Masonry Society (June 1998): p 49-54.
61. Hughes, D., M. Kazemi, K. Marler, R. Zoughi, J. Myers and A. Nanni. "Microwave Detection of Delamination between Fiber Reinforced Polymer (FRP) Composite and Hardened Cement Paste." *Review of Progress in Quantitative Nondestructive Evaluation* [Brunswick, ME, July-August 2001]. Vol. 21A. Melville, NY: American Institute of Physics (2002): p 512-519.

Wire Rope Bibliography

Geller, L.B., F. Kitzinger and K. Leung. "NDT of Wire Ropes with a Novel, Computer-Controlled, Dual Function Electromagnetic Wire Rope Tester." *Materials Evaluation.* Vol. 56, No. 3. Columbus, OH: American Society for Nondestructive Testing (March 1988): p 403-408, 410.

Geller, L.B., K. Leung and F. Kitzinger. "Computerized Operational Control of an Electro-Magnetic Wire Rope Tester." *Materials Evaluation.* Vol. 53, No. 9. Columbus, OH: American Society for Nondestructive Testing (September 1995): p 1002-1006.

Mironenko, A.S. and V.V. Sukhorukov. "Non-Destructive Testing of Steel Wire Ropes in Russia." *Insight.* Vol. 40, No. 6. Northhampton, United Kingdom: British Institute of Non-Destructive Testing (June 1998): p 395-397.

Offshore Technology Report OTO 2000 064, *Wire Rope Non-Destructive Testing — Survey of Instrument Manufacturers.* London, United Kingdom: Health and Safety Executive (2000).

Poffenroth, D.N. "Nondestructive Testing of Elevator Suspension and Governor Ropes." *Elevator World.* Vol. 44, No. 4. Mobile, AL: Elevator World (April 1996): p 73-75.

Singh, S.D. and B. Ghara. "Steel Wire Rope Condition Monitoring by Non-Destructive Investigation and Evaluation While on Installation/Service." *15th World Conference on NDT Proceedings* [Rome, Italy, October 2000]. Brescia, Italy: Italian Society for Nondestructive Testing and Monitoring Diagnostics (2000).

Weischedel, H.R. United States Patent 4 659 991, *Method and Apparatus for Magnetically Inspecting Elongated Objects for Structural Defects* (April 1987).

18

CHAPTER

Aerospace Applications of Eddy Current Testing

Donald J. Hagemaier, Huntington Beach, California (Parts 1 to 6)

Craig W. Benson, General Dynamics, Dayton, Ohio (Part 7)

Part 1. Introduction to Eddy Current Testing of Aircraft[1]

Typical Applications

Applications of eddy current tests in the aerospace industry include the following: (1) measurement of metal and coating thickness; (2) measurement of effects that alloying elements, heat treatment, test object dimensions, cladding, liftoff and temperature have on eddy current conductivity values; (3) sorting of mixed aluminum alloys; (4) evaluation of overaging or heat damage to aluminum; (5) detection of alpha case hardening on titanium alloys and detection of titanium aluminide in aluminum brazed titanium honeycomb; (6) eddy current testing of bolt holes in aircraft structures; (7) analysis and applications using impedance plane techniques to measure metal thinning and metal spacing and to detect exfoliation corrosion and surface cracks in aluminum aircraft structures; and (8) low frequency testing of aircraft structures for the detection of subsurface cracks.

Testing of Aluminum Alloys

The widespread use of aluminum alloys in critical aerospace structures requires careful control during manufacture, to ensure (1) that each test object is made from the specified aluminum alloy and (2) that the properties resulting from the composition and heat treatment of these alloys are those required for the specific application. The electrical conductivity of aluminum alloys depends on their alloy content and the effects of prior processing. Thus, eddy current conductivity tests provide unique versatility in their capabilities for detecting mixed alloys or alloys having properties changed by thermal processing.

Eddy current tests can supplement or in some cases replace liquid penetrant tests for detection of surface connected discontinuities and can also be used to measure the presence of coating or cladding on substrates. In addition, eddy current tests are often used to test aerospace structures during maintenance overhauls and to detect damage incurred in service. Such tests can be used to detect fatigue cracks (including those near fasteners) resulting from cyclic loading during flight, takeoff or landing. They can also be used to indicate the extent of corrosion damage or to identify portions of aircraft structures damaged by fire. In this way, it is possible to identify damaged components that need to be removed and replaced to restore the craft to airworthiness.

PART 2. Eddy Current Tests of Metal and Coating Thickness[1]

Coating Thickness on Metals

In the aerospace industry, numerous metal parts are coated or plated to obtain special surface properties such as corrosion resistance, wear resistance or improved appearance. These coatings are prepared by various means, including electrodeposition, hot dipping, cladding and spraying. For controlling the thickness of the deposit and also for testing purposes, reliable and rapid measurements of coatings are necessary. There are many methods available, including magnetic, mechanical, optical, radiographic, ultrasonic and electromagnetic testing. Each of these methods has applications and more than one method may be used for the same application.

Classification of Coatings and Base Materials

Four general types of coatings and base materials lend themselves to eddy current testing. The operating procedure used with the test instrument is determined by the specific combination of materials. These material combinations are classified as follows: (1) metal foil and sheet or metal coating on a nonconductive base material, such as metallic film on glass, ceramics or plastics; (2) metal cladding with a higher conductivity than the base metal, such as copper, zinc or cadmium on steel or pure aluminum on aluminum alloy; (3) metal cladding with a lower conductivity than the base metal, such as nickel on aluminum; and (4) nonconductive coatings on a metallic base material, such as anodic film or paint on aluminum or such as other organic coatings on metals.

Reference Standards for Thickness

For eddy current thickness tests, at least three specimens with known metal thicknesses are needed as reference standards to calibrate the equipment. One reference standard represents the minimum acceptable thickness, the second represents the maximum acceptable thickness and the third is from the middle of the range. In some cases, a bare metal specimen is necessary. All reference standards must have the same conductivity, permeability, substrate thickness and geometry as the test objects.

Metal Thickness

Tests for measuring metal thickness are generally used on chemically milled sheet stock, thin walled tubing, metal foil bonded to nonmetallic materials and any parts that may experience thinning from corrosion.

Null Balancing

Phase sensitive equipment permits plotting metal thickness profiles on the impedance plane. In Fig. 1, the operating point has a value of inductive reactance $X = 800$ and resistance $R = 900$. This point was purposely chosen so that the voltage vectors are about 45 degrees from the locus for change in thickness and 90 degrees, or normal, to the liftoff curve. When this is done, thickness changes also result in maximum instrument response. This happens because the vector to the thin metal is shorter than the vector to the thick metal, producing a large change between the two. However, the vectors to

FIGURE 1. Impedance plane thickness plot for Unified Numbering System A97075 wrought aluminum alloy, temper 6, at 10 kHz.

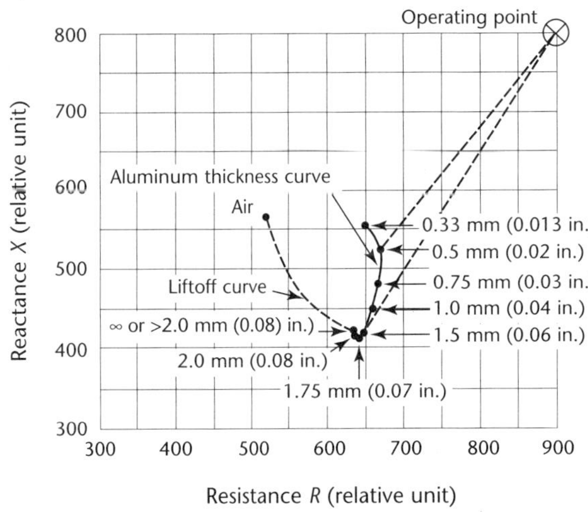

the liftoff curve are equal, resulting in no change in instrument amplitude response.

A technique called *null balancing* can be used with any phase sensitive metered instrument. The operating point should be chosen so that the signal to be suppressed is normal to a line connecting it with the operating point; the variable to be measured is parallel or at least less than 45 degrees to the point. Note that in Fig. 1 the short vector registers as low values on the meter and the long vector registers as large values.

Null balancing is very useful for eliminating or reducing unwanted effects. Increasing or decreasing the signal gain control, to bring the meter reading back on scale with the signal level control, permits larger or smaller spreads of thickness to be displayed on the panel meter. Null balancing is discussed below, in the section on impedance plane analysis.

Conductive Coating Thickness

Cladding thickness measurement is used in two types of situations where the conductivities of two metal layers are very different. The first situation, in which the cladding is a better conductor than the base material, pertains to copper, zinc or cadmium coatings on steel base materials. The second situation, in which the cladding is the poorer conductor of the two metals, pertains to lead coatings on copper or to nickel coatings on aluminum. The conducting coating thickness may be experimentally determined for a particular combination of coating metal, eddy current probe and test frequency by increasing the cladding thickness in steps until a given increase in thickness does not change the indicated meter reading (Fig. 2).

Figure 2. Relative instrument readings for various thicknesses of copper plate on brass base.

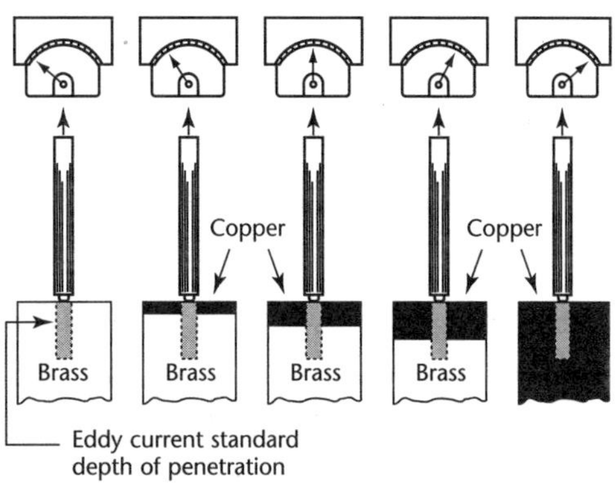

The layered examples in Fig. 3 result from matching arbitrary combinations of metals to illustrate how these combinations behave on the impedance

Figure 3. Impedance graphs showing coating thickness effects: (a) low conductivity on high conductivity material; (b) high conductivity coating on low conductivity material; (c) aluminum coating on steel at 25 kHz, in units of 25 µm (0.001 in.).[2]

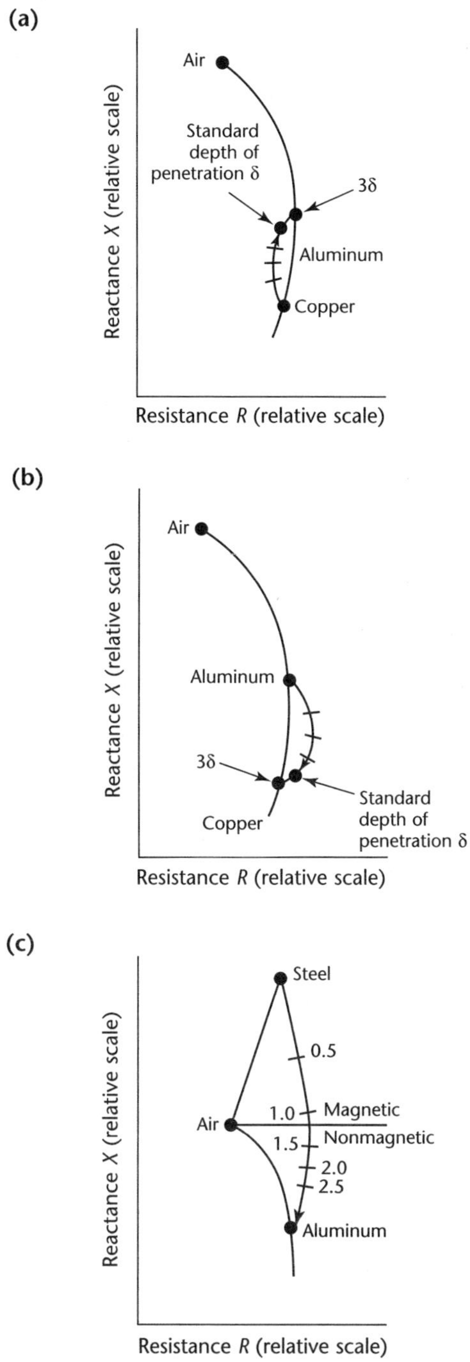

plane.[2] The principles illustrated have practical applications.

The impedance point of the thick copper base material would be a point on the conductivity locus as shown in Fig. 3a. If the copper base material is plated with aluminum, the probe coil impedance moves upward and to the left (clockwise) on the conductivity locus. The impedance on the conductivity locus approaches 3δ, corresponding to thick aluminum.

If copper of various thicknesses is added on top of the aluminum, the probe coil impedance follows a locus downward and to the right of the conductivity locus. The movement of the impedance point (with increasing thickness) is not proportional to the thickness; the deflections for equal thickness changes become less as the thickness increases. The impedance point corresponding to a thickness about equal to one standard depth of penetration δ is shown in Fig. 3b. The impedance point finally converges rapidly on the copper conductivity locus at 3δ.

The important fact illustrated in Figs. 3a and 3b is that the curves spiral with increasing curvature in a clockwise direction. The start of the spiral in each case approximates a straight line but as the thickness increases the spiral becomes more accentuated. The degree of spiraling and the shape of the curve will vary as the operating frequency is changed.

A further example is the case of nonmagnetic but conductive coating on a magnetic substrate or base material, as illustrated in Fig. 3c for aluminum on steel. In this example, the inductive reactance decreases in a nonlinear fashion. The locus of points is to the right of the liftoff locus line. The locus crosses from the magnetic domain to the nonmagnetic domain on the impedance plane at rather thin values of aluminum. For plating on steel alloys, similar results would be obtained for cadmium at 14.5 MS·m^{-1} (25 percent of the International Annealed Copper Standard) and for chromium at 4.6 MS·m^{-1} (8 percent of the International Annealed Copper Standard).

Nonconductive Coating Thickness

The measurement of nonconductive coating thicknesses on metallic base materials is basically a liftoff measurement. The measurement is applicable to nonconductive coatings such as anodic coatings, paint or plastic on a conductive base. Normally, magnetic attraction gages are used to measure nonmagnetic coatings (conductive or nonconductive) on steel and an eddy current test is used to measure nonconductive coatings on nonmagnetic base materials.

The spacing from probe to test object is referred to as *liftoff*. Figure 4 shows the impedance plane response that occurs when liftoff is increased at 100 kHz. The upper portion of the impedance plane is the magnetic domain where responses occur from ferromagnetic materials. The lower portion is the domain where responses are obtained from an aluminum alloy tested. Note the nonlinear changes along the liftoff locus for equal increments of spacing.

Feasibility Criterion for Coating Thickness Tests

For the coating material and the base material, the product of conductivity and permeability ($\sigma_1\mu_1$) for one of the materials should be at least 1.5 times the product of conductivity and permeability ($\sigma_2\mu_2$) for the other material. Nonconductors such as air, organics and paint have a conductivity value of zero. Nonferromagnetic materials have a relative magnetic permeability equal to 1.0. When measuring coatings on a metallic base, the total thickness of the base material plus the coating (if conductive) should be at least three standard depths of eddy current penetration, that is, should be ≥3δ. For eddy current tests of materials with various permeability and conductivity product values, standard depths of

FIGURE 4. Impedance graph showing liftoff effect at 100 kHz. Liftoff is spacing of probe to part in units of 25 µm (0.001 in.).

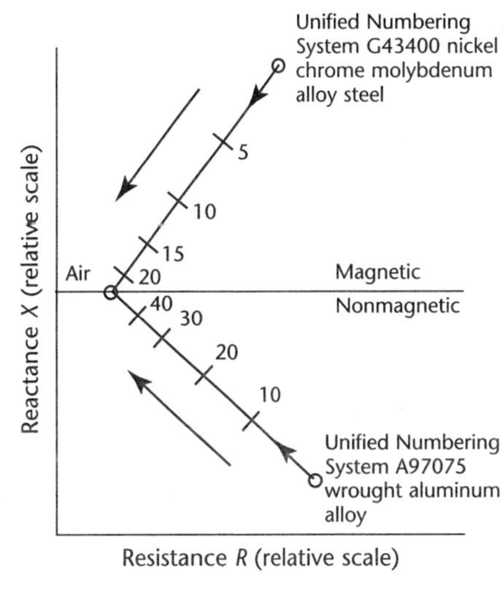

penetration as a function of frequency can be predicted by using Fig. 5.

Metal Spacing[2]

There are times when a gap separates two metal sheets. The gap may be filled with a nonconductive adhesive layer or a nonmetallic shim or may be purposely produced as a fixed dimension. If it is desired to measure the gap or spacing, an operating frequency must be chosen so that eddy currents will be generated in the second (subsurface) layer. The frequency chosen should produce eddy currents that penetrate both metals to a depth of 3δ.

When the gap is zero (both metals touching), the combination appears infinitely thick and produces a vector point on the conductivity locus. As the gap increases, the inductive reactance increases and the phasors produce a locus of points (in the vertical direction) that cross the conductivity locus at some point. As the gap increases, the locus of points will also intersect the thinning locus at a point that represents the thickness of the upper metal specimen (Fig. 6). At this point, the gap has reached a value where eddy currents are being generated only in the upper layer. Hence, any further increase in the gap between the two metal specimens will not cause any further change in impedance. If a larger gap needs to be measured, the operating frequency must be reduced.

FIGURE 5. Chart for determining standard depth of penetration from relative magnetic permeability μ_r (henry per meter) and conductivity σ (siemens per meter) as percentage of International Annealed Copper Standard (%IACS).

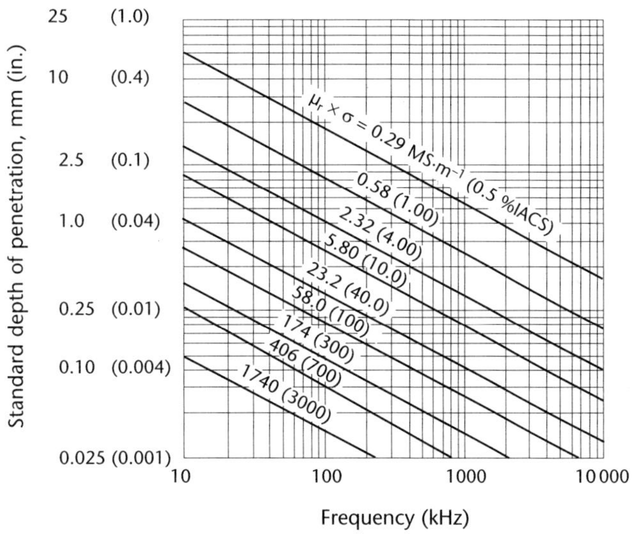

Gap or spacing measurements are best made with the probe located over the thin member; this reduces the problem with depth of penetration. If both members are equally thick but one has a lower conductivity, the probe should be placed on the low conductivity specimen.

FIGURE 6. Probe coil impedance curve showing effect of metal spacing between two flat, parallel, metallic conducting sheets.

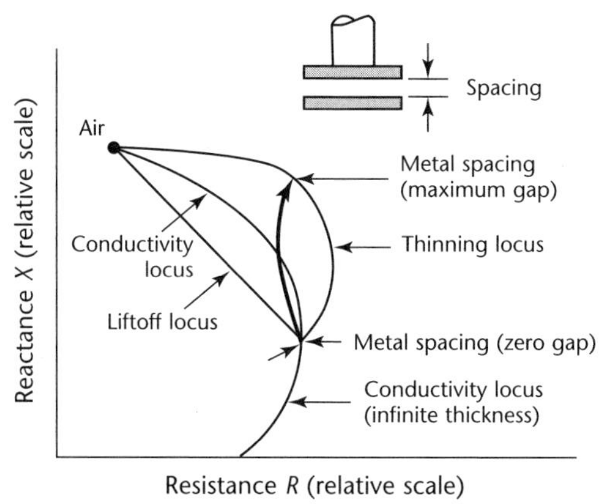

PART 3. Eddy Current Tests of Metal Conductivity[1]

For extending the use of eddy current testing in the control of metal alloys and their heat treatment, the most important prerequisite is the establishment of suitable documentation, test procedures and limits.

Control of Tests

Important considerations in eddy current conductivity testing are (1) selection of suitable and reliable instruments, (2) careful laboratory evaluation for defining test requirements and (3) selection of suitable reference standards. There must be no doubt about the accuracy and reproducibility of instrument performance from one test to the next or between two or more instruments of the same type and from the same manufacturer.

Conductivity Instruments

Eddy current conductivity meters usually differ with respect to operating frequency, liftoff compensation, temperature compensation, sensitivity, probe size or means of presenting test results. Most conductivity meters can provide results as percentages of the International Annealed Copper Standard. Other instruments can provide conductivity measurements in siemens per meter or as customized displays of signal amplitude or phase angle. Whatever type of instrument is used, reference standards must be used to calibrate the instrument, permitting qualitative judgments on all future test results.

Laboratory Evaluations

Laboratory evaluations are an important step in establishing eddy current test procedures. A suitable laboratory evaluation typically involves analysis of test object composition, heat treating times and temperatures and mechanical processing. The laboratory evaluations serve to establish a conductivity value for an alloy, as illustrated in Fig. 7.

Conductivity Reference Standards

Nondestructive test calibration reference standards should be based on some measurable physical characteristic related to the critical properties being measured. This characteristic must be not only measurable but also reproducible. For conductivity testing, reference standards are usually made from a sample of metal with resistivity that is known and thickness that can be considered infinite for purposes of the test.

Physical reference standards for conductivity may be obtained from the National Institute of Standards and Technology or may be fabricated.

The reference standards are identified by the material's conductivity, a reciprocal of its resistivity. When testing aluminum alloys, it is necessary to maintain two or more aluminum alloy reference standards. Typically, one of these reference standards has conductivity in the range of 14.5 to 18.6 $MS \cdot m^{-1}$ (25 to 32 percent of the International Annealed Copper Standard) and the other standard is in the range of 22.0 to 29.0 $MS \cdot m^{-1}$ (38 to 50 percent of the International Annealed Copper Standard). Such reference standards are certified and their accuracy is documented.[3] Nonferrous conductivity

FIGURE 7. Relative conductivity of metals and alloys shown by eddy current meter readings. Unified Numbering System (UNS) is used here for wrought aluminum alloys.

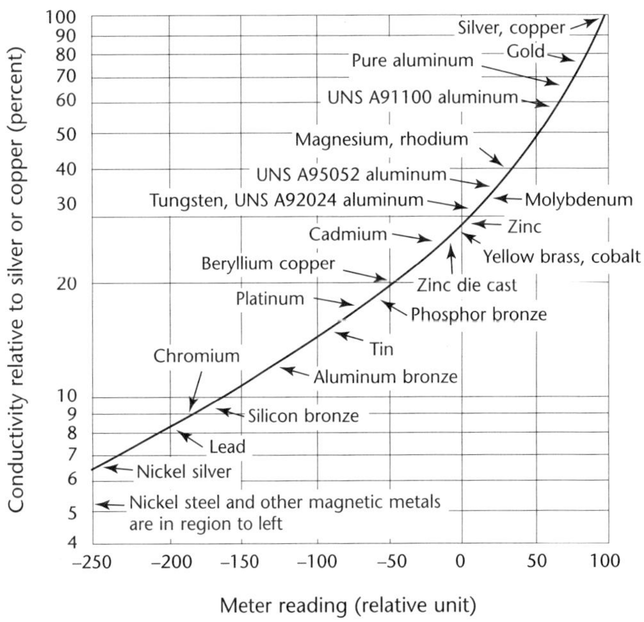

reference standards can be fabricated for instrument calibration.[4] Conductivity standards in the United States are usually referenced to the National Institute of Standards and Technology.

Factors Affecting Conductivity

Alloying Elements

If a pure metal were tested in the annealed or unstrained condition, a single conductivity value would be obtained, such as the 100 percent value arbitrarily assigned to pure annealed copper. However, most structural metals are alloys consisting of a base metal with several alloying elements added to obtain specific properties. The addition of alloying elements to a pure metal alters its conductivity. Usually, the alloying element provides centers of interference to electron flow. This reduces the alloyed metal's conductivity below that of the pure metal. Although the conductivity of copper is usually given as 58.0 $MS \cdot m^{-1}$ (100 percent of the International Annealed Copper Standard), care must be exercised when testing copper, because the conductivity is influenced by small additions of impurities or alloying elements. Conductivity is also affected by the metal's temperature at the time and place of eddy current testing. Touching the reference standard with fingers can change its temperature and apparent conductivity. The specific resistivity of standard annealed copper changes by 0.4 percent for each kelvin or degree celsius of temperature change in tests near room temperature, 20 °C (68 °F).

Heat Treatment and Mechanical Stresses

The conductivity of a particular alloy is further influenced by heat treatment and by stresses introduced in the metal during fabrication. Heat treatment alters crystal or grain structure and the distribution of the alloying elements, both of which affect electron flow. Stress introduced during fabrication provides centers of interference to electron flow. The combined effects of alloying elements (with their permissible tolerances), of heat treatment (with its permissible time and temperature ranges) and of fabrication stress establish a conductivity range instead of a specific conductivity value for any metal alloy and heat treatment condition. It is the function of laboratory evaluations to define these conductivity ranges and their variations under various test conditions.

Laboratory Evaluation of Aluminum Alloys

The following are practical considerations for selecting instruments, reference standards and test samples for a laboratory evaluation to establish the conductivity ranges of an aluminum alloy.

First, eddy current instruments, probes and reference standards will be required. Instrument performance should be checked for such effects as drift, including that caused by power supply fluctuations and temperature, and for reproducibility of calibration and instrument readings. Calibration standards of known conductivity should be obtained for the approximate conductivity range expected.

Test samples should be flat, unclad and uncoated material of sufficient size to eliminate edge effects and of sufficient thickness to eliminate thickness effects. Surface finish roughness up to 6.3 µm (2.5×10^{-4} in.) or oxide coating thicknesses normally encountered on aluminum sheet, plate or bar will not significantly interfere with the eddy current readings, provided the instrument has liftoff compensation up to 0.08 mm (0.003 in.). Some knowledge of variables that influence eddy current measurements helps ensure proper selection of test samples for establishing conductivity values.

Factors Affecting Conductivity Tests

Any variation from the condition for which eddy current conductivity meters are calibrated may have a significant effect on the accuracy of their measurements. The basic test condition variables have been studied to determine their effects on measurements of true base material conductivities. Six basic test condition variables are (1) curved surfaces, (2) edge effect, (3) material thickness, (4) aluminum cladding thickness, (5) liftoff or nonconductive coating thickness and (6) temperature.

Curved Surfaces

Eddy current tests on surfaces having small radii of curvature are to be avoided. A curved surface changes the effective probe coil liftoff and the true material mass subjected to eddy current influence. Figure 8 indicates the effect of curved surfaces on typical eddy current measurements.[5] Eddy current laboratory evaluations for establishing conductivities should be made on test objects with flat surfaces. For convex surfaces, a V block is used to maintain parallelism between the probe face and the curved surface. Curved surface standards are made by machining a thick, flat specimen to various radii of curvature.

Edges

Edge effects can be demonstrated by moving the eddy current probe toward the edge of a test sample and observing the change in instrument reading that results. Figure 9 is an example of instrument change resulting from edge effect.[5] Test samples must be large enough to prevent this edge interference or probes must be shielded to collimate the field. In this particular case, an edge distance of 5 mm (0.2 in.) or more must be maintained to avoid errors. If parts having a narrow width must be tested, a centering jig or holder should be used to maintain the probe on center. Edge distance curves are used to apply correction factors to conductivity readings on production parts.

FIGURE 8. Influence of cylindrical curved surfaces on conductivity measurements: (a) probe on convex curvature; (b) probe on concave curvature; (c) signals.[5]

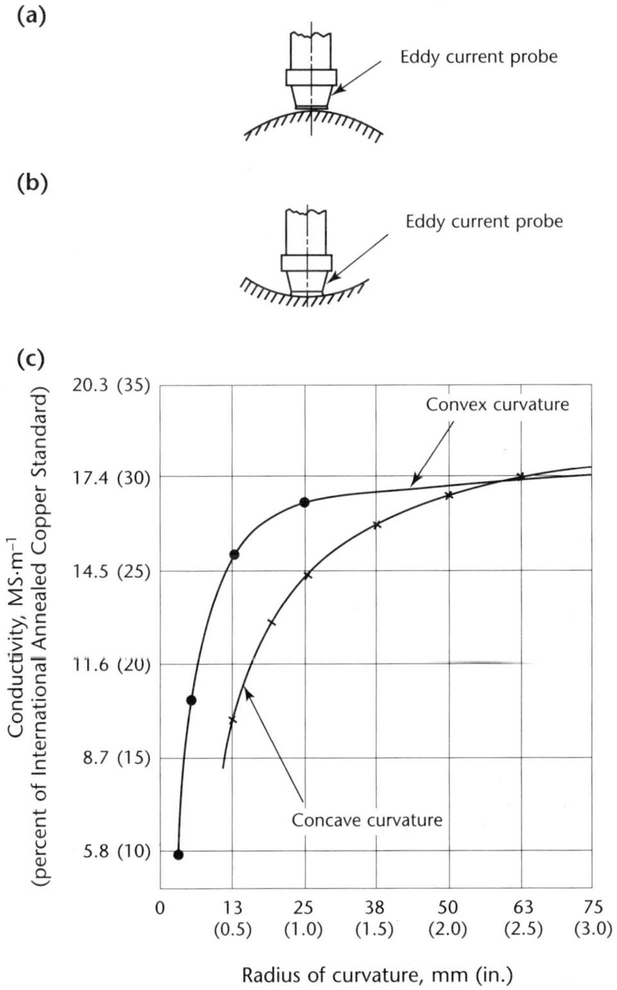

Material Thickness

Material thickness will affect test results when the depth of eddy current penetration is greater than the material thickness. The eddy currents will penetrate the test object to an effective depth of penetration 3δ. Various mathematical equations are documented for determining the standard depth of penetration δ, as described elsewhere in this volume and other publications.[1,6]

The skin effect (described elsewhere in this volume) causes the currents to be concentrated near the surface next to the excitation coil. The effect increases, that is, the depth decreases, with increasing test operating frequency, test object electrical conductivity and test object magnetic permeability. The currents decrease exponentially or almost exponentially with depth, depending on test object shape and thickness.

In addition, the phase angle of the current becomes increasingly lagging as the depth increases. The phase angle lag θ (radian) uses the phase angle of the

FIGURE 9. Influence of edge effect on conductivity measured by eddy current probe: (a) probe at edge; (b) signal.

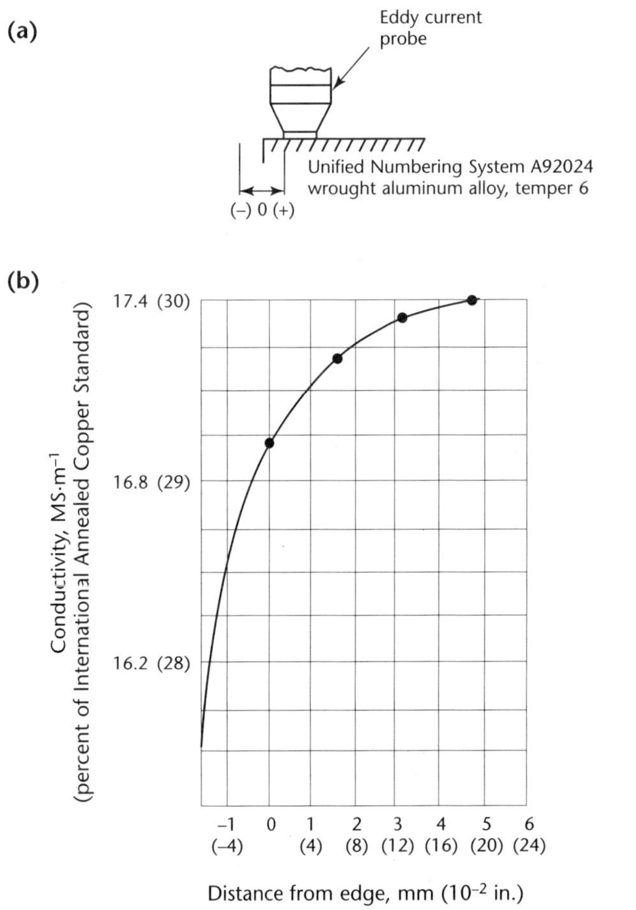

current density at the surface as a reference angle:[6]

$$\theta = x\sqrt{\pi\mu\sigma} = \frac{x}{\delta} \tag{1}$$

where x is distance (meter) below the surface, δ is the standard depth of penetration (meter), μ is magnetic permeability (henry per meter) and σ is electric conductivity (in Eq. 1, as a percentage of the International Annealed Copper Standard). The calculated values at 18.6 $MS \cdot m^{-1}$ (32 percent of the International Annealed Copper Standard, corresponding to a resistivity of 54 $n\Omega \cdot m$) and 60 kHz yield a value of $\delta = 0.48$ mm (0.019 in.). These values agree with those obtained using a commercial eddy current slide rule.[7]

Figure 10 shows the general effect on the coil impedance from changes in conductivity, thickness and liftoff. The included angle is defined as the angle, on the impedance plane, between the locus of liftoff changes and the locus of metal thickness changes. This angle increases with plate thickness and becomes important when impedance analysis is used to minimize the effects of liftoff.

To avoid thickness effects, the effective depth of penetration 3δ may be used:[6]

$$3\delta = \frac{3}{\sqrt{\pi f \mu \sigma}} \tag{2}$$

where f is frequency (hertz) and, for nonmagnetic materials, $\mu = 4\pi \times 10^{-7}\ H \cdot m^{-1} = 1.26\ \mu H \cdot m^{-1}$.

Whatever the test material is, it must be remembered that the standard depth of eddy current penetration δ is a function of material conductivity, operating frequency and magnetic permeability. Because all these values are in the denominator of the depth of penetration equations, an increase in any of these values will result in a shallower depth of penetration. With fixed frequency instruments, not much can be done, except for stacking, to eliminate the influence of thickness variations on the test results. However, if variable frequency instruments are available, then the operating frequency can be altered to match the effective depth of penetration 3δ value and eliminate errors caused by thickness variations.

Cladding Thickness on Aluminum Alloy Sheet

Cladding with commercially pure, high conductivity aluminum on aluminum alloy base materials (which have lower electrical conductivities) causes a greater positive deviation in apparent conductivity as the cladding thickness is increased. Figure 11 shows typical readings of the thickness of commercially pure aluminum cladding on base sheets of Unified Numbering System A97075 and A92024 wrought aluminum alloys. The ordinate (vertical) scale of Fig. 11 shows the change in apparent conductivity of the base metal, resulting from the higher conductivity of the aluminum cladding layer.[8] In typical cases, the cladding adds only about 5 percent to the total sheet thickness but results in an apparent conductivity change of 2.3 to 4.1 $MS \cdot m^{-1}$ (4 to 7 percent of the International Annealed Copper Standard).

FIGURE 10. Conductivity variations with changes of material thickness for 18.6 $MS \cdot m^{-1}$ (32 percent of International Annealed Copper Standard) aluminum tested at 60 kHz: (a) impedance diagram; (b) conductivity versus thickness. Included angles β and θ correspond to phase angles in instrument display.

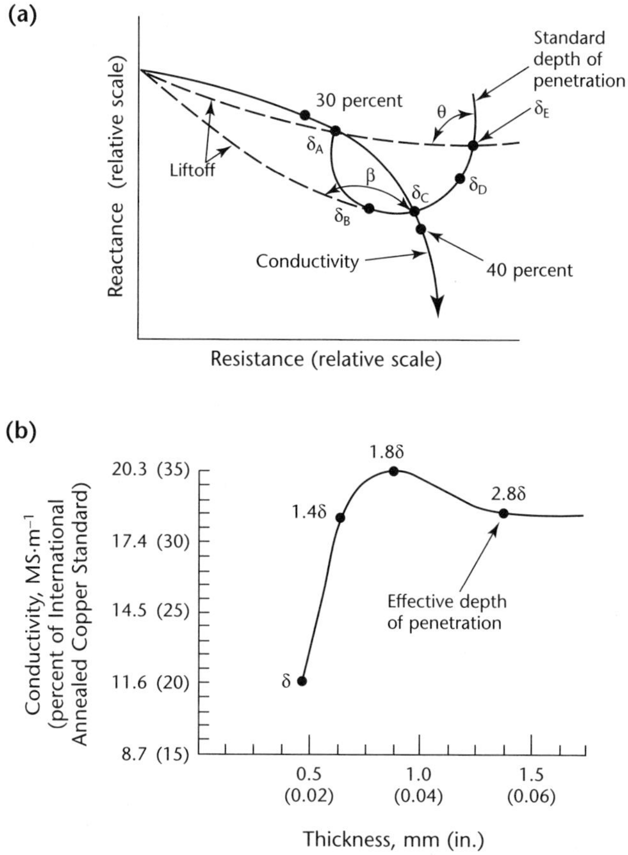

Legend

β = included angle difficult to interpret
δ = standard depth of penetration
δ_A = 1.40 mm (0.055 in.)
δ_B = 0.89 mm (0.035 in.)
δ_C = 0.66 mm (0.026 in.)
δ_D = 0.48 mm (0.019 in.)
δ_E = 0.36 mm (0.014 in.)
θ = included angle easy to interpret

Eddy current conductivity testing (at 60 kHz) for acceptance of aluminum clad material is usually restricted to material less than 2.3 mm (0.09 in.) in total thickness. Testing of thicker materials requires removal of most of the aluminum cladding layer by spot facing a small area and taking a reading there. This reading is taken on a small percentage of parts in the lot. A correction factor is determined between the spot faced areas and those where the cladding has not been removed. The remaining parts are checked without spot facing and the correction factor is applied to the readings.

FIGURE 11. Conductivity change as function of aluminum alloy cladding thickness at 60 kHz: (a) Unified Numbering System A97075 wrought aluminum alloy base material; (b) Unified Numbering System A92024 wrought aluminum alloy base material.[8]

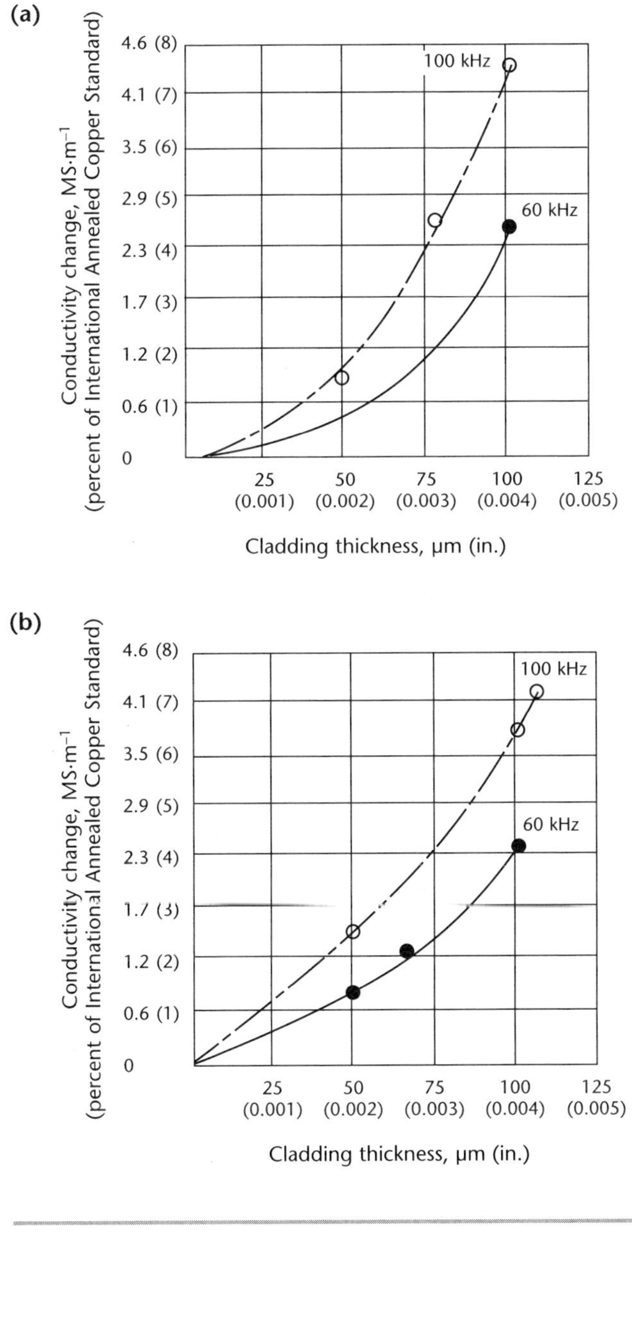

Compensation for Probe Coil Liftoff on Coated Metals

The *liftoff* is the distance that the conductivity meter probe coil is held away from the metal surface by a nonconductive film. The liftoff setting of the test instrument is made by the liftoff control available on some eddy current instruments. This adjustment is made so that true conductivity readings for the base metal are the same as the readings made on a base metal covered by a thin, nonconductive film. Thus, when the liftoff control is set for a specific film thickness on metal with a conductivity of 17.7 $MS \cdot m^{-1}$ (30.5 percent of the International Annealed Copper Standard), the conductivity of a metal covered with a paint layer of that thickness will be read as 17.7 $MS \cdot m^{-1}$ (30.5 percent of the International Annealed Copper Standard). During tests of alloys with different conductivities and with varying thicknesses of nonconductive coatings, it would be necessary to correct the apparent conductivity readings for the deviations caused by liftoff effects to obtain true readings for the base metal conductivities.

Test Material Temperatures

Temperature is an important test variable, particularly when eddy currents are used to establish a basic conductivity range for an alloy. Consideration must be given to (1) the temperature of the test material, (2) the difference in temperature between the test sample and reference standards used for calibration and (3) the type of eddy current instrument being used. The influence of temperature on the resistivity of a metal can be determined from Eq. 3:

$$R_t = R_0 \left(1 + \alpha \Delta T\right) \quad (3)$$

where R_t is resistivity (ohm meter) of the metal at the test temperature, R_0 is resistivity (ohm meter) of the metal at standard temperature, ΔT is the difference between the standard and test temperatures (kelvin) and α is the resistivity temperature coefficient (rarely negative).

From Eq. 3 it can be seen that if the temperature is increased, resistivity increases and conductivity decreases from their ambient temperature levels. Conversely, if temperature is decreased, the resistivity decreases and conductivity increases.

PART 4. Eddy Current Testing of Bolt Holes[1]

Small cracks in material next to fastener holes may go undetected until the cracks grow to a size that allows detection at a surface not covered by the head of the fastener or nut. Unfortunately, in many highly stressed areas, a crack of this magnitude may continue to grow to failure at a rapid rate. Eddy current testing is a reliable way to detect cracks in material adjacent to fastener holes. Eddy current tests are well known for detecting small fatigue cracks, particularly in aluminum fastener holes.[9,10] Automated bolt hole scanners provide reliable and repeatable test results.

Finding small cracks at fastener holes sometimes requires removing the fastener and performing an eddy current check. Automatic scanning[11] may be used to detect cracks as small as 0.13 mm (0.005 in.). Figure 12 illustrates an eddy current bolt hole probe. The probe coil axis is perpendicular to the material adjacent to the hole. An adjustable collar is used to position and locate the coil a desired distance inside the hole. Usually, testing is started with the coil just below the test material surface and the probe is manually rotated 360 degrees. After each rotation, the probe is advanced in increments to ensure adequate overlap and coverage. The spherical end of the probe is often split in the middle so that a small rubber or plastic V wedge may be inserted to make the probe snug in oversized holes. These wedges are not generally used for probes less than 6.3 mm (0.25 in.) in diameter.

FIGURE 12. Eddy current hole probe.

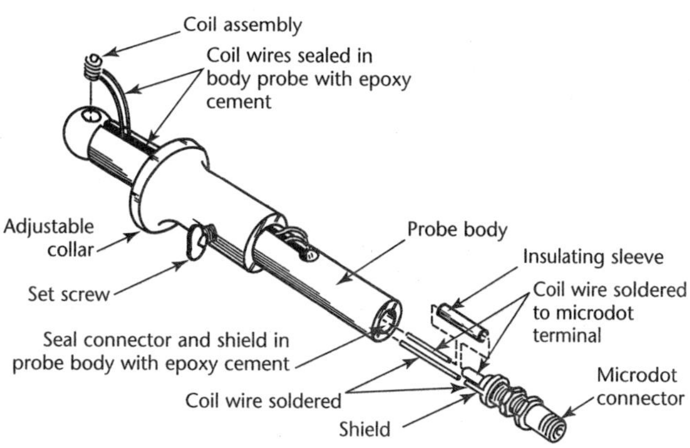

During bolt hole testing, the absolute technique is generally used. The absolute technique uses one coil on a ferrite core; measurements are made while the material is in direct contact with the probe coil. The differential technique uses two coils and compares measurements obtained from the (uncracked) material under one coil with the (cracked) material under the other coil. A bridge circuit is unbalanced if one coil is over sound material and the second coil is over a discontinuity. Figure 13 shows crack

FIGURE 13. Normalized impedance graph showing crack responses for absolute and differential bolt hole probes: (a) absolute probe crack response at 100 kHz; (b) differential probe crack response at 500 kHz.

(a)

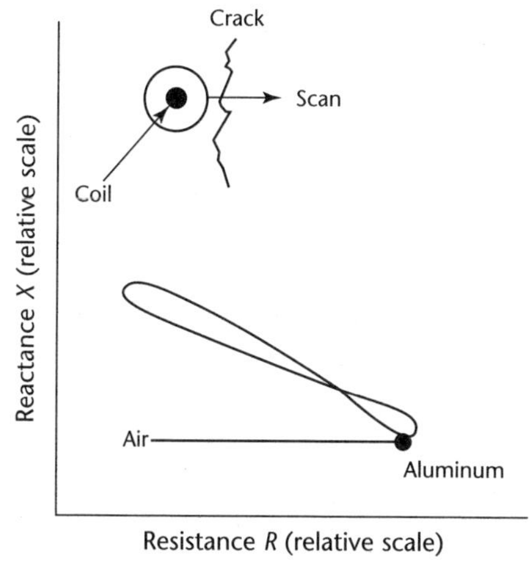

(b)

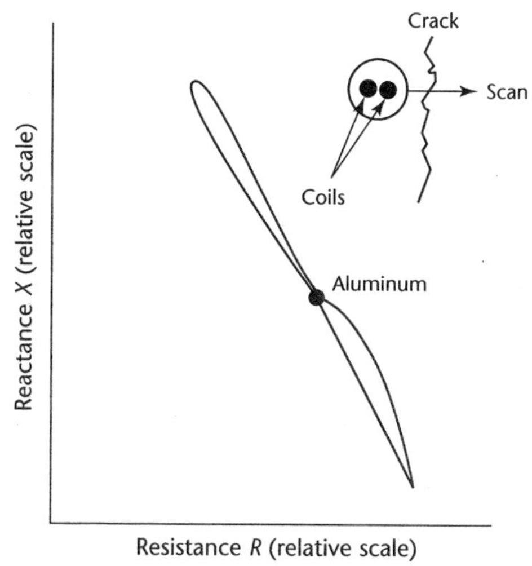

responses typical of bolt hole probes inserted in an aluminum sample: a single-loop response from an absolute bolt hole probe (Fig. 13a) and a double-loop response from a differential probe (Fig. 13b).

Reference Standards for Bolt Hole Testing

Reference standards aid in calibrating or adjusting instrument controls and are used to ensure that cracks will be detected with a predetermined sensitivity, based on depth, length and location (see Fig. 14). Reference standards have been fabricated by simulating cracks with electric discharge machined radial slots of various depths and lengths (Fig. 15). This type of reference standard works well if the notches are less than 0.13 mm (0.005 in.) wide. However, simulated discontinuities do not respond exactly the same as real fatigue cracks.[10] Some investigators have produced cracks by introducing an electric discharge machined slot in the hole and then subjecting the specimen to fatigue loading. A crack initiates and grows outward from the machined slot. The depth and length of the crack are qualitatively controlled by the stress and the number of cycles. After the crack has grown to the approximate size, the electric discharge machined slot is removed by drilling the hole to a sufficient, predetermined oversize. The depth of the resulting crack is estimated; the true depth can be determined only by fracturing the reference standard.

Procedure for Bolt Hole Testing

A typical procedure for performing eddy current bolt hole tests is listed below.

1. Determine the diameter and depth of holes to be tested from engineering drawings or other applicable documents.
2. Select sizes of probes and of reference standards.
3. Calibrate the instrument response to the reference standards.
4. Visually check holes for excess corrosion or galling. Remove material from a hole by manually passing a reamer through the hole or drilling out to next oversize diameter.
5. Scan for discontinuities by rotating the probe at constant depths, at intervals not exceeding 1.5 mm (0.06 in.) deep. Conduct scanning in intervals of about one revolution each. If suspect areas are noted near the ends of scans, retest these areas so that the suspect areas are in the middle of the second scans.

Fatigue crack depths may be estimated from the amplitudes of eddy current readings. When observing meter deflections, make sure the deflection is not caused by scratches, surface irregularities, out-of-roundness of the hole or tilting of the probe.

Bolt Hole Probes

Bolt hole probes are generally available from eddy current equipment manufacturers. It must be remembered that different types of probes are used for testing straight or tapered holes and that different collars are used for flat, countersunk or curved outer surfaces.

Strip Chart Recorders

Although many bolt hole tests are performed manually (where the operator directly observes the meter deflection or screen presentation), strip charts or computers may be used to process, record or interpret the test results. Signals

FIGURE 14. Typical location of cracks adjacent to fastener holes: (a) at top; (b) at middle of hole; (c) at bottom; (d) on beveled surface.

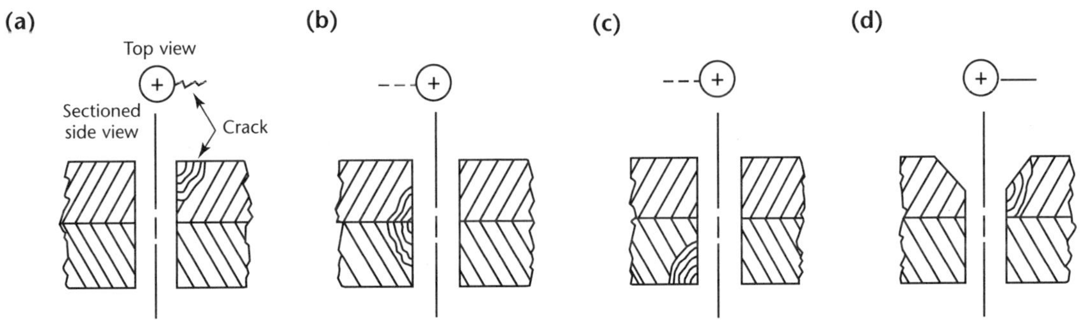

produced by various discontinuities are shown in Fig. 16. The trace has been rotated to make the air-to-aluminum line horizontal as in Fig. 13a. The vertical output is then recorded by a slow running strip chart recorder.

Automated Bolt Hole Testing

When numerous bolt holes need to be tested, it is often desirable to eliminate manual scanning and use a device that automatically rotates the probe in the hole. Eddy current equipment manufacturers have developed such systems. A system has been designed for the testing of shallow holes and tube ends.[12] The probe is attached to the end of a short spindle and is rotated in a hand held, rotating head.

The United States Air Force has also developed bolt hole tests. An automatic eddy current discontinuity detection system detects and records discontinuities on conductive metal surfaces and in structural fastener holes, reducing the errors of conventional hand scanning.[11] The system can detect cracks as small as 0.25 mm (0.010 in.) long by 0.13 mm (0.005 in.) deep in aluminum.[11] Scan depths are adjustable from 6.3 to 3.8 mm (0.25 to 1.50 in.). The scanner's speed can be varied from 0 to 2.45 Hz (0 to 150 rotations per minute), providing simultaneous linear and rotational (spiral) motion. Either filtered or unfiltered signals from the scanner can be displayed on a chart recording or oscilloscope. The analog recorder provides a detailed profile of discontinuities for analysis. The system is fully portable and all equipment is contained in a protective case.

Other rotating eddy current test systems are available for rapidly detecting small cracks in aircraft bolt holes. They use small, light, hand held scanners and various sizes of bolt hole probes. These instruments are simplified, easy to use and portable. They generally operate at 500 kHz and use differential probes rotating at 50 cycles per second (3000 rotations per minute).

The resolution of the differential scanning coil system follows the physical laws for general eddy current testing; when the probe moves on an orthogonal path over a crack, the crack depth is the principal factor affecting the amplitude, provided its length is equal to or greater than the coil diameter.

The main application is the rapid testing of fastener holes for cracks. The high revolution speed of 50 cycles per second (3000 rotations per minute) of the rotating probe and the clear display on the screen permit a high test sensitivity. The saving of time compared to the hand held technique is about 80 percent. In principle, all metals can be tested, including titanium, aluminum and steel. The detection sensitivity depends on the roughness of the surface and begins at

FIGURE 15. Calibration block for bolt hole and surface probe tests.[11]

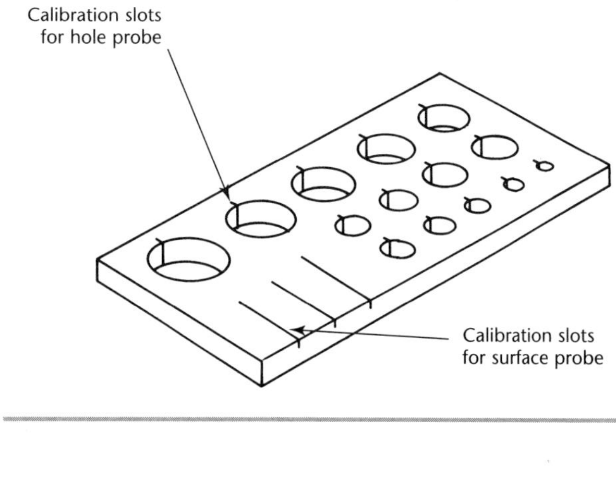

FIGURE 16. Strip chart indications obtained from typical discontinuities during eddy current bolt hole test: (a) scratch; (b) crack or corrosion; (c) corrosion; (d) conductivity variation; (e) nonperpendicular hole axis. Deflections to left indicate downscale meter deflections.

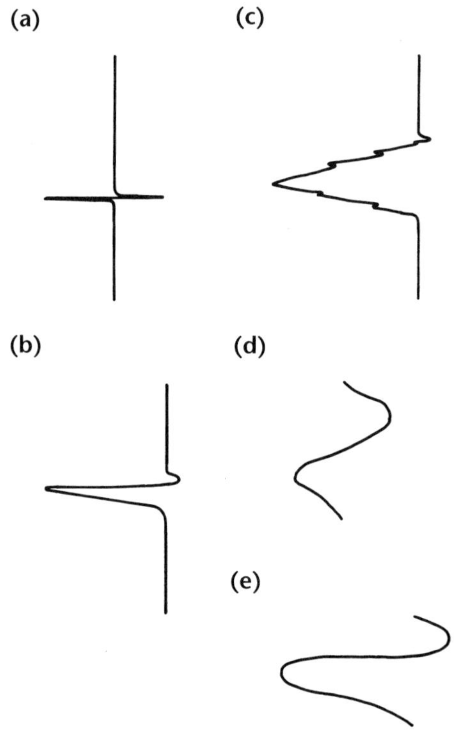

about 0.20 mm (0.008 in.) depth. Two different kinds of display may be used: Y,*t* with one sweep per revolution (Fig. 17a) and the X,Y vector for differentiation of the signal in phase and amplitude (Fig. 17b). The lightness and small size of the detection probe allows the instrument to be used in confined areas.

FIGURE 17. Cathode ray tube displays: (a) amplitude versus time; (b) lissajous phase analysis presentation of X,Y components on impedance plane.

(a)

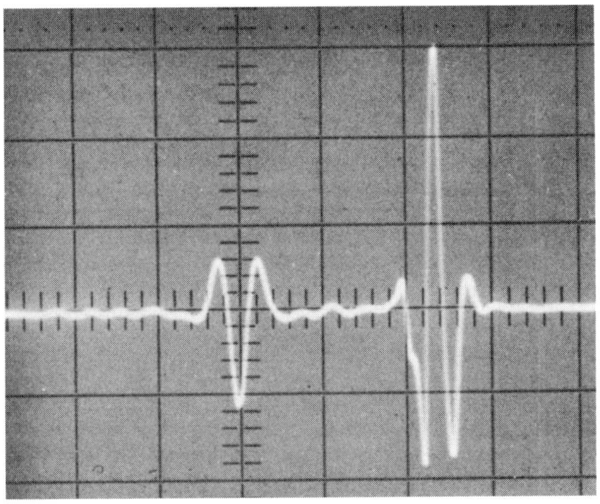

(b)

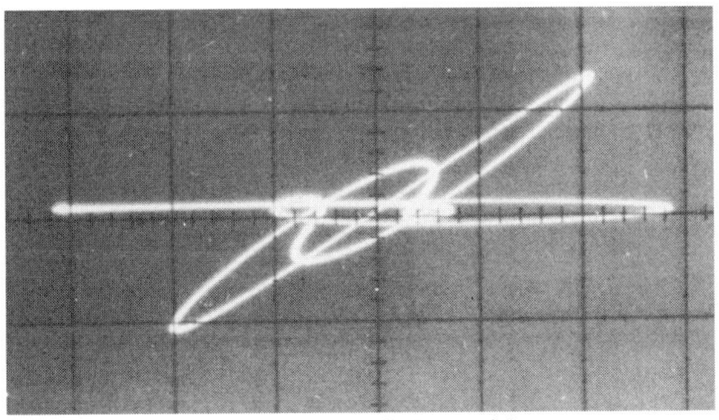

PART 5. Impedance Plane Analysis of Typical Aerospace Material Tests[1]

Impedance Plane[2]

Eddy current probes typically consist of one or more wire wound coils that behave electromagnetically as inductors. Thus, the probe coils possess both an inductive reactance and a resistance. These two values combine to produce the overall electrical impedance of the probe. Instruments that measure impedance provide more complete test data during eddy current testing.

Because electrical impedance contains two components, it can be represented as a two-dimensional graph such as that shown in Fig. 18. This graph is called the *impedance plane*. The coil resistance R is plotted horizontally and its inductive reactance X is plotted vertically. During eddy current testing, a relative impedance point (material test point) is obtained by placing the probe against the sample and setting the bridge controls for X and R. These X and R values (phasors) are then traced on the instrument screen display.

Each eddy current measurement of a test variable generally possesses a unique location on the normalized impedance graph. Different material conditions — such as conductivity, permeability, discontinuity, liftoff, spacing and thinning — produce their own unique impedance plane plots. The diversity of plots allows rapid, significant impedance plane analysis in a wide variety of applications.

Continuously changing the magnitude of a given variable traces out an impedance curve (locus of points) that is a characteristic signature for that variable. The amount of movement and the position of a test point along the curve is related to the magnitude of the impedance change. This analysis of impedance components is the basis for eddy current signal interpretation.

Eddy current phase analysis instruments enable the operator to produce impedance plane traces automatically on the integral X,Y storage oscilloscope. These instruments operate from 60 Hz to 6 MHz, allowing the operator to choose the best frequency for a given material and test.

For the eddy current test response to different conductivities of various nonmagnetic alloys, the material points trace out a characteristic comma shaped curve with conductivity increasing in a clockwise direction. The figure also shows probe liftoff impedance curves for two of the alloys (liftoff is the spacing between the probe and the bare metal surface).

FIGURE 18. Impedance changes in relation to one another for surface probes of 6.3 mm (0.25 in.) diameter.[2]

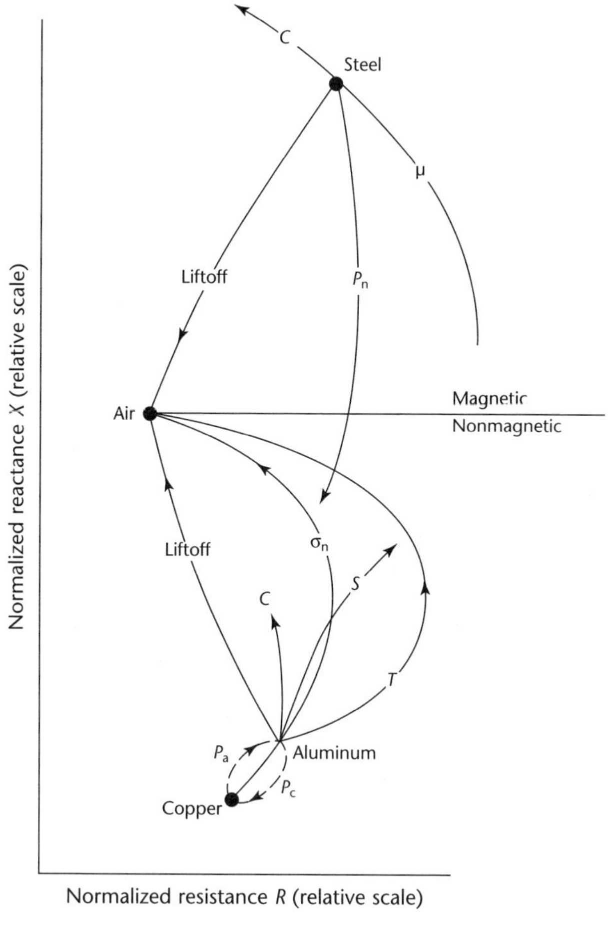

Legend

C = crack in aluminum
P_a = plating aluminum on copper
P_c = plating copper on aluminum
P_n = plating nonmagnetic
S = spacing between aluminum layers
T = thinning in aluminum
μ = permeability
σ_n = conductivity nonmagnetic materials

The material test points move clockwise around all impedance curves in a nonlinear fashion as test material conductivity and test frequency are increased.

Figure 18 shows impedance plane responses for several other types of common variables. Examples of metal thinning are shown in Figs. 1 and 10. An impedance chart of cladding thickness is shown in Fig. 3 and an impedance chart of liftoff, for both magnetic and nonmagnetic materials, is shown in Fig. 4. The impedance plane response for separation (spacing) between two aluminum sheets is shown in Fig. 6. The impedance response for surface cracks in aluminum is shown in Fig. 13.

Test Frequency Selection

One of the most important considerations in eddy current testing is the choice of test frequency. The type of alloy involved and the variables to be measured or suppressed determine the best frequency. The depth of eddy current penetration within test materials is strongly affected by test frequency. For a given alloy, higher frequencies normally limit the eddy current test to inspection of the excited metal surface nearest the magnetizing coil winding. Lower frequencies permit deeper eddy current penetration. A given test frequency will allow eddy currents to penetrate deeper in lower conductivity alloys than in higher conductivity alloys.

High test frequencies are normally used for detecting small surface cracks or surface contamination and for gaging thin coatings. Medium frequencies are useful for conductivity measurements such as alloy sorting. Low test frequencies are usually required for testing thicker materials (for opposite side corrosion, for example), for thickness gaging and for penetrating into magnetic materials.

Penetration depth, however, is only part of the process for selection of optimum eddy current test frequencies. The geometric relationship between the impedance curves for the variable to be measured and the variable to be suppressed must also be considered. Furthermore, the position of the variable magnitudes (different conductivity points or liftoff points) along their respective impedance curves is important.

Test frequency affects the conductivity and liftoff curves of nonmagnetic alloys. Changes in frequency shift the points along the curves in a nonlinear fashion. This phenomenon, also true for other impedance curves (including magnetic materials), can be advantageously used because it allows the material points to be positioned for optimum response to desired variables and for suppression of undesired variables. Specifically, a frequency should be chosen that causes the impedance curve points (material points) for the variable to be measured to move in a substantially different direction than those points for the variable to be suppressed.

At low frequencies, the angle between the liftoff curve and the region of the conductivity curve for the titanium point is quite small. Thus, it becomes more difficult to obtain liftoff suppression without a noticeable expense to the response for conductivity variations or surface cracks in titanium. By selecting a higher frequency, the angle between titanium test points and liftoff is quite large, which allows liftoff suppression and good sensitivity to conductivity variations in the titanium.

Conductivity is measured to identify alloys and heat treatment conditions. For conductivity measurements and for surface crack detection, a frequency in the conductivity curve should be chosen that places the material test point just below the point of maximum coil resistance and eddy current losses (the region of highest sensitivity to changes in all test variables). This location for the material test point creates a desirably large angle between the liftoff and conductivity responses, allows good liftoff suppression and provides good sensitivity to conductivity variations.

The material points for all other variables (metal thickness, plating thickness and others) are also shifted clockwise around their impedance curves by frequency increases. Therefore, a rationale similar to that discussed above applies to all variables.

Applications of Impedance Plane Analysis[13]

Examples of metal thinning caused by corrosion include the following: (1) metal spacing between an aluminum and titanium skin in an engine nose cowl, (2) localized metal thinning of a chemically milled step and (3) surface crack detection in aircraft structure.

Metal Thinning from Corrosion[13]

Moisture entrapment between an aluminum fuselage belly skin and a titanium doubler causes corrosion thinning in the skin. To detect the corrosion and determine metal thinning, the inspector can be given a sketch showing the location and outline of the internal doubler. The skin material and temper can be identified along with the electrical conductivity values. An

FIGURE 19. Screen display calibration at 20 kHz for 1.3 mm (0.05 in.) thick skins: (a) aluminum taper gage; (b) taper gage signal at 1.3 mm (0.05 in.) thickness; (c) taper gage signal at 0.2 mm (0.01 in.) thickness; (d) completed thickness calibration using taper gage. Voltage measurement increment is assigned during calibration.

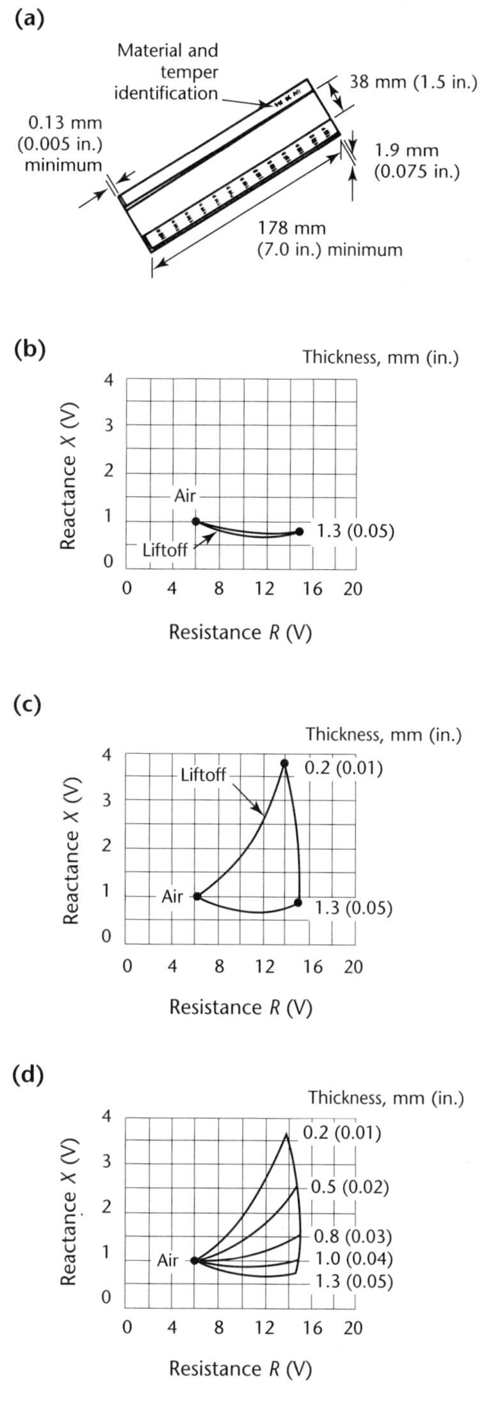

aluminum thickness taper gage (Fig. 19a) of alloy similar to the skin material is used to calibrate the eddy current instrument for metal thinning. The probe is placed on the taper gage at the 1.3 mm (0.05 in.) thickness and the response shown in Fig. 19b is obtained. The probe is then slowly scanned to the thin end of the gage and is lifted off at 0.2 mm (0.01 in.) thickness, yielding the response shown in Fig. 19c. Thickness calibrations are performed by rescanning the taper gage and lifting off the probe at various thickness increments as shown in Fig. 19d. The instrument is now calibrated for the test.

The operator looks for corrosion thinning while the aluminum external skin is scanned over the area of the internal doubler. If there is no corrosion, the response shown in Fig. 20b will be obtained. Severe corrosion will yield a response similar to that shown in Fig. 20c. The depth of the corrosion is determined by using the taper gage to measure the depth from maximum amplitude response as shown in Fig. 20d. In this case, the

FIGURE 20. Impedance plane screen display presentations for exfoliation corrosion metal thinning: (a) diagram; (b) screen display presentation for no corrosion; (c) screen display presentation for severe corrosion; (d) method for determining depth of corrosion; (e) surface crack response in corroded area.

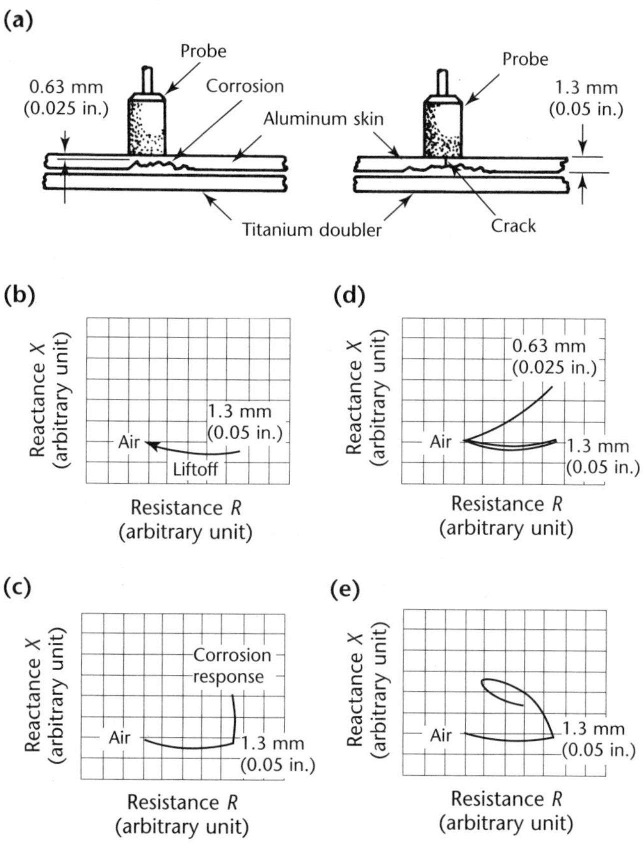

corrosion thinning is 0.63 mm (0.025 in.), which represents 50 percent of the skin thickness. Cracks may be associated with severe corrosion and, if they exist, the response shown in Fig. 20e will be obtained.

Metal Spacing[13]

During fabrication of engine nose cowl inlet ducts, problems have been encountered in maintaining a gap between the outer aluminum skin and inner titanium skin for the purpose of hot air deicing. An eddy current technique was developed and used to evaluate the gap at various locations around the nose cowl. Results indicated that a consistent spacing was not being maintained and that the gap spacing was below minimum tolerance in some areas of the duct. Because of this, aluminum button spacers were installed between the aluminum and titanium skins to make the gap consistent.

Before taking gap thickness measurements, a reference standard was fabricated from a plate of 1.5 mm (0.06 in.) thick Unified Numbering System A92219 wrought aluminum alloy (Fig. 21) to represent the outer skin of the cowl. The 0.5 mm (0.02 in.) thick commercially pure titanium shims represent the internal skin of the cowl. The nonconductive acrylic step wedge represents the air gap or space between the inner and outer skins. The instrument was calibrated by placing the probe on the low conductivity (titanium) side of the standard at the zero and 2.5 mm (0.10 in.) gap positions. The frequency and instrument settings varied until a maximum spread was obtained on the screen display.

FIGURE 21. Eddy current reference standard for spacing measurement: (a) diagram of use; (b) top view; (c) side view.[15]

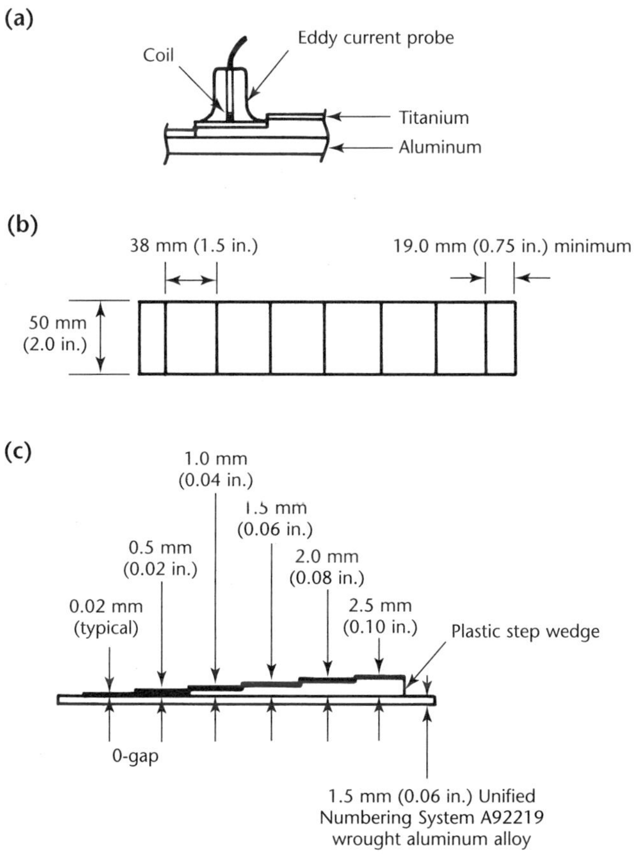

Localized Metal Thinning[13]

In one case, a thinning of a chemically milled step in a bonded honeycomb wing tip was discovered. A test was needed to both detect and measure the thickness at the discrepant area of the Unified Numbering System A97075, temper 6, wrought aluminum alloy skin. The skin thickness ranged from 0.45 to 0.70 mm (0.018 to 0.028 in.) in the chemically milled area. Reference standards 0.41, 0.46 and 0.51 mm (0.016, 0.018 and 0.020 in.) thick were used to calibrate the instrument at 20 kHz. The liftoff was rotated on the screen display by a phase adjustment so that it was displayed in the horizontal direction. The wing tips were measured by placing the probe at the inboard or thin part of the bonded panel and scanning outboard across the chemically milled step (Fig. 22). When the test object had a normal chemically milled step, the screen display trace shown in Fig. 23a was obtained. When the part had local undercut in the chemically milled step, then the trace shown in Fig. 23b was obtained. By marking the screen display with the minimum 0.46 mm (0.018 in.) thickness during calibration, the depth of the undercut could be determined.

FIGURE 22. Eddy current impedance plane testing of bonded honeycomb panel in wing tip.

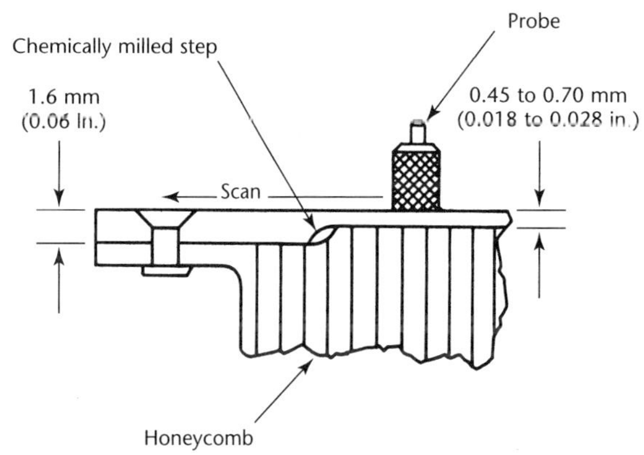

Numerous test objects were evaluated and test objects showing less than minimum thickness were rejected and sectioned and the eddy current results were verified.

Exfoliation Corrosion around Wing Skin Fastener Holes[13]

Although contact between galvanically dissimilar metals, such as steel and aluminum, is known to be a cause of corrosion, the design of aircraft structures and systems occasionally requires that such metals be joined together. When dissimilar metals must be joined, it is a design requirement that the contacting surfaces be electrically insulated with organic paint or sealant or that one of the surfaces be coated with a metallic coating galvanically similar to the other surface. For example, cadmium plated steel bolts are used on many aircraft. The cadmium plating protects the steel bolt from corrosion, provides a surface galvanically similar to aluminum and reduces the problem of accelerated corrosion. If the cadmium is depleted or a crevice where moisture can collect exists between the fastener head and the aluminum skin, pitting and intergranular corrosion may occur.

FIGURE 23. Eddy current impedance plane results from inspection of wing tip: (a) results from normal chemically milled step; (b) results from undercut chemically milled step.

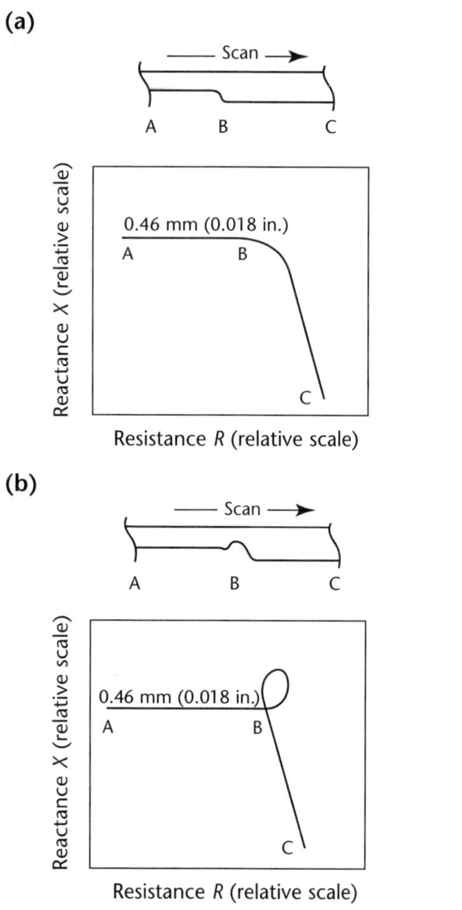

Intergranular corrosion occurs along aluminum grain boundaries, which in sheet and plate are oriented parallel to the surface of the material as a result of the rolling process. Intergranular corrosion, called *exfoliation corrosion* in its more severe form, is characterized by delamination of thin layers of aluminum parallel to the sheet surface, with white corrosion products between the layers. Where fasteners are used, the corrosion extends outward from the fastener hole, either from the entire circumference of the hole or in one direction from a segment of the hole. In severe cases, the surface bulges upward. In less severe cases, there may be no telltale bulging: corrosion may be detected only by nondestructive testing.

Eddy current tests for exfoliation corrosion are shown in Fig. 24. A phase sensitive instrument providing an impedance plane display was operated at 20 and 50 kHz. The depth of penetration of the eddy currents is a function of operating frequency and material conductivity. The effective depth of penetration in the skins, clad with Unified Numbering System A97075 wrought aluminum alloy, temper 6, was about 1.3 mm (0.05 in.) at 50 kHz and 2.3 mm (0.09 in.) at 20 kHz. The conductivity of the coated aluminum skins ranged from 26.1 to 29.0 $MS{\cdot}m^{-1}$ (45 to 50 percent of the International Annealed Copper Standard).

The phase sensitive eddy current instrument was calibrated to yield a maximum response difference between a corroded and uncorroded area. Figure 24 shows the screen display response from an uncorroded area (Fig. 24b), from a corroded area with 20 kHz and a 13 mm (0.5 in.) diameter probe (Fig. 24c) and from a corroded area with a 50 kHz, 6.3 mm (0.25 in.) diameter probe (Fig. 24d). Identical responses were obtained with and without a steel fastener in the hole.

The phase angle between the liftoff and corrosion response was about 90 degrees at 50 kHz and 45 degrees at 20 kHz (Fig. 24). Because of the greater phase angle response at 50 kHz, remaining tests were conducted at this frequency. The test was performed by scanning 360 degrees around the periphery of each hole. Corroded holes were marked and the extent of the corrosion was outlined on the skin surface with a marking pen.

If corroded samples are not available to calibrate the instrument before testing, a taper gage reference standard (Fig. 19a) may be made from similar material. The instrument is null balanced with the probe on the thick portion of the standard. After calibration, a corroded area will yield a response similar to that shown in Fig. 24d. A circle template is generally used to guide the probe as it is scanned 360 degrees around each fastener.

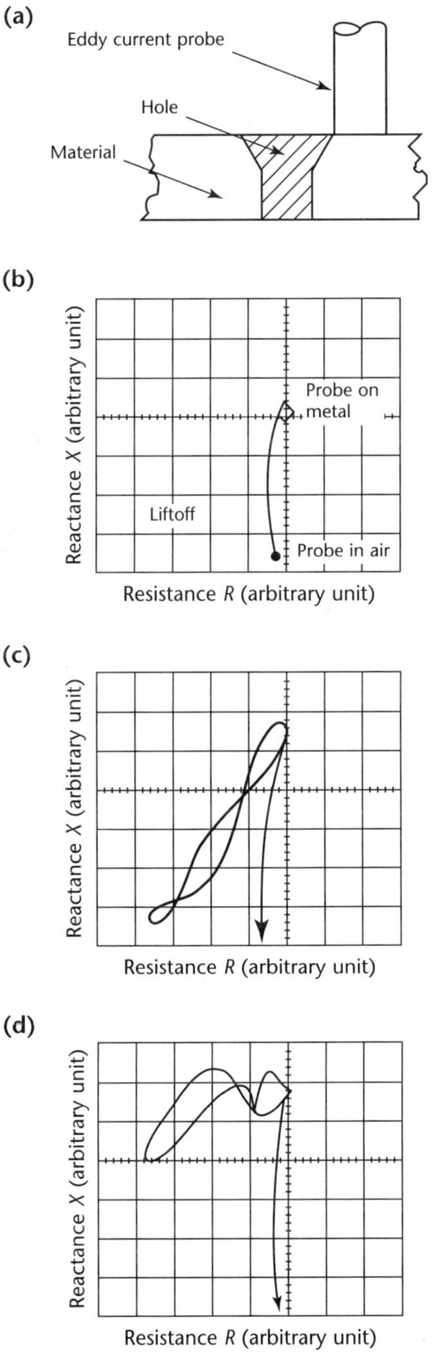

FIGURE 24. Eddy current cathode ray tube responses for exfoliation corrosion around fastener holes in wing skins: (a) diagram; (b) eddy current response, no corrosion; (c) eddy current response from corrosion using 20 kHz and 13 mm (0.5 in.) diameter probe; (d) eddy current response from corrosion using 50 kHz and 6.3 mm (0.25 in.) diameter probe.

Surface Crack Detection[2]

Eddy current testing is used to find cracks in aircraft structures during inservice maintenance checks. Impedance plane analysis provides a better understanding of crack response loci with other variables on the impedance plane. The value of this approach is evident in Fig. 25a where the crack response locus is 90 degrees out of phase with the liftoff locus for Unified Numbering System G43400 nickel chrome molybdenum alloy steel. Because of the wide separation angle between these two loci, cracks in steel are fairly easy to detect. The lower amplitudes and phase angles for electric discharge machined

FIGURE 25. Notches and cracks at 200 kHz on impedance plane screen display: (a) electric discharge machined notches and crack in steel; (b) electric discharge machined notches and crack in aluminum.

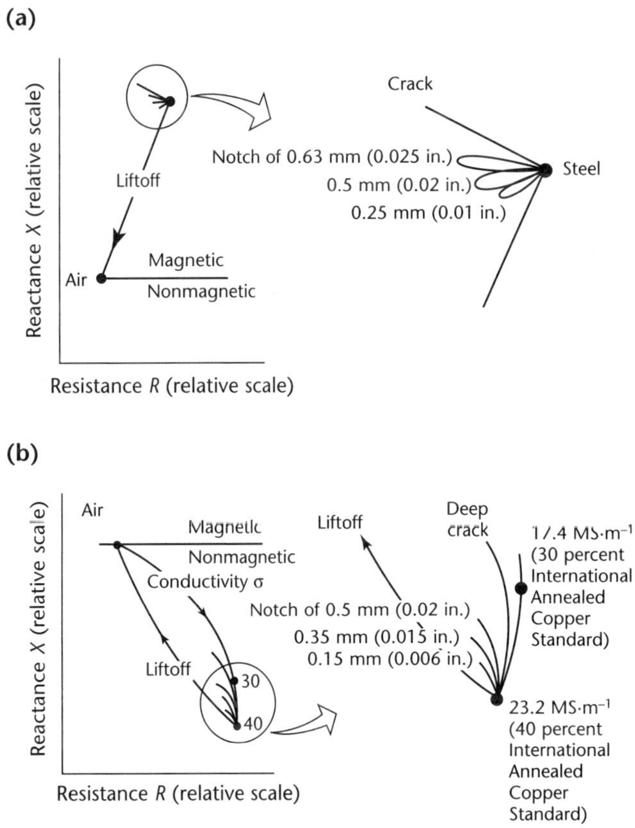

notches in the steel sample are also shown. Figure 25b shows that for aluminum alloys, the electric discharge machined notch and crack response lie between the liftoff and conductivity loci. The smaller notch response almost parallels the liftoff response but the angle of separation and amplitude increases as the notch depth increases. A fairly good response is obtained from a deep crack in a similar aluminum specimen. It can be seen in Fig. 25b that the crack response is almost parallel to the conductivity locus. This characteristic is true for all nonmagnetic and magnetic materials, as illustrated in Fig. 26.

Skin cracks, which initiated at fastener holes in a fuselage splice, grew by fatigue in a circumferential direction. Flush head aluminum rivets joined the fuselage skin to the frames and longerons. Detection of the surface cracks was possible with standard eddy current crack depth meters. These instruments required scanning between each row of fasteners, which was somewhat time consuming. To speed up the test, a procedure (operating at 5 kHz) was developed for phase analysis screen display instrumentation. Using this type of equipment, the operator can scan in a direct circumferential line over the installed fasteners as shown in Fig. 27a.

In Fig. 26, crack responses are in the same direction as the conductivity locus and therefore should be in the same direction as the change in conductivity caused by the rivets. Figure 27b shows the response obtained by scanning over the rivets with no cracks in the skin. In Fig. 27c, the crack response is in the same direction but has a much greater impedance, resulting in a clearly discernible increased amplitude signal.

FIGURE 26. Direction of surface cracks on impedance plane cathode ray tube.[2]

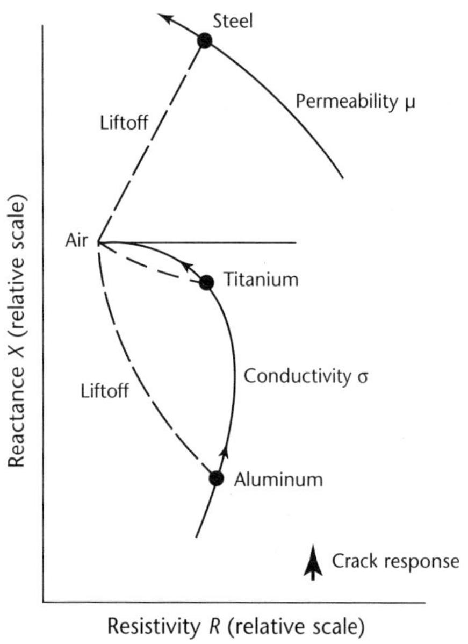

Gain Adjustment to Enhance Indications[2]

With phase analysis instruments, the technique called *null balancing* described earlier makes the test results easier to interpret.[14] Null balancing entails rotating the impedance place to help interpret signals of interest. This technique is sometimes difficult with screen display

FIGURE 27. Cracks in first layer initiating at aluminum fastener holes: (a) looking inboard, left side shown, right side opposite; (b) scan over rivets with no crack in skin at 5 kHz using 9.6 mm (0.38 in.) diameter probe; (c) scan over rivets with crack in skin.

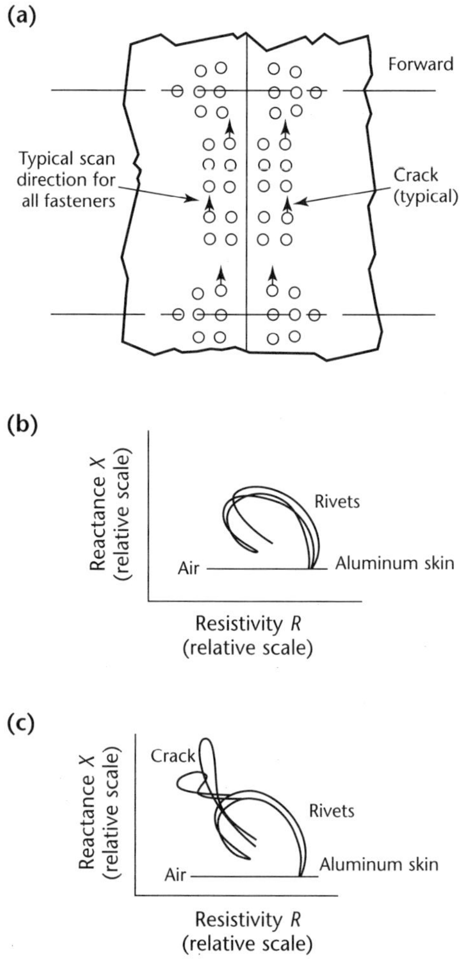

FIGURE 28. Improving test results at 50 kHz by changing screen display vertical gain: (a) normal impedance plane response from crack in aluminum; (b) crack responses at various changes in vertical gain; (c) normal impedance plane response from crack in steel; (d) crack responses at various changes in vertical gain.[2]

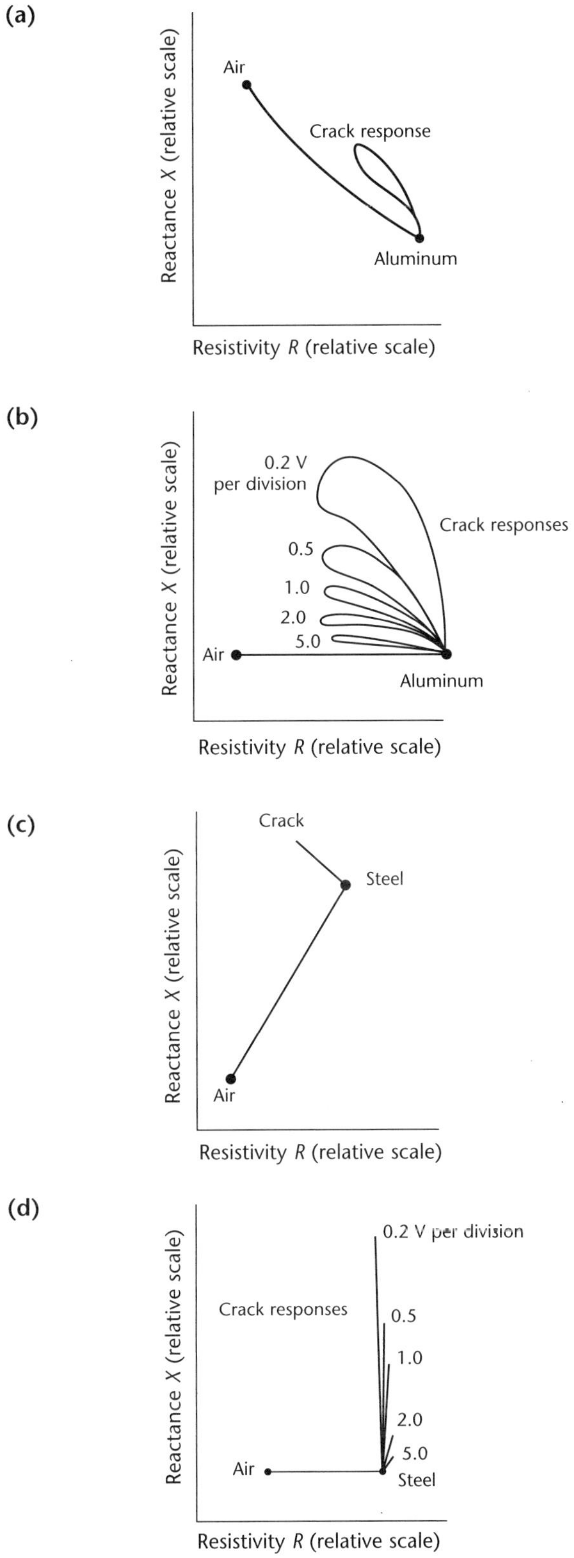

instruments because the flying dot impedance point will in many cases be offscreen.

The onscreen results of null balancing can sometimes be improved by adjusting the gain. As the optimum operating frequency is selected by rotating the signal traces on the impedance plane, the vertical or horizontal gains on the screen display can be increased or decreased. The traces are rotated by turning the phase control on the face of the instrument. At 5 V per division, the gain is low on the screen display; at 0.2 V per division, the gain is high on the screen display. Two examples are shown in Fig. 28.

The first example shows the normal impedance plane response for a crack in an aluminum sample (Fig. 28a). The liftoff locus is rotated about 45 degrees to a horizontal position as shown in Fig. 28b. Because the liftoff is an unwanted response, the gain in the horizontal direction is reduced by using a setting of 2 V per division on the screen display. To amplify the crack response in the vertical direction, the vertical gain is increased. This change in gain amplifies both the separation angle (between liftoff and crack response) and the amplitude of the crack response. Figure 28b shows the response that can be expected with differential gain settings superimposed for the same frequency response.

The second example is for a crack in steel as normally presented on the impedance plane (Fig. 28c). Here again, the liftoff response is rotated to a horizontal direction with the crack response in the vertical direction. The horizontal gain is set at 2 V per division and the vertical gain is increased to 0.2 V per division (see Fig. 28d). In this case, the phase angle does not change appreciably but the signal amplitude is increased considerably at the 0.2 V per division.

These two examples are typical of situations where null balancing reduces the irrelevant signal amplitude while increasing the signal amplitude of interest.

Subsurface Crack Detection[2]

Engineers for two aircraft manufacturers investigated subsurface crack detection in aircraft structures using low frequency eddy current techniques. Figure 29 shows how cracks through the thickness of second and third layer aluminum structures appear on the impedance plane at 1 kHz. The surface crack response in the first layer runs almost parallel to the liftoff locus. The second layer crack runs parallel to the conductivity locus and the third layer crack response is to the right of the conductivity locus. The phase lag causes the phase shift of the crack signals as the crack depth below the surface increases.

FIGURE 29. Direction of surface and subsurface crack indications in aluminum on impedance plane: (a) test setup at 1 kHz; (b) impedance plane with enlargement of circled area.

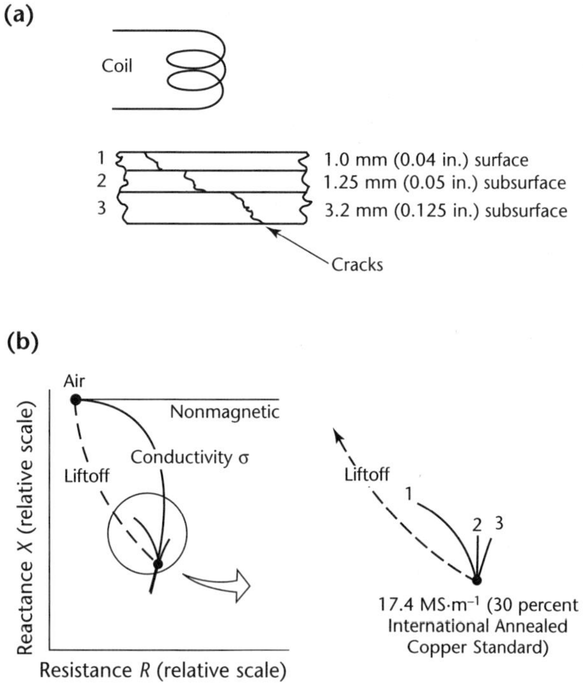

Part 6. Low Frequency Eddy Current Testing of Aircraft Structure[1,15]

Airlines and airframe manufacturers have been using eddy current crack detection techniques since the early 1960s. Most of the eddy current equipment and procedures were designed to operate at high frequencies (10 kHz to 1 MHz) to detect very small surface cracks. Around 1973, it was determined by damage tolerance studies that eddy current testing was more sensitive than radiography for detecting fatigue cracks.[16,17] It then became desirable to use eddy current testing instead of radiography to detect subsurface cracks. Unfortunately, at that time, instruments operating at low frequencies (100 Hz to 10 kHz) were scarce and, because the probes were large in diameter, they lacked the sensitivity to detect small cracks.

Low frequency eddy current testing became reliable with the introduction of phase analysis instruments. In addition, instrument manufacturers lowered the operating frequency to 60 Hz and reduced the size of the probes.

Depth of Penetration

The depth of penetration of eddy currents below the test surface is influenced by operating frequency, material conductivity and magnetic permeability. These three variables are used to define the standard depth of penetration δ, the depth at which the eddy current strength or intensity is reduced to 37 percent of its surface value.

The relative magnetic permeability is 1.0 for nonmagnetic materials and the conductivity of a particular material is generally known and constant. Hence, the depth of penetration is controlled by the operating frequency.

The δ value is used when the eddy currents need to penetrate a faying surface layer to detect anomalies in a second layer. If the inspector does not want the thickness variations of the test object to influence the test, the effective depth of penetration 3δ is used. The effective depth of penetration is the point at which the eddy current density is reduced to about 5 percent of its value at the surface (and thickness effects are no longer noticed). Figure 30 is a plot of standard depth of penetration δ versus operating frequency for (1) titanium, (2) Unified Numbering System S30400 austenitic chromium nickel stainless steel, (3) Unified Numbering System A97075 wrought aluminum alloy and (4) high alloy steel. By using these data, δ values or operating frequencies may be obtained for each application. (A chart as in Fig. 5 may also be used.)

Applications for Low Frequency Testing[15]

Low frequency eddy current techniques have been used successfully for detecting hidden corrosion (see Fig. 31). The areas that are most prone to corrosion are those that collect moisture either by leakage or by condensation. Once the corrosion starts, it becomes intergranular and causes continuous exfoliation until the corrosion propagates all the way through the member. If the corrosion products collect between faying surfaces, a blister may form and cause interface separation or spacing. Thinning or spacing effects can be detected and measured by low frequency eddy current testing.

Low frequency eddy current techniques have detected corrosion in jet aircraft.[18] An operating frequency of 4.5 kHz was used, producing a standard depth of

Figure 30. Eddy current standard depth of penetration for various conductivity materials at different frequencies.

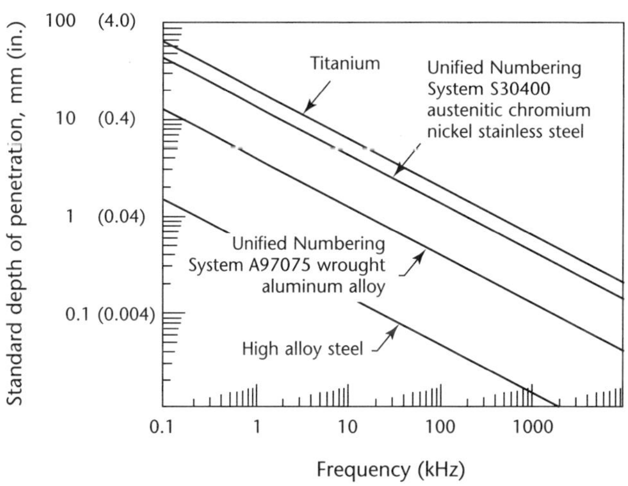

penetration of 1.8 mm (0.07 in.), sufficient to guarantee penetration into the second member. A shielded 9 mm (0.35 in.) diameter probe was designed to minimize edge effects. Reference standards to simulate spacing or thinning were produced by end milling flat bottom holes in an aluminum skin. A spacing of 0.5 mm (0.02 in.) produced a full scale vertical movement of the flying dot and the 0.13 mm (0.005 in.) gap caused a proportionally smaller movement. Gaps greater than 0.5 mm (0.02 in.) were considered serious corrosion and gaps between 0.25 mm (0.010 in.) and 0.51 mm (0.020 in.) were considered suspect. This example is similar to one discussed above (Fig. 20).

Low frequency eddy current equipment operating at 500 Hz has been used[19] to detect 1.0 mm (0.04 in.) subsurface cracks in aluminum structure through aluminum layers 5 mm (0.2 in.) thick. This test technique has been effective for detecting cracks in second layer structure and has detected cracks missed by radiography. Although the low frequency, 500 Hz eddy current technique is effective, it has several limitations.

1. It cannot be used to test through outer layers thicker than 6.3 mm (0.25 in.).
2. It requires multiple positions around the circumference of a hole for crack detection.
3. Larger probes prohibit testing between fasteners separated by a distance less than the probe diameter.

For application on an aluminum structure 15 mm (0.6 in.) thick containing steel fasteners, a 100 Hz encircling probe (Fig. 31d) has been developed.[19] The alternating current coils produce an electromagnetic field that penetrates both the aluminum and steel fastener. In addition, the steel fastener acts as a core and tends to concentrate the electromagnetic field. Cracks 13 mm (0.5 in.) long were detected through outer layers of aluminum 8.9 mm (0.35 in.) thick and longer cracks may be detected through even thicker outer layers. The encircling probe can be used where the fastener head spacing is 6.3 mm (0.25 in.) or greater. The major disadvantage of the encircling probe is the absolute measurement system. Changes in fastener

FIGURE 31. Typical applications for low frequency eddy current testing: (a) interstitial corrosion; (b) thinning; (c) second layer cracking; (d) cracking in fastener hole.

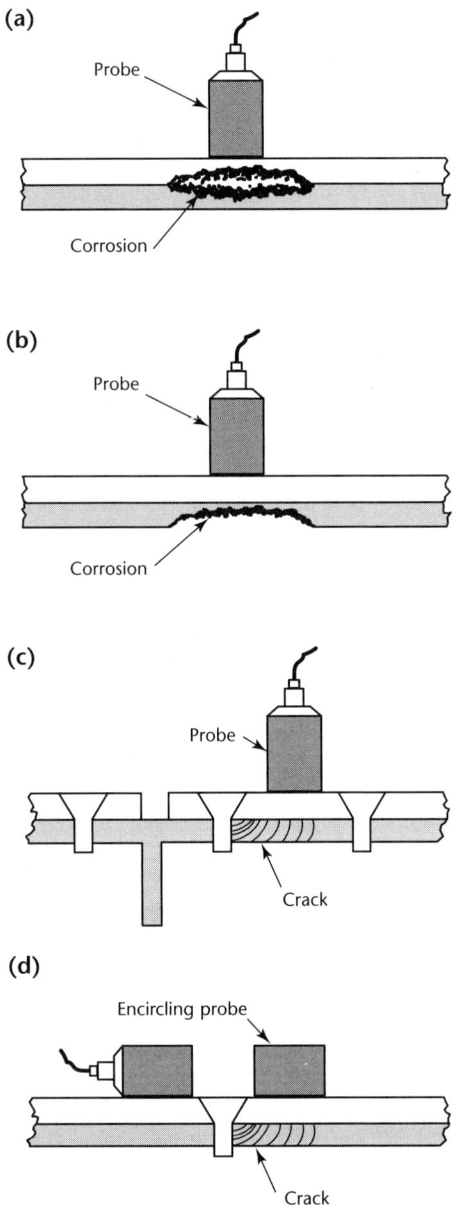

FIGURE 32. Instrument response at 500 Hz related to crack length and upper layer thickness in typical aircraft structure.[20]

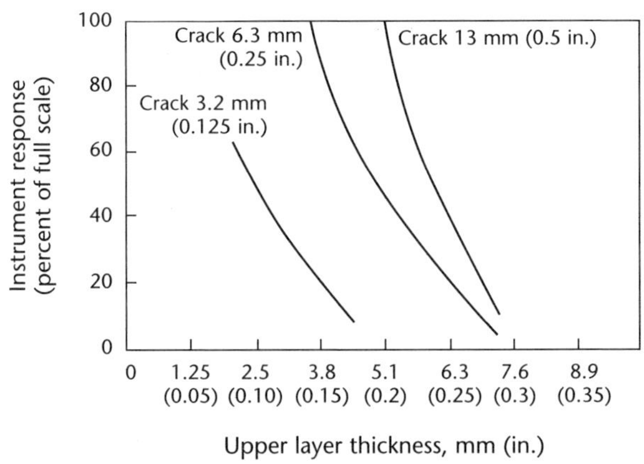

length, fastener permeability, aluminum thickness and conductivity may cause signal variations that interfere with the test.

Crack detection sensitivity for laboratory conditions (Fig. 32) has been reported.[20] The actual crack detection capability in the field is reduced by signal noise from temperature variations, by separations between the layers and by variation in probe placement. Field experience with a 500 Hz probe confirms that cracks can be reliably detected — 13 mm (0.5 in.) long cracks under a 5 mm (0.2 in.) thick upper layer and 6.3 mm (0.25 in.) long cracks under a 2.5 mm (0.10 in.) thick upper layer. Hundreds of test procedures have been developed for crack detection in various aircraft components and structures.

Aircraft Window Belt Splice Cracks[13]

Cracks were detected in the first row of rivets above the longitudinal belt splices of aircraft windows. The cracks initiated at fastener holes in the internal (second layer) skin and grew in a longitudinal direction (Fig. 33).

Using a meter or screen display instrument, the test may be performed by scanning between fasteners with a 10 mm (0.4 in.) probe at 2 kHz (Fig. 34). The crack response is shown in Fig. 34c. To speed up the test, a procedure was developed for scanning in line with and across the aluminum fasteners by using a 15 mm (0.6 in.) diameter probe operating at 1 kHz (Fig. 35). The crack response is shown in Fig. 35c. Both test techniques produce easily interpreted crack responses on a screen display.

Fatigue cracks caused by pressurization cycles were detected in the window belt panels on an airplane with many hours in service (Fig. 36). The window belt was removed from the aircraft and sent to the manufacturer for evaluation. A section of the cracked panel was used to develop a nondestructive test. The low frequency eddy current test was performed at a frequency range of 500 Hz to 1 kHz using a 13 mm (0.5 in.) diameter probe. A reference standard was made, simulating the window belt splice joint. It contained a 25 mm (1.0 in.) electric discharge machined notch at the change of thickness area in the window belt panel. A nonmetallic straight edge was used to guide the center of the probe directly over the center of the splice. For instruments equipped with meters, the liftoff and phase angles were set so that liftoff drove the meter up scale and passing the probe over the electric discharge machined notch drove the meter down scale. For

FIGURE 33. Cracks in second layer initiating at aluminum fastener holes: (a) outboard side; (b) inboard side.

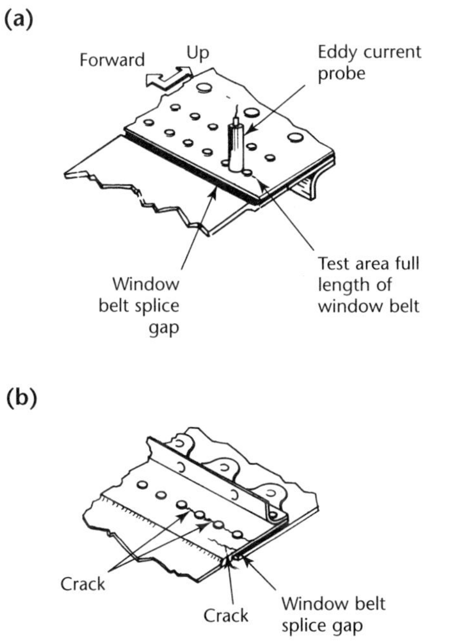

FIGURE 34. Detection of crack in second layer by scanning between fasteners with 10 mm (0.4 in.) probe at 2 kHz: (a) scanning procedure; (b) no crack response; (c) crack response.

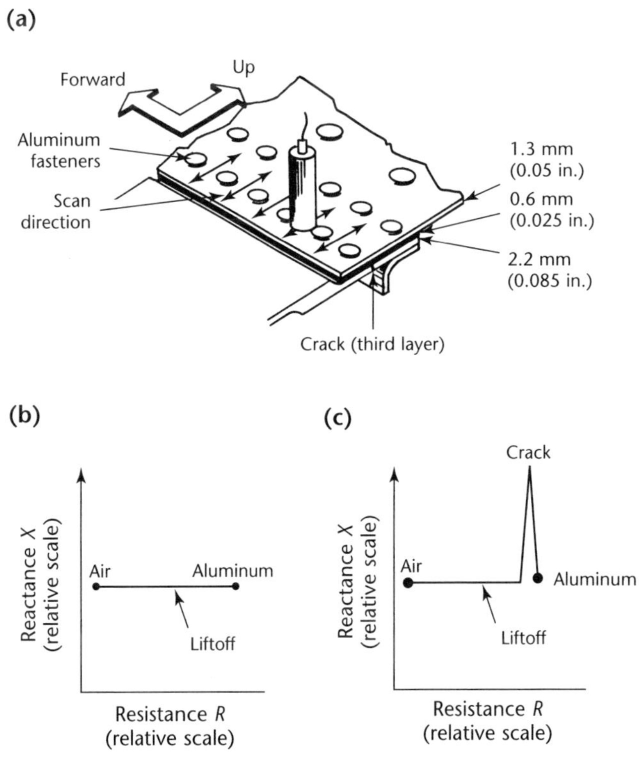

instruments equipped with oscilloscopes, the phase rotation was null balanced to produce a crack response like that in Fig. 34c.

When a crack signal is obtained, using the low frequency eddy current procedure on the outside of the aircraft, verification requires removal of the interior seats, lining and insulation to conduct a visual or high frequency eddy current check of the internal surface of the window belt panel.

Automated Fastener Hole Testing

Inservice test techniques using a driver and receiver sliding probe eddy current system have been developed for improved crack detection in aging aircraft fastener hole tests.[13,21] The basic idea of this test system is to slide the probe along a line of fasteners without removing them and to observe the screen display pattern for crack responses. The technique has used a circular probe centered over the installed fastener and the screen display is observed for a crack response. Unfortunately, the results with circular probes are affected by fastener permeability, fastener length, fastener spacing, ambient temperature and probe centering. Flat surface low frequency eddy current probes offset some of these problems. Sliding probe systems, however, offer better crack sensitivity and reliability as well as a wide frequency range and good penetration.

Scanning is done in the direction of crack extension (but it will detect cracks within a 90 degrees angle or less). The crack direction is normally known by the direction of the stress in a given test object. The probes are reflection type and must be used with appropriate instruments. When an aluminum specimen with subsurface cracks is scanned, the fastener response is almost parallel to the liftoff and the crack response has an increasing clockwise phase angle response.

FIGURE 35. Detection of crack in second layer by scanning over fasteners with a 15 mm (0.6 in.) probe at 1 kHz: (a) scanning procedure; (b) no crack response; (c) crack response.

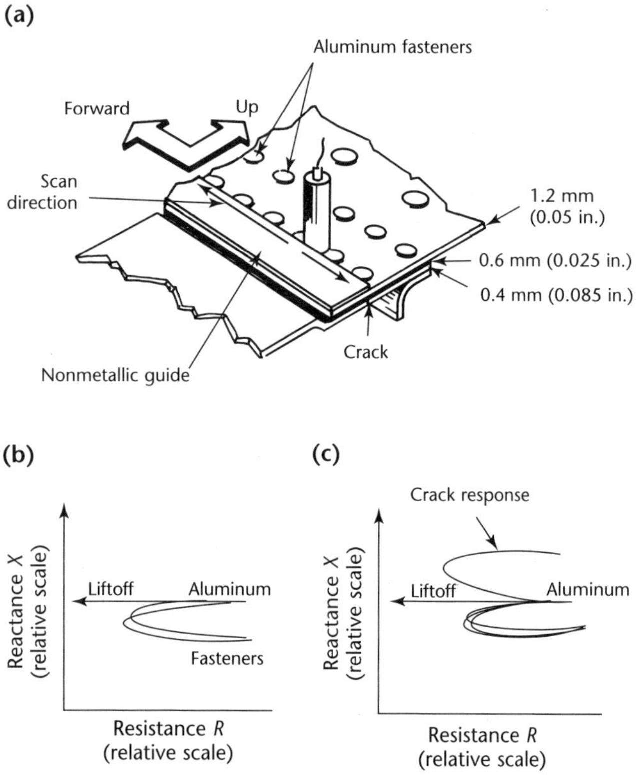

Conclusion

Eddy current testing is a primary test technique for detecting cracks and corrosion of aircraft structure and engine hardware during maintenance overhaul. These tests are performed in accordance with procedures developed by the manufacturer. Such tests can be used to detect fatigue cracks resulting from cyclic loading during flight, takeoff or landing. They can also be used to detect stress corrosion cracks to indicate the extent of corrosion damage or to identify portions of aircraft structures damaged by fire. In

FIGURE 36. Window belt crack location: (a) area; (b) close view.[15]

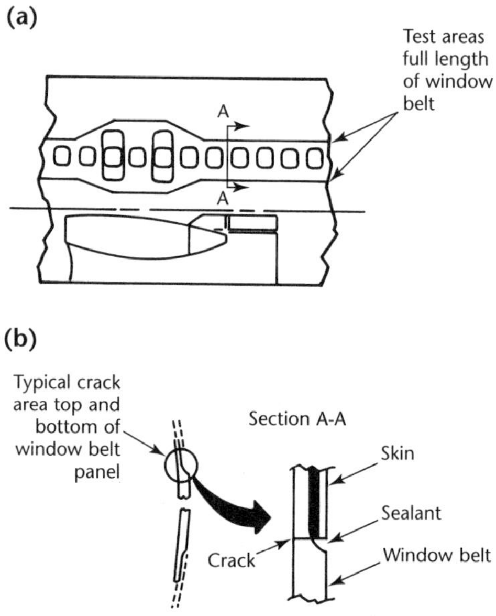

this way, it is possible to identify damaged components that need to be replaced to restore the craft to airworthiness.

Fracture mechanics and nondestructive testing programs have consistently shown that eddy current tests are very reliable in detecting cracks that would be missed by radiography or that require careful cleaning or coating removal to be detected by liquid penetrant or magnetic particle testing. In addition, these studies also reveal that very small cracks can be detected by high frequency eddy current tests. As a consequence, aircraft and engine inspections must use eddy current testing to ensure product reliability. Fortunately, this demand for eddy current testing has generated considerable research and development by aircraft manufacturers, airlines, military and eddy current equipment manufacturers. Development has resulted in advances in instruments, probes and other aspects of the technology.

PART 7. Eddy Current Testing of Jet Engines

Eddy current testing is a well established technique for the detection of surface discontinuities in conductive test parts. The United States Air Force and other organizations have used automated eddy current testing to ensure that critical engine components are free of rejectable surface discontinuities.

Eddy Current Techniques for Jet Engines

Eddy current tests of military aircraft engines can be categorized as manufacturing, depot or field tests. When aircraft engine components are initially produced, eddy current tests of the manufactured parts are often required by specification to ensure that the new engine components are free of undesirable material properties or machining discontinuities. Inservice parts similarly require depot tests at predetermined intervals to ensure that parts are free of small discontinuities in critical, predetermined zones. Manufacturing tests and depot tests are performed as preventive screening.

Structural integrity requirements are based on fracture mechanics analysis using mission critical parameters for each engine component.[22] For a helicopter engine, conservative service life limits and critical discontinuity sizes are determined by using minimum material properties and other worst case operating conditions.[23] Aerodynamic, heat transfer, stress and life analysis models are used along with the criticality of a failure to determine test zones and discontinuity requirements for critical engine components. The aviation community recognizes that eddy current tests can reliably detect small discontinuities and locate critical discontinuities.[23]

Eddy current tests are typically required for discontinuities smaller than 1.8 mm (0.07 in.) long whereas fluorescent liquid penetrant testing is often sufficient for discontinuities over 1.8 mm (0.07 in.) long. Although eddy current requirements have been as small as 0.13 mm (0.005 in.) deep by 0.25 mm (0.010 in.) long, surface requirements are typically in the range of 0.25 × 0.50 mm to 0.75 × 1.50 mm (0.010 × 0.020 in. to 0.030 × 0.060 in.). The required discontinuity sizes, orientations, test areas and surface conditions dictate the test technique. Parameters such as coil type, test frequency, scanning, indexing and filtering are chosen to optimize the ratio of signal to noise for each test.

Probability of Detection

To demonstrate a technique's reliability after it is developed, data for probability of detection are collected from fatigue crack specimens of the same material and by using the same parameters. The term *probability of detection* refers to the statistical protocols used to quantify the effects of test variables on test reliability, specifically on the identification of rejectable discontinuities. Probability of detection data analysis is used to quantify a system's response and to determine the required test thresholds.

Many aircraft manufacturers require potential eddy current test service agencies to submit a test matrix to assess potential test variables and their impact on test sensitivity. Such variables include repeat runs, different operators, different machines and different serial number probes of the same probe type. Variability test matrices for an automated eddy current test station (for retirement for cause) have identified probe variability to be the most significant cause of variation in the probability of detection data for the test station. Therefore, a minimum of two or three probes with the same specifications are used to collect the probability of detection data. Although a greater number of probes would be desirable, cost and schedule constraints typically prohibit additional collections. After the probability of detection data are collected, a statistical fit is performed by using the broadest analysis range that is statistically acceptable. The analysis range must contain all required crack lengths for a given combination of technique and test object.[24] Slope, intercept and scatter parameters are then obtained from the data analysis. These parameters are used to calculate required limits and thresholds necessary to detect required crack sizes with, for example, 90 percent probability at a 95 percent mean confidence level. This requirement is known as the *90/95 requirement.*

Sizing information can also be obtained from the probability of detection results to convert recorded signal amplitude into estimated crack size. It is important to note that estimated crack sizes are inaccurate outside of the probability of detection data analysis range because of instrument saturation, signal rolloff (where discontinuity sizes exceed settings for signal maxima) and other nonlinearities (Figs. 37 and 38). Therefore, it is crucial that all required discontinuity sizes are sufficiently bounded by the probability of detection data analysis range and that no extrapolations are made for required discontinuity sizes. Although indications with amplitudes outside the probability of detection analysis range will not be accurately sized, they will still be detected and the tested region will receive a reject status.

Standard versus Complex Geometries

Within the manufacturing and depot categories, eddy current tests can be further subdivided into standard and complex geometry categories. Standard tests typically include part configurations that lack significant geometry signals or other significant noise. Examples of standard geometries include continuous surfaces, bolt holes and scallops (Fig. 39).

Although surface tests are typically free of strong geometry signals, test zones can sometimes run close enough to bolt holes or other features to pick up a geometry signal. Surface tests can also require thresholds so small that additional work is needed to differentiate discontinuity responses.

Although bolt hole and scallop geometries have significant edge geometries, coil configuration and scan techniques can typically be selected so no appreciable edge signal is present during the tests. However, significant geometry signals can occur if a bolt hole probe is not accurately centered on a bolt hole or precise following of scallops cannot be accomplished. These geometry signals can render the test of such features impractical. Lack of sophisticated probe positioning apparatus can permit unwanted geometry signals. Bolt hole and scallop geometries with very low test requirements force standard geometries to use complex test techniques in order to extract very small discontinuity signals out of the background noise.

Complex geometries include broach or dovetail slots, oil drain holes, shaped holes, antirotation windows, knife edges, air foils and other geometries that produce large unwanted edge signals

FIGURE 37. Probability of detection data for three similar split D shaped (Figs. 40 to 42a) reflection probes on bolt hole.

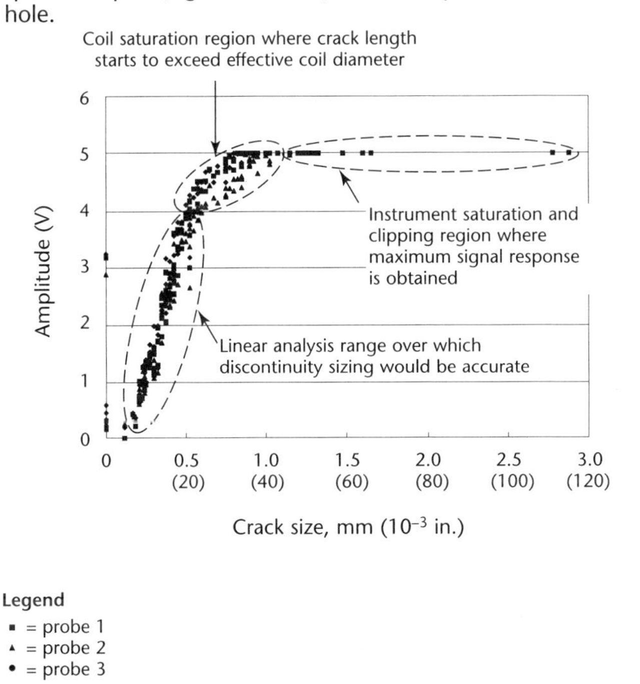

FIGURE 38. Typical probability of detection response for three similar wide field probes on flat plate specimens.

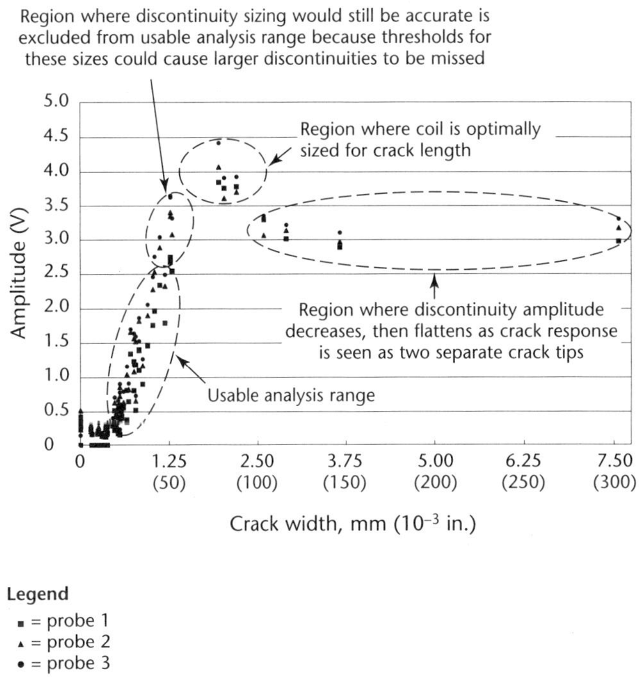

(Fig. 39). These geometries require special techniques, sophisticated signal processing or both to suppress the large unwanted geometry response.

The development of tests for complex geometries typically requires extremely accurate probe centering to obtain a very consistent edge response combined with a signal removal or differencing technique. These edge responses can easily be one hundred times greater than the desired discontinuity response. Often, differencing techniques alone are insufficient for suppressing strong geometry signals and signal processing techniques are required.

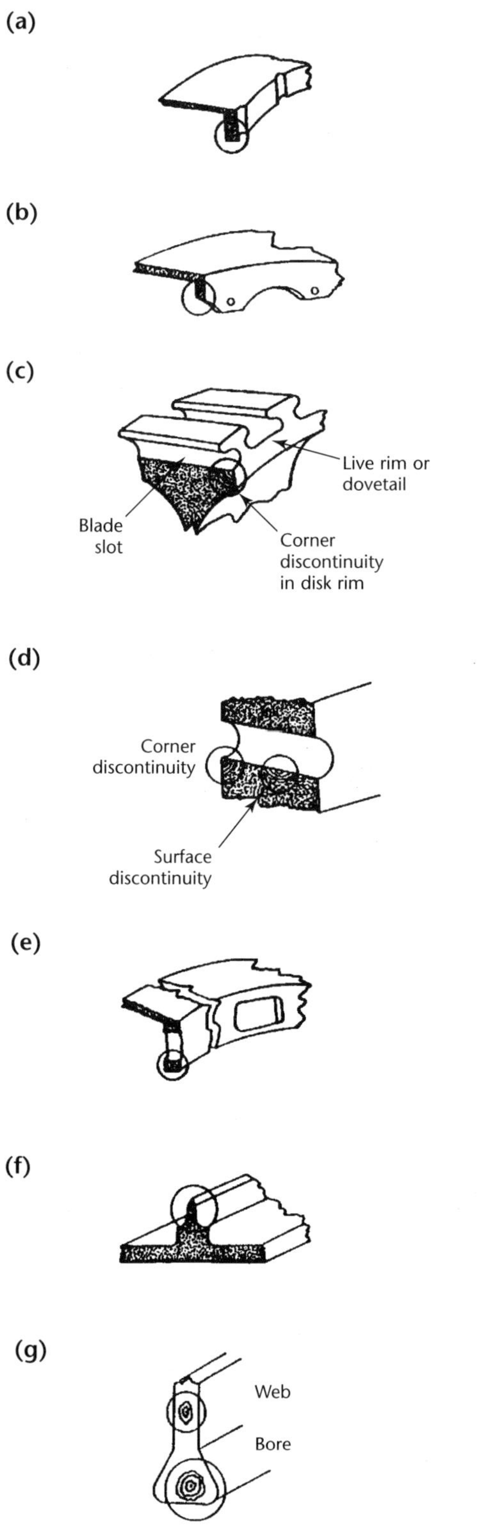

FIGURE 39. Discontinuity locations for fighter engines: (a) corner of cut for balance or drain slot; (b) corner of balance flange scallop; (c) disk rim; (d) air holes for oil drain, fastener or cooling; (e) corner of antirotation window; (f) through knife edge seal; (g) blade interior.

Signal Processing

Signal processing techniques include parametric modeling of edge responses, pattern recognition, frequency discrimination, digital filtering and phase discrimination.

1. Parametric modeling involves adjusting slope, amplitude, direct current offset, phase or other parameters to obtain a best fit to some predetermined model to significantly reduce edge variations.
2. Imaging and pattern recognition techniques can also be used to reduce edge variations.
3. Frequency discrimination techniques can be as simple as selecting the proper filter settings to eliminate unwanted noise but available instrument filter settings are often insufficient to discriminate between an indication and noise.
4. Digital filters are used to supplement instrument filters and improve the ratio of signal to noise.
5. Phase discrimination techniques can be used to extract discontinuity signals if there is enough phase separation between the discontinuity signal and unwanted signals.[25]

Regardless of the signal processing technique applied to a given test, it is important that test parameters are equivalent to those used to collect the probability of detection data. Without equivalent test parameters, it is difficult to ensure that test thresholds and discontinuity sizing are accurate.

Eddy Current Probes for Jet Engines

Because probe and coil variations were identified as a significant variable in the automated eddy current test station probability of detection data, great care is taken to control slight variations in the probes. Probes have been studied to assess their parameters, optimize their design and minimize variations from probe to probe.

Split Core Coils

Split core coils use two separate receiving coils wound in opposite directions on a split cylinder of ferrite whose cross section resembles two D shapes, one D being the mirror image of the other. The two Ds are back to back so that their cross section looks like a bifurcated circle. These D shaped receive coils are typically bound together with a drive coil wound around them both.

For the study of these coils, probe footprints were collected by indexing a probe across a small cross section of a 150 µm (0.006 in.) gage wire in a nonconductive plate. These footprints illustrate some probe parameters such as coil orientation, coil size and coil tilt (Figs. 40 and 41).[26] With careful alignment, coil positioning can also be easily visualized and quantified. In addition to orientation, size and tilt, the footprint assessment can also provide insight into a probe's working sensitivity. Results have shown a correlation between a reduced probe footprint and a reduction in the probe's response to larger fatigue cracks.

Many standard techniques for aircraft engines use a split D shaped coil with the coil axis running normal to the test object (Fig. 42a). A differential coil is used to exclude environmental noise such as temperature variations and bulk material properties. To further increase the coil sensitivity, a separate driving coil is often wound around the two receiving coils. In this configuration, the probe is classified as a *differential reflection probe*. Probes are generally specified according to coil diameter.

One of the D shaped coils has a 0.51 mm (0.020 in.) diameter core. Coils with 1.27 mm (0.050 in.) and 1.52 mm (0.060 in.) diameter cores are some of the larger D shaped coils used for the detection of relatively small discontinuities. The smaller diameter coils provide a greater ratio of signal to noise but require a greater number of indexes to adequately cover the same amount of test

FIGURE 40. Axial alignment of probe having a 4 to 5 mV, split D shaped coil: (a) tilt perpendicular to split; (b) acceptable alignment; (c) tilt in split axis.

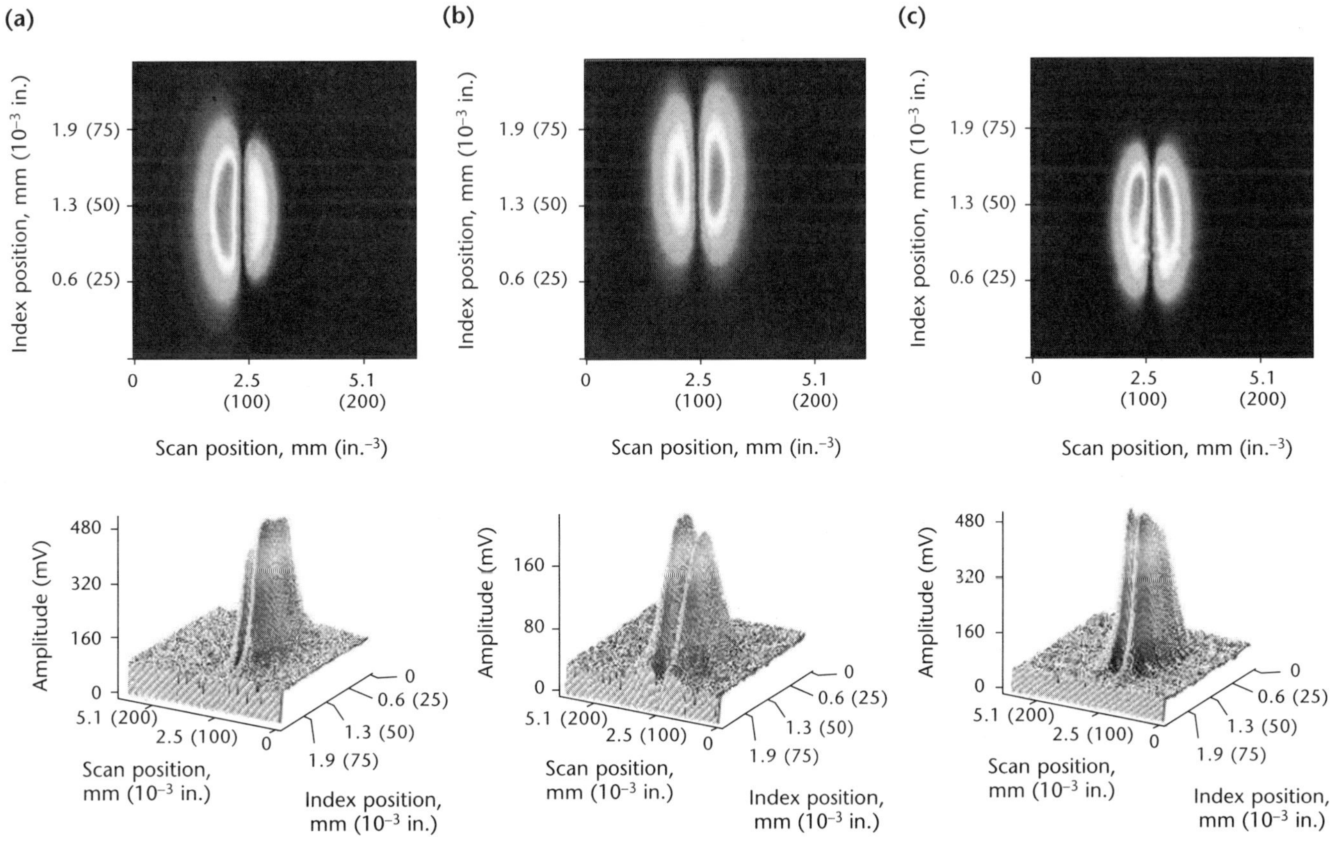

area. Several techniques have been used to increase test throughput.

Wide Field Probes

The term *wide field probes* typically refers to probes with circumferentially wound coils whose axes are parallel to the test surface to achieve a wider field of view. Two receiving coils are abutted (so that a circular end of one cylindrical coil is next to the end of the other) such that their currents flow in opposing directions. These two receiving coils are often supplemented with a separate drive coil in the same probe. When the probe is applied with the axis parallel to the test surface, the discontinuity frequency no longer depends on the coil diameter and the coil can be larger so that the field of view is as wide as the contact area between the coil and the test surface. Because part noise is integrated over the contact area of the coil, the practical size of the wide field probe is determined by the required discontinuity detection size and the anticipated noise of typical test surfaces.

Wide field probes can be further categorized as bobbin coils, shaped coils and flex coils.

1. Bobbin coils generally refer to coils whose circumferential windings are circular or nearly circular (Fig. 42b). Bobbin coils have cores of air, ferrite, annealed steel or other materials.

FIGURE 41. Footprints of eddy current probes having split D shaped coil: (a) first probe; (b) second probe; (c) third probe.

(a)

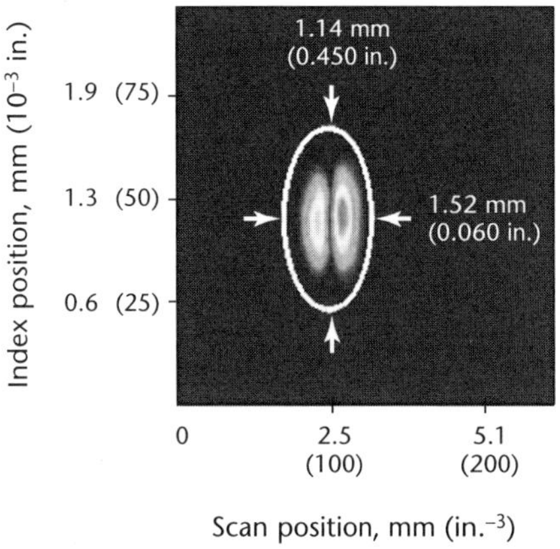

(b)

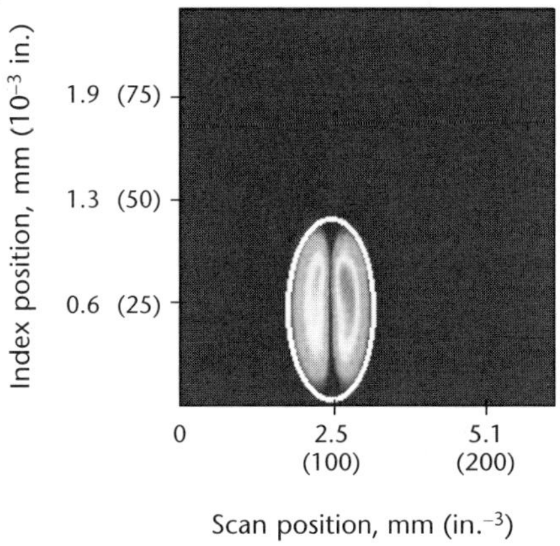

(c)

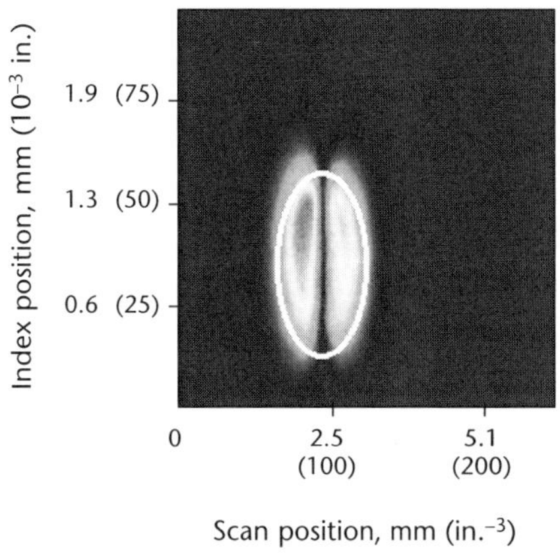

FIGURE 42. Eddy current probes: (a) split, D shaped reflection coil; (b) wide field bobbin coils; (c) wide field flex coil.

(a)

(b)

(c)

2. Shaped coils are a slight derivative from the bobbin coils in that the coils are shaped to fit a distinct geometry.[27]
3. Flex coils differ from bobbin and shaped coils in that each coil is housed in a flexible protective jacket that lets the coil be pressed into a geometry to conform to the surface (Fig. 42c).

The main advantage of the wide field probe is the increased throughput provided by a single-pass test. However, the increased throughput comes at a cost.

The disadvantages of the wide field coils include (1) a greatly diminished signal response at a nonoptimal coil orientation, (2) greater rolloff (where a large discontinuity signal amplitude exceeds the threshold of rejectable indications) and (3) an inability to identify the circumferential location of an indication about a coil.

1. Because the required discontinuity orientation is typically ideal for the wide field coil technique, orientation sensitivity is generally not an issue.
2. Although rolloff at larger discontinuity sizes is of greater concern than rolloff of the D shaped reflection probe, the amplitudes of larger discontinuity sizes are generally much greater than the threshold required for the test. Although discontinuities beyond the rolloff range will be reported at a reduced size, they will still exceed the rejectability threshold and will need to be addressed before acceptance of a part. A typical response related to rolloff is illustrated in Fig. 38 for a wide field probe.
3. Although knowing the circumferential location of an indication can be useful in locating a discontinuity, the axial location of the discontinuity is usually sufficient in approximating and locating a discontinuity. In applications where knowing the circumferential position of a discontinuity is critical, an overlapping test with a coil can be used to help determine the approximate circumferential location of the discontinuity.

Array Probes

A second approach to improving test throughput involves an array of eddy current coils. Increasing throughput comes at a cost. Eddy current arrays require more channels. Although multiplexing can be performed to significantly reduce the costs associated with parallel processing of multiple signals, the costs will still be greater than that of a system with a single probe. In addition to the increased equipment costs, sensor and probe costs typically increase as more sensors are added.

Although printed circuit board eddy current coil arrays can be used to reduce sensor costs and provide a flexible coil array, these arrays typically come with a decrease in coil sensitivity because of the single-turn nature of these coils. In addition to the sensor production costs, there are also costs to normalize coils and ensure their proper sensitivity. After these costs and problems are addressed, the eddy current array approach can offer advantages over single-sensor testing.

Conclusion

The testing of critical components of aircraft engines requires careful attention to ensure that test parameters are carefully set to meet test requirements. Automated equipment is essential to precisely position and scan sensors on geometries where slight inaccuracies in coil positioning can produce indications that lead to rejection of an acceptable part. To further improve ratios of signal to noise with very stringent discontinuity requirements, sophisticated signal processing techniques are often applied. Great care is also taken to minimize system variations.

Although certain techniques are used to increase test throughput, the driving factors for scheduled tests of critical engine component tests are adherence to the discontinuity requirements and test zones specified by the manufacturer.

References

1. Hagemaier, D.J. Section 14, "Applications of Eddy Current Testing to Airframes." *Nondestructive Testing Handbook,* second edition: Vol. 4, *Electromagnetic Testing*. Columbus, OH: American Society for Nondestructive Testing (1986): p 369-421.
2. Hagemaier, D.J. "Eddy Current Impedance Plane Analysis." *Materials Evaluation*. Vol. 41, No. 2. Columbus, OH: American Society for Nondestructive Testing (February 1983): p 211-218.
3. ASTM B 193, *Standard Test Method for Resistivity of Electrical Conductor Materials*. West Conshohocken, PA: ASTM International (2002).
4. Jones, A.R., G.S. Tsiang and T.M. Barber. "Semiautomated Conductivity Standards Calibration." *Materials Evaluation*. Vol. 41, No. 10. Columbus, OH: American Society for Nondestructive Testing (September 1983): p 1194-1200.
5. Raatz, C.F., Jr. *Eddy Current Testing in Determining Metal Alloy Type and Heat Treatment*. Seattle, WA: Boeing Company (August 1966).
6. Libby, H. *Introduction to Electromagnetic Nondestructive Test Methods*. New York, NY: Wiley Interscience (1971).
7. Brown, R.L. "The Eddy Current Slide Rule." *Materials Evaluation*. Vol. 26, No. 6. Columbus, OH: American Society for Nondestructive Testing (June 1969): p 120-123.
8. Hagemaier, D.J. "Evaluation of Heat Damage to Aluminum Aircraft Structures." *Materials Evaluation*. Vol. 40, No. 9. Columbus, OH: American Society for Nondestructive Testing (September 1982): p 962-969.
9. Padilla, V.E. and J.W. Parks. "Definition of Fatigue Crack Geometry by Eddy Current Techniques." *Seventh Symposium on Nondestructive Evaluation of Components and Materials in Aerospace, Weapon Systems and Nuclear Applications* [San Antonio, TX]. San Antonio, TX: Southwest Research Institute (April 1969): p 79-94.
10. Parks, J.W. "Detection and Definition of Environmental Damage (Fatigue, Stress Corrosion, etc.) Using Eddy Current Techniques." *Proceedings of the Ninth Symposium on Nondestructive Evaluation American Society of Nondestructive Testing* [San Antonio, TX, April 1973]. San Antonio, TX: Southwest Research Institute (1973): p 291-297.
11. Rogel, A.P. and J.J. Scalese. "Automatic Eddy Current Bolt-Hole Scanning System." *Materials Evaluation*. Vol. 41, No. 7. Columbus, OH: American Society for Nondestructive Testing (June 1983): p 839-843.
12. *Eddy Current Characterization of Materials and Structures*. Special Technical Publication 722. West Conshohocken, PA: ASTM International (1979).
13. Hagemaier, D.J. "Application of Eddy Current Impedance Plane Testing." *Materials Evaluation*. Vol. 42, No. 8. Columbus, OH: American Society for Nondestructive Testing (July 1984): p 1035-1040.
14. [Botsko,] R.J. *Instruction Manual Vector III*. Huntington Beach, CA: NDT Systems (1974).
15. Hagemaier, D.J. and A.P. Steinberg. "Low-Frequency Eddy Current Inspection of Aircraft Structure." *Materials Evaluation*. Vol. 40, No. 2. Columbus, OH: American Society for Nondestructive Testing (February 1982): p 206-210.
16. Southworth, H. AFMLTR-1, *Practical Sensitivity Limits of Production Nondestructive Testing Methods in Aluminum and Steel*. Wright Patterson Air Force Base, OH: United States Air Force (November 1973).
17. Anderson, R.T., T.J. DeLacy and R.C. Steward. Final Report NASA-CR-128946 (GDCA-DBG73-002), *Detection of Fatigue Cracks by Nondestructive Testing Methods*. Springfield, VA: National Technical Information Service (1973).
18. McFarlan, T.W. et al. "Low-Frequency Eddy Current Test for Detection of Corrosion beneath Aircraft Skin Surfaces." Presented at the Air Transport Association Nondestructive Testing Forum [Hartford, CT, 1977].

19. Ansley, G. and R. Neufeld. "Application and Development Work in Low-Frequency Eddy Current — An Update." Presented at the Air Transport Association Nondestructive Testing Forum [Tulsa, OK, 1976].
20. Neufeld, R.M. "Detection of Cracks in Second Layer Structure with Low-Frequency Eddy Current." *36th National Fall Conference* [Houston, TX]. Columbus, OH: American Society for Nondestructive Testing (September 1976): p 46-65.
21. Pellicer, J. "Sliding Probe Eddy Current System for Improved Fastener Hole Inspection." Presented at the 1983 Air Transport Association Nondestructive Testing Forum [Kansas City, MO, August 1983].
22. MIL-HDBK-1783B, *Department of Defense Handbook, Engine Structural Integrity Program (ENSIP)*. Arlington, VA: United States Department of Defense (2002).
23. Caraway, J. and C. Benson. "Automated Eddy Current Inspections for Turbine Engine Component Life Extensions." Paper 30. *RTO Meeting Proceedings 34*. Brussels, Belgium: North Atlantic Treaty Organization, Research and Technology Organization (September 2000): p 1-4.
24. Berens, A.P. "NDE Reliability Data Analysis." *ASM Handbook*, ninth edition: Vol. 17, *Nondestructive Evaluation and Quality Control*. Materials Park, OH: ASM International (1992): p 689-701.
25. Stepinski, T. and N. Mszi. "Conjugate Spectrum Filters for Eddy Current Signal Processing." *Materials Evaluation*. Vol. 51, No. 7. Columbus, OH: American Society for Nondestructive Testing (July 1993): p 839-844.
26. Benson, C.W. "'Footprint' Assessment of Eddy Current Probes." *Review of Progress in Quantitative Nondestructive Evaluation* [Ames, IA, July 2000]. Vol. 20B. Melville, NY: American Institute of Physics (2001): p 1995-2000.
27. Concordia, M., P. Ballard, L.S. Price, J. Stawarz, R. Stone, B. Arispe and T.J. Braun. "Extended Field ECI Applications Progress in Air Force Depot Level Jet Engine Inspections." *ASNT Fall Conference and Quality Testing Show Paper Summaries* [Nashville, TN]. Columbus, OH: American Society for Nondestructive Testing (October 1998): p 166-171.

Bibliography

Aluminum Standards and Data 2003: Metric SI. Washington, DC: Aluminum Association (2003).

Chao, J. and J. Amos. "Assessment of Eddy Current Inspection Development through Numerical Simulation." *ASNT Fall Conference Quality Testing Show Paper Summaries* [Pittsburgh, PA]. Columbus, OH: American Society for Nondestructive Testing (October 1997): p 126.

DIN EN ISO 21968, *Non-Magnetic Metallic Coatings on Metallic and Non-Metallic Basis Materials — Measurement of Coating Thickness — Phase Sensitive Eddy Current Method*. Berlin, Germany: Deutsches Institut für Normung (2003).

DIN EN ISO 2360, *Non-Conductive Coatings on Non-Magnetic Electrically Conductive Basis Materials — Measurement of Coating Thickness — Amplitude Sensitive Eddy Current Method*. Berlin, Germany: Deutsches Institut für Normung (2000).

IEC 60028, *International Standard of Resistance for Copper*. Geneva, Switzerland: International Electrotechnical Commission (2001).

Eua-Anant, N., X. Cai, L. Udpa, J. Chao and I. Elshafiey. "Crack Detection in Eddy Current Images of Jet Engine Disks." *Review of Progress in Quantitative Nondestructive Evaluation* [Montreal, Canada, July 1999]. Vol. 19A. Melville, NY: American Institute of Physics (May 2000): p 773-780.

Franklin, E.M. "Eddy Current Inspection." *Materials Evaluation*. Vol. 40, No. 10. Columbus, OH: American Society for Nondestructive Testing (September 1982): p 1008, 1010.

Goldfine, N., V. Zilbertstein, K. Walrath, E. Hill and C. Paraizaman. "Inspection of Gas Turbine Components Using Conformable MWM Eddy-Current Sensors." *ASNT Fall Conference and Quality Testing Show — 2000: Paper Summaries Book* [Indianapolis, IN]. Columbus, OH: American Society for Nondestructive Testing (November 2000): p 29-31.

Hagemaier, D.J. "Eddy Current Standard Depth of Penetration." *Materials Evaluation*. Vol. 43, No. 11. Columbus, OH: American Society for Nondestructive Testing (October 1985): p 1438–1442, 1454.

Hagemaier, D.J. "Factors Influencing Eddy Current PoD in the Field Environment." *ASNT Spring Conference and 8th Annual Research Symposium Paper Summaries.* Columbus, OH: American Society for Nondestructive Testing (1999): p 64-66.

Hagemaier, D.J. "Nondestructive Testing of Aging Aircraft." *FAA Aging Aircraft Workshop* [Valley Forge, PA, October 1989]. Columbus, OH: American Society for Nondestructive Testing (1990): p 4-12.

Hagemaier, D.J. and G. Klark. " Eddy Current Detection of Short Cracks under Installed Fasteners." *Materials Evaluation.* Vol. 55, No. 1. Columbus, OH: American Society for Nondestructive Testing (January 1997): p 25-30.

Hagemaier, D.J. and K. Nguyen. "Automated Eddy Current Scanning of Aircraft for Corrosion Detection." *Materials Evaluation.* Vol. 52, No. 1. Columbus, OH: American Society for Nondestructive Testing (January 1994): p 91-95.

Irvine, A.W. "Quick Eddy Current Inspection of Aircraft Wheels." *Materials Evaluation.* Vol. 55, No. 5. Columbus, OH: American Society for Nondestructive Testing (May 1997): p 573.

KSA W 1212, *Electrical Conductivity Test for Measurement of Heat Treatment of Aluminum Alloys, Eddy Current Method, for Aerospace Use.* Seoul, Republic of Korea: Korean Standards Association (1988).

Kramer, S.R. and J.F. Harmon. "Eddy Current Testing to Detect Cracks and Corrosion in the P-3C Orion Vertical Stabilizer Spar Caps." *Materials Evaluation.* Vol. 59, No. 2. Columbus, OH: American Society for Nondestructive Testing (February 2002): p 166-169.

Leclerc, R. and R. Samson. "Eddy Current Array Probes for Aerospace Applications." *ASNT Fall Conference and Quality Testing Show — 2000: Paper Summaries Book* [Indianapolis, IN]. Columbus, OH: American Society for Nondestructive Testing (November 2000): p 40-43.

Lepine, B., D. Forsyth, S. Guiguere and S. Dubois. "Comparison of Pulsed Eddy Current NDT to Conventional Eddy Current Testing." *ASNT Spring Conference and 8th Annual Research Symposium Paper Summaries* [Orlando, FL]. Columbus, OH: American Society for Nondestructive Testing (March 1999): p 124.

MIL-STD-1537C, *Electrical Conductivity Test for Verification of Heat Treatment of Aluminum Alloys, Eddy Current Method.* Arlington, VA: United States Department of Defense (2002).

Moore, D. and M. Hutchinson. "Effect of Conductivity between Fasteners and Aluminum Skin on Eddy Current Specimens." *ASNT's 1997 Spring Conference and Sixth Annual Research Symposium* [Houston, TX]. Columbus, OH: American Society for Nondestructive Testing (1997): p 79-87.

Moore, D., J. Mihelic and J.D. Barnes. "Crack Detection on HC-130H Aircraft Using Low Frequency Eddy Current." *The 9th Asia-Pacific Conference on Nondestructive Testing in Conjunction with ASNT's 1998 Spring Conference and 7th Annual Research Symposium* [Anaheim, CA]. Columbus, OH: American Society for Nondestructive Testing (March 1998): p 202-205.

Nath, S., M. Mina and Y. Sun. "New Eddy Current Sensor Development for Detecting Defects in Thick Aluminum Structures." *ASNT's 1997 Spring Conference and Sixth Annual Research Symposium* [Houston, TX]. Columbus, OH: American Society for Nondestructive Testing (March 1997): p 59-61.

Rempt, R. "Eddy Current C-Scans of Subsurface Flaws in Aircraft Structure Using Magnetoresistive Sensors." *ASNT Spring Conference and 9th Annual Research Symposium Abstracts* [Birmingham, AL]. Columbus, OH: American Society for Nondestructive Testing (March 2000): p 41.

Rempt, R. and G. Geithman. "Eddy Current Detection of Subsurface Flaws with Magnetoresistive Sensors." *ASNT Spring Conference and 8th Annual Research Symposium Paper Summaries* [Orlando, FL]. Columbus, OH: American Society for Nondestructive Testing (March 1999): p 128.

SAE ARP 891A, *Determination of Aluminum Alloy Tempers through Electrical Conductivity Measurements (Eddy Current).* Warrendale, PA: SAE International (2001).

SAE ARP 4402, *Eddy Current Inspection of Open Fastener Holes in Aluminum Aircraft Structure.* Warrendale, PA: SAE International (April 1992).

SAE AMS 4957-C, *Titanium Alloy, Round Bar and Wire 3Al 8V 6Cr 4Mo 4Zr Consumable Electrode Melted Solution Heat Treated and Cold Drawn.* Warrendale, PA: SAE International (2003).

SAE AS 4787, *Eddy Current Inspection of Circular Holes in Nonferrous Metallic Aircraft Engine Hardware.* Warrendale, PA: SAE International (October 1992).

SAE J 457, *Chemical Compositions, Mechanical Property Limits, and Dimensional Tolerances of SAE Wrought Aluminum Alloys.* Warrendale, PA: SAE International (1991).

SAE MAM 2771, *Heat Treatment of Aluminum Alloy Castings.* Warrendale, PA: SAE International (2003).

Safizadeh, M.-S., D.S. Forsyth, Z. Liu, B.A. Lepine and M. Liao. Document 2003-01-2916, "Pulsed Eddy Current Inspections of Aircraft Structures in Support of Holistic Damage Tolerance." *Aerospace Manufacturing Technology Conference and Exposition* [Montreal, Canada, September 2003]. Warrendale, PA: SAE International (2003).

Sheppard, W.R. and O.V. Manning. "Low Frequency Eddy Current Array Assessment at the FAA NDI Validation Center." *Materials Evaluation.* Vol. 53, No. 7. Columbus, OH: American Society for Nondestructive Testing (July 1995): p 844-847.

Sun, Y.S., S. Udpa, W. Lord and D. Cooley. "A Remote Field Eddy Current NDT Probe for the Inspection of Metallic Plates." *Materials Evaluation.* Vol. 54, No. 4. Columbus, OH: American Society for Nondestructive Testing (April 1996): p 510-512.

Vernon, S.N. "Eddy Current Measurement of Char Depth in the Advanced Solid Rocket Motor Nozzle." *Materials Evaluation.* Vol. 53, No. 2. Columbus, OH: American Society for Nondestructive Testing (February 1995): p 127-128, 130-131.

19

CHAPTER

Electromagnetic Testing Glossary

Paul M. Gammel, Gammel Applied Technologies,
Exmore, Virginia

Introduction

Most of the definitions in this glossary are adapted from the second edition of the *Nondestructive Testing Handbook*.[1-10] The definitions in this glossary have been modified to satisfy peer review and editorial style. For these reasons, references in this glossary should be considered not attributions but rather acknowledgments and suggestions for further reading.

The definitions in this *Nondestructive Testing Handbook* volume should not be referenced for tests performed according to standards or specifications or in fulfillment of contracts. Standards writing bodies take great pains to ensure that their standards are definitive in wording and technical accuracy. People working to written contracts or procedures should consult definitions referenced in standards when appropriate.

This glossary is provided for instructional purposes. No other use is intended.

Terms

A

absolute coil: Coil that responds to the electromagnetic properties of that region of the test part within the magnetic field of the coil, without comparison to the response of a second coil at a different location on the same or similar material.[4]

absolute measurement: (1) Measurement made with an absolute coil.[4] (2) Measurement of a property without reference to another measurement of that property.

acceptance criterion: Benchmark against which test results are to be compared for purposes of establishing the functional acceptability of a part or system being examined.

acceptance level: Measured value or values above or below which test specimens are acceptable in contrast to *rejection level*.[4,11]

acceptance limit: Test signal value used in electromagnetic testing, establishing the group to which a material under test belongs.[4,11]

acceptance standard: Specimen, similar to the product to be tested, containing natural or artificial discontinuities that are well defined and similar in size or extent to the maximum acceptable in the product.[4] Acceptance standards are available also for material properties such as conductivity and hardness.

algorithm: Set of well defined rules or processes that prescribe the solution of a problem in a finite number of steps.[4,12]

alternating current: Electrical current that reverses its direction at regular intervals.[6,10]

alternating current field: Varying magnetic field produced around a conductor by alternating current flowing in the conductor.[6,10]

alternating current magnetization: Magnetization by a magnetic field generated when alternating current is flowing.[6,10,13]

ampere (A): SI unit of electric current.[6,10]

ampere per meter ($A \cdot m^{-1}$): SI compound unit for magnetic field intensity. The measurement 1 $A \cdot m^{-1}$, for example, describes a current of 1 A flowing through a coil of 1 m diameter.[6,10,13]

amplitude response: Property of a test system whereby the amplitude of the detected signal is measured without regard to phase. See also *phase analysis*.[4,11]

analog-to-digital converter: Circuit whose input is information in analog form and whose output is essentially the same information in digital form.[4,12]

annular coil: See *encircling coil.*

annular coil clearance: Mean radial distance between the inner diameter of an encircling coil assembly and test object surface in electromagnetic testing. See *fill factor*.[4,10,11]

anomaly: Variation from normal material or product quality.[4]

argand diagram: Graphical representation of a vector quantity on the *complex plane*.

artifact: In nondestructive testing, an indication that may be interpreted erroneously as a discontinuity.[10,15]

artificial discontinuity: See *discontinuity, artificial*.

artificial discontinuity standard: See *acceptance standard.*

ASNT: American Society for Nondestructive Testing.

***ASNT Recommended Practice No. SNT-TC-1A*:** See *Recommended Practice No. SNT-TC-1A.*

attenuation: Decrease in signal amplitude over distance, often called *loss*; can be expressed in decibels or as a scalar ratio of the input magnitude to the output magnitude.[4,12]

automated system: Acting mechanism that performs required tasks at a determined time and in a fixed sequence in response to certain conditions.[10]

B

band pass filter: Frequency filter that has a single transmission band between two cutoff frequencies, neither of the cutoff frequencies being zero or infinity.[4,10,12]

bandwidth: Difference between the cutoff frequencies of a bandpass filter.[14]

blister: Discontinuity in metal, on or near the surface, resulting from the expansion of gas in a subsurface zone. Very small blisters are called *pinheads* or *pepper blisters*.[10]

blowhole: Hole in a casting or a weld caused by gas entrapped during solidification.[10]

bobbin coil: Cylindrically wound coil.

brittleness: Characteristic of a material that leads to crack propagation without appreciable plastic deformation.[10]

bucking coil: See *differential coils.*

C

calibration, instrument: Adjustment of instrument readings to known reference standard.

casing: Many strings of pipe that are used to line the hole during and after drilling of a water, gas or oil well.[10]

casing string: Tubular structure on the outer perimeter of a water, gas or oil well hole. The casing string is a permanent part of the well and many casing strings are cemented into the formation.[10]

central conductor: Electric conductor passed through the opening in a part with an aperture, or through a hole in a test object, for the purpose of creating a circular magnetic field in the object.[10]

certification: With respect to nondestructive test personnel, process of providing written testimony that an individual is qualified. See also *certified* and *qualified.*

certified: With respect to nondestructive test personnel, having written testimony of qualification. See also *certification* and *qualification.*

circular magnetization: Magnetization in an object resulting from current passed longitudinally through the object itself or through an inserted central conductor.[10,15]

circumferential coil: See *encircling coil.*

coil: One or more loops of a conducting material; a single coil may be an exciter and induce currents in the material or it may be a detector or both simultaneously.[4]

coil clearance: See *liftoff.*

coil spacing: In electromagnetic testing, the axial distance between two encircling or inside coils of a differential or remote field test system.[4,11]

comparative measurement: In electromagnetic testing, a measurement based on the imbalance in a system and using comparator coils in contrast to differential and absolute measurements. See also *comparator coils*.[4,11]

comparator coils: In electromagnetic testing, two or more coils electrically connected in series opposition and arranged so that there is no mutual induction (coupling) between them. Any electromagnetic condition that is not common to the test specimen and the standard will produce an imbalance in the system and thereby yield an indication. See also *differential coils*.[4,11]

complex plane: Plane defined by two perpendicular reference axes, used for plotting a complex variable (such as impedance) or functions of this variable (such as a transfer function).[4,12]

complex plane diagram: Graphical presentation of complex quantities where the real and imaginary components are represented along the horizontal and vertical axes, respectively.[4] Types of complex plane diagram include *impedance plane diagram*, voltage plane diagram and phase amplitude diagram.

conductance (G): Transmission of electric current through material. Measured in siemens (S). Inversely related to *resistance* R (ohm).

$$G = \frac{1}{R}$$

conductivity (σ): Ability of material to transmit electric current. Measured in siemens per meter. Inversely related to *resistivity* ρ.

$$\sigma = \frac{1}{\rho}$$

contact head: Electrode assembly used to clamp and support an object to facilitate passage of electric current through the object for circular magnetization.[10,15]

coupled: (1) Of two electric circuits, having an impedance in common so that a current in one causes a voltage in the other.[10,11] (2) Of two coils, sharing parts of their magnetic flux paths.

coupling: Percentage of magnetic flux from a primary circuit that links a secondary circuit; effectiveness of a coil in inducing eddy currents in the test object.[4]

coupling coefficient: Fraction of magnetic flux from one circuit (test coil) that threads a second circuit (test object); the ratio of impedance of the coupling to the square root of the product of the total impedances of similar elements in the two meshes.[4,12]

crack: (1) Break, fissure or rupture, sometimes V shaped and relatively narrow and deep. Discontinuity that has a relatively large cross section in one direction and a small or negligible cross section when viewed in a direction perpendicular to the first. (2) Propagating discontinuity caused by stresses such as heat treating or grinding. Difficult to detect unaided because of fineness of line and pattern (may have a radial or latticed appearance).[10]

crack, cold: Crack that occurs in a casting after solidification, because of excessive stress generally resulting from nonuniform cooling.[10]

crack, cooling: Crack in bars of alloy or tool steels resulting from uneven cooling after heating or hot rolling. Cooling cracks are usually deep and lie in a longitudinal direction but are usually not straight.[10]

crack, fatigue: Progressive crack that develops on the surface and is caused by the repeated loading and unloading of the object.[10]

crack, forging: Crack developed in the forging operation because of forging at too low a temperature, resulting in rupturing of the material.[10]

crack, hot: Crack that develops before the casting has completely cooled, as contrasted with cold cracks, that develop after solidification.[10]

crack, longitudinal: Crack parallel to the length of the test object.[10]

crack, quenching: During quenching of hot metal, rupture produced by more rapid cooling and contraction of one portion of a test object than occurs in adjacent portions.[10]

crack, transverse: Crack at right angle to the length of the test object.[10]

current flow technique: Magnetizing by passing current through an object using prods or contact heads. The current may be alternating current or rectified alternating current.[10,15]

current induction technique: Magnetization in which a circulating current is induced in a ring component by a fluctuating magnetic field.[10,15]

cycle: Single period of a waveform or other variable. See *period.*

D

defect: Discontinuity whose size, shape, orientation or location make it detrimental to the useful service of its host object or which exceeds the accept/reject criteria of an applicable specification.[10,16] Note that some discontinuities may not exceed specifications and are therefore not defects. Compare *discontinuity* and *indication.*[10]

demodulation: Process wherein a carrier frequency modulated with a signal of lower frequency than the carrier frequency is converted to a close representation of the original modulating signal.[14] See *modulation.*

depth of penetration: See *skin effect* and *standard depth of penetration.*

detector coil: See *sensing coil.*

differential amplifier: Amplifier whose output signal is proportional to the algebraic difference between two input signals.[4,12]

differential coils: Two or more physically adjacent and mutually coupled coils connected in series opposition such that an imbalance between them, causing a signal, will be produced only when the electromagnetic conditions are different in the regions beneath two of the coils. In contrast, *comparator coils* are not adjacent or mutually coupled.

differential measurement: In electromagnetic testing, the measurement of system imbalance by using differential coils, in contrast to absolute and comparative measurements.[4,11]

differentiated signal: In electromagnetic testing, an output signal proportional to the input signal's rate of change.[4,11]

direct current: Electric current flowing continually in one direction without variation in amplitude through a conductor.[10,16] See also *full-wave rectified direct current* and *half-wave direct current.*

direct current field: Active magnetic field produced by direct current flowing in a conductor or coil.[10,16]

discontinuity: Interruption in the physical structure or configuration of a test object.[10,17] After nondestructive testing, unintentional discontinuities interpreted as detrimental to the serviceability of the host object may be called *flaws* or *defects.*[10] Compare *defect* and *indication.*

discontinuity, artificial: Reference discontinuity such as hole, indentation, crack, groove or notch introduced into a reference standard to provide accurately reproducible indications for determining sensitivity levels.[10]

discontinuity inversion: Technique for measuring some dimension(s) of a discontinuity by the application of a mathematical algorithm to the measured test data.[4]

discontinuity resolution: Property of a test system that enables the separation of indications due to discontinuities near each other in a test specimen.[4,11]

E

eddy current: Electrical current induced in a conductor by a time varying magnetic field.[4]

eddy current testing: Nondestructive test technique in which eddy current flow is induced in the test object. Changes in the flow caused by variations in the specimen are reflected into a nearby coil, coils, hall effect device or other magnetic flux sensor for subsequent analysis by suitable instrumentation and techniques.[4,11]

edge effect: In electromagnetic testing, the disturbance of the magnetic field and eddy currents because of the proximity of an abrupt change in geometry, such as an edge of the test object. Sometimes called *end effect.* The effect generally results in the masking of discontinuities within the affected region.[4,11]

effective depth of penetration: In electromagnetic testing, the minimum depth beyond which a test system can no longer practically detect a further increase in specimen thickness.

electric field: Vector field of either the electric field intensity (volt per meter) or of the electric flux density (coulomb per meter squared).

electrical center: Center established by the electromagnetic field distribution within a test coil. A constant intensity signal, irrespective of the circumferential position of a discontinuity, is indicative of electrical centering. The electrical center may be different from the physical center of the test coil.[4,11]

electrode: Conductor by which a current passes into or out of a test object.[10,13]

electromagnet: Ferromagnetic core surrounded by a coil of wire that temporarily becomes a magnet when an electric current flows through the wire.[10,15]

electromagnetic acoustic transducer (EMAT): Electromagnetic device using lorentz forces and magnetostriction in conductive and ferromagnetic materials to generate and receive acoustic signals for ultrasonic nondestructive tests.[10]

electromagnetic testing (ET): Nondestructive test method for materials, including magnetic materials, that uses electromagnetic energy, either alternating or direct current, to yield information regarding the quality and characteristics of the tested material.[10,11]

EMAT: *Electromagnetic acoustic transducer.*

encircling coil: In electromagnetic testing, a coil or coil assembly that surrounds the test object. Such a coil is also called an *annular coil, circumferential coil* or *feed-through coil.*[10,11]

end effect: In bar and tube testing, *edge effect.*

ET: Electromagnetic testing.

evaluation: Review following interpretation of indications, to determine whether they meet specified acceptance criteria.

excitation coil: Coil that carries the excitation current. Also called *primary coil* or *winding.* See *sensing coil.*[10]

external discontinuities: Discontinuities on the outside or exposed surface of a test object.[10]

F

false indication: Test indication that could be interpreted as originating from a discontinuity but which actually originates where no discontinuity exists in the test object. Distinct from nonrelevant indication. Compare *defect.*[10]

feed-through coil: See *encircling coil.*

ferrite: Any of several magnetic substances that consist essentially of an iron oxide combined with one or more metals (such as manganese, nickel or zinc) having high magnetic permeability and high electrical resistivity.[6]

ferromagnetic material: Material such as iron, nickel or cobalt whose relative permeability is considerably greater than unity and depends on the magnetizing force and often exhibits hysteresis.[10,12] Materials that are most strongly affected by magnetism are called *ferromagnetic.*[10]

fill factor: For encircling coil electromagnetic testing, the ratio of the cross sectional area of the test object to the effective cross sectional core area of the primary encircling coil (outside diameter of coil form, not inside diameter that is adjacent to the object).[10,11,13] For internal probe electromagnetic testing, the ratio of the effective cross sectional area of the primary internal probe coil to the cross sectional area of the tube interior.[10,11]

fill factor effect: Effect of fill factor on coupling between coil and test object. See *coupling coefficient.*[4]

filter: Network that leaves a signal unaffected over a prescribed range of frequencies and attenuates signal components at all other frequencies.[4,11]

finite element analysis: Numerical technique for the analysis of a continuous system whereby that system is decomposed into a collection of finite sized elements.[4]

flaw: Rejectable or unintentional anomaly. See also *defect* and *discontinuity.*[10]

flaw inversion: See *discontinuity inversion.*

flux density: See *magnetic flux density.*

flux leakage: See *magnetic flux leakage field; magnetic flux leakage technique; magnetic flux meter.*

flux meter: See *magnetic flux meter.*

full-wave rectified direct current: Single-phase or three-phase alternating current rectified to produce unidirectional current. The rectified current contains ripple.

G

gauss (G): Obsolete unit of magnetic flux density, replaced in SI by *tesla* (T). 1 G = 0.1 mT.[10]

gauss meter: Gage that measures magnetic flux density in gauss (or tesla).[10]

general examination: In personnel qualification, test or examination of a person's knowledge, typically (in the case of nondestructive testing personnel qualification) a written test on the basic principles of a nondestructive testing method and general knowledge of basic equipment used in the method. (According to ASNT's guidelines, the general examination should not address knowledge of specific equipment, codes, standards and procedures pertaining to a particular application.)[10]

grinding crack: Shallow crack formed in the surface of relatively hard materials because of excessive grinding heat or the high sensitivity of the material. Grinding cracks typically are 90 degrees to the direction of grinding.[10]

H

half-wave direct current: Single-phase alternating current half-wave rectified to produce a pulsating unidirectional current. Also called *half-wave current.*[10,15]

hall detector: Semiconductor element that produces an output electromotive force proportional to the product of the magnetic field intensity and a biasing current.[10]

hall effect: Potential difference developed across a conductor at right angles to the direction of both the magnetic field and the electric current. Produced when a current flows along a rectangular conductor subjected to a transverse magnetic field.[10,13]

hardness: Resistance of metal to plastic deformation, usually by indentation. However, the term may also refer to stiffness or temper or to resistance to scratching, abrasion or cutting.[10]

heat affected zone (HAZ): Base metal that was not melted during brazing, cutting or welding but whose microstructure and physical properties were altered by the heat.[10]

hertz: Measurement unit of frequency, equivalent to one cycle per second.[10,12]

horseshoe coil: Probe coil in which the ferrite core of the coil is horseshoe shaped. Also called a *U shaped coil.*[4]

hysteresis: Apparent lagging of the magnetic effect when the magnetizing force acting on a ferromagnetic body is changed; phenomenon exhibited by a magnetic system wherein its state is influenced by its previous history.[10]

hysteresis loop: Curve showing flux density B plotted as a function of magnetizing force H as magnetizing force is increased to the saturation point in both negative and positive directions sequentially. The curve forms a characteristic shaped loop.

I

IACS: *International Annealed Copper Standard.*

impedance: Opposition that a circuit presents to the flow of an alternating current, specifically the complex quotient of voltage divided by current.[10,11]

impedance analysis: In electromagnetic testing, an analytical technique that consists of correlating changes in the amplitude, phase, quadrature components or all of these of a complex test signal voltage to the condition of the test specimen.[10,11]

impedance plane diagram: Graphical representation of the locus of points indicating the variations in the impedance of a test coil as a function of a parameter, such as *conductivity* or *liftoff.*

incremental permeability: Ratio of the change in magnetic induction to the corresponding change in magnetizing force.

indication: Nondestructive test equipment response to a discontinuity that requires interpretation to determine its relevance. Compare *defect, discontinuity* and *false indication.*[10]

indication, discontinuity: Visible evidence of a material discontinuity. Subsequent interpretation is required to determine the significance of an indication.[10]

indication, false: See *false indication.*

indication, nonrelevant: Indication due to misapplied or improper testing. May also be an indication caused by an actual discontinuity that does not affect the usability of the test object (a change of section, for instance).[10]

indication, relevant: Indication from a discontinuity (as opposed to a nonrelevant indication) requiring evaluation by a qualified inspector, typically with reference to an acceptance standard, by virtue of the discontinuity's size, shape, orientation or location.[10,17]

induced current technique: See *current induction technique.*

inductor: Device consisting of one or more associated windings, with or without a magnetic core, which impedes the flow of current.

initial permeability: Slope of the induction curve at zero magnetizing force as the test specimen begins to be magnetized from a demagnetized condition (slope at the origin of the *B,H* curve before hysteresis is observed).

inserted coil: See *inside diameter coil.*

inside coil: See *inside diameter coil.*

inside diameter coil: Coil or coil assembly used for electromagnetic testing by insertion into the test piece, as with an inside probe for tubing. Also called *inserted coil.*[4,11]

intergranular stress corrosion cracking (IGSCC): Anomaly caused by intergranular corrosion as a result of sensitized material, stress and corrosive environment (typical in the heat affected zone of stainless steel welds).

International Annealed Copper Standard (IACS): Conductivity measurement system in which the conductivity of annealed, unalloyed copper is arbitrarily rated at 100 percent and in which the conductivities of other materials are expressed as percentages of this standard. See also *conductivity* and *percent International Annealed Copper Standard.*

interpretation: Determination of the significance of test indications from the standpoint of their relevance or irrelevance, that is, from the standpoint of whether they are detrimental or inconsequential.[10]

inversion, discontinuity: See *discontinuity inversion.*

L

leakage flux: (1) Magnetic flux of the coil that does not link with the test object. (2) Magnetic flux that leaves a saturated or nearly saturated specimen at a discontinuity.[4]

level, acceptance: See *acceptance level.*

level, rejection: See *rejection level.*

liftoff: Distance between the probe coil and the test object.[4]

liftoff effect: In an electromagnetic test system output, the effect observed due to a change in coupling between a test object and a probe whenever the distance between them is varied.[10,11]

longitudinal magnetic field: Magnetic field wherein the flux lines traverse the component in a direction essentially parallel with its longitudinal axis.[10,15]

M

magnetic circuit: Closed path that allows magnetic flux to flow.

magnetic field: Distribution of a vector quantity that is a measure of an exerted magnetic force.

magnetic field indicator: Device used to locate or determine relative intensity of a flux leakage field.[10,15]

magnetic field intensity: Strength of a magnetic field at a specific point. Measured in ampere per meter.[10]

magnetic flux density: Normal magnetic flux per unit area, measured in *tesla* (T).[10,13]

magnetic flux leakage field: Magnetic field that leaves or enters the surface of an object.[10,15]

magnetic flux leakage technique: Electromagnetic test technique for the detection and analysis of a surface discontinuity or near surface discontinuity using the flux that leaves a magnetically saturated, or nearly saturated, test object at a discontinuity.[10]

magnetic flux meter: Electronic device for measuring magnetic flux leakage.[10]

magnetic flux leakage: Excursion of magnetic lines of force from the surface of a test specimen.[4,11]

magnetic particle testing (MT): Nondestructive test method using magnetic leakage fields and indication materials to disclose surface and near surface discontinuities.[10,15]

magnetic saturation: That degree of magnetization where a further increase in magnetizing force produces no significant increase in magnetic flux density in an object.[10,11]

magnetometer: Device for measuring the strength of magnets or the intensity of magnetic fields.[10,16]

magnitude: Absolute value of a complex quantity (number) without reference to the phase of the quantity.[4]

Maxwell's equations: Fundamental equations of electromagnetic field theory:

$$\nabla \times \boldsymbol{E} = -\frac{\partial \boldsymbol{B}}{\partial t}$$

$$\nabla \times \boldsymbol{H} = \frac{\partial \boldsymbol{D}}{\partial t} + \boldsymbol{J}$$

$$\nabla \bullet \boldsymbol{B} = 0$$

$$\nabla \bullet \boldsymbol{D} = \rho$$

where $\boldsymbol{B}$ = magnetic flux density, $\boldsymbol{D}$ = electric flux density, $\boldsymbol{E}$ = electric field intensity, $\boldsymbol{H}$ = magnetic field strength, $\boldsymbol{J}$ = current density and ρ = volume charge density.[4]

microwave testing: Nondestructive testing method that uses, for its probing energy, electromagnetic radiation at radio frequencies — from 0.3 to 300 GHz, with wavelengths from 1 mm to 1 m.[10]

model, analytical: Mathematical representation of a process or phenomenon.

modulation: Process of imparting information to a carrier signal by the introduction of amplitude or phase perturbation.[14]

MT: Magnetic particle testing.

multifrequency: Two or more frequencies applied sequentially or simultaneously to the test coil.

multifrequency technique: Use of the response of a test specimen to more than one frequency, usually to separate effects that would be indistinguishable at a single frequency.[4]

multiparameter: Of or pertaining to a test system having many parameters that affect the response. These parameters can often be distinguished with a multifrequency technique.[4]

multivariable: See *multiparameter*.[4]

mutual inductance: Property of two electrical circuits whereby a voltage is induced in one circuit by a change of current in the other circuit.[4,12]

N

NDC: *Nondestructive characterization.*

NDE: (1) *Nondestructive evaluation.* (2) *Nondestructive examination.*

NDI: *Nondestructive inspection.*

NDT: *Nondestructive testing.*

noise: In electromagnetic testing, any nonrelevant signal that tends to interfere with the normal reception or processing of a desired discontinuity signal. Such noise signals may be due to an extraneous source or generated by heterogeneities in the test part that are not detrimental to the use of the part.[4,11]

nondestructive characterization (NDC): Branch of nondestructive testing concerned with the description and prediction of material properties and behaviors of components and systems.

nondestructive evaluation (NDE): Another term for *nondestructive testing.* In research and academic communities, the word *evaluation* is sometimes preferred because it implies interpretation by knowledgeable personnel or systems.[10]

nondestructive examination (NDE): Another term for *nondestructive testing.* In the utilities and nuclear industry, *examination* is sometimes preferred because *testing* can imply performance trials of pressure containment or power generation systems.[10]

nondestructive inspection (NDI): Another term for *nondestructive testing.* In some industries (utilities, aviation), the word *inspection* often implies maintenance for a component that has been in service.[10]

nondestructive testing (NDT): Determination of the physical condition of an object without affecting that object's ability to fulfill its intended function. Nondestructive test methods typically use an appropriate form of energy to determine material properties or to indicate the presence of material discontinuities (surface, internal or concealed).[10]

nonferromagnetic material: Material not magnetizable and essentially not affected by magnetic fields.[4,11]

normalized impedance diagram: Diagram in which the impedance of the probe in air is a reference value to which impedance values in other conditions are compared. Usually the plotted data are (1) the measured reactance divided by the reactance of the coil in air versus (2) the measured resistance less the resistance in air divided by the coil reactance in air.

null: To adjust a bridge circuit so that the test sample and reference arms produce equal and opposite currents through the detector.[4]

null signal: Fixed component of the test coil signal that is subtracted from the output signal leaving only that part of the signal that varies with test object conditions; it reduces dynamic range requirements.[4]

numerical analysis: Technique to generate numbers as the solution to a mathematical model of a physical system; used in place of a closed form analytic expression; usually requires digital computation.[4]

O

OCTG: Oil country tubular goods.[10]

oersted (Oe): Obsolete measurement unit of magnetic field intensity, replaced in SI by ampere per meter ($A \cdot m^{-1}$). 1 Oe = 79.57747 $A \cdot m^{-1}$.

ohm (Ω): Measurement unit of electrical resistance.

oil country tubular goods: Hollow cylindrical components used to convey petroleum and related products.[10]

optimum frequency: In electromagnetic testing, that frequency that provides the largest signal-to-noise ratio obtainable for the detection of an individual material property.

P

pancake coil: Probe coil whose axis is normal to the surface of the test material and whose length is not larger than the radius.[4]

paramagnetic material: In electromagnetic testing, a material that has a relative permeability slightly greater than unity and is practically independent of the magnetizing force.[10,11]

percent International Annealed Copper Standard (%IACS): Measurement of *conductivity* as a percentage of the conductivity of pure copper, arbitrarily rated at 100 percent. See also *International Annealed Copper Standard.*

period: Absolute value of the minimum interval after which the same characteristics of a periodic waveform or a periodic feature repeat.[4]

permeability: Ratio of magnetic induction to magnetizing force. This relationship is either (1) absolute permeability, in general the quotient of magnetic induction divided by the magnetizing force, or (2) *relative permeability* (or specific permeability), a pure number that is the same in all unit systems. The value and dimension of absolute permeability depend on the system of units used. In anisotropic media, permeability is a matrix.[4,12]

phase analysis: Analytical technique that discriminates between variables in a part undergoing electromagnetic testing by the different phase angle and amplitude changes that these conditions produce in the test signal. See also *phase detection.*[4,11]

phase angle: Angular equivalent of the time displacement between corresponding points on two sine waves of the same frequency.[4,11]

phase detection: Derivation of a signal whose amplitude is a function of the phase angle between two alternating currents, one of which is used as a reference.[4,11]

phase sensitive system: System whose output signal depends on the phase relationship between the voltage returned from a pickup or sensing coil and a reference voltage.[4,11]

phase shift: Change in the phase relationship between two alternating quantities of the same frequency.[4,11]

phasor: Complex number that represents the amplitude and phase of a quantity that varies sinusoidally with time. A phasor is not a vector, because the orientation of a vector represents direction.

physical properties: Nonmechanical properties such as density, electrical conductivity, heat conductivity and thermal expansion.[10]

probe coil clearance: Perpendicular distance between adjacent surfaces of the probe and test part. See *liftoff.*[4,11]

pulse technique: Multifrequency technique in which a broadband excitation such as an impulse is used. Either the frequency components are extracted and analyzed or the interpretation is based directly on characteristics of the time domain waveform.[4]

Q

Q of a coil: *Quality factor* of a coil; related to the ratio of maximum energy stored to the total energy lost per period.

quadrature: Relation between two periodic functions when the phase difference between them is one fourth of a period.[4,12]

qualification: Process of demonstrating that an individual has the required amount and the required type of training, experience, knowledge and capabilities.[10]

qualified: Having demonstrated the required amount and the required type of training, experience, knowledge and abilities.[10]

quality: Ability of a process or product to meet specifications or to meet the expectations of its users in terms of efficiency, appearance, longevity and ergonomics.[10]

quality assurance: Administrative actions that specify, enforce and verify a quality program.[10]

quality control: Physical and administrative actions required to ensure compliance with the quality assurance program. Quality control may include nondestructive testing in the manufacturing cycle.[10]

quality factor: Of a coil, the ratio of reactance to resistance defined at the operating frequency.

R

recommended practice: Set of guidelines or recommendations.[10]

Recommended Practice No. SNT-TC-1A: Set of guidelines published by the American Society for Nondestructive Testing, for employers to establish and conduct a nondestructive testing personnel qualification and certification program.[10]

recovery time: Time required for a test system to return to its original state after overload or signal reception.

reference coil: In electromagnetic testing, the section of the coil assembly that excites or detects the electromagnetic field in the reference standard of a comparative system.[4,11]

reference number: Number associated with the impedance of a coil adjacent to a test sample.

reference standard: Reference used as a basis for comparison or calibration. In tube testing, a tube with artificial discontinuities used for establishing the test sensitivity setting and for periodically checking and adjusting the sensitivity setting as required. See also *acceptance standard.*[4,11]

reflection probe: Coil system that uses both an excitation and a detection or sensing coil on the same side of the sample.[4,11]

rejection level: Value established for a test signal above or below which test specimens are rejectable or otherwise distinguished from the remaining specimens. This level is different from the *rejection level* as defined for ultrasonic and other test systems.[4,11,20]

relative permeability: Ratio of the permeability of the material to the permeability of vacuum. See also *permeability.*[4]

resistance, electrical (*R*): Opposition to transmission of electric current through material; ratio of voltage to current. Measured in ohm (Ω). Inversely related to *conductance:*

$$R = \frac{1}{G} = \frac{\rho L}{A}$$

where A is the conductor's cross sectional area (square meter), G is conductance (siemens), L is the length of the conductor (meter) and ρ is resistivity (ohm meter).

resistivity (ρ): Ability of material to resist electric current. Measured in ohm meter (Ω·m), which is the resistance of a cube made of the material whose dimensions are 1 m on each side. Inversely related to *conductivity* σ (siemens per meter):

$$\rho = \frac{1}{\sigma}$$

response function: Ratio of response to excitation, both expressed as functions of the complex impedance.[4,12]

S

scalar: Quantity completely specified by a single number and unit.[4,12]

search coil: Detection coil, usually smaller than the excitation coil.[4]

secondary magnetic flux: Magnetic flux due to induced flow of eddy currents.[4]

selectivity: Characteristic of a test system, a measure of the extent to which an instrument can differentiate between the desired signal and disturbances of other frequencies or phases.[4,11]

self-inductance: Property of an electric circuit whereby an electromotive force is induced in that circuit by a change of current in the circuit.[4,12]

sensing coil: Coil that detects changes in the flow of eddy currents induced by an excitation coil; sensing and excitation coils can be one and the same.[4] Also called *detector coil.*

shielding: Conducting or magnetic material (or a combination of both) placed so as to decrease susceptibility to interference.[4]

SI (International System of Units): Universal, coherent system of measurement in which the following seven units are considered basic: meter, mole, kilogram, second, ampere, kelvin and candela.[4,12]

siemens per meter ($S{\cdot}m^{-1}$): SI unit of conductivity.

signal: Physical quantity, such as electrical voltage, that contains relevant information.[4,11]

signal-to-noise ratio: Ratio of signal values (responses that contain relevant information) to baseline noise values (responses that contain nonrelevant information). See *noise*.[10,11]

skin depth: *Standard depth of penetration.* See also *skin effect.*

skin effect: Phenomenon wherein the depth of penetration of electrical currents into a conductor decreases as the frequency of the current is increased. At very high frequencies, the current flow is restricted to an extremely thin outer layer of the conductor. See standard *depth of penetration*.[10,11]

SNT-TC-1A: See *Recommended Practice No. SNT-TC-1A.*

specification: Set of instructions or standards invoked by a specific customer to govern the results or performance of a specific set of tasks or products.[10]

spectrum: Signal aspect showing the distribution of the various frequency components of the signal.[14] Also called *fourier spectrum.*

SQUID: Superconducting quantum interference device, a sensitive detector of magnetic fields using quantum effect.[4]

standard: (1) Physical object with known material characteristics used as a basis for comparison or calibration; *reference standard.* (2) Concept established by authority, custom or agreement to serve as a model or rule in the measurement of quantity or the establishment of a practice or procedure. (3) Document to control and govern practices in an industry or application, applied on a national or international basis and usually produced by consensus. See also *acceptance standard* and *reference standard*.[10,11,18]

standard depth of penetration: In electromagnetic testing, the depth at which the magnetic field intensity or intensity of induced eddy currents has decreased to 37 percent of its surface value. The square of the depth of penetration is inversely proportional to the frequency of the signal, the conductivity of the material and the permeability of the material. See also *skin effect*.[10,11]

standardization, instrument: Adjustment of instrument readout before use to an arbitrary reference value.

stationary: Of a signal, having statistical properties such as mean and variance that do not vary with time.[14]

T

tangential magnetic field: Magnetic field at an object's surface parallel to the surface. The tangential field is continuous (equal on either side) with the interface of material to air. Measurement can be influenced by external fields.[10]

tape head probe: Head of a tape recorder used as an eddy current coil; a type of horseshoe coil.[4]

tesla (T): SI unit of measure for magnetic flux density. $1\ T = 1\ Wb{\cdot}m^{-2} = 10\,000\ G$.[10]

tesla meter: Gage that measures magnetic flux density in tesla.[10]

test coil: Section of a coil assembly that excites or detects the magnetic field in the material under electromagnetic test.[4,11]

test frequency: In electromagnetic testing, the number of complete cycles per unit time of the alternating current applied to the primary test coil.[4,11]

test quality level: See *rejection level.*

text information: Information stored on recording medium to support recorded eddy current data.

three-way sort: Electromagnetic sort based on a test object signal response above or below two levels established by three or more calibration standards.[4,11]

threshold level: Setting of an instrument that causes it to register only those changes in response greater or less than a specified magnitude.[4,11]

through-transmission: Of or pertaining to electromagnetic techniques where the excitation field penetrates the test object so that the detected signal is responsive to features external to or on the opposite surface.

toroidal field: Induced magnetic field occurring in a ring test object when current is induced. See *current induction technique*.[10]

trace: Line formed by an electron beam scanning from left to right on a video or computer screen to generate an image.[10]

transducer: Device by means of which energy can flow from one or more transmission systems or media to one or more other transmission systems or media; sensor or probe.

tubing string: Pipe with which oil or gas has contact as it is brought to the earth's surface.[10]

two-way sort: Electromagnetic sort based on a test object signal response above or below a level established by two or more calibration standards.[4,11]

U

U shaped coil: See *horseshoe coil.*

Unified Numbering System: Alphanumeric system for identifying alloys according to a registry maintained by ASTM International and SAE International.[21]

unit of data storage: Discrete physical recording medium on which text information is stored.

V

vector quantity: Any physical quantity whose specification involves both magnitude and direction and that obeys the parallelogram law of addition.[4,12]

volt (V): Measurement unit of electric potential.

W

wobble: In electromagnetic testing, an effect that produces variations in an output signal of a test system and arises from coil spacing (operational liftoff) variations due to lateral motion of the test specimen in passing through an encircling coil or of a bobbin coil passing through a cylindrical test object.[4,11]

Y

yoke: Magnet that induces a magnetic field in the area of a part that lies between its poles. Yokes may be permanent magnets or either alternating current or direct current electromagnets.[4,11]

References

1. *Nondestructive Testing Handbook,* second edition: Vol. 1, *Leak Testing.* Columbus, OH: American Society for Nondestructive Testing (1982).
2. *Nondestructive Testing Handbook,* second edition: Vol. 2, *Liquid Penetrant Tests.* Columbus, OH: American Society for Nondestructive Testing (1982).
3. *Nondestructive Testing Handbook,* second edition: Vol. 3, *Radiography and Radiation Testing.* Columbus, OH: American Society for Nondestructive Testing (1985).
4. *Nondestructive Testing Handbook,* second edition: Vol. 4, *Electromagnetic Testing.* Columbus, OH: American Society for Nondestructive Testing (1986).
5. *Nondestructive Testing Handbook,* second edition: Vol. 5, *Acoustic Emission Testing.* Columbus, OH: American Society for Nondestructive Testing (1987).
6. *Nondestructive Testing Handbook,* second edition: Vol. 6, *Magnetic Particle Testing.* Columbus, OH: American Society for Nondestructive Testing (1989).
7. *Nondestructive Testing Handbook,* second edition: Vol. 7, *Ultrasonic Testing.* Columbus, OH: American Society for Nondestructive Testing (1991).
8. *Nondestructive Testing Handbook,* second edition: Vol. 8, *Visual and Optical Testing.* Columbus, OH: American Society for Nondestructive Testing (1993).
9. *Nondestructive Testing Handbook,* second edition: Vol. 9, *Special Nondestructive Testing Methods.* Columbus, OH: American Society for Nondestructive Testing (1995).
10. *Nondestructive Testing Handbook,* second edition: Vol. 10, *Nondestructive Testing Overview.* Columbus, OH: American Society for Nondestructive Testing (1996).
11. E 268-81, *Definitions Approved for Use by Agencies of the Department of Defense as Part of Federal Test Method Standard No. 151b and for Listing in the DoD Index of Specifications and Standards.* West Conshohocken, PA: ASTM International (1981).
12. *IEEE Standard Dictionary of Electrical and Electronic Terms.* New York, NY: Wiley-Interscience, for the Institute of Electrical and Electronics Engineers (1984).
13. *Glossary of Terms Used in Nondestructive Testing, Part 2.* London, United Kingdom: British Standards Institute (November 1984).
14. Couch, L.W. *Digital and Analog Communication Systems.* Upper Saddle River, NJ: Prentice-Hall (1997).
15. E 269-89, *Standard Definitions of Terms Relating to Magnetic Particle Examination.* West Conshohocken, PA: ASTM International (1989).
16. API RP5A5, *Recommended Practice for Field Inspection of New Casing, Tubing and Plain End Drill Pipe,* third edition. Washington, DC: American Petroleum Institute (1987).
17. *Annual Book of ASTM Standards:* Section 3, *Metals Test Methods and Analytical Procedures.* Vol. 03.03, *Nondestructive Testing.* West Conshohocken, PA: ASTM International (2001).
18. *Nondestructive Testing Methods.* TO33B-1-1 (NAVAIR 01-1A-16) TM43-0103. Washington, DC: Department of Defense, United States Air Force (June 1984): p 1.25.
19. Libby, H. *Introduction to Electromagnetic Nondestructive Test Methods.* New York, NY: Wiley-Interscience (1971).
20. *Webster's Ninth New Collegiate Dictionary.* Springfield, MA: Merriam-Webster Incorporated (1990).
21. ASTM DS-56H [SAE HS-1086], *Metals and Alloys in the Unified Numbering System,* ninth edition. Warrendale, PA: SAE International (2001).

Index

Note

Readers are encouraged to consult this volume's glossary; glossary entries are not entered in this index.

A

B

C

D

E

F

G

H

I

J

K

L

M

N

O

P

Q

R

S

T

U

V

W

X

Y

Z

Figure Sources

Chapter 1. Introduction to Electromagnetic Testing

Figure 3b — Zetec, Incorporated, Issaquah, WA.

Chapter 2. History of Electromagnetic Testing

Figure 1 — Stanford University, Palo Alto, CA.
Figures 2a, 3a, 4 — Burndy Library, Cambridge, MA.
Figure 6 — General Electric Research and Development Center, Schenectady, NY.
Figure 8 — LTV Steel, Cleveland, OH.

Chapter 5. Probes for Electromagnetic Testing

Figures 3-10 — Plenum Press, New York, NY.
Figure 38 — Honeywell, Microswitch Division, Richardson, TX.

Chapter 13. Electromagnetic Techniques for Material Identification

Figures 7-8 — Boeing Company, Long Beach, CA.
Figure 15 — NDT Instruments, Huntington Beach, CA.
Figures 16-17, 20-22 — ASM International, Materials Park, OH.
Figures 18-19 — Institut Dr. Förster, Reutlingen, Germany.
Figure 23 — Sensor Corporation, Scottdale, NY.
Figure 24 — John Wiley and Sons, New York, NY.
Figures 27-32 — Jentek Sensors, Waltham, MA.

Chapter 18. Aerospace Applications of Eddy Current Testing

Figures 1, 3-16, 18-30, 32-36 — Boeing Company, Long Beach, CA.
Figure 31 — Hocking NDT Limited, Saint Albans, United Kingdom.